are both in $\mathbf{R}^n$. Then the dot product of $\mathbf{u}$ and $\mathbf{v}$ is given by

$$\mathbf{u} \cdot \mathbf{v} = u_1 v_1 + \cdots + u_n v_n$$

The dot product can also be expressed $\mathbf{u} \cdot \mathbf{v} = \mathbf{u}^T \mathbf{v}$. (Sect. 8.1)

Eigenspace Let A be a square matrix with eigenvalue λ. The subspace of all eigenvectors associated with λ, together with the zero vector, is the eigenspace of λ. Each distinct eigenvalue of A has its own associated eigenspace. (Sect. 6.1)

Eigenvalue Let A be an $n \times n$ matrix. Suppose that λ is a scalar and $\mathbf{u} \neq \mathbf{0}$ is a vector satisfying

$$A\mathbf{u} = \lambda\mathbf{u}$$

The scalar λ is called an eigenvalue of A. (Sect. 6.1)

Eigenvector Let A be an $n \times n$ matrix. Suppose that λ is a scalar and $\mathbf{u} \neq \mathbf{0}$ is a vector satisfying

$$A\mathbf{u} = \lambda\mathbf{u}$$

Then $\mathbf{u}$ is called an eigenvector of A. (Sect. 6.1)

Elementary row operations Three row operations that can be performed on an augmented matrix that do not change the set of solutions to the corresponding linear system. They are interchanging two rows; multiplying a row by a nonzero constant (1), (2), and (3) adding a multiple of one row to another. (Sect. 1.2)

Equivalent matrices Two matrices are equivalent if one can be transformed into the other through a sequence of elementary row operations. If the matrices in question are augmented matrices, then the corresponding linear systems have the same set of solutions. (Sect. 1.2)

Equivalent systems Two linear systems are equivalent if they have the same set of solutions. (Sect. 1.2)

Free variable Any variable in a linear system in echelon form that is not a leading variable. (Sect. 1.1)

Gauss–Jordan elimination An algorithm that extends Gaussian elimination, applying row operations in a manner that will transform a matrix to reduced echelon form. (Sect. 1.2)

Gaussian elimination An algorithm for applying row operations in a manner that will transform a matrix to echelon form. (Sect. 1.2)

General solution A description of the set of all solutions to a linear equation or linear system. (Sect. 1.1)

Homogeneous system A linear system is homogeneous if it has the form

$$\begin{aligned}
a_{11}x_1 + a_{12}x_2 + a_{13}x_3 + \cdots + a_{1n}x_n &= 0 \\
a_{21}x_1 + a_{22}x_2 + a_{23}x_3 + \cdots + a_{2n}x_n &= 0 \\
&\vdots \\
a_{m1}x_1 + a_{m2}x_2 + a_{m3}x_3 + \cdots + a_{mn}x_n &= 0
\end{aligned}$$

Such systems always have the trivial solution, so are consistent. This system can also be expressed by $A\mathbf{x} = \mathbf{0}$. (Sect. 1.2; see also Sect. 2.3)

Identity matrix The $n \times n$ identity matrix is given by

$$I_n = \begin{bmatrix} \mathbf{e}_1 & \mathbf{e}_2 & \cdots & \mathbf{e}_m \end{bmatrix} = \begin{bmatrix} 1 & 0 & 0 & \cdots & 0 \\ 0 & 1 & 0 & \cdots & 0 \\ 0 & 0 & 1 & \cdots & 0 \\ \vdots & \vdots & \vdots & \ddots & \vdots \\ 0 & 0 & 0 & \cdots & 1 \end{bmatrix}$$

(Sect. 3.2)

Inconsistent linear system A linear system that has no solutions. (Sect. 1.1)

Inner product (real) Let $\mathbf{u}$, $\mathbf{v}$, and $\mathbf{w}$ be elements of a vector space V, and let c be a scalar. An inner product on V is a function that takes two vectors in V as input and produces a scalar as output. An inner product function is denoted by $\langle \mathbf{u}, \mathbf{v} \rangle$, and satisfies the following conditions:

(a) $\langle \mathbf{u}, \mathbf{v} \rangle = \langle \mathbf{v}, \mathbf{u} \rangle$

(b) $\langle \mathbf{u} + \mathbf{v}, \mathbf{w} \rangle = \langle \mathbf{u}, \mathbf{w} \rangle + \langle \mathbf{v}, \mathbf{w} \rangle$

(c) $\langle c\mathbf{u}, \mathbf{v} \rangle = \langle \mathbf{u}, c\mathbf{v} \rangle = c\langle \mathbf{u}, \mathbf{v} \rangle$

(d) $\langle \mathbf{u}, \mathbf{u} \rangle \geq 0$, and $\langle \mathbf{u}, \mathbf{u} \rangle = 0$ only when $\mathbf{u} = \mathbf{0}$

(Sect. 10.1)

Inner product space A vector space V with an inner product defined on it. (Sect. 10.1)

Invertible matrix An $n \times n$ matrix A is invertible if there exists an $n \times n$ matrix B such that $AB = I_n$. The matrix B is called the inverse of A and is denoted A^{-1}. (Sect. 3.3)

Isomorphic vector spaces V and W are isomorphic vector spaces if there exists a isomorphism $T : V \rightarrow W$. (Sect. 9.2)

(Continues on the inside of the back cover.)

LINEAR
ALGEBRA
WITH
APPLICATIONS

JEFFREY HOLT

University of Virginia

 W. H. FREEMAN AND COMPANY
New York

Senior Publisher: Ruth Baruth

Executive Editor: Terri Ward

Developmental Editor: Leslie Lahr

Senior Media Editor: Laura Judge

Associate Editor: Jorge Amaral

Director of Market Research and Development: Steve Rigolosi

Associate Media Editor: Catriona Kaplan

Associate Media Editor: Courtney M. Elezovic

Editorial Assistant: Liam Ferguson

Marketing Manager: Steve Thomas

Marketing Assistant: Alissa Nigro

Photo Editor: Ted Szczepanski

Cover and Text Designer: Vicki Tomaselli

Senior Project Editor: Vivien Weiss, MPS Ltd.

Illustrations: MPS Ltd.

Illustration Coordinator: Bill Page

Production Coordinator: Julia DeRosa

Composition: MPS Ltd.

Printing and Binding: Quad Graphics

Library of Congress Control Number: 2012950275

ISBN-13: 978-0-7167-8667-2
ISBN-10: 0-7167-8667-2

Printed in the United States of America

First printing

W. H. Freeman and Company
41 Madison Avenue
New York, NY 10010
Houndmills, Basingstoke RG21 6XS, England
www.whfreeman.com

To my family, Mike, Laura, Tom, and Kathy.

CONTENTS

---

*Sections with an asterisk can be omitted without loss of continuity but may be required for later optional sections. See each indicated section for dependency information.

PREFACE

Welcome to *Linear Algebra with Applications*. This book is designed for use in a standard linear algebra course for an applied audience, usually populated by sophomores and juniors. While the majority of students in this type of course are majoring in engineering, some also come from the sciences, economics, and other disciplines. To accommodate a broad audience, applications covering a variety of topics are included.

Although this book is targeted toward an applied audience, full development of the theoretical side of linear algebra is included, so this textbook can also be used as an introductory course for mathematics majors. I have designed this book so that instructors can teach from it at a conceptual level that is appropriate for their individual course.

There is a collection of core topics that appear in virtually all linear algebra texts, and these are included in this text. In particular, the core topics recommended by the Linear Algebra Curriculum Study Group are covered here. The organization of these core topics varies from text to text, with the recent trend being to introduce more of the "abstract" material earlier rather than later. The organization here reflects this trend, with the chapters (approximately) alternating between computational and conceptual topics.

Text Features

Early Presentation of Key Concepts. Traditional linear algebra texts initially focus on computational topics, then treat more conceptual subjects soon after introducing abstract vector spaces. As a result, at the point abstract vector spaces are introduced, students face two simultaneous challenges:

(a) A change in mode of thinking, from largely mechanical and computational (solving systems of equations, performing matrix arithmetic) to wrestling with conceptual topics (span, linear independence).

(b) A change in context from the familiar and concrete $\mathbf{R}^n$ to abstract vector spaces.

Many students cannot effectively meet both these challenges at the same time. The organization of topics in this book is designed to address this significant problem.

In *Linear Algebra with Applications*, we first address challenge (a). Conceptual topics are explored early and often, blended in with topics that are more computational. This spreads out the impact of conceptual topics, giving students more time to digest them. The first six chapters are presented solely in the context of Euclidean space, which is relatively familiar to students. This defers challenge (b) and also allows for a treatment of eigenvalues and eigenvectors that comes earlier than in other texts.

Challenge (b) is taken up in Chapter 7, where abstract vector spaces are introduced. Here, many of the conceptual topics explored in the context of Euclidean space are revisited in this more general setting. Definitions and theorems presented are similar to those given earlier (with explicit references to reinforce connections), so students have less trouble grasping them and can focus more attention on the new concept of an abstract vector space.

From a mathematical standpoint, there is a certain amount of redundancy in this book. Quite a lot of the material in Chapters 7, 9, and 10, where the majority of the development of abstract vector spaces resides, has close analogs in earlier chapters. This is a deliberate part of the book design, to give students a second pass through key ideas to reinforce understanding and promote success.

Topics Introduced and Motivated Through Applications. To provide understanding of why a topic is of interest, when it makes sense I use applications to introduce and motivate new topics, definitions, and concepts. In particular, many sections open with an application. Applications are also distributed in other places, including the exercises. In a few instances, entire sections are devoted to applications.

Extensive Exercise Sets. Recently, I read a review of a text in which the reviewer stated, in essence, "This text has a very nice collection of exercises, which is the only thing I care about. When will textbook authors learn that the most important consideration is the exercises?" Although this sentiment might be extreme, most instructors share it to some degree, and it is certainly true that a text with inadequate problem sets can be frustrating. *Linear Algebra with Applications* contains over 2600 exercises, covering a wide range of types (computational to conceptual to proofs) and diffculty levels.

Ample Instructional Examples. For many students, a primary use of a mathematics text is to learn by studying examples. Besides those examples used to introduce new topics, this text contains a large number of additional representative examples. Perhaps the number one complaint from students about mathematics texts is that there are not enough examples. I have tried to address that in this text.

Support for Theory and Proofs. Many students in a first linear algebra course are usually not math majors, and many have limited experience with proofs. Proofs of most theorems are supplied in this book, but it is possible for a course instructor to vary the level of emphasis given to proofs through choice of lecture topics and homework exercises.

Throughout the book, the goal of proofs is to help students understand why a statement is true. Thus, proofs are presented in different ways. Sometimes a theorem might be proved for a special case, when it is clear that no additional understanding results from presenting the more general case (especially if the general case is more notationally messy). If a proof is diffcult and will not help students understand why the theorem is true, then it might be given at the end of the section or omitted entirely. If it provides a source of motivation for the theorem, the proof might come before the statement of the theorem. I have also written an appendix containing an overview of how to read and write proofs to assist those with limited experience. (See the text website at **www.whfreeman.com/holt.**)

Most linear algebra texts handle theorems and proofs in similar ways, although there is some variety in the level of rigor. However, it seems that often there is not enough concern for whether or not the proof is conveying why the theorem is true, with the goal instead being to keep the proof as short as possible. Sometimes it is worth taking a bit of extra time to give a complete explanation. For example, in Section 1.1, a system of equations is reduced to "$0 = 8$." At this point, most texts would state that this shows the system has no solutions, and it is likely that most students would agree. However, it is also possible that many students will not know why the system has no solutions, so a brief explanation is included.

Organization of Material

Roughly speaking, the chapters alternate between computational and conceptual topics. This is deliberate, in order to spread out the challenge of the conceptual topics and to give students more time to digest them. The material in Chapters 1–6 and 8 is exclusively in the context of Euclidean space and includes the core topics recommended by the Linear Algebra Curriculum Study Group. Chapters 7, 9, and 10 cover topics in the context of abstract vector space, and Chapter 11 contains a collection of optional topics that can be included at the end of a course.

Those sections marked with an asterisk (∗) can be omitted without loss of continuity, but in some cases they may be assumed in optional sections that come later. See the start of each optional section for dependency information.

1. Systems of Linear Equations

1.1 Lines and Linear Equations

1.2 Linear Equations and Matrices

1.3 Numerical Solutions*

1.4 Applications of Linear Systems*

Chapter 1 is fairly computational, providing a comprehensive introduction to systems of linear equations and their solutions. Iterative solutions to systems are also treated. The chapter closes with a section containing in-depth descriptions of several applications of linear systems. By the end of this chapter, students should be proficient in using augmented matrices and row operations to find the set of solutions to a linear system.

2. Euclidean Space

2.1 Vectors

2.2 Span

2.3 Linear Independence

Chapter 2 shifts from mechanical to conceptual material. This chapter is devoted to introducing vectors and the important concepts of span and linear independence, all in the concrete context of $\mathbf{R}^n$. These topics appear early so that students have more time to absorb these important concepts.

3. Matrices

3.1 Linear Transformations

3.2 Matrix Algebra

3.3 Inverses

3.4 LU Factorization*

3.5 Markov Chains*

Chapter 3 shifts from conceptual back to (mostly) mechanical material, starting with a treatment of linear transformations from $\mathbf{R}^n$ to $\mathbf{R}^m$. This is used to motivate the definition of matrix multiplication, which is covered in the next section along with other matrix arithmetic. This is followed by a section on computing the inverse of a matrix, motivated by finding the inverse of a linear transformation. Matrix factorizations, arguably related to numerical methods, provide an alternate way of organizing computations. The chapter closes with Markov chains, a topic not typically covered until after discussing eigenvalues and eigenvectors. But this subject easily can be covered earlier, and as there are a number of interesting applications of Markov chains, they are included here.

4. Subspaces

4.1 Introduction to Subspaces

4.2 Basis and Dimension

4.3 Row and Column Spaces

In Chapter 4, we again shift back to a more conceptual topic, subspaces in $\mathbf{R}^n$. The first section provides the definition of subspace along with examples. The second section develops the notion of basis and dimension for subspaces in $\mathbf{R}^n$, and the last section thoroughly treats row and column spaces. By the end of this chapter, students will have been exposed to many of the central conceptual topics typically covered in a linear algebra course. These are revisited (and eventually generalized) throughout the remainder of the book.

5. Determinants

5.1 The Determinant Function
5.2 Properties of the Determinant*
5.3 Applications of the Determinant*

Chapter 5 develops the usual properties of determinants. This topic has moved around in texts in recent years. For some time, the trend was to reduce the emphasis on determinants, but lately they have made something of a comeback. This chapter is relatively short and is introduced at this point in the text to support the introduction of eigenvalues and eigenvectors in the next chapter. Those who want only enough of determinants for eigenvalues can cover only Section 5.1.

6. Eigenvalues and Eigenvectors

6.1 Eigenvalues and Eigenvectors
6.2 Approximation Methods*
6.3 Change of Basis
6.4 Diagonalization
6.5 Complex Eigenvalues*
6.6 Systems of Differential Equations*

Chapter 6 provides a treatment of eigenvalues and eigenvectors that comes earlier than in most books. Section 6.2 covers numerical methods for approximating eigenvalues and eigenvectors and can be deferred until later or omitted entirely. Diagonalization is presented as a special type of change of basis and is revisited for symmetric matrices in Chapter 8.

7. Vector Spaces

7.1 Vector Spaces and Subspaces
7.2 Span and Linear Independence
7.3 Basis and Dimension

Abstract vector spaces are first introduced in Chapter 7. This relatively late introduction allows students time to internalize key concepts such as span, linear independence, and subspaces before being presented with the challenge of abstract vector spaces. To further smooth this transition, definitions and theorems in this chapter typically include specific references to analogs in earlier chapters to reinforce connections. Since most proofs are similar to those given in Euclidean space, many are left as homework exercises. Making the parallels between Euclidean space and abstract vector spaces very explicit helps students more easily assimilate this material.

The order of Chapter 7 and Chapter 8 can be reversed, so if time is limited, Chapter 8 can be covered immediately after Chapter 6. However, if both Chapters 7 and 8 are going to be covered, it is recommended that Chapter 7 be covered first so that this new, more abstract material is not appearing at the end of the course.

8. Orthogonality

8.1 Dot Products and Orthogonal Sets

8.2 Projection and the Gram-Schmidt Process

8.3 Diagonalizing Symmetric Matrices and QR Factorization

8.4 The Singular Value Decomposition*

8.5 Least Squares Regression*

In Chapter 8, the context shifts back to Euclidean space and treats topics that are more computational than conceptual. Chapter 7 is placed before Chapter 8 to allow for an introduction to abstract vector spaces that does not come at the end of the course, and to a degree preserves the chapter alternation between computational and conceptual. However, the two chapters are interchangeable.

9. Linear Transformations

9.1 Definition and Properties

9.2 Isomorphisms

9.3 The Matrix of a Linear Transformation

9.4 Similarity

The focus of Chapter 9 shifts back to abstract vector spaces, with a general development of linear transformations. As in Chapter 7, there is some deliberate redundancy between the material in Chapter 9 and that presented in earlier chapters. Explicit references to earlier analogous definitions and theorems are provided to reinforce connections and improve understanding.

10. Inner Product Spaces

10.1 Inner Products

10.2 The Gram–Schmidt Process Revisited

10.3 Applications of Inner Products*

Chapter 10 is in the context of abstract vector spaces. The content is somewhat parallel to the first two sections of Chapter 8, with explicit analogs noted. The first section defines the inner product and inner product spaces and gives numerous examples of each. The second section generalizes the notion of projection and the Gram–Schmidt process to inner product spaces, and the last section provides applications of inner products. For the most part, Chapter 10 is independent of Chapter 9 (except for a small number of exercise references to linear transformations), so Chapter 10 can be covered without covering Chapter 9.

11. Additional Topics and Applications*

11.1 Quadratic Forms

11.2 Positive Definite Matrices

11.3 Constrained Optimization

11.4 Complex Vector Spaces

11.5 Hermitian Matrices

Chapter 11 provides a collection of topics and applications that most instructors consider optional but that are nonetheless important and interesting. These can be inserted at the end of a course as desired.

Course Coverage

Most schools teach linear algebra as a semester-long course that meets 3 hours per week. This does not allow enough time to cover everything in this book, so decisions about coverage are required.

The dependencies among chapters are fairly straightforward.

- The first six chapters are designed to be covered in order, although there are some optional sections (flagged in the table of contents) that can be skipped.
- The order of Chapter 7 and Chapter 8 can be interchanged.
- The order of Chapter 9 and Chapter 10 can be interchanged (except for a small number of exercises in Chapter 10 that use linear transformations).
- Chapter 9 assumes Chapter 7, and Chapter 10 assumes Chapter 7 and Chapter 8.

Below are a few options for course coverage. Note that some sections or even subsections can be omitted to fine-tune the course to local needs.

- **Modest Pace:** Chapters 1–8. This course covers all key concepts in the context of Euclidean space and provides an introduction to abstract vector spaces.
- **Intermediate Pace:** Chapters 1–9 or Chapters 1–8 and 10. This includes everything from the Modest Pace course and either linear transformations on abstract vector spaces (Chapter 9) or inner product spaces (Chapter 10).
- **Brisk Pace:** Chapters 1–10. This will include everything from the Modest Pace course, as well as both linear transformations on abstract vector spaces and inner product spaces. This is roughly the syllabus we follow at here at the University of Virginia, although we omit a few optional sections and we give exams in the evening, which makes available more lecture time. (A detailed list of sections that we cover is available on request from the author.)

Chapter Transitions

Each chapter opens with the picture of a bridge and a brief description. These are included for a number of reasons. One reason is to provide a metaphor: linear algebra provides a bridge to higher understanding. Another reason is the clear engineering component. Even if the mathematics behind bridge building is not discussed, it is implicitly present. Finally, we discovered that many of the text reviewers are fans of bridges, with a number willing to nominate favorites for inclusion here. We assume that they are not alone.

Supplements for Instructors

Instructor's Solutions Manual

Instructor's Resource Manual

Test Bank

PowerPoint Slides

Matlab Manual

Maple Manual

Mathematica Manual

TI Manual

Supplements for Students

Student Solutions Manual
Matlab Manual
Maple Manual
Mathematica Manual
TI Manual

Media

www.yourmathportal.com MathPortal combines a fully customizable e-Book, exceptional student and instructor resources, and an online homework assignment center. Included are algorithmically generated exercises, as well as diagnostic quizzes; interactive applets; student solutions; online quizzes; Mathematica®, Maple™, and MathLab® technology manuals; and homework management tools—all in one affordable, easy-to-use, fully customizable learning space.

WeBWorK http://webwork.maa.org W. H. Freeman offers algorithmically generated questions (with full solutions) through this free open source online homework system developed at the University of Rochester.

Additional Media Resources

Online e-Book. In addition to being integrated into MathPortal, the e-Book for Holt's *Linear Algebra with Applications* is available as a stand-alone resource to be used with, or instead of, the printed text. Access can also be packaged with the text at a substantial discount.

Online Study Center. The Online Study Center helps students pinpoint where to focus their study and provides a variety of resources tied to the text. The features include:

- *Personalized Study Plan:* Students take preliminary quizzes and are directed to specific text sections and resources to review the questions they missed.

- *Premium Resources*, including:
  - Interactive Applets
  - Mathematica Manual
  - Maple Manual
  - MathLab Manual
  - Student Solutions Manual
  - Instructor Resources

SolutionMaster With this innovative online tool, instructors can provide selected secure solutions for any assignment from the textbook to their students.

i>clicker The hassle-free solution created for educators by educators. For more information, visit **www.iclicker.com.**

Acknowledgments

A large number of experienced mathematics faculty were generous in sharing their thoughts as this book was developed. Their input contributed in countless ways to the improvement of this book. I gratefully acknowledge the comments and suggestions from the following individuals.

Lowell Abrams, *The George Washington University*

Maria Theodora Acosta, *Texas State University-San Marcos*

Ulrich Albrecht, *Auburn University*

John Alongi, *Northwestern University*

Paolo Aluffi, *Florida State University*

Dorothy C. Attaway, *Boston University*

Chris Bernhardt, *Fairfield University*

Eddie Boyd Jr., *University of Maryland Eastern Shore*

Natasha Bozovic, *San Jose State University*

Mary E. Bradley, *University of Louisville*

Terry J. Bridgman, *Colorado School of Mines*

Fernando Burgos, Ph.D., *University of South Florida*

Robert E. Byerly, *Texas Tech University*

Nancy Childress, *Arizona State University*

Peter Cholak, *University of Notre Dame*

Matthew T. Clay, *University of Arkansas*

Adam Coffman, *Indiana University-Purdue Fort Wayne*

Ray E. Collings, *Georgia Perimeter College*

Ben W. Crain, *George Mason University*

Curtis Crawford, *Columbia Basin College*

Xianzhe Dai, *University of California, Santa Barbara*

Alain D'Amour, *Southern Connecticut State University*

James W. Daniel, *University of Texas at Austin*

Gregory Daubenmire, *Las Positas College*

Donald Davis, *Lehigh University*

Tristan Denley, *Austin Peay State University*

Yssa DeWoody, *Texas Tech University*

Caren Diefenderfer, *Hollins University*

Javid Dizgam, *University of Illinois Urbana-Champaign*

Edward Tauscher Dobson, *Mississippi State University*

Neil Malcolm Donaldson, *University of California, Irvine*

Alina Raluca Dumitru, *University of North Florida*

Della Duncan-Schnell, *California State University, Fresno*

Alexander Dynin, *The Ohio State University*

Daniel J. Endres, *University of Central Oklahoma*

Alex Feingold, *SUNY Binghamton*

John Fink, *Kalamazoo College*

Timothy J. Flaherty, *Carnegie Mellon University*

Bill Fleissner, *University of Kansas*

Chris Francisco, *Oklahoma State University*

Natalie P. Frank, *Vassar College*

Chris Frenzen, *Naval Postgraduate School*

Anda Gadidov, *Kennesaw State University*

Scott Glasgow, *Brigham Young University*

Jay Gopalakrishnan, *Portland State University*

Anton Gorodetski, *University of California-Irvine*

John Goulet, *Worcester Polytechnic Institute*

Barry Griffiths, *Central Florida University*

William Hager, *University of Florida*

Patricia Hale, *California State Polytechnic University, Pomona*

Chungsim Han, *Baldwin-Wallace College*

James Hartsman, *Colorado School of Mines*

Willy Hereman, *Colorado School of Mines*

Konrad J. Heuvers, *Michigan Technological University*

Allen Hibbard, *Central College*

Rudy Horne, *Morehouse College*

Mark Hunacek, *Iowa State University*

Kevin James, *Clemson University*

Bin Jiang, *Portland State University*

Naihuan Jing, *North Carolina State University*

Raymond L. Johnson, *Rice University*

Thomas W. Judson, *Stephen F. Austin State University*

Steven Kahan, *CUNY Queens College*

Jennifer D. Key, *Clemson University*

In-Jae Kim, *Minnesota State University*

Alan Koch, *Agnes Scott College*

Joseph D. Lakey, *New Mexico State University*

Namyong Lee, *Minnesota State University*

Luen-Chau Li, *Penn State University*

Lucy Lifschitz, *University of Oklahoma*

Roger Lipsett, *Brandeis University*

Xinfeng Liu, *University of South Carolina*

Satyagopal Mandal, *University of Kansas*

Aldo J. Manfroi, *University of Illinois at Urbana-Champaign*

Judith McDonald, *Washington State University*

Douglas Bradley Meade, *University of South Carolina*

Valentin Milanov, *Fayetteville State University*

Mona Mocanasu, *Northwestern University*

Mariana Montiel, *Georgia State University*

Carrie Muir, *University of Colorado at Boulder*

Shashikant Mulay, *The University of Tennessee*

Bruno Nachtergaele, *University of California*

Ralph Oberste-Vorth, *Marshall University*

Timothy E. Olson, *University of Florida*

Boon Wee Ong, *Penn State Erie, The Behrend College*

Seth F. Oppenheimer, *Mississippi State University*
Bonsu M. Osei, *Eastern Connecticut State University*
Allison Pacelli, *Williams College*
Richard O. Pellerin, *Northern Virginia Community College*
Jack Porter, *University of Kansas*
Chuanxi Qian, *Mississippi State University*
Ernest F. Ratliff, *Texas State University*
David Richter, *West Michigan University*
John Rossi, *Virginia Tech*
Matthew Saltzman, *Clemson University*
Alicia Sevilla, *Moravian College*
Alexander Shibakov, *Tennessee Tech University*
Rick L. Smith, *University of Florida*
Katherine F. Stevenson, *California State University, Northridge*

Allan Struthers, *Michigan Technological University*
Alexey Sukhinin, *University of New Mexico*
Gnana Bhaskar Tenali, *Florida Institute of Technology*
Magdalena Toda, *Texas Tech University*
Mark Tomforde, *University of Houston*
Douglas Torrance, *University of Idaho*
Michael Tsatsomeros, *Washington State University*
Haiyan Wang, *Arizona State University*
Tamas Wiandt, *Rochester Institute of Technology*
Scott Wilson, *CUNY Queens College*
Amy Yielding, *Eastern Oregon University*
Jeong-Mi Yoon, *UH-Downtown*
John Zerger, *Catawba College*
Jianqiang Zhao, *Eckerd College*

A large round of thanks are due to all of the people associated with W. H. Freeman for their assistance and guidance during this project. These include Laura Judge, Bruce Kaplan, Tony Palermino, Frank Purcell, Steve Rigolosi, Leslie Lahr, Jorge Amaral, and Liam Ferguson. I particularly want to thank Terri Ward, who has been with this project since the beginning and displayed a remarkable level of patience with my consistently inconsistent progress.

I gratefully acknowledge the support of the University of Virginia, where I class tested portions of this book. I also thank Simon Fraser University and the IRMACS Centre, for their warm hospitality during my sabbatical visit while I completed the final draft.

Last but certainly not least, I thank my family for their ongoing and unconditional support as I wrote this book.

Systems of Linear Equations

The New River Gorge Bridge near Fayetteville, West Virginia is the world's third longest steel single-span arch bridge, and one of the highest vehicular bridges at 876 feet above the ravine floor below. Like all arch bridges, the New River Gorge Bridge transfers its weight and loads onto a horizontal thrust restrained by the abutments on both sides. Before it was completed, travelers faced a 45-minute drive along a winding road to get from one side of the New River Gorge to the other. Now it takes less than a minute. The bridge is commemorated on West Virginia's state quarter as a monumental achievement in engineering.

Bridge suggested by Matt Clay, Allegheny College (Pat & Chuck Blackley/Alamy)

There are endless applications of linear algebra in the sciences, social sciences, and business, and many are included throughout this book. Chapter 1 begins our tour of linear algebra in territory that may be familiar, systems of linear equations. In the first two sections, we develop a systematic method for finding the set of solutions to a linear system. This method can be impractical for large linear systems, so in Section 1.3 we consider numerical methods for approximating solutions that can be applied to large systems. Section 1.4 focuses on applications.

1.1 Lines and Linear Equations

The goal of this section is to provide an introduction to systems of linear equations. The following example is a good place to start. Although not complicated, it contains the essential elements of other applications and also serves as a gateway to our treatment of more general systems of linear equations.

EXAMPLE 1 Fran is designing a solar hot water system for her home. The system works by circulating a mixture of water and propylene glycol through rooftop solar panels to absorb heat, and then through a heat exchanger to heat household water (Figure 1). The glycol is included in the mixture to prevent freezing during cold weather. Table 1 shows the percentage of glycol required for various minimum temperatures.

Active, Closed-Loop Solar Water Heater

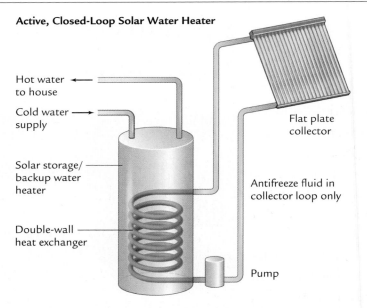

Hot water → to house

Cold water → supply

Flat plate collector

Solar storage/ backup water heater

Antifreeze fluid in collector loop only

Double-wall heat exchanger

Pump

Figure 1 Schematic of a solar hot water system. (Source: U.S. Dept. of Energy).

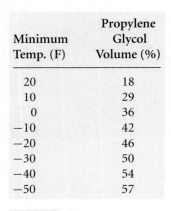

Minimum Temp. (F)	Propylene Glycol Volume (%)
20	18
10	29
0	36
−10	42
−20	46
−30	50
−40	54
−50	57

Table 1 Percentage of Glycol Required to Prevent Freezing

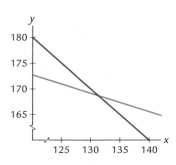

Figure 2 Graphs of $x + y = 300$ (blue) and $0.18x + 0.50y = 108$ (red) from Example 1.

The lowest the temperature ever gets at Fran's house is $0°F$. Fran can purchase solutions of water and glycol that contain either 18% glycol or 50% glycol, which she will combine for her 300-liter system. How much of each type of solution is required?

Solution To solve this problem, we start by translating the given information into equations. Let x denote the required number of liters of the 18% solution, and y the required number of liters of the 50% solution. Since the system requires a total of 300 liters, it follows that

$$x + y = 300$$

To prevent freezing at $0°F$, we must determine how much of each solution is needed for the mixture. From Table 1, we see that we need a 36% glycol mixture. Thus the total amount of glycol in the system must be $0.36(300) = 108$ liters. We will get $0.18x$ liters of glycol from the 18% solution and $0.50y$ liters of glycol from the 50% solution. This leads to a second equation,

$$0.18x + 0.50y = 108$$

Both $x + y = 300$ and $0.18x + 0.50y = 108$ are equations of lines. Figure 2 shows their graphs on the same set of axes. In our problem, we are looking for values of x and y that satisfy *both* equations, which means that the point with coordinates (x, y) will lie on the graph of *both* lines—that is, at the point of intersection of the two lines.

Instead of trying to determine the exact point of intersection from the graph, we use algebraic methods. Here are the two equations again,

$$\begin{aligned} x + \quad y &= 300 \\ 0.18x + 0.50y &= 108 \end{aligned} \qquad (1)$$

We can "eliminate" x by multiplying the first equation by -0.18 and then adding it to the second equation,

$$\begin{array}{r} -0.18x - 0.18y = -54 \\ + \quad (\quad 0.18x + 0.50y = 108) \\ \hline \Rightarrow \qquad\qquad 0.32y = 54 \end{array}$$

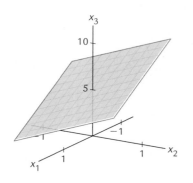

Figure 3 Graph of the solutions to $3x_1 - 2x_2 + x_3 = 5$.

Hence $y = 54/0.32 = 168.75$. Next, we substitute this back into the top equation in (1) to find x. Plugging in $y = 168.75$ gives

$$x + 168.75 = 300$$

which simplifies to $x = 131.25$. Writing our solution in the form (x, y), we have (131.25, 168.75). Referring back to Figure 2, we see that this looks like a plausible candidate for the point of intersection. We can check our answer by substituting the values $x = 131.25$ and $y = 168.75$ into the original pair of equations, to confirm that

$$131.25 + 168.75 = 300$$

and

$$0.18(131.25) + 0.50(168.75) = 108$$

This verifies that a combination of 131.25 liters of the 18% solution and 168.75 liters of the 50% solution should be used in the solar system. ∎

Systems of Linear Equations

Definition **Linear Equation**

The equations in the preceding problem are examples of **linear equations**. In general, a linear equation has the form

$$a_1 x_1 + a_2 x_2 + a_3 x_3 + \cdots + a_n x_n = b \tag{2}$$

Definition **Solution of Linear Equation**

where $a_1, a_2, \cdots, a_n$ and b are constants and $x_1, x_2, \cdots, x_n$ are variables or unknowns. A **solution** $(s_1, s_2, \cdots, s_n)$ to (2) is an ordered set of n numbers (sometimes called an *n-tuple*) such that if we set $x_1 = s_1, x_2 = s_2, \cdots, x_n = s_n$, then (2) is satisfied. That is, $(s_1, s_2, \cdots, s_n)$ is a solution to (2) if

$$a_1 s_1 + a_2 s_2 + a_3 s_3 + \cdots + a_n s_n = b$$

For example, $(-2, 5, 1, 13)$ is a solution to $3x_1 + 4x_2 - 7x_3 - 2x_4 = -19$, because

$$3(-2) + 4(5) - 7(1) - 2(13) = -19$$

Definition **Solution Set**

The **solution set** for a linear equation such as (2) consists of the set of all solutions to the equation. When the equation has two variables, the graph of the solution set is a line. In three variables, the graph of a solution set is a plane. (See Figure 3 for an example.) If $n \geq 4$, then the solution set of all points that satisfy equation (2) is called a **hyperplane**.

Definition **Hyperplane**

The set of two linear equations in (1) is an example of a *system of linear equations*. Other examples of systems of linear equations are

$$\begin{array}{rcrcrcr} -3x_1 & + & 5x_2 & - & x_3 & = & 4 \\ & - & x_2 & - & 9x_3 & = & -4 \\ 6x_1 & + & 4x_2 & - & 8x_3 & = & 11 \\ -5x_1 & - & 9x_2 & & & = & 0 \end{array} \quad \text{and} \quad \begin{array}{rcrcrcrcr} 4x_1 & - & 2x_2 & - & 8x_3 & + & 5x_4 & = & -1 \\ -x_1 & + & 7x_2 & & & + & 2x_4 & = & 13 \\ & & & & x_3 & - & 2x_4 & = & 5 \end{array} \tag{3}$$

Our standard practice is to write all systems of linear equations as shown above, aligning the variables vertically and with $x_1, x_2, \ldots$ appearing in order from left to right.

DEFINITION 1.1

Definition System of Linear Equations

▶ For brevity, we sometimes use "linear system" or "system" when referring to a system of linear equations.

A **system of linear equations** is a collection of equations of the form

$$
\begin{aligned}
a_{11}x_1 + a_{12}x_2 + a_{13}x_3 + \cdots + a_{1n}x_n &= b_1 \\
a_{21}x_1 + a_{22}x_2 + a_{23}x_3 + \cdots + a_{2n}x_n &= b_2 \\
a_{31}x_1 + a_{32}x_2 + a_{33}x_3 + \cdots + a_{3n}x_n &= b_3 \\
\vdots \qquad\qquad \vdots \qquad\qquad \vdots \qquad\quad \vdots \\
a_{m1}x_1 + a_{m2}x_2 + a_{m3}x_3 + \cdots + a_{mn}x_n &= b_m
\end{aligned}
\tag{4}
$$

When reading the coefficients, for a_{32} we say "a-three-two" instead of "a-thirty-two" because the "32" indicates that a_{32} is the coefficient from the third equation that is multiplied by x_2. For example, in the system on the right of (3) we have $a_{14} = 5$, $a_{22} = 7$, $a_{34} = -2$, and $a_{32} = 0$. Here $a_{32} = 0$ because there is no x_2 term in the third equation.

Definition Solution for Linear System, Solution Set for a Linear System

The system (4) has m equations with n unknowns. It is possible for m to be greater than, equal to, or less than n, and we will encounter all three cases. A **solution** to the linear system (4) is an n-tuple $(s_1, s_2, \cdots, s_n)$ that satisfies every equation in the system. The collection of all solutions to a linear system is called the **solution set** for the system.

In Example 1, there was exactly one solution to the linear system. This is not always the case.

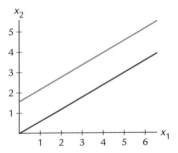

Figure 4 Graphs of $6x_1 - 10x_2 = 0$ (blue) and $-3x_1 + 5x_2 = 8$ (red) from Example 2.

EXAMPLE 2   Find all solutions to the system of linear equations

$$
\begin{aligned}
6x_1 - 10x_2 &= 0 \\
-3x_1 + 5x_2 &= 8
\end{aligned}
\tag{5}
$$

Solution   We will proceed as we did in Example 1, by eliminating a variable. This time we multiply the first equation by $\frac{1}{2}$ and then add it to the second,

$$
\begin{array}{r}
3x_1 - 5x_2 = 0 \\
+ \quad (-3x_1 + 5x_2 = 8\ ) \\
\hline
\Rightarrow \quad 0 = 8
\end{array}
$$

▶ This explanation gives an example of a mathematical proof technique called "proof by contradiction." You can read about this and other methods of proof in the appendix "Reading and writing proofs" posted on the text website. (See the Preface for the Web address.)

The equation $0 = 8$ tells us that there are *no* solutions to the system. Why? Because if there were values of x_1 and x_2 that satisfied both the equations in (5), then we could plug them in, work through the above algebraic steps with these values in place, and *prove* that $0 = 8$, which we know is not true. So, it must be that our original assumption that there are values of x_1 and x_2 that satisfy (5) is false, and therefore we can conclude that the system has no solutions.   ■

The graphs of the two equations in Example 2 are parallel lines (see Figure 4). Since the lines do not have any points in common, there cannot be values that satisfy both equations, confirming what we discovered algebraically.

Definition Consistent Linear System, Inconsistent Linear System

If a linear system has at least one solution, then we say that it is **consistent**. If not (as in Example 2), then it is **inconsistent**.

EXAMPLE 3   Find all solutions to the system of linear equations

$$
\begin{aligned}
4x_1 + 10x_2 &= 14 \\
-6x_1 - 15x_2 &= -21
\end{aligned}
\tag{6}
$$

Solution This time, we multiply the first equation by $\frac{3}{2}$ and then add,

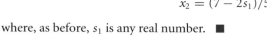

$$\begin{array}{r} 6x_1 + 15x_2 = 21 \\ +\quad (-6x_1 - 15x_2 = -21\,) \\ \hline \Rightarrow 0 = 0 \end{array}$$

Unlike Example 2, where we ended up with an equation that had no solutions, here we find ourselves with the equation $0 = 0$ that is satisfied by *any* choices of x_1 and x_2. This tells us that the relationship between x_1 and x_2 is the same in both equations. In this case we select one of the equations (either will work) and solve for x_1 in terms of x_2, which gives us

$$x_1 = \frac{7 - 5x_2}{2}$$

For *every* choice of x_2 there will be a corresponding choice of x_1 that satisfies the original system (6). Therefore there are infinitely many solutions. To avoid confusing variables with values satisfying the linear system, we describe the solutions to (6) by

$$\begin{aligned} x_1 &= (7 - 5s_1)/2 \\ x_2 &= s_1 \end{aligned} \tag{7}$$

where s_1 is called a **free parameter** and can be any real number. This is known as the **general solution** because it gives all solutions to the system of equations.

We note that (7) is not the only way to describe the solutions. If we solve for x_2 instead of x_1, then we arrive at the formulation of the general solution

$$\begin{aligned} x_1 &= s_1 \\ x_2 &= (7 - 2s_1)/5 \end{aligned}$$

where, as before, s_1 is any real number. ∎

Figure 5 shows the graphs of the two equations in (6). It looks like something is missing, but there is only one line because the two equations have the same graph. Since the graphs coincide, they have infinitely many points in common, which is consistent with our algebraic conclusion that there are infinitely many solutions to the system of linear equations.

In Examples 1–3, we have seen that a linear system can have a single solution, no solutions, or infinitely many solutions.

Figure 6 shows that our examples illustrate all possibilities for two lines, which are intersecting in exactly one point, being parallel and having no points in common, or coinciding and having infinitely many points of intersection. Thus it follows that a system of two linear equations with two variables can have zero, one, or infinitely many solutions.

Definition Free Parameter, General Solution

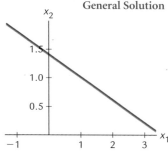

Figure 5 Graphs of $4x_1 + 10x_2 = 14$ (blue) and $-6x_1 - 15x_2 = -21$ (red) from Example 3.

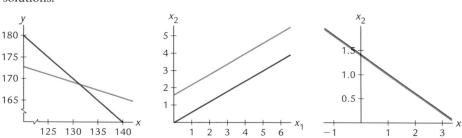

Figure 6 Graphs of equations in Examples 1–3.

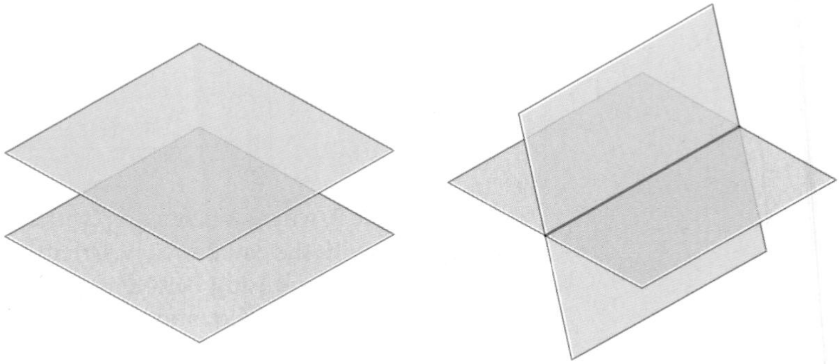

(a) Parallel planes, no points in common.

(b) Planes intersect, infinitely many points in common.

Figure 7 Graphs of systems of two equations with three variables.

Now consider systems of linear equations with three variables. Recall that the graph of the solutions for each equation is a plane. To explore the solutions such a system can have, you can experiment by using a few pieces of cardboard to represent planes.

Starting with two pieces, you will quickly discover that the only two possibilities for the number of points of intersection is either none or infinitely many. (See Figure 7.)

This geometric observation is equivalent to the algebraic statement that a system of two linear equations in three variables has either no solutions or infinitely many solutions.

Now try three pieces of cardboard. There are more possible configurations, some shown in Figure 8.

This time, we see that the number of points of intersection can be zero, one, or infinitely many. (Note that this also held for a pair of lines.) In fact, this turns out to be true in general, not only for planes but also for solution sets in higher dimensions. The equivalent statement for systems of linear equations is contained in Theorem 1.2.

▶ A **theorem** is a mathematical statement that has been rigorously proved to be true. As we progress through this book, theorems will serve to organize our expanding body of linear algebra knowledge.

THEOREM 1.2

A system of linear equations has no solutions, exactly one solution, or infinitely many solutions.

We will prove this theorem at the end of the next section.

Finding Solutions: Triangular Systems

▶ In Section 1.2 we show how to generalize the results given here to find the solutions to *any* linear system.

Now that we know how many solutions a linear system can have, we turn to the problem of *finding* the solutions. For the remainder of this section, we concentrate on special types of linear systems.

Consider the two systems below. Although not obvious, these systems have exactly the same solution set.

$$\begin{aligned}
-2x_1 + 4x_2 + 11x_3 - 4x_4 &= 4 \\
3x_1 - 6x_2 - 15x_3 + 10x_4 &= 11 \\
2x_1 - 4x_2 - 10x_3 + 6x_4 &= 4 \\
-3x_1 + 7x_2 + 18x_3 - 13x_4 &= 1
\end{aligned}
\qquad
\begin{aligned}
x_1 - 2x_2 - 5x_3 + 3x_4 &= 2 \\
x_2 + 3x_3 - 4x_4 &= 7 \\
x_3 + 2x_4 &= -4 \\
x_4 &= 5
\end{aligned}$$

The one on the right looks easier to solve, so let's find its solutions.

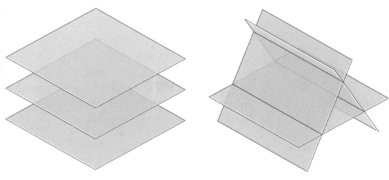

(a) Parallel planes, no points
in common to all three.

(b) Planes intersect in pairs, no
points in common to all three.

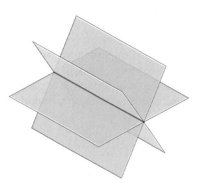

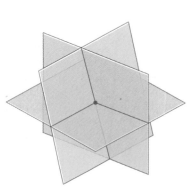

(c) Planes intersect in a line,
infinitely many points in common.

(d) Planes intersect at a point,
unique point in common.

Figure 8 Graphs of systems of three equations with three variables.

EXAMPLE 4   Find all solutions to the system of linear equations

$$\begin{aligned}
x_1 - 2x_2 - 5x_3 + 3x_4 &= 2 \\
x_2 + 3x_3 - 4x_4 &= 7 \\
x_3 + 2x_4 &= -4 \\
x_4 &= 5
\end{aligned} \tag{8}$$

Definition Back Substitution

Solution The method that we use here is called **back substitution**. Looking at the system, we see that the easiest place to start is at the bottom. Since $x_4 = 5$, substituting this back (hence the name for the method) into the next equation up gives us

$$x_3 + 2(5) = -4$$

which simplifies to $x_3 = -14$. Now we know the values of x_3 and x_4. Substituting these back into the next equation up (second from the top) gives

$$x_2 + 3(-14) - 4(5) = 7$$

so that $x_2 = 69$. Finally, we substitute the values of x_2, x_3, and x_4 back into the top equation to get

$$x_1 - 2(69) - 5(-14) + 3(5) = 2$$

which simplifies to $x_1 = 55$. Thus this system of linear equations has one solution,

$$x_1 = 55, \quad x_2 = 69, \quad x_3 = -14, \quad x_4 = 5$$

∎

Definition Leading Variable

In Example 4, each variable x_1, x_2, x_3, and x_4 appears as the first term of an equation. In a system of linear equations, a variable that appears as the first term in at least one equation is called a **leading variable**. Thus in Example 4 each of x_1, x_2, x_3, and x_4 is a leading variable. In the system

$$
\begin{aligned}
-4x_1 + 2x_2 - x_3 \qquad\quad + 3x_5 &= 7 \\
- 3x_4 + 4x_5 &= -7 \\
x_4 - 2x_5 &= 1 \\
7x_5 &= 2
\end{aligned}
\tag{9}
$$

x_1, x_4, and x_5 are leading variables, while x_2 and x_3 are not.

A key reason why the system in Example 4 is easy to solve is that every variable is a leading variable in exactly one equation. This feature is useful because as we back substitute from the bottom equation upward, at each step we are working with an equation that has only one remaining unknown variable.

The system in Example 4 is said to be in *triangular form*, with the name suggested by the triangular shape of the left side of the system. In general, a linear system is in **triangular form** (and is said to be a **triangular system**) if it has the form

Definition Triangular Form, Triangular System

$$
\begin{aligned}
a_{11}x_1 + a_{12}x_2 + a_{13}x_3 + \cdots + a_{1n}x_n &= b_1 \\
a_{22}x_2 + a_{23}x_3 + \cdots + a_{2n}x_n &= b_2 \\
a_{33}x_3 + \cdots + a_{3n}x_n &= b_3 \\
\ddots \qquad \vdots \quad &\ \ \vdots \\
a_{nn}x_n &= b_n
\end{aligned}
$$

where $a_{11}, a_{22}, \ldots, a_{nn}$ are all nonzero. It is straightforward to verify that triangular systems have the following properties.

PROPERTIES OF TRIANGULAR SYSTEMS

(a) Every variable of a triangular system is the leading variable of exactly one equation.

(b) A triangular system has the same number of equations as variables.

(c) A triangular system has exactly one solution.

Figure 9 Golden Gate Bridge.
(Photo taken by John Holt.)

EXAMPLE 5 A bowling ball dropped off the Golden Gate bridge has height H (in meters) above the water at time t (in seconds after release time) given by $H(t) = at^2 + bt + c$, where a, b, and c are constants. Using ideas from calculus, it follows that the velocity is $V(t) = 2at + b$ and the acceleration is $A(t) = 2a$. At $t = 2$, it is known that the ball has height 47.4 m, velocity -19.6 m/s, and acceleration -9.8 m/s^2. (The velocity and acceleration are negative because the ball is moving in the negative direction.) What is the height of the bridge and when does the ball hit the water?

Solution We need to find the values of a, b, and c in order to answer these questions. At time $t = 2$, we have

$$47.4 = H(2) = 4a + 2b + c, \quad -19.6 = V(2) = 4a + b, \quad -9.8 = A(2) = 2a$$

This gives us the linear system

$$\begin{array}{rrrr} 4a + 2b + c = & 47.4 \\ 4a + b = & -19.6 \\ 2a = & -9.8 \end{array} \qquad (10)$$

▶ Our model ignores forces other than gravity. For falling objects, wind resistance can be significant. We chose to drop a bowling ball to reduce the effects of wind resistance to make the model more accurate.

Back substituting as usual, we find that

$$a = -4.9, \qquad b = 0, \qquad c = 67$$

so that the height function is $H(t) = -4.9t^2 + 67$. At time $t = 0$ the ball is just starting its descent, so the bridge has height $H(0) = 67$ meters. The ball strikes the water when $H(t) = 0$, which leads to the equation

$$-4.9t^2 + 67 = 0$$

The solution is $t = \sqrt{67/4.9} \approx 3.7$ seconds after the ball is released. ■

Finding Solutions: Echelon Systems

In the next example, we consider a linear system where each variable is a leading variable for *at most* one equation. Although this system is not quite triangular, it is close enough that the solutions still can be found using back substitution.

EXAMPLE 6 Find all solutions to the system of linear equations

$$\begin{array}{rrrrr} 2x_1 - 4x_2 + 2x_3 + & x_4 = 11 \\ x_2 - & x_3 + 2x_4 = & 5 \\ 3x_4 = & 9 \end{array} \qquad (11)$$

Solution We find the solutions by back substituting, just like with a triangular system. Starting with the bottom equation yields $x_4 = 3$.

The middle equation has x_2 as the leading variable, but we do not yet have a value for x_3. We address this by setting $x_3 = s_1$, where s_1 is a free parameter. We now have values for both x_3 and x_4, which we substitute into the middle equation, giving us

$$x_2 - s_1 + 2(3) = 5$$

Thus $x_2 = -1 + s_1$. Substituting our values for x_2, x_3, and x_4 into the top equation, we have

$$2x_1 - 4(-1 + s_1) + 2s_1 + 3 = 11$$

which simplifies to $x_1 = 2 + s_1$. Therefore the general solution is

$$\begin{array}{l} x_1 = 2 + s_1 \\ x_2 = -1 + s_1 \\ x_3 = s_1 \\ x_4 = 3 \end{array}$$

Definition Echelon Form, Echelon System

where s_1 can be any real number. Note that each distinct choice for s_1 gives a new solution, so the system has infinitely many solutions. ■

▶ Note that all triangular systems are in echelon form.

The system (11) in Example 6 is in *echelon form* and is said to be an *echelon system*. Such systems have the properties given in the definition below.

DEFINITION 1.3

A linear system is in **echelon form** (and is called an **echelon system**) if

(a) Every variable is the leading variable of *at most* one equation.

(b) The system is organized in a descending "stair step" pattern so that the index of the leading variables increases from the top to bottom.

(c) Every equation has a leading variable.

For example, the systems (11) and (12) are in echelon form, but the system (9) is not, because x_4 is the leading variable of two equations.

Definition Free Variable

For a system in echelon form, any variable that is not a leading variable is called a **free variable**. For instance, x_3 is a free variable in Example 6. To find the general solution to a system in echelon form, we use the following two-step algorithm.

▶ For a system in echelon form, the total number of variables is equal to the number of leading variables plus the number of free variables.

(a) Set each free variable equal to a distinct free parameter.

(b) Back substitute to solve for the leading variables.

EXAMPLE 7 Find all solutions to the system of linear equations

$$\begin{aligned}
x_1 + 2x_2 - x_3 \qquad\quad + 3x_5 &= 7 \\
x_2 - 4x_3 \qquad + x_5 &= -2 \\
x_4 - 2x_5 &= 1
\end{aligned} \qquad (12)$$

Solution In this system x_3 and x_5 are free variables, so we set each equal to a free parameter

$$x_3 = s_1 \quad \text{and} \quad x_5 = s_2$$

It remains to determine the values of the leading variables. Substituting x_5 into the bottom equation, we have

$$x_4 - 2s_2 = 1$$

so that $x_4 = 1 + 2s_2$. Substituting our values for x_3 and x_5 into the next equation up gives

$$x_2 - 4s_1 + s_2 = -2$$

so that $x_2 = -2 + 4s_1 - s_2$. Finally, substituting in for x_2, x_3, and x_5 in the top equation, we have

$$x_1 + 2(-2 + 4s_1 - s_2) - s_1 + 3s_2 = 7$$

Hence $x_1 = 11 - 7s_1 - s_2$. Therefore the general solution is

$$\begin{aligned}
x_1 &= 11 - 7s_1 - s_2 \\
x_2 &= -2 + 4s_1 - s_2 \\
x_3 &= s_1 \\
x_4 &= 1 + 2s_2 \\
x_5 &= s_2
\end{aligned}$$

where s_1 and s_2 can be any real numbers. ∎

EXAMPLE 8   Find all solutions to the system of linear equations

$$x_1 - 4x_2 + x_3 + 5x_4 - x_5 = -3$$
$$- x_3 + 4x_4 + 3x_5 = 8$$

(13)

Solution We see that x_2, x_4, and x_5 are free variables, so we set $x_2 = s_1$, $x_4 = s_2$, and $x_5 = s_3$, where s_1, s_2, and s_3 are free parameters.

Turning to the bottom equation, we substitute in our values for x_4 and x_5, yielding the equation

$$-x_3 + 4s_2 + 3s_3 = 8$$

so that $x_3 = -8 + 4s_2 + 3s_3$. Back substituting into the top equation gives us

$$x_1 - 4s_1 + (-8 + 4s_2 + 3s_3) + 5s_2 - s_3 = -3$$

which simplifies to $x_1 = 5 + 4s_1 - 9s_2 - 2s_3$. Therefore the general solution is

$$x_1 = 5 + 4s_1 - 9s_2 - 2s_3$$
$$x_2 = s_1$$
$$x_3 = -8 + 4s_2 + 3s_3$$
$$x_4 = s_2$$
$$x_5 = s_3$$

where s_1, s_2, and s_3 can be any real numbers. ■

To sum up, there are two possibilities for a linear system in echelon form.

1. The system has no free variables. In this case, the system is also triangular and there is exactly one solution.

2. The system has at least one free variable. In this case, the general solution has free parameters and there are infinitely many solutions.

| EXERCISES |

In each exercise set, problems marked with Ⓒ are designed to be solved using a programmable calculator or computer algebra system.

1. Determine which of the points $(1, -2), (-3, -3)$, and $(-2, -3)$ lie on the line $2x_1 - 5x_2 = 9$.

2. Determine which of the points $(1, -2, 0)$, $(4, 2, 1)$, and $(2, -5, 1)$ lie in the plane $x_1 - 3x_2 + 4x_3 = 7$.

3. Determine which of the points $(-1, 2), (-2, 5)$, and $(1, -5)$ lie on both the lines $3x_1 + x_2 = -1$ and $-5x_1 + 2x_2 = 20$.

4. Determine which of the points $(3, 1)$, $(2, -4)$, and $(-4, 5)$ lie on both the lines $2x_1 - 5x_2 = 1$ and $-4x_1 + 10x_2 = -2$.

5. Determine which of the points $(1, 2, 3)$, $(1, -1, 1)$, and $(-1, -2, -6)$ satisfy the linear system

$$-2x_1 + 9x_2 - x_3 = -10$$
$$x_1 - 5x_2 + 2x_3 = 4$$

6. Determine which of the points $(1, -2, -1, 3), (-1, 0, 2, 1)$, and $(-2, -1, 4, -3)$ satisfy the linear system

$$3x_1 - x_2 + 2x_3 = 1$$
$$2x_1 + 3x_2 - x_4 = -3$$

In Exercises 7–8, determine which of (a)–(d) form a solution to the given system for any choice of the free parameter(s). (HINT: All parameters of a solution must cancel completely when substituted into each equation.)

7. $-2x_1 + 3x_2 + 2x_3 = 6$

Note: This system has only one equation.

(a) $(-3 + s_1 + s_2, s_1, s_2)$

(b) $(-3 + 3s_1 + s_2, 2s_1, s_2)$

(c) $(3s_1 + s_2, 2s_1 + 2, s_2)$

(d) $(s_1, s_2, 3 - 3s_2/2 + s_1)$

8. $3x_1 + 8x_2 - 14x_3 = 6$
$$x_1 + 3x_2 - 4x_3 = 1$$

(a) $(5 - 2s_1, 7 + 3s_1, s_1)$

(b) $(-5 - 5s_1, s_1, -(3 + s_1)/2)$

(c) $(10 + 10s_1, -3 - 2s_1, s_1)$

(d) $((6 - 4s_1)/3, s_1, -(5 - s_1)/4)$

In Exercises 9–14, find all solutions to the given system by eliminating one of the variables.

9. $3x_1 + 5x_2 = 4$
$\quad 2x_1 - 7x_2 = 13$

10. $-3x_1 + 2x_2 = 1$
$\quad 5x_1 + x_2 = -4$

11. $-10x_1 + 4x_2 = 2$
$\quad 15x_1 - 6x_2 = -3$

12. $-3x_1 + 4x_2 = 0$
$\quad 9x_1 - 12x_2 = -2$

13. $7x_1 - 3x_2 = -1$
$\quad -5x_1 + 8x_2 = 0$

14. $6x_1 - 3x_2 = 5$
$\quad -8x_1 + 4x_2 = 1$

In Exercises 15–22, determine if the given linear system is in echelon form. If so, identify the leading variables and the free variables. If not, explain why not.

15. $x_1 - x_2 = 7$
$\quad 7x_2 = 0$

16. $6x_1 - 5x_2 = 12$
$\quad -2x_1 + 7x_2 = 0$

17. $-7x_1 - x_2 + 2x_3 = 11$
$\quad 6x_3 = -1$

18. $3x_1 + 2x_2 + 7x_3 = 0$
$\quad -3x_3 = -3$
$\quad -x_2 - 4x_3 = 13$

19. $4x_1 + 3x_2 - 9x_3 + 2x_4 = 3$
$\quad 6x_2 + x_3 = -2$
$\quad -5x_2 - 8x_3 + x_4 = -4$

20. $2x_1 + 2x_3 = 12$
$\quad 12x_2 - 5x_4 = -19$
$\quad 3x_3 + 11x_4 = 14$
$\quad -x_4 = 3$

21. $-2x_1 - 3x_2 + x_3 - 13x_4 = 2$
$\quad 2x_3 = -7$

22. $-7x_1 + 3x_2 + 8x_4 - 2x_5 + 13x_6 = -6$
$\quad -5x_3 - x_4 + 6x_5 + 3x_6 = 0$
$\quad 2x_4 + 5x_5 = 1$

In Exercises 23–30, find the set of solutions for the given linear system. Note that some systems have only one equation.

23. $-5x_1 - 3x_2 = 4$
$\quad 2x_2 = 10$

24. $x_1 + 4x_2 - 7x_3 = -3$
$\quad -x_2 + 4x_3 = 1$
$\quad 3x_3 = -9$

25. $-3x_1 + 4x_2 = 2$

26. $3x_1 - 2x_2 + x_3 = 4$
$\quad -6x_3 = -12$

27. $x_1 + 5x_2 - 2x_3 = 0$
$\quad -2x_2 + x_3 - x_4 = -1$
$\quad x_4 = 5$

28. $2x_1 - x_2 + 6x_3 = -3$

29. $-2x_1 + x_2 + 2x_3 = 1$
$\quad -3x_3 + x_4 = -4$

30. $-7x_1 + 3x_2 + 8x_4 - 2x_5 + 13x_6 = -6$
$\quad -5x_3 - x_4 + 6x_5 + 3x_6 = 0$
$\quad 2x_4 + 5x_5 = 1$

In Exercises 31–34, each linear system is not in echelon form but can be put in echelon form by reordering the equations. Write the system in echelon form, and then find the set of solutions.

31. $-5x_2 = 4$
$\quad 3x_1 + 2x_2 = 1$

32. $-3x_3 = -3$
$\quad -x_2 - 4x_3 = 13$
$\quad 3x_1 + 2x_2 + 7x_3 = 0$

33. $2x_2 + x_3 - 5x_4 = 0$
$\quad x_1 + 3x_2 - 2x_3 + 2x_4 = -1$

34. $x_2 - 4x_3 + 3x_4 = 2$
$\quad x_1 - 5x_2 - 6x_3 + 3x_4 = 3$
$\quad -3x_4 = 15$
$\quad 5x_3 - 4x_4 = 10$

35. For what value(s) of k is the linear system consistent?

$$6x_1 - 5x_2 = 4$$
$$9x_1 + kx_2 = -1$$

36. For what value(s) of h is the linear system consistent?

$$6x_1 - 8x_2 = h$$
$$-9x_1 + 12x_2 = -1$$

37. Find values of h and k so that the linear system has no solutions.

$$2x_1 + 5x_2 = -1$$
$$hx_1 + 5x_2 = k$$

38. For what values of h and k does the linear system have infinitely many solutions?

$$2x_1 + 5x_2 = -1$$
$$hx_1 + kx_2 = 3$$

39. A system of linear equations is in echelon form. If there are four free variables and five leading variables, how many variables are there? Justify your answer.

40. Suppose that a system of five equations with eight unknowns is in echelon form. How many free variables are there? Justify your answer.

41. Suppose that a system of seven equations with thirteen unknowns is in echelon form. How many leading variables are there? Justify your answer.

42. A linear system is in echelon form. There are a total of nine variables, of which four are free variables. How many equations does the system have? Justify your answer.

FIND AN EXAMPLE For Exercises 43–50, find an example that meets the given specifications.

43. A linear system with three equations and three variables that has exactly one solution.

44. A linear system with three equations and three variables that has infinitely many solutions.

45. A linear system with four equations and three variables that has infinitely many solutions.

46. A linear system with three equations and four variables that has no solutions.

47. Come up with an application that has a solution found by solving an echelon linear system. Then solve the system to find the solution.

48. A linear system with two equations and two variables that has $x_1 = -1$ and $x_2 = 3$ as the only solution.

49. A linear system with two equations and three variables that has solutions $x_1 = 1$, $x_2 = 4$, $x_3 = -1$ and $x_1 = 2$, $x_2 = 5$, $x_3 = 2$.

50. A linear system with two equations and two variables that has the line $x_1 = 2x_2$ for solutions.

TRUE OR FALSE   For Exercises 51–60, determine if the statement is true or false, and justify your answer.

51. A linear system with three equations and two variables must be inconsistent.

52. A linear system with three equations and five variables must be consistent.

53. There is only one way to express the general solution for a linear system.

54. A triangular system always has exactly one solution.

55. All triangular systems are in echelon form.

56. All systems in echelon form are also triangular systems.

57. A system in echelon form can be inconsistent.

58. A system in echelon form can have more equations than variables.

59. If a triangular system has integer coefficients (including the constant terms), then the solution consists of rational numbers.

60. A system in echelon form can have more variables than equations.

61. Referring to Example 1, suppose that the minimum outside temperature is $10°$F. In this case, how much of each type of solution is required?

62. Referring to Example 1, suppose that the minimum outside temperature is $-20°$F. In this case, how much of each type of solution is required?

63. A total of 385 people attend the premiere of a new movie. Ticket prices are $11 for adults and $8 for children. If the total revenue is $3974, how many adults and children attended?

64. For tax and accounting purposes, corporations depreciate the value of equipment each year. One method used is called "linear depreciation," where the value decreases over time in a linear manner. Suppose that two years after purchase, an industrial milling machine is worth $800,000, and five years after purchase, the machine is worth $440,000. Find a formula for the machine value at time $t \geq 0$ after purchase.

65. (Calculus required) Suppose that $f(x) = a_1 e^{2x} + a_2 e^{-3x}$ is a solution to a differential equation. If we know that $f(0) = 5$ and $f'(0) = -1$ (these are the *initial conditions*), what are the values of a_1 and a_2? (HINT: $f'(x) = 2a_1 e^{2x} - 3a_2 e^{-3x}$.)

66. An investor has $100,000 and can invest in any combination of two types of bonds, one that is safe and pays 3% annually, and one that carries risk and pays 9% annually. The investor would like to keep risk as low as possible while realizing a 7% annual return. How much should be invested in each type of bond?

67. Degrees Fahrenheit (F) and Celsius (C) are related by a linear equation $C = aF + b$. Pure water freezes at $32°$F and $0°$C, and boils at $212°$F and $100°$C. Use this information to find a and b.

68. A 60-gallon bathtub is to be filled with water that is exactly $100°$F. Unfortunately, the two sources of water available are $125°$F and $60°$F. When mixed, the temperature will be a weighted average based on the amount of each water source in the mix. How much of each should be used to fill the tub as specified?

69. This problem requires about 8 nickels, 8 quarters, and a sheet of 8.5-by-11-inch paper. The goal is to estimate the diameter of each type of coin as follows: Using trial and error, find a combination of nickels and quarters that, when placed side by side, extend the height (long side) of the paper. Then do the same along the width (short side) of the paper. Use the information obtained to write two linear equations involving the unknown diameters of each type of coin, then solve the resulting system to find the diameter for each type of coin.

70. The Bixby Creek Bridge is located along California's Big Sur coast and has been featured in numerous television commercials. Suppose that a bag of concrete is projected downward from the bridge deck at an initial rate of 5 meters per second. After 3 seconds, the bag is 25.9 meters from the Bixby Creek, has a velocity of -34.4 m/s, and has an acceleration of -9.8 m/s^2. Use the model in Example 5 to find a formula for $H(t)$, the height at time t.

Bixby Creek Bridge. (Dennis Frates/Alamy)

Ⓒ In Exercises 71–76, use computational assistance to find the set of solutions to the linear system.

71. $-4x_1 + 7x_2 = -13$
$3x_1 - 5x_2 = 11$

72. $\begin{aligned} 3x_1 - 5x_2 &= 0 \\ -7x_1 - 2x_2 &= -2 \end{aligned}$

73. $\begin{aligned} 2x_1 - 5x_2 + 3x_3 &= 10 \\ 4x_2 - 9x_3 &= -7 \end{aligned}$

74. $\begin{aligned} -x_1 + 4x_2 + 7x_3 &= 6 \\ -3x_2 &= 1 \end{aligned}$

75. $\begin{aligned} -2x_1 - x_2 + 5x_3 + x_4 &= 20 \\ 3x_2 + 6x_4 &= 13 \\ -4x_3 + 7x_4 &= -6 \end{aligned}$

76. $\begin{aligned} 3x_1 + 5x_2 - x_3 - x_4 &= 17 \\ -x_2 - 6x_3 + 11x_4 &= 5 \\ 2x_3 + x_4 &= 11 \end{aligned}$

1.2 Linear Systems and Matrices

Systems of linear equations arise naturally in many applications, but the systems rarely are in echelon form. For instance, consider the following projectile motion problem. Suppose that a cannon sits on a hill and fires a ball across a flat field below. The path of the ball is known to be approximately parabolic and so can be modeled by a quadratic function $E(x) = ax^2 + bx + c$, where E is the elevation (in feet) over position x, and a, b, and c are constants.

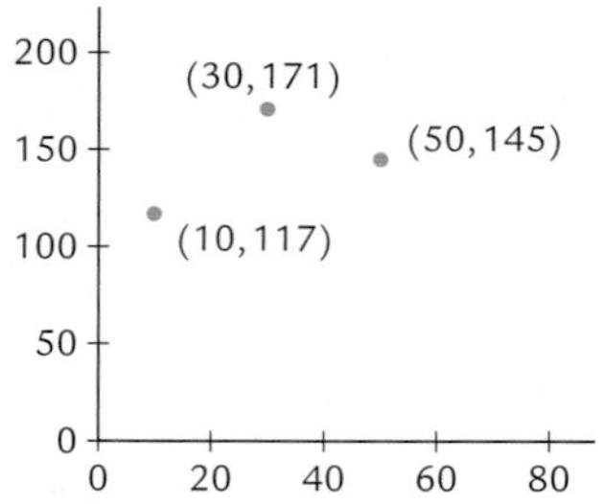

Figure 1 shows the elevation of the ball at three separate places. Since every point on its path is given by $(x, E(x))$, the data can be converted into three linear equations

$$\begin{aligned} 100a + 10b + c &= 117 \\ 900a + 30b + c &= 171 \\ 2500a + 50b + c &= 145 \end{aligned} \qquad (1)$$

Figure 1 Positions and elevations $(x, E(x))$ of an airborne cannonball.

This system is not in echelon form, so back substitution is not easy to use here. We will return to this system shortly, after developing the tools to find a solution.

The primary goal of this section is to develop a systematic procedure for transforming *any* linear system into a system that is in echelon form. The key feature of our transformation procedure is that it produces a new linear system that is in echelon form (hence solvable using back substitution) and has exactly the same set of solutions as the original system.

▶ Two linear systems are said to be **equivalent** if they have the same set of solutions.

Elementary Operations

We can transform a linear system using a sequence of **elementary operations**. Each operation produces a new system that is equivalent to the old one, so the solution set is unchanged. There are three types of elementary operations.

Definition Elementary Operations

1. Interchange the position of two equations.

▶ The symbol $\sim$ indicates the transformation from one linear system to an equivalent linear system.

This amounts to nothing more than rewriting the system of equations. For example, we exchange the places of the first and second equations in the following system.

$$\begin{aligned} 3x_1 - 5x_2 - 8x_3 &= -4 \\ x_1 + 2x_2 - 4x_3 &= 5 \\ -2x_1 + 6x_2 + x_3 &= 3 \end{aligned} \sim \begin{aligned} x_1 + 2x_2 - 4x_3 &= 5 \\ 3x_1 - 5x_2 - 8x_3 &= -4 \\ -2x_1 + 6x_2 + x_3 &= 3 \end{aligned}$$

2. Multiply an equation by a nonzero constant.

▶ Verifying that each elementary operation produces an equivalent linear system is left as Exercise 56.

For example, here we multiply the third equation by -2.

$$\begin{aligned} x_1 + 2x_2 - 4x_3 &= 5 \\ 3x_1 - 5x_2 - 8x_3 &= -4 \\ -2x_1 + 6x_2 + x_3 &= 3 \end{aligned} \sim \begin{aligned} x_1 + 2x_2 - 4x_3 &= 5 \\ 3x_1 - 5x_2 - 8x_3 &= -4 \\ 4x_1 - 12x_2 - 2x_3 &= -6 \end{aligned}$$

3. Add a multiple of one equation to another.

For this operation, we multiply one of the equations by a constant and then add it to another equation, replacing the latter with the result. For example, below we multiply the top equation by -4 and add it to the bottom equation, replacing the bottom equation with the result.

$$
\begin{aligned}
x_1 + 2x_2 - 4x_3 &= 5 \\
3x_1 - 5x_2 - 8x_3 &= -4 \\
4x_1 - 12x_2 - 2x_3 &= -6
\end{aligned}
\quad \sim \quad
\begin{aligned}
x_1 + 2x_2 - 4x_3 &= 5 \\
3x_1 - 5x_2 - 8x_3 &= -4 \\
-20x_2 + 14x_3 &= -26
\end{aligned}
$$

The third operation may look familiar. It is similar to the method used in the first three examples of Section 1.1 to eliminate a variable. Note that this is exactly what happened here, with the lower left coefficient becoming zero, transforming the system closer to echelon form. This illustrates a single step of our basic strategy for transforming any linear system into a system that is in echelon form.

▶ Generic linear system

$$
\begin{aligned}
a_{11}x_1 + a_{12}x_2 + \cdots + a_{1n}x_n &= b_1 \\
a_{21}x_1 + a_{22}x_2 + \cdots + a_{2n}x_n &= b_2 \\
a_{31}x_1 + a_{32}x_2 + \cdots + a_{3n}x_n &= b_3 \\
\vdots \qquad \vdots \qquad\qquad \vdots \qquad \vdots \\
a_{m1}x_1 + a_{m2}x_2 + \cdots + a_{mn}x_n &= b_m
\end{aligned}
$$

EXAMPLE 1   Find the set of solutions to the system of linear equations

$$
\begin{aligned}
x_1 - 3x_2 + 2x_3 &= -1 \\
2x_1 - 5x_2 - x_3 &= 2 \\
-4x_1 + 13x_2 - 12x_3 &= 11
\end{aligned}
$$

Solution We begin by focusing on the variable x_1 in each equation. Our goal is to transform the system to echelon form, so we want to eliminate the x_1 terms in the second and third equations. This will leave x_1 as the leading variable in only the top equation.

NOTE: Going forward, we identify coefficients using the notation for a generic system of equations introduced in Section 1.1 and shown again in the margin.

- **Add a multiple of one equation to another** (focus on x_1).

We need to transform a_{21} and a_{31} to 0. We do this in two parts. Since $a_{21} = 2$, if we take -2 times the first equation and add it to the second, then the resulting coefficient on x_1 will be $(-2) \cdot 1 + 2 = 0$, which is what we want.

$$
\begin{aligned}
x_1 - 3x_2 + 2x_3 &= -1 \\
2x_1 - 5x_2 - x_3 &= 2 \\
-4x_1 + 13x_2 - 12x_3 &= 11
\end{aligned}
\quad \sim \quad
\begin{aligned}
x_1 - 3x_2 + 2x_3 &= -1 \\
x_2 - 5x_3 &= 4 \\
-4x_1 + 13x_2 - 12x_3 &= 11
\end{aligned}
$$

The second part is similar. This time, since $(4) \cdot 1 - 4 = 0$, we multiply 4 times the first equation and add it to the third.

$$
\begin{aligned}
x_1 - 3x_2 + 2x_3 &= -1 \\
x_2 - 5x_3 &= 4 \\
-4x_1 + 13x_2 - 12x_3 &= 11
\end{aligned}
\quad \sim \quad
\begin{aligned}
x_1 - 3x_2 + 2x_3 &= -1 \\
x_2 - 5x_3 &= 4 \\
x_2 - 4x_3 &= 7
\end{aligned}
$$

With these steps complete, the x_1 terms in the second and third equations are gone, exactly as we want.

Next, we focus on the x_2 coefficients. Since our goal is to reach echelon form, we do not care about the coefficient on x_2 in the top equation, so we concentrate on the second and third equations.

- **Add a multiple of one equation to another** (focus on x_2).

  Here we need to transform a_{32} to 0. Since $(-1) \cdot 1 + 1 = 0$, we multiply -1 times the second equation and add the result to the third equation.

▶ Using only the second and third equations avoids reintroducing x_1 into the third equation.

$$
\begin{aligned}
x_1 - 3x_2 + 2x_3 &= -1 \\
x_2 - 5x_3 &= 4 \\
x_2 - 4x_3 &= 7
\end{aligned}
\quad \sim \quad
\begin{aligned}
x_1 - 3x_2 + 2x_3 &= -1 \\
x_2 - 5x_3 &= 4 \\
x_3 &= 3
\end{aligned}
$$

The system is now in echelon (indeed, triangular) form, and using back substitution we can easily show that the solution (we know there is only one) is $x_1 = 50$, $x_2 = 19$, and $x_3 = 3$. To check our solution, we plug these values into the original system.

$$
\begin{aligned}
1(50) - 3(19) + 2(3) &= -1 \\
2(50) - 5(19) - 1(3) &= 2 \\
-4(50) + 13(19) - 12(3) &= 11
\end{aligned}
$$

∎

EXAMPLE 2 Find the set of solutions to the linear system 1 from the start of the section,

$$
\begin{aligned}
100a + 10b + c &= 117 \\
900a + 30b + c &= 171 \\
2500a + 50b + c &= 145
\end{aligned}
$$

Solution We follow the same procedure as in the previous example.

- **Add a multiple of one equation to another** (focus on x_1).

  We need to transform a_{21} and a_{31} to 0. Since $a_{21} = 900$, we multiply the first equation by -9 and add it to the second, so that

$$
\begin{aligned}
100a + 10b + c &= 117 \\
900a + 30b + c &= 171 \\
2500a + 50b + c &= 145
\end{aligned}
\quad \sim \quad
\begin{aligned}
100a + 10b + c &= 117 \\
-60b - 8c &= -882 \\
2500a + 50b + c &= 145
\end{aligned}
$$

The second part is similar. We multiply the first equation by -25 and add it to the third.

$$
\begin{aligned}
100a + 10b + c &= 117 \\
-60b - 8c &= -882 \\
2500a + 50b + c &= 145
\end{aligned}
\quad \sim \quad
\begin{aligned}
100a + 10b + c &= 117 \\
-60b - 8c &= -882 \\
-200b - 24c &= -2780
\end{aligned}
$$

- **Multiply an equation by a nonzero constant** (focus on x_2).

  Here we multiply the third equation by -0.3, so that the coefficients on b match up (other than sign).

$$
\begin{aligned}
100a + 10b + c &= 117 \\
-60b - 8c &= -882 \\
-200b - 24c &= -2780
\end{aligned}
\quad \sim \quad
\begin{aligned}
100a + 10b + c &= 117 \\
-60b - 8c &= -882 \\
60b + 7.2c &= 834
\end{aligned}
$$

- **Add a multiple of one equation to another** (focus on x_2).

  Thanks to the previous step, we need only add the second equation to the third to transform a_{32} to 0.

$$
\begin{aligned}
100a + 10b + c &= 117 \\
-60b - 8c &= -882 \\
60b + 7.2c &= 834
\end{aligned}
\quad \sim \quad
\begin{aligned}
100a + 10b + c &= 117 \\
-60b - 8c &= -882 \\
-0.8c &= -48
\end{aligned}
$$

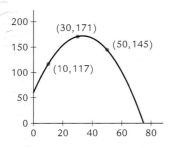

Figure 2 Cannonball data and the graph of the model.

The system is now in triangular form. Using back substitution, we can show that the solution is $a = 0.1$, $b = 6.7$, and $c = 60$, which gives us $E(x) = 0.1x^2 + 6.7x + 60$. Figure 2 shows a graph of the model together with the known points. ■

Matrices and the Augmented Matrix

In the preceding examples, as we manipulated the equations the variables just served as placeholders. One way to simplify our work is by transferring the coefficients to a **matrix**, which for the moment we can think of as a rectangular table of numbers. When a matrix contains all the coefficients of a linear system, including the constant terms on the right side of each equation, it is called an **augmented matrix**. For instance, the system in Example 1 is transferred to an augmented matrix by

Definition Matrix

Definition Augmented Matrix

Linear System

$$\begin{aligned} x_1 - 3x_2 + 2x_3 &= -1 \\ 2x_1 - 5x_2 - x_3 &= 2 \\ -4x_1 + 13x_2 - 12x_3 &= 11 \end{aligned}$$

Augmented Matrix

$$\begin{bmatrix} 1 & -3 & 2 & -1 \\ 2 & -5 & -1 & 2 \\ -4 & 13 & -12 & 11 \end{bmatrix}$$

Definition Elementary Row Operations

The three elementary operations that we performed on equations can be translated into equivalent **elementary row operations** for matrices.[1]

ELEMENTARY ROW OPERATIONS

1. Interchange two rows.
2. Multiply a row by a nonzero constant.
3. Add a multiple of one row to another.

Definition Equivalent Matrices

Borrowing from the terminology for systems of equations, we say that two matrices are **equivalent** if one can be obtained from the other through a sequence of elementary row operations. Hence equivalent augmented matrices correspond to equivalent linear systems.

When discussing matrices, the rows are numbered from top to bottom, and the columns are numbered from left to right. A **zero row** is a row consisting entirely of zeros, and a **nonzero row** contains at least one nonzero entry. The terms **zero column** and **nonzero column** are similarly defined.

Definition Zero Row, Zero Column

In the examples that follow, we transfer the system of equations to an augmented matrix, but our goal is the same as before, to find an equivalent system in echelon form.

EXAMPLE 3   Find all solutions to the system of linear equations

$$\begin{aligned} 2x_1 - 3x_2 + 10x_3 &= -2 \\ x_1 - 2x_2 + 3x_3 &= -2 \\ -x_1 + 3x_2 + x_3 &= 4 \end{aligned}$$

Solution   We begin by converting the system to an augmented matrix:

$$\begin{bmatrix} 2 & -3 & 10 & -2 \\ 1 & -2 & 3 & -2 \\ -1 & 3 & 1 & 4 \end{bmatrix}$$

---

[1]The plural of matrix is matrices.

► As with linear systems, we use the symbol $\sim$ to indicate that two matrices are equivalent.

- **Interchange rows** (focus on column 1).

    We focus on the first column of the matrix, which contains the coefficients of x_1. Although this step is not required, exchanging Row 1 and Row 2 will move a 1 into the upper left position and avoid the early introduction of fractions.

► We express this operation compactly as $R_1 \Leftrightarrow R_2$.

$$\begin{bmatrix} 2 & -3 & 10 & -2 \\ 1 & -2 & 3 & -2 \\ -1 & 3 & 1 & 4 \end{bmatrix} \sim \begin{bmatrix} 1 & -2 & 3 & -2 \\ 2 & -3 & 10 & -2 \\ -1 & 3 & 1 & 4 \end{bmatrix}$$

- **Add a multiple of one row to another** (focus on column 1).

    To transform the system to echelon form, we need to introduce zeros in the first column below Row 1. This requires two operations. Focusing first on Row 2, since $(-2)(1) + 2 = 0$, we add -2 times Row 1 to Row 2 and replace Row 2 with the result.

► We express this operation compactly as $-2R_1 + R_2 \Rightarrow R_2$.

$$\begin{bmatrix} 1 & -2 & 3 & -2 \\ 2 & -3 & 10 & -2 \\ -1 & 3 & 1 & 4 \end{bmatrix} \sim \begin{bmatrix} 1 & -2 & 3 & -2 \\ 0 & 1 & 4 & 2 \\ -1 & 3 & 1 & 4 \end{bmatrix}$$

Focusing now on Row 3, since $(1)(1) + (-1) = 0$ we add 1 times Row 1 to Row 3 and replace Row 3 with the result.

► We express this operation compactly as $R_1 + R_3 \Rightarrow R_3$.

$$\begin{bmatrix} 1 & -2 & 3 & -2 \\ 0 & 1 & 4 & 2 \\ -1 & 3 & 1 & 4 \end{bmatrix} \sim \begin{bmatrix} 1 & -2 & 3 & -2 \\ 0 & 1 & 4 & 2 \\ 0 & 1 & 4 & 2 \end{bmatrix}$$

- **Add a multiple of one row to another** (focus on column 2).

    With the first column complete, we move down to the second row and to the right to the second column. Since $(-1)(1) + (1) = 0$, we add -1 times Row 2 to Row 3 and replace Row 3 with the result.

► We express this operation compactly as $-R_2 + R_3 \Rightarrow R_3$.

$$\begin{bmatrix} 1 & -2 & 3 & -2 \\ 0 & 1 & 4 & 2 \\ 0 & 1 & 4 & 2 \end{bmatrix} \sim \begin{bmatrix} 1 & -2 & 3 & -2 \\ 0 & 1 & 4 & 2 \\ 0 & 0 & 0 & 0 \end{bmatrix}$$

We now extract the transformed system of equations from the matrix. The row of zeros indicates that one of the equations in the transformed system is $0 = 0$. Since any choice of values for the variables will satisfy $0 = 0$, this equation contributes no information about the solution set and so can be ignored. The new equivalent system is therefore

$$\begin{aligned} x_1 - 2x_2 + 3x_3 &= -2 \\ x_2 + 4x_3 &= 2 \end{aligned}$$

Back substitution can be used to show that the general solution is

$$\begin{aligned} x_1 &= 2 - 11s_2 \\ x_2 &= 2 - 4s_1 \\ x_3 &= s_1 \end{aligned}$$

where s_1 can be any real number. We can substitute into the original system to verify our solution.

$$\begin{aligned} 2(2 - 11s_1) - 3(2 - 4s_1) + 10s_1 &= 4 - 22s_1 - 6 + 12s_1 + 10s_1 = -2, \\ (2 - 11s_1) - 2(2 - 4s_1) + 3s_1 &= 2 - 11s_1 - 4 + 8s_1 + 3s_1 = -2, \\ -(2 - 11s_1) + 3(2 - 4s_1) + s_1 &= -2 + 11s_1 + 6 - 12s_1 + s_1 = 4 \end{aligned}$$ ∎

Gaussian Elimination

Definition Gaussian
Elimination
Definition Echelon Form
Definition Leading Term

The procedure that we used in Example 3 is known as **Gaussian elimination**. The resulting matrix is said to be in **echelon form** (or **row echelon form**) and will have the properties given in Definition 1.4 below. In the definition, the **leading term** of a row is the leftmost nonzero term in that row, and a row of all zeros has no leading term.

DEFINITION 1.4

A matrix is in **echelon form** if

(a) Every leading term is in a column to the left of the leading term of the row below it.

(b) Any zero rows are at the bottom of the matrix.

▶ Gaussian elimination was originally discovered by Chinese mathematicians over 2000 years ago. It is named in honor of German mathematician Carl Friedrich Gauss, who independently discovered the method and introduced it to the West in the nineteenth century.

Note that the first condition in the definition implies that a matrix in echelon form will have zeros filling out the column below each of the leading terms. Examples of matrices in echelon form are

$$\begin{bmatrix} 5 & 1 & -4 & 0 & 9 & 2 \\ 0 & 2 & -3 & -6 & 7 & 31 \\ 0 & 0 & 0 & -2 & 4 & 9 \end{bmatrix} \quad \text{and} \quad \begin{bmatrix} -1 & 2 & 3 & -2 & 17 & 9 & 7 \\ 0 & 0 & 9 & -6 & 26 & 3 & -6 \\ 0 & 0 & 0 & 0 & 0 & -3 & 0 \\ 0 & 0 & 0 & 0 & 0 & 0 & 4 \\ 0 & 0 & 0 & 0 & 0 & 0 & 0 \end{bmatrix} \quad (2)$$

Definition Pivot Positions

Definition Pivot Columns, Pivot

For a matrix in echelon form, the **pivot positions** are those that contain a leading term. The entries in the pivot positions for the matrices in (2) are shown in boldface. The **pivot columns** are the columns that contain pivot positions, and a **pivot** is a nonzero number in a pivot position that is used during row operations to produce zeros.

In what follows, it will be handy to have a general matrix to refer to when talking about entries in specific positions. We adopt a notation similar to that for a general system of equations given in (4) of Section 1.1,

$$\begin{bmatrix} a_{11} & a_{12} & a_{13} & \cdots & a_{1n} & b_1 \\ a_{21} & a_{22} & a_{23} & \cdots & a_{2n} & b_2 \\ a_{31} & a_{32} & a_{33} & \cdots & a_{3n} & b_3 \\ \vdots & \vdots & \vdots & & \vdots & \vdots \\ a_{m1} & a_{m2} & a_{m3} & \cdots & a_{mn} & b_m \end{bmatrix}$$

In the previous example, the linear system had the same number of equations and variables. However, this is not required.

EXAMPLE 4 Find all solutions to the system of linear equations

$$\begin{aligned} 6x_3 \quad\quad + 19x_5 + 11x_6 &= -27 \\ 3x_1 + 12x_2 + 9x_3 - 6x_4 + 26x_5 + 31x_6 &= -63 \\ x_1 + 4x_2 + 3x_3 - 2x_4 + 10x_5 + 9x_6 &= -17 \\ -x_1 - 4x_2 - 4x_3 + 2x_4 - 13x_5 - 11x_6 &= 22 \end{aligned}$$

Solution The augmented matrix for this system is

$$\begin{bmatrix} 0 & 0 & 6 & 0 & 19 & 11 & -27 \\ 3 & 12 & 9 & -6 & 26 & 31 & -63 \\ 1 & 4 & 3 & -2 & 10 & 9 & -17 \\ -1 & -4 & -4 & 2 & -13 & -11 & 22 \end{bmatrix}$$

- **Identify pivot position for Row 1.**

    Starting with the first column, we see that $a_{11} = 0$, which will not work for a pivot. However, there are nonzero terms down the first column, so we interchange Row 1 and Row 3 to place a 1 in the pivot position.

▶ The operation is $R_1 \Leftrightarrow R_3$.

$$\begin{bmatrix} 0 & 0 & 6 & 0 & 19 & 11 & -27 \\ 3 & 12 & 9 & -6 & 26 & 31 & -63 \\ 1 & 4 & 3 & -2 & 10 & 9 & -17 \\ -1 & -4 & -4 & 2 & -13 & -11 & 22 \end{bmatrix} \sim \begin{bmatrix} 1 & 4 & 3 & -2 & 10 & 9 & -17 \\ 3 & 12 & 9 & -6 & 26 & 31 & -63 \\ 0 & 0 & 6 & 0 & 19 & 11 & -27 \\ -1 & -4 & -4 & 2 & -13 & -11 & 22 \end{bmatrix}$$

- **Elimination.**

    Next, we need zeros down the first column below the pivot position. We already have $a_{31} = 0$, and we arrange for $a_{21} = 0$ and $a_{41} = 0$ by using the operations shown in the margin.

▶ The *elimination* steps are used to "eliminate" coefficients by transforming them to zero.

▶ The operations are
$$-3\,R_1 + R_2 \Rightarrow R_2$$
$$R_1 + R_4 \Rightarrow R_4$$

$$\begin{bmatrix} 1 & 4 & 3 & -2 & 10 & 9 & -17 \\ 3 & 12 & 9 & -6 & 26 & 31 & -63 \\ 0 & 0 & 6 & 0 & 19 & 11 & -27 \\ -1 & -4 & -4 & 2 & -13 & -11 & 22 \end{bmatrix} \sim \begin{bmatrix} 1 & 4 & 3 & -2 & 10 & 9 & -17 \\ 0 & 0 & 0 & 0 & -4 & 4 & -12 \\ 0 & 0 & 6 & 0 & 19 & 11 & -27 \\ 0 & 0 & -1 & 0 & -3 & -2 & 5 \end{bmatrix}$$

- **Identify pivot position for Row 2.**

    Moving down one row and to the right one column from a_{11}, we find $a_{22} = 0$. Since all the entries below a_{22} are also zero, interchanging with lower rows will not put a nonzero term in the a_{22} position. Thus a_{22} cannot be a pivot position, so we move to the right to the third column to determine if a_{23} is a suitable pivot position. Although a_{23} is also zero, there are nonzero terms below, so we interchange Row 2 and Row 4, putting a -1 in the pivot position.

▶ Do not be tempted to perform the operation $R_1 \Leftrightarrow R_2$. This will undo the zeros in the first column.

▶ The operation is $R_2 \Leftrightarrow R_4$.

$$\begin{bmatrix} 1 & 4 & 3 & -2 & 10 & 9 & -17 \\ 0 & 0 & 0 & 0 & -4 & 4 & -12 \\ 0 & 0 & 6 & 0 & 19 & 11 & -27 \\ 0 & 0 & -1 & 0 & -3 & -2 & 5 \end{bmatrix} \sim \begin{bmatrix} 1 & 4 & 3 & -2 & 10 & 9 & -17 \\ 0 & 0 & -1 & 0 & -3 & -2 & 5 \\ 0 & 0 & 6 & 0 & 19 & 11 & -27 \\ 0 & 0 & 0 & 0 & -4 & 4 & -12 \end{bmatrix}$$

- **Elimination.**

    Down the remainder of the third column, we already have $a_{43} = 0$, so we need only introduce a zero at a_{33} by using the operation shown in the margin.

▶ The operation is $6\,R_2 + R_3 \Rightarrow R_3$.

$$\begin{bmatrix} 1 & 4 & 3 & -2 & 10 & 9 & -17 \\ 0 & 0 & -1 & 0 & -3 & -2 & 5 \\ 0 & 0 & 6 & 0 & 19 & 11 & -27 \\ 0 & 0 & 0 & 0 & -4 & 4 & -12 \end{bmatrix} \sim \begin{bmatrix} 1 & 4 & 3 & -2 & 10 & 9 & -17 \\ 0 & 0 & -1 & 0 & -3 & -2 & 5 \\ 0 & 0 & 0 & 0 & 1 & -1 & 3 \\ 0 & 0 & 0 & 0 & -4 & 4 & -12 \end{bmatrix}$$

- **Identify pivot position for Row 3.**

    From the Row 2 pivot position, we move down one row and to the right one column to a_{34}. This entry is 0, as is the entry below, so interchanging rows will not yield an acceptable pivot. As we did before, we move one column to the right. Since $a_{35} = 1$ is nonzero, this becomes the pivot for Row 3.

- **Elimination.**

    We introduce a zero in the a_{45} position by using the operation shown in the margin.

▶ The operation is $4R_3 + R_4 \Rightarrow$ R_4.

$$\begin{bmatrix} 1 & 4 & 3 & -2 & 10 & 9 & -17 \\ 0 & 0 & -1 & 0 & -3 & -2 & 5 \\ 0 & 0 & 0 & 0 & 1 & -1 & 3 \\ 0 & 0 & 0 & 0 & -4 & 4 & -12 \end{bmatrix} \sim \begin{bmatrix} 1 & 4 & 3 & -2 & 10 & 9 & -17 \\ 0 & 0 & -1 & 0 & -3 & -2 & 5 \\ 0 & 0 & 0 & 0 & 1 & -1 & 3 \\ 0 & 0 & 0 & 0 & 0 & 0 & 0 \end{bmatrix}$$

- **Identify pivot position for Row 4.**

    Since Row 4 is the only remaining row and consists entirely of zeros, it has no pivot position. The matrix is now in echelon form, so no additional row operations are required. Converting the augmented matrix back to a linear system gives us

$$\begin{aligned} x_1 + 4x_2 + 3x_3 - 2x_4 + 10x_5 + 9x_6 &= -17 \\ - x_3 \quad\quad - 3x_5 - 2x_6 &= 5 \\ x_5 - x_6 &= 3 \end{aligned}$$

Using back substitution, we arrive at the general solution

$$\begin{aligned} x_1 &= -5 - 4s_1 + 2s_2 - 4s_3 \\ x_2 &= s_1 \\ x_3 &= -14 - 5s_3 \\ x_4 &= s_2 \\ x_5 &= 3 + s_3 \\ x_6 &= s_3 \end{aligned}$$

where s_1, s_2, and s_3 can be any real numbers.   ■

Gaussian elimination can be applied to any matrix to find an equivalent matrix that is in echelon form. If matrix A is equivalent to matrix B that is in echelon form, we say that B is an echelon form of A. Different sequences of row operations can produce different echelon forms of the same starting matrix, but all echelon forms of a given matrix will have the same pivot positions.

EXAMPLE 5   Use Gaussian elimination to find all solutions to the system of linear equations

$$\begin{aligned} x_1 + 4x_2 - 3x_3 &= 2 \\ 3x_1 - 2x_2 - x_3 &= -1 \\ -x_1 + 10x_2 - 5x_3 &= 3 \end{aligned}$$

Solution   The augmented matrix for this system is

$$\begin{bmatrix} 1 & 4 & -3 & 2 \\ 3 & -2 & -1 & -1 \\ -1 & 10 & -5 & 3 \end{bmatrix}$$

- **Identify pivot position for Row 1, then elimination.**

    We have $a_{11} = 1$, so this is the pivot position for Row 1. We introduce zeros down the first column with the row operations shown in the margin.

▶ The operations are
$-3R_1 + R_2 \Rightarrow R_2$
$R_1 + R_3 \Rightarrow R_3$

$$\begin{bmatrix} 1 & 4 & -3 & 2 \\ 3 & -2 & -1 & -1 \\ -1 & 10 & -5 & 3 \end{bmatrix} \sim \begin{bmatrix} 1 & 4 & -3 & 2 \\ 0 & -14 & 8 & -7 \\ 0 & 14 & -8 & 5 \end{bmatrix}$$ ■

- **Identify pivot position for Row 2, then elimination.**

▶ The operation is $R_2 + R_3 \Rightarrow R_3$.

$$\begin{bmatrix} 1 & 4 & -3 & 2 \\ 0 & -14 & 8 & -7 \\ 0 & 0 & 0 & -2 \end{bmatrix}$$

Before continuing, let's consider what we have. We find ourselves with a matrix in echelon form, but when we translate the last row back into an equation, we get $0 = -2$, which clearly has no solutions. Thus this system has no solutions, and so is inconsistent. ■

The preceding example illustrates a general principle. When applying row operations to an augmented matrix, if at any point in the process the matrix has a row of the form

$$\begin{bmatrix} 0 & 0 & 0 & \cdots & 0 & c \end{bmatrix} \tag{3}$$

where c is nonzero, then stop. The system is inconsistent.

Gauss–Jordan Elimination

▶ Gauss–Jordan elimination is named for the previously encountered C. F. Gauss, and Wilhelm Jordan (1842–1899), a German engineer who popularized this method for finding solutions to linear systems in his book on geodesy (the science of measuring earth shapes).

Let's return to the echelon form of the augmented matrix from Example 4,

$$\begin{bmatrix} 1 & 4 & 3 & -2 & 10 & 9 & -17 \\ 0 & 0 & -1 & 0 & -3 & -2 & 5 \\ 0 & 0 & 0 & 0 & 1 & -1 & 3 \\ 0 & 0 & 0 & 0 & 0 & 0 & 0 \end{bmatrix}$$

After extracting the linear system from this matrix, we back substituted and simplified to find the general solution. We can make it easier to find the general solution by performing additional row operations on the matrix. Specifically, we do the following:

1. Multiply each nonzero row by the reciprocal of the pivot so that we end up with a 1 as the leading term in each nonzero row.

2. Use row operations to introduce zeros in the entries *above* each pivot position.

Picking up with our matrix, we see that the first and third rows already have a 1 in the pivot position. Multiplying the second row by -1 takes care of the remaining nonzero row.

▶ The operation is $-R_2 \Rightarrow R_2$.

$$\begin{bmatrix} 1 & 4 & 3 & -2 & 10 & 9 & -17 \\ 0 & 0 & -1 & 0 & -3 & -2 & 5 \\ 0 & 0 & 0 & 0 & 1 & -1 & 3 \\ 0 & 0 & 0 & 0 & 0 & 0 & 0 \end{bmatrix} \sim \begin{bmatrix} 1 & 4 & 3 & -2 & 10 & 9 & -17 \\ 0 & 0 & 1 & 0 & 3 & 2 & -5 \\ 0 & 0 & 0 & 0 & 1 & -1 & 3 \\ 0 & 0 & 0 & 0 & 0 & 0 & 0 \end{bmatrix}$$

When implementing Gaussian elimination, we worked from left to right. To put zeros above pivot positions, we work from right to left, starting with the rightmost pivot, which in this case appears in the fifth column. Two row operations are required to introduce zeros above this pivot.

▶ The operations are

$$-3R_3 + R_2 \Rightarrow R_2$$
$$-10R_3 + R_1 \Rightarrow R_1$$

$$\begin{bmatrix} 1 & 4 & 3 & -2 & 10 & 9 & -17 \\ 0 & 0 & 1 & 0 & 3 & 2 & -5 \\ 0 & 0 & 0 & 0 & 1 & -1 & 3 \\ 0 & 0 & 0 & 0 & 0 & 0 & 0 \end{bmatrix} \sim \begin{bmatrix} 1 & 4 & 3 & -2 & 0 & 19 & -47 \\ 0 & 0 & 1 & 0 & 0 & 5 & -14 \\ 0 & 0 & 0 & 0 & 1 & -1 & 3 \\ 0 & 0 & 0 & 0 & 0 & 0 & 0 \end{bmatrix}$$

Next, we move up to the pivot in the second row, located in the third column. One row operation is required to introduce a zero in the a_{13} position.

▶ The operation is
$$-3R_2 + R_1 \Rightarrow R_1$$

$$\begin{bmatrix} 1 & 4 & 3 & -2 & 0 & 19 & -47 \\ 0 & 0 & 1 & 0 & 0 & 5 & -14 \\ 0 & 0 & 0 & 0 & 1 & -1 & 3 \\ 0 & 0 & 0 & 0 & 0 & 0 & 0 \end{bmatrix} \sim \begin{bmatrix} 1 & 4 & 0 & -2 & 0 & 4 & -5 \\ 0 & 0 & 1 & 0 & 0 & 5 & -14 \\ 0 & 0 & 0 & 0 & 1 & -1 & 3 \\ 0 & 0 & 0 & 0 & 0 & 0 & 0 \end{bmatrix} \qquad (4)$$

Naturally there are no rows above the pivot position in the first row, so we are done. Now when we extract the linear system, it has the form

$$\begin{aligned} x_1 + 4x_2 \quad - 2x_4 \quad + 4x_6 &= -5 \\ x_3 \qquad + 5x_6 &= -14 \\ x_5 - x_6 &= 3 \end{aligned}$$

▶ When using Gaussian and Gauss-Jordan elimination, do not yield to the temptation to alter the order of the row operations. Changing the order can result in a "circular" sequence of operations that lead nowhere.

Note that when the system is expressed in this form, the leading variables appear *only* in the equation that they lead. Thus during back substitution we need only plug in free parameters and then subtract to solve for the leading variables, simplifying the process considerably.

The matrix on the right in (4) is said to be in *reduced echelon form*.

DEFINITION 1.5

Definition Reduced Echelon Form

A matrix is in **reduced echelon form** (or **reduced row echelon form**) if

(a) It is in echelon form.

(b) All pivot positions contain a 1.

(c) The only nonzero term in a pivot column is in the pivot position.

Examples of matrices in reduced echelon form include

$$\begin{bmatrix} 0 & 1 & 0 & -2 & 0 & 0 & 17 \\ 0 & 0 & 1 & -6 & 0 & 3 & -6 \\ 0 & 0 & 0 & 0 & 1 & -2 & 5 \\ 0 & 0 & 0 & 0 & 0 & 0 & 0 \end{bmatrix} \quad \text{and} \quad \begin{bmatrix} 1 & -3 & 0 & 0 & -7 & 21 \\ 0 & 0 & 1 & 0 & 2 & 13 \\ 0 & 0 & 0 & 1 & 5 & -9 \end{bmatrix}.$$

Definition Forward Phase, Backward Phase

Definition Gauss–Jordan Elimination

Transforming a matrix to reduced echelon form can be viewed as having two parts: The **forward phase** is Gaussian elimination, transforming the matrix to echelon form, and the **backward phase**, which completes the transformation to reduced echelon form. The combination of the forward and backward phases is referred to as **Gauss–Jordan elimination**. Although a given matrix can be equivalent to many different echelon form matrices, the same is not true of reduced echelon form matrices.

THEOREM 1.6

A given matrix is equivalent to a unique matrix that is in reduced echelon form.

The proof of this theorem is omitted.

EXAMPLE 6   Use Gauss–Jordan elimination to find all solutions to the system of linear equations

▶ Going forward, we omit detailed explanations and instead just show the row operations in the order performed.

$$\begin{aligned} x_1 - 2x_2 - 3x_3 &= -1 \\ x_1 - x_2 - 2x_3 &= 1 \\ -x_1 + 3x_2 + 5x_3 &= 2 \end{aligned}$$

Solution The augmented matrix and row operations are shown below.

$$\begin{bmatrix} 1 & -2 & -3 & -1 \\ 1 & -1 & -2 & 1 \\ -1 & 3 & 5 & 2 \end{bmatrix} \begin{array}{l} -R_1+R_2 \Rightarrow R_2 \\ R_1+R_3 \Rightarrow R_3 \\ \sim \end{array} \begin{bmatrix} 1 & -2 & -3 & -1 \\ 0 & 1 & 1 & 2 \\ 0 & 1 & 2 & 1 \end{bmatrix}$$

$$\begin{array}{l} -R_2+R_3 \Rightarrow R_3 \\ \sim \end{array} \begin{bmatrix} 1 & -2 & -3 & -1 \\ 0 & 1 & 1 & 2 \\ 0 & 0 & 1 & -1 \end{bmatrix}$$

That completes the forward phase, yielding a matrix in echelon form. Next, we implement the backward phase to transform the matrix to reduced echelon form.

$$\begin{bmatrix} 1 & -2 & -3 & -1 \\ 0 & 1 & 1 & 2 \\ 0 & 0 & 1 & -1 \end{bmatrix} \begin{array}{l} -R_3+R_2 \Rightarrow R_2 \\ 3R_3+R_1 \Rightarrow R_1 \\ \sim \end{array} \begin{bmatrix} 1 & -2 & 0 & -4 \\ 0 & 1 & 0 & 3 \\ 0 & 0 & 1 & -1 \end{bmatrix}$$

$$\begin{array}{l} 2R_2+R_1 \Rightarrow R_1 \\ \sim \end{array} \begin{bmatrix} 1 & 0 & 0 & 2 \\ 0 & 1 & 0 & 3 \\ 0 & 0 & 1 & -1 \end{bmatrix}$$

The reduced echelon form is equivalent to the linear system

$$\begin{aligned} x_1 \quad\quad &= \quad 2 \\ x_2 \quad &= \quad 3 \\ x_3 &= -1 \end{aligned}$$

We see immediately that the system has unique solution $x_1 = 2, x_2 = 3,$ and $x_3 = -1.$ ∎

Homogeneous Linear Systems

A linear equation is **homogeneous** if it has the form

$$a_1 x_1 + a_2 x_2 + \cdots + a_n x_n = 0$$

Definition Homogeneous Equation, Homogeneous System

Homogeneous linear systems are an important class of systems that are made up of homogeneous linear equations.

$$\begin{aligned} a_{11} x_1 + a_{12} x_2 + a_{13} x_3 + \cdots + a_{1n} x_n &= 0 \\ a_{21} x_1 + a_{22} x_2 + a_{23} x_3 + \cdots + a_{2n} x_n &= 0 \\ \vdots \quad\quad \vdots \quad\quad\quad \vdots \quad\quad \vdots \\ a_{m1} x_1 + a_{m2} x_2 + a_{m3} x_3 + \cdots + a_{mn} x_n &= 0 \end{aligned}$$

Note that all homogeneous systems are consistent, because there is always one easy solution, namely,

$$x_1 = 0, \quad x_2 = 0, \quad \ldots, \quad x_n = 0$$

Definition Trivial Solution, Nontrivial Solution

This is called the **trivial solution**. If there are additional solutions, they are called **nontrivial solutions**. We determine if there are nontrivial solutions in the usual way, using elimination methods.

EXAMPLE 7   Use Gauss–Jordan elimination to find all solutions to the homogeneous system of linear equations

$$2x_1 - 6x_2 - x_3 + 8x_4 = 0$$
$$x_1 - 3x_2 - x_3 + 6x_4 = 0$$
$$-x_1 + 3x_2 - x_3 + 2x_4 = 0$$

Solution   As the system is homogeneous, we know that it has the trivial solution. To find the other solutions, we load the system into an augmented matrix and transform to reduced echelon form.

$$\begin{bmatrix} 2 & -6 & -1 & 8 & 0 \\ 1 & -3 & -1 & 6 & 0 \\ -1 & 3 & -1 & 2 & 0 \end{bmatrix} \quad \overset{R_1 \Leftrightarrow R_2}{\sim} \quad \begin{bmatrix} 1 & -3 & -1 & 6 & 0 \\ 2 & -6 & -1 & 8 & 0 \\ -1 & 3 & -1 & 2 & 0 \end{bmatrix}$$

$$\overset{-2R_1 + R_2 \Rightarrow R_2}{\underset{\sim}{R_1 + R_3 \Rightarrow R_3}} \quad \begin{bmatrix} 1 & -3 & -1 & 6 & 0 \\ 0 & 0 & 1 & -4 & 0 \\ 0 & 0 & -2 & 8 & 0 \end{bmatrix}$$

$$\overset{2R_2 + R_3 \Rightarrow R_3}{\sim} \quad \begin{bmatrix} 1 & -3 & -1 & 6 & 0 \\ 0 & 0 & 1 & -4 & 0 \\ 0 & 0 & 0 & 0 & 0 \end{bmatrix}$$

$$\overset{R_2 + R_1 \Rightarrow R_1}{\sim} \quad \begin{bmatrix} 1 & -3 & 0 & 2 & 0 \\ 0 & 0 & 1 & -4 & 0 \\ 0 & 0 & 0 & 0 & 0 \end{bmatrix}$$

The last matrix is in reduced echelon form. The corresponding linear system is

$$x_1 - 3x_2 \quad\; + 2x_4 = 0$$
$$x_3 - 4x_4 = 0$$

Back substituting yields the general solution

$$x_1 = 3s_1 - 2s_2$$
$$x_2 = s_1$$
$$x_3 = 4s_2$$
$$x_4 = s_2$$

where s_1 and s_2 can be any real numbers.   ■

Proof of Theorem 1.2

We are now in a position to revisit and prove Theorem 1.2 from Section 1.1. Recall the statement of the theorem.

THEOREM 1.2

A system of linear equations has no solutions, exactly one solution, or infinitely many solutions.

Proof   We can take any linear system, form the augmented matrix, use Gaussian elimination to reduce to echelon form, and extract the transformed system. There are three possible outcomes from this process:

(a)   The system has an equation of the form $0 = c$ for $c \neq 0$. In this case, the system has no solutions.

If (a) does not occur, then one of (b) or (c) must:

(b) The transformed system is triangular, and thus has no free variables and hence exactly one solution.

(c) The transformed system is not triangular, and so has one or more free variables and hence infinitely many solutions.

Homogeneous linear systems are even simpler. Since all such systems have the trivial solution, (a) cannot happen. Therefore a homogeneous linear system has either a unique solution or infinitely many solutions. ■

Computational Comments

▶ There are various similar definitions for what constitutes a "flop." Here we take a "flop" to be one arithmetic operation, either addition or multiplication. Counting flops gives a measure of algorithm efficiency.

- We can find the solutions to any system by using either Gaussian elimination or Gauss–Jordan elimination. Which is better? For a system of n equations with n unknowns, Gaussian elimination requires approximately $\frac{2}{3}n^3$ flops (i.e., arithmetic operations) and Gauss–Jordan requires about n^3 flops. Back substitution is slightly more complicated for Gaussian elimination than for Gauss–Jordan, but overall Gaussian elimination is more efficient and is the method that is usually implemented in computer software.

- When elimination methods are implemented on computers, to control round-off error they typically include an extra step called "partial pivoting," which involves selecting the entry having the largest absolute value to serve as the pivot. When performing row operations by hand, partial pivoting tends to introduce fractions and leads to messy calculations, so we avoided the topic. However, it is discussed in the next section.

| EXERCISES |

In each exercise set, problems marked with Ⓒ are designed to be solved using a programmable calculator or computer algebra system.

In Exercises 1–4, convert the given augmented matrix to the equivalent linear system.

1. $\begin{bmatrix} 4 & 2 & -1 & 2 \\ -1 & 0 & 5 & 7 \end{bmatrix}$

2. $\begin{bmatrix} -2 & 1 & 0 \\ 13 & -3 & 6 \\ -11 & 7 & -5 \end{bmatrix}$

3. $\begin{bmatrix} 0 & 12 & -3 & -9 & 17 \\ -12 & 5 & -3 & 11 & 0 \\ 6 & 8 & 2 & 10 & -8 \\ 17 & 0 & 0 & 13 & -1 \end{bmatrix}$

4. $\begin{bmatrix} -1 & 2 \\ 5 & -7 \\ 3 & 0 \end{bmatrix}$

In Exercises 5–10, determine those matrices that are in echelon form, and those that are also in reduced echelon form.

5. $\begin{bmatrix} 1 & 3 & -2 \\ 0 & 2 & 6 \\ 0 & 0 & 0 \end{bmatrix}$

6. $\begin{bmatrix} 1 & 3 & 0 & 6 & -2 \\ 0 & 0 & 1 & 5 & 3 \\ 0 & 0 & 0 & 0 & 0 \end{bmatrix}$

7. $\begin{bmatrix} 3 & -3 & 1 & 1 & 0 \\ 0 & 0 & -2 & 4 & 0 \\ 0 & 0 & 1 & 0 & 0 \end{bmatrix}$

8. $\begin{bmatrix} 1 & -3 & 1 & -7 \\ 0 & 1 & -2 & 4 \\ 0 & 0 & 1 & -2 \\ 0 & 0 & 0 & 2 \end{bmatrix}$

9. $\begin{bmatrix} 1 & 0 & 0 & 5 & -1 \\ 0 & 2 & 0 & -2 & 0 \\ 0 & 0 & 1 & -3 & 2 \end{bmatrix}$

10. $\begin{bmatrix} 1 & -1 & 0 & 9 & 0 \\ 0 & 0 & 1 & 8 & 0 \\ 0 & 0 & 0 & 0 & 1 \end{bmatrix}$

In Exercises 11–14, the matrix on the right results after performing a single row operation on the matrix on the left. Identify the row operation.

11. $\begin{bmatrix} -2 & 1 & 0 \\ 13 & -3 & 6 \\ -11 & 7 & -5 \end{bmatrix} \sim \begin{bmatrix} 4 & -2 & 0 \\ 13 & -3 & 6 \\ -11 & 7 & -5 \end{bmatrix}$

12. $\begin{bmatrix} 4 & 2 & -1 & 2 \\ -1 & 0 & 5 & 7 \end{bmatrix} \sim \begin{bmatrix} 1 & 2 & 14 & 23 \\ -1 & 0 & 5 & 7 \end{bmatrix}$

13. $\begin{bmatrix} 2 & -1 & 3 & 0 \\ 4 & 9 & -2 & 3 \\ 6 & 7 & 5 & -1 \end{bmatrix} \sim \begin{bmatrix} 2 & -1 & 3 & 0 \\ 4 & 9 & -2 & 3 \\ -2 & -11 & 9 & -7 \end{bmatrix}$

14. $\begin{bmatrix} 2 & -1 & 3 & 0 \\ 4 & 9 & -2 & 3 \\ 6 & 7 & 5 & -1 \end{bmatrix} \sim \begin{bmatrix} 6 & 7 & 5 & -1 \\ 4 & 9 & -2 & 3 \\ 2 & -1 & 3 & 0 \end{bmatrix}$

In Exercises 15–18, a single row operation was performed on the matrix on the left to produce the matrix on the right. Unfortunately, an error was made when performing the row operation. Identify the operation and fix the error.

15. $\begin{bmatrix} 3 & 7 & -2 \\ -1 & 4 & 3 \\ 5 & 0 & -3 \end{bmatrix} \sim \begin{bmatrix} -1 & 4 & -3 \\ 3 & 7 & -2 \\ 5 & 0 & -3 \end{bmatrix}$

16. $\begin{bmatrix} -2 & -2 & 1 & 6 \\ 4 & -1 & 0 & -5 \end{bmatrix} \sim \begin{bmatrix} -2 & -2 & 1 & 6 \\ 0 & 5 & 2 & 7 \end{bmatrix}$

17. $\begin{bmatrix} 0 & 3 & -1 & 2 \\ -1 & -9 & 4 & 1 \\ 5 & 0 & 7 & 2 \end{bmatrix} \sim \begin{bmatrix} 2 & 6 & -2 & 4 \\ -1 & -9 & 4 & 1 \\ 5 & 0 & 7 & 2 \end{bmatrix}$

18. $\begin{bmatrix} 1 & 7 & 2 & 0 \\ 0 & 4 & -8 & -3 \\ 3 & 0 & 0 & 1 \end{bmatrix} \sim \begin{bmatrix} 1 & 7 & 2 & 0 \\ 0 & 4 & -8 & -3 \\ 1 & -14 & 0 & 1 \end{bmatrix}$

In Exercises 19–26, convert the given system to an augmented matrix and then find all solutions by reducing the system to echelon form and back substituting.

19. $\begin{aligned} 2x_1 + x_2 &= 1 \\ -4x_1 - x_2 &= 3 \end{aligned}$

20. $\begin{aligned} 3x_1 - 7x_2 &= 0 \\ x_1 + 4x_2 &= 0 \end{aligned}$

21. $\begin{aligned} -2x_1 + 5x_2 - 10x_3 &= 4 \\ x_1 - 2x_2 + 3x_3 &= -1 \\ 7x_1 - 17x_2 + 34x_3 &= -16 \end{aligned}$

22. $\begin{aligned} 2x_1 + 8x_2 - 4x_3 &= -10 \\ -x_1 - 3x_2 + 5x_3 &= 4 \end{aligned}$

23. $\begin{aligned} 2x_1 + 2x_2 - x_3 &= 8 \\ -x_1 - x_2 &= -3 \\ 3x_1 + 3x_2 + x_3 &= 7 \end{aligned}$

24. $\begin{aligned} -5x_1 + 9x_2 &= 13 \\ 3x_1 - 5x_2 &= -9 \\ x_1 - 2x_2 &= -2 \end{aligned}$

25. $\begin{aligned} 2x_1 + 6x_2 - 9x_3 - 4x_4 &= 0 \\ -3x_1 - 11x_2 + 9x_3 - x_4 &= 0 \\ x_1 + 4x_2 - 2x_3 + x_4 &= 0 \end{aligned}$

26. $\begin{aligned} x_1 - x_2 - 3x_3 - x_4 &= -1 \\ -2x_1 + 2x_2 + 6x_3 + 2x_4 &= -1 \\ -3x_1 - 3x_2 + 10x_3 &= 5 \end{aligned}$

In Exercises 27–30, convert the given system to an augmented matrix and then find all solutions by transforming the system to reduced echelon form and back substituting.

27. $\begin{aligned} -2x_1 - 5x_2 &= 0 \\ x_1 + 3x_2 &= 1 \end{aligned}$

28. $\begin{aligned} -4x_1 + 2x_2 - 2x_3 &= 10 \\ x_1 + x_3 &= -3 \\ 3x_1 - x_2 + x_3 &= -8 \end{aligned}$

29. $\begin{aligned} 2x_1 + x_2 &= 2 \\ -x_1 - x_2 - x_3 &= 1 \end{aligned}$

30. $\begin{aligned} -3x_1 + 2x_2 - x_3 + 6x_4 &= -7 \\ 7x_1 - 3x_2 + 2x_3 - 11x_4 &= 14 \\ x_1 - x_4 &= 1 \end{aligned}$

For each of Exercises 31–36, suppose that the given row operation is used to transform a matrix. Which row operation will transform the matrix back to its original form?

31. $5R_1 \implies R_1$

32. $-2R_3 \implies R_3$

33. $R_1 \iff R_3$

34. $R_4 \iff R_1$

35. $-5R_2 + R_6 \implies R_6$

36. $-3R_1 + R_3 \implies R_3$

FIND AN EXAMPLE For Exercises 37–42, find an example that meets the given specifications.

37. A matrix with three rows and five columns that is in echelon form, but not in reduced echelon form.

38. A matrix with six rows and four columns that is in echelon form, but not in reduced echelon form.

39. An augmented matrix for an inconsistent linear system that has four equations and three variables.

40. An augmented matrix for an inconsistent linear system that has three equations and four variables.

41. A homogeneous linear system with three equations, four variables, and infinitely many solutions.

42. Two matrices that are distinct yet equivalent.

TRUE OR FALSE For Exercises 43–50, determine if the statement is true or false, and justify your answer.

43. If two matrices are equivalent, then one can be transformed into the other with a sequence of elementary row operations.

44. Different sequences of row operations can lead to different echelon forms for the same matrix.

45. Different sequences of row operations can lead to different reduced echelon forms for the same matrix.

46. If a linear system has four equations and seven variables, then it must have infinitely many solutions.

47. If a linear system has seven equations and four variables, then it must be inconsistent.

48. Every linear system with free variables has infinitely many solutions.

49. Any linear system with more variables than equations cannot have a unique solution.

50. If a linear system has the same number of equations and variables, then it must have a unique solution.

51. Suppose that the echelon form of an augmented matrix has a pivot position in every column except the rightmost one. How many solutions does the associated linear system have? Justify your answer.

52. Suppose that the echelon form of an augmented matrix has a pivot position in every column. How many solutions does the associated linear system have? Justify your answer.

53. Show that if a linear system has two different solutions, then it must have infinitely many solutions.

54. Show that if a matrix has more rows than columns and is in echelon form, then it must have at least one row of zeros at the bottom.

55. Show that a homogeneous linear system with more variables than equations must have an infinite number of solutions.

56. Show that each of the elementary operations on linear systems (see pages 14–15) produces an equivalent linear system. (Recall two linear systems are equivalent if they have the same solution set.)

(a) Interchange the position of two equations.

(b) Multiply an equation by a nonzero constant.

(c) Add a multiple of one equation to another.

Ⓒ In Exercises 57–58 you are asked to find an *interpolating polynomial*, which is used to fit a function to a set of data.

57. Figure 3 shows the plot of the points $(1, 4)$, $(2, 7)$, and $(3, 14)$. Find a polynomial of degree 2 of the form $f(x) = ax^2 + bx + c$ whose graph passes through these points.

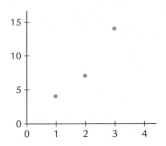

Figure 3 Exercise 57 data.

58. Figure 4 shows the plot of the points $(1, 8)$, $(2, 3)$, $(3, 9)$, $(5, 1)$, and $(7, 7)$. Find a polynomial of degree 4 of the form $f(x) = ax^4 + bx^3 + cx^2 + dx + e$ whose graph passes through these points.

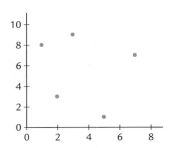

Figure 4 Exercise 58 data.

Ⓒ Exercises 59–60 refer to the cannonball scenario described at the start of the section. For each problem, the three ordered pairs are $(x, E(x))$, where x is the distance on the ground from the position of the cannon and $E(x)$ is the elevation of the ball. Find a model for the elevation of the ball, and use the model to determine where it hits the ground.

59. $(20, 288)$, $(40, 364)$, $(60, 360)$

60. $(40, 814)$, $(80, 1218)$, $(110, 1311)$

Ⓒ In Exercises 61–68, the given matrix is the augmented matrix for a linear system. Use technology to perform the row operations needed to transform the matrix to reduced echelon form, and then find all solutions to the system.

61. $\begin{bmatrix} 2 & 7 & -3 & 0 \\ -3 & 0 & 5 & 1 \\ -2 & 6 & -5 & 4 \end{bmatrix}$

62. $\begin{bmatrix} 11 & -5 & 0 & 0 \\ 2 & -3 & 8 & 0 \\ 7 & 3 & 3 & 0 \end{bmatrix}$

63. $\begin{bmatrix} 5 & -2 & 0 & 3 & 9 \\ 7 & 1 & 6 & 2 & -2 \\ 2 & 0 & -3 & 5 & 4 \end{bmatrix}$

64. $\begin{bmatrix} 9 & -2 & 0 & -4 & 6 \\ 0 & 7 & -1 & -1 & 3 \\ 8 & 12 & -6 & 5 & -8 \end{bmatrix}$

65. $\begin{bmatrix} 8 & -8 & 0 & -1 \\ 6 & 2 & -1 & 0 \\ 5 & 6 & -3 & 10 \\ -2 & 0 & -1 & -4 \end{bmatrix}$

66. $\begin{bmatrix} 5 & 3 & 7 & 5 \\ 4 & -3 & -2 & 0 \\ 0 & 3 & 17 & -2 \\ 4 & 7 & 8 & 12 \end{bmatrix}$

67. $\begin{bmatrix} 6 & 5 & 1 & 0 & -3 & 0 \\ 3 & -2 & -1 & 8 & 12 & 0 \\ -7 & 1 & 3 & 0 & 11 & 0 \\ 13 & 2 & 0 & -2 & -7 & 0 \end{bmatrix}$

68. $\begin{bmatrix} 2 & 1 & 0 & 0 & 3 & -5 & 7 \\ 0 & 5 & -1 & 8 & -1 & 4 & 0 \\ 3 & 11 & -9 & 1 & 6 & 0 & 13 \\ 7 & 0 & 5 & 5 & -3 & 2 & 11 \end{bmatrix}$

▶ This section is optional and can be omitted without loss of continuity.

1.3 Numerical Solutions

In theory, the elimination methods developed in Section 1.2 can be used to find the solutions to *any* system of linear equations. And in practice, elimination methods work fine as long as the system is not too large. However, when implemented on a computer, elimination methods can lead to the wrong answer due to round-off error. Furthermore, for very large systems elimination methods may not be efficient enough to be practical. In this section we consider some shortcomings of elimination methods, and develop alternative solution methods.

▶ "In theory, there is no difference between theory and practice. In practice, there is." —*Yogi Berra* (*Also attributed to computer scientist Jan L. A. van de Snepscheut and physicist Albert Einstein.*)

Round-off Error

No sensible person spends their day solving complicated systems of linear equations by hand—they use a computer. But while computers are fast, they have drawbacks, one being the round-off errors that can arise when using floating-point representations for numbers.

For example, suppose that we have a simple computer that has only four digits of accuracy. Using this computer and Gauss–Jordan elimination to solve the system

$$\begin{aligned} 7x_1 - 3x_2 + 2x_3 + 6x_4 &= 13 \\ -3x_1 + 9x_2 + 5x_3 - 2x_4 &= 9 \\ x_1 - 13x_2 - 3x_3 + 8x_4 &= -13 \\ 2x_1 \quad\quad - x_3 + 3x_4 &= -6 \end{aligned}$$

yields the solution $x_1 = 2$, $x_2 = -0.999$, $x_3 = 3.998$, and $x_4 = -1.999$. This differs from the exact solution $x_1 = 2$, $x_2 = -1$, $x_3 = 4$, and $x_4 = -2$ because of round-off error occurring while performing row operations. The degree of error here is not too large, but this is a small system. Elimination methods applied to larger systems will require many more arithmetic operations, which can result in accumulation of round-off errors, even on a high-precision computer.

If the right combination of conditions exists, even small systems can generate significant round-off errors.

EXAMPLE 1   Suppose that we are using a computer with four digits of accuracy. Apply Gaussian elimination to find the solution to the system

▶ The choice of four digits of accuracy does not restrict us to numbers less than 10,000. For instance, a number such as 973,400 can be represented as 9.734×10^5.

$$\begin{aligned} 3x_1 + 1000x_2 &= 7006 \\ 42x_1 - 36x_2 &= -168 \end{aligned} \qquad (1)$$

Solution We need only one row operation to put the system in triangular form. The exact computations are

$$\begin{bmatrix} 3 & 1000 & 7006 \\ 42 & -36 & -168 \end{bmatrix} \xrightarrow{-14R_1 + R_2 \Rightarrow R_2} \begin{bmatrix} 3 & 1000 & 7006 \\ 0 & -14036 & -98252 \end{bmatrix}$$

Since our computer only carries four digits of accuracy, the number $-14,036$ is rounded to $-14,040$ and $-98,252$ is rounded to $-98,250$. Thus the triangular system we end up with is

$$\begin{aligned} 3x_1 + 1000x_2 &= 7006 \\ -14,040x_2 &= -98,250 \end{aligned}$$

▶ The notation $:\approx$ means that rounding has occurred, and that the value on the right is being assigned to the indicated variable.

Solving for x_2, we get

$$x_2 = \frac{-98,250}{-14,040} :\approx 6.998$$

Back substituting to solve for x_1 gives us

$$x_1 = \frac{7006 - 1000(6.998)}{3} \; :\approx 2.667$$

The exact solution to the system is $x_1 = 2$ and $x_2 = 7$. Although the approximation for x_2 is fairly good, the approximation for x_1 is off by quite a bit. The source of the problem is that the coefficients in the equation

$$3x_1 + 1000x_2 = 7006$$

differ dramatically in size. During back substitution into this equation, the error in x_2 is magnified by the coefficient 1000 and only can be compensated for by the $3x_1$ term. But since the coefficient on this term is so much smaller, the error in x_1 is forced to be large. ∎

Definition Partial Pivoting

One way to combat round-off error is to use **partial pivoting**, which adds a step to the usual elimination algorithms. With partial pivoting, when starting on a new column we first switch the row with the largest leading entry (compared using absolute values) to the pivot position before beginning the elimination process.

For instance, with the system (1) we interchange the position of the two rows because because $|42| > |3|$, which gives us

$$\begin{array}{rcr} 42x_1 - & 36x_2 = & -168 \\ 3x_1 + & 1000x_2 = & 7006 \end{array}$$

This time the single elimination step is (shown with four digits of accuracy)

$$\begin{bmatrix} 42 & -36 & -168 \\ 3 & 1000 & 7006 \end{bmatrix} \overset{-\frac{1}{14}R_1 + R_2 \Rightarrow R_2}{\sim} \begin{bmatrix} 42 & -36 & -168 \\ 0 & 1003 & 7018 \end{bmatrix}$$

From this we have $x_2 = 7018/1003 = 6.997$, which is slightly less accurate than before. However, when we back substitute this value of x_2 into the equation

$$42x_1 - 36x_2 = -168$$

we get $x_1 = 1.997$, a much better approximation for the exact value of x_1.

Definition Full Pivoting

When Gaussian and Gauss–Jordan elimination are implemented in computer software, partial pivoting is often used to help control round-off errors. It is also possible to implement **full pivoting**, where both rows and columns are interchanged to arrange for the largest possible leading coefficient. However, full pivoting is slower and so is employed less frequently than partial pivoting.

Jacobi Iteration

It is not at all unusual for an application to yield a system of linear equations with thousands of equations and variables. In such a case, even if round-off error is controlled, elimination methods may not be efficient enough to be practical.

Here we turn our attention to a pair of related *iterative methods* that attempt to find the solution to a system of equations through a sequence of approximations. These methods do not suffer from the round-off problems described earlier, and in many cases they are faster than elimination methods. However, they only work on systems where the number of equations equals the number of variables, and sometimes they **diverge**—that is, they fail to reach the solution. In the cases where a solution is found, we say that the method **converges**.

Definition Diverge, Converge

Our first approximation method is called **Jacobi iteration**. We illustrate this method by using it to find the solution to the system

▶ Jacobi iteration is named for German mathematician Karl Gustav Jacobi (1804–1851).

$$10x_1 + 4x_2 - x_3 = 3$$
$$2x_1 + 10x_2 + x_3 = -19 \qquad (2)$$
$$x_1 - x_2 + 5x_3 = -2$$

Step 1: Solve the first equation of the system for x_1, the second equation for x_2, and so on,

$$x_1 = 0.3 - 0.4x_2 + 0.1x_3$$
$$x_2 = -1.9 - 0.2x_1 - 0.1x_3 \qquad (3)$$
$$x_3 = -0.4 - 0.2x_1 + 0.2x_2$$

Step 2: Make a guess at the values of x_1, x_2, and x_3 that satisfy the system. If we have no idea about the solution, then set each equal to 0,

$$x_1 = 0, \qquad x_2 = 0, \qquad x_3 = 0$$

Step 3: Substitute the values for x_1, x_2, and x_3 into (3). This is Iteration 1, and it gives the updated values:

Iteration 1:
$$x_1 = 0.3 - 0.4(0) + 0.1(0) = 0.3$$
$$x_2 = -1.9 - 0.2(0) - 0.1(0) = -1.9$$
$$x_3 = -0.4 - 0.2(0) + 0.2(0) = -0.4$$

Now repeat the process, substituting the new values for x_1, x_2, and x_3 into the equations in Step 1.

Iteration 2:
$$x_1 = 0.3 - 0.4(-1.9) + 0.1(-0.4) = 1.02$$
$$x_2 = -1.9 - 0.2(0.3) - 0.1(-0.4) = -1.92$$
$$x_3 = -0.4 - 0.2(0.3) + 0.2(-1.9) = -0.84$$

We keep repeating this procedure until we have two consecutive iterations where each value differs from its predecessor by no more than the accuracy desired. Table 1 shows the outcome from the first nine iterations, each rounded to four decimal places. We see that the values have converged to $x_1 = 1$, $x_2 = -2$, and $x_3 = -1$, which is the exact solution to the system.

▶ Table values are rounded to four decimal places, and the rounded values are carried to the next iteration.

n	x_1	x_2	x_3
0	0	0	0
1	0.3000	−1.9000	−0.4000
2	1.0200	−1.9200	−0.8400
3	0.9840	−2.0200	−0.9880
4	1.0092	−1.9980	−1.0008
5	0.9991	−2.0018	−1.0014
6	1.0006	−1.9997	−1.0002
7	0.9999	−2.0001	−1.0001
8	1.0000	−2.0000	−1.0000
9	1.0000	−2.0000	−1.0000

Table 1 Jacobi Iterations (n is the iteration number)

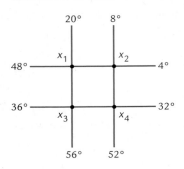

Figure 1 Grid Temperatures for Example 2.

EXAMPLE 2   Figure 1 gives a diagram of a piece of heavy wire mesh. Each of the eight wire ends has temperature held fixed as shown. When the temperature of the mesh reaches equilibrium, the temperature at each connecting point will be the average of the temperatures of the adjacent points and fixed ends. Determine the equilibrium temperature at the connecting points x_1, x_2, x_3, and x_4.

Solution   The temperature of each connecting point depends in part on the temperature of other connecting points. For instance, since x_1 is adjacent to x_2, x_3, and the ends held fixed at 48° and 20°, its temperature at equilibrium will be the average

$$x_1 = \frac{x_2 + x_3 + 48 + 20}{4} = 0.25x_2 + 0.25x_3 + 17$$

Similarly, for the other connecting points we have the equations (after simplifying)

$$x_2 = 0.25x_1 + 0.25x_4 + 3$$
$$x_3 = 0.25x_1 + 0.25x_4 + 23$$
$$x_4 = 0.25x_2 + 0.25x_3 + 21$$

Our four equations could be reorganized into the usual form of a linear equation and solved using elimination methods. But since each equation has one variable written in terms of the other variables, the problem sets up perfectly for Jacobi iteration. Starting with initial choices $x_1 = x_2 = x_3 = x_4 = 0$, the first two iterations are

Iteration 1: $x_1 = 0.25(0) + 0.25(0) + 17 = 17$
$x_2 = 0.25(0) + 0.25(0) + 3 = 3$
$x_3 = 0.25(0) + 0.25(0) + 23 = 23$
$x_4 = 0.25(0) + 0.25(0) + 21 = 21$

Iteration 2: $x_1 = 0.25(3) + 0.25(23) + 17 = 23.5$
$x_2 = 0.25(17) + 0.25(21) + 3 = 12.5$
$x_3 = 0.25(17) + 0.25(21) + 23 = 32.5$
$x_4 = 0.25(3) + 0.25(23) + 21 = 27.5$

Table 2 shows additional Jacobi iterations.

▶ To save space, only every fourth iteration is given in Table 2.

n	x_1	x_2	x_3	x_4
4	29.8750	18.1250	38.1250	33.8750
8	31.8672	19.8828	39.8828	35.8672
12	31.9917	19.9927	39.9927	35.9917
16	31.9995	19.9995	39.9995	35.9995
20	32.0000	20.0000	40.0000	36.0000
24	32.0000	20.0000	40.0000	36.0000

Table 2   Jacobi Iterations for Example 2

▶ We encountered C. F. Gauss earlier. Ludwig Philipp von Seidel (1821–1896) was a German mathematician. Interestingly, Gauss discovered the method long before Seidel but discarded it as worthless. Nonetheless, Gauss's name was attached to the algorithm along with that of Seidel, who independently discovered and published it after Gauss died.

This suggests equilibrium temperatures of $x_1 = 32$, $x_2 = 20$, $x_3 = 40$, and $x_4 = 36$. Substituting these values into our four equations confirms that this is correct. ∎

Gauss–Seidel Iteration

At each step of Jacobi iteration we take the values from the previous step and plug them into the set of equations, updating the values of all variables at the same time. We modify this approach with a variant of Jacobi iteration called Gauss–Seidel iteration. With this method, we always use the current value of each variable.

To illustrate how Gauss–Seidel works, we use the system (2) considered before.

Step 1: As with Jacobi, we start by solving for x_1, x_2, and x_3,

$$
\begin{aligned}
x_1 &= 0.3 - 0.4x_2 + 0.1x_3 \\
x_2 &= -1.9 - 0.2x_1 - 0.1x_3 \\
x_3 &= -0.4 - 0.2x_1 + 0.2x_2
\end{aligned}
$$

Step 2: We set initial values for x_1, x_2, and x_3. In the absence of any approximation for the solution, we use

$$
x_1 = 0, \qquad x_2 = 0, \qquad x_3 = 0
$$

Step 3: For the first part of Iteration 1, we have (again as with Jacobi)

$$
x_1 = 0.3 - 0.4(0) + 0.1(0) = 0.3
$$

At this point the Jacobi and Gauss–Seidel methods begin to differ. To calculate the updated value of x_2, we use the most current variable values, which are $x_1 = 0.3$ and $x_3 = 0$.

$$
x_2 = -1.9 - 0.2(0.3) - 0.1(0) = -1.96
$$

We finish this iteration by updating the value of x_3, using the current values $x_1 = 0.3$ and $x_2 = -1.96$, so that

$$
x_3 = -0.4 - 0.2(0.3) + 0.2(-1.96) = -0.852
$$

Subsequent iterations proceed in the same way, always incorporating the most current variable values. The second iteration is

$$
\begin{aligned}
\text{Iteration 2:} \quad x_1 &= 0.3 - 0.4(-1.96) + 0.1(-0.852) = 0.9988 \\
x_2 &= -1.9 - 0.2(0.9988) - 0.1(-0.852) :\approx -2.0146 \\
x_3 &= -0.4 - 0.2(0.9988) + 0.2(-2.01456) :\approx -1.0027
\end{aligned}
$$

As with Jacobi, we continue until reaching the point where two consecutive iterations yield values sufficiently close together. Table 3 gives the first six iterations of Gauss–Seidel applied to our system.

Note that Gauss–Seidel converged to the solution faster than Jacobi. Since Gauss–Seidel immediately incorporates new values into the computations, it seems reasonable to expect that it would converge faster than Jacobi. Most of the time this is true, but surprisingly not always—there are systems where Jacobi iteration converges more rapidly.

n	x_1	x_2	x_3
0	0	0	0
1	0.3000	−1.9600	−0.8520
2	0.9988	−2.0146	−1.0027
3	1.0056	−2.0008	−1.0013
4	1.0002	−1.9999	−1.0000
5	1.0000	−2.0000	−1.0000
6	1.0000	−2.0000	−1.0000

Table 3 Gauss–Seidel Iterations

▶ Example 3 is based on the work of Nobel Prize–winning economist Wassily Leontief (1906–1999). He divided the economy into 500 sectors in developing his input-output model.

EXAMPLE 3   Imagine a simple economy that consists of consumers and just three industries, which we refer to as A, B, and C. These industries have annual consumer sales of 60, 75, and 40 (in billions of dollars), respectively. In addition, for every dollar of goods A sells, A requires 10 cents of goods from B and 15 cents of goods from C to support production. (For instance, maybe B sells electricity and C sells shipping services.) Similarly, each dollar of goods B sells requires 20 cents of goods from A and 5 cents of goods from C, and each dollar of goods C sells requires 25 cents of goods from A and 15 cents of goods from B. What output from each industry will satisfy both consumer and between-industry demand?

Solution   Let a, b, and c denote the total output from each of A, B, and C, respectively. The entire output for A is 60 for consumers, $0.20b$ for B, and $0.25c$ for C. Totaling this up yields the equation

$$a = 60 + 0.20b + 0.25c$$

Similar reasoning applied to industries B and C yields the equations

$$b = 75 + 0.10a + 0.15c$$
$$c = 40 + 0.15a + 0.05b$$

Here we apply Gauss–Seidel iteration to find a solution. Since we are given the consumer demand for each industry, we take that as our starting point, initially setting $a = 60$, $b = 75$, and $c = 40$. For the first two iterations, we have

Iteration 1: $a = 60 + 0.20(75) + 0.25(40) = 85$
$b = 75 + 0.10(85) + 0.15(40) = 89.5$
$c = 40 + 0.15(85) + 0.05(89.5) = 57.225$

Iteration 2: $a = 60 + 0.20(89.5) + 0.25(57.225) :\approx 92.2063$
$b = 75 + 0.10(92.2063) + 0.15(57.225) :\approx 92.8044$
$c = 40 + 0.15(92.2063) + 0.05(92.8044) :\approx 58.4712$

n	a	b	c
0	60	75	40
1	85	89.5	57.225
2	92.2063	92.8044	58.4712
3	93.1787	93.0885	58.6312
4	93.2755	93.1222	58.6474
5	93.2863	93.1257	58.6492
6	93.2875	93.1261	58.6494
7	93.2876	93.1262	58.6494
8	93.2876	93.1262	58.6494

Table 4 **Gauss–Seidel Iterations for Example 3**

Additional iterations are shown in Table 4, and suggest convergence to $a = 93.2876$, $b = 93.1262$, and $c = 58.6494$. These match the exact solution to four decimal places. ∎

Convergence

Gaussian and Gauss–Jordan elimination are called *direct methods*, because they will always yield the solution in a finite number of steps (ignoring the potential problems brought about by round-off). On the other hand, as noted earlier, Jacobi and Gauss–Seidel are

n	x_1	x_2
0	0	0
1	6	11
2	-27	-55
3	171	341
4	-1017	-2035
5	6111	12221

Table 5   Gauss–Seidel
Iterations

Definition **Diagonally
Dominant**

iterative methods and do not converge to a solution in all cases. For instance, applying Gauss–Seidel iteration starting at $x_1 = x_2 = 0$ to the system

$$\begin{aligned} x_1 + 3x_2 &= 6 \\ 2x_1 - x_2 &= 1 \end{aligned} \qquad (4)$$

yields the sequence shown in Table 5. The values grow quickly in absolute value, and do not converge.

One case where we are guaranteed convergence is if the coefficients of the system are **diagonally dominant**. This means that for each equation of the system, the coefficient a_{ii} (in equation i) along the diagonal has absolute value larger than the sum of the absolute values of the other coefficients in the equation. For example, the system

$$\begin{aligned} 7x_1 - 3x_2 + 2x_3 &= 6 \\ x_1 + 5x_2 - 2x_3 &= 1 \\ -3x_1 + x_2 - 6x_3 &= -4 \end{aligned} \qquad (5)$$

is diagonally dominant because

$$\begin{aligned} |7| &> |-3| + |2| \\ |5| &> |1| + |-2| \\ |-6| &> |-3| + |1| \end{aligned}$$

On the other hand, the system

▶ Diagonal dominance is not required in order for the iterative methods to converge. There are instances where convergence occurs without diagonal dominance.

$$\begin{aligned} -2x_1 + x_2 - 9x_3 &= 0 \\ 6x_1 - x_2 + 4x_3 &= -12 \\ -x_1 + 4x_2 - x_3 &= 3 \end{aligned} \qquad (6)$$

is not diagonally dominant as expressed, but reordering the equations to

$$\begin{aligned} 6x_1 - x_2 + 4x_3 &= -12 \\ -x_1 + 4x_2 - x_3 &= 3 \\ -2x_1 + x_2 - 9x_3 &= 0 \end{aligned} \qquad (7)$$

makes it diagonally dominant.

EXAMPLE 4   Reverse the order of the equations in 4 to make the system diagonally dominant, and then find the solution using Gauss–Seidel iteration.

Solution   Reversing the order of the equations gives us

$$\begin{aligned} 2x_1 - x_2 &= 1 \\ x_1 + 3x_2 &= 6 \end{aligned}$$

which is diagonally dominant. Next, we solve for x_1 and x_2 (rounded to four decimal places),

$$\begin{aligned} x_1 &= 0.5 + 0.5x_2 \\ x_2 &= 2 - 0.3333x_1 \end{aligned}$$

Starting with $x_1 = 0$ and $x_2 = 0$, we have

Iteration 1: $x_1 = 0.5 + 0.5(0) = 0.5$
$x_2 = 2 - 0.3333(0.5) :\approx 1.8333$

Iteration 2: $x_1 = 0.5 + 0.5(1.8333) :\approx 1.4167$
$x_2 = 2 - 0.3333(1.4167) :\approx 1.5278$

n	x_1	x_2
0	0	0
1	0.5	1.8333
2	1.4167	1.5278
3	1.2639	1.5787
4	1.2894	1.5702
5	1.2851	1.5716
6	1.2858	1.5714
7	1.2857	1.5714
8	1.2857	1.5714

Table 6 Gauss–Seidel Iterations for Example 4

The first eight iterations are shown in Table 6, which shows convergence to $x_1 = 1.2857$ and $x_2 = 1.5714$. These match the exact solutions, which are $x_1 = 9/7$ and $x_2 = 11/7$. ∎

Computational Comments

- Our iterative methods do not suffer from the round-off errors that can afflict elimination methods. Since the values from one iteration can be thought of as an "initial guess" for the next, there is no accumulation of errors. For the same reason, if a computation error is made, the result still can be used in the next iteration. By contrast, if a computation error is made when using elimination methods, the end result is almost always wrong.

- For a system of n equations with n unknowns, Jacobi and Gauss–Seidel both require about $2n^2$ flops per iteration. As mentioned earlier, Gauss–Seidel usually converges in fewer iterations than Jacobi, so Gauss–Seidel is typically the preferred method.

- If we ignore potential round-off issues and go solely by the number of flops, then Gaussian elimination requires about $2n^3/3$ flops, versus $2n^2$ flops per iteration for Gauss–Seidel. Hence, as long as Gauss–Seidel converges in fewer than $n/3$ iterations, this will be the more efficient method.

▶ See *Matrix Computations* by G. Golub and C. Van Loan for a more extensive discussion of iterative methods and an explanation of why those described here work.

- The rate of convergence of our iterative methods is influenced by the degree of diagonal dominance of the system. If the diagonal terms are much larger than the others, then iterative methods generally will converge relatively quickly. If the diagonal terms are only slightly dominant, then although iterative methods eventually will converge, they can be too slow to be practical. There are other iterative methods besides those presented here that are designed to have better convergence properties.

Definition Sparse System, Sparse Matrix

- Iterative methods are particularly useful for solving **sparse systems**, which are linear systems where most of the coefficients are zero. The augmented matrix of such a system has mostly zero entries and is said to be a **sparse matrix**. Elimination methods applied to sparse systems have a tendency to change the zeros to nonzero terms, removing the sparseness.

| EXERCISES |

In each exercise set, problems marked with Ⓒ are designed to be solved using a programmable calculator or computer algebra system.

In Exercises 1–4, use partial pivoting with Gaussian elimination to find the solutions to the given system.

1. $-2x_1 + 3x_2 = 4$
$5x_1 - 2x_2 = 1$

2. $x_1 - 2x_2 = -1$
$-3x_1 + 7x_2 = 5$

3. $x_1 + x_2 - 2x_3 = -3$
$3x_1 - 2x_2 + 2x_3 = 9$
$6x_1 - 7x_2 - x_3 = 4$

4. $x_1 - 3x_2 + 2x_3 = 4$
$-2x_1 + 7x_2 - 2x_3 = -7$
$4x_1 - 13x_2 + 7x_3 = 12$

Ⓒ In Exercises 5–8, solve the system as given with Gaussian elimination with three significant digits of accuracy. Then solve the system again, incorporating partial pivoting.

5. $2x_1 + 975x_2 = 41$
$53x_1 - 82x_2 = -13$

6. $3x_1 - 813x_2 = 32$
$71x_1 - 93x_2 = -5$

7. $3x_1 - 7x_2 + 639x_3 = 12$
$-2x_1 + 5x_2 + 803x_3 = 7$
$56x_1 - 41x_2 + 79x_3 = 10$

8. $2x_1 - 5x_2 + 802x_3 = -1$
$-x_1 + 3x_2 - 789x_3 = -8$
$40x_1 + 34x_2 + 51x_3 = 19$

Ⓒ In Exercises 9–12, compute the first three Jacobi iterations for the given system, using 0 as the initial value for each variable. Then find the exact solution and compare.

9. $-5x_1 + 2x_2 = 6$
$3x_1 + 10x_2 = 2$

10. $2x_1 - x_2 = -4$
$-4x_1 + 5x_2 = 11$

11. $20x_1 + 3x_2 + 5x_3 = -26$
$-2x_1 - 10x_2 + 3x_3 = -23$
$x_1 - 2x_2 - 5x_3 = -13$

12. $-2x_1 \quad + \quad x_3 = 5$
$-x_1 + 5x_2 - \quad x_3 = 8$
$2x_1 - 6x_2 + 10x_3 = 16$

C In Exercises 13–16, compute the first three Gauss–Seidel iterations for the given system, using 0 as the initial value for each variable. Then find the exact solution and compare.

13. The system given in Exercise 9.

14. The system given in Exercise 10.

15. The system given in Exercise 11.

16. The system given in Exercise 12.

In Exercises 17–20, determine if the given system is diagonally dominant. If not, then (if possible) rewrite the system so that it is diagonally dominant.

17. $2x_1 - 5x_2 = 7$
$3x_1 + 7x_2 = 4$

18. $4x_1 + 2x_2 - \quad x_3 = 13$
$-2x_1 + 7x_2 + 2x_3 = -9$
$x_1 + 3x_2 - 5x_3 = 6$

19. $3x_1 + 6x_2 - \quad x_3 = 0$
$-x_1 - 2x_2 + 4x_3 = -1$
$7x_1 + 5x_2 - 3x_3 = 3$

20. $-2x_1 + 6x_2 = 12$
$5x_1 - \quad x_2 = -4$

C In Exercises 21–24, compute the first four Jacobi iterations for the system as written, with the initial value of each variable set equal to 0. Then rewrite the system so that it is diagonally dominant, set the value of each variable to 0, and again compute 4 Jacobi iterations.

21. $x_1 - 2x_2 = -1$
$2x_1 - \quad x_2 = 1$

22. $x_1 - 3x_2 = -2$
$3x_1 - \quad x_2 = 2$

23. $x_1 - 2x_2 + 5x_3 = -1$
$5x_1 + \quad x_2 - 2x_3 = 8$
$2x_1 - 10x_2 + 3x_3 = -1$

24. $2x_1 + 4x_2 - 10x_3 = -3$
$3x_1 - \quad x_2 + \quad x_3 = 7$
$-x_1 + 6x_2 - 2x_3 = -6$

C In Exercises 25–28, compute the first four Gauss–Seidel iterations for the system as written, with the initial value of each variable set equal to 0. Then rewrite the system so that it is diagonally dominant, set the value of each variable to 0, and again compute four Gauss–Seidel iterations.

25. The system given in Exercise 21.

26. The system given in Exercise 22.

27. The system given in Exercise 23.

28. The system given in Exercise 24.

C In Exercises 29–30, the values from the first few Jacobi iterations are given for an unknown system. Find the values for the next iteration.

29.

n	x_1	x_2
0	0	0
1	1	-2
2	5	2
3	?	?

30.

n	x_1	x_2	x_3
0	0	0	0
1	-2	-1	1
2	-4	-4	5
3	-11	-4	5
4	?	?	?

C In Exercises 31–32, the values from the first few Gauss–Seidel iterations are given for an unknown system. Find the values for the next iteration.

31.

n	x_1	x_2
0	0	0
1	3	4
2	-5	-12
3	?	?

32.

n	x_1	x_2	x_3
0	0	0	0
1	3	4	12
2	7	-24	-76
3	-25	176	556
4	?	?	?

▶ This section is optional. Although some applications presented here are referred to later, they can be reviewed as needed.

1.4 Applications of Linear Systems

In this section we consider in depth some applications of linear systems. These are but a few of the many different possible applications that exist.

Traffic Flow

Arcata, on the northern coast of California, is a small college town with a central plaza (Figure 1). Figure 2 shows the streets surrounding and adjacent to the town's central plaza. As indicated by the arrows, all streets in the vicinity of the plaza are one-way.

Figure 1 The Arcata plaza. (Photo taken by Terrence McNally of Arcata Photo.)

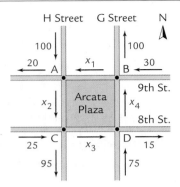

Figure 2 Traffic volumes around the Arcata plaza.

Traffic flows north and south on G and H streets, respectively, and east and west on 8th and 9th streets, respectively. The number of cars flowing on and off the plaza during a typical 15-minute period on a Saturday morning is also shown. Our goal is to find x_1, x_2, x_3, and x_4, the volume of traffic along each side of the plaza.

The four intersections are labeled A, B, C, and D. At each intersection, the number of cars entering the intersection must equal the number leaving. For example, the number of cars entering A is $100 + x_1$ and the number exiting is $20 + x_2$. Since these must be equal, we end up with the equation

$$\text{A:} \quad 100 + x_1 = 20 + x_2$$

Applying the same reasoning to intersections B, C, and D, we arrive at three more equations,

$$\text{B:} \quad x_4 + 30 = x_1 + 100$$
$$\text{C:} \quad x_2 + 25 = x_3 + 95$$
$$\text{D:} \quad x_3 + 75 = x_4 + 15$$

Rewriting the equations in the usual form, we obtain the system

$$\begin{aligned}
x_1 - x_2 &&&= -80 \\
x_1 && - x_4 &= -70 \\
x_2 - x_3 &&&= 70 \\
x_3 - x_4 &&&= -60
\end{aligned}$$

To solve the system, we populate an augmented matrix and transform to echelon form.

$$\begin{bmatrix} 1 & -1 & 0 & 0 & -80 \\ 1 & 0 & 0 & -1 & -70 \\ 0 & 1 & -1 & 0 & 70 \\ 0 & 0 & 1 & -1 & -60 \end{bmatrix} \overset{-R_1+R_2 \Rightarrow R_2}{\underset{\sim}{}} \begin{bmatrix} 1 & -1 & 0 & 0 & -80 \\ 0 & 1 & 0 & -1 & 10 \\ 0 & 1 & -1 & 0 & 70 \\ 0 & 0 & 1 & -1 & -60 \end{bmatrix}$$

$$\overset{-R_2+R_3 \Rightarrow R_3}{\underset{\sim}{}} \begin{bmatrix} 1 & -1 & 0 & 0 & -80 \\ 0 & 1 & 0 & -1 & 10 \\ 0 & 0 & -1 & 1 & 60 \\ 0 & 0 & 1 & -1 & -60 \end{bmatrix}$$

$$\overset{R_3+R_4 \Rightarrow R_4}{\underset{\sim}{}} \begin{bmatrix} 1 & -1 & 0 & 0 & -80 \\ 0 & 1 & 0 & -1 & 10 \\ 0 & 0 & -1 & 1 & 60 \\ 0 & 0 & 0 & 0 & 0 \end{bmatrix}$$

Back substitution yields the general solution

$$x_1 = -70 + s_1, \quad x_2 = 10 + s_1, \quad x_3 = -60 + s_1, \quad x_4 = s_1$$

where s_1 is a free parameter.

A moment's thought reveals why it makes sense that this system has infinitely many solutions. There can be an arbitrary number of cars simply circling the plaza, perhaps looking for a parking space. Note also that since each of x_1, x_2, x_3, and x_4 must be nonnegative, it follows that the parameter $s_1 \geq 70$.

The analysis performed here can be carried over to much more complex traffic questions, or to other similar settings, such as computer networks.

The BCS Ranking System

The BCS (Bowl Championship Series) is a system for ranking college football teams. The two teams ranked highest at the end of the regular season get to play in the national championship game. A "BCS Index" is calculated to rank the teams. The BCS Index is calculated by combining three team rankings from different sources:

USA: A survey of 62 college football coaches compiled by *USA Today*.

Harris: A survey of 114 college football experts compiled by *Harris Interactive*.

Computer: An average of computer-based rankings from various sources.

Table 1 gives the points awarded by each source to determine individual rankings together with the BCS Index. The higher the BCS Index, the higher the BCS ranking.

> ▶ A *conjecture* is the mathematical equivalent of an educated guess.

The information in Table 1 appeared frequently in the media, but the formula for computing the BCS Index rarely did. Our goal here is to deduce the BCS Index formula using linear algebra. To find our formula, we start by conjecturing that the BCS Index is a linear combination of the three components, so that each team's data will satisfy

$$x_1(\text{USA}) + x_2(\text{Harris}) + x_3(\text{Computer}) = \text{BCS Index}$$

for the right choice of x_1, x_2, and x_3. For example, using the data for Oklahoma gives us the equation

$$1482x_1 + 2699x_2 + 100x_3 = 0.9757$$

> ▶ This system can be solved using the methods presented in Section 1.2 or Section 1.3, but here a computer algebra system was used. There is a small amount of rounding in the solution.

Data from other schools can be used in the same way to obtain additional equations. Since we have three unknowns, we need three equations in order to find the values of x_1, x_2, and x_3. Taking the top three schools, we arrive at the linear system

$$1482x_1 + 2699x_2 + 100x_3 = 0.9757$$
$$1481x_1 + 2776x_2 + 89x_3 = 0.9479$$
$$1408x_1 + 2616x_2 + 94x_3 = 0.9298$$

Rank	Team	USA	Harris	Computer	BCS Index
1	Oklahoma	1482	2699	100	0.9757
2	Florida	1481	2776	89	0.9479
3	Texas	1408	2616	94	0.9298
4	Alabama	1309	2442	81	0.8443
5	Southern Cal	1309	2413	75	0.8208
6	Penn State	1193	2186	66	0.7387

Table 1 The 2008 Final Regular Season BCS Ranks

This system has unique solution

$$x_1 = 0.0002151, \quad x_2 = 0.0001195, \quad x_3 = 0.003344$$

This gives us the formula

$$\text{BCS Index} = 0.0002151(\text{USA}) + 0.0001195(\text{Harris}) + 0.003344(\text{Computer})$$

To check if we have the right formula, let's test it out on some schools not represented in our system, say, Alabama, Southern Cal, and Penn State.

Alabama:	$0.0002151(1309) + 0.0001195(2442) + 0.003344(81) = 0.8443$
Southern Cal:	$0.0002151(1309) + 0.0001195(2413) + 0.003344(75) = 0.8207$
Penn State:	$0.0002151(1193) + 0.0001195(2186) + 0.003344(66) = 0.7386$

Other than small differences due to rounding, the formula checks out. Thus it is reasonable to assume that we have found the correct formula for the BCS index.

▶ At the time of this writing, a limited playoff system has been approved.

Finally, we note that the BCS index formula has been changed several times in the past, so it may no longer have this form. (See Exercise 32 for an older version that was more complicated.) However, if you obtain a version of Table 1 for the most recent college football season, you can probably perform the same analysis as we have here to find the current formula for the BCS index.

Planetary Orbital Periods

Most people are aware that the planets that are closer to the sun take a shorter amount of time to make one orbit around the sun than those that are farther out. Table 2 gives the average distance from the sun and the number of Earth days required to make one orbit.

Our goal here is to develop an equation that describes the relationship between the distance from the sun and the length of the orbital period. As a starting point, consider the scatter plot of the data given in Figure 3.

There seems to be a pattern to the data. The points do not lie on a line, but the curved shape suggests that for constants a and b, the data may come close to satisfying the equation

$$p = ad^b \tag{1}$$

where p is the orbital period and d is the distance from the sun. Here we proceed as in the BCS example, substituting data to create a system of equations to solve. However, before doing that, we note equation (1) is not linear in a and b, but it can be if we apply

Planet	Distance from Sun ($\times 10^6$ km)	Orbital Period (days)
Mercury	57.9	88
Venus	108.2	224.7
Earth	149.6	365.2
Mars	227.9	687
Jupiter	778.6	4331
Saturn	1433.5	10747
Uranus	2872.5	30589
Neptune	4495.1	59800

Table 2 Planetary Orbital Distances and Periods

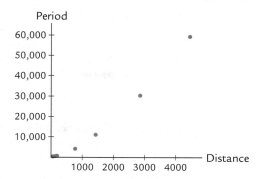

Figure 3 Orbital Distance vs. Orbit Period.

the logarithm function to both sides. This gives us

$$\ln(p) = \ln(ad^b)$$
$$= \ln(a) + b\ln(d)$$

If we let $a_1 = \ln(a)$ and substitute the data from Mercury and Venus, we get the system of two equations and two unknowns

$$a_1 + b\ln(57.9) = \ln(88)$$
$$a_1 + b\ln(108.2) = \ln(224.7)$$

The solution to this system is $a_1 \approx -1.60771$ and $b \approx 1.49925$. Since $a_1 = \ln(a)$, we have $a \approx e^{-1.60771} = 0.200346$, yielding the formula

$$p = (0.200346)d^{1.49925}$$

Table 3 gives the actual and predicted (using the above formula) orbital period for each planet.

Planet	Distance	Actual Period	Predicted Period
Mercury	57.9	88	87.9988
Venus	108.2	224.7	224.696
Earth	149.6	365.2	365.214
Mars	227.9	687	686.482
Jupiter	778.6	4331	4330.96
Saturn	1433.5	10,747	10,814.6
Uranus	2872.5	30,589	30,644.3
Neptune	4495.1	59,800	59,999.8

Table 3 **Planetary Orbital Distances and Periods**

The predictions are fairly good, suggesting that our formula is on the right track. However, the predictions become less accurate for those planets farther from the sun. Because we used the data for Mercury and Venus to develop our formula, perhaps this is not surprising. If instead we use Uranus and Neptune, we arrive at the formula

$$p = (0.20349)d^{1.497}$$

Table 4 shows this formula produces better predictions.

Planet	Predicted Period
Mercury	88.6
Venus	225.8
Earth	366.8
Mars	688.8
Jupiter	4334
Saturn	10,806
Uranus	30,589
Neptune	59,799

Table 4 **Predicted Orbital Periods**

A natural idea is to incorporate more data into our formula, by using more planets to generate a larger system of equations. Unfortunately, if we use more than two planets, we end up with a system that has no solutions. (Try it for yourself.) Thus there are limitations to what we can do with the tools we currently have available. In Chapter 8 we develop a more sophisticated method that allows us to use all of our data simultaneously to come up with a formula that provides a good estimate for a range of distances from the sun.

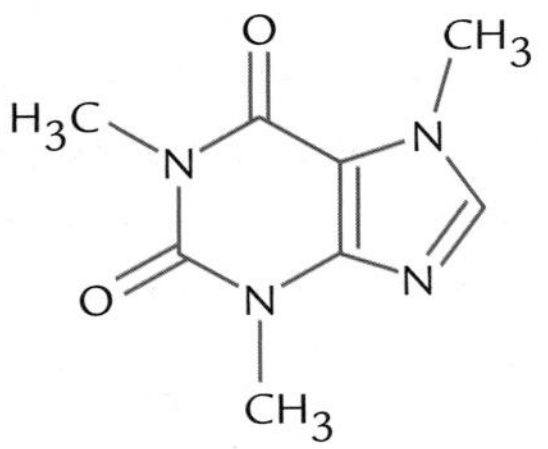

Figure 4 The caffeine molecule. (Source: Wikimedia Commons; Author: NEUROtiker)

Balancing Chemical Equations

A popular chemical among college students is caffeine, which has chemical composition $C_8H_{10}N_4O_2$. When heated and combined with oxygen (O_2), the ensuing reaction produces carbon dioxide (CO_2), water (H_2O), and nitrogen dioxide (NO_2). This chemical reaction is indicated using the notation

$$x_1 C_8H_{10}N_4O_2 + x_2 O_2 \rightarrow x_3 CO_2 + x_4 H_2O + x_5 NO_2 \tag{2}$$

where the subscripts on the elements indicate the number of atoms. (No subscript indicates one atom.) Balancing the equation involves finding values for $x_1, x_2, x_3, x_4,$ and x_5 so that the number of atoms of each element is the same before and after the reaction. Most chemistry texts describe a method of solution that is best described as trial and error. However, there is no need for a haphazard approach—we can use linear algebra.

In the reaction in 2, let's start with carbon. On the left side there are $8x_1$ carbon atoms, while on the right side there are x_3 carbon atoms. This yields the equation

$$8x_1 = x_3$$

For oxygen, we see that there are $2x_1 + 2x_2$ atoms on the left side and $2x_3 + x_4 + 2x_5$ on the right, producing another equation,

$$2x_1 + 2x_2 = 2x_3 + x_4 + 2x_5$$

Similar analysis on nitrogen and hydrogen results in two additional equations,

$$4x_1 = x_5 \quad \text{and} \quad 10x_1 = 2x_4$$

To balance the chemical equation, we must find a solution that satisfies all four equations. That is, we need to find the solution set to the linear system

$$
\begin{aligned}
2x_1 + 2x_2 - 2x_3 - x_4 - 2x_5 &= 0 \\
4x_1 \qquad\qquad\qquad - x_5 &= 0 \\
8x_1 \qquad - x_3 \qquad\qquad &= 0 \\
10x_1 \qquad\qquad - 2x_4 \qquad &= 0
\end{aligned}
$$

The augmented matrix and row operations are

$$
\begin{bmatrix}
2 & 2 & -2 & -1 & -2 & 0 \\
4 & 0 & 0 & 0 & -1 & 0 \\
8 & 0 & -1 & 0 & 0 & 0 \\
10 & 0 & 0 & -2 & 0 & 0
\end{bmatrix}
\begin{array}{l} -2R_1+R_2 \Rightarrow R_2 \\ -4R_1+R_3 \Rightarrow R_3 \\ -5R_1+R_4 \Rightarrow R_4 \\ \sim \end{array}
\begin{bmatrix}
2 & 2 & -2 & -1 & -2 & 0 \\
0 & -4 & 4 & 2 & 3 & 0 \\
0 & -8 & 7 & 4 & 8 & 0 \\
0 & -10 & 10 & 3 & 10 & 0
\end{bmatrix}
$$

$$
\begin{array}{l} -2R_2+R_3 \Rightarrow R_3 \\ -\frac{5}{2}R_2+R_4 \Rightarrow R_4 \\ \sim \end{array}
\begin{bmatrix}
2 & 2 & -2 & -1 & -2 & 0 \\
0 & -4 & 4 & 2 & 3 & 0 \\
0 & 0 & -1 & 0 & 2 & 0 \\
0 & 0 & 0 & -2 & \frac{5}{2} & 0
\end{bmatrix}
$$

Back substituting and scaling the free parameter gives the general solution

$$x_1 = 2s_1$$
$$x_2 = 27s_1$$
$$x_3 = 16s_1$$
$$x_4 = 10s_1$$
$$x_5 = 8s_1$$

where s_1 can be any real number. Any choice of s_1 yields constants that balance our chemical equation, but it is customary to select the specific solution that makes each of the coefficients x_1, x_2, x_3, x_4, and x_5 integers that have no common factors. Setting $s_1 = 1$ accomplishes this, yielding the balanced equation

$$2C_8H_{10}N_4O_2 + 27O_2 \rightarrow 16CO_2 + 10H_2O + 8NO_2$$

| EXERCISES |

In each exercise set, problems marked with $\boxed{C}$ are designed to be solved using a programmable calculator or computer algebra system.

1. The volume of traffic for a collection of intersections is shown in the figure below. Find all possible values for x_1, x_2, and x_3. What is the minimum volume of traffic from C to A?

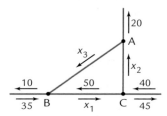

2. $\boxed{C}$ The volume of traffic for a collection of intersections is shown in the figure below. Find all possible values for x_1, x_2, x_3, and x_4. What is the minimum volume of traffic from C to D?

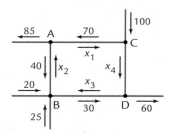

3. $\boxed{C}$ The volume of traffic for a collection of intersections is shown in the figure below. Find all possible values for x_1, x_2, x_3, and x_4. What is the minimum volume of traffic from C to A?

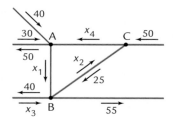

4. $\boxed{C}$ The volume of traffic for a collection of intersections is shown in the figure below. Find all possible values for x_1, x_2, x_3, x_4, x_5, and x_6.

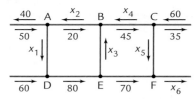

In Exercises 5–12, balance the given chemical equations.

5. Hydrogen burned in oxygen forms steam:
$$H_2 + O \longrightarrow H_2O$$

6. Hydrogen and nitrogen combine to form ammonia:
$$H_2 + N_2 \longrightarrow NH_3$$

7. Iron and oxygen combine to form iron oxide:
$$Fe + O_2 \longrightarrow Fe_2O_3$$

8. Sodium and water react to form sodium hydroxide (lye) and hydrogen:
$$Na + H_2O \longrightarrow NaOH + H_2$$

9. When propane burns in oxygen, it produces carbon dioxide and water:
$$C_3H_8 + O_2 \longrightarrow CO_2 + H_2O$$

10. When acetylene burns in oxygen, it produces carbon dioxide and water:
$$C_2H_2 + O_2 \longrightarrow CO_2 + H_2O$$

11. Potassium superoxide and carbon dioxide react to form potassium carbonate and oxygen:
$$KO_2 + CO_2 \longrightarrow K_2CO_3 + O_2$$

12. Manganese dioxide and hydrochloric acid combine to form manganese chloride, water, and chlorine gas:
$$MnO_2 + HCl \longrightarrow MnCl_2 + H_2O + Cl_2$$

In Exercises 13–16, find a model for planetary orbital period using the data for the given planets.

13. Earth and Mars.

14. Mercury and Uranus.

15. Venus and Neptune.

16. Jupiter and Saturn.

In Exercises 17–18, the data given provides the distance required for a particular type of car to stop when traveling at a variety of speeds. A reasonable model for braking distance is $d = as^k$, where d is distance, s is speed, and a and k are constants. Use the data in the table to find values for a and k, and test your model. (HINT: Methods similar to those used to find a model for planetary orbital periods can be applied here.)

17.

Speed (MPH)	10	20	30	40
Distance (Feet)	4.5	18	40.5	72

18.

Speed (MPH)	10	20	30	40
Distance (Feet)	20	80	180	320

When using partial fractions to find antiderivatives in calculus, we decompose complicated rational expressions into the sum of simpler expressions that can be integrated individually. In Exercises 19–22, the required decomposition is given. Find the values of the missing constants.

19. $\dfrac{1}{x(x + 1)} = \dfrac{A}{x} + \dfrac{B}{x + 1}$

20. $\dfrac{3x - 1}{(x - 1)(x + 1)} = \dfrac{A}{x - 1} + \dfrac{B}{x + 1}$

21. $\dfrac{1}{x^2(x - 1)} = \dfrac{A}{x} + \dfrac{B}{x^2} + \dfrac{C}{x - 1}$

22. $\dfrac{1}{x(x^2 + 1)} = \dfrac{A}{x} + \dfrac{Bx + C}{x^2 + 1}$

23. The points $(1, 3)$ and $(-2, 6)$ lie on a line. Where does the line cross the x-axis?

24. The points $(2, -1, -2), (1, 3, 12),$ and $(4, 2, 3)$ lie on a unique plane. Where does this plane cross the z-axis?

25. The equation for a parabola has the form $y = ax^2 + bx + c$, where a, b, and c are constants and $a \neq 0$. Find an equation for the parabola that passes through the points $(-1, -10), (1, -4),$ and $(2, -7)$.

26. ⓒ Find a polynomial of the form

$$f(x) = ax^3 + bx^2 + cx + d$$

such that $f(0) = -3$, $f(1) = 2$, $f(3) = 5$, and $f(4) = 0$.

27. ⓒ Find a polynomial of the form

$$g(x) = ax^4 + bx^3 + cx^2 + dx + e$$

such that $g(-2) = -17$, $g(-1) = 6$, $g(0) = 5$, $g(1) = 4$, and $g(2) = 3$.

ⓒ (Calculus required) In Exercises 28–29, find the values of the coefficients a, b, and c so the given conditions for the function f and its derivatives are met. (This type of problem arises in the study of differential equations.)

28. $f(x) = ae^x + be^{2x} + ce^{-3x}$; $f(0) = 2$, $f'(0) = 1$, and $f''(0) = 19$.

29. $f(x) = ae^{-2x} + be^x + cxe^x$; $f(0) = -1$, $f'(0) = -2$, and $f''(0) = 3$.

ⓒ In Exercises 30–31, a new "LAI" (for Linear Algebra Index) formula has been used to rank the eight college football teams shown. The new formula uses the same components as the 2008 BCS formula described earlier. Determine the formula for the LAI, and test it to be sure it is correct.

30.

Rank	Team	LAI
1	Oklahoma	0.9655
2	Florida	0.9652
3	Texas	0.9237
4	Alabama	0.8538
5	Southern Cal	0.8436
6	Penn State	0.7646
7	Utah	0.7560
8	Texas Tech	0.7522

31.

Rank	Team	LAI
1	Oklahoma	0.9895
2	Texas	0.9364
3	Florida	0.9204
4	Texas Tech	0.8285
5	Alabama	0.8284
6	Utah	0.8236
7	Southern Cal	0.7866
8	Penn State	0.7004

32. ⓒ The BCS ranking system was more complicated in 2001 than in 2008. The table below gives the BCS rankings at the end of the regular season. (A lower BCS Index gave a higher rank.)

Table headings:

- AP and USA gives the rank of each team in the two opinion polls of writers and coaches, respectively.

- SS stands for strength of schedule ranking, with 1 being the most challenging.

- L is the number of losses during the season.

- CA (Computer Average) is the average of computer rankings from various sources.

- QW (Quality Wins) gives a measurement of the number of victories over highly ranked teams.

Rank	Team	AP	USA	SS	L	CA	QW	BCS Index
1	Miami	1	1	18	0	1.00	0.1	2.62
2	Nebraska	4	4	14	1	2.17	0.5	7.23
3	Colorado	3	3	2	2	4.50	2.3	7.28
4	Oregon	2	2	31	1	4.83	0.4	8.67
5	Florida	5	5	19	2	5.83	0.5	13.09
6	Tennessee	8	8	3	2	6.17	1.6	14.69
7	Texas	9	9	33	2	6.67	1.2	17.79
8	Illinois	7	7	37	1	9.83	0.0	19.31
9	Stanford	11	11	22	2	7.83	1.3	20.41
10	Maryland	6	6	78	1	11.17	0.0	21.29
11	Oklahoma	10	10	36	2	9.00	0.9	21.54
12	Washington St	13	13	42	2	10.83	0.6	26.91
13	LSU	12	12	10	3	13.33	1.0	27.73
14	South Carolina	14	14	40	3	19.17	0.0	37.77
15	Washington	21	20	21	3	14.83	1.0	38.17

Find the 2001 BCS ranking formula. Test it for three schools not used to develop your formula to check for correctness. (HINT: To avoid a system with infinitely many solutions, include Washington among the schools used to develop the formula. Explain why including Washington will accomplish this.)

Euclidean Space

The Willis Family Bridge provides a walkway from the main campus of Indiana University–Purdue University Fort Wayne to the Waterfield Campus Student Housing. The bridge was designed by Kurt Heidenreich and dedicated in 2003. The triangular design accommodates the need to cross Crescent Ave. as well an area of uneven terrain. The two angled pylons and four support cables suggest a diagram for the addition of force vectors; the ellipse at the top of the triangle is formed by the intersection of two circular cylinders.

Bridge suggested by Adam Coffman, Indiana University – Purdue University Fort Wayne (James E. Whitcraft)

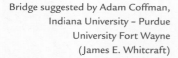

We can think of algebra as the study of the properties of arithmetic on the real numbers. In linear algebra, we study the properties of arithmetic performed on objects called *vectors*. As we shall see shortly, one use of vectors is to compactly describe the set of solutions of a linear system, but vectors have many other applications as well. Section 2.1 gives an introduction to vectors, arithmetic with vectors, and the geometry of vectors. Section 2.2 and Section 2.3 describe important properties of sets of vectors.

2.1 Vectors

In this section we introduce vectors, which for the moment we can think of as a list of numbers. We start with a specific example of vectors that occur in plain sight, on packages of plant fertilizer. Fertilizer is sold in bags labelled with three numbers that indicate the amount of nitrogen (N), phosphoric acid (P_2O_5), and potash (K_2O) present. The mixture of these nutrients varies from one type of fertilizer to the next. For example, a bag of *Vigoro Ultra Turf* has the numbers "29–3–4," which means that 100 pounds of this fertilizer contains 29 pounds of nitrogen, 3 pounds of phosphoric acid, and 4 pounds of potash. Organizing these quantities vertically in a matrix, we have

$$\begin{bmatrix} 29 \\ 3 \\ 4 \end{bmatrix}$$

This representation is an example of a *vector*. Using a vector provides a convenient way to record the amounts of each nutrient and also lends itself to compact forms of algebraic operations that arise naturally. For instance, if we want to know the amount of nitrogen, phosphoric acid, and potash contained in a ton (2000 pounds) of Ultra Turf, we just multiply each vector entry by 20. If we think of this as multiplying the vector by 20, then it is reasonable to represent this operation by

$$20 \begin{bmatrix} 29 \\ 3 \\ 4 \end{bmatrix}$$

so that we have

$$20 \begin{bmatrix} 29 \\ 3 \\ 4 \end{bmatrix} = \begin{bmatrix} 20 \cdot 29 \\ 20 \cdot 3 \\ 20 \cdot 4 \end{bmatrix} = \begin{bmatrix} 580 \\ 60 \\ 80 \end{bmatrix}$$

Note that the "=" sign between the vectors means that the entries in corresponding positions are equal.

Another type of fertilizer, *Parker's Premium Starter*, has 18 pounds of nitrogen, 25 pounds of phosphoric acid, and 6 pounds of potash per 100 pounds, which is represented in vector form by

$$\begin{bmatrix} 18 \\ 25 \\ 6 \end{bmatrix}$$

If we mix together 100 pounds of each type of fertilizer, then we can find the total amount of each nutrient in the mixture by adding entries in each of the vectors. Thinking of this as adding the vectors, we have

$$\begin{bmatrix} 29 \\ 3 \\ 4 \end{bmatrix} + \begin{bmatrix} 18 \\ 25 \\ 6 \end{bmatrix} = \begin{bmatrix} 29 + 18 \\ 3 + 25 \\ 4 + 6 \end{bmatrix} = \begin{bmatrix} 47 \\ 28 \\ 10 \end{bmatrix}$$

Vectors and $\mathbf{R}^n$

We formalize our notion of vector with the following definition.

Definition Vector

DEFINITION 2.1

A **vector** is an ordered list of real numbers $u_1, u_2, \ldots, u_n$ expressed as

$$\mathbf{u} = \begin{bmatrix} u_1 \\ u_2 \\ \vdots \\ u_n \end{bmatrix}$$

Definition $\mathbf{R}^n$

or as $\mathbf{u} = (u_1, u_2, \ldots, u_n)$. The set of all vectors with n entries is denoted by $\mathbf{R}^n$.

Our convention will be to denote vectors using boldface, as in $\mathbf{u}$. Each of the entries $u_1, u_2, \ldots, u_n$ is called a **component** of the vector. A vector expressed in the vertical form is also called a **column vector**, and a vector expressed in horizontal form is also called a **row vector**. It is customary to express vectors in column form, but we will occasionally use row form to save space.

Definition Component
Definition Column Vector
Definition Row Vector

The fertilizer discussion provides a good model for how vector arithmetic works. Here we formalize the definitions.

DEFINITION 2.2

Definition Vector Arithmetic,
Scalar, Euclidean Space

Let **u** and **v** be vectors in $\mathbf{R}^n$ given by

$$\mathbf{u} = \begin{bmatrix} u_1 \\ u_2 \\ \vdots \\ u_n \end{bmatrix} \quad \text{and} \quad \mathbf{v} = \begin{bmatrix} v_1 \\ v_2 \\ \vdots \\ v_n \end{bmatrix}$$

Suppose that c is a real number, called a **scalar** in this context. Then we have the following definitions:

Equality: $\mathbf{u} = \mathbf{v}$ if and only if $u_1 = v_1, u_2 = v_2, \ldots, u_n = v_n$.

Addition: $\mathbf{u} + \mathbf{v} = \begin{bmatrix} u_1 \\ u_2 \\ \vdots \\ u_n \end{bmatrix} + \begin{bmatrix} v_1 \\ v_2 \\ \vdots \\ v_n \end{bmatrix} = \begin{bmatrix} u_1 + v_1 \\ u_2 + v_2 \\ \vdots \\ u_n + v_n \end{bmatrix}$

Scalar Multiplication: $c\mathbf{u} = c \begin{bmatrix} u_1 \\ u_2 \\ \vdots \\ u_n \end{bmatrix} = \begin{bmatrix} c \cdot u_1 \\ c \cdot u_2 \\ \vdots \\ c \cdot u_n \end{bmatrix}$

The set of all vectors in $\mathbf{R}^n$, taken together with these definitions of addition and scalar multiplication, is called **Euclidean space**.

▶ Euclidean space is named for the Greek mathematician Euclid, the father of geometry. Euclidean space is an example of a *vector space*, discussed in Chapter 7.

Although vectors with negative components and negative scalars do not make sense in the fertilizer discussion, they do in other contexts and are included in Definition 2.2.

EXAMPLE 1   Suppose that

$$\mathbf{u} = \begin{bmatrix} 2 \\ -3 \\ 0 \\ -1 \end{bmatrix} \quad \text{and} \quad \mathbf{v} = \begin{bmatrix} -4 \\ 6 \\ -2 \\ 7 \end{bmatrix}$$

Find $\mathbf{u} + \mathbf{v}$, $-4\mathbf{v}$, and $2\mathbf{u} - 3\mathbf{v}$.

▶ Two vectors can be equal only if they have the same number of components. Similarly, there is no way to add two vectors that have a different number of components.

Solution The solutions to the first two parts are

$$\mathbf{u} + \mathbf{v} = \begin{bmatrix} 2 \\ -3 \\ 0 \\ -1 \end{bmatrix} + \begin{bmatrix} -4 \\ 6 \\ -2 \\ 7 \end{bmatrix} = \begin{bmatrix} 2 - 4 \\ -3 + 6 \\ 0 - 2 \\ -1 + 7 \end{bmatrix} = \begin{bmatrix} -2 \\ 3 \\ -2 \\ 6 \end{bmatrix}$$

$$-4\mathbf{v} = -4 \begin{bmatrix} -4 \\ 6 \\ -2 \\ 7 \end{bmatrix} = \begin{bmatrix} -4(-4) \\ -4(\ 6) \\ -4(-2) \\ -4(\ 7) \end{bmatrix} = \begin{bmatrix} 16 \\ -24 \\ 8 \\ -28 \end{bmatrix}$$

The third computation has a slight twist because we have not yet defined the difference of two vectors. But subtraction works exactly as we would expect and follows from the natural interpretation that $2\mathbf{u} - 3\mathbf{v} = 2\mathbf{u} + (-3)\mathbf{v}$.

$$2\mathbf{u} - 3\mathbf{v} = 2 \begin{bmatrix} 2 \\ -3 \\ 0 \\ -1 \end{bmatrix} - 3 \begin{bmatrix} -4 \\ 6 \\ -2 \\ 7 \end{bmatrix} = \begin{bmatrix} 2(\ 2) - 3(-4) \\ 2(-3) - 3(\ 6) \\ 2(\ 0) - 3(-2) \\ 2(-1) - 3(\ 7) \end{bmatrix} = \begin{bmatrix} 16 \\ -24 \\ 6 \\ -23 \end{bmatrix}$$ ∎

Many of the properties of arithmetic of real numbers, such as the commutative, distributive, and associative laws, carry over as properties of vector arithmetic. These are summarized in the next theorem.

THEOREM 2.3

▶ The **zero vector** is given by

$$\mathbf{0} = \begin{bmatrix} 0 \\ 0 \\ \vdots \\ 0 \end{bmatrix}$$

and $-\mathbf{u} = (-1)\mathbf{u}$.

(ALGEBRAIC PROPERTIES OF VECTORS)   Let a and b be scalars, and $\mathbf{u}$, $\mathbf{v}$, and $\mathbf{w}$ be vectors in $\mathbf{R}^n$. Then

(a) $\mathbf{u} + \mathbf{v} = \mathbf{v} + \mathbf{u}$

(b) $a(\mathbf{u} + \mathbf{v}) = a\mathbf{u} + a\mathbf{v}$

(c) $(a + b)\mathbf{u} = a\mathbf{u} + b\mathbf{u}$

(d) $(\mathbf{u} + \mathbf{v}) + \mathbf{w} = \mathbf{u} + (\mathbf{v} + \mathbf{w})$

(e) $a(b\mathbf{u}) = (ab)\mathbf{u}$

(f) $\mathbf{u} + (-\mathbf{u}) = \mathbf{0}$

(g) $\mathbf{u} + \mathbf{0} = \mathbf{0} + \mathbf{u} = \mathbf{u}$

(h) $1\mathbf{u} = \mathbf{u}$

Proof   Let

$$\mathbf{u} = \begin{bmatrix} u_1 \\ u_2 \\ \vdots \\ u_n \end{bmatrix} \quad \text{and} \quad \mathbf{v} = \begin{bmatrix} v_1 \\ v_2 \\ \vdots \\ v_n \end{bmatrix}$$

Since the components of each vector are real numbers, we have

$$u_1 + v_1 = v_1 + u_1, \quad \ldots \quad u_n + v_n = v_n + u_n$$

Hence

$$\mathbf{u} + \mathbf{v} = \begin{bmatrix} u_1 + v_1 \\ u_2 + v_2 \\ \vdots \\ u_n + v_n \end{bmatrix} = \begin{bmatrix} v_1 + u_1 \\ v_2 + u_2 \\ \vdots \\ v_n + u_n \end{bmatrix} = \mathbf{v} + \mathbf{u}$$

which proves (a). For (b), suppose that a is a scalar. Since

$$a(u_1 + v_1) = au_1 + av_1, \quad \ldots \quad a(u_n + v_n) = au_n + av_n$$

it follows that

$$a(\mathbf{u} + \mathbf{v}) = a \begin{bmatrix} u_1 + v_1 \\ u_2 + v_2 \\ \vdots \\ u_n + v_n \end{bmatrix} = \begin{bmatrix} a(u_1 + v_1) \\ a(u_2 + v_2) \\ \vdots \\ a(u_n + v_n) \end{bmatrix} = \begin{bmatrix} au_1 + av_1 \\ au_2 + av_2 \\ \vdots \\ au_n + av_n \end{bmatrix} = a\mathbf{v} + a\mathbf{u}$$

Therefore (b) is true. Proofs of the remaining properties are left as exercises.   ∎

Vectors and Systems of Equations

Let's return to the fertilizer example from the beginning of this section. We have two different kinds, Vigoro and Parker's, with nutrient vectors given by

$$\text{Vigoro: } \mathbf{v} = \begin{bmatrix} 29 \\ 3 \\ 4 \end{bmatrix} \quad \text{Parker's: } \mathbf{p} = \begin{bmatrix} 18 \\ 25 \\ 6 \end{bmatrix}$$

By using vector arithmetic, we can find the nutrient vector for combinations of the two fertilizers. For example, if 500 pounds of Vigoro and 300 pounds of Parker's are mixed, then the total amount of each nutrient is given by

$$5\mathbf{v} + 3\mathbf{p} = 5 \begin{bmatrix} 29 \\ 3 \\ 4 \end{bmatrix} + 3 \begin{bmatrix} 18 \\ 25 \\ 6 \end{bmatrix} = \begin{bmatrix} 145 + 54 \\ 15 + 75 \\ 20 + 18 \end{bmatrix} = \begin{bmatrix} 199 \\ 90 \\ 38 \end{bmatrix}$$

The sum $5\mathbf{v} + 3\mathbf{p}$ is an example of a *linear combination* of vectors.

DEFINITION 2.4

If $\mathbf{u}_1, \mathbf{u}_2, \ldots, \mathbf{u}_m$ are vectors and $c_1, c_2, \ldots, c_m$ are scalars, then

$$c_1\mathbf{u}_1 + c_2\mathbf{u}_2 + \cdots + c_m\mathbf{u}_m$$

is a **linear combination** of the vectors. Note that it is possible for scalars to be negative or equal to zero.

Definition **Linear Combination**

EXAMPLE 2   If possible, find the amount of Vigoro and Parker's required to create a mixture containing 148 pounds of nitrogen, 131 pounds of phosphoric acid, and 38 pounds of potash.

Solution We formulate the problem in terms of a linear combination. Specifically, we need to find scalars x_1 and x_2 such that

► (1) is an example of a **vector equation**.

$$x_1 \begin{bmatrix} 29 \\ 3 \\ 4 \end{bmatrix} + x_2 \begin{bmatrix} 18 \\ 25 \\ 6 \end{bmatrix} = \begin{bmatrix} 148 \\ 131 \\ 38 \end{bmatrix} \tag{1}$$

Taking one component at a time, we see that this vector equation is equivalent to the system of equations

$$\begin{aligned} 29x_1 + 18x_2 &= 148 \\ 3x_1 + 25x_2 &= 131 \\ 4x_1 + 6x_2 &= 38 \end{aligned}$$

► The row operations used in (2) are (in order performed):

$$\begin{aligned} R_1 &\Leftrightarrow R_3 \\ -(3/4)R_1 + R_2 &\Rightarrow R_2 \\ -(29/4)R_1 + R_3 &\Rightarrow R_3 \\ (51/41)R_2 + R_3 &\Rightarrow R_3 \\ (2/41)R_2 &\Rightarrow R_3 \end{aligned}$$

The augmented matrix and echelon form are

$$\begin{bmatrix} 29 & 18 & 148 \\ 3 & 25 & 131 \\ 4 & 6 & 38 \end{bmatrix} \sim \begin{bmatrix} 4 & 6 & 38 \\ 0 & 1 & 5 \\ 0 & 0 & 0 \end{bmatrix} \tag{2}$$

Back substitution gives the solution $x_1 = 2$ and $x_2 = 5$. ∎

We were lucky in the mixture of components in Example 2. Had we needed 40 pounds of potash instead of 38 pounds, there is no combination that works (see Exercise 37).

Solutions as Linear Combinations

The solution to Example 2 can be expressed in the form of a vector,

$$\mathbf{x} = \begin{bmatrix} x_1 \\ x_2 \end{bmatrix} = \begin{bmatrix} 2 \\ 5 \end{bmatrix}$$

Definition **Vector Form**

In fact, the general solution to any system of linear equations can be expressed as a linear combination of vectors, called the **vector form** of the general solution.

EXAMPLE 3   Express the general solution to the linear system

$$\begin{array}{rcrcrcr} 2x_1 & - & 3x_2 & + & 10x_3 & = & -2 \\ x_1 & - & 2x_2 & + & 3x_3 & = & -2 \\ -x_1 & + & 3x_2 & + & x_3 & = & 4 \end{array}$$

in vector form.

Solution In Example 3 of Section 1.2, we found the general solution to this system. Separating the general solution into the constant term and the term multiplied by the parameter s_1, we have

$$\begin{array}{rclcl} x_1 & = & 2 - 11s_1 & = & 2 - 11s_1 \\ x_2 & = & 2 - 4s_1 & = & 2 - 4s_1 \\ x_3 & = & s_1 & = & 0 + 1s_1 \end{array}$$

Thus the vector form of the general solution is

$$\mathbf{x} = \begin{bmatrix} x_1 \\ x_2 \\ x_3 \end{bmatrix} = \begin{bmatrix} 2 \\ 2 \\ 0 \end{bmatrix} + s_1 \begin{bmatrix} -11 \\ -4 \\ 1 \end{bmatrix}$$

where s_1 can be any real number. ∎

A more complicated general solution arises in Example 4 of Section 1.2. There we found the general solution

$$\begin{array}{rclcl} x_1 & = & -5 - 4s_1 + 2s_2 - 4s_3 & = & -5 - 4s_1 + 2s_2 - 4s_3 \\ x_2 & = & s_1 & = & 0 + 1s_1 + 0s_2 + 0s_3 \\ x_3 & = & -14 - 5s_3 & = & -14 + 0s_1 + 0s_2 - 5s_3 \\ x_4 & = & s_2 & = & 0 + 0s_1 + 1s_2 + 0s_3 \\ x_5 & = & 3 + s_3 & = & 3 + 0s_1 + 0s_2 + 1s_3 \\ x_6 & = & s_3 & = & 0 + 0s_1 + 0s_2 + 1s_3 \end{array}$$

where s_1, s_2, and s_3 can be any real numbers. In vector form, the general solution is given by

$$\mathbf{x} = \begin{bmatrix} x_1 \\ x_2 \\ x_3 \\ x_4 \\ x_5 \\ x_6 \end{bmatrix} = \begin{bmatrix} -5 \\ 0 \\ -14 \\ 0 \\ 3 \\ 0 \end{bmatrix} + s_1 \begin{bmatrix} -4 \\ 1 \\ 0 \\ 0 \\ 0 \\ 0 \end{bmatrix} + s_2 \begin{bmatrix} 2 \\ 0 \\ 0 \\ 1 \\ 0 \\ 0 \end{bmatrix} + s_3 \begin{bmatrix} -4 \\ 0 \\ -5 \\ 0 \\ 1 \\ 1 \end{bmatrix}$$

Geometry of Vectors

Vectors have a geometric interpretation that is most easily understood in $\mathbf{R}^2$. We plot the vector $\begin{bmatrix} x_1 \\ x_2 \end{bmatrix}$ by drawing an arrow from the origin to the point (x_1, x_2) in the plane. For example, the vectors $(2, 3)$ and $(3, -1)$ are illustrated in Figure 1. Using an arrow to denote a vector suggests a direction, which is a common interpretation in physics and other sciences, and will frequently be useful for us as well. We call the end of the vector with the arrow the **tip**, and the end at the origin the **tail**.

Note that the ordered pair for the point (x_1, x_2) looks the same as the row vector (x_1, x_2). The difference between the two is that vectors have an algebraic and geometric

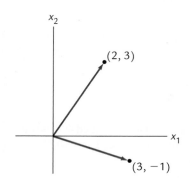

Figure 1 Vectors in $\mathbf{R}^2$.

Definition Tip, Tail of Vector

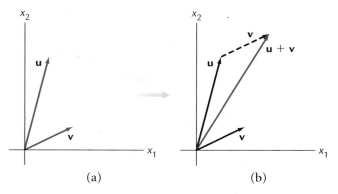

Figure 2 (a) The vectors **u** and **v**. (b) The vector **u** + **v**.

structure that is not associated with points. Most of the time we focus on vectors, so use that interpretation unless the alternative is clearly appropriate.

There are two related geometric procedures for adding vectors.

    1. The Tip-to-Tail Rule: Let **u** and **v** be two vectors. Translate the graph of **v**, preserving direction, so that its tail is at the tip of **u**. Then the tip of the translated **v** is at the tip of **u** + **v**.

Figure 2(a) shows vectors **u** and **v**, and Figure 2(b) shows **u**, the translated **v** (dashed), and **u** + **v**.

The Tip-to-Tail Rule makes sense from an algebraic standpoint. When we add **v** to **u**, we add each component of **v** to the corresponding component of **u**, which is exactly what we are doing geometrically. We also see in Figure 2(b) that we get to the same place if we translate **u** instead of **v**.

The second rule follows easily from the first.

    2. The Parallelogram Rule: Let vectors **u** and **v** form two adjacent sides of a parallelogram with vertices at the origin, the tip of **u**, and the tip of **v**. Then the tip of **u** + **v** is at the fourth vertex.

Figure 3 illustrates the Parallelogram Rule. It is evident that the third and fourth sides of the parallelogram are translated copies of **u** and **v**, which shows the connection to the Tip-to-Tail Rule.

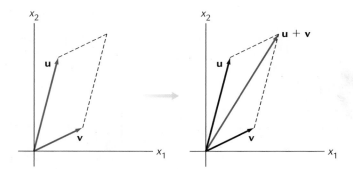

Figure 3 The Parallelogram Rule for vector addition.

Scalar multiplication and subtraction also have nice geometric interpretations.

    Scalar Multiplication: If a vector **u** is multiplied by a scalar c, then the new vector $c\mathbf{u}$ points in the same direction as **u** when $c > 0$ and in the opposite direction when $c < 0$. The length of $c\mathbf{u}$ is equal to the length of **u** multiplied by $|c|$.

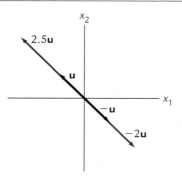

Figure 4 Scalar multiples of the vector **u**. **Figure 5** Subtracting vectors.

For example, $-2\mathbf{u}$ points in the opposite direction of **u** and is twice as long. (We will consider how to find the length of a vector later in the book.) A few examples of scalar multiples, starting with $\mathbf{u} = (-2, 3)$, are shown in Figure 4.

Subtraction: Draw a vector **w** from the tip of **v** to the tip of **u**. Then translate **w**, preserving direction and placing the tail at the origin. The resulting vector is $\mathbf{u} - \mathbf{v}$.

The subtraction procedure is illustrated in Figure 5; it is considered in more detail in Exercise 74.

EXERCISES

For Exercises 1–6, let

$$\mathbf{u} = \begin{bmatrix} 3 \\ -2 \\ 0 \end{bmatrix} \quad \mathbf{v} = \begin{bmatrix} -4 \\ 1 \\ 5 \end{bmatrix} \quad \text{and} \quad \mathbf{w} = \begin{bmatrix} 2 \\ -7 \\ -1 \end{bmatrix}$$

1. Compute $\mathbf{u} - \mathbf{v}$.

2. Compute $-5\mathbf{u}$.

3. Compute $\mathbf{w} + 3\mathbf{v}$.

4. Compute $4\mathbf{w} - \mathbf{u}$.

5. Compute $-\mathbf{u} + \mathbf{v} + \mathbf{w}$.

6. Compute $3\mathbf{u} - 2\mathbf{v} + 5\mathbf{w}$.

In Exercises 7–10, express the given vector equation as a system of linear equations.

7. $x_1 \begin{bmatrix} 3 \\ 2 \end{bmatrix} + x_2 \begin{bmatrix} -1 \\ 5 \end{bmatrix} = \begin{bmatrix} 8 \\ 13 \end{bmatrix}$

8. $x_1 \begin{bmatrix} -1 \\ 6 \\ -4 \end{bmatrix} + x_2 \begin{bmatrix} 9 \\ -5 \\ 0 \end{bmatrix} = \begin{bmatrix} -7 \\ -11 \\ 3 \end{bmatrix}$

9. $x_1 \begin{bmatrix} -6 \\ 5 \end{bmatrix} + x_2 \begin{bmatrix} 5 \\ -3 \end{bmatrix} + x_3 \begin{bmatrix} 0 \\ 2 \end{bmatrix} = \begin{bmatrix} 4 \\ 16 \end{bmatrix}$

10. $x_1 \begin{bmatrix} 2 \\ 7 \\ 8 \\ 3 \end{bmatrix} + x_2 \begin{bmatrix} 0 \\ 2 \\ 4 \\ 2 \end{bmatrix} + x_3 \begin{bmatrix} 5 \\ 1 \\ 6 \\ 1 \end{bmatrix} + x_4 \begin{bmatrix} 4 \\ 5 \\ 7 \\ 0 \end{bmatrix} = \begin{bmatrix} 0 \\ 4 \\ 3 \\ 5 \end{bmatrix}$

In Exercises 11–14, express the given system of linear equations as a vector equation.

11. $\begin{aligned} 2x_1 + 8x_2 - 4x_3 &= -10 \\ -x_1 - 3x_2 + 5x_3 &= 4 \end{aligned}$

12. $\begin{aligned} -2x_1 + 5x_2 - 10x_3 &= 4 \\ x_1 - 2x_2 + 3x_3 &= -1 \\ 7x_1 - 17x_2 + 34x_3 &= -16 \end{aligned}$

13. $\begin{aligned} x_1 - x_2 - 3x_3 - x_4 &= -1 \\ -2x_1 + 2x_2 + 6x_3 + 2x_4 &= -1 \\ -3x_1 - 3x_2 + 10x_3 &= 5 \end{aligned}$

14. $\begin{aligned} -5x_1 + 9x_2 &= 13 \\ 3x_1 - 5x_2 &= -9 \\ x_1 - 2x_2 &= -2 \end{aligned}$

In Exercises 15–18, the general solution to a linear system is given. Express this as a linear combination of vectors.

15. $\begin{aligned} x_1 &= -4 + 3s_1 \\ x_2 &= s_1 \end{aligned}$

16. $\begin{aligned} x_1 &= 7 - 2s_1 \\ x_2 &= -3 \\ x_3 &= s_1 \end{aligned}$

17. $\begin{aligned} x_1 &= 4 + 6s_1 - 5s_2 \\ x_2 &= s_2 \\ x_3 &= -9 + 3s_1 \\ x_4 &= s_1 \end{aligned}$

18. $\begin{aligned} x_1 &= 1 - 7s_1 + 14s_2 - s_3 \\ x_2 &= s_3 \\ x_3 &= s_2 \\ x_4 &= -12 + s_1 \\ x_5 &= s_1 \end{aligned}$

In Exercises 19–22, find three different vectors that are a linear combination of the given vectors.

19. $\mathbf{u} = \begin{bmatrix} 3 \\ -2 \end{bmatrix}, \quad \mathbf{v} = \begin{bmatrix} -1 \\ -4 \end{bmatrix}$

20. $\mathbf{u} = \begin{bmatrix} 7 \\ 1 \\ -13 \end{bmatrix}$, $\mathbf{v} = \begin{bmatrix} 5 \\ -3 \\ 2 \end{bmatrix}$

21. $\mathbf{u} = \begin{bmatrix} -4 \\ 0 \\ -3 \end{bmatrix}$, $\mathbf{v} = \begin{bmatrix} -2 \\ -1 \\ 5 \end{bmatrix}$, $\mathbf{w} = \begin{bmatrix} 9 \\ 6 \\ 11 \end{bmatrix}$

22. $\mathbf{u} = \begin{bmatrix} 1 \\ 8 \\ 2 \\ 2 \end{bmatrix}$, $\mathbf{v} = \begin{bmatrix} 4 \\ -2 \\ 5 \\ -5 \end{bmatrix}$, $\mathbf{w} = \begin{bmatrix} 9 \\ 9 \\ 0 \\ 1 \end{bmatrix}$

In Exercises 23–26, a vector equation is given with some unknown entries. Find the unknowns.

23. $-3 \begin{bmatrix} a \\ 3 \end{bmatrix} + 4 \begin{bmatrix} -1 \\ b \end{bmatrix} = \begin{bmatrix} -10 \\ 19 \end{bmatrix}$

24. $4 \begin{bmatrix} 4 \\ a \end{bmatrix} + 3 \begin{bmatrix} -3 \\ 5 \end{bmatrix} - 2 \begin{bmatrix} b \\ 8 \end{bmatrix} = \begin{bmatrix} -1 \\ 7 \end{bmatrix}$

25. $- \begin{bmatrix} -1 \\ a \\ 2 \end{bmatrix} + 2 \begin{bmatrix} 3 \\ -2 \\ b \end{bmatrix} = \begin{bmatrix} c \\ -7 \\ 8 \end{bmatrix}$

26. $- \begin{bmatrix} a \\ 4 \\ -2 \\ -1 \end{bmatrix} + 2 \begin{bmatrix} 5 \\ 1 \\ b \\ 3 \end{bmatrix} - \begin{bmatrix} 2 \\ c \\ -3 \\ -6 \end{bmatrix} = \begin{bmatrix} 11 \\ -4 \\ 3 \\ d \end{bmatrix}$

In Exercises 27–30, determine if **b** is a linear combination of the other vectors. If so, write **b** as a linear combination.

27. $\mathbf{a}_1 = \begin{bmatrix} -2 \\ 5 \end{bmatrix}$, $\mathbf{a}_2 = \begin{bmatrix} 7 \\ -3 \end{bmatrix}$, $\mathbf{b} = \begin{bmatrix} 8 \\ 9 \end{bmatrix}$

28. $\mathbf{a}_1 = \begin{bmatrix} 2 \\ -3 \\ 1 \end{bmatrix}$, $\mathbf{a}_2 = \begin{bmatrix} 0 \\ 3 \\ -3 \end{bmatrix}$, $\mathbf{b} = \begin{bmatrix} 1 \\ -5 \\ -2 \end{bmatrix}$

29. $\mathbf{a}_1 = \begin{bmatrix} 2 \\ -3 \\ 1 \end{bmatrix}$, $\mathbf{a}_2 = \begin{bmatrix} 0 \\ 3 \\ -3 \end{bmatrix}$, $\mathbf{b} = \begin{bmatrix} 6 \\ 3 \\ -9 \end{bmatrix}$

30. $\mathbf{a}_1 = \begin{bmatrix} 2 \\ -3 \\ 1 \end{bmatrix}$, $\mathbf{a}_2 = \begin{bmatrix} 0 \\ 3 \\ -3 \end{bmatrix}$, $\mathbf{a}_3 = \begin{bmatrix} -2 \\ -1 \\ 3 \end{bmatrix}$, $\mathbf{b} = \begin{bmatrix} 2 \\ -4 \\ 5 \end{bmatrix}$

Exercises 31–32 refer to Vigoro and Parker's fertilizers described at the start of the section. Determine the total amount of nitrogen, phosphoric acid, and potash in the given mixture.

31. 200 pounds of Vigoro, 100 pounds of Parker's.

32. 400 pounds of Vigoro, 700 pounds of Parker's.

Exercises 33–36 refer to Vigoro and Parker's fertilizers described at the start of the section. Determine the amount of each type required to yield a mixture containing the given amounts of nitrogen, phosphoric acid, and potash.

33. 112 pounds of nitrogen, 81 pounds of phosphoric acid, and 26 pounds of potash.

34. 285 pounds of nitrogen, 284 pounds of phosphoric acid, and 78 pounds of potash.

35. 123 pounds of nitrogen, 59 pounds of phosphoric acid, and 24 pounds of potash.

36. 159 pounds of nitrogen, 109 pounds of phosphoric acid, and 36 pounds of potash.

Exercises 37–40 refer to Vigoro and Parker's fertilizers described at the start of the section. Show that it is not possible to combine Vigoro and Parker's to obtain the specified mixture of nitrogen, phosphoric acid, and potash.

37. 148 pounds of nitrogen, 131 pounds of phosphoric acid, and 40 pounds of potash.

38. 100 pounds of nitrogen, 120 pounds of phosphoric acid, and 40 pounds of potash.

39. 25 pounds of nitrogen, 72 pounds of phosphoric acid, and 14 pounds of potash.

40. 301 pounds of nitrogen, 8 pounds of phosphoric acid, and 38 pounds of potash.

One 8.3 ounce can of Red Bull contains energy in two forms: 27 grams of sugar and 80 milligrams of caffeine. One 23.5 ounce can of Jolt Cola contains 94 grams of sugar and 280 milligrams of caffeine. In Exercises 41–44, determine the number of cans of each drink that when combined will contain the specified nerve-jangling combination of sugar and caffeine.

41. 148 grams sugar, 440 milligrams caffeine.

42. 309 grams sugar, 920 milligrams caffeine.

43. 242 grams sugar, 720 milligrams caffeine.

44. 457 grams sugar, 1360 milligrams caffeine.

One serving of Lucky Charms contains 10% of the percent daily values (PDV) for calcium, 25% of the PDV for iron, and 25% of the PDV for zinc. One serving of Raisin Bran contains 2% of the PDV for calcium, 25% of the PDV for iron, and 10% of the PDV for zinc. In Exercises 45–48, determine the number of servings of each cereal required to get the given mix of nutrients.

45. 40% of the PDV for calcium, 200% of the PDV for iron, and 125% of the PDV for zinc.

46. 34% of the PDV for calcium, 125% of the PDV for iron, and 95% of the PDV for zinc.

47. 26% of the PDV for calcium, 125% of the PDV for iron, and 80% of the PDV for zinc.

48. 38% of the PDV for calcium, 175% of the PDV for iron, and 115% of the PDV for zinc.

49. An electronics company has two production facilities, A and B. During an average week, facility A produces 2000 computer monitors and 8000 flat panel televisions, and facility B produces 3000 computer monitors and 10,000 flat panel televisions.

(a) Give vectors **a** and **b** that give the weekly production amounts at *A* and *B*, respectively.

(b) Compute 8**b**, and then describe what the entries tell us.

(c) Determine the combined output from A and B over a 6-week period.

(d) Determine the number of weeks of production from A and B required to produce 24,000 monitors and 92,000 televisions.

50. An industrial chemical company has three facilities A, B, and C. Each facility produces polyethylene (PE), polyvinyl chloride (PVC), and polystyrene (PS). The table below gives the daily production output (in metric tons) for each facility:

	Facility		
Product	A	B	C
PE	10	20	40
PVC	20	30	70
PS	10	40	50

(a) Give vectors **a**, **b**, and **c** that give the daily production amounts at each facility.

(b) Compute 20**c**, and describe what the entries tell us.

(c) Determine the combined output from all three facilities over a 2-week period. (Note: The facility does not operate on weekends.)

(d) Determine the number of days of production from each facility required to produce 240 metric tons of polyethylene, 420 metric tons of polyvinyl chloride and 320 metric tons of polystyrene.

Exercises 51–54 refer to the following: Let $\mathbf{v}_1, \ldots, \mathbf{v}_k$ be vectors, and suppose that a *point mass* of $m_1, \ldots, m_k$ is located at the tip of each vector. The *center of mass* for this set of point masses is

$$\bar{\mathbf{v}} = \frac{m_1 \mathbf{v}_1 + \cdots + m_k \mathbf{v}_k}{m}$$

where $m = m_1 + \cdots + m_k$.

51. Let $\mathbf{u}_1 = (3, 2)$ have mass 5kg, $\mathbf{u}_2 = (-1, 4)$ have mass 3kg, and $\mathbf{u}_3 = (2, 5)$ have mass 2kg. Graph the vectors, and then determine the center of mass.

52. Determine the center of mass for the vectors $\mathbf{u}_1 = (-1, 0, 2)$ (mass 4kg), $\mathbf{u}_2 = (2, 1, -3)$ (mass 1kg), $\mathbf{u}_3 = (0, 4, 3)$ (mass 2kg), and $\mathbf{u}_4 = (5, 2, 0)$ (mass 5kg).

53. Determine how to divide a total mass of 11kg among the vectors $\mathbf{u}_1 = (-1, 3)$, $\mathbf{u}_2 = (3, -2)$, and $\mathbf{u}_3 = (5, 2)$ so that the center of mass is $(13/11, 16/11)$.

54. Determine how to divide a total mass of 11kg among the vectors $\mathbf{u}_1 = (1, 1, 2)$, $\mathbf{u}_2 = (2, -1, 0)$, $\mathbf{u}_3 = (0, 3, 2)$, and $\mathbf{u}_4 = (-1, 0, 1)$ so that the center of mass is $(4/11, 5/11, 12/11)$.

FIND AN EXAMPLE For Exercises 55–62, find an example that meets the given specifications.

55. Two nonzero vectors **u** and **v** in $\mathbf{R}^3$ such that $\mathbf{u} + \mathbf{v} = (3, 2, -1)$.

56. Two nonzero vectors **u** and **v** in $\mathbf{R}^4$ such that $\mathbf{u} - \mathbf{v} = (4, -2, 0, -1)$.

57. Three nonzero vectors in $\mathbf{R}^3$ whose sum is the zero vector.

58. Three nonzero vectors in $\mathbf{R}^4$ whose sum is the zero vector.

59. Two vectors **u** and **v** in $\mathbf{R}^2$ that point in the same direction.

60. Two vectors **u** and **v** in $\mathbf{R}^2$ that point in opposite directions.

61. A linear system with two equations and two variables that has $\mathbf{x} = \begin{bmatrix} 3 \\ -2 \end{bmatrix}$ as the only solution.

62. A linear system with two equations and three variables that has $\mathbf{x} = \begin{bmatrix} 1 \\ 0 \\ 1 \end{bmatrix} + s \begin{bmatrix} 2 \\ 1 \\ -1 \end{bmatrix}$ as the general solution.

TRUE OR FALSE For Exercises 63–72, determine if the statement is true or false, and justify your answer.

63. If $\mathbf{u} = \begin{bmatrix} -3 \\ 5 \end{bmatrix}$, then $-2\mathbf{u} = \begin{bmatrix} 6 \\ -10 \end{bmatrix}$.

64. If $\mathbf{u} = \begin{bmatrix} 1 \\ 3 \end{bmatrix}$ and $\mathbf{v} = \begin{bmatrix} -4 \\ 2 \end{bmatrix}$, then $\mathbf{u} - \mathbf{v} = \begin{bmatrix} -3 \\ 1 \end{bmatrix}$.

65. If **u** and **v** are vectors and c is a scalar, then $c(\mathbf{u} + \mathbf{v}) = c\mathbf{u} + c\mathbf{v}$.

66. A vector can have positive or negative components, but a scalar must be positive.

67. If c_1 and c_2 are scalars and **u** is a vector, then $(c_1 + \mathbf{u})c_2 = c_1 c_2 + c_2 \mathbf{u}$.

68. The vectors $\begin{bmatrix} 1 \\ -2 \\ 4 \end{bmatrix}$ and $\begin{bmatrix} -2 \\ 4 \\ 8 \end{bmatrix}$ point in opposite directions.

69. $\begin{bmatrix} -2 \\ 1 \end{bmatrix}$ and $(-2, 1)$ are the same when both are considered as vectors.

70. The vector $2\mathbf{u}$ is longer than the vector $-3\mathbf{u}$.

71. The parallelogram rule for adding vectors only works in the first quadrant.

72. The difference $\mathbf{u} - \mathbf{v}$ is found by adding $-\mathbf{u}$ to **v**.

73. Prove each of the following parts of Theorem 2.3:

(a) Part (c).   **(b)** Part (d).   **(c)** Part (e).

(d) Part (f).   **(e)** Part (g).   **(f)** Part (h).

74. In this exercise we verify the geometric subtraction rule shown in Figure 5 by combining the identity $\mathbf{u} - \mathbf{v} = \mathbf{u} + (-\mathbf{v})$ and the Tip-to-Tail rule for addition. Draw a set of coordinate axes, and then sketch and label each of the following:

(a) Vectors **u** and **v** of your choosing.

(b) The vector $-\mathbf{v}$.

(c) The translation of $-\mathbf{v}$ so that its tail is at the tip of **u**.

(d) Using the Tip-to-Tail rule, the vector $\mathbf{u} + (-\mathbf{v})$.

Explain why the vector you get is the same as the one obtained using the subtraction rule shown in Figure 5.

In Exercises 75–76, sketch the graph of the vectors **u** and **v** and then use the Tip-to-Tail Rule to sketch the graph of $\mathbf{u} + \mathbf{v}$.

75. $\mathbf{u} = \begin{bmatrix} -2 \\ 3 \end{bmatrix}$, $\mathbf{v} = \begin{bmatrix} 1 \\ 4 \end{bmatrix}$

76. $\mathbf{u} = \begin{bmatrix} -1 \\ -2 \end{bmatrix}$, $\mathbf{v} = \begin{bmatrix} 3 \\ 1 \end{bmatrix}$

In Exercises 77–78, sketch the graph of the vectors **u** and **v** and then use the Parallelogram Rule to sketch the graph of **u** + **v**.

77. $\mathbf{u} = \begin{bmatrix} 0 \\ -3 \end{bmatrix}$, $\mathbf{v} = \begin{bmatrix} 2 \\ 2 \end{bmatrix}$

78. $\mathbf{u} = \begin{bmatrix} 4 \\ 2 \end{bmatrix}$, $\mathbf{v} = \begin{bmatrix} 2 \\ 0 \end{bmatrix}$

In Exercises 79–80, sketch the graph of the vectors **u** and **v**, and then use the subtraction procedure shown in Figure 5 to sketch the graph of **u** − **v**.

79. $\mathbf{u} = \begin{bmatrix} 3 \\ 2 \end{bmatrix}$, $\mathbf{v} = \begin{bmatrix} 1 \\ -1 \end{bmatrix}$

80. $\mathbf{u} = \begin{bmatrix} 1 \\ 3 \end{bmatrix}$, $\mathbf{v} = \begin{bmatrix} 2 \\ -3 \end{bmatrix}$

C In Exercises 81–82, find the solutions to the vector equation.

81. $x_1 \begin{bmatrix} 2 \\ 7 \\ 3 \end{bmatrix} + x_2 \begin{bmatrix} 2 \\ 4 \\ 2 \end{bmatrix} + x_3 \begin{bmatrix} 5 \\ 1 \\ 6 \end{bmatrix} = \begin{bmatrix} 0 \\ 3 \\ 5 \end{bmatrix}$

82. $x_1 \begin{bmatrix} 1 \\ -3 \\ 2 \\ 0 \end{bmatrix} + x_2 \begin{bmatrix} 4 \\ 3 \\ 2 \\ 1 \end{bmatrix} + x_3 \begin{bmatrix} -4 \\ 2 \\ -3 \\ 1 \end{bmatrix} + x_4 \begin{bmatrix} 5 \\ 2 \\ -4 \\ 0 \end{bmatrix} = \begin{bmatrix} 1 \\ 7 \\ 2 \\ -6 \end{bmatrix}$

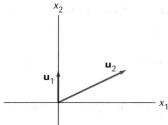

Figure 1 The VecMobile II vectors.

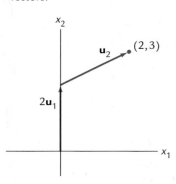

Figure 2 $2\mathbf{u}_1 + \mathbf{u}_2 = (2, 3)$.

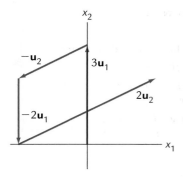

Figure 3 $3\mathbf{u}_1 - \mathbf{u}_2 - 2\mathbf{u}_1 + 2\mathbf{u}_2 = (2, 2)$.

2.2 Span

We open this section with a fictitious hypothetical situation. Imagine that you live in the two-dimensional plane $\mathbf{R}^2$ and have just purchased a new car, the VecMobile II. The VecMobile II is delivered at the origin $(0, 0)$ and is a fairly simple vehicle. At any given time, it can be pointed in the direction of

$$\mathbf{u}_1 = \begin{bmatrix} 0 \\ 1 \end{bmatrix} \quad \text{or} \quad \mathbf{u}_2 = \begin{bmatrix} 2 \\ 1 \end{bmatrix}$$

both shown in Figure 1. The VecMobile II can also go in forward or reverse.

Despite its simplicity, there are many places in $\mathbf{R}^2$ that we can go in the VecMobile II. For instance, the point $(2, 3)$ can be reached by first traversing $2\mathbf{u}_1$, changing direction, and then traversing $\mathbf{u}_2$, as shown in Figure 2. The trip also could be made in the reverse order, first taking $\mathbf{u}_2$ and then $2\mathbf{u}_1$. Since we are traversing vectors in a "tip-to-tail" manner, the entire trip can be summarized by the sum

$$2\mathbf{u}_1 + \mathbf{u}_2 = 2\begin{bmatrix} 0 \\ 1 \end{bmatrix} + \begin{bmatrix} 2 \\ 1 \end{bmatrix} = \begin{bmatrix} 2 \\ 3 \end{bmatrix}$$

Figure 3 depicts a more complicated path that arises from traversing $3\mathbf{u}_1$, then $-\mathbf{u}_2$, then $-2\mathbf{u}_1$, and finally $2\mathbf{u}_2$. This simplifies algebraically to

$$3\mathbf{u}_1 - \mathbf{u}_2 - 2\mathbf{u}_1 + 2\mathbf{u}_2 = \mathbf{u}_1 + \mathbf{u}_2$$

Any path taken in the VecMobile II can be similarly simplified, so that the set of all possible destinations can be expressed as

$$x_1\mathbf{u}_1 + x_2\mathbf{u}_2$$

where x_1 and x_2 can be any real numbers. This set of linear combinations is called the *span* of the vectors $\mathbf{u}_1$ and $\mathbf{u}_2$.

Although the VecMobile II is simple, we can go anywhere within $\mathbf{R}^2$ by first selecting the multiple (positive or negative) of $\mathbf{u}_2$ to reach the horizontal position we desire and then using a multiple of $\mathbf{u}_1$ to adjust the vertical position. For example, to reach $(6, 5)$, we start with $3\mathbf{u}_2 = \begin{bmatrix} 6 \\ 3 \end{bmatrix}$ and then add $2\mathbf{u}_1 = \begin{bmatrix} 0 \\ 2 \end{bmatrix}$.

$$3\mathbf{u}_2 + 2\mathbf{u}_1 = 3\begin{bmatrix} 2 \\ 1 \end{bmatrix} + 2\begin{bmatrix} 0 \\ 1 \end{bmatrix} = \begin{bmatrix} 6 \\ 5 \end{bmatrix}$$

EXAMPLE 1 Show algebraically that the VecMobile II can reach any position in $\mathbf{R}^2$.

Solution Suppose that we want to reach an arbitrary point (a, b). To do so, we need to find scalars x_1 and x_2 such that

$$x_1 \begin{bmatrix} 0 \\ 1 \end{bmatrix} + x_2 \begin{bmatrix} 2 \\ 1 \end{bmatrix} = \begin{bmatrix} a \\ b \end{bmatrix}$$

This vector equation translates into the system of equations

$$\begin{aligned} 2x_2 &= a \\ x_1 + x_2 &= b \end{aligned}$$

which has the unique solution $x_1 = b - a/2$ and $x_2 = a/2$. We now know exactly how to find the scalars x_1 and x_2 required to reach any point (a, b), and so we can conclude that the VecMobile II can get anywhere in $\mathbf{R}^2$. ∎

The notion of span generalizes to sets of vectors in $\mathbf{R}^n$.

DEFINITION 2.5

Definition Span

Let $\{\mathbf{u}_1, \mathbf{u}_2, \ldots, \mathbf{u}_m\}$ be a set of vectors in $\mathbf{R}^n$. The **span** of this set is denoted $\text{span}\{\mathbf{u}_1, \mathbf{u}_2, \ldots, \mathbf{u}_m\}$ and is defined to be the set of all linear combinations

$$x_1 \mathbf{u}_1 + x_2 \mathbf{u}_2 + \cdots + x_m \mathbf{u}_m$$

where $x_1, x_2, \ldots, x_m$ can be any real numbers.

If $\text{span}\{\mathbf{u}_1, \mathbf{u}_2, \ldots, \mathbf{u}_m\} = \mathbf{R}^n$, then we say that the set $\{\mathbf{u}_1, \mathbf{u}_2, \ldots, \mathbf{u}_m\}$ spans $\mathbf{R}^n$.

The VecMobile II in $\mathbf{R}^3$

Suppose that our world has expanded from $\mathbf{R}^2$ to $\mathbf{R}^3$. Happily, a VecMobile II model exists in $\mathbf{R}^3$. Like the $\mathbf{R}^2$ version, this vehicle only can move in two directions—in this case, that of the vectors

$$\mathbf{u}_1 = \begin{bmatrix} 2 \\ 1 \\ 1 \end{bmatrix} \quad \text{or} \quad \mathbf{u}_2 = \begin{bmatrix} 1 \\ 2 \\ 3 \end{bmatrix}$$

shown in Figure 4. Following the reasoning given earlier, we know that it is possible to get to any location that can be described as a linear combination of the form

$$x_1 \mathbf{u}_1 + x_2 \mathbf{u}_2$$

That is, we can get anywhere within $\text{span}\{\mathbf{u}_1, \mathbf{u}_2\}$. This covers a lot of territory, but not all of $\mathbf{R}^3$.

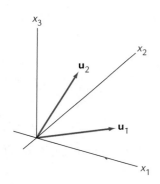

Figure 4 The VecMobile II (in $\mathbf{R}^3$) vectors.

EXAMPLE 2 Show that the VecMobile II cannot reach the point $(1, 0, 0)$.

Solution In order for the VecMobile II to reach $(1, 0, 0)$, there need to be scalars x_1 and x_2 that satisfy the equation

$$x_1 \begin{bmatrix} 2 \\ 1 \\ 1 \end{bmatrix} + x_2 \begin{bmatrix} 1 \\ 2 \\ 3 \end{bmatrix} = \begin{bmatrix} 1 \\ 0 \\ 0 \end{bmatrix}$$

▶ The row operations used in (1) are (in order performed):

$$R_1 \Leftrightarrow R_2$$
$$-2R_1 + R_2 \Rightarrow R_2$$
$$-R_1 + R_3 \Rightarrow R_3$$
$$R_2 \Leftrightarrow R_3$$
$$-3R_2 + R_3 \Rightarrow R_3$$

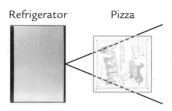

Figure 5 String vectors and pizza box.

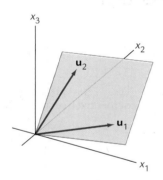

Figure 6 The plane is equal to span$\{\mathbf{u}_1, \mathbf{u}_2\}$.

This is equivalent to the linear system

$$2x_1 + x_2 = 1$$
$$x_1 + 2x_2 = 0$$
$$x_1 + 3x_2 = 0$$

Transferring to an augmented matrix and reducing, we find

$$\begin{bmatrix} 2 & 1 & 1 \\ 1 & 2 & 0 \\ 1 & 3 & 0 \end{bmatrix} \sim \begin{bmatrix} 1 & 2 & 0 \\ 0 & 1 & 0 \\ 0 & 0 & 1 \end{bmatrix} \tag{1}$$

The third row of the reduced matrix corresponds to the equation $0 = 1$. Thus the system has no solutions and hence the VecMobile II cannot reach the point $(1, 0, 0)$. ■

Here is another way to visualize the span of two vectors in $\mathbf{R}^3$. Get two pieces of string, about 3 feet long each, and tie them both to some solid object (like a refrigerator). Get a friend to pull the strings tight and in different directions. These are your vectors.

Next get a light-weight flat surface (a pizza box works well) and gently rest it on the strings (see Figure 5). Think of the surface as representing a plane. Then the span of the two "string" vectors is the set of all vectors that lie within the plane. Note that no matter the angle of the strings, if you are doing this correctly it is possible to rest the surface on them.

Figure 6 shows a plane resting on $\mathbf{u}_1$ and $\mathbf{u}_2$. The span$\{\mathbf{u}_1, \mathbf{u}_2\}$ consists exactly of those vectors that are contained in the plane. Therefore, if $\mathbf{u}_3$ is contained in span$\{\mathbf{u}_1, \mathbf{u}_2\}$, then $\mathbf{u}_3$ will lie in the plane, as shown in Figure 7(a). On the other hand, if $\mathbf{u}_3$ is not contained in span$\{\mathbf{u}_1, \mathbf{u}_2\}$, then $\mathbf{u}_3$ will be outside the plane, as in Figure 7(b).

The VecMobile III in $\mathbf{R}^3$

Continuing with our hypothetical line of cars, consumers disappointed by the limitations of the VecMobile II in $\mathbf{R}^3$ can spend more money and buy the VecMobile III, which can be pointed in the directions given by vectors

$$\mathbf{u}_1 = \begin{bmatrix} 2 \\ 1 \\ 1 \end{bmatrix}, \quad \mathbf{u}_2 = \begin{bmatrix} 1 \\ 2 \\ 3 \end{bmatrix}, \quad \text{or} \quad \mathbf{u}_3 = \begin{bmatrix} 1 \\ 0 \\ 0 \end{bmatrix}$$

With the addition of $\mathbf{u}_3$, it is clear that the point $(1, 0, 0)$ can now be reached. But can we get to all points in $\mathbf{R}^3$?

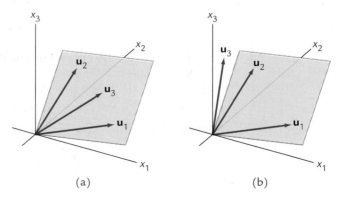

(a)   (b)

Figure 7 In (a), $\mathbf{u}_3$ is in span$\{\mathbf{u}_1, \mathbf{u}_2\}$. In (b), $\mathbf{u}_3$ is *not* in span$\{\mathbf{u}_1, \mathbf{u}_2\}$.

EXAMPLE 3 Show that the VecMobile III can reach any point in $\mathbf{R}^3$.

Solution To reach an arbitrary point (a, b, c), we need to find scalars x_1, x_2, and x_3 such that

$$x_1 \begin{bmatrix} 2 \\ 1 \\ 1 \end{bmatrix} + x_2 \begin{bmatrix} 1 \\ 2 \\ 3 \end{bmatrix} + x_3 \begin{bmatrix} 1 \\ 0 \\ 0 \end{bmatrix} = \begin{bmatrix} a \\ b \\ c \end{bmatrix} \tag{2}$$

The augmented matrix and equivalent reduced echelon form are

$$\begin{bmatrix} 2 & 1 & 1 & a \\ 1 & 2 & 0 & b \\ 1 & 3 & 0 & c \end{bmatrix} \sim \begin{bmatrix} 1 & 0 & 0 & (\quad 3b - 2c) \\ 0 & 1 & 0 & (\quad -b + c) \\ 0 & 0 & 1 & (a - 5b + 3c) \end{bmatrix} \tag{3}$$

▶ The row operations used in (3) are (in order performed):

$$R_1 \Leftrightarrow R_2$$
$$-2R_1 + R_2 \Rightarrow R_2$$
$$-R_1 + R_3 \Rightarrow R_3$$
$$R_2 \Leftrightarrow R_3$$
$$3R_2 + R_3 \Rightarrow R_3$$
$$-2R_2 + R_1 \Rightarrow R_1$$

We see from the reduced echelon form that the solution is

$$x_1 = \quad 3b - 2c$$
$$x_2 = \quad -b + c$$
$$x_3 = a - 5b + 3c$$

This shows that we can reach any point (a, b, c), and so $span\{\mathbf{u}_1, \mathbf{u}_2, \mathbf{u}_3\} = \mathbf{R}^3$. ■

The solution also shows how to get to any point. For example, if we want to get to the point $(-3, 1, 4)$, we substitute in $a = -3$, $b = 1$, and $c = 4$, which gives $x_1 = -5$, $x_2 = 3$, and $x_3 = 4$. Hence the linear combination we need is

$$-5\mathbf{u}_1 + 3\mathbf{u}_2 + 4\mathbf{u}_3$$

In Example 3, the vectors in (2) become the columns in the augmented matrix in (3),

$$\mathbf{u}_1 = \begin{bmatrix} 2 \\ 1 \\ 1 \end{bmatrix}, \mathbf{u}_2 = \begin{bmatrix} 1 \\ 2 \\ 3 \end{bmatrix}, \mathbf{u}_3 = \begin{bmatrix} 1 \\ 0 \\ 0 \end{bmatrix}, \mathbf{v} = \begin{bmatrix} a \\ b \\ c \end{bmatrix} \Rightarrow \begin{bmatrix} 2 & 1 & 1 & a \\ 1 & 2 & 0 & b \\ 1 & 3 & 0 & c \end{bmatrix} = \begin{bmatrix} \mathbf{u}_1 & \mathbf{u}_2 & \mathbf{u}_3 & \mathbf{v} \end{bmatrix}$$
$$\qquad\qquad\qquad\qquad\qquad\qquad\qquad\qquad\qquad\qquad\qquad\quad \uparrow \quad \uparrow \quad \uparrow \quad \uparrow$$
$$\qquad\qquad\qquad\qquad\qquad\qquad\qquad\qquad\qquad\qquad\quad \mathbf{u}_1 \ \mathbf{u}_2 \ \mathbf{u}_3 \ \mathbf{v}$$

The same is true in general. Given the vector equation

$$x_1\mathbf{u}_1 + x_2\mathbf{u}_2 + \cdots + x_m\mathbf{u}_m = \mathbf{v}$$

the augmented matrix for the corresponding linear system is

$$\begin{bmatrix} \mathbf{u}_1 & \mathbf{u}_2 & \cdots & \mathbf{u}_m & \mathbf{v} \end{bmatrix}$$

where the columns are given by the vectors $\mathbf{u}_1, \ldots, \mathbf{u}_m$, and $\mathbf{v}$.

THEOREM 2.6

Let $\mathbf{u}_1, \mathbf{u}_2, \ldots, \mathbf{u}_m$ and $\mathbf{v}$ be vectors in $\mathbf{R}^n$. Then $\mathbf{v}$ is an element of $span\{\mathbf{u}_1, \mathbf{u}_2, \ldots, \mathbf{u}_m\}$ if and only if the linear system represented by the augmented matrix

$$\begin{bmatrix} \mathbf{u}_1 & \mathbf{u}_2 & \cdots & \mathbf{u}_m & \mathbf{v} \end{bmatrix} \tag{4}$$

has a solution.

Proof The vector $\mathbf{v}$ is in $span\{\mathbf{u}_1, \mathbf{u}_2, \ldots, \mathbf{u}_m\}$ if and only if there exist scalars $x_1, x_2, \ldots, x_m$ that satisfy

$$x_1\mathbf{u}_1 + x_2\mathbf{u}_2 + \cdots + x_m\mathbf{u}_m = \mathbf{v}$$

This is true if and only if the corresponding linear system has a solution. As noted above, the linear system is equivalent to the augmented matrix (4), so the proof is complete. ■

How Many Vectors Are Needed to Span $\mathbf{R}^n$?

In the examples we have seen, a set of two vectors spanned $\mathbf{R}^2$ and a set of three vectors spanned $\mathbf{R}^3$. This suggests the following question.

EXAMPLE 4   Is it always true that a set of n vectors will span $\mathbf{R}^n$?

Solution Not always. For example, the span of the vectors

$$\mathbf{v}_1 = \begin{bmatrix} 1 \\ 1 \end{bmatrix}, \qquad \mathbf{v}_2 = \begin{bmatrix} 2 \\ 2 \end{bmatrix}$$

is a line in $\mathbf{R}^2$ (shown in Figure 8) because $\mathbf{v}_2 = 2\mathbf{v}_1$, so is not all of $\mathbf{R}^2$.

A more subtle example is given by the set of vectors in $\mathbf{R}^3$,

$$\mathbf{u}_1 = \begin{bmatrix} 2 \\ 1 \\ 1 \end{bmatrix}, \qquad \mathbf{u}_2 = \begin{bmatrix} 1 \\ 2 \\ 3 \end{bmatrix}, \quad \text{and} \quad \mathbf{u}_3 = \begin{bmatrix} 1 \\ -4 \\ -7 \end{bmatrix}$$

Note that $\mathbf{u}_1$ and $\mathbf{u}_2$ are the vectors from the VecMobile II in $\mathbf{R}^3$. It is straightforward to verify that $\mathbf{u}_3$ is given by the linear combination

$$\mathbf{u}_3 = 2\mathbf{u}_1 - 3\mathbf{u}_2$$

Thus any vector that is a linear combination of $\mathbf{u}_1$, $\mathbf{u}_2$, and $\mathbf{u}_3$ can be written as a linear combination of just $\mathbf{u}_1$ and $\mathbf{u}_2$ by substituting

$$x_1\mathbf{u}_1 + x_2\mathbf{u}_2 + x_3\mathbf{u}_3 = x_1\mathbf{u}_1 + x_2\mathbf{u}_2 + x_3(2\mathbf{u}_1 - 3\mathbf{u}_2)$$
$$= (x_1 + 2x_3)\mathbf{u}_1 + (x_2 - 3x_3)\mathbf{u}_2$$

Therefore span$\{\mathbf{u}_1, \mathbf{u}_2, \mathbf{u}_3\}$ = span$\{\mathbf{u}_1, \mathbf{u}_2\}$, and since we have already shown (in Example 2) that span$\{\mathbf{u}_1, \mathbf{u}_2\} \neq \mathbf{R}^3$, it follows that span$\{\mathbf{u}_1, \mathbf{u}_2, \mathbf{u}_3\} \neq \mathbf{R}^3$. ■

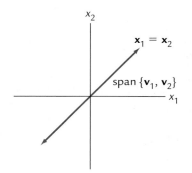

Figure 8 The span of $\{\mathbf{v}_1, \mathbf{v}_2\}$ in $\mathbf{R}^2$ in Example 4.

The preceding argument serves as a model for proving the next theorem.

THEOREM 2.7

Let $\mathbf{u}_1, \mathbf{u}_2, \ldots, \mathbf{u}_m$ and $\mathbf{u}$ be vectors in $\mathbf{R}^n$. If $\mathbf{u}$ is in $span\{\mathbf{u}_1, \mathbf{u}_2, \ldots, \mathbf{u}_m\}$, then

$$span\{\mathbf{u}, \mathbf{u}_1, \mathbf{u}_2, \ldots, \mathbf{u}_m\} = span\{\mathbf{u}_1, \mathbf{u}_2, \ldots, \mathbf{u}_m\}.$$

Proof   Let $S_0 = $ span$\{\mathbf{u}, \mathbf{u}_1, \mathbf{u}_2, \ldots, \mathbf{u}_m\}$ and $S_1 = $ span$\{\mathbf{u}_1, \mathbf{u}_2, \ldots, \mathbf{u}_m\}$. We need to show that sets $S_0 = S_1$, which we do by showing that each is a subset of the other. First suppose that a vector $\mathbf{v}$ is in S_1. Then there exist scalars $a_1, \ldots, a_m$ such that

$$\mathbf{v} = a_1\mathbf{u}_1 + a_2\mathbf{u}_2 + \cdots + a_m\mathbf{u}_m = 0\mathbf{u} + a_1\mathbf{u}_1 + a_2\mathbf{u}_2 + \cdots + a_m\mathbf{u}_m$$

Hence $\mathbf{v}$ is also in S_0, so S_1 is a subset of S_0.

Now suppose that $\mathbf{v}$ is in S_0. Then there exist scalars $b_0, b_1, \ldots, b_m$ such that $\mathbf{v} = b_0\mathbf{u} + b_1\mathbf{u}_1 + \cdots + b_m\mathbf{u}_m$. Since $\mathbf{u}$ is in S_1, there also exist scalars $c_1, \ldots, c_m$ such that $\mathbf{u} = c_1\mathbf{u}_1 + \cdots + c_m\mathbf{u}_m$. Then

$$\mathbf{v} = b_0(c_1\mathbf{u}_1 + c_2\mathbf{u}_2 + \cdots + c_m\mathbf{u}_m) + b_1\mathbf{u}_1 + b_2\mathbf{u}_2 + \cdots + b_m\mathbf{u}_m$$
$$= (b_0c_1 + b_1)\mathbf{u}_1 + (b_0c_2 + b_2)\mathbf{u}_2 + \cdots + (b_0c_m + b_m)\mathbf{u}_m$$

Hence $\mathbf{v}$ is in S_1, so S_0 is a subset of S_1. Since S_0 and S_1 are subsets of each other, it follows that $S_0 = S_1$. ■

EXAMPLE 5 Suppose that $\mathbf{u}_1, \mathbf{u}_2, \ldots, \mathbf{u}_m$ are vectors in $\mathbf{R}^n$. If $m < n$, can $\{\mathbf{u}_1, \mathbf{u}_2, \ldots, \mathbf{u}_m\}$ span $\mathbf{R}^n$?

Solution The answer is no. If $m < n$, we can always construct $\mathbf{b}$ in $\mathbf{R}^n$ that is not in the span of the given vectors $\mathbf{u}_1, \mathbf{u}_2, \ldots, \mathbf{u}_m$. We illustrate the procedure for vectors

$$
\mathbf{u}_1 = \begin{bmatrix} 1 \\ -2 \\ -1 \\ 2 \end{bmatrix}, \quad \mathbf{u}_2 = \begin{bmatrix} -3 \\ 7 \\ 4 \\ -6 \end{bmatrix}, \quad \text{and} \quad \mathbf{u}_3 = \begin{bmatrix} 2 \\ 0 \\ 2 \\ 5 \end{bmatrix}
$$

in $\mathbf{R}^4$, but it will work in general. Start by forming the matrix with only our vectors as columns,

$$
\begin{bmatrix} \mathbf{u}_1 & \mathbf{u}_2 & \mathbf{u}_3 \end{bmatrix} = \begin{bmatrix} 1 & -3 & 2 \\ -2 & 7 & 0 \\ -1 & 4 & 2 \\ 2 & -6 & 5 \end{bmatrix}
$$

Now perform the usual row operations needed to transform the matrix to echelon form, recording each operation along the way. We need only perform enough operations to introduce a row of zeroes on the bottom of the matrix, which must be possible because there are more rows than columns (see Exercise 54 in Section 1.2). The "Forward Operations" shown in the margin yields the transformation

▶ Forward Operations:

$$2R_1 + R_2 \Rightarrow R_2$$
$$R_1 + R_3 \Rightarrow R_3$$
$$-2R_1 + R_4 \Rightarrow R_4$$
$$-R_2 + R_3 \Rightarrow R_3$$
$$R_3 \Leftrightarrow R_4$$

$$
\begin{bmatrix} 1 & -3 & 2 \\ -2 & 7 & 0 \\ -1 & 4 & 2 \\ 2 & -6 & 5 \end{bmatrix} \sim \begin{bmatrix} 1 & -3 & 2 \\ 0 & 1 & 4 \\ 0 & 0 & 1 \\ 0 & 0 & 0 \end{bmatrix}
$$

The next step is to append a column to the right side of the echelon matrix,

$$
\begin{bmatrix} 1 & -3 & 2 & 0 \\ 0 & 1 & 4 & 0 \\ 0 & 0 & 1 & 0 \\ 0 & 0 & 0 & 1 \end{bmatrix} \tag{5}
$$

Viewing this new matrix as an augmented matrix, we see that the bottom row is equivalent to the equation $0 = 1$, so that the associated linear system has no solutions. If we now reverse the row operations used previously (as shown in the margin), the first three columns of the augmented matrix will be returned to their original form. This gives us

▶ Reverse Operations:

$$R_3 \Leftrightarrow R_4$$
$$R_2 + R_3 \Rightarrow R_3$$
$$2R_1 + R_4 \Rightarrow R_4$$
$$-R_1 + R_3 \Rightarrow R_3$$
$$-2R_1 + R_2 \Rightarrow R_2$$

$$
\begin{bmatrix} 1 & -3 & 2 & 0 \\ 0 & 1 & 4 & 0 \\ 0 & 0 & 1 & 0 \\ 0 & 0 & 0 & 1 \end{bmatrix} \sim \begin{bmatrix} 1 & -3 & 2 & 0 \\ -2 & 7 & 0 & 0 \\ -1 & 4 & 2 & 1 \\ 2 & -6 & 5 & 0 \end{bmatrix} \tag{6}
$$

Since the system associated with the augmented matrix (5) has no solutions, the system associated with the equivalent augmented matrix (6) also has no solutions. But this system is represented by the vector equation

$$
x_1\mathbf{u}_1 + x_2\mathbf{u}_2 + x_3\mathbf{u}_3 = \mathbf{b} \quad \text{where} \quad \mathbf{b} = \begin{bmatrix} 0 \\ 0 \\ 1 \\ 0 \end{bmatrix}
$$

Thus this vector equation can have no solutions, and so it follows that **b** is *not* in the span of $\{\mathbf{u}_1, \mathbf{u}_2, \mathbf{u}_3\}$. Therefore this set of vectors does not span $\mathbf{R}^4$. ■

As long as $m < n$, the argument described above generalizes to any set of vectors $\{\mathbf{u}_1, \mathbf{u}_2, \ldots, \mathbf{u}_m\}$ in $\mathbf{R}^n$. Theorem 2.8 summarizes the main results of this subsection.

THEOREM 2.8

Let $\{\mathbf{u}_1, \mathbf{u}_2, \ldots, \mathbf{u}_m\}$ be a set of vectors in $\mathbf{R}^n$. If $m < n$, then this set does not span $\mathbf{R}^n$. If $m \geq n$, then the set might span $\mathbf{R}^n$ or it might not. In this case, we cannot say more without additional information about the vectors.

The proof of this theorem is left as an exercise.

The Equation $A\mathbf{x} = \mathbf{b}$

By now we are comfortable with translating back and forth between vector equations and linear systems. Here we give new notation that will be used for a variety of purposes, including expressing linear systems in a compact form.

$$\text{Let } A \text{ be the matrix with columns } \mathbf{a}_1 = \begin{bmatrix} 10 \\ 5 \\ 7 \end{bmatrix} \text{ and } \mathbf{a}_2 = \begin{bmatrix} 8 \\ 6 \\ -1 \end{bmatrix}. \text{ That is,}$$

$$A = \begin{bmatrix} \mathbf{a}_1 & \mathbf{a}_2 \end{bmatrix} = \begin{bmatrix} 10 & 8 \\ 5 & 6 \\ 7 & -1 \end{bmatrix}$$

Also, let $\mathbf{x} = \begin{bmatrix} x_1 \\ x_2 \end{bmatrix}$. Then the product of the matrix A and the vector $\mathbf{x}$ is defined to be

$$A\mathbf{x} = \begin{bmatrix} \mathbf{a}_1 & \mathbf{a}_2 \end{bmatrix} \begin{bmatrix} x_1 \\ x_2 \end{bmatrix} = x_1 \mathbf{a}_1 + x_2 \mathbf{a}_2$$

Thus $A\mathbf{x}$ is a linear combination of the columns of A, with the scalars given by the components of $\mathbf{x}$. Now take it a step farther and let

$$\mathbf{b} = \begin{bmatrix} 18 \\ 31 \\ 3 \end{bmatrix}$$

Then $A\mathbf{x} = \mathbf{b}$ is a compact form of the vector equation $x_1\mathbf{a}_1 + x_2\mathbf{a}_2 = \mathbf{b}$, which in turn is equivalent to the linear system

$$\begin{array}{rcr} 10x_1 + 8x_2 &=& 18 \\ 5x_1 + 6x_2 &=& 31 \\ 7x_1 - x_2 &=& 3 \end{array}$$

Below is the general formula for multiplying a matrix by a vector.

DEFINITION 2.9

Let $\mathbf{a}_1, \mathbf{a}_2, \ldots, \mathbf{a}_m$ be vectors in $\mathbf{R}^n$. If

$$A = \begin{bmatrix} \mathbf{a}_1 & \mathbf{a}_2 & \cdots & \mathbf{a}_m \end{bmatrix} \quad \text{and} \quad \mathbf{x} = \begin{bmatrix} x_1 \\ x_2 \\ \vdots \\ x_m \end{bmatrix} \tag{7}$$

then $A\mathbf{x} = x_1\mathbf{a}_1 + x_2\mathbf{a}_2 + \cdots + x_m\mathbf{a}_m$.

▶ Remember: The product $A\mathbf{x}$ only is defined when the number of columns of A equals the number of components (entries) of $\mathbf{x}$.

EXAMPLE 6   Find A, $\mathbf{x}$, and $\mathbf{b}$ so that the equation $A\mathbf{x} = \mathbf{b}$ corresponds to the system of equations

$$\begin{array}{rcrcrcrcr} 4x_1 & - & 3x_2 & + & 7x_3 & - & x_4 & = & 13 \\ -x_1 & + & 2x_2 & & & + & 6x_4 & = & -2 \\ & & x_2 & - & 3x_3 & - & 5x_4 & = & 29 \end{array}$$

Solution Translating the system to the form $A\mathbf{x} = \mathbf{b}$, the matrix A will contain the coefficients of the system, the vector $\mathbf{x}$ has the variables, and the vector $\mathbf{b}$ will contain the constant terms. Thus we have

$$A = \begin{bmatrix} 4 & -3 & 7 & -1 \\ -1 & 2 & 0 & 6 \\ 0 & 1 & -3 & -5 \end{bmatrix}, \quad \mathbf{x} = \begin{bmatrix} x_1 \\ x_2 \\ x_3 \\ x_4 \end{bmatrix}, \quad \mathbf{b} = \begin{bmatrix} 13 \\ -2 \\ 29 \end{bmatrix} \qquad \blacksquare$$

EXAMPLE 7   Suppose that

$$\mathbf{a}_1 = \begin{bmatrix} 1 \\ -2 \\ 0 \end{bmatrix}, \quad \mathbf{a}_2 = \begin{bmatrix} 3 \\ 1 \\ -1 \end{bmatrix}, \quad \mathbf{v}_1 = \begin{bmatrix} 7 \\ 5 \\ -2 \end{bmatrix}, \quad \mathbf{v}_2 = \begin{bmatrix} -6 \\ 4 \end{bmatrix}, \quad \mathbf{x} = \begin{bmatrix} x_1 \\ x_2 \end{bmatrix}, \quad \mathbf{b} = \begin{bmatrix} -1 \\ 2 \\ 5 \end{bmatrix}$$

Let $A = \begin{bmatrix} \mathbf{a}_1 & \mathbf{a}_2 \end{bmatrix}$. Find the following (if they exist):

(a) $A\mathbf{v}_1$ and $A\mathbf{v}_2$.

(b) The system of equations corresponding to $A\mathbf{x} = \mathbf{b}$ and $A\mathbf{x} = \mathbf{v}_2$.

Solution

(a) In order for $A\mathbf{v}_1$ to exist, the number of columns of A must equal the number of components of $\mathbf{v}_1$. Since this is *not* the case, $A\mathbf{v}_1$ does not exist.

On the other hand, $\mathbf{v}_2$ has two components, so $A\mathbf{v}_2$ exists. We have

$$A\mathbf{v}_2 = -6\mathbf{a}_1 + 4\mathbf{a}_2 = -6 \begin{bmatrix} 1 \\ -2 \\ 0 \end{bmatrix} + 4 \begin{bmatrix} 3 \\ 1 \\ -1 \end{bmatrix} = \begin{bmatrix} 6 \\ 16 \\ -4 \end{bmatrix}$$

(b) Since $\mathbf{x}$ has two components, $A\mathbf{x}$ exists. Moreover, $A\mathbf{x}$ and $\mathbf{b}$ both have three components, so the equation $A\mathbf{x} = \mathbf{b}$ is defined and corresponds to the system

$$\begin{array}{rcrcr} x_1 & + & 3x_2 & = & -1 \\ -2x_1 & + & x_2 & = & 2 \\ & - & x_2 & = & 5 \end{array}$$

For the second part, $A\mathbf{x}$ exists and has three components, but $\mathbf{v}_2$ has only two components, so $A\mathbf{x} = \mathbf{v}_2$ is undefined. $\qquad \blacksquare$

We close this section with a theorem that ties together several closely related ideas.

THEOREM 2.10

Let $\mathbf{a}_1, \mathbf{a}_2, \ldots, \mathbf{a}_m$ and $\mathbf{b}$ be vectors in $\mathbf{R}^n$. Then the following statements are equivalent. That is, if one is true, then so are the others, and if one is false, then so are the others.

(a) $\mathbf{b}$ is in span$\{\mathbf{a}_1, \mathbf{a}_2, \ldots, \mathbf{a}_m\}$.

(b) The vector equation $x_1\mathbf{a}_1 + x_2\mathbf{a}_2 + \cdots + x_m\mathbf{a}_m = \mathbf{b}$ has at least one solution.

(c) The linear system corresponding to $\begin{bmatrix} \mathbf{a}_1 & \mathbf{a}_2 & \cdots & \mathbf{a}_m & \mathbf{b} \end{bmatrix}$ has at least one solution.

(d) The equation $A\mathbf{x} = \mathbf{b}$, with A and $\mathbf{x}$ given as in (7), has at least one solution.

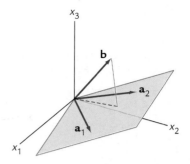

Figure 9 span{$\mathbf{a}_1$, $\mathbf{a}_2$} includes **b**, so (8) has a solution.

Figure 10 span{$\mathbf{a}_1$, $\mathbf{a}_2$} does not include **b**, so (8) does not have a solution.

The theorem follows directly from the definitions, so the proof is left as an exercise. Although this result is not hard to arrive at, it is important because it explicitly states the connection between these different formulations of the same basic idea.

As a quick application, note that the vector **b** in Figure 9 is in span{$\mathbf{a}_1$, $\mathbf{a}_2$}, so by Theorem 2.10 it follows that

$$x_1\mathbf{a}_1 + x_2\mathbf{a}_2 = \mathbf{b} \tag{8}$$

has at least one solution. On the other hand, if $\mathbf{a}_1$, $\mathbf{a}_2$, and **b** are as shown in Figure 10, then (8) has no solutions.

EXERCISES

For Exercises 1–6, find three vectors that are in the span of the given vectors.

1. $\mathbf{u}_1 = \begin{bmatrix} 2 \\ 6 \end{bmatrix}$; $\mathbf{u}_2 = \begin{bmatrix} 9 \\ 15 \end{bmatrix}$

2. $\mathbf{u}_1 = \begin{bmatrix} -2 \\ 7 \end{bmatrix}$, $\mathbf{u}_2 = \begin{bmatrix} -3 \\ 4 \end{bmatrix}$

3. $\mathbf{u}_1 = \begin{bmatrix} 2 \\ 5 \\ -3 \end{bmatrix}$, $\mathbf{u}_2 = \begin{bmatrix} 1 \\ 0 \\ 4 \end{bmatrix}$

4. $\mathbf{u}_1 = \begin{bmatrix} 0 \\ 5 \\ -2 \end{bmatrix}$, $\mathbf{u}_2 = \begin{bmatrix} 1 \\ 2 \\ 6 \end{bmatrix}$, $\mathbf{u}_3 = \begin{bmatrix} -6 \\ 7 \\ 2 \end{bmatrix}$

5. $\mathbf{u}_1 = \begin{bmatrix} 2 \\ 0 \\ 0 \end{bmatrix}$, $\mathbf{u}_2 = \begin{bmatrix} 4 \\ 1 \\ 6 \end{bmatrix}$, $\mathbf{u}_3 = \begin{bmatrix} -4 \\ 0 \\ 7 \end{bmatrix}$

6. $\mathbf{u}_1 = \begin{bmatrix} 0 \\ 1 \\ 3 \\ 0 \end{bmatrix}$, $\mathbf{u}_2 = \begin{bmatrix} -1 \\ 8 \\ -5 \\ 2 \end{bmatrix}$, $\mathbf{u}_3 = \begin{bmatrix} 12 \\ -1 \\ 1 \\ 0 \end{bmatrix}$

For Exercises 7–12, determine if **b** is in the span of the other given vectors. If so, write **b** as a linear combination of the other vectors.

7. $\mathbf{a}_1 = \begin{bmatrix} 3 \\ 5 \end{bmatrix}$, $\mathbf{b} = \begin{bmatrix} 9 \\ -15 \end{bmatrix}$

8. $\mathbf{a}_1 = \begin{bmatrix} 10 \\ -15 \end{bmatrix}$, $\mathbf{b} = \begin{bmatrix} -30 \\ 45 \end{bmatrix}$

9. $\mathbf{a}_1 = \begin{bmatrix} 4 \\ -2 \\ 10 \end{bmatrix}$, $\mathbf{b} = \begin{bmatrix} 2 \\ -1 \\ -5 \end{bmatrix}$

10. $\mathbf{a}_1 = \begin{bmatrix} -1 \\ 3 \\ -1 \end{bmatrix}$, $\mathbf{a}_2 = \begin{bmatrix} -2 \\ -3 \\ 6 \end{bmatrix}$, $\mathbf{b} = \begin{bmatrix} -6 \\ 9 \\ 2 \end{bmatrix}$

11. $\mathbf{a}_1 = \begin{bmatrix} -1 \\ 4 \\ -3 \end{bmatrix}$, $\mathbf{a}_2 = \begin{bmatrix} 2 \\ 8 \\ -7 \end{bmatrix}$, $\mathbf{b} = \begin{bmatrix} -10 \\ -8 \\ 7 \end{bmatrix}$

12. $\mathbf{a}_1 = \begin{bmatrix} 3 \\ 1 \\ -2 \\ -1 \end{bmatrix}$, $\mathbf{a}_2 = \begin{bmatrix} -4 \\ 2 \\ 3 \\ 3 \end{bmatrix}$, $\mathbf{b} = \begin{bmatrix} 0 \\ 10 \\ 1 \\ 5 \end{bmatrix}$

In Exercises 13–16, find A, **x**, and **b** such that $A\mathbf{x} = \mathbf{b}$ corresponds to the given linear system.

13. $\begin{aligned} 2x_1 + 8x_2 - 4x_3 &= -10 \\ -x_1 - 3x_2 + 5x_3 &= 4 \end{aligned}$

14. $\begin{aligned} -2x_1 + 5x_2 - 10x_3 &= 4 \\ x_1 - 2x_2 + 3x_3 &= -1 \\ 7x_1 - 17x_2 + 34x_3 &= -16 \end{aligned}$

15. $\begin{aligned} x_1 - x_2 - 3x_3 - x_4 &= -1 \\ -2x_1 + 2x_2 + 6x_3 + 2x_4 &= -1 \\ -3x_1 - 3x_2 + 10x_3 &= 5 \end{aligned}$

16. $\begin{aligned} -5x_1 + 9x_2 &= 13 \\ 3x_1 - 5x_2 &= -9 \\ x_1 - 2x_2 &= -2 \end{aligned}$

In Exercises 17–20, find an equation involving vectors that corresponds to the given linear system.

17. $\begin{aligned} 5x_1 + 7x_2 - 2x_3 &= 9 \\ x_1 - 5x_2 - 4x_3 &= 2 \end{aligned}$

18. $\begin{aligned} 4x_1 - 5x_2 - 3x_3 &= 0 \\ 3x_1 + 4x_2 + 2x_3 &= 1 \\ 6x_1 - 13x_2 + 7x_3 &= 2 \end{aligned}$

19. $\begin{aligned} 4x_1 - 2x_2 - 3x_3 + 5x_4 &= 12 \\ - 5x_2 + 7x_3 + 3x_4 &= 6 \\ 3x_1 + 8x_2 + 2x_3 - x_4 &= 2 \end{aligned}$

20. $\begin{aligned} 4x_1 - 9x_2 &= 11 \\ 2x_1 + 4x_2 &= 9 \\ x_1 - 7x_2 &= 2 \end{aligned}$

In Exercises 21–24, determine if the columns of the given matrix span $\mathbf{R}^2$.

21. $\begin{bmatrix} 15 & -6 \\ -5 & 2 \end{bmatrix}$

22. $\begin{bmatrix} 4 & -12 \\ 2 & 6 \end{bmatrix}$

23. $\begin{bmatrix} 2 & 1 & 0 \\ 6 & -3 & -1 \end{bmatrix}$

24. $\begin{bmatrix} 1 & 0 & 5 \\ -2 & 2 & 7 \end{bmatrix}$

In Exercises 25–28, determine if the columns of the given matrix span $\mathbf{R}^3$.

25. $\begin{bmatrix} 3 & 1 & 0 \\ 5 & -2 & -1 \\ 4 & -4 & -3 \end{bmatrix}$

26. $\begin{bmatrix} 1 & 2 & 8 \\ -2 & 3 & 7 \\ 3 & -1 & 1 \end{bmatrix}$

27. $\begin{bmatrix} 2 & 1 & -3 & 5 \\ 1 & 4 & 2 & 6 \\ 0 & 3 & 3 & 3 \end{bmatrix}$

28. $\begin{bmatrix} -4 & -7 & 1 & 2 \\ 0 & 0 & 3 & 8 \\ 5 & -1 & 1 & -4 \end{bmatrix}$

In Exercises 29–34, a matrix A is given. Determine if the system $A\mathbf{x} = \mathbf{b}$ (where $\mathbf{x}$ and $\mathbf{b}$ have the appropriate number of components) has a solution for all choices of $\mathbf{b}$.

29. $A = \begin{bmatrix} 3 & -4 \\ 4 & 2 \end{bmatrix}$

30. $A = \begin{bmatrix} -9 & 21 \\ 6 & -14 \end{bmatrix}$

31. $A = \begin{bmatrix} 8 & 1 \\ 0 & -1 \\ -3 & 2 \end{bmatrix}$

32. $A = \begin{bmatrix} 1 & -1 & 2 \\ -2 & 3 & -1 \\ 1 & 0 & 5 \end{bmatrix}$

33. $A = \begin{bmatrix} -3 & 2 & 1 \\ 1 & -1 & -1 \\ 5 & -4 & -3 \end{bmatrix}$

34. $A = \begin{bmatrix} 2 & -3 & 0 \\ 0 & 1 & 2 \\ -5 & 3 & -9 \\ 3 & 0 & 9 \end{bmatrix}$

For Exercises 35–38, find a vector of matching dimension that is *not* in the given span.

35. span $\left\{ \begin{bmatrix} 1 \\ -2 \end{bmatrix}, \begin{bmatrix} -3 \\ 6 \end{bmatrix} \right\}$

36. span $\left\{ \begin{bmatrix} 3 \\ 1 \end{bmatrix}, \begin{bmatrix} 6 \\ 2 \end{bmatrix} \right\}$

37. span $\left\{ \begin{bmatrix} 1 \\ 3 \\ -2 \end{bmatrix}, \begin{bmatrix} 2 \\ -1 \\ 1 \end{bmatrix} \right\}$

38. span $\left\{ \begin{bmatrix} 1 \\ 2 \\ 1 \end{bmatrix}, \begin{bmatrix} 3 \\ -1 \\ 1 \end{bmatrix}, \begin{bmatrix} -1 \\ 5 \\ 1 \end{bmatrix} \right\}$

39. Find all values of h such that the vectors $\{\mathbf{a}_1, \mathbf{a}_2\}$ span $\mathbf{R}^2$, where

$$\mathbf{a}_1 = \begin{bmatrix} 2 \\ 4 \end{bmatrix}, \quad \mathbf{a}_2 = \begin{bmatrix} h \\ 6 \end{bmatrix}$$

40. Find all values of h such that the vectors $\{\mathbf{a}_1, \mathbf{a}_2\}$ span $\mathbf{R}^2$, where

$$\mathbf{a}_1 = \begin{bmatrix} -3 \\ h \end{bmatrix}, \quad \mathbf{a}_2 = \begin{bmatrix} 5 \\ -4 \end{bmatrix}$$

41. Find all values of h such that the vectors $\{\mathbf{a}_1, \mathbf{a}_2, \mathbf{a}_3\}$ span $\mathbf{R}^3$, where

$$\mathbf{a}_1 = \begin{bmatrix} 2 \\ 4 \\ 5 \end{bmatrix}, \quad \mathbf{a}_2 = \begin{bmatrix} h \\ 8 \\ 10 \end{bmatrix}, \quad \mathbf{a}_3 = \begin{bmatrix} 1 \\ 2 \\ 6 \end{bmatrix}$$

42. Find all values of h such that the vectors $\{\mathbf{a}_1, \mathbf{a}_2, \mathbf{a}_3\}$ span $\mathbf{R}^3$, where

$$\mathbf{a}_1 = \begin{bmatrix} -1 \\ h \\ 7 \end{bmatrix}, \quad \mathbf{a}_2 = \begin{bmatrix} 4 \\ -2 \\ 5 \end{bmatrix}, \quad \mathbf{a}_3 = \begin{bmatrix} 1 \\ -3 \\ 2 \end{bmatrix}$$

FIND AN EXAMPLE For Exercises 43–50, find an example that meets the given specifications.

43. Four distinct nonzero vectors that span $\mathbf{R}^3$.

44. Four distinct nonzero vectors that span $\mathbf{R}^4$.

45. Four distinct nonzero vectors that do not span $\mathbf{R}^3$.

46. Four distinct nonzero vectors that do not span $\mathbf{R}^4$.

47. Two vectors $\mathbf{u}_1$ and $\mathbf{u}_2$ in $\mathbf{R}^3$ that span the set of all vectors of the form $\mathbf{v} = (v_1, v_2, 0)$.

48. Three vectors $\mathbf{u}_1, \mathbf{u}_2,$ and $\mathbf{u}_3$ in $\mathbf{R}^4$ that span the set of all vectors of the form $\mathbf{v} = (0, v_2, v_3, v_4)$.

49. Two vectors $\mathbf{u}_1$ and $\mathbf{u}_2$ in $\mathbf{R}^3$ that span the set of all vectors of the form $\mathbf{v} = (v_1, v_2, v_3)$ where $v_1 + v_2 + v_3 = 0$.

50. Three vectors $\mathbf{u}_1, \mathbf{u}_2,$ and $\mathbf{u}_3$ in $\mathbf{R}^4$ that span the set of all vectors of the form $\mathbf{v} = (v_1, v_2, v_3, v_4)$ where $v_1 + v_2 + v_3 + v_4 = 0$.

TRUE OR FALSE For Exercises 51–64, determine if the statement is true or false, and justify your answer.

51. If $m < n$, then a set of m vectors cannot span $\mathbf{R}^n$.

52. If a set of vectors includes $\mathbf{0}$, then it cannot span $\mathbf{R}^n$.

53. Suppose A is a matrix with n rows and m columns. If $n < m$, then the columns of A span $\mathbf{R}^n$.

54. Suppose A is a matrix with n rows and m columns. If $m < n$, then the columns of A span $\mathbf{R}^n$.

55. If A is a matrix with columns that span $\mathbf{R}^n$, then $A\mathbf{x} = \mathbf{0}$ has nontrivial solutions.

56. If A is a matrix with columns that span $\mathbf{R}^n$, then $A\mathbf{x} = \mathbf{b}$ has a solution for all $\mathbf{b}$ in $\mathbf{R}^n$.

57. If $\{\mathbf{u}_1, \mathbf{u}_2, \mathbf{u}_3\}$ spans $\mathbf{R}^3$, then so does $\{\mathbf{u}_1, \mathbf{u}_2, \mathbf{u}_3, \mathbf{u}_4\}$.

58. If $\{\mathbf{u}_1, \mathbf{u}_2, \mathbf{u}_3\}$ does not span $\mathbf{R}^3$, then neither does $\{\mathbf{u}_1, \mathbf{u}_2, \mathbf{u}_3, \mathbf{u}_4\}$.

59. If $\{\mathbf{u}_1, \mathbf{u}_2, \mathbf{u}_3, \mathbf{u}_4\}$ spans $\mathbf{R}^3$, then so does $\{\mathbf{u}_1, \mathbf{u}_2, \mathbf{u}_3\}$.

60. If $\{\mathbf{u}_1, \mathbf{u}_2, \mathbf{u}_3, \mathbf{u}_4\}$ does not span $\mathbf{R}^3$, then neither does $\{\mathbf{u}_1, \mathbf{u}_2, \mathbf{u}_3\}$.

61. If $\mathbf{u}_4$ is a linear combination of $\{\mathbf{u}_1, \mathbf{u}_2, \mathbf{u}_3\}$, then
$$\text{span}\{\mathbf{u}_1, \mathbf{u}_2, \mathbf{u}_3, \mathbf{u}_4\} = \text{span}\{\mathbf{u}_1, \mathbf{u}_2, \mathbf{u}_3\}.$$

62. If $\mathbf{u}_4$ is a linear combination of $\{\mathbf{u}_1, \mathbf{u}_2, \mathbf{u}_3\}$, then
$$\text{span}\{\mathbf{u}_1, \mathbf{u}_2, \mathbf{u}_3, \mathbf{u}_4\} \neq \text{span}\{\mathbf{u}_1, \mathbf{u}_2, \mathbf{u}_3\}.$$

63. If $\mathbf{u}_4$ is *not* a linear combination of $\{\mathbf{u}_1, \mathbf{u}_2, \mathbf{u}_3\}$, then
$$\text{span}\{\mathbf{u}_1, \mathbf{u}_2, \mathbf{u}_3, \mathbf{u}_4\} = \text{span}\{\mathbf{u}_1, \mathbf{u}_2, \mathbf{u}_3\}.$$

64. If $\mathbf{u}_4$ is *not* a linear combination of $\{\mathbf{u}_1, \mathbf{u}_2, \mathbf{u}_3\}$, then
$$\text{span}\{\mathbf{u}_1, \mathbf{u}_2, \mathbf{u}_3, \mathbf{u}_4\} \neq \text{span}\{\mathbf{u}_1, \mathbf{u}_2, \mathbf{u}_3\}.$$

65. Which of the following sets of vectors in $\mathbf{R}^3$ can possibly span $\mathbf{R}^3$? Justify your answer.

(a) $\{\mathbf{u}_1\}$

(b) $\{\mathbf{u}_1, \mathbf{u}_2\}$

(c) $\{\mathbf{u}_1, \mathbf{u}_2, \mathbf{u}_3\}$

(d) $\{\mathbf{u}_1, \mathbf{u}_2, \mathbf{u}_3, \mathbf{u}_4\}$

66. Which of the following sets of vectors in $\mathbf{R}^3$ cannot possibly span $\mathbf{R}^3$? Justify your answer.

(a) $\{\mathbf{u}_1\}$

(b) $\{\mathbf{u}_1, \mathbf{u}_2\}$

(c) $\{\mathbf{u}_1, \mathbf{u}_2, \mathbf{u}_3\}$

(d) $\{\mathbf{u}_1, \mathbf{u}_2, \mathbf{u}_3, \mathbf{u}_4\}$

67. Prove that if c is a nonzero scalar, then $\text{span}\{\mathbf{u}\} = \text{span}\{c\mathbf{u}\}$.

68. Prove that if c_1 and c_2 are nonzero scalars, then $\text{span}\{\mathbf{u}_1, \mathbf{u}_2\} = \text{span}\{c_1\mathbf{u}_1, c_2\mathbf{u}_2\}$.

69. Suppose that S_1 are S_2 are two finite sets of vectors, and that S_1 is a subset of S_2. Prove that the span of S_1 is a subset of the span of S_2.

70. Prove that if $\text{span}\{\mathbf{u}_1, \mathbf{u}_2\} = \mathbf{R}^2$, then $\text{span}\{\mathbf{u}_1 + \mathbf{u}_2, \mathbf{u}_1 - \mathbf{u}_2\} = \mathbf{R}^2$.

71. Prove that if $\text{span}\{\mathbf{u}_1, \mathbf{u}_2, \mathbf{u}_3\} = \mathbf{R}^3$, then $\text{span}\{\mathbf{u}_1 + \mathbf{u}_2, \mathbf{u}_1 + \mathbf{u}_3, \mathbf{u}_2 + \mathbf{u}_3\} = \mathbf{R}^3$.

72. Suppose that $\{\mathbf{u}_1, \dots, \mathbf{u}_m\}$ is a subset of $\mathbf{R}^n$, with $m > n$. Prove that if $\mathbf{b}$ is in $\text{span}\{\mathbf{u}_1, \dots, \mathbf{u}_m\}$, then there are infinitely many ways to express $\mathbf{b}$ as a linear combination of $\{\mathbf{u}_1, \dots, \mathbf{u}_m\}$.

73. Prove Theorem 2.8.

74. Prove Theorem 2.10.

Ⓒ For Exercises 75–78, determine if the claimed equality is true or false.

75. $\text{span}\left\{ \begin{bmatrix} 3 \\ 2 \\ 1 \end{bmatrix}, \begin{bmatrix} 1 \\ 5 \\ 4 \end{bmatrix}, \begin{bmatrix} -2 \\ 3 \\ 0 \end{bmatrix} \right\} = \mathbf{R}^3$

76. $\text{span}\left\{ \begin{bmatrix} 1 \\ 1 \\ 2 \end{bmatrix}, \begin{bmatrix} 3 \\ 4 \\ -1 \end{bmatrix}, \begin{bmatrix} 4 \\ 6 \\ -6 \end{bmatrix} \right\} = \mathbf{R}^3$

77. $\text{span}\left\{ \begin{bmatrix} 4 \\ 0 \\ 2 \\ 3 \end{bmatrix}, \begin{bmatrix} 7 \\ -4 \\ 6 \\ 7 \end{bmatrix}, \begin{bmatrix} 1 \\ 3 \\ -2 \\ 1 \end{bmatrix}, \begin{bmatrix} 3 \\ 2 \\ 0 \\ 2 \end{bmatrix} \right\} = \mathbf{R}^4$

78. $\text{span}\left\{ \begin{bmatrix} 3 \\ -2 \\ 1 \\ 0 \end{bmatrix}, \begin{bmatrix} 8 \\ 5 \\ -9 \\ 7 \end{bmatrix}, \begin{bmatrix} 1 \\ 6 \\ 2 \\ -3 \end{bmatrix}, \begin{bmatrix} 2 \\ -1 \\ 3 \\ 5 \end{bmatrix} \right\} = \mathbf{R}^4$

2.3 Linear Independence

The myology clinic at a university research hospital helps patients recover muscle mass lost due to illness. After a full evaluation, patients receive exercise training and each is given a nutritional powder that has the exact balance of protein, fat, and carbohydrates required to meet his or her needs. The nutritional powders are created by combining

	Brand			
	A	B	C	D
Protein	16	22	18	18
Fat	2	4	0	2
Carbohydrates	8	4	4	6

Table 1 Nutritional Powder Brand Components (grams per serving)

some or all of four powder brands that the clinic keeps in stock. The components for brands A, B, C, and D (in grams per serving) are shown in Table 1.

Stocking all four brands is expensive, as they have a limited shelf life and take up valuable storage space. The clinic would like to eliminate unnecessary brands, but it does not want to sacrifice any flexibility to create specialized combinations. Are all four brands needed, or can they get by with fewer?

We can solve this problem using vectors, but first we need to develop some additional concepts. Recall that in the previous section we noted the set of vectors

$$\mathbf{u}_1 = \begin{bmatrix} 2 \\ 1 \\ 1 \end{bmatrix}, \quad \mathbf{u}_2 = \begin{bmatrix} 1 \\ 2 \\ 3 \end{bmatrix}, \quad \mathbf{u}_3 = \begin{bmatrix} 1 \\ -4 \\ -7 \end{bmatrix} \tag{1}$$

are such that the third is a linear combination of the first two, with

$$\mathbf{u}_3 = 2\mathbf{u}_1 - 3\mathbf{u}_2 \tag{2}$$

Thus, in a sense, $\mathbf{u}_3$ depends on $\mathbf{u}_1$ and $\mathbf{u}_2$. We can also solve (2) for $\mathbf{u}_1$ or $\mathbf{u}_2$,

$$\mathbf{u}_1 = \frac{3}{2}\mathbf{u}_2 + \frac{1}{2}\mathbf{u}_3 \quad \text{or} \quad \mathbf{u}_2 = \frac{2}{3}\mathbf{u}_1 - \frac{1}{3}\mathbf{u}_3$$

so each of the vectors is "dependent" on the others. Rather than separating out one particular vector, we can move all terms to one side of the equation, giving us

$$2\mathbf{u}_1 - 3\mathbf{u}_2 - \mathbf{u}_3 = \mathbf{0}$$

Definition Linear Independence   This brings us to the following important definition.

DEFINITION 2.11

Let $\{\mathbf{u}_1, \mathbf{u}_2, \ldots, \mathbf{u}_m\}$ be a set of vectors in $\mathbf{R}^n$. If the only solution to the vector equation

$$x_1\mathbf{u}_1 + x_2\mathbf{u}_2 + \cdots + x_m\mathbf{u}_m = \mathbf{0}$$

is the trivial solution given by $x_1 = x_2 = \cdots = x_m = 0$, then the set $\{\mathbf{u}_1, \mathbf{u}_2, \ldots, \mathbf{u}_m\}$ is **linearly independent**. If there are nontrivial solutions, then the set is **linearly dependent**.

▶ To determine if a set of vectors is linearly dependent or independent, we almost always use the method illustrated in Example 1: Set the linear combination equal to **0** and find the solutions.

EXAMPLE 1 Determine if the set

$$\mathbf{u}_1 = \begin{bmatrix} -1 \\ 4 \\ -2 \\ -3 \end{bmatrix}, \quad \mathbf{u}_2 = \begin{bmatrix} 3 \\ -13 \\ 7 \\ 7 \end{bmatrix}, \quad \mathbf{u}_3 = \begin{bmatrix} -2 \\ 1 \\ 9 \\ -5 \end{bmatrix}$$

is linearly dependent or linearly independent.

Solution To determine if the set $\{\mathbf{u}_1, \mathbf{u}_2, \mathbf{u}_3\}$ is linearly dependent or linearly independent, we need to find the solutions of the vector equation

$$x_1\mathbf{u}_1 + x_2\mathbf{u}_2 + x_3\mathbf{u}_3 = \mathbf{0}$$

This is equivalent to the linear system

$$
\begin{aligned}
-x_1 + 3x_2 - 2x_3 &= 0 \\
4x_1 - 13x_2 + x_3 &= 0 \\
-2x_1 + 7x_2 + 9x_3 &= 0 \\
-3x_1 + 7x_2 - 5x_3 &= 0
\end{aligned}
$$

The corresponding augmented matrix and echelon form are

◀ The row operations used in (3) are (in order performed):

$$
\begin{aligned}
4R_1 + R_2 &\Rightarrow R_2 \\
-2R_1 + R_3 &\Rightarrow R_3 \\
-3R_1 + R_4 &\Rightarrow R_4 \\
R_2 + R_3 &\Rightarrow R_3 \\
-2R_2 + R_4 &\Rightarrow R_4 \\
(-5/2)R_3 + R_4 &\Rightarrow R_4
\end{aligned}
$$

$$
\begin{bmatrix}
-1 & 3 & -2 & 0 \\
4 & -13 & 1 & 0 \\
-2 & 7 & 9 & 0 \\
-3 & 7 & -5 & 0
\end{bmatrix}
\sim
\begin{bmatrix}
-1 & 3 & -2 & 0 \\
0 & -1 & -7 & 0 \\
0 & 0 & 6 & 0 \\
0 & 0 & 0 & 0
\end{bmatrix}
\tag{3}
$$

Back substitution shows that the only solution is the trivial one, $x_1 = x_2 = x_3 = 0$. Hence the set $\{\mathbf{u}_1, \mathbf{u}_2, \mathbf{u}_3\}$ is linearly independent. ■

EXAMPLE 2 Determine if the myology clinic described earlier can eliminate any of the nutritional powder brands with components given in Table 1.

Solution We start by determining if the nutrient vectors for the four brands

$$
\mathbf{a} = \begin{bmatrix} 16 \\ 2 \\ 8 \end{bmatrix}, \quad
\mathbf{b} = \begin{bmatrix} 22 \\ 4 \\ 4 \end{bmatrix}, \quad
\mathbf{c} = \begin{bmatrix} 18 \\ 0 \\ 4 \end{bmatrix}, \quad
\mathbf{d} = \begin{bmatrix} 18 \\ 2 \\ 6 \end{bmatrix}
$$

◀ The row operations used in (4) are (in order performed):

$$
\begin{aligned}
R_1 &\Leftrightarrow R_2 \\
-8R_1 + R_2 &\Rightarrow R_2 \\
-4R_1 + R_3 &\Rightarrow R_3 \\
(-6/5)R_2 + R_3 &\Rightarrow R_3 \\
(-5/22)R_3 &\Rightarrow R_3
\end{aligned}
$$

are linearly independent. To do so, we must find the solutions to the vector equation $x_1\mathbf{a} + x_2\mathbf{b} + x_3\mathbf{c} + x_4\mathbf{d} = \mathbf{0}$. The augmented matrix of the equivalent linear system and the echelon form are

$$
\begin{bmatrix}
16 & 22 & 18 & 18 & 0 \\
2 & 4 & 0 & 2 & 0 \\
8 & 4 & 4 & 6 & 0
\end{bmatrix}
\sim
\begin{bmatrix}
2 & 4 & 0 & 2 & 0 \\
0 & -10 & 18 & 2 & 0 \\
0 & 0 & 4 & 1 & 0
\end{bmatrix}
\tag{4}
$$

Back substitution leads to the general solution

$$
x_1 = -\tfrac{1}{2}s, \qquad x_2 = -\tfrac{1}{4}s, \qquad x_3 = -\tfrac{1}{4}s, \qquad x_4 = s
$$

which holds for all choices of s. Thus there exist nontrivial solutions, so the set of vectors is linearly dependent. Letting $s = 1$, we have $x_1 = -\tfrac{1}{2}$, $x_2 = -\tfrac{1}{4}$, $x_3 = -\tfrac{1}{4}$, and $x_4 = 1$, which gives us

$$
-\tfrac{1}{2}\mathbf{a} - \tfrac{1}{4}\mathbf{b} - \tfrac{1}{4}\mathbf{c} + \mathbf{d} = \mathbf{0} \quad \Longrightarrow \quad \mathbf{d} = \tfrac{1}{2}\mathbf{a} + \tfrac{1}{4}\mathbf{b} + \tfrac{1}{4}\mathbf{c}
$$

Thus we obtain a serving of brand D by combining a $\tfrac{1}{2}$ serving of A, a $\tfrac{1}{4}$ serving of B, and a $\tfrac{1}{4}$ serving of C. Hence there is no need to stock brand D. ■

When working with a new concept, it can be helpful to start with simple cases. In this spirit, suppose that we have a set with one vector, $\{\mathbf{u}_1\}$. Is this set linearly independent? To check, we need to determine the solutions to the equation

$$x_1\mathbf{u}_1 = \mathbf{0} \tag{5}$$

At first glance, it seems that the only solution is the trivial one $x_1 = 0$, and for most choices of $\mathbf{u}_1$ that is true. Specifically, as long as $\mathbf{u}_1 \neq \mathbf{0}$, then the only solution is $x_1 = 0$ and the set $\{\mathbf{u}_1\}$ is linearly independent. But if it happens that $\mathbf{u}_1 = \mathbf{0}$, then the set is linearly dependent, because now there are nontrivial solutions to (5), such as

$$3\mathbf{u}_1 = \mathbf{0}$$

In fact, having $\mathbf{0}$ in *any* set of vectors always guarantees that the set will be linearly dependent.

THEOREM 2.12

Suppose that $\{\mathbf{0}, \mathbf{u}_1, \mathbf{u}_2, \ldots, \mathbf{u}_m\}$ is a set of vectors in $\mathbf{R}^n$. Then the set is linearly dependent.

Proof   We need to determine if the vector equation

$$x_0\mathbf{0} + x_1\mathbf{u}_1 + x_2\mathbf{u}_2 + \cdots + x_m\mathbf{u}_m = \mathbf{0}$$

has any nontrivial solutions. Regardless of the values of $\mathbf{u}_1, \mathbf{u}_2, \ldots, \mathbf{u}_m$, setting $x_0 = 1$ and $x_1 = x_2 = \cdots = x_m = 0$ gives us an easy (but legitimate) nontrivial solution, so that the set is linearly dependent. ∎

For a set of two vectors $\{\mathbf{u}_1, \mathbf{u}_2\}$, we already know what happens if one of these is $\mathbf{0}$. Let's assume that both vectors are nonzero and ask the same question as above: Is this set linearly independent? As usual, we need to determine the nature of the solutions to

$$x_1\mathbf{u}_1 + x_2\mathbf{u}_2 = \mathbf{0} \tag{6}$$

If there is a nontrivial solution, then it must be that both x_1 and x_2 are nonzero. (Why?) In this case, we can solve (6) for $\mathbf{u}_1$, giving us

$$\mathbf{u}_1 = -\frac{x_2}{x_1}\mathbf{u}_2$$

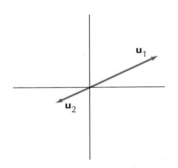

Figure 1 $\mathbf{u}_1$ and $\mathbf{u}_2$ are linearly dependent vectors.

Thus we see that the set $\{\mathbf{u}_1, \mathbf{u}_2\}$ is linearly dependent if and only if $\mathbf{u}_1$ is a scalar multiple of $\mathbf{u}_2$. Geometrically, the set is linearly dependent if and only if the two vectors point in the same (or opposite) direction. (See Figure 1.)

When trying to determine if a set $\{\mathbf{u}_1, \mathbf{u}_2, \ldots, \mathbf{u}_m\}$ of three or more vectors is linearly dependent or independent, in general we have to find the solutions to

$$x_1\mathbf{u}_1 + x_2\mathbf{u}_2 + \cdots + x_m\mathbf{u}_m = \mathbf{0}$$

However, there is a special case where virtually no work is required.

THEOREM 2.13

Suppose that $\{\mathbf{u}_1, \mathbf{u}_2, \ldots, \mathbf{u}_m\}$ is a set of vectors in $\mathbf{R}^n$. If $n < m$, then the set is linearly dependent.

In other words, if the number of vectors m exceeds the number of components n, then the set is linearly dependent.

Proof   As usual when testing for linear independence, we start with the vector equation

$$x_1\mathbf{u}_1 + x_2\mathbf{u}_2 + \cdots + x_m\mathbf{u}_m = \mathbf{0} \tag{7}$$

Since $\mathbf{u}_1, \ldots, \mathbf{u}_m$ each have n components, this is equivalent to a homogeneous linear system with n equations and m unknowns. Because $n < m$, this system has more variables than equations, so there are infinitely many solutions (see Exercise 55 in Section 1.2). Therefore (7) has nontrivial solutions, and hence the set $\{\mathbf{u}_1, \mathbf{u}_2, \ldots, \mathbf{u}_m\}$ is linearly dependent.   ∎

This theorem immediately tells us that the myology clinic's set of four nutritional powder brands must not all be needed, because we can represent them by four vectors $(\mathbf{a}, \mathbf{b}, \mathbf{c}, \mathbf{d})$ in $\mathbf{R}^3$ (protein, fat, carbohydrates). However, the theorem does not tell us which brand can be eliminated.

Note also that Theorem 2.13 tells us nothing about the case when $n \geq m$. In this instance, it is possible for the set to be linearly dependent, as in (1), where $n = 3$ and $m = 3$. Or the set can be linearly independent, as in Example 1, where $n = 4$ and $m = 3$.

Span and Linear Independence

It is common to confuse span and linear independence, because although they are different concepts, they are related. To see the connection, let's return to the earlier discussion about the set of two vectors $\{\mathbf{u}_1, \mathbf{u}_2\}$. This set is linearly dependent exactly when $\mathbf{u}_1$ is a multiple of $\mathbf{u}_2$—that is, exactly when $\mathbf{u}_1$ is in span$\{\mathbf{u}_2\}$. This connection between span and linear independence holds more generally.

THEOREM 2.14

Let $\{\mathbf{u}_1, \mathbf{u}_2, \ldots, \mathbf{u}_m\}$ be a set of vectors in $\mathbf{R}^n$. Then the set is linearly dependent if and only if one of the vectors in the set is in the span of the other vectors.

Proof   Suppose first that the set is linearly dependent. Then there exist scalars $c_1, \ldots, c_m$, not all zero, such that

$$c_1\mathbf{u}_1 + c_2\mathbf{u}_2 + \cdots + c_m\mathbf{u}_m = \mathbf{0}$$

To simplify notation, assume that $c_1 \neq 0$. Then we can solve for $\mathbf{u}_1$,

$$\mathbf{u}_1 = -\frac{c_2}{c_1}\mathbf{u}_2 - \cdots - \frac{c_m}{c_1}\mathbf{u}_m$$

which shows that $\mathbf{u}_1$ is in span$\{\mathbf{u}_2, \ldots, \mathbf{u}_m\}$. This completes the "forward" direction of the proof.

Now suppose that one of the vectors in the set is in the span of the remaining vectors—say, $\mathbf{u}_1$ is in span$\{\mathbf{u}_2, \ldots, \mathbf{u}_m\}$. Then there exist scalars $c_2, c_3, \ldots, c_m$ such that

$$\mathbf{u}_1 = c_2\mathbf{u}_2 + \cdots + c_m\mathbf{u}_m \tag{8}$$

Moving all terms to the left side in (8), we have

$$\mathbf{u}_1 - c_2\mathbf{u}_2 - \cdots - c_m\mathbf{u}_m = \mathbf{0}$$

Since the coefficient on $\mathbf{u}_1$ is 1, this shows the set is linearly dependent. This completes the "backward" direction, finishing the proof.   ∎

EXAMPLE 3   Give a linearly dependent set of vectors such that one vector is not a linear combination of the others. Explain why this does not contradict Theorem 2.14.

Solution   Theorem 2.14 tells us that in a linearly dependent set, at least one vector is a linear combination of the other vectors. However, it does not say that *every* vector is a linear combination of the others, so what we seek does not contradict Theorem 2.14.

Let

$$\mathbf{u}_1 = \begin{bmatrix} -1 \\ 0 \\ -2 \end{bmatrix}, \quad \mathbf{u}_2 = \begin{bmatrix} 3 \\ -2 \\ 2 \end{bmatrix}, \quad \mathbf{u}_3 = \begin{bmatrix} 5 \\ -2 \\ 6 \end{bmatrix}, \quad \mathbf{u}_4 = \begin{bmatrix} 1 \\ 2 \\ 3 \end{bmatrix}$$

By Theorem 2.13, this set must be linearly dependent because there are four vectors with three components. Theorem 2.14 says that one of the vectors is a linear combination of the others, and indeed we have $\mathbf{u}_3 = -2\mathbf{u}_1 + \mathbf{u}_2$. On the other hand, the equation

$$x_1\mathbf{u}_1 + x_2\mathbf{u}_2 + x_3\mathbf{u}_3 = \mathbf{u}_4 \tag{9}$$

has corresponding augmented matrix and echelon form

$$\begin{bmatrix} -1 & 3 & 5 & 1 \\ 0 & -2 & -2 & 2 \\ -2 & 2 & 6 & 3 \end{bmatrix} \sim \begin{bmatrix} -1 & 3 & 5 & 1 \\ 0 & -2 & -2 & 2 \\ 0 & 0 & 0 & -3 \end{bmatrix} \tag{10}$$

▶ The row operations used in (10) are (in order performed):

$-2R_1 + R_3 \Rightarrow R_3$
$-2R_2 + R_3 \Rightarrow R_3$

From the echelon form it follows that (9) has no solutions, so that $\mathbf{u}_4$ is not a linear combination of the other vectors in the set. ∎

Homogeneous Systems

In Section 2.2, we introduced the notation

$$A\mathbf{x} = x_1\mathbf{a}_1 + x_2\mathbf{a}_2 + \cdots + x_m\mathbf{a}_m$$

where $A = \begin{bmatrix} \mathbf{a}_1 & \mathbf{a}_2 & \dots & \mathbf{a}_m \end{bmatrix}$ and $\mathbf{x} = (x_1, x_2, \ldots, x_m)$, and noted that any linear system can be expressed in the compact form

$$A\mathbf{x} = \mathbf{b}$$

The system $A\mathbf{x} = \mathbf{0}$ is a **homogeneous** linear system, introduced in Section 1.2. There we showed that homogeneous linear systems have either one solution (the trivial solution) or infinitely many solutions.

The next theorem shows that there is a direct connection between the number of solutions to $A\mathbf{x} = \mathbf{0}$ and whether the columns of A are linearly independent.

THEOREM 2.15

Let $A = \begin{bmatrix} \mathbf{a}_1 & \mathbf{a}_2 & \dots & \mathbf{a}_m \end{bmatrix}$ and $\mathbf{x} = (x_1, x_2, \ldots, x_m)$. The set $\{\mathbf{a}_1, \mathbf{a}_2, \ldots, \mathbf{a}_m\}$ is linearly independent if and only if the homogeneous linear system

$$A\mathbf{x} = \mathbf{0}$$

has only the trivial solution.

Proof   Written as a vector equation, the system $A\mathbf{x} = \mathbf{0}$ has the form

$$x_1\mathbf{a}_1 + x_2\mathbf{a}_2 + \cdots + x_m\mathbf{a}_m = \mathbf{0} \tag{11}$$

Thus if $A\mathbf{x} = \mathbf{0}$ has only the trivial solution, then so does (11) and the columns of A are linearly independent. On the other hand, if $A\mathbf{x} = \mathbf{0}$ has nontrivial solutions, then so does (11) and the columns of A are linearly dependent. ∎

Definition Nonhomogeneous, Associated Homogeneous System

If $\mathbf{b} \neq \mathbf{0}$, then the system $A\mathbf{x} = \mathbf{b}$ is **nonhomogeneous**, and the **associated homogeneous system** is $A\mathbf{x} = \mathbf{0}$. There is a close connection between the set of solutions to a nonhomogeneous system $A\mathbf{x} = \mathbf{b}$ and the associated homogeneous system $A\mathbf{x} = \mathbf{0}$, illustrated in the following example.

EXAMPLE 4   Find the general solution for the linear system

$$
\begin{aligned}
2x_1 - 6x_2 - x_3 + 8x_4 &= 7 \\
x_1 - 3x_2 - x_3 + 6x_4 &= 6 \\
-x_1 + 3x_2 - x_3 + 2x_4 &= 4
\end{aligned}
\tag{12}
$$

and the general solution for the associated homogeneous system.

Solution   Applying our usual matrix and row reduction methods, we find that the general solution to (12) is

$$
\mathbf{x} = \begin{bmatrix} 1 \\ 0 \\ -5 \\ 0 \end{bmatrix} + s_1 \begin{bmatrix} 3 \\ 1 \\ 0 \\ 0 \end{bmatrix} + s_2 \begin{bmatrix} -2 \\ 0 \\ 4 \\ 1 \end{bmatrix}
$$

The general solution to the associated homogeneous system was found in Example 7 of Section 1.2. It is

$$
\mathbf{x} = s_1 \begin{bmatrix} 3 \\ 1 \\ 0 \\ 0 \end{bmatrix} + s_2 \begin{bmatrix} -2 \\ 0 \\ 4 \\ 1 \end{bmatrix}
$$

For both solutions, s_1 and s_2 can be any real numbers.   ■

Comparing the preceding solutions, we see that the only difference is the "constant" vector

$$
\begin{bmatrix} 1 \\ 0 \\ -5 \\ 0 \end{bmatrix}
$$

This type of relationship between general solutions occurs in all such cases. To see why, it is helpful to have the following result showing that the product $A\mathbf{x}$ obeys the distributive law.

THEOREM 2.16

Suppose that $A = \begin{bmatrix} \mathbf{a}_1 & \mathbf{a}_2 & \cdots & \mathbf{a}_m \end{bmatrix}$, and let $\mathbf{x} = (x_1, x_2, \ldots, x_m)$ and $\mathbf{y} = (y_1, y_2, \ldots, y_m)$. Then

(a) $A(\mathbf{x} + \mathbf{y}) = A\mathbf{x} + A\mathbf{y}$

(b) $A(\mathbf{x} - \mathbf{y}) = A\mathbf{x} - A\mathbf{y}$

Proof   The results follow from the definition of the product and a bit of algebra. Starting with (a), we have

$$
\mathbf{x} + \mathbf{y} = \begin{bmatrix} x_1 + y_1 \\ x_2 + y_2 \\ \vdots \\ x_m + y_m \end{bmatrix}
$$

so that

$$A(\mathbf{x} + \mathbf{y}) = (x_1 + y_1)\mathbf{a}_1 + (x_2 + y_2)\mathbf{a}_2 + \cdots + (x_m + y_m)\mathbf{a}_m$$
$$= (x_1\mathbf{a}_1 + x_2\mathbf{a}_2 + \cdots + x_m\mathbf{a}_m) + (y_1\mathbf{a}_1 + y_2\mathbf{a}_2 + \cdots + y_m\mathbf{a}_m)$$
$$= A\mathbf{x} + A\mathbf{y}$$

The proof of (b) is similar and left as an exercise. ∎

Definition Particular Solution

Now let $\mathbf{x}_p$ be any solution to $A\mathbf{x} = \mathbf{b}$. We call $\mathbf{x}_p$ a **particular solution** to the system, and it can be thought of as any fixed solution to the system.

THEOREM 2.17

Let $\mathbf{x}_p$ be a particular solution to

$$A\mathbf{x} = \mathbf{b} \tag{13}$$

Then all solutions $\mathbf{x}_g$ to (13) have the form $\mathbf{x}_g = \mathbf{x}_p + \mathbf{x}_h$, where $\mathbf{x}_h$ is a solution to the associated homogeneous system $A\mathbf{x} = \mathbf{0}$.

Proof Let $\mathbf{x}_p$ be a particular solution to (13), and suppose that $\mathbf{x}_g$ is any solution to the same system. Then

$$A(\mathbf{x}_g - \mathbf{x}_p) = A\mathbf{x}_g - A\mathbf{x}_p = \mathbf{b} - \mathbf{b} = \mathbf{0}$$

Hence if we let $\mathbf{x}_h = \mathbf{x}_g - \mathbf{x}_p$, then $\mathbf{x}_h$ is a solution to the associated homogeneous system $A\mathbf{x} = \mathbf{0}$. Solving for $\mathbf{x}_g$, we have

$$\mathbf{x}_g = \mathbf{x}_p + \mathbf{x}_h$$

so that $\mathbf{x}_g$ has the form claimed. ∎

EXAMPLE 5 Find the general solution and solution to the associated homogeneous system for

$$\begin{aligned} 4x_1 - 6x_2 &= -14 \\ -6x_1 + 9x_2 &= 21 \end{aligned} \tag{14}$$

Solution Applying our standard solution procedures yields the vector form of the general solution to (14)

$$\mathbf{x}_g = \begin{bmatrix} 1 \\ 3 \end{bmatrix} + s \begin{bmatrix} 3 \\ 2 \end{bmatrix}$$

which is a line when graphed in $\mathbf{R}^2$. The solutions to the associated homogeneous system are

$$\mathbf{x}_h = s \begin{bmatrix} 3 \\ 2 \end{bmatrix}$$

which is also a line when graphed in $\mathbf{R}^2$. If we let $\mathbf{x}_p = \begin{bmatrix} 1 \\ 3 \end{bmatrix}$, then every value of $\mathbf{x}_g$ can be expressed as the sum of $\mathbf{x}_p$ and one of the homogeneous solutions $\mathbf{x}_h$. Thus the general solution $\mathbf{x}_g$ is a translation by $\mathbf{x}_p$ of the general solution $\mathbf{x}_h$ of the associated homogeneous system. The graphs are shown in Figure 2. ∎

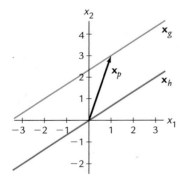

Figure 2 Graphs of $\mathbf{x}_g$, $\mathbf{x}_h$, and $\mathbf{x}_p$ from Example 5.

At the end of Section 2.2, Theorem 2.10 linked span with solutions to linear systems. Theorem 2.18 is similar in spirit, this time linking linear independence with solutions to linear systems.

THEOREM 2.18

Let $\mathbf{a}_1, \mathbf{a}_2, \ldots, \mathbf{a}_m$ and $\mathbf{b}$ be vectors in $\mathbf{R}^n$. Then the following statements are equivalent. That is, if one is true, then so are the others, and if one is false, then so are the others.

(a) The set $\{\mathbf{a}_1, \mathbf{a}_2, \ldots, \mathbf{a}_m\}$ is linearly independent.

(b) The vector equation $x_1\mathbf{a}_1 + x_2\mathbf{a}_2 + \cdots + x_m\mathbf{a}_m = \mathbf{b}$ has at most one solution.

(c) The linear system corresponding to $\begin{bmatrix} \mathbf{a}_1 & \mathbf{a}_2 & \cdots & \mathbf{a}_m & \mathbf{b} \end{bmatrix}$ has at most one solution.

(d) The equation $A\mathbf{x} = \mathbf{b}$, with $A = \begin{bmatrix} \mathbf{a}_1 & \mathbf{a}_2 & \cdots & \mathbf{a}_m \end{bmatrix}$, has at most one solution.

Proof   The equivalence of (b), (c), and (d) is immediate from the definitions, so all that is needed to complete the proof is show that (a) and (b) are equivalent.

We start by showing that (a) implies (b). Let $\{\mathbf{a}_1, \mathbf{a}_2, \ldots, \mathbf{a}_m\}$ be linearly independent, and suppose to the contrary that

$$x_1\mathbf{a}_1 + x_2\mathbf{a}_2 + \cdots + x_m\mathbf{a}_m = \mathbf{b}$$

has more than one solution. Then there exist scalars $r_1, \ldots, r_m$ and $s_1, \ldots, s_m$ such that

$$r_1\mathbf{a}_1 + r_2\mathbf{a}_2 + \cdots + r_m\mathbf{a}_m = \mathbf{b}$$
$$s_1\mathbf{a}_1 + s_2\mathbf{a}_2 + \cdots + s_m\mathbf{a}_m = \mathbf{b}$$

and so

$$r_1\mathbf{a}_1 + r_2\mathbf{a}_2 + \cdots + r_m\mathbf{a}_m = s_1\mathbf{a}_1 + s_2\mathbf{a}_2 + \cdots + s_m\mathbf{a}_m$$

Moving all terms to one side and regrouping yields

$$(r_1 - s_1)\mathbf{a}_1 + (r_2 - s_2)\mathbf{a}_2 + \cdots + (r_m - s_m)\mathbf{a}_m = \mathbf{0}$$

Since $\{\mathbf{a}_1, \mathbf{a}_2, \ldots, \mathbf{a}_m\}$ is linearly independent, each coefficient must be 0. Hence $r_1 = s_1$, $\ldots, r_m = s_m$, so there is just one solution to $x_1\mathbf{a}_1 + \cdots + x_m\mathbf{a}_m = \mathbf{b}$.

Proving that (b) implies (a) is easier. Since $\mathbf{b}$ can be any vector, we can set $\mathbf{b} = \mathbf{0}$. By (b) there is at most one solution to

$$x_1\mathbf{a}_1 + x_2\mathbf{a}_2 + \cdots + x_m\mathbf{a}_m = \mathbf{0}$$

Of course, there is the trivial solution $x_1 = \cdots = x_m = 0$, so this must be the only solution. Hence the set $\{\mathbf{a}_1, \mathbf{a}_2, \ldots, \mathbf{a}_m\}$ is linearly independent, and (a) follows. ■

The Big Theorem – Version 1

Next, we present the first version of the Big Theorem. This theorem is Big for two reasons: (a) it is Big—as in important—because it will serve to tie together and unify many of the ideas that we shall be developing, and (b) subsequent versions will include more and more equivalent statements, making it Big in size.

THEOREM 2.19

▶ In all versions of the Big Theorem, n is both the number of vectors in $\mathcal{A}$ and the number of components in each vector. Thus A has n rows and n columns.

THE BIG THEOREM — VERSION 1 Let $\mathcal{A} = \{\mathbf{a}_1, \ldots, \mathbf{a}_n\}$ be a set of n vectors in $\mathbf{R}^n$, and let $A = \begin{bmatrix} \mathbf{a}_1 & \cdots & \mathbf{a}_n \end{bmatrix}$. Then the following are equivalent:

(a) $\mathcal{A}$ spans $\mathbf{R}^n$.

(b) $\mathcal{A}$ is linearly independent.

(c) $A\mathbf{x} = \mathbf{b}$ has a unique solution for all $\mathbf{b}$ in $\mathbf{R}^n$.

Proof We start by showing that (a) and (b) are equivalent. First suppose that $\mathcal{A}$ spans $\mathbf{R}^n$. If $\mathcal{A}$ is linearly dependent, then one of $\mathbf{a}_1, \ldots, \mathbf{a}_n$—say, $\mathbf{a}_1$—is a linear combination of the others. Then by Theorem 2.7, it follows that

$$\text{span}\{\mathbf{a}_1, \ldots, \mathbf{a}_n\} = \text{span}\{\mathbf{a}_2, \ldots, \mathbf{a}_n\}.$$

But this implies that $\mathbf{R}^n = \text{span}\{\mathbf{a}_2, \ldots, \mathbf{a}_n\}$, contradicting Theorem 2.8. Hence it must be that $\mathcal{A}$ is linearly independent. This shows that (a) implies (b).

To show that (b) implies (a), we assume that $\mathcal{A}$ is linearly independent. Now, if $\mathcal{A}$ does not span $\mathbf{R}^n$, then there exists a vector $\mathbf{a}$ that is not a linear combination of $\mathbf{a}_1, \ldots, \mathbf{a}_n$. Since $\mathcal{A}$ is linearly independent, it follows that the set $\{\mathbf{a}, \mathbf{a}_1, \ldots, \mathbf{a}_n\}$ of $n + 1$ vectors is also linearly independent, contradicting Theorem 2.13. Hence $\mathcal{A}$ must span $\mathbf{R}^n$. Thus (b) implies (a), and therefore (a) is equivalent to (b).

Now suppose that (a) and (b) are both true. Then by Theorem 2.10, (a) implies that $A\mathbf{x} = \mathbf{b}$ has *at least* one solution for every $\mathbf{b}$ in $\mathbf{R}^n$. On the other hand, from Theorem 2.18 we know that (b) implies that $A\mathbf{x} = \mathbf{b}$ has *at most* one solution for every $\mathbf{b}$ in $\mathbf{R}^n$. This leaves us with only one possibility, that $A\mathbf{x} = \mathbf{b}$ has exactly one solution for every $\mathbf{b}$ in $\mathbf{R}^n$, confirming (c) is true.

Finally, suppose that (c) is true. Since $A\mathbf{x} = \mathbf{b}$ has a unique solution for every $\mathbf{b}$ in $\mathbf{R}^n$, then in particular $A\mathbf{x} = \mathbf{0}$ has only the trivial solution. Appealing to Theorem 2.15, we conclude that $\mathcal{A}$ must be linearly independent and hence also spans $\mathbf{R}^n$. ■

EXAMPLE 6 Suppose that

$$\mathbf{a}_1 = \begin{bmatrix} 1 \\ 7 \\ -2 \end{bmatrix}, \quad \mathbf{a}_2 = \begin{bmatrix} 3 \\ 0 \\ 1 \end{bmatrix}, \quad \mathbf{a}_3 = \begin{bmatrix} 5 \\ 2 \\ -6 \end{bmatrix}, \quad \text{and} \quad A = \begin{bmatrix} \mathbf{a}_1 & \mathbf{a}_2 & \mathbf{a}_3 \end{bmatrix}$$

Show that the columns of A are linearly independent and span $\mathbf{R}^3$, and that $A\mathbf{x} = \mathbf{b}$ has a unique solution for every $\mathbf{b}$ in $\mathbf{R}^3$.

Solution We start with linear independence, so we need to find the solutions to

$$x_1\mathbf{a}_1 + x_2\mathbf{a}_2 + x_3\mathbf{a}_3 = \mathbf{0} \tag{15}$$

The corresponding augmented matrix and echelon form are

▶ The row operations used in (16) are (in order performed):

$$-7R_1 + R_2 \Rightarrow R_2$$
$$2R_1 + R_3 \Rightarrow R_3$$
$$R_2 \Leftrightarrow R_3$$
$$3R_2 + R_3 \Rightarrow R_3$$

$$\begin{bmatrix} 1 & 3 & 5 & 0 \\ 7 & 0 & 2 & 0 \\ -2 & 1 & -6 & 0 \end{bmatrix} \sim \begin{bmatrix} 1 & 3 & 5 & 0 \\ 0 & 7 & 4 & 0 \\ 0 & 0 & -21 & 0 \end{bmatrix} \tag{16}$$

From the echelon form it follows that (15) has only the trivial solution, so the columns of A are linearly independent.

Because we have three vectors and each has three components, the other questions follow immediately from the Big Theorem. Specifically, since $\{\mathbf{a}_1, \mathbf{a}_2, \mathbf{a}_3\}$ is linearly independent, the set must also span $\mathbf{R}^3$ and there is exactly one solution to $A\mathbf{x} = \mathbf{b}$ for any $\mathbf{b}$ in $\mathbf{R}^3$. The Big Theorem and its successors to come are all very powerful. ■

For one more application of Theorem 2.19, let's return to the nutritional powder problem described at the beginning of the section.

EXAMPLE 7   Use the Big Theorem to show that stocking powder brands A, B, and C is efficient.

Solution We previously determined that the nutrient vectors **a**, **b**, **c**, and **d** are linearly dependent, and we concluded that brand D can be eliminated. For the remaining three brands, it can be verified that

$$x_1\mathbf{a} + x_2\mathbf{b} + x_3\mathbf{c} = \mathbf{0}$$

has only the trivial solution, which tells us that **a**, **b**, and **c** are linearly independent. By Theorem 2.19, we can conclude the following:

- The vectors **a**, **b**, and **c** span all of $\mathbf{R}^3$. Therefore *every* vector in $\mathbf{R}^3$ can be expressed as a linear combination of these three vectors. (Note, though, that some combinations will require negative values of x_1, x_2, and x_3, which is not physically possible when combining powders.)

- Item (c) of the Big Theorem tells us that there is exactly one way to combine brands A, B, and C to create any blend with a specific combination of protein, fat, and carbohydrates. Thus stocking brands A, B, and C is efficient, in that there is no redundancy. ∎

EXERCISES

For Exercises 1–6, determine if the given vectors are linearly independent.

1. $\mathbf{u} = \begin{bmatrix} 3 \\ -2 \end{bmatrix}, \quad \mathbf{v} = \begin{bmatrix} -1 \\ -4 \end{bmatrix}$

2. $\mathbf{u} = \begin{bmatrix} 6 \\ -15 \end{bmatrix}, \quad \mathbf{v} = \begin{bmatrix} -4 \\ -10 \end{bmatrix}$

3. $\mathbf{u} = \begin{bmatrix} 7 \\ 1 \\ -13 \end{bmatrix}, \quad \mathbf{v} = \begin{bmatrix} 5 \\ -3 \\ 2 \end{bmatrix}$

4. $\mathbf{u} = \begin{bmatrix} -4 \\ 0 \\ -3 \end{bmatrix}, \quad \mathbf{v} = \begin{bmatrix} -2 \\ -1 \\ 5 \end{bmatrix}, \quad \mathbf{w} = \begin{bmatrix} -8 \\ 2 \\ -19 \end{bmatrix}$

5. $\mathbf{u} = \begin{bmatrix} 3 \\ -1 \\ 2 \end{bmatrix}, \quad \mathbf{v} = \begin{bmatrix} 0 \\ 4 \\ 1 \end{bmatrix}, \quad \mathbf{w} = \begin{bmatrix} 2 \\ 4 \\ 7 \end{bmatrix}$

6. $\mathbf{u} = \begin{bmatrix} 1 \\ 8 \\ 3 \\ 3 \end{bmatrix}, \quad \mathbf{v} = \begin{bmatrix} 4 \\ -2 \\ 5 \\ -5 \end{bmatrix}, \quad \mathbf{w} = \begin{bmatrix} -1 \\ 2 \\ 0 \\ 1 \end{bmatrix}$

In Exercises 7–12, determine if the columns of the given matrix are linearly independent.

7. $\begin{bmatrix} 15 & -6 \\ -5 & 2 \end{bmatrix}$

8. $\begin{bmatrix} 4 & -12 \\ 2 & 6 \end{bmatrix}$

9. $\begin{bmatrix} 1 & 0 \\ -2 & 2 \\ 5 & -7 \end{bmatrix}$

10. $\begin{bmatrix} 1 & -1 & 2 \\ -4 & 5 & -5 \\ -1 & 2 & 1 \end{bmatrix}$

11. $\begin{bmatrix} 3 & 1 & 0 \\ 5 & -2 & -1 \\ 4 & -4 & -3 \end{bmatrix}$

12. $\begin{bmatrix} -4 & -7 & 1 \\ 0 & 0 & 3 \\ 5 & -1 & 1 \\ 8 & 2 & -4 \end{bmatrix}$

In Exercises 13–18, a matrix A is given. Determine if the homogeneous system $A\mathbf{x} = \mathbf{0}$ (where $\mathbf{x}$ and $\mathbf{0}$ have the appropriate number of components) has any nontrivial solutions.

13. $A = \begin{bmatrix} -3 & 5 \\ 4 & 1 \end{bmatrix}$

14. $A = \begin{bmatrix} 12 & 10 \\ 6 & 5 \end{bmatrix}$

15. $A = \begin{bmatrix} 8 & 1 \\ 0 & -1 \\ -3 & 2 \end{bmatrix}$

16. $A = \begin{bmatrix} -3 & 2 & 1 \\ 1 & -1 & -1 \\ 5 & -4 & -3 \end{bmatrix}$

17. $A = \begin{bmatrix} -1 & 3 & 1 \\ 4 & -3 & -1 \\ 3 & 0 & 5 \end{bmatrix}$

18. $A = \begin{bmatrix} 2 & -3 & 0 \\ 0 & 1 & 2 \\ -5 & 3 & -9 \\ 3 & 0 & 9 \end{bmatrix}$

In Exercises 19–24, determine by inspection (that is, with only minimal calculations) if the given vectors form a linearly dependent or linearly independent set. Justify your answer.

19. $\mathbf{u} = \begin{bmatrix} 14 \\ -6 \end{bmatrix}$, $\mathbf{v} = \begin{bmatrix} 7 \\ -3 \end{bmatrix}$

20. $\mathbf{u} = \begin{bmatrix} 2 \\ 1 \end{bmatrix}$, $\mathbf{v} = \begin{bmatrix} 5 \\ 3 \end{bmatrix}$

21. $\mathbf{u} = \begin{bmatrix} 3 \\ -1 \end{bmatrix}$, $\mathbf{v} = \begin{bmatrix} 6 \\ -5 \end{bmatrix}$, $\mathbf{w} = \begin{bmatrix} 1 \\ 4 \end{bmatrix}$

22. $\mathbf{u} = \begin{bmatrix} 6 \\ -4 \\ 2 \end{bmatrix}$, $\mathbf{v} = \begin{bmatrix} 3 \\ -2 \\ -1 \end{bmatrix}$

23. $\mathbf{u} = \begin{bmatrix} 1 \\ -8 \\ 3 \end{bmatrix}$, $\mathbf{v} = \begin{bmatrix} 0 \\ 0 \\ 0 \end{bmatrix}$, $\mathbf{w} = \begin{bmatrix} -7 \\ 1 \\ 12 \end{bmatrix}$

24. $\mathbf{u} = \begin{bmatrix} 1 \\ 2 \\ 3 \\ 4 \end{bmatrix}$, $\mathbf{v} = \begin{bmatrix} 1 \\ 2 \\ 3 \\ 4 \end{bmatrix}$, $\mathbf{w} = \begin{bmatrix} 4 \\ 3 \\ 2 \\ 1 \end{bmatrix}$

In Exercises 25–28, determine if one of the given vectors is in the span of the other vectors. (HINT: Check to see if the vectors are linearly dependent, and then appeal to Theorem 2.14.)

25. $\mathbf{u} = \begin{bmatrix} 6 \\ 2 \\ -5 \end{bmatrix}$, $\mathbf{v} = \begin{bmatrix} 1 \\ 7 \\ 0 \end{bmatrix}$

26. $\mathbf{u} = \begin{bmatrix} 2 \\ 7 \\ -1 \end{bmatrix}$, $\mathbf{v} = \begin{bmatrix} 1 \\ 1 \\ 6 \end{bmatrix}$, $\mathbf{w} = \begin{bmatrix} 1 \\ 3 \\ 0 \end{bmatrix}$

27. $\mathbf{u} = \begin{bmatrix} 4 \\ -1 \\ 3 \end{bmatrix}$, $\mathbf{v} = \begin{bmatrix} 3 \\ 5 \\ -2 \end{bmatrix}$, $\mathbf{w} = \begin{bmatrix} -5 \\ 7 \\ -7 \end{bmatrix}$

28. $\mathbf{u} = \begin{bmatrix} 1 \\ 7 \\ 8 \\ 4 \end{bmatrix}$, $\mathbf{v} = \begin{bmatrix} -1 \\ 3 \\ 5 \\ 2 \end{bmatrix}$, $\mathbf{w} = \begin{bmatrix} 3 \\ 1 \\ -2 \\ 0 \end{bmatrix}$

For each matrix A given in Exercises 29–32, determine if $A\mathbf{x} = \mathbf{b}$ has a unique solution for every $\mathbf{b}$ in $\mathbf{R}^3$. (HINT: the Big Theorem is helpful here.)

29. $A = \begin{bmatrix} 2 & -1 & 0 \\ 1 & 0 & 1 \\ -3 & 4 & 5 \end{bmatrix}$

30. $A = \begin{bmatrix} 3 & 4 & 7 \\ 7 & -1 & 6 \\ -2 & 0 & 2 \end{bmatrix}$

31. $A = \begin{bmatrix} 3 & -2 & 1 \\ -4 & 1 & 0 \\ -5 & 0 & 1 \end{bmatrix}$

32. $A = \begin{bmatrix} 1 & -3 & -2 \\ 0 & 1 & 1 \\ 2 & 4 & 7 \end{bmatrix}$

FIND AN EXAMPLE For Exercises 33–38, find an example that meets the given specifications.

33. Three distinct nonzero linearly dependent vectors in $\mathbf{R}^4$.

34. Three linearly independent vectors in $\mathbf{R}^5$.

35. Three distinct nonzero linearly dependent vectors in $\mathbf{R}^2$ that do not span $\mathbf{R}^2$.

36. Three distinct nonzero vectors in $\mathbf{R}^2$ such that any pair is linearly independent.

37. Three distinct nonzero linearly dependent vectors in $\mathbf{R}^3$ such that each vector is in the span of the other two vectors.

38. Four vectors in $\mathbf{R}^3$ such that no vector is a nontrivial linear combination of the other three. (Explain why this does not contradict Theorem 2.14.)

TRUE OR FALSE For Exercises 39–52, determine if the statement is true or false, and justify your answer.

39. If a set of vectors in $\mathbf{R}^n$ is linearly dependent, then the set must span $\mathbf{R}^n$.

40. If $m > n$, then a set of m vectors in $\mathbf{R}^n$ is linearly dependent.

41. If A is a matrix with more rows than columns, then the columns of A are linearly independent.

42. If A is a matrix with more columns than rows, then the columns of A are linearly independent.

43. If A is a matrix with linearly independent columns, then $A\mathbf{x} = \mathbf{0}$ has nontrivial solutions.

44. If A is a matrix with linearly independent columns, then $A\mathbf{x} = \mathbf{b}$ has a solution for all $\mathbf{b}$.

45. If $\{\mathbf{u}_1, \mathbf{u}_2, \mathbf{u}_3\}$ is linearly independent, then so is $\{\mathbf{u}_1, \mathbf{u}_2, \mathbf{u}_3, \mathbf{u}_4\}$.

46. If $\{\mathbf{u}_1, \mathbf{u}_2, \mathbf{u}_3\}$ is linearly dependent, then so is $\{\mathbf{u}_1, \mathbf{u}_2, \mathbf{u}_3, \mathbf{u}_4\}$.

47. If $\{\mathbf{u}_1, \mathbf{u}_2, \mathbf{u}_3, \mathbf{u}_4\}$ is linearly independent, then so is $\{\mathbf{u}_1, \mathbf{u}_2, \mathbf{u}_3\}$.

48. If $\{\mathbf{u}_1, \mathbf{u}_2, \mathbf{u}_3, \mathbf{u}_4\}$ is linearly dependent, then so is $\{\mathbf{u}_1, \mathbf{u}_2, \mathbf{u}_3\}$.

49. If $\mathbf{u}_4$ is a linear combination of $\{\mathbf{u}_1, \mathbf{u}_2, \mathbf{u}_3\}$, then $\{\mathbf{u}_1, \mathbf{u}_2, \mathbf{u}_3, \mathbf{u}_4\}$ is linearly independent.

50. If $\mathbf{u}_4$ is a linear combination of $\{\mathbf{u}_1, \mathbf{u}_2, \mathbf{u}_3\}$, then $\{\mathbf{u}_1, \mathbf{u}_2, \mathbf{u}_3, \mathbf{u}_4\}$ is linearly dependent.

51. If $\mathbf{u}_4$ is *not* a linear combination of $\{\mathbf{u}_1, \mathbf{u}_2, \mathbf{u}_3\}$, then $\{\mathbf{u}_1, \mathbf{u}_2, \mathbf{u}_3, \mathbf{u}_4\}$ is linearly independent.

52. If $\mathbf{u}_4$ is *not* a linear combination of $\{\mathbf{u}_1, \mathbf{u}_2, \mathbf{u}_3\}$, then $\{\mathbf{u}_1, \mathbf{u}_2, \mathbf{u}_3, \mathbf{u}_4\}$ is linearly dependent.

53. Which of the following sets of vectors in $\mathbf{R}^3$ could possibly be linearly independent? Justify your answer.

(a) $\{\mathbf{u}_1\}$

(b) $\{\mathbf{u}_1, \mathbf{u}_2\}$

(c) $\{\mathbf{u}_1, \mathbf{u}_2, \mathbf{u}_3\}$

(d) $\{\mathbf{u}_1, \mathbf{u}_2, \mathbf{u}_3, \mathbf{u}_4\}$

54. Which of the following sets of vectors in $\mathbf{R}^3$ could possibly be linearly independent *and* span $\mathbf{R}^3$? Justify your answer.

(a) $\{\mathbf{u}_1\}$

(b) $\{\mathbf{u}_1, \mathbf{u}_2\}$

(c) $\{\mathbf{u}_1, \mathbf{u}_2, \mathbf{u}_3\}$

(d) $\{\mathbf{u}_1, \mathbf{u}_2, \mathbf{u}_3, \mathbf{u}_4\}$

55. Prove that if c_1, c_2, and c_3 are nonzero scalars and $\{\mathbf{u}_1, \mathbf{u}_2, \mathbf{u}_3\}$ is a linearly independent set of vectors, then so is $\{c_1\mathbf{u}_1, c_2\mathbf{u}_2, c_3\mathbf{u}_3\}$.

56. Prove that if $\mathbf{u}$ and $\mathbf{v}$ are linearly independent vectors, then so are $\mathbf{u} + \mathbf{v}$ and $\mathbf{u} - \mathbf{v}$.

57. Prove that if $\{\mathbf{u}_1, \mathbf{u}_2, \mathbf{u}_3\}$ is a linearly independent set of vectors, then so is $\{\mathbf{u}_1 + \mathbf{u}_2, \mathbf{u}_1 + \mathbf{u}_3, \mathbf{u}_2 + \mathbf{u}_3\}$.

58. Prove that if $U = \{\mathbf{u}_1, \ldots, \mathbf{u}_m\}$ is linearly independent, then any nonempty subset of U is also linearly independent.

59. Prove that if a set of vectors is linearly dependent, then adding additional vectors to the set will create a new set that is still linearly dependent.

60. Prove that if $\mathbf{u}$ and $\mathbf{v}$ are linearly independent and the set $\{\mathbf{u}, \mathbf{v}, \mathbf{w}\}$ is a linearly dependent set, then $\mathbf{w}$ is in span$\{\mathbf{u}, \mathbf{v}\}$.

61. Prove that two nonzero vectors $\mathbf{u}$ and $\mathbf{v}$ are linearly dependent if and only if $\mathbf{u} = c\mathbf{v}$ for some scalar c.

62. Let A be an $n \times m$ matrix that is in echelon form. Prove that the nonzero rows of A, when considered as vectors in $\mathbf{R}^m$, are a linearly independent set.

63. Prove part (*b*) of Theorem 2.16.

64. Let $\{\mathbf{u}_1, \ldots, \mathbf{u}_m\}$ be a linearly dependent set of nonzero vectors. Prove that some vector in the set can be written as a linear combination of a linearly independent subset of the remaining vectors, with the set of coefficients all nonzero and unique for the given subset. (HINT: Start with Theorem 2.14.)

In Exercises 65–66, suppose that the given vectors are direction vectors for a model of the VecMobile III (discussed in Section 2.2).

Determine if there is any redundancy in the vectors and if it is possible to reach every point in $\mathbf{R}^3$.

65. $\mathbf{u}_1 = \begin{bmatrix} 1 \\ -2 \\ 5 \end{bmatrix}$, $\mathbf{u}_2 = \begin{bmatrix} 4 \\ 2 \\ 0 \end{bmatrix}$, $\mathbf{u}_3 = \begin{bmatrix} 2 \\ 6 \\ 3 \end{bmatrix}$

66. $\mathbf{u}_1 = \begin{bmatrix} 2 \\ -5 \\ 1 \end{bmatrix}$, $\mathbf{u}_2 = \begin{bmatrix} 1 \\ 3 \\ -4 \end{bmatrix}$, $\mathbf{u}_3 = \begin{bmatrix} -5 \\ 7 \\ 2 \end{bmatrix}$

Ⓒ In Exercises 67–70, determine if the given vectors form a linearly dependent or linearly independent set.

67. $\begin{bmatrix} 2 \\ -3 \\ 5 \end{bmatrix}$, $\begin{bmatrix} 3 \\ -4 \\ 2 \end{bmatrix}$, $\begin{bmatrix} -1 \\ 1 \\ 7 \end{bmatrix}$

68. $\begin{bmatrix} -4 \\ 2 \\ 3 \end{bmatrix}$, $\begin{bmatrix} 1 \\ 3 \\ 1 \end{bmatrix}$, $\begin{bmatrix} -3 \\ 5 \\ 4 \end{bmatrix}$

69. $\begin{bmatrix} 2 \\ 0 \\ 1 \\ -1 \end{bmatrix}$, $\begin{bmatrix} -3 \\ 2 \\ 5 \\ 6 \end{bmatrix}$, $\begin{bmatrix} 6 \\ 7 \\ 0 \\ -5 \end{bmatrix}$, $\begin{bmatrix} 5 \\ -3 \\ 7 \\ -3 \end{bmatrix}$

70. $\begin{bmatrix} 3 \\ 5 \\ -2 \\ -4 \end{bmatrix}$, $\begin{bmatrix} 2 \\ -4 \\ 3 \\ -1 \end{bmatrix}$, $\begin{bmatrix} -4 \\ 6 \\ 6 \\ 2 \end{bmatrix}$, $\begin{bmatrix} -7 \\ 2 \\ 2 \\ 6 \end{bmatrix}$

Ⓒ In Exercises 71–72, determine if $A\mathbf{x} = \mathbf{b}$ has a unique solution for every $\mathbf{b}$ in $\mathbf{R}^3$.

71. $A = \begin{bmatrix} 1 & -2 & 4 \\ 5 & -3 & -1 \\ -3 & -7 & -9 \end{bmatrix}$, $\mathbf{x} = \begin{bmatrix} x_1 \\ x_2 \\ x_3 \end{bmatrix}$

72. $A = \begin{bmatrix} 3 & -2 & 5 \\ 2 & 0 & -4 \\ -2 & 7 & 1 \end{bmatrix}$, $\mathbf{x} = \begin{bmatrix} x_1 \\ x_2 \\ x_3 \end{bmatrix}$

Ⓒ In Exercises 73–74, determine if $A\mathbf{x} = \mathbf{b}$ has a unique solution for every $\mathbf{b}$ in $\mathbf{R}^4$.

73. $A = \begin{bmatrix} 2 & 5 & -3 & 6 \\ -1 & 0 & 1 & -1 \\ 5 & 2 & -3 & 9 \\ 3 & -4 & 6 & 8 \end{bmatrix}$, $\mathbf{x} = \begin{bmatrix} x_1 \\ x_2 \\ x_3 \\ x_4 \end{bmatrix}$

74. $A = \begin{bmatrix} 5 & 1 & 0 & 8 \\ -2 & 4 & 3 & 11 \\ -3 & 8 & 2 & 5 \\ 0 & 3 & -1 & 8 \end{bmatrix}$, $\mathbf{x} = \begin{bmatrix} x_1 \\ x_2 \\ x_3 \\ x_4 \end{bmatrix}$

Matrices

The Brooklyn Bridge, completed in 1883, was the first major steel-wire suspension bridge in the world. It was designed by John Roebling, though the plan was ultimately carried out by his son Washington Roebling and daughter-in-law Emily upon John's death in 1869. The bridge employs a balance of tensile force in the supporting cables, which take on a parabolic shape, and compressive force in the towers through which the cables pass.

Bridge suggested by Mark Hunacek, Iowa State University (Reflexstock)

In this chapter we expand our development of matrices. Thus far we have used matrices to solve systems of equations and have defined how to multiply a matrix times a vector. In Section 3.1 we use this multiplication to define an important type of function called a linear transformation and investigate its properties and applications. Section 3.2 and Section 3.3 focus on the algebra of matrices. In Section 3.4 we develop a factorization method that uses matrix multiplication to efficiently find solutions to linear systems. Section 3.5 is about Markov Chains, an application of matrix multiplication that arises in a variety of contexts.

3.1 Linear Transformations

In this section we consider an important class of functions called *linear transformations*. These functions arise in many fields and are defined naturally in terms of matrix multiplication. The following example gives a sense of how one might encounter linear transformations.

A consumer electronics company makes three different types of MP3 players, the J8 (8 GB), the J40 (40 GB), and the J80 (80 GB). The manufacturing cost includes labor, materials, and overhead (facilities, etc.). The company's costs (in dollars) per unit for each type are summarized in Table 1.

	J8	J40	J80
Labor	15	46	65
Materials	14	43	61
Overhead	20	60	81

Table 1 **MP3 Manufacturing Costs**

This table can be organized into three cost vectors,

$$\text{J8: } \mathbf{j}_8 = \begin{bmatrix} 15 \\ 14 \\ 20 \end{bmatrix}, \quad \text{J40: } \mathbf{j}_{40} = \begin{bmatrix} 46 \\ 43 \\ 60 \end{bmatrix}, \quad \text{J80: } \mathbf{j}_{80} = \begin{bmatrix} 65 \\ 61 \\ 81 \end{bmatrix}$$

To determine costs for different manufacturing levels, we compute linear combinations of these vectors. For instance, the cost vector for producing 12 J8's, 10 J40's, and 6 J80's is

$$12\mathbf{j}_8 + 10\mathbf{j}_{40} + 6\mathbf{j}_{80} = 12 \begin{bmatrix} 15 \\ 14 \\ 20 \end{bmatrix} + 10 \begin{bmatrix} 46 \\ 43 \\ 60 \end{bmatrix} + 6 \begin{bmatrix} 65 \\ 61 \\ 81 \end{bmatrix} = \begin{bmatrix} 1030 \\ 964 \\ 1326 \end{bmatrix}$$

Thus, for this production mix, the company will incur costs of \$1030 for labor, \$964 for materials, and \$1326 for overhead. More generally, let

$$\mathbf{x} = \begin{bmatrix} x_1 \\ x_2 \\ x_3 \end{bmatrix}$$

where $x_1, x_2,$ and x_3 indicate desired production levels for J8's, J40's, and J80's, respectively. Let T be the function that takes the production vector $\mathbf{x}$ as input and produces the corresponding total cost vector as output. Using the individual cost vectors, we have

$$T(\mathbf{x}) = x_1 \begin{bmatrix} 15 \\ 14 \\ 20 \end{bmatrix} + x_2 \begin{bmatrix} 46 \\ 43 \\ 60 \end{bmatrix} + x_3 \begin{bmatrix} 65 \\ 61 \\ 81 \end{bmatrix}$$

Now suppose that we define the 3×3 matrix

$$A = \begin{bmatrix} \mathbf{j}_8 & \mathbf{j}_{40} & \mathbf{j}_{80} \end{bmatrix} = \begin{bmatrix} 15 & 46 & 65 \\ 14 & 43 & 61 \\ 20 & 60 & 81 \end{bmatrix}$$

Recalling the formula for multiplying a matrix by a vector (see Definition 2.9 in Section 2.2), we see that we can write the cost function compactly as

$$T(\mathbf{x}) = \begin{bmatrix} 15 & 46 & 65 \\ 14 & 43 & 61 \\ 20 & 60 & 81 \end{bmatrix} \begin{bmatrix} x_1 \\ x_2 \\ x_3 \end{bmatrix} = A\mathbf{x}$$

Thus, for example, a production level of 10 J8's, 20 J40's, and 36 J80's will have a cost vector

$$T\left(\begin{bmatrix} 10 \\ 20 \\ 36 \end{bmatrix} \right) = \begin{bmatrix} 15 & 46 & 65 \\ 14 & 43 & 61 \\ 20 & 60 & 81 \end{bmatrix} \begin{bmatrix} 10 \\ 20 \\ 36 \end{bmatrix} = \begin{bmatrix} 3410 \\ 3196 \\ 4316 \end{bmatrix}$$

Linear Transformations

The MP3 cost function is an example of an important class of functions that we now describe, starting with some notation and terminology. Let

$$T : \mathbf{R}^m \to \mathbf{R}^n$$

Definition Domain, Codomain, Image, Range

denote a function T that takes vectors in $\mathbf{R}^m$ as input and produces vectors in $\mathbf{R}^n$ as output. Put another way, the **domain** of T is $\mathbf{R}^m$ and the **codomain** is $\mathbf{R}^n$. For every vector $\mathbf{u}$ in $\mathbf{R}^m$, the vector $T(\mathbf{u})$ is called the **image of u under** T. The set of all images of vectors $\mathbf{u}$ in $\mathbf{R}^m$ under T is called the **range** of T, denoted range(T). Thus the range of T is a subset of the codomain of T.

DEFINITION 3.1

Definition Linear Transformation

A function $T : \mathbf{R}^m \to \mathbf{R}^n$ is a **linear transformation** if for all vectors $\mathbf{u}$ and $\mathbf{v}$ in $\mathbf{R}^m$ and all scalars r we have

(a) $T(\mathbf{u} + \mathbf{v}) = T(\mathbf{u}) + T(\mathbf{v})$

(b) $T(r\mathbf{u}) = r\,T(\mathbf{u})$

The MP3 cost function T is an example of a linear transformation. This claim follows immediately from the next theorem, but before getting to that we need to introduce new terminology.

Definition Matrix Dimensions, Square Matrix

Suppose that A is a matrix with n rows and m columns. Then we say that A is an **$n \times m$ matrix** and that A has **dimensions** $n \times m$. If $n = m$, then A is a **square matrix**. For instance, if

$$A = \begin{bmatrix} 1 & 4 & -2 & 0 & 9 \\ 3 & 0 & 1 & 9 & 11 \\ 2 & -1 & 7 & 5 & 8 \end{bmatrix} \quad \text{and} \quad B = \begin{bmatrix} 0 & 4 & 1 & 3 \\ 2 & 0 & 7 & 8 \\ -1 & 3 & 5 & 9 \\ 8 & 6 & -4 & 1 \end{bmatrix}$$

▶ We say that A is a "3-by-5" matrix and B is a "4-by-4" matrix.

then A is a 3×5 matrix and B is a 4×4 (square) matrix.

THEOREM 3.2

Let A be an $n \times m$ matrix, and define $T(\mathbf{x}) = A\mathbf{x}$. Then $T : \mathbf{R}^m \to \mathbf{R}^n$ is a linear transformation.

Proof To show that T is a linear transformation, we must verify that the two conditions in Definition 3.1 both hold. Starting with condition (a), given vectors $\mathbf{u}$ and $\mathbf{v}$, we have

$$\begin{aligned} T(\mathbf{u} + \mathbf{v}) &= A(\mathbf{u} + \mathbf{v}) \\ &= A\mathbf{u} + A\mathbf{v} \quad \text{(by Theorem 2.16)} \\ &= T(\mathbf{u}) + T(\mathbf{v}) \end{aligned}$$

That shows (a) holds. Condition (b) is covered in Exercise 57. Verifying the two conditions completes the proof. ■

Since our MP3 cost function is $T(\mathbf{x}) = A\mathbf{x}$, it follows immediately from Theorem 3.2 that T is a linear transformation. Furthermore, since A is a 3×3 matrix, the domain and codomain of T are both $\mathbf{R}^3$.

It turns out that *all* linear transformations $T : \mathbf{R}^m \to \mathbf{R}^n$ are of the form $T(\mathbf{x}) = A\mathbf{x}$ for some $n \times m$ matrix A. The proof is given later in this section.

EXAMPLE 1 Let

$$A = \begin{bmatrix} 1 & -2 & 4 \\ 3 & 0 & -5 \end{bmatrix}, \quad \mathbf{u} = \begin{bmatrix} 1 \\ 2 \\ 1 \end{bmatrix}, \quad \mathbf{v} = \begin{bmatrix} 0 \\ 0 \\ 0 \end{bmatrix}, \quad \text{and} \quad \mathbf{w} = \begin{bmatrix} 3 \\ 4 \end{bmatrix} \tag{1}$$

Suppose that $T : \mathbf{R}^3 \to \mathbf{R}^2$ with $T(\mathbf{x}) = A\mathbf{x}$. Compute $T(\mathbf{u})$ and $T(\mathbf{v})$, and determine if $\mathbf{w}$ is in the range of T.

Solution We start by computing

$$T(\mathbf{u}) = A\mathbf{u} = \begin{bmatrix} 1 & -2 & 4 \\ 3 & 0 & -5 \end{bmatrix} \begin{bmatrix} 1 \\ 2 \\ 1 \end{bmatrix} = \begin{bmatrix} 1 \\ -2 \end{bmatrix}$$

$$T(\mathbf{v}) = A\mathbf{v} = \begin{bmatrix} 1 & -2 & 4 \\ 3 & 0 & -5 \end{bmatrix} \begin{bmatrix} 0 \\ 0 \\ 0 \end{bmatrix} = \begin{bmatrix} 0 \\ 0 \end{bmatrix}$$

Next, in order for $\mathbf{w}$ to be in the range of T, there must exist a solution to $T(\mathbf{x}) = \mathbf{w}$, which is equivalent to the linear system $A\mathbf{x} = \mathbf{w}$. The corresponding augmented matrix and echelon form are

▶ Going forward, we usually will not include the row operations.

$$\begin{bmatrix} 1 & -2 & 4 & 3 \\ 3 & 0 & -5 & 4 \end{bmatrix} \sim \begin{bmatrix} 1 & -2 & 4 & 3 \\ 0 & 6 & -17 & -5 \end{bmatrix}$$

Back substitution yields the general solution

$$\mathbf{x} = \begin{bmatrix} 4/3 \\ -5/6 \\ 0 \end{bmatrix} + s \begin{bmatrix} 5/3 \\ 17/6 \\ 1 \end{bmatrix}$$

where s can be any real number. Thus $\mathbf{w}$ is the image of many different vectors and hence is in the range of T. ∎

THEOREM 3.3

Let $A = \begin{bmatrix} \mathbf{a}_1 & \mathbf{a}_2 & \cdots & \mathbf{a}_m \end{bmatrix}$ be an $n \times m$ matrix, and let $T : \mathbf{R}^m \to \mathbf{R}^n$ with $T(\mathbf{x}) = A\mathbf{x}$ be a linear transformation. Then

(a) The vector $\mathbf{w}$ is in the range of T if and only if $A\mathbf{x} = \mathbf{w}$ is a consistent linear system.

(b) range$(T) = \text{span}\{\mathbf{a}_1, \ldots, \mathbf{a}_m\}$.

Proof A vector $\mathbf{w}$ is in the range of T if and only if there exists a vector $\mathbf{u}$ such that $T(\mathbf{u}) = \mathbf{w}$. As $T(\mathbf{x}) = A\mathbf{x}$, this is equivalent to $A\mathbf{u} = \mathbf{w}$, which is true if and only if the linear system $A\mathbf{x} = \mathbf{w}$ is consistent.

For part (b), from Theorem 2.10 it follows that $A\mathbf{x} = \mathbf{w}$ is consistent if and only if $\mathbf{w}$ is in the span of the columns of A. Therefore range$(T) = \text{span}\{\mathbf{a}_1, \ldots, \mathbf{a}_m\}$. ∎

The next example shows that not all functions $T : \mathbf{R}^m \to \mathbf{R}^n$ are linear transformations.

EXAMPLE 2 Show that $T : \mathbf{R}^3 \to \mathbf{R}^2$ defined by

$$T \left(\begin{bmatrix} x_1 \\ x_2 \\ x_3 \end{bmatrix} \right) = \begin{bmatrix} x_1 - 3x_2 \\ x_2 x_3^2 \end{bmatrix}$$

is *not* a linear transformation.

Solution We address this question by appealing to Definition 3.1. We only have to find a single specific example where one of the required conditions does not hold. One possibility is to let

$$\mathbf{x} = \begin{bmatrix} 1 \\ 2 \\ 3 \end{bmatrix} \quad \text{and} \quad r = 2$$

Then we have

$$T(r\mathbf{x}) = T\left(\begin{bmatrix} 2 \\ 4 \\ 6 \end{bmatrix}\right) = \begin{bmatrix} 2 - 3(4) \\ (4)(6^2) \end{bmatrix} = \begin{bmatrix} -10 \\ 144 \end{bmatrix}$$

and

$$r\,T(\mathbf{x}) = 2T\left(\begin{bmatrix} 1 \\ 2 \\ 3 \end{bmatrix}\right) = 2\begin{bmatrix} 1 - 3(2) \\ (2)(3^2) \end{bmatrix} = \begin{bmatrix} -10 \\ 36 \end{bmatrix}$$

Since $T(r\mathbf{x}) \neq r\,T(\mathbf{x})$, it follows that T is not a linear transformation. ∎

Conditions (a) and (b) of Definition 3.1 can be combined into a single condition

$$T(r\mathbf{u} + s\mathbf{v}) = r\,T(\mathbf{u}) + s\,T(\mathbf{v}) \tag{2}$$

for all vectors $\mathbf{u}$ and $\mathbf{v}$ and all scalars r and s. The proof that these are equivalent is covered in Exercise 58. The following example shows that this is true for a specific case.

EXAMPLE 3 Let

$$A = \begin{bmatrix} 4 & -1 \\ -2 & 2 \\ 0 & 3 \end{bmatrix}, \quad \mathbf{x} = \begin{bmatrix} -1 \\ -3 \end{bmatrix}, \quad \text{and} \quad \mathbf{y} = \begin{bmatrix} 2 \\ 5 \end{bmatrix}$$

Let $r = 2$ and $s = -1$. Verify that 2 holds for $T(\mathbf{x}) = A\mathbf{x}$ in this case.

Solution We have

$$r\mathbf{x} + s\mathbf{y} = 2\begin{bmatrix} -1 \\ -3 \end{bmatrix} - \begin{bmatrix} 2 \\ 5 \end{bmatrix} = \begin{bmatrix} -4 \\ -11 \end{bmatrix}$$

so that $T(r\mathbf{x} + s\mathbf{y}) = A\begin{bmatrix} -4 \\ -11 \end{bmatrix} = \begin{bmatrix} -5 \\ -14 \\ -33 \end{bmatrix}$.

On the other hand, we also have

$$r\,T(\mathbf{x}) + s\,T(\mathbf{y}) = 2A\begin{bmatrix} -1 \\ -3 \end{bmatrix} - A\begin{bmatrix} 2 \\ 5 \end{bmatrix}$$

$$= 2\begin{bmatrix} -1 \\ -4 \\ -9 \end{bmatrix} - \begin{bmatrix} 3 \\ 6 \\ 15 \end{bmatrix} = \begin{bmatrix} -5 \\ -14 \\ -33 \end{bmatrix}$$

Thus $T(r\mathbf{x} + s\mathbf{y}) = r\,T(\mathbf{x}) + s\,T(\mathbf{y})$. ∎

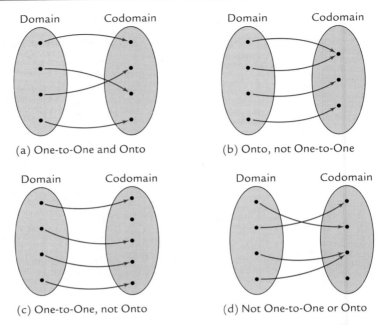

Figure 1 Graphical depiction of various combinations of one-to-one and onto.

One-to-One and Onto Linear Transformations

Here we consider two special types of linear transformations.

DEFINITION 3.4

Definition One-to-One, Onto

Let $T : \mathbf{R}^m \to \mathbf{R}^n$ be a linear transformation. Then

(a) T is **one-to-one** if for every vector $\mathbf{w}$ in $\mathbf{R}^n$ there exists *at most* one vector $\mathbf{u}$ in $\mathbf{R}^m$ such that $T(\mathbf{u}) = \mathbf{w}$.

(b) T is **onto** if for every vector $\mathbf{w}$ in $\mathbf{R}^n$ there exists *at least* one vector $\mathbf{u}$ in $\mathbf{R}^m$ such that $T(\mathbf{u}) = \mathbf{w}$.

Put another way, T is one-to-one if every vector in the domain of T is sent to its "own" unique vector in the range (see Figure 1(a) and 1(c)). T is onto if the range is equal to the codomain (see Figure 1(a) and 1(b)).

An equivalent formulation (see Exercise 59) for the definition of one-to-one given below.

DEFINITION

(ALTERNATE) A linear transformation T is **one-to-one** if $T(\mathbf{u}) = T(\mathbf{v})$ implies that $\mathbf{u} = \mathbf{v}$.

This version of one-to-one often is more convenient for proofs. It is used in the proof of the next theorem, which provides an easy way to determine if a linear transformation is one-to-one.

THEOREM 3.5

Let T be a linear transformation. Then T is one-to-one if and only if $T(\mathbf{x}) = \mathbf{0}$ has only the trivial solution $\mathbf{x} = \mathbf{0}$.

Proof First suppose that T is one-to-one. Then there is at most one solution to $T(\mathbf{x}) = \mathbf{0}$. Moreover, since T is a linear transformation, it follows that $T(\mathbf{0}) = \mathbf{0}$ (see Exercise 55), so that $T(\mathbf{x}) = \mathbf{0}$ has only the trivial solution.

Now suppose that $T(\mathbf{x}) = \mathbf{0}$ has only the trivial solution. If $T(\mathbf{u}) = T(\mathbf{v})$, then

$$T(\mathbf{u}) - T(\mathbf{v}) = \mathbf{0} \quad \Longrightarrow \quad T(\mathbf{u} - \mathbf{v}) = \mathbf{0} \quad (T \text{ is a linear transformation})$$

Since $T(\mathbf{x}) = \mathbf{0}$ has only the trivial solution, it follows that $\mathbf{u} - \mathbf{v} = \mathbf{0}$ and hence $\mathbf{u} = \mathbf{v}$. Therefore T is one-to-one. ■

EXAMPLE 4   Let A be as given in Example 3. Determine if $T(\mathbf{x}) = A\mathbf{x}$ one-to-one.

Solution   By Theorem 3.5, we need only find the solutions to $T(\mathbf{x}) = \mathbf{0}$, which is equivalent to solving $A\mathbf{x} = \mathbf{0}$. Populating the augmented matrix and reducing to echelon form gives

$$\begin{bmatrix} 4 & -1 & 0 \\ -2 & 2 & 0 \\ 0 & 3 & 0 \end{bmatrix} \sim \begin{bmatrix} -2 & 2 & 0 \\ 0 & 3 & 0 \\ 0 & 0 & 0 \end{bmatrix}$$

From the echelon form we can see that $A\mathbf{x} = \mathbf{0}$ has only the trivial solution. Thus $T(\mathbf{x}) = \mathbf{0}$ has only the trivial solution and hence T is one-to-one. ■

The next two theorems follow from results on span and linear independence developed in Section 2.2 and Section 2.3.

THEOREM 3.6

Let A be an $n \times m$ matrix and define $T : \mathbf{R}^m \to \mathbf{R}^n$ by $T(\mathbf{x}) = A\mathbf{x}$. Then

(a) T is one-to-one if and only if the columns of A are linearly independent.

(b) If $n < m$, then T is *not* one-to-one.

▶ This proof is carried out by applying results we have already proved, and illustrates a few of the many interconnections in linear algebra.

Proof   To prove part (a), note that by Theorem 3.5 T is one-to-one if and only if $T(\mathbf{x}) = \mathbf{0}$ has only the trivial solution. By Theorem 2.15, $T(\mathbf{x}) = \mathbf{0}$ has only the trivial solution if and only if the columns of A are linearly independent.

For part (b), if A has more columns than rows, then by Theorem 2.13 the columns are linearly dependent. Hence, by part (a), T is not one-to-one. ■

Returning to Example 4, we see that the two columns of A are linearly independent. (Why?) Hence we can also conclude from Theorem 3.6 that the linear transformation T is one-to-one.

The next theorem is the counterpart to Theorem 3.6 that shows when a linear transformation is onto.

THEOREM 3.7

Let A be an $n \times m$ matrix and define $T : \mathbf{R}^m \to \mathbf{R}^n$ by $T(\mathbf{x}) = A\mathbf{x}$. Then

(a) T is onto if and only if the columns of A span the codomain R^n.

(b) If $n > m$, then T is *not* onto.

Proof   For part (a), if T is onto then range$(T) = \mathbf{R}^n$. By Theorem 3.3, range(T) equals the span of the columns of A. Hence T is onto if and only if the columns of A span $\mathbf{R}^n$.

For part (b), by Theorem 2.8 if $n > m$ then the columns of A cannot span $\mathbf{R}^n$. Thus, by part (a), T is not onto. ■

EXAMPLE 5   Suppose that A is the matrix given in Example 3. Determine if $T(\mathbf{x}) = A\mathbf{x}$ is onto.

Solution   Since A is a 3×2 matrix, by Theorem 3.7(b) it follows that T is not onto. ■

EXAMPLE 6 Suppose that

$$A = \begin{bmatrix} 2 & 1 & 1 \\ 1 & 2 & 0 \\ 1 & 3 & 0 \end{bmatrix}$$

Determine if the linear transformation $T(\mathbf{x}) = A\mathbf{x}$ is onto.

Solution In Section 2.2 we showed that the set of vectors

$$\left\{ \begin{bmatrix} 2 \\ 1 \\ 1 \end{bmatrix}, \begin{bmatrix} 1 \\ 2 \\ 3 \end{bmatrix}, \begin{bmatrix} 1 \\ 0 \\ 0 \end{bmatrix} \right\}$$

spans $\mathbf{R}^3$. Thus, by Theorem 3.7(a), T is onto. ∎

In Theorem 3.2, we showed that a function $T : \mathbf{R}^m \rightarrow \mathbf{R}^n$ of the form $T(\mathbf{x}) = A\mathbf{x}$ must be a linear transformation. The next theorem combines Theorem 3.2 with its converse.

THEOREM 3.8

Let $T : \mathbf{R}^m \rightarrow \mathbf{R}^n$. Then $T(\mathbf{x}) = A\mathbf{x}$, where A is an $n \times m$ matrix, if and only if T is a linear transformation.

Proof One direction of this theorem is proved in Theorem 3.2. For the other direction, suppose that T is a linear transformation. Let

$$\mathbf{e}_1 = \begin{bmatrix} 1 \\ 0 \\ 0 \\ \vdots \\ 0 \end{bmatrix}, \quad \mathbf{e}_2 = \begin{bmatrix} 0 \\ 1 \\ 0 \\ \vdots \\ 0 \end{bmatrix}, \quad \mathbf{e}_3 = \begin{bmatrix} 0 \\ 0 \\ 1 \\ \vdots \\ 0 \end{bmatrix}, \quad \cdots, \quad \mathbf{e}_m = \begin{bmatrix} 0 \\ 0 \\ 0 \\ \vdots \\ 1 \end{bmatrix} \tag{3}$$

be vectors in $\mathbf{R}^m$, and then let A be the $n \times m$ matrix with columns $T(\mathbf{e}_1)$, $T(\mathbf{e}_2)$, ..., $T(\mathbf{e}_m)$,

$$A = \begin{bmatrix} T(\mathbf{e}_1) & T(\mathbf{e}_2) & \cdots & T(\mathbf{e}_m) \end{bmatrix}$$

Note that any vector $\mathbf{x}$ in $\mathbf{R}^m$ can be written as a linear combination of $\mathbf{e}_1, \mathbf{e}_2, \ldots, \mathbf{e}_m$, by

$$\mathbf{x} = \begin{bmatrix} x_1 \\ x_2 \\ x_3 \\ \vdots \\ x_m \end{bmatrix} = x_1 \begin{bmatrix} 1 \\ 0 \\ 0 \\ \vdots \\ 0 \end{bmatrix} + x_2 \begin{bmatrix} 0 \\ 1 \\ 0 \\ \vdots \\ 0 \end{bmatrix} + \cdots + x_m \begin{bmatrix} 0 \\ 0 \\ 0 \\ \vdots \\ 1 \end{bmatrix} = x_1 \mathbf{e}_1 + x_2 \mathbf{e}_2 + \cdots + x_m \mathbf{e}_m$$

From the properties of linear transformations, we have

$$\begin{aligned} T(\mathbf{x}) &= T(x_1 \mathbf{e}_1 + x_2 \mathbf{e}_2 + \cdots + x_m \mathbf{e}_m) \\ &= x_1 T(\mathbf{e}_1) + x_2 T(\mathbf{e}_2) + \cdots + x_m T(\mathbf{e}_m) \\ &= A\mathbf{x} \end{aligned}$$

Thus T has the required form and the proof is complete. ∎

EXAMPLE 7 Suppose that $T : \mathbf{R}^3 \to \mathbf{R}^4$ is defined by

$$T\left(\begin{bmatrix} x_1 \\ x_2 \\ x_3 \end{bmatrix}\right) = \begin{bmatrix} 2x_1 + x_3 \\ -x_1 + 2x_2 \\ x_1 - 3x_2 + 5x_3 \\ 4x_2 \end{bmatrix}$$

Show that T is a linear transformation.

Solution We could solve this by directly appealing to Definition 3.1. But instead, let's apply Theorem 3.8 by finding the matrix A such that $T(\mathbf{x}) = A\mathbf{x}$. We start by noting that

$$T\left(\begin{bmatrix} x_1 \\ x_2 \\ x_3 \end{bmatrix}\right) = \begin{bmatrix} 2x_1 + x_3 \\ -x_1 + 2x_2 \\ x_1 - 3x_2 + 5x_3 \\ 4x_2 \end{bmatrix} = \begin{bmatrix} 2x_1 + 0x_2 + 1x_3 \\ -1x_1 + 2x_2 + 0x_3 \\ 1x_1 - 3x_2 + 5x_3 \\ 0x_1 + 4x_2 + 0x_3 \end{bmatrix} = \begin{bmatrix} 2 & 0 & 1 \\ -1 & 2 & 0 \\ 1 & -3 & 5 \\ 0 & 4 & 0 \end{bmatrix} \begin{bmatrix} x_1 \\ x_2 \\ x_3 \end{bmatrix}$$

Thus, if

$$A = \begin{bmatrix} 2 & 0 & 1 \\ -1 & 2 & 0 \\ 1 & -3 & 5 \\ 0 & 4 & 0 \end{bmatrix}$$

then $T(\mathbf{x}) = A\mathbf{x}$. Hence, by Theorem 3.8, T is a linear transformation. ∎

The Big Theorem, Version 2

Incorporating the preceding work, let us add two new conditions to the Big Theorem, Version 1 (Theorem 2.19) that we proved in Section 2.3.

THEOREM 3.9

(THE BIG THEOREM, VERSION 2) Let $\mathcal{A} = \{\mathbf{a}_1, \ldots, \mathbf{a}_n\}$ be a set of n vectors in $\mathbf{R}^n$, let $A = \begin{bmatrix} \mathbf{a}_1 & \cdots & \mathbf{a}_n \end{bmatrix}$, and let $T : \mathbf{R}^n \to \mathbf{R}^n$ be given by $T(\mathbf{x}) = A\mathbf{x}$. Then the following are equivalent:

(a) $\mathcal{A}$ spans $\mathbf{R}^n$.

(b) $\mathcal{A}$ is linearly independent.

(c) $A\mathbf{x} = \mathbf{b}$ has a unique solution for all $\mathbf{b}$ in $\mathbf{R}^n$.

(d) T is onto.

(e) T is one-to-one.

Proof From the Big Theorem, Version 1, we know that (a), (b), and (c) are equivalent. It follows from Theorem 3.7 that (a) and (d) are equivalent, and it follows from Theorem 3.6 that (b) and (e) are equivalent. Thus all five conditions are equivalent to each other. ∎

Geometry of Linear Transformations

One reason for the name of linear transformations is because they (usually) transform lines in the domain to lines in the range. To see why, recall that (see Exercise 64) the line segment from $\mathbf{u}$ to $\mathbf{v}$ can be parameterized by

$$(1 - s)\mathbf{u} + s\mathbf{v}, \quad 0 \leq s \leq 1$$

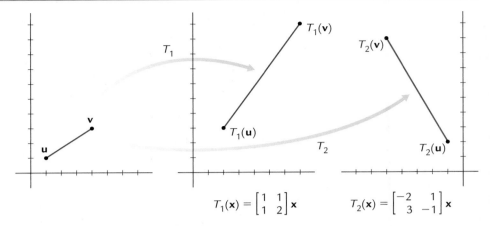

$$T_1(\mathbf{x}) = \begin{bmatrix} 1 & 1 \\ 1 & 2 \end{bmatrix} \mathbf{x} \qquad T_2(\mathbf{x}) = \begin{bmatrix} -2 & 1 \\ 3 & -1 \end{bmatrix} \mathbf{x}$$

Figure 2 Image of the line segment between **u** and **v** under linear transformations $T_1(\mathbf{x})$ and $T_2(\mathbf{x})$.

Applying a linear transformation T to this, we find that

$$T\big[(1-s)\mathbf{u}+s\mathbf{v}\big] = (1-s)T(\mathbf{u}) + sT(\mathbf{v}), \quad 0 \le s \le 1$$

which is the parameterization of a line in the range of T. Figure 2 illustrates this for two different linear transformations, $T_1 : \mathbf{R}^2 \to \mathbf{R}^2$ and $T_2 : \mathbf{R}^2 \to \mathbf{R}^2$.

What effect do linear transformations have on regions? Figure 3 shows the effect of applying three different linear transformations to the unit square S in the first quadrant.

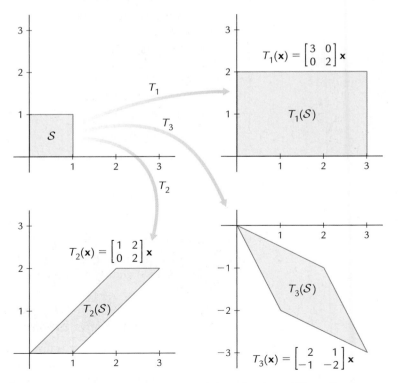

Figure 3 The image of the unit square S (*upper left*) under three different linear transformations.

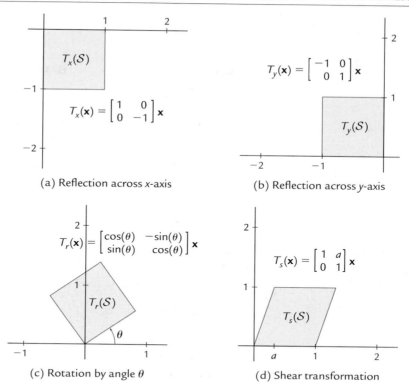

$$T_x(\mathbf{x}) = \begin{bmatrix} 1 & 0 \\ 0 & -1 \end{bmatrix} \mathbf{x}$$

(a) Reflection across *x*-axis

$$T_y(\mathbf{x}) = \begin{bmatrix} -1 & 0 \\ 0 & 1 \end{bmatrix} \mathbf{x}$$

(b) Reflection across *y*-axis

$$T_r(\mathbf{x}) = \begin{bmatrix} \cos(\theta) & -\sin(\theta) \\ \sin(\theta) & \cos(\theta) \end{bmatrix} \mathbf{x}$$

(c) Rotation by angle θ

$$T_s(\mathbf{x}) = \begin{bmatrix} 1 & a \\ 0 & 1 \end{bmatrix} \mathbf{x}$$

(d) Shear transformation

Figure 4 The image of the unit square S under reflection, rotation, and shear linear transformations used in computer graphics.

Figure 5 Karl Jacobi (Source: Smithsonian Institution Libraries).

Since linear transformations map lines to lines, the boundaries of S are mapped to lines. When the linear transformation is one-to-one, the unit square S is mapped to a parallelogram.

Linear transformations are used extensively in computer graphics. Figure 4 shows the results of linear transformations that reflect, rotate, and shear the unit square S in Figure 3.

In Section 1.3 we encountered Karl Jacobi, whose picture is shown in Figure 5. Reflection, rotation, and shear transformations applied to that picture are shown in Figure 6.

(a) Reflection across *x*-axis   (b) Reflection across *y*-axis   (c) Rotation by $\theta = 30°$ (d) Shear, $a = 1/3$

Figure 6 The linear transformations from Figure 4 applied to an image of Karl Jacobi. (Source: Smithsonian Institution Libraries).

| EXERCISES |

In Exercises 1–4, let $T(\mathbf{x}) = A\mathbf{x}$ for the given matrix A, and find $T(\mathbf{u}_1)$ and $T(\mathbf{u}_2)$ for the given $\mathbf{u}_1$ and $\mathbf{u}_2$.

1. $A = \begin{bmatrix} 2 & 1 \\ -3 & 5 \end{bmatrix}$, $\mathbf{u}_1 = \begin{bmatrix} -4 \\ -2 \end{bmatrix}$, $\mathbf{u}_2 = \begin{bmatrix} 1 \\ -6 \end{bmatrix}$

2. $A = \begin{bmatrix} 1 & 0 \\ 2 & -4 \\ 3 & 3 \end{bmatrix}$, $\mathbf{u}_1 = \begin{bmatrix} 1 \\ 2 \end{bmatrix}$, $\mathbf{u}_2 = \begin{bmatrix} -5 \\ 0 \end{bmatrix}$

3. $A = \begin{bmatrix} 0 & -4 & 2 \\ 3 & 1 & -2 \end{bmatrix}$, $\mathbf{u}_1 = \begin{bmatrix} 3 \\ 2 \\ 1 \end{bmatrix}$, $\mathbf{u}_2 = \begin{bmatrix} 4 \\ -5 \\ -2 \end{bmatrix}$

4. $A = \begin{bmatrix} -2 & 5 & -2 \\ 0 & -1 & -2 \\ 0 & -1 & -1 \end{bmatrix}$, $\mathbf{u}_1 = \begin{bmatrix} 0 \\ 7 \\ -2 \end{bmatrix}$, $\mathbf{u}_2 = \begin{bmatrix} 3 \\ 5 \\ -1 \end{bmatrix}$

In Exercises 5–8, determine if the given vector is in the range of $T(\mathbf{x}) = A\mathbf{x}$, where

$$A = \begin{bmatrix} 1 & -2 & 0 \\ 3 & 2 & 1 \end{bmatrix}$$

5. $\mathbf{y} = \begin{bmatrix} -3 \\ 6 \end{bmatrix}$

6. $\mathbf{y} = \begin{bmatrix} 1 \\ -4 \end{bmatrix}$

7. $\mathbf{y} = \begin{bmatrix} 2 \\ 7 \end{bmatrix}$

8. $\mathbf{y} = \begin{bmatrix} 4 \\ 5 \end{bmatrix}$

9. Suppose that a linear transformation T satisfies

$$T(\mathbf{u}_1) = \begin{bmatrix} 2 \\ 1 \end{bmatrix}, \quad T(\mathbf{u}_2) = \begin{bmatrix} -3 \\ 2 \end{bmatrix}$$

Find $T(-2\mathbf{u}_1 + 3\mathbf{u}_2)$.

10. Suppose that a linear transformation T satisfies

$$T(\mathbf{u}_1) = \begin{bmatrix} 3 \\ -1 \\ -2 \end{bmatrix}, \quad T(\mathbf{u}_2) = \begin{bmatrix} 1 \\ 1 \\ 4 \end{bmatrix}$$

Find $T(3\mathbf{u}_1 - 2\mathbf{u}_2)$.

11. Suppose that a linear transformation T satisfies

$$T(\mathbf{u}_1) = \begin{bmatrix} -3 \\ 0 \end{bmatrix}, \quad T(\mathbf{u}_2) = \begin{bmatrix} 2 \\ -1 \end{bmatrix}, \quad T(\mathbf{u}_3) = \begin{bmatrix} 0 \\ 5 \end{bmatrix}$$

Find $T(-\mathbf{u}_1 + 4\mathbf{u}_2 - 3\mathbf{u}_3)$.

12. Suppose that a linear transformation T satisfies

$$T(\mathbf{u}_1) = \begin{bmatrix} 3 \\ -1 \\ -2 \end{bmatrix}, \quad T(\mathbf{u}_2) = \begin{bmatrix} 1 \\ 1 \\ 4 \end{bmatrix}, \quad T(\mathbf{u}_3) = \begin{bmatrix} 6 \\ 0 \\ 0 \end{bmatrix}$$

Find $T(\mathbf{u}_1 + 4\mathbf{u}_2 - 2\mathbf{u}_3)$.

In Exercises 13–20, determine if the given function is a linear transformation. If so, identify the matrix A such that $T(\mathbf{x}) = A\mathbf{x}$. If not, explain why not.

13. $T(x_1, x_2) = (3x_1 + x_2, -2x_1 + 4x_2)$

14. $T(x_1, x_2) = (x_1 - x_2, x_1 x_2)$

15. $T(x_1, x_2, x_3) = (2\cos(x_2), 3\sin(x_3), x_1)$

16. $T(x_1, x_2, x_3) = (-5x_2, 7x_3)$

17. $T(x_1, x_2, x_3) = (-4x_1 + x_3, 6x_1 + 5x_2)$

18. $T(x_1, x_2, x_3) = (-x_1 + 3x_2 + x_3, 2x_1 + 7x_2 + 4, 3x_3)$

19. $T(x_1, x_2) = (x_2 \sin(\pi/4), x_1 \ln(2))$

20. $T(x_1, x_2) = (3x_2, -x_1 + 5|x_2|, 2x_1)$

For Exercises 21–28, let $T(\mathbf{x}) = A\mathbf{x}$ for the given matrix A. Determine if T is one-to-one and if T is onto.

21. $A = \begin{bmatrix} 1 & -3 \\ -2 & 5 \end{bmatrix}$

22. $A = \begin{bmatrix} 3 & 2 \\ 9 & 6 \end{bmatrix}$

23. $A = \begin{bmatrix} 5 & 4 & -2 \\ 3 & -1 & 0 \end{bmatrix}$

24. $A = \begin{bmatrix} -1 & 3 & 2 \\ 4 & -12 & -8 \end{bmatrix}$

25. $A = \begin{bmatrix} 1 & -2 \\ -3 & 5 \\ 2 & -7 \end{bmatrix}$

26. $A = \begin{bmatrix} 2 & -4 \\ 5 & -10 \\ -4 & 8 \end{bmatrix}$

27. $A = \begin{bmatrix} 2 & 8 & 4 \\ 3 & 2 & 3 \\ 1 & 14 & 5 \end{bmatrix}$

28. $A = \begin{bmatrix} 1 & 2 & -5 \\ 3 & 7 & -8 \\ -2 & -4 & 6 \end{bmatrix}$

For Exercises 29–32, suppose that $T(\mathbf{x}) = A\mathbf{x}$ for the given A. Sketch a graph of the image under T of the unit square in the first quadrant of $\mathbf{R}^2$.

29. $A = \begin{bmatrix} 3 & 0 \\ 0 & 3 \end{bmatrix}$

30. $A = \begin{bmatrix} -2 & 0 \\ 0 & 4 \end{bmatrix}$

31. $A = \begin{bmatrix} 1 & -2 \\ 3 & 1 \end{bmatrix}$

32. $A = \begin{bmatrix} -3 & 1 \\ 6 & -2 \end{bmatrix}$

FIND AN EXAMPLE For Exercises 33–38, find an example that meets the given specifications.

33. A linear transformation $T : \mathbf{R}^2 \to \mathbf{R}^2$ such that

$$T\left(\begin{bmatrix} 1 \\ 0 \end{bmatrix}\right) = \begin{bmatrix} 2 \\ 3 \end{bmatrix}$$

34. A linear transformation $T : \mathbf{R}^2 \to \mathbf{R}^3$ such that

$$T\left(\begin{bmatrix} 0 \\ 1 \end{bmatrix}\right) = \begin{bmatrix} 1 \\ 4 \\ 5 \end{bmatrix}$$

35. A linear transformation $T : \mathbf{R}^2 \to \mathbf{R}^2$ such that

$$T\left(\begin{bmatrix} 3 \\ 1 \end{bmatrix}\right) = \begin{bmatrix} 7 \\ 0 \end{bmatrix}$$

36. A linear transformation $T : \mathbf{R}^3 \to \mathbf{R}^2$ such that

$$T\left(\begin{bmatrix} 3 \\ 1 \\ -2 \end{bmatrix}\right) = \begin{bmatrix} 7 \\ -1 \end{bmatrix}$$

37. A linear transformation $T : \mathbf{R}^2 \to \mathbf{R}^2$ such that

$$T\left(\begin{bmatrix} 2 \\ 1 \end{bmatrix}\right) = \begin{bmatrix} 0 \\ 7 \end{bmatrix} \quad \text{and} \quad T\left(\begin{bmatrix} 1 \\ 3 \end{bmatrix}\right) = \begin{bmatrix} -5 \\ 6 \end{bmatrix}$$

38. A linear transformation $T : \mathbf{R}^2 \to \mathbf{R}^2$ such that

$$T\left(\begin{bmatrix} -1 \\ 2 \end{bmatrix}\right) = \begin{bmatrix} -1 \\ 8 \end{bmatrix} \quad \text{and} \quad T\left(\begin{bmatrix} 2 \\ -3 \end{bmatrix}\right) = \begin{bmatrix} 2 \\ -13 \end{bmatrix}$$

TRUE OR FALSE For Exercises 39–48, determine if the statement is true or false, and justify your answer.

39. The codomain of a linear transformation is a subset of the range.

40. The range of a linear transformation must be a subset of the domain.

41. If T is a linear transformation and $\mathbf{v}$ is in range(T), then there is at least one $\mathbf{u}$ in the domain such that $T(\mathbf{u}) = \mathbf{v}$.

42. If $T(\mathbf{x})$ is not a linear transformation, then $T(r\mathbf{x}) \neq r T(\mathbf{x})$ for all r and $\mathbf{x}$.

43. The function $T(\mathbf{x}) = A\mathbf{x} + \mathbf{b}$ is a linear transformation only when $\mathbf{b} = \mathbf{0}$.

44. If $T : \mathbf{R}^2 \longrightarrow \mathbf{R}^2$ is a linear transformation, then the image under T of the unit square in the first quadrant will be a parallelogram.

45. If $T_1(\mathbf{x})$ and $T_2(\mathbf{x})$ are one-to-one linear transformations from $\mathbf{R}^n$ to $\mathbf{R}^m$, then so is $W(\mathbf{x}) = T_1(\mathbf{x}) + T_2(\mathbf{x})$.

46. If $T_1(\mathbf{x})$ and $T_2(\mathbf{x})$ are onto linear transformations from $\mathbf{R}^n$ to $\mathbf{R}^m$, then so is $W(\mathbf{x}) = T_1(\mathbf{x}) + T_2(\mathbf{x})$.

47. If a linear transformation $T : \mathbf{R}^4 \to \mathbf{R}^4$ is one-to-one, then $T(\mathbf{x}) = \mathbf{0}$ has nontrivial solutions.

48. If a linear transformation $T : \mathbf{R}^3 \to \mathbf{R}^3$ is one-to-one, then T also must be onto.

49. A linear transformation $T : \mathbf{R}^2 \to \mathbf{R}^2$ is called a *dilation* if $T(\mathbf{x}) = r\mathbf{x}$ for $r > 1$. (It is called a *contraction* if $0 < r < 1$.)

(a) Find the matrix A such that $T(\mathbf{x}) = A\mathbf{x}$.

(b) Let $r = 2$, and then sketch the graphs of $\mathbf{x} = \begin{bmatrix} 2 \\ -1 \end{bmatrix}$ and $T(\mathbf{x})$.

50. Suppose that $T : \mathbf{R}^3 \to \mathbf{R}^2$ is given by

$$T\left(\begin{bmatrix} x_1 \\ x_2 \\ x_3 \end{bmatrix}\right) = \begin{bmatrix} x_1 \\ x_2 \end{bmatrix}$$

The T is called a *projection transformation* because it projects vectors in $\mathbf{R}^3$ onto $\mathbf{R}^2$.

(a) Prove that T is a linear transformation.

(b) Find the matrix A such that $T(\mathbf{x}) = A\mathbf{x}$.

(c) Describe the set of vectors in $\mathbf{R}^3$ such that $T(\mathbf{x}) = \mathbf{0}$.

51. Suppose that $\mathbf{x} = (x_1, \dots, x_n)$ and $\mathbf{y} = (y_1, \dots, y_n)$ are vectors in $\mathbf{R}^n$. Then the *dot product* of $\mathbf{x}$ and $\mathbf{y}$ is given by

$$\mathbf{x} \cdot \mathbf{y} = x_1 y_1 + \cdots + x_n y_n$$

Now let $\mathbf{u}$ be a fixed vector in $\mathbf{R}^n$, and define $T(\mathbf{x}) = \mathbf{u} \cdot \mathbf{x}$. Show that T is a linear transformation.

52. Suppose that $\mathbf{x} = (x_1, x_2, x_3)$ and $\mathbf{y} = (y_1, y_2, y_3)$ are vectors in $\mathbf{R}^3$. Then the *cross product* of $\mathbf{x}$ and $\mathbf{y}$ is given by

$$\mathbf{x} \times \mathbf{y} = \begin{bmatrix} x_2 y_3 - x_3 y_2 \\ x_3 y_1 - x_1 y_3 \\ x_1 y_2 - x_2 y_1 \end{bmatrix}$$

Now let $\mathbf{u}$ be a fixed vector in $\mathbf{R}^n$, and define $T(\mathbf{x}) = \mathbf{u} \times \mathbf{x}$. Show that T is a linear transformation.

53. Suppose that $T : \mathbf{R}^3 \to \mathbf{R}^2$ is a linear transformation. Prove that T is not one-to-one.

54. Suppose that $T : \mathbf{R}^2 \to \mathbf{R}^4$ is a linear transformation. Prove that T is not onto.

55. Suppose that T is a linear transformation. Show that $T(\mathbf{0}) = \mathbf{0}$.

56. Suppose that $T(\mathbf{x}) = A\mathbf{x}$ is a linear transformation and that there exists $\mathbf{u} \neq \mathbf{0}$ such that $T(\mathbf{u}) = \mathbf{0}$. Show that the columns of A must be linearly dependent.

57. Suppose that $T(\mathbf{x}) = A\mathbf{x}$. Show that

$$T(r\mathbf{u}) = r T(\mathbf{u})$$

for all scalars r and all vectors $\mathbf{u}$.

58. Suppose that $T : \mathbf{R}^m \to \mathbf{R}^n$.

(a) Show that if T is a linear transformation, then

$$T(r\mathbf{x} + s\mathbf{y}) = r T(\mathbf{x}) + s T(\mathbf{y})$$

for all scalars r and s and all vectors $\mathbf{x}$ and $\mathbf{y}$.

(b) Now show the converse: If

$$T(r\mathbf{x} + s\mathbf{y}) = r T(\mathbf{x}) + s T(\mathbf{y})$$

for all scalars r and s and all vectors $\mathbf{x}$ and $\mathbf{y}$, then T is a linear transformation.

59. Prove that a linear transformation T is one-to-one if and only if $T(\mathbf{u}) = T(\mathbf{v})$ implies that $\mathbf{u} = \mathbf{v}$.

60. Suppose that T is a linear transformation and that $\mathbf{u}_1$ and $\mathbf{u}_2$ are linearly dependent. Prove that $T(\mathbf{u}_1)$ and $T(\mathbf{u}_2)$ are also linearly dependent.

61. Suppose that T is a linear transformation and that $T(\mathbf{u}_1)$ and $T(\mathbf{u}_2)$ are linearly independent. Prove that $\mathbf{u}_1$ and $\mathbf{u}_2$ must be linearly independent.

62. Suppose that T is a linear transformation and that $\mathbf{u}_1$ and $\mathbf{u}_2$ are linearly independent. Show that $T(\mathbf{u}_1)$ and $T(\mathbf{u}_2)$ need not be linearly independent.

63. Suppose that $\mathbf{y}$ is in the range of a linear transformation T and that there exists a vector $\mathbf{u} \neq \mathbf{0}$ such that $T(\mathbf{u}) = \mathbf{0}$. Show that there are infinitely many solutions $\mathbf{x}$ to $T(\mathbf{x}) = \mathbf{y}$. (HINT: First, explain why $T(r\mathbf{u}) = \mathbf{0}$ for any scalar r, and then show that $T(\mathbf{x} + r\mathbf{u}) = \mathbf{y}$ when $T(\mathbf{x}) = \mathbf{y}$.)

64. Let $\mathbf{u}$ and $\mathbf{v}$ be two distinct vectors in $\mathbf{R}^2$. Show that the set of points on the line segment connecting $\mathbf{u}$ and $\mathbf{v}$ is the same as the set of points

$$(1 - s)\mathbf{u} + s\mathbf{v}, \quad 0 \le s \le 1$$

65. Let $T : \mathbf{R}^2 \to \mathbf{R}^2$ be a linear transformation with $T(\mathbf{x}) = A\mathbf{x}$. Prove that the image of the unit square in the first quadrant is a parallelogram if the columns of A are linearly independent, and a line segment (possibly of zero length) if the columns of A are linearly dependent.

66. In graph theory, an *adjacency matrix* A has an entry of 1 at a_{ij} if there is an edge connecting node i with node j, and a zero otherwise. (Such matrices come up in network analysis.) Suppose that a graph with five nodes has adjacency matrix

$$A = \begin{bmatrix} 0 & 1 & 0 & 1 & 1 \\ 1 & 0 & 1 & 0 & 1 \\ 0 & 1 & 0 & 1 & 0 \\ 1 & 0 & 1 & 0 & 0 \\ 1 & 1 & 0 & 0 & 0 \end{bmatrix}$$

Let $T : \mathbf{R}^5 \to \mathbf{R}^5$ be given by $T(\mathbf{x}) = A\mathbf{x}$.

(a) Describe how to use $T(\mathbf{x})$ to determine the number of edges connected to node j.

(b) How can one use $T(\mathbf{x})$ to help determine the total number of graph edges?

67. (Calculus required) Suppose that for each polynomial of degree 2 or less, we identify the coefficients with a vector in $\mathbf{R}^3$ by

$$ax^2 + bx + c \quad \leftrightarrow \quad \begin{bmatrix} a \\ b \\ c \end{bmatrix}$$

(a) Show that addition of polynomials corresponds to vector addition and that multiplication of a polynomial by a constant corresponds to scalar multiplication of a vector.

(b) Let $T : \mathbf{R}^3 \to \mathbf{R}^3$ be the function that takes a polynomial vector as input and produces the vector of the derivative as output. Prove that T is a linear transformation.

(c) Find the matrix A such that $T(\mathbf{x}) = A\mathbf{x}$.

(d) Is T one-to-one? Onto? Give a proof or counter-example for each.

68. (Calculus required) Complete Exercise 67 for polynomials of degree 3 or less identified with vectors in $\mathbf{R}^4$.

(Calculus required) In Chapter 9 we extend the concept of a linear transformation by observing that the two conditions given in Definition 3.1 exist for other types of mathematical operations. Exercises 69–70 provide a sneak preview. Assume that $f(x)$ and $g(x)$ are in the set of functions $C^\infty(\mathbf{R})$ that have infinitely many derivatives on $\mathbf{R}$ and that r is a real number.

69. Let $T : C^\infty(\mathbf{R}) \to C^\infty(\mathbf{R})$ be defined by $T\big(f(x)\big) = f'(x)$.

(a) Evaluate $T\big(x^2 + \sin(x)\big)$.

(b) Prove that T satisfies conditions analogous to those given in Definition 3.1:

  i. $T\big(f(x) + g(x)\big) = T\big(f(x)\big) + T\big(g(x)\big)$

  ii. $T\big(rf(x)\big) = rT\big(f(x)\big)$

70. Let $T : C^\infty(\mathbf{R}) \to \mathbf{R}$ be defined by

$$T\big(f(x)\big) = \int_0^1 f(x)\, dx$$

(a) Evaluate $T\big(4x^3 - 6x^2 + 1\big)$.

(b) Prove that T satisfies conditions analogous to those given in Definition 3.1:

  i. $T\big(f(x) + g(x)\big) = T\big(f(x)\big) + T\big(g(x)\big)$

  ii. $T\big(rf(x)\big) = rT\big(f(x)\big)$

Ⓒ In Exercises 71–74, refer to the MP3 scenario given at the beginning of the section. Use the linear transformation T to determine the cost vector that results from producing the specified number of J8's, J40's, and J80's.

71. 5 J8's, 3 J40's, and 6 J80's.

72. 6 J8's, 4 J40's, and 10 J80's.

73. 14 J8's, 10 J40's, and 9 J80's.

74. 16 J8's, 20 J40's, and 18 J80's.

Ⓒ For Exercises 75–80, let $T(\mathbf{x}) = A\mathbf{x}$ for the given matrix A. Determine if T is one-to-one and if T is onto.

75. $A = \begin{bmatrix} 4 & 2 & -5 & 2 & 6 \\ 7 & -2 & 0 & -4 & 1 \\ 0 & 3 & -5 & 7 & -1 \end{bmatrix}$

76. $A = \begin{bmatrix} 4 & -2 & 5 & 2 & 1 \\ 5 & 14 & 4 & -5 & 8 \\ -1 & 6 & -2 & -3 & 2 \end{bmatrix}$

77. $A = \begin{bmatrix} 2 & -1 & 4 & 0 \\ 3 & -3 & 1 & 1 \\ 1 & -1 & 8 & 3 \\ 0 & -2 & 1 & 4 \end{bmatrix}$

78. $A = \begin{bmatrix} 3 & 2 & 0 & 5 \\ 0 & 1 & 2 & -3 \\ -2 & -1 & 3 & 1 \\ 4 & -2 & 3 & -1 \end{bmatrix}$

79. $A = \begin{bmatrix} 2 & -3 & 5 & 1 \\ 6 & 0 & 3 & -2 \\ -4 & 2 & 1 & 1 \\ 8 & 2 & 3 & -4 \\ -1 & 2 & 5 & -3 \end{bmatrix}$

80. $A = \begin{bmatrix} 4 & 3 & -2 & 9 \\ -1 & 0 & 1 & -1 \\ 3 & 0 & -2 & 4 \\ 2 & -4 & 3 & 3 \\ 5 & -7 & 0 & 3 \end{bmatrix}$

3.2 Matrix Algebra

In this section, we develop the algebra of matrices. This algebraic structure has many things in common with the algebra of the real numbers, but there are also some important differences.

To get us started, consider a hypothetical natural foods store that sells organic chicken eggs that come from two suppliers, *The Happy Coop* and *Eggspeditious*. Each provides both white and brown eggs by the dozen in medium, large, and extra large sizes. The store's current inventory of 12–egg cartons is given in the following two tables.

The Happy Coop			Eggspeditious		
	White	Brown		White	Brown
Medium	5	3	Medium	8	5
Large	11	6	Large	3	6
XLarge	4	6	XLarge	8	10

Table 1 Egg Inventories at a Natural Foods Store

This information can be transferred into a pair of matrices

$$H = \begin{bmatrix} 5 & 3 \\ 11 & 6 \\ 4 & 6 \end{bmatrix} \quad \text{and} \quad E = \begin{bmatrix} 8 & 5 \\ 3 & 6 \\ 8 & 10 \end{bmatrix}$$

If we want to know the total number of egg cartons of each type in stock, we just add the corresponding terms in each matrix, giving

$$\begin{bmatrix} (5+8) & (3+5) \\ (11+3) & (6+6) \\ (4+8) & (6+10) \end{bmatrix} = \begin{bmatrix} 13 & 8 \\ 14 & 12 \\ 12 & 16 \end{bmatrix}$$

Thus, for instance, there are 13 cartons of medium white eggs in stock. If instead we want to find the total number of each type of Eggspeditious eggs, we multiply each term in E by 12:

$$\begin{bmatrix} (12 \cdot 8) & (12 \cdot 5) \\ (12 \cdot 3) & (12 \cdot 6) \\ (12 \cdot 8) & (12 \cdot 10) \end{bmatrix} = \begin{bmatrix} 96 & 60 \\ 36 & 72 \\ 96 & 120 \end{bmatrix}$$

Definition Equal Matrices   These computations illustrate addition and scalar multiplication of matrices, and are analogous to those for vectors. As these operations suggest, two matrices are **equal** if they have the same dimensions and if their corresponding entries are equal.

This example serves as a model for a formal definition of addition and scalar multiplication.

DEFINITION 3.10

Let c be a scalar, and let

$$A = \begin{bmatrix} a_{11} & a_{12} & \cdots & a_{1m} \\ a_{21} & a_{22} & \cdots & a_{2m} \\ \vdots & \vdots & & \vdots \\ a_{n1} & a_{n2} & \cdots & a_{nm} \end{bmatrix} \quad \text{and} \quad B = \begin{bmatrix} b_{11} & b_{12} & \cdots & b_{1m} \\ b_{21} & b_{22} & \cdots & b_{2m} \\ \vdots & \vdots & & \vdots \\ b_{n1} & b_{n2} & \cdots & b_{nm} \end{bmatrix}$$

be $n \times m$ matrices. Then addition and scalar multiplication of matrices are defined as follows:

Definition Addition, Scalar Multiplication of Matrices

(a) **Addition:** $A + B = \begin{bmatrix} (a_{11} + b_{11}) & (a_{12} + b_{12}) & \cdots & (a_{1m} + b_{1m}) \\ (a_{21} + b_{21}) & (a_{22} + b_{22}) & \cdots & (a_{2m} + b_{2m}) \\ \vdots & \vdots & & \vdots \\ (a_{n1} + b_{n1}) & (a_{n2} + b_{n2}) & \cdots & (a_{nm} + a_{nm}) \end{bmatrix}$

(b) **Scalar Multiplication:** $cA = \begin{bmatrix} ca_{11} & ca_{12} & \cdots & ca_{1m} \\ ca_{21} & ca_{22} & \cdots & ca_{2m} \\ \vdots & \vdots & & \vdots \\ ca_{n1} & ca_{n2} & \cdots & ca_{nm} \end{bmatrix}$

EXAMPLE 1 Let

$$A = \begin{bmatrix} 4 & -1 \\ 2 & -3 \\ 7 & 0 \end{bmatrix} \quad \text{and} \quad B = \begin{bmatrix} 3 & -1 \\ 5 & 0 \\ 0 & 2 \end{bmatrix}$$

Find $3A$ and $A - 2B$.

Solution We have

$$3A = 3 \begin{bmatrix} 4 & -1 \\ 2 & -3 \\ 7 & 0 \end{bmatrix} = \begin{bmatrix} 3(4) & 3(-1) \\ 3(2) & 3(-3) \\ 3(7) & 3(0) \end{bmatrix} = \begin{bmatrix} 12 & -3 \\ 6 & -9 \\ 21 & 0 \end{bmatrix}$$

and

$$A - 2B = \begin{bmatrix} 4 & -1 \\ 2 & -3 \\ 7 & 0 \end{bmatrix} - 2 \begin{bmatrix} 3 & -1 \\ 5 & 0 \\ 0 & 2 \end{bmatrix} = \begin{bmatrix} 4 & -1 \\ 2 & -3 \\ 7 & 0 \end{bmatrix} - \begin{bmatrix} 6 & -2 \\ 10 & 0 \\ 0 & 4 \end{bmatrix} = \begin{bmatrix} -2 & 1 \\ -8 & -3 \\ 7 & -4 \end{bmatrix} \blacksquare$$

Suppose that r, s, and t are real numbers. Recall the following laws for arithmetic:

(a) $r + s = s + r$ (Commutative)

(b) $(r + s) + t = r + (s + t)$ (Associative)

(c) $r(s + t) = rs + rt$ (Distributive)

(d) $r + 0 = r$ (Additive Identity)

Similar laws hold for addition and scalar multiplication of matrices.

THEOREM 3.11

Let s and t be scalars, A, B, and C be matrices of dimension $n \times m$, and 0_{nm} be the $n \times m$ matrix with all zero entries. Then

(a) $A + B = B + A$

(b) $s(A + B) = sA + sB$

(c) $(s + t)A = sA + tA$

(d) $(A + B) + C = A + (B + C)$

(e) $(st)A = s(tA)$

(f) $A + 0_{nm} = A$

Proof   The proof of each part follows from the analogous laws for the real numbers. We prove part (e) here and leave the rest as exercises.

Let

$$
A = \begin{bmatrix}
a_{11} & a_{12} & \cdots & a_{1m} \\
a_{21} & a_{22} & \cdots & a_{2m} \\
\vdots & \vdots & & \vdots \\
a_{n1} & a_{n2} & \cdots & a_{nm}
\end{bmatrix}
$$

Then we have

$$
(st)A = (st)\begin{bmatrix}
a_{11} & a_{12} & \cdots & a_{1m} \\
a_{21} & a_{22} & \cdots & a_{2m} \\
\vdots & \vdots & & \vdots \\
a_{n1} & a_{n2} & \cdots & a_{nm}
\end{bmatrix} = \begin{bmatrix}
sta_{11} & sta_{12} & \cdots & sta_{1m} \\
sta_{21} & sta_{22} & \cdots & sta_{2m} \\
\vdots & \vdots & & \vdots \\
sta_{n1} & sta_{n2} & \cdots & sta_{nm}
\end{bmatrix}
$$

On the other hand,

$$
s(tA) = s\begin{bmatrix}
ta_{11} & ta_{12} & \cdots & ta_{1m} \\
ta_{21} & ta_{22} & \cdots & ta_{2m} \\
\vdots & \vdots & & \vdots \\
ta_{n1} & ta_{n2} & \cdots & ta_{nm}
\end{bmatrix} = \begin{bmatrix}
sta_{11} & sta_{12} & \cdots & sta_{1m} \\
sta_{21} & sta_{22} & \cdots & sta_{2m} \\
\vdots & \vdots & & \vdots \\
sta_{n1} & sta_{n2} & \cdots & sta_{nm}
\end{bmatrix}
$$

Both are the same, so $(st)A = s(tA)$. ∎

Matrix Multiplication

It is tempting to expect that matrix multiplication would be defined in the same term-by-term manner as addition, so that (for example) we would have

Warning: This is not correct!
$$
\begin{bmatrix} 1 & 2 \\ 3 & 4 \end{bmatrix}\begin{bmatrix} 5 & 6 \\ 7 & 8 \end{bmatrix} = \begin{bmatrix} 1 \cdot 5 & 2 \cdot 6 \\ 3 \cdot 7 & 4 \cdot 8 \end{bmatrix} = \begin{bmatrix} 5 & 12 \\ 21 & 32 \end{bmatrix}
$$

However, defining multiplication in this way is not helpful in most applications. Instead, it is more useful to define matrix multiplication so that it is consistent with the composition of linear transformations.

Let $S : \mathbf{R}^m \rightarrow \mathbf{R}^k$ and $T : \mathbf{R}^k \rightarrow \mathbf{R}^n$ be linear transformations. By Theorem 3.8, there exists a $k \times m$ matrix $B = \begin{bmatrix} \mathbf{b}_1 & \cdots & \mathbf{b}_m \end{bmatrix}$ and an $n \times k$ matrix A such that

$$
S(\mathbf{x}) = B\mathbf{x} \quad \text{and} \quad T(\mathbf{y}) = A\mathbf{y}
$$

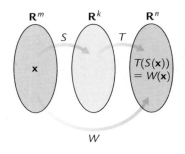

Figure 1 The composition $W(\mathbf{x}) = T\big(S(\mathbf{x})\big)$.

where $\mathbf{x}$ is in $\mathbf{R}^m$ and $\mathbf{y}$ is in $\mathbf{R}^k$. Now let

$$W(\mathbf{x}) = T\big(S(\mathbf{x})\big)$$

be the composition of T with S (see Figure 1). Then $W : \mathbf{R}^m \to \mathbf{R}^n$ and

$$
\begin{aligned}
W(\mathbf{x}) &= T\big(S(\mathbf{x})\big) \\
&= A(B\mathbf{x}) \\
&= A\big(x_1\mathbf{b}_1 + x_2\mathbf{b}_2 + \cdots + x_m\mathbf{b}_m\big) \\
&= x_1(A\mathbf{b}_1) + x_2(A\mathbf{b}_2) + \cdots + x_m(A\mathbf{b}_m) \\
&= \underbrace{\begin{bmatrix} A\mathbf{b}_1 & A\mathbf{b}_2 & \cdots & A\mathbf{b}_m \end{bmatrix}}_{n \times m \text{ matrix}} \begin{bmatrix} x_1 \\ \vdots \\ x_m \end{bmatrix}
\end{aligned}
$$

Thus, if

$$C = \begin{bmatrix} A\mathbf{b}_1 & A\mathbf{b}_2 & \cdots & A\mathbf{b}_m \end{bmatrix}$$

then C is an $n \times m$ matrix and $W(\mathbf{x}) = C\mathbf{x}$. Therefore, by Theorem 3.8, W is a linear transformation. This also shows how to define the matrix product to make it consistent with composition of linear transformations.

DEFINITION 3.12

Definition Matrix Multiplication

▶ For AB to exist, the number of columns of A must equal the number of rows of B.

Let A be an $n \times k$ matrix and B a $k \times m$ matrix. Then the **product** AB is an $n \times m$ matrix given by

$$AB = \begin{bmatrix} A\mathbf{b}_1 & A\mathbf{b}_2 & \cdots & A\mathbf{b}_m \end{bmatrix}$$

In other words, the product AB is the product of the matrix A and each of the columns of B. Note also that if $S(\mathbf{x}) = B\mathbf{x}$ and $T(\mathbf{y}) = A\mathbf{y}$, then $T\big(S(\mathbf{x})\big) = AB\mathbf{x}$.

EXAMPLE 2 Let

$$A = \begin{bmatrix} 3 & 1 \\ -2 & 0 \end{bmatrix} \quad \text{and} \quad B = \begin{bmatrix} -1 & 0 & 2 \\ 4 & -3 & -1 \end{bmatrix}$$

Find (if they exist) AB and BA.

Solution Since A is a 2×2 matrix and B is a 2×3 matrix, it follows that $C = AB$ exists and is a 2×3 matrix, with columns

$$\mathbf{c}_1 = A\mathbf{b}_1 = -1\begin{bmatrix} 3 \\ -2 \end{bmatrix} + 4\begin{bmatrix} 1 \\ 0 \end{bmatrix} = \begin{bmatrix} 1 \\ 2 \end{bmatrix}$$

$$\mathbf{c}_2 = A\mathbf{b}_2 = 0\begin{bmatrix} 3 \\ -2 \end{bmatrix} - 3\begin{bmatrix} 1 \\ 0 \end{bmatrix} = \begin{bmatrix} -3 \\ 0 \end{bmatrix}$$

$$\mathbf{c}_3 = A\mathbf{b}_3 = 2\begin{bmatrix} 3 \\ -2 \end{bmatrix} - 1\begin{bmatrix} 1 \\ 0 \end{bmatrix} = \begin{bmatrix} 5 \\ -4 \end{bmatrix}$$

Thus

$$AB = \begin{bmatrix} 1 & -3 & 5 \\ 2 & 0 & -4 \end{bmatrix}$$

Turning to BA, since B is 2×3 and A is 2×2, B has three columns but A has two rows. These do not match, so the product BA is not defined. ∎

The preceding example shows that even when AB is defined, BA still might not be. Example 3 shows that even if both AB and BA are defined, they typically are not equal.

EXAMPLE 3 Let

$$A = \begin{bmatrix} 2 & -1 \\ 1 & 3 \end{bmatrix} \quad \text{and} \quad B = \begin{bmatrix} 4 & -2 \\ -1 & 1 \end{bmatrix}$$

Find (if they exist) AB and BA.

Solution Since A and B both are 2×2 matrices, AB and BA are defined. If $C = AB$, then

$$\mathbf{c}_1 = A\mathbf{b}_1 = 4 \begin{bmatrix} 2 \\ 1 \end{bmatrix} - 1 \begin{bmatrix} -1 \\ 3 \end{bmatrix} = \begin{bmatrix} 9 \\ 1 \end{bmatrix}$$

$$\mathbf{c}_2 = A\mathbf{b}_2 = -2 \begin{bmatrix} 2 \\ 1 \end{bmatrix} + 1 \begin{bmatrix} -1 \\ 3 \end{bmatrix} = \begin{bmatrix} -5 \\ 1 \end{bmatrix}$$

so that

$$AB = \begin{bmatrix} 9 & -5 \\ 1 & 1 \end{bmatrix}$$

Reversing A and B, if $D = BA$, then

$$\mathbf{d}_1 = B\mathbf{a}_1 = 2 \begin{bmatrix} 4 \\ -1 \end{bmatrix} + 1 \begin{bmatrix} -2 \\ 1 \end{bmatrix} = \begin{bmatrix} 6 \\ -1 \end{bmatrix}$$

$$\mathbf{d}_2 = B\mathbf{a}_2 = -1 \begin{bmatrix} 4 \\ -1 \end{bmatrix} + 3 \begin{bmatrix} -2 \\ 1 \end{bmatrix} = \begin{bmatrix} -10 \\ 4 \end{bmatrix}$$

and so

$$BA = \begin{bmatrix} 6 & -10 \\ -1 & 4 \end{bmatrix}$$

Thus we see that $AB \neq BA$. ■

▶ Example 3 shows that matrix multiplication is not commutative.

In some applications, we need only a single entry of the matrix product AB. Assume that A is an $n \times k$ matrix and B is a $k \times m$ matrix. Then $C = AB$ is defined, with jth column

$$\mathbf{c}_j = A\mathbf{b}_j = b_{1j}\mathbf{a}_1 + b_{2j}\mathbf{a}_2 + \cdots + b_{kj}\mathbf{a}_k$$

for $j = 1, \ldots, m$. Selecting the ith component from each of the vectors $\mathbf{a}_1, \ldots, \mathbf{a}_k$ gives the formula for individual entries of C,

$$c_{ij} = a_{i1}b_{1j} + a_{i2}b_{2j} + \cdots + a_{ik}b_{kj} \tag{1}$$

Note that we are multiplying each term from row i of A times the corresponding term in column j of B and then adding the resulting products. (This is called the *dot product*, which we will study later.) Figure 2 shows a graphical depiction of computing an entry.

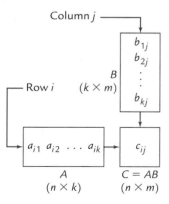

$$c_{ij} = a_{i1}b_{1j} + a_{i2}b_{2j} + \ldots + a_{ik}b_{kj}$$

Figure 2 Computing an entry in matrix multiplication.

EXAMPLE 4 Use 1 to compute the entries of $C = AB$, where

$$A = \begin{bmatrix} 2 & -3 \\ 0 & -1 \\ 1 & 4 \end{bmatrix} \quad \text{and} \quad B = \begin{bmatrix} 2 & 3 & -4 & 2 \\ 4 & -2 & 5 & -3 \end{bmatrix}$$

Solution Let's start with c_{11}. To compute this, we need the entries in row 1 of A and column 1 of B. Setting $i = 1$ and $j = 1$, our formula 1 gives us

$$c_{11} = a_{11}b_{11} + a_{12}b_{21} = (2)(2) + (-3)(4) = -8$$

All entries of C are found in a similar manner, with

$$c_{11} = (2)(2) + (-3)(4) = -8 \qquad\qquad c_{13} = (2)(-4) + (-3)(5) = -23$$
$$c_{21} = (0)(2) + (-1)(4) = -4 \qquad\qquad c_{23} = (0)(-4) + (-1)(5) = -5$$
$$c_{31} = (1)(2) + (4)(4) = 18 \qquad\qquad c_{33} = (1)(-4) + (4)(5) = 16$$
$$c_{12} = (2)(3) + (-3)(-2) = 12 \qquad\qquad c_{14} = (2)(2) + (-3)(-3) = 13$$
$$c_{22} = (0)(3) + (-1)(-2) = 2 \qquad\qquad c_{24} = (0)(2) + (-1)(-3) = 3$$
$$c_{32} = (1)(3) + (4)(-2) = -5 \qquad\qquad c_{34} = (1)(2) + (4)(-3) = -10$$

This gives the product

$$AB = C = \begin{bmatrix} -8 & 12 & -23 & 13 \\ -4 & 2 & -5 & 3 \\ 18 & -5 & 16 & -10 \end{bmatrix}$$

Some find that the method illustrated above is easier for computing products by hand than that shown in Example 2 and Example 3. Both methods are perfectly fine, so use whichever you prefer.

The Identity Matrix

Given an $n \times m$ matrix $A = \begin{bmatrix} \mathbf{a}_1 & \cdots & \mathbf{a}_m \end{bmatrix}$, we can see that the zero matrix 0_{nm} satisfies

$$A + 0_{nm} = A$$

Definition Additive Identity

For this reason, the matrix 0_{nm} is called the **additive identity**. It is less apparent which matrix plays the role of the multiplicative identity—that is, the matrix I such that

$$A = AI$$

for all $n \times m$ matrices A. But we can deduce I. For AI to be defined and equal to A, I must be an $m \times m$ matrix, so let $I_m = \begin{bmatrix} \mathbf{i}_1 & \mathbf{i}_2 & \cdots & \mathbf{i}_m \end{bmatrix}$. The first columns of A and AI must be the same, so that

$$\mathbf{a}_1 = A\mathbf{i}_1 = i_{11}\mathbf{a}_1 + i_{21}\mathbf{a}_2 + \cdots + i_{m1}\mathbf{a}_m$$

The only way that this will hold for every possible matrix A is if

$$\mathbf{i}_1 = \begin{bmatrix} 1 \\ 0 \\ 0 \\ \vdots \\ 0 \end{bmatrix}$$

Taking each of the other columns in turn, similar arguments show that

$$
\mathbf{i}_2 = \begin{bmatrix} 0 \\ 1 \\ 0 \\ \vdots \\ 0 \end{bmatrix}, \qquad
\mathbf{i}_3 = \begin{bmatrix} 0 \\ 0 \\ 1 \\ \vdots \\ 0 \end{bmatrix}, \qquad \cdots \qquad
\mathbf{i}_m = \begin{bmatrix} 0 \\ 0 \\ 0 \\ \vdots \\ 1 \end{bmatrix}
$$

Note that $\mathbf{i}_1 = \mathbf{e}_1, \ldots, \mathbf{i}_m = \mathbf{e}_m$, where $\mathbf{e}_1, \ldots, \mathbf{e}_m$ are defined in (3) of Section 3.1. Thus we find that if I_m is the $m \times m$ matrix

$$
I_m = \begin{bmatrix} \mathbf{e}_1 & \mathbf{e}_2 & \cdots & \mathbf{e}_m \end{bmatrix} = \begin{bmatrix}
1 & 0 & 0 & \cdots & 0 \\
0 & 1 & 0 & \cdots & 0 \\
0 & 0 & 1 & \cdots & 0 \\
\vdots & \vdots & \vdots & \ddots & \vdots \\
0 & 0 & 0 & \cdots & 1
\end{bmatrix}
$$

then $A = AI_m$ for all $n \times m$ matrices A. Similarly, we have

$$
A = I_n A \tag{2}
$$

Definition Identity Matrix

for all $n \times m$ matrices A. (Verification of (2) is left as an exercise.) The matrices I_n for $n = 1, 2, \ldots$ are referred to as **identity matrices**.

Properties of Matrix Algebra

As we have already seen, addition and scalar multiplication of matrices share many of the same properties as addition and multiplication of real numbers. However, not all properties of multiplication of real numbers carry over to multiplication of matrices. Some of those that do are given in Theorem 3.13.

THEOREM 3.13

Let s be a scalar, and let A, B, and C be matrices. Then each of the following holds in the cases where the indicated operations are defined:

(a) $A(BC) = (AB)C$

(b) $A(B + C) = AB + AC$

(c) $(A + B)C = AC + BC$

(d) $s(AB) = (sA)B = A(sB)$

(e) $AI = A$

(f) $IA = A$

Here I denotes an identity matrix of appropriate dimension.

▶ $D = [d_{ij}]$ is shorthand notation for

$$
D = \begin{bmatrix}
d_{11} & d_{12} & \cdots & d_{1m} \\
d_{21} & d_{22} & \cdots & d_{2m} \\
\vdots & \vdots & & \vdots \\
d_{n1} & d_{n2} & \cdots & d_{nm}
\end{bmatrix}
$$

Proof We have already supplied a proof for part (e). Here we prove part (c) and leave the rest as exercises.

Suppose that $A = [a_{ij}]$ and $B = [b_{ij}]$ are $n \times m$ matrices and that $C = [c_{ij}]$ is an $m \times k$ matrix. Let $F = [f_{ij}] = (A + B)C$ and $G = [g_{ij}] = AC + BC$. It is straightforward to verify that F and G both have dimension $n \times k$. Using the formula for calculating product entries given in 1, we find that

$$
\begin{aligned}
f_{ij} &= (a_{i1} + b_{i1})c_{1j} + (a_{i2} + b_{i2})c_{2j} + \cdots + (a_{im} + b_{im})c_{mj} \\
&= (a_{i1}c_{1j} + a_{i2}c_{2j} + \cdots + a_{im}c_{mj}) + (b_{i1}c_{1j} + b_{i2}c_{2j} + \cdots + b_{im}c_{mj}) \\
&= g_{ij}
\end{aligned}
$$

which shows that $(A + B)C = AC + BC$ as claimed. ∎

EXAMPLE 5 Let

$$A = \begin{bmatrix} 2 & -3 \\ -1 & 5 \end{bmatrix}, \quad B = \begin{bmatrix} 0 & 7 \\ 4 & -2 \end{bmatrix}, \quad \text{and} \quad C = \begin{bmatrix} -3 & -4 \\ 0 & -1 \end{bmatrix}$$

Verify that $A(BC) = (AB)C$ and $A(B + C) = AB + AC$.

Solution We have

$$A(BC) = \begin{bmatrix} 2 & -3 \\ -1 & 5 \end{bmatrix} \left(\begin{bmatrix} 0 & 7 \\ 4 & -2 \end{bmatrix} \begin{bmatrix} -3 & -4 \\ 0 & -1 \end{bmatrix} \right)$$

$$= \begin{bmatrix} 2 & -3 \\ -1 & 5 \end{bmatrix} \begin{bmatrix} 0 & -7 \\ -12 & -14 \end{bmatrix} = \begin{bmatrix} 36 & 28 \\ -60 & -63 \end{bmatrix}$$

and

$$(AB)C = \left(\begin{bmatrix} 2 & -3 \\ -1 & 5 \end{bmatrix} \begin{bmatrix} 0 & 7 \\ 4 & -2 \end{bmatrix} \right) \begin{bmatrix} -3 & -4 \\ 0 & -1 \end{bmatrix}$$

$$= \begin{bmatrix} -12 & 20 \\ 20 & -17 \end{bmatrix} \begin{bmatrix} 0 & -7 \\ -12 & -14 \end{bmatrix} = \begin{bmatrix} 36 & 28 \\ -60 & -63 \end{bmatrix}$$

so that $A(BC) = (AB)C$. We also have

$$A(B + C) = \begin{bmatrix} 2 & -3 \\ -1 & 5 \end{bmatrix} \left(\begin{bmatrix} 0 & 7 \\ 4 & -2 \end{bmatrix} + \begin{bmatrix} -3 & -4 \\ 0 & -1 \end{bmatrix} \right)$$

$$= \begin{bmatrix} 2 & -3 \\ -1 & 5 \end{bmatrix} \begin{bmatrix} -3 & 3 \\ 4 & -3 \end{bmatrix} = \begin{bmatrix} -18 & 15 \\ 23 & -18 \end{bmatrix}$$

and

$$AB + AC = \begin{bmatrix} 2 & -3 \\ -1 & 5 \end{bmatrix} \begin{bmatrix} 0 & 7 \\ 4 & -2 \end{bmatrix} + \begin{bmatrix} 2 & -3 \\ -1 & 5 \end{bmatrix} \begin{bmatrix} -3 & -4 \\ 0 & -1 \end{bmatrix}$$

$$= \begin{bmatrix} -12 & 20 \\ 20 & -17 \end{bmatrix} \begin{bmatrix} -6 & -5 \\ 3 & -1 \end{bmatrix} = \begin{bmatrix} -18 & 15 \\ 23 & -18 \end{bmatrix}$$

Thus $A(B + C) = AB + AC$. ∎

Although many of the rules for the algebra of matrices are the same as the rules for the algebra of real numbers, there are important differences. For instance, if a and b are any real numbers, then $ab = ba$. The same is not true of matrices, where multiplication is not generally commutative (see Example 3). This and two other properties of real numbers that do not carry over to matrices are given in the next theorem.

THEOREM 3.14

▶ Here 0 represents the zero matrix of appropriate dimension.

Let A, B, and C be nonzero matrices.

(a) It is possible that $AB \neq BA$.

(b) $AB = 0$ does not imply that $A = 0$ or $B = 0$.

(c) $AC = BC$ does not imply that $A = B$ or $C = 0$.

Proof Part (a) follows from Example 3. For part (b), let

$$A = \begin{bmatrix} 1 & 2 \\ 3 & 6 \end{bmatrix} \text{ and } B = \begin{bmatrix} -4 & 6 \\ 2 & -3 \end{bmatrix} \implies AB = \begin{bmatrix} 0 & 0 \\ 0 & 0 \end{bmatrix}$$

Thus $AB = 0_{22}$ even though A and B are nonzero matrices. For (c), let

$$A = \begin{bmatrix} -3 & 3 \\ 11 & -3 \end{bmatrix}, \qquad B = \begin{bmatrix} -1 & 2 \\ 3 & 1 \end{bmatrix}, \qquad C = \begin{bmatrix} 1 & 3 \\ 2 & 6 \end{bmatrix}$$

Although $A \neq B$ and C is a nonzero matrix, we have

$$AC = \begin{bmatrix} 3 & 9 \\ 5 & 15 \end{bmatrix} \quad \text{and} \quad BC = \begin{bmatrix} 3 & 9 \\ 5 & 15 \end{bmatrix} \quad \Longrightarrow \quad AC = BC \quad \blacksquare$$

Because of the results in Theorem 3.14, we must take care when performing algebra with matrices.

Definition Transpose

Transpose of a Matrix

The **transpose** of a matrix A is denoted by A^T and results from interchanging the rows and columns of A. For example,

$$A = \begin{bmatrix} 1 & 2 & 3 & 4 \\ 5 & 6 & 7 & 8 \\ 9 & 10 & 11 & 12 \end{bmatrix} \quad \Longrightarrow \quad A^T = \begin{bmatrix} 1 & 5 & 9 \\ 2 & 6 & 10 \\ 3 & 7 & 11 \\ 4 & 8 & 12 \end{bmatrix}$$

Focusing on individual entries, the entry in row i and column j of A becomes the entry in row j and column i of A^T.

A few properties of matrix transposes are given in the next theorem.

THEOREM 3.15

▶ Theorem 3.15(c) says that the transpose of a product is the product of transposes with the order reversed.

Let A and B be $n \times m$ matrices, C an $m \times k$ matrix, and s a scalar. Then

(a) $(A + B)^T = A^T + B^T$

(b) $(sA)^T = sA^T$

(c) $(AC)^T = C^T A^T$

The proofs of parts (a) and (b) are straightforward and left as exercises. A general proof of part (c) is not difficult but is notationally messy and is omitted. The next example illustrates Theorem 3.15(c) for a specific pair of matrices.

EXAMPLE 6 Show that $(AC)^T = C^T A^T$ for

$$A = \begin{bmatrix} 1 & -2 & 0 \\ 3 & 1 & -4 \end{bmatrix} \quad \text{and} \quad C = \begin{bmatrix} 5 & 0 \\ -1 & 2 \\ 0 & 3 \end{bmatrix}$$

Solution We have

$$AC = \begin{bmatrix} 7 & -4 \\ 14 & -10 \end{bmatrix} \quad \Longrightarrow \quad (AC)^T = \begin{bmatrix} 7 & 14 \\ -4 & -10 \end{bmatrix}$$

On the other hand,

$$C^T A^T = \begin{bmatrix} 5 & -1 & 0 \\ 0 & 2 & 3 \end{bmatrix} \begin{bmatrix} 1 & 3 \\ -2 & 1 \\ 0 & -4 \end{bmatrix} = \begin{bmatrix} 7 & 14 \\ -4 & -10 \end{bmatrix}$$

Thus $(AC)^T = C^T A^T$. $\blacksquare$

A special class of square matrices are those such that $A = A^T$. Matrices with this property are said to be **symmetric**. For instance, if

$$A = \begin{bmatrix} 1 & 2 & 4 \\ 2 & 2 & -1 \\ 4 & -1 & -3 \end{bmatrix} \implies A^T = \begin{bmatrix} 1 & 2 & 4 \\ 2 & 2 & -1 \\ 4 & -1 & -3 \end{bmatrix}$$

then A is a symmetric matrix.

Powers of a Matrix

Suppose that A is the 2×2 matrix

$$\begin{bmatrix} 2 & 2 \\ -3 & -1 \end{bmatrix}$$

Then we let A^2 denote $A \cdot A$, so that

$$A^2 = \begin{bmatrix} 2 & 2 \\ -3 & -1 \end{bmatrix} \begin{bmatrix} 2 & 2 \\ -3 & -1 \end{bmatrix} = \begin{bmatrix} -2 & 2 \\ -3 & -5 \end{bmatrix}$$

Similarly, $A^3 = A \cdot A \cdot A$. By the associative law, we have $A \cdot (A \cdot A) = (A \cdot A) \cdot A$, so that we can interpret $A^3 = A \cdot A^2$ or $A^3 = A^2 \cdot A$—either way, we get the same result. For our matrix, we have

$$A^3 = A \cdot A^2 = \begin{bmatrix} 2 & 2 \\ -3 & -1 \end{bmatrix} \begin{bmatrix} -2 & 2 \\ -3 & -5 \end{bmatrix} = \begin{bmatrix} -10 & -6 \\ 9 & -1 \end{bmatrix}$$

In general, if A is an $n \times n$ matrix, then

$$A^k = \underbrace{A \cdot A \cdots A}_{k \text{ terms}}$$

As with A^3, the associative law ensures that we get the same result regardless of how we organize the products.

EXAMPLE 7 In a small town there are 10,000 homes. When it comes to television viewing, the residents have three choices: they can subscribe to cable, they can pay for satellite service, or they watch no TV. (The town is sufficiently remote so that an antenna does not work.) In a given year, 80% of the cable customers stick with cable, 10% switch to satellite, and 10% get totally disgusted and quit watching TV. Over the same time period, 90% of satellite viewers continue with satellite service, 5% switch to cable, and 5% quit watching TV. And of those people who start the year not watching TV, 85% continue not watching, 5% subscribe to cable, and 10% get satellite service (see Figure 3). If the current distribution is 6000 homes with cable, 2500 with satellite service, and 1500 with no TV, how many of each type will there be a year from now? How about two years from now? Three years from now?

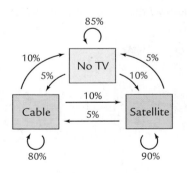

Figure 3 Customer transition percentages between cable, satellite, and no television.

Solution The information given is summarized in Table 2. Reading down each column, we see the percentage of viewers in a given group that switches to one of the other groups. At the start of the year, 6000 homes have cable, 2500 have satellite service, and 1500 have no TV. From our table we see that at the end of the year, the number of homes with cable is

$$0.80\,(6000) + 0.05\,(2500) + 0.05\,(1500) = 5000 \tag{3}$$

		Start of Year		
		Cable	Satellite	No TV
	Cable	80%	5%	5%
End of Year	Satellite	10%	90%	10%
	No TV	10%	5%	85%

Table 2 Rates of Customer Transitions

Similar calculations can be performed to determine the number of satellite customers and the number with no TV at year's end. We can simplify these calculations by letting A be the matrix formed from the values in our table (converted from percentages to proportions) and $\mathbf{x}$ be the vector containing the initial number of people in each category,

$$A = \begin{bmatrix} 0.80 & 0.05 & 0.05 \\ 0.10 & 0.90 & 0.10 \\ 0.10 & 0.05 & 0.85 \end{bmatrix} \quad \text{and} \quad \mathbf{x} = \begin{bmatrix} 6000 \\ 2500 \\ 1500 \end{bmatrix}$$

Note that the top entry of $A\mathbf{x}$ is the same as the left side of (3) and that in general $A\mathbf{x}$ gives the number of people in each category after a year has passed. We have

$$A\mathbf{x} = \begin{bmatrix} 0.80 & 0.05 & 0.05 \\ 0.10 & 0.90 & 0.10 \\ 0.10 & 0.05 & 0.85 \end{bmatrix} \begin{bmatrix} 6000 \\ 2500 \\ 1500 \end{bmatrix} = \begin{bmatrix} 5000 \\ 3000 \\ 2000 \end{bmatrix}$$

which shows that after one year the town will have 5000 cable subscribers, 3000 receiving satellite service, and 2000 with no TV. If the proportion of homes switching among categories remains unchanged, after two years the number of households in each group will be

$$A(A\mathbf{x}) = A^2\mathbf{x} = \begin{bmatrix} 0.65 & 0.0875 & 0.0875 \\ 0.18 & 0.82 & 0.18 \\ 0.17 & 0.0925 & 0.7325 \end{bmatrix} \begin{bmatrix} 6000 \\ 2500 \\ 1500 \end{bmatrix} = \begin{bmatrix} 4250 \\ 3400 \\ 2350 \end{bmatrix}$$

Similarly, after three years, the number of homes in each group is

▶ Here we have rounded the entries to the nearest integer.

$$A^3\mathbf{x} = \begin{bmatrix} 0.655375 & 0.115625 & 0.115625 \\ 0.244 & 0.756 & 0.244 \\ 0.2185 & 0.128375 & 0.640375 \end{bmatrix} \begin{bmatrix} 6000 \\ 2500 \\ 1500 \end{bmatrix} \approx \begin{bmatrix} 3688 \\ 3720 \\ 2593 \end{bmatrix}$$

More generally, the number of households in each category after n years is given by $A^n\mathbf{x}$. These results suggest that cable is losing customers to both satellite and no TV, but eventually the populations stabilize. In Section 3.5, we see what happens as $n \to \infty$. ∎

Definition Diagonal Matrix

There are two types of matrices that retain their form when raised to powers. The first of these is the **diagonal matrix**, which has the form

$$A = \begin{bmatrix} a_{11} & 0 & 0 & \cdots & 0 \\ 0 & a_{22} & 0 & \cdots & 0 \\ 0 & 0 & a_{33} & \cdots & 0 \\ \vdots & \vdots & \vdots & \ddots & \vdots \\ 0 & 0 & 0 & \cdots & a_{nn} \end{bmatrix} \tag{4}$$

The **diagonal** of A consists of the entries $a_{11}, \ldots, a_{nn}$, each of which can be zero or nonzero. For instance,

$$\begin{bmatrix} 1 & 0 & 0 \\ 0 & -3 & 0 \\ 0 & 0 & 4 \end{bmatrix} \quad \text{and} \quad \begin{bmatrix} 2 & 0 & 0 & 0 \\ 0 & 0 & 0 & 0 \\ 0 & 0 & 7 & 0 \\ 0 & 0 & 0 & 5 \end{bmatrix}$$

are both diagonal matrices.

EXAMPLE 8   Compute A^2 and A^3 for the diagonal matrix

$$A = \begin{bmatrix} 2 & 0 & 0 \\ 0 & -3 & 0 \\ 0 & 0 & 5 \end{bmatrix}$$

Solution We have

$$A^2 = \begin{bmatrix} 2 & 0 & 0 \\ 0 & -3 & 0 \\ 0 & 0 & 5 \end{bmatrix} \begin{bmatrix} 2 & 0 & 0 \\ 0 & -3 & 0 \\ 0 & 0 & 5 \end{bmatrix} = \begin{bmatrix} 2^2 & 0 & 0 \\ 0 & (-3)^2 & 0 \\ 0 & 0 & 5^2 \end{bmatrix} = \begin{bmatrix} 4 & 0 & 0 \\ 0 & 9 & 0 \\ 0 & 0 & 25 \end{bmatrix}$$

and

$$A^3 = A^2 \cdot A = \begin{bmatrix} 2^2 & 0 & 0 \\ 0 & (-3)^2 & 0 \\ 0 & 0 & 5^2 \end{bmatrix} \begin{bmatrix} 2 & 0 & 0 \\ 0 & -3 & 0 \\ 0 & 0 & 5 \end{bmatrix}$$

$$= \begin{bmatrix} 2^3 & 0 & 0 \\ 0 & (-3)^3 & 0 \\ 0 & 0 & 5^3 \end{bmatrix} = \begin{bmatrix} 8 & 0 & 0 \\ 0 & -27 & 0 \\ 0 & 0 & 125 \end{bmatrix} \qquad ■$$

In Example 8 the powers of the diagonal matrix A are just the powers of the diagonal entries. This is true for any diagonal matrix.

THEOREM 3.16

If A is the diagonal matrix in (4), then for each integer $k \geq 1$,

$$A^k = \begin{bmatrix} a_{11}^k & 0 & 0 & \cdots & 0 \\ 0 & a_{22}^k & 0 & \cdots & 0 \\ 0 & 0 & a_{33}^k & \cdots & 0 \\ \vdots & \vdots & \vdots & \ddots & \vdots \\ 0 & 0 & 0 & \cdots & a_{nn}^k \end{bmatrix}$$

Proof   We use induction for the proof. First, if $k = 1$, then $A^k = A$ so that A^k clearly has the form shown. Next is the induction hypothesis, which states that the theorem is true for exponent $k - 1$, so that A^{k-1} is diagonal and given by

$$A^{k-1} = \begin{bmatrix} a_{11}^{k-1} & 0 & 0 & \cdots & 0 \\ 0 & a_{22}^{k-1} & 0 & \cdots & 0 \\ 0 & 0 & a_{33}^{k-1} & \cdots & 0 \\ \vdots & \vdots & \vdots & \ddots & \vdots \\ 0 & 0 & 0 & \cdots & a_{nn}^{k-1} \end{bmatrix}$$

Therefore

$$A^k = A^{k-1} \cdot A = \begin{bmatrix} a_{11}^{k-1} & 0 & 0 & \cdots & 0 \\ 0 & a_{22}^{k-1} & 0 & \cdots & 0 \\ 0 & 0 & a_{33}^{k-1} & \cdots & 0 \\ \vdots & \vdots & \vdots & \ddots & \vdots \\ 0 & 0 & 0 & \cdots & a_{nn}^{k-1} \end{bmatrix} \begin{bmatrix} a_{11} & 0 & 0 & \cdots & 0 \\ 0 & a_{22} & 0 & \cdots & 0 \\ 0 & 0 & a_{33} & \cdots & 0 \\ \vdots & \vdots & \vdots & \ddots & \vdots \\ 0 & 0 & 0 & \cdots & a_{nn} \end{bmatrix}$$

$$= \begin{bmatrix} a_{11}^{k} & 0 & 0 & \cdots & 0 \\ 0 & a_{22}^{k} & 0 & \cdots & 0 \\ 0 & 0 & a_{33}^{k} & \cdots & 0 \\ \vdots & \vdots & \vdots & \ddots & \vdots \\ 0 & 0 & 0 & \cdots & a_{nn}^{k} \end{bmatrix}$$

Hence A^k has the claimed form. ∎

Definition Upper Triangular Matrix

A second class of matrices whose form is unchanged when raised to a power are triangular matrices. An $n \times n$ matrix A is **upper triangular** if it has the form

$$A = \begin{bmatrix} a_{11} & a_{12} & a_{13} & \cdots & a_{1n} \\ 0 & a_{22} & a_{23} & \cdots & a_{2n} \\ 0 & 0 & a_{33} & \cdots & a_{3n} \\ \vdots & \vdots & \vdots & \ddots & \vdots \\ 0 & 0 & 0 & \cdots & a_{nn} \end{bmatrix}$$

Definition Lower Triangular Matrix

That is, A is upper triangular if the entries below the diagonal are all zero. Similarly, an $n \times n$ matrix A is **lower triangular** if the terms above the diagonal are all zero,

$$A = \begin{bmatrix} a_{11} & 0 & 0 & \cdots & 0 \\ a_{21} & a_{22} & 0 & \cdots & 0 \\ a_{31} & a_{32} & a_{33} & \cdots & 0 \\ \vdots & \vdots & \vdots & \ddots & \vdots \\ a_{n1} & a_{n2} & a_{n3} & \cdots & a_{nn} \end{bmatrix}$$

▶ As with diagonal matrices, triangular matrices can have entries equal to zero along the diagonal. Note that this is different than triangular linear systems, where the leading (diagonal) coefficients must be nonzero.

A matrix is **triangular** if it is either upper or lower triangular. (Diagonal matrices are both.) For example, the matrix

$$A = \begin{bmatrix} -1 & 0 & 1 \\ 0 & -2 & 4 \\ 0 & 0 & 3 \end{bmatrix}$$

is upper triangular, as are the matrix powers

$$A^2 = \begin{bmatrix} 1 & 0 & 2 \\ 0 & 4 & 4 \\ 0 & 0 & 9 \end{bmatrix} \quad \text{and} \quad A^3 = \begin{bmatrix} -1 & 0 & 7 \\ 0 & -8 & 28 \\ 0 & 0 & 27 \end{bmatrix}$$

In fact, powers of upper (or lower) triangular matrices are also upper (or lower) triangular.

THEOREM 3.17

Let A be an $n \times n$ upper (lower) triangular matrix and $k \geq 1$ an integer. Then A^k is also an upper (lower) triangular.

The proof makes use of the fact that the product of two upper (lower) triangular matrices is again an upper (lower) triangular matrix (see Exercises 57–58), and is left as an exercise.

Partitioned Matrices

▶ The material on partitioned matrices is optional.

▶ For instance, the Google Page Rank search algorithm uses a matrix with several billion rows and columns.

Some applications require working with really, really big matrices. (Think tens of millions of entries.) In such situations, we can divide the matrices into smaller submatrices that are more manageable.

Let's start with a concrete example, say,

$$A = \begin{bmatrix} 2 & 0 & -3 & 1 & 7 \\ -1 & 4 & 2 & 0 & 4 \\ 6 & -1 & 1 & 8 & -3 \\ 0 & 2 & 7 & -3 & 3 \\ 2 & 0 & -6 & 9 & 0 \\ 1 & -1 & 8 & 5 & -1 \\ 4 & 6 & 9 & 7 & 8 \end{bmatrix}$$

Definition Partitioned Matrix, Blocks

Below, A is shown **partitioned** into six different submatrices (called **blocks**),

$$A = \begin{bmatrix} A_{11} & A_{12} \\ A_{21} & A_{22} \\ A_{31} & A_{32} \end{bmatrix}$$

where

$$A_{11} = \begin{bmatrix} 2 & 0 & -3 \\ -1 & 4 & 2 \\ 6 & -1 & 1 \end{bmatrix} \qquad A_{12} = \begin{bmatrix} 1 & 7 \\ 0 & 4 \\ 8 & -3 \end{bmatrix}$$

$$A_{21} = \begin{bmatrix} 0 & 2 & 7 \\ 2 & 0 & -6 \end{bmatrix} \qquad A_{22} = \begin{bmatrix} -3 & 3 \\ 9 & 0 \end{bmatrix}$$

$$A_{31} = \begin{bmatrix} 1 & -1 & 8 \\ 4 & 6 & 9 \end{bmatrix} \qquad A_{32} = \begin{bmatrix} 5 & -1 \\ 7 & 8 \end{bmatrix}$$

Matrices can be partitioned in any manner desired. The advantage of working with partitioned matrices is that we can do arithmetic on a few blocks at a time to make better use of computer memory. In addition, if numerous processors are available, computations can be distributed across them and simultaneously performed in parallel.

Of the arithmetic operations that can be performed on partitioned matrices, addition and scalar multiplication are the easiest to understand. Suppose that A and B are $n \times m$ matrices, partitioned into blocks as shown:

$$A = \begin{bmatrix} A_{11} & A_{12} & A_{13} \\ A_{21} & A_{22} & A_{23} \end{bmatrix} \begin{matrix} n_1 \\ n_2 \end{matrix} \qquad B = \begin{bmatrix} B_{11} & B_{12} & B_{13} \\ B_{21} & B_{22} & B_{23} \end{bmatrix} \begin{matrix} n_1 \\ n_2 \end{matrix}$$
$$\quad\ \ m_1 \quad\ m_2 \quad\ m_3 \qquad\qquad\quad m_1 \quad\ m_2 \quad\ m_3$$

The notation around each matrix indicates the block dimensions. For example, A_{21} and B_{21} are both $n_2 \times m_1$ submatrices. Since the corresponding blocks have the same dimensions, they can be added together in the usual manner, yielding

$$A + B = \begin{bmatrix} (A_{11} + B_{11}) & (A_{12} + B_{12}) & (A_{13} + B_{13}) \\ (A_{21} + B_{21}) & (A_{22} + B_{22}) & (A_{23} + B_{23}) \end{bmatrix}$$

The formula for scalar multiplication of partitioned matrices is quite natural,

$$r A = \begin{bmatrix} r\,A_{11} & r\,A_{12} & r\,A_{13} \\ r\,A_{21} & r\,A_{22} & r\,A_{23} \end{bmatrix}$$

EXAMPLE 9 Suppose that A and B are two matrices partitioned into blocks,

$$A = \begin{bmatrix} 1 & -2 & 0 & 3 & 4 \\ 4 & 2 & 1 & 1 & -2 \\ 7 & -3 & 5 & 0 & -1 \end{bmatrix} \quad \text{and} \quad B = \begin{bmatrix} 6 & 2 & 4 & 1 & 0 \\ 3 & 7 & -1 & 2 & 2 \\ 4 & 5 & 0 & 1 & 6 \end{bmatrix}$$

Use addition and scalar multiplication of partitioned matrices to find $A + B$ and $-4B$.

Solution We have

$$A_{11} + B_{11} = \begin{bmatrix} 1 & -2 \\ 4 & 2 \end{bmatrix} + \begin{bmatrix} 6 & 2 \\ 3 & 7 \end{bmatrix} = \begin{bmatrix} 7 & 0 \\ 7 & 9 \end{bmatrix}$$

$$A_{12} + B_{12} = \begin{bmatrix} 0 & 3 & 4 \\ 1 & 1 & -2 \end{bmatrix} + \begin{bmatrix} 4 & 1 & 0 \\ -1 & 2 & 2 \end{bmatrix} = \begin{bmatrix} 4 & 4 & 4 \\ 0 & 3 & 0 \end{bmatrix}$$

$$A_{21} + B_{21} = \begin{bmatrix} 7 & -3 \end{bmatrix} + \begin{bmatrix} 4 & 5 \end{bmatrix} = \begin{bmatrix} 11 & 2 \end{bmatrix}$$

$$A_{22} + B_{22} = \begin{bmatrix} 5 & 0 & -1 \end{bmatrix} + \begin{bmatrix} 0 & 1 & 6 \end{bmatrix} = \begin{bmatrix} 5 & 1 & 5 \end{bmatrix}$$

Pulling the block sums back together gives

$$A + B = \begin{bmatrix} (A_{11} + B_{11}) & (A_{12} + B_{12}) \\ (A_{21} + B_{21}) & (A_{22} + B_{22}) \end{bmatrix} = \begin{bmatrix} 7 & 0 & 4 & 4 & 4 \\ 7 & 9 & 0 & 3 & 0 \\ 11 & 2 & 5 & 1 & 5 \end{bmatrix}$$

To find $-4B$, we need

$$-4B_{11} = -4 \begin{bmatrix} 6 & 2 \\ 3 & 7 \end{bmatrix} = \begin{bmatrix} -24 & -8 \\ -12 & -28 \end{bmatrix}$$

$$-4B_{12} = -4 \begin{bmatrix} 4 & 1 & 0 \\ -1 & 2 & 2 \end{bmatrix} = \begin{bmatrix} -16 & -4 & 0 \\ 4 & -8 & -8 \end{bmatrix}$$

$$-4B_{21} = -4 \begin{bmatrix} 4 & 5 \end{bmatrix} = \begin{bmatrix} -16 & -20 \end{bmatrix}$$

$$-4B_{22} = -4 \begin{bmatrix} 0 & 1 & 6 \end{bmatrix} = \begin{bmatrix} 0 & -4 & -24 \end{bmatrix}$$

Putting everything back together yields

$$-4B = \begin{bmatrix} -4B_{11} & -4B_{12} \\ -4B_{21} & -4B_{22} \end{bmatrix} = \begin{bmatrix} -24 & -8 & -16 & -4 & 0 \\ -12 & -28 & 4 & -8 & -8 \\ -16 & -20 & 0 & -4 & -24 \end{bmatrix} \quad\blacksquare$$

Multiplication of partitioned matrices nicely mimics the usual multiplication of matrices. Suppose that A is an $n \times k$ matrix and B is a $k \times m$ matrix, so that AB is defined. Partition A and B as shown:

$$A = \begin{bmatrix} A_{11} & A_{12} & \cdots & A_{1i} \\ A_{21} & A_{22} & \cdots & A_{2i} \\ \vdots & \vdots & \cdots & \vdots \\ A_{j1} & A_{j2} & \cdots & A_{ji} \end{bmatrix} \begin{matrix} n_1 \\ n_2 \\ \vdots \\ n_j \end{matrix} \qquad B = \begin{bmatrix} B_{11} & B_{12} & \cdots & B_{1l} \\ B_{21} & B_{22} & \cdots & B_{2l} \\ \vdots & \vdots & \cdots & \vdots \\ B_{i1} & B_{i2} & \cdots & B_{il} \end{bmatrix} \begin{matrix} k_1 \\ k_2 \\ \vdots \\ k_i \end{matrix}$$

$$\begin{matrix} k_1 & k_2 & \cdots & k_i \end{matrix} \qquad\qquad \begin{matrix} m_1 & m_2 & \cdots & m_l \end{matrix}$$

The only requirement when setting the size of the partitions is that the column sizes for A must match the row sizes for B. (Here both are $k_1, k_2, \ldots, k_i$.) We can compute AB using the blocks in exactly the same manner as we do with regular matrix multiplication. For instance, the upper left block of AB is given by

$$A_{11} B_{11} + A_{12} B_{21} + \cdots + A_{1i} B_{i1}$$

Note that each product in this sum is an $n_1 \times m_1$ matrix, so that the upper left block of AB is also an $n_1 \times m_1$ matrix.

EXAMPLE 10 Suppose that

$$A = \left[\begin{array}{cc|ccc} 3 & -1 & 2 & 4 & 0 \\ 0 & 2 & 1 & -3 & 1 \\ \hline 2 & 3 & -4 & 0 & -4 \\ 1 & 6 & 0 & 2 & -2 \end{array}\right] = \begin{bmatrix} A_{11} & A_{12} \\ A_{21} & A_{22} \end{bmatrix}$$

$$B = \left[\begin{array}{c|cc} 3 & 1 & 2 \\ 4 & 0 & 3 \\ \hline -1 & 7 & 0 \\ 2 & 4 & 1 \\ 0 & -1 & -1 \end{array}\right] = \begin{bmatrix} B_{11} & B_{12} & B_{13} \\ B_{21} & B_{22} & B_{23} \end{bmatrix}$$

Find AB using block multiplication with the given partitions.

Solution First, the block multiplication yields

$$AB = \begin{bmatrix} A_{11} & A_{12} \\ A_{21} & A_{22} \end{bmatrix} \begin{bmatrix} B_{11} & B_{12} & B_{13} \\ B_{21} & B_{22} & B_{23} \end{bmatrix}$$

$$= \begin{bmatrix} (A_{11} B_{11} + A_{12} B_{21}) & (A_{11} B_{12} + A_{12} B_{22}) & (A_{11} B_{13} + A_{12} B_{23}) \\ (A_{21} B_{11} + A_{22} B_{21}) & (A_{21} B_{12} + A_{22} B_{22}) & (A_{21} B_{13} + A_{22} B_{23}) \end{bmatrix}$$

For the upper left block we need the products

$$A_{11} B_{11} = \begin{bmatrix} 3 & -1 \\ 0 & 2 \\ 2 & 3 \end{bmatrix} \begin{bmatrix} 3 \\ 4 \end{bmatrix} = \begin{bmatrix} 5 \\ 8 \\ 18 \end{bmatrix} \quad \text{and} \quad A_{12} B_{21} = \begin{bmatrix} 2 & 4 & 0 \\ 1 & -3 & 1 \\ -4 & 0 & -4 \end{bmatrix} \begin{bmatrix} -1 \\ 2 \\ 0 \end{bmatrix} = \begin{bmatrix} 6 \\ -7 \\ 4 \end{bmatrix}$$

Thus the upper left block of AB is given by

$$A_{11} B_{11} + A_{12} B_{21} = \begin{bmatrix} 5 \\ 8 \\ 18 \end{bmatrix} + \begin{bmatrix} 6 \\ -7 \\ 4 \end{bmatrix} = \begin{bmatrix} 11 \\ 1 \\ 22 \end{bmatrix}$$

Similar computations give us

$$A_{11}B_{12} + A_{12}B_{22} = \begin{bmatrix} 33 \\ -4 \\ -30 \end{bmatrix}, \qquad A_{11}B_{13} + A_{12}B_{23} = \begin{bmatrix} 7 \\ 4 \\ 9 \end{bmatrix},$$

$$A_{21}B_{11} + A_{22}B_{21} = \begin{bmatrix} 31 \end{bmatrix}, \qquad A_{21}B_{12} + A_{22}B_{22} = \begin{bmatrix} 11 \end{bmatrix}, \qquad A_{21}B_{13} + A_{22}B_{23} = \begin{bmatrix} 24 \end{bmatrix}$$

Finally, we pull everything together to arrive at

$$AB = \left[\begin{array}{c|cc} 11 & 33 & 7 \\ 1 & -4 & 4 \\ \hline 22 & -30 & 9 \\ 31 & 11 & 24 \end{array} \right] \quad \blacksquare$$

EXERCISES

In Exercises 1–6, perform the indicated computations when possible, using the matrices given below. If a computation is not possible, explain why.

$$A = \begin{bmatrix} -3 & 1 \\ 2 & -1 \end{bmatrix}, \quad B = \begin{bmatrix} 0 & 4 \\ -2 & 5 \end{bmatrix}, \quad C = \begin{bmatrix} 5 & 0 \\ -1 & 4 \\ 3 & 3 \end{bmatrix}$$

$$D = \begin{bmatrix} 1 & 0 & -3 \\ -2 & 5 & -1 \end{bmatrix}, \quad E = \begin{bmatrix} 1 & 4 & -5 \\ -2 & 1 & -3 \\ 0 & 2 & 6 \end{bmatrix}$$

1. (a) $A + B$, (b) $AB + I_2$, (c) $A + C$

2. (a) AC, (b) $C + D^T$, (c) $CB + I_2$

3. (a) $(AB)^T$, (b) CE, (c) $(A - B)D$

4. (a) A^3, (b) BC^T, (c) $EC + I_3$

5. (a) $(C + E)B$, (b) $B(C^T + D)$, (c) $E + CD$

6. (a) $AD - C^T$, (b) $AB - DC$, (c) $DE + CB$

In Exercises 7–10, find the missing values in the given matrix equation.

7. $\begin{bmatrix} 2 & a \\ 3 & -2 \end{bmatrix} \begin{bmatrix} b & -3 \\ -1 & 2 \end{bmatrix} = \begin{bmatrix} 3 & -8 \\ 5 & c \end{bmatrix}$

8. $\begin{bmatrix} 1 & 4 \\ a & 7 \end{bmatrix} \begin{bmatrix} 2 & -1 \\ b & 3 \end{bmatrix} = \begin{bmatrix} 6 & d \\ 11 & c \end{bmatrix}$

9. $\begin{bmatrix} a & 3 & -2 \\ 3 & -2 & 4 \end{bmatrix} \begin{bmatrix} 2 & -1 \\ 0 & b \\ c & 1 \end{bmatrix} = \begin{bmatrix} 4 & d \\ -6 & -5 \end{bmatrix}$

10. $\begin{bmatrix} 1 & a \\ 0 & -2 \\ 5 & b \end{bmatrix} \begin{bmatrix} 3 & c & d \\ -2 & 1 & 2 \end{bmatrix} = \begin{bmatrix} -3 & 3 & 7 \\ e & -2 & -4 \\ f & -2 & 1 \end{bmatrix}$

11. Find all values of a such that $A^2 = A$ for

$$A = \begin{bmatrix} 5 & -10 \\ a & -4 \end{bmatrix}$$

12. Find all values of a such that $A^3 = 2A$ for

$$A = \begin{bmatrix} -2 & 2 \\ -1 & a \end{bmatrix}$$

13. Let T_1 and T_2 be linear transformations given by

$$T_1\left(\begin{bmatrix} x_1 \\ x_2 \end{bmatrix}\right) = \begin{bmatrix} 3x_1 + 5x_2 \\ -2x_1 + 7x_2 \end{bmatrix}$$

$$T_2\left(\begin{bmatrix} x_1 \\ x_2 \end{bmatrix}\right) = \begin{bmatrix} -2x_1 + 9x_2 \\ 5x_2 \end{bmatrix}$$

Find the matrix A such that

(a) $T_1(T_2(\mathbf{x})) = A\mathbf{x}$

(b) $T_2(T_1(\mathbf{x})) = A\mathbf{x}$

(c) $T_1(T_1(\mathbf{x})) = A\mathbf{x}$

(d) $T_2(T_2(\mathbf{x})) = A\mathbf{x}$

14. Let T_1 and T_2 be linear transformations given by

$$T_1\left(\begin{bmatrix} x_1 \\ x_2 \end{bmatrix}\right) = \begin{bmatrix} -2x_1 + 3x_2 \\ x_1 + 6x_2 \end{bmatrix}$$

$$T_2\left(\begin{bmatrix} x_1 \\ x_2 \end{bmatrix}\right) = \begin{bmatrix} 4x_1 - 5x_2 \\ x_1 + 5x_2 \end{bmatrix}$$

Find the matrix A such that

(a) $T_1(T_2(\mathbf{x})) = A\mathbf{x}$

(b) $T_2(T_1(\mathbf{x})) = A\mathbf{x}$

(c) $T_1(T_1(\mathbf{x})) = A\mathbf{x}$

(d) $T_2(T_2(\mathbf{x})) = A\mathbf{x}$

In Exercises 15–18, expand each of the given matrix expressions and combine as many terms as possible. Assume that all matrices are $n \times n$.

15. $(A + I)(A - I)$

16. $(A + I)(A^2 + A)$

17. $(A + B^2)(BA - A)$

18. $A(A + B) + B(B - A)$

In Exercises 19–22, the given matrix equation is *not* true in general. Explain why. Assume that all matrices are $n \times n$.

19. $(A + B)^2 = A^2 + 2AB + B^2$

20. $(A - B)^2 = A^2 - 2AB + B^2$

21. $A^2 - B^2 = (A - B)(A + B)$

22. $A^3 + B^3 = (A + B)(A^2 - AB + B^2)$

23. Suppose that A has four rows and B has five columns. If AB is defined, what are its dimensions?

24. Suppose that A has four rows and B has five columns. If BA is defined, what are its dimensions?

In Exercises 25–28,

$$A = \begin{bmatrix} 1 & -2 & -1 & 3 \\ -2 & 0 & 1 & 4 \\ -1 & 2 & -2 & 0 \\ 0 & 1 & 2 & 1 \end{bmatrix}, \quad B = \begin{bmatrix} 2 & 0 & -1 & 1 \\ -3 & 1 & 2 & 1 \\ 0 & -1 & -2 & 3 \\ 2 & 2 & -1 & -2 \end{bmatrix}$$

25. Partition A and B into four 2×2 blocks, and then use them to compute each of the following:

(a) $A - B$

(b) AB

(c) BA

26. Partition A and B into four blocks, with the upper left of each a 3×3 matrix, and then use them to compute each of the following:

(a) $A + B$

(b) AB

(c) BA

27. Partition A and B into four blocks, with the lower left of each a 3×3 matrix, and then use them to compute each of the following:

(a) $B - A$

(b) AB

(c) $BA + A$

28. Partition A and B into four blocks, with the lower right of each a 3×3 matrix, and then use them to compute each of the following:

(a) $A + B$

(b) AB

(c) BA

29. Suppose that A is a 3×3 matrix. Find a 3×3 matrix E such that the product EA is equal to A with

(a) the first and second rows interchanged.

(b) the first and third rows interchanged.

(c) the second row multiplied by -2.

30. Suppose that A is a 4×3 matrix. Find a 4×4 matrix E such that the product EA is equal to A with

(a) the first and fourth rows interchanged.

(b) the second and third rows interchanged.

(c) the third row multiplied by -2.

FIND AN EXAMPLE For Exercises 31–38, find an example that meets the given specifications.

31. 3×3 matrices A and B such that $AB \neq BA$.

32. 3×3 matrices A and B such that $AB = BA$.

33. 2×2 nonzero matrices A and B (other than those given earlier) such that $AB = 0_{22}$.

34. 3×3 nonzero matrices A and B such that $AB = 0_{33}$.

35. 2×2 matrices A and B (other than those given earlier) that have *no* zero entries and yet $AB = 0_{22}$.

36. 3×3 matrices A and B that have *no* zero entries and yet $AB = 0_{33}$.

37. 2×2 matrices A, B, and C (other than those given earlier) that are nonzero, where $A \neq B$ but $AC = BC$.

38. 3×3 matrices A, B, and C that are nonzero, where $A \neq B$ but $AC = BC$.

TRUE OR FALSE For Exercises 39–48, determine if the statement is true or false, and justify your answer. You may assume that A, B, and C are $n \times n$ matrices.

39. If A and B are nonzero (that is, not equal to 0_{nn}), then so is $A + B$.

40. If A and B are diagonal matrices, then so is $A - B$.

41. If A is upper triangular, then A^T is lower triangular.

42. $AB \neq BA$

43. $C + I_n = C$

44. If A is symmetric, then so is $A + I_n$.

45. $(ABC)^T = C^T B^T A^T$

46. If $AB = BA$, then either $A = I_n$ or $B = I_n$.

47. $(AB + C)^T = C^T + B^T A^T$

48. $(AB)^2 = A^2 B^2$

49. Prove the remaining unproven parts of Theorem 3.11.

(a) $A + B = B + A$

(b) $s(A + B) = sA + sB$

(c) $(s + t)A = sA + tA$

(d) $(A + B) + C = A + (B + C)$

(f) $A + 0_{nm} = A$

50. Prove the remaining unproven parts of Theorem 3.13.

(a) $A(BC) = (AB)C$

(b) $A(B + C) = AB + AC$

(d) $s(AB) = (sA)B = A(sB)$

(f) $IA = A$

51. Prove the remaining unproven parts of Theorem 3.15.

(a) $(A + B)^T = A^T + B^T$

(b) $(sA)^T = sA^T$

52. Verify Equation (2): If A is an $n \times m$ matrix and I_n is the $n \times n$ identity matrix, then $A = I_n A$.

53. Show that if A and B are symmetric matrices and $AB = BA$, then AB is also a symmetric matrix.

54. Let A and D be $n \times n$ matrices, and suppose that the only nonzero terms of D are along the diagonal. Must $AD = DA$? If so, prove it. If not, give a counter-example.

55. Let A be an $n \times m$ matrix.

(a) What are the dimensions of $A^T A$?

(b) Show that $A^T A$ is symmetric.

56. Suppose that A and B are both $n \times n$ diagonal matrices. Prove that AB is also an $n \times n$ diagonal matrix. (HINT: The formula given in (1) can be helpful here.)

57. Suppose that A and B are both $n \times n$ upper triangular matrices. Prove that AB is also an $n \times n$ upper triangular matrix. (HINT: The formula given in (1) can be helpful here.)

58. Suppose that A and B are both $n \times n$ lower triangular matrices. Prove that AB is also an $n \times n$ lower triangular matrix. (HINT: The formula given in (1) can be helpful here.)

59. Prove Theorem 3.17: If A is an upper (lower) triangular matrix and $k \geq 1$ is an integer, then A^k is also an upper (lower) triangular matrix.

60. If A is a square matrix, show that $A + A^T$ is symmetric.

61. A square matrix A is **skew symmetric** if $A^T = -A$.

(a) Find a 3×3 skew symmetric matrix.

(b) Show that the same numbers must be on the diagonal of all skew symmetric matrices.

62. A square matrix A is **idempotent** if $A^2 = A$.

(a) Find a 2×2 matrix, not equal to 0_{22} or I, that is idempotent.

(b) Show that if A is idempotent, then so is $I - A$.

63. If A is a square matrix, show that $(A^T)^T = A$.

64. The **trace** of a square matrix A is the sum of the diagonal terms of A and is denoted by tr(A).

(a) Find a 3×3 matrix A with nonzero entries such that tr(A) = 0.

(b) If A and B are both $n \times n$ matrices, show that tr($A + B$) = tr(A) + tr(A).

(c) Show that tr(A) = tr(A^T).

(d) Select two nonzero 2×2 matrices A and B of your choosing, and check if tr(AB) = tr(A)tr(B).

65. ⓒ In Example 7, suppose that the current distribution is 8000 homes with cable, 1500 homes with satellite, and 500 homes with no TV. Find the distribution one year, two years, three years, and four years from now.

66. ⓒ In Example 7, suppose that the current distribution is 5000 homes with cable, 3000 homes with satellite, and 2000 homes with no TV. Find the distribution one year, two years, three years, and four years from now.

67. ⓒ In an office complex of 1000 employees, on any given day some are at work and the rest are absent. It is known that if an employee is at work today, there is an 85% chance that she will be at work tomorrow, and if the employee is absent today, there is a 60% chance that she will be absent tomorrow. Suppose that today there are 760 employees at work. Predict the number that will be at work tomorrow, the following day, and the day after that.

68. ⓒ The star quarterback of a university football team has decided to return for one more season. He tells one person, who in turn tells someone else, and so on, with each person talking to someone who has not heard the news. At each step in this chain, if the message heard is "yes" (he is returning) then there is a 10% chance it will be changed to "no," and if the message heard is "no," then there is a 15% chance that it will be changed to "yes." Determine the probability that the fourth person in the chain hears the correct news.

ⓒ In Exercises 69–74, perform the indicated computations when possible, using the matrices given below. If a computation is not possible, explain why.

$$A = \begin{bmatrix} 2 & -1 & 0 & 4 \\ 0 & 3 & 3 & -1 \\ 6 & 8 & 1 & 1 \\ 5 & -3 & 1 & -2 \end{bmatrix}, \quad B = \begin{bmatrix} -6 & 2 & -3 & 1 \\ -5 & 2 & 0 & 3 \\ 0 & 3 & -1 & 4 \\ 8 & 5 & -2 & 0 \end{bmatrix}$$

$$C = \begin{bmatrix} 2 & 0 & 1 & 1 & 1 \\ 5 & 1 & 2 & 4 & 3 \\ 6 & 2 & 4 & 0 & 8 \\ 7 & 3 & 3 & 3 & 2 \end{bmatrix}, \quad D = \begin{bmatrix} 5 & 2 & 0 & 0 \\ 2 & 5 & 1 & 3 \\ 0 & 7 & 1 & 4 \\ 3 & 6 & 9 & 2 \\ 1 & 4 & 7 & 1 \end{bmatrix}$$

69. (a) $A + B$, (b) $BA - I_4$, (c) $D + C$

70. (a) AC, (b) $C^T - D^T$, (c) $CB + I_2$

71. (a) AB, (b) CD, (c) $(A - B)C^T$

72. (a) B^4, (b) BC^T, (c) $D + I_4$

73. (a) $(C + A)B$, (b) $C(C^T + D)$, (c) $A + CD$

74. (a) $AB - D^T$, (b) $AB - DC$, (c) $D + CB$

3.3 Inverses

In Section 3.1, we defined the linear transformation and developed the properties of this type of function. In this section we consider the problem of "reversing" a linear transformation. An application of this can be found in encoding messages so that they cannot be read by anyone besides the intended recipient. The history of secret codes is long, going back at least as far as Julius Caesar. Here we give a brief description of an encoding method that uses linear transformations.

▶ We could also have numerical equivalencies for spaces, punctuation, and uppercase letters, but these are not needed here.

We start by establishing numerical equivalencies between letters and numbers,

$$a = 1, \quad b = 2, \quad c = 3, \quad \ldots \quad z = 26 \tag{1}$$

One way to encode messages is to map each number $1, 2, \ldots, 26$ to some other number, such as

$$1 \to 4, \quad 2 \to 5, \quad 3 \to 6, \quad \ldots \quad 26 \to 29$$

▶ This particular shift cipher is called the *Caesar cipher* because it is said to have been invented by Julius Caesar.

This is called a *shift cipher* and gives encodings such as

$$\text{``linear''} = \underbrace{\{12, 9, 14, 5, 1, 18\}}_{\text{numerical equiv.}} \longrightarrow \underbrace{\{15, 12, 17, 8, 4, 21\}}_{\text{encoded message}}$$

We could convert the encoded message back to letters before transmitting the message, but that is not necessary — the string of encoded numbers can be sent. This type of encoding system is easy to implement but not very secure. In particular, it is vulnerable to *frequency analysis*, which involves breaking the code by matching up the numbers that occur most frequently in the encoded message with the letters that occur most frequently in the language of the original message (Table 1).

One way to deter frequency analysis is by encoding letters in groups called *blocks*. Although there are only 26 letters in English, there are $26^3 = 17{,}576$ possible blocks of three letters, making frequency analysis more difficult. We start by converting each letter of a block using the equivalences in (1) and placing them in a vector in $\mathbf{R}^3$. For example, we would have

$$\text{``the''} = \begin{bmatrix} 20 \\ 8 \\ 5 \end{bmatrix} \quad \text{and} \quad \text{``dog''} = \begin{bmatrix} 4 \\ 15 \\ 7 \end{bmatrix}$$

Letter	Frequency
e	12.70%
t	9.06%
a	8.17%
o	7.51%
i	6.97%
n	6.75%
s	6.33%
h	6.09%
r	5.99%
d	4.53%

Table 1 Relative Frequency of Letters in English

We then apply a linear transformation to encode the vector. For instance, we could use $T(\mathbf{x}) = A\mathbf{x}$, where

$$A = \begin{bmatrix} 1 & 3 & 2 \\ -2 & -7 & 2 \\ 2 & 6 & 5 \end{bmatrix}$$

For example, we encode "linear" by splitting it into blocks and then encoding as shown.

$$\text{``linear''} = \left\{ \underbrace{\begin{bmatrix} 12 \\ 9 \\ 14 \end{bmatrix}}_{\text{``lin''}}, \underbrace{\begin{bmatrix} 5 \\ 1 \\ 18 \end{bmatrix}}_{\text{``ear''}} \right\} \longrightarrow \left\{ T\left(\begin{bmatrix} 12 \\ 9 \\ 14 \end{bmatrix} \right), T\left(\begin{bmatrix} 5 \\ 1 \\ 18 \end{bmatrix} \right) \right\} = \underbrace{\left\{ \begin{bmatrix} 67 \\ -59 \\ 148 \end{bmatrix}, \begin{bmatrix} 44 \\ 19 \\ 106 \end{bmatrix} \right\}}_{\text{encoded message}}$$

Of course, it does no good to encode a message if it cannot be decoded. To see how we decode, suppose that we have received the encoded block

$$\mathbf{y} = \begin{bmatrix} 109 \\ -95 \\ 241 \end{bmatrix}$$

To decode, we need to find the corresponding vector $\mathbf{x}$ such that $T(\mathbf{x}) = \mathbf{y}$, or equivalently, $A\mathbf{x} = \mathbf{y}$. Expressed as a linear system, we have

$$\begin{aligned} x_1 + 3x_2 + 2x_3 &= 109 \\ -2x_1 - 7x_2 + 2x_3 &= -95 \\ 2x_1 + 6x_2 + 5x_3 &= 241 \end{aligned} \tag{2}$$

Our usual solution methods can be used to show that $x_1 = 18$, $x_2 = 15$, and $x_3 = 23$, so that

$$\mathbf{x} = \begin{bmatrix} 18 \\ 15 \\ 23 \end{bmatrix} = \text{"row"}$$

In solving for x_1, x_2, and x_3, we have computed the *inverse* of T for the vector $\mathbf{y}$. The notation for the inverse function is T^{-1}, and we have the relationship

$$T(\mathbf{x}) = \mathbf{y} \iff T^{-1}(\mathbf{y}) = \mathbf{x}$$

A typical encoded message will consist of many blocks. Instead of repeatedly solving (2) with only changes to the right-hand numbers, it is more efficient to find a formula for $T^{-1}(\mathbf{y})$ for a generic vector $\mathbf{y} = (y_1, y_2, y_3)$. To find T^{-1}, we need to solve

$$\begin{aligned} x_1 + 3x_2 + 2x_3 &= y_1 \\ -2x_1 - 7x_2 + 2x_3 &= y_2 \\ 2x_1 + 6x_2 + 5x_3 &= y_3 \end{aligned} \qquad (3)$$

for x_1, x_2, and x_3 in terms of y_1, y_2, and y_3. Transferring (3) to an augmented matrix and transforming to reduced echelon form yields the solution

$$\begin{aligned} x_1 &= 47y_1 + 3y_2 - 20y_3 \\ x_2 &= -14y_1 - y_2 + 6y_3 \\ x_3 &= -2y_1 + y_3 \end{aligned}$$

Note that if we set

$$B = \begin{bmatrix} 47 & 3 & -20 \\ -14 & -1 & 6 \\ -2 & 0 & 1 \end{bmatrix}$$

then we have $T^{-1}(\mathbf{y}) = B\mathbf{y}$. Therefore $T(\mathbf{x})$ and $T^{-1}(\mathbf{y})$ are both linear transformations. As a quick application of $T^{-1}(\mathbf{y}) = B\mathbf{y}$, we have

$$\underbrace{\left\{ \begin{bmatrix} 85 \\ -132 \\ 179 \end{bmatrix}, \begin{bmatrix} 74 \\ -4 \\ 173 \end{bmatrix} \right\}}_{\text{encoded message}} \longrightarrow \left\{ T^{-1}\left(\begin{bmatrix} 85 \\ -132 \\ 179 \end{bmatrix} \right), T^{-1}\left(\begin{bmatrix} 74 \\ -4 \\ 173 \end{bmatrix} \right) \right\} = \left\{ \underbrace{\begin{bmatrix} 19 \\ 16 \\ 9 \end{bmatrix}}_{\text{"spi"}}, \underbrace{\begin{bmatrix} 6 \\ 6 \\ 25 \end{bmatrix}}_{\text{"ffy"}} \right\} = \text{"spiffy"}$$

Larger blocks and corresponding encoding matrices can be used if needed. More can be learned about encoding messages in *Cryptological Mathematics* by Robert Lewand (Mathematical Association of America Textbooks).

Inverse Linear Transformations

A linear transformation $T : \mathbf{R}^m \to \mathbf{R}^n$ that is one-to-one and onto pairs up each vector $\mathbf{x}$ in $\mathbf{R}^m$ with a unique vector $\mathbf{y} = T(\mathbf{x})$ in $\mathbf{R}^n$. The *inverse* T^{-1} creates the same pairs but in reverse, so that $T^{-1}(\mathbf{y}) = \mathbf{x}$ (see Figure 1). Hence we can think of T^{-1} as reversing T. Here we put the notion of inverse on a firmer footing with a definition of an inverse linear transformation.

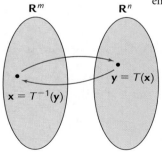

Figure 1 The relationship between T and T^{-1}.

DEFINITION 3.18

Definition Inverse, Invertible

A linear transformation $T : \mathbf{R}^m \to \mathbf{R}^n$ is **invertible** if T is one-to-one and onto. When T is invertible, the **inverse** function $T^{-1} : \mathbf{R}^n \to \mathbf{R}^m$ is defined by

$$T^{-1}(\mathbf{y}) = \mathbf{x} \quad \text{if and only if} \quad T(\mathbf{x}) = \mathbf{y}$$

If T is an invertible linear transformation, then the inverse T^{-1} is unique and satisfies

$$T\big(T^{-1}(\mathbf{y})\big) = \mathbf{y} \quad \text{and} \quad T^{-1}\big(T(\mathbf{x})\big) = \mathbf{x}$$

In our secret-code example, the encoding linear transformation mapped vectors in $\mathbf{R}^3$ to other vectors in $\mathbf{R}^3$. We also observed that the inverse function for decoding was another linear transformation. The next theorem gives a required condition for a linear transformation to be invertible and tells us that the inverse of a linear transformation is also a linear transformation.

THEOREM 3.19

Let $T : \mathbf{R}^m \to \mathbf{R}^n$ be a linear transformation. Then

(a) T has an inverse only if $m = n$.

(b) If T is invertible, then T^{-1} is also a linear transformation.

Proof For (a), note that Theorem 3.6 in Section 3.1 tells us that the only way T can be one-to-one is if $n \geq m$. Moreover, Theorem 3.7 in Section 3.1 tells us the only way that T can be onto is if $n \leq m$. Thus the only way that T can be one-to-one and onto, and hence invertible, is if $m = n$.

For (b), let $\mathbf{y}_1$ and $\mathbf{y}_2$ be vectors in $\mathbf{R}^n$. If T is invertible, then T is onto, so there exist vectors $\mathbf{x}_1$ and $\mathbf{x}_2$ in $\mathbf{R}^n$ such that $T(\mathbf{x}_1) = \mathbf{y}_1$ and $T(\mathbf{x}_2) = \mathbf{y}_2$. Hence

$$\begin{aligned} T^{-1}(\mathbf{y}_1 + \mathbf{y}_2) &= T^{-1}\big(T(\mathbf{x}_1) + T(\mathbf{x}_2)\big) \\ &= T^{-1}\big(T(\mathbf{x}_1 + \mathbf{x}_2)\big) \qquad (T \text{ is a linear transformation}) \\ &= \mathbf{x}_1 + \mathbf{x}_2 \\ &= T^{-1}(\mathbf{y}_1) + T^{-1}(\mathbf{y}_2) \end{aligned}$$

A similar argument can be used to show that $T^{-1}(r\mathbf{y}) = r\,T^{-1}(\mathbf{y})$ (see Exercise 69). Therefore T^{-1} is a linear transformation. ∎

Note that having $m = n$ does not guarantee that T will have an inverse. For instance, if

$$A = \begin{bmatrix} 1 & 2 \\ 3 & 6 \end{bmatrix}$$

and $T(\mathbf{x}) = A\mathbf{x}$, then $T : \mathbf{R}^2 \to \mathbf{R}^2$ but is not invertible, because T is neither one-to-one nor onto.

If $T : \mathbf{R}^n \to \mathbf{R}^n$ is an invertible linear transformation, then $T^{-1} : \mathbf{R}^n \to \mathbf{R}^n$ is also a linear transformation. Hence there exist $n \times n$ matrices A and B such that $T(\mathbf{x}) = A\mathbf{x}$ and $T^{-1}(\mathbf{x}) = B\mathbf{x}$. Furthermore, for each $\mathbf{x}$ in $\mathbf{R}^n$ we have

$$\mathbf{x} = T\big(T^{-1}(\mathbf{x})\big) = T\big(B\mathbf{x}\big) = AB\mathbf{x}$$

Since $AB\mathbf{x} = \mathbf{x}$ for all $\mathbf{x}$ in $\mathbf{R}^n$, it follows that $AB = I_n$. We use this relationship to characterize what it means for a matrix to be invertible.

DEFINITION 3.20

Definition Invertible Matrix

An $n \times n$ matrix A is **invertible** if there exists an $n \times n$ matrix B such that $AB = I_n$.

A matrix A is invertible precisely when the associated linear transformation $T(\mathbf{x}) = A\mathbf{x}$ is invertible.

EXAMPLE 1   Let

$$A = \begin{bmatrix} 1 & 2 & -1 \\ 2 & 5 & -1 \\ 1 & 2 & 0 \end{bmatrix} \quad \text{and} \quad B = \begin{bmatrix} 2 & -2 & 3 \\ -1 & 1 & -1 \\ -1 & 0 & 1 \end{bmatrix}$$

Prove that A is invertible by showing $AB = I_3$.

Solution   We have

$$AB = \begin{bmatrix} 1 & 2 & -1 \\ 2 & 5 & -1 \\ 1 & 2 & 0 \end{bmatrix} \begin{bmatrix} 2 & -2 & 3 \\ -1 & 1 & -1 \\ -1 & 0 & 1 \end{bmatrix} = \begin{bmatrix} 1 & 0 & 0 \\ 0 & 1 & 0 \\ 0 & 0 & 1 \end{bmatrix} = I_3$$

so A is invertible.   ■

In Section 3.2 we showed that sometimes $AB \neq BA$ because, in general, matrix multiplication is not commutative. However, somewhat surprisingly, if $AB = I_n$, then $BA = AB = I_n$.

THEOREM 3.21

Suppose that A is an invertible matrix with $AB = I_n$. Then $BA = I_n$, and the matrix B such that $AB = BA = I_n$ is unique.

Proof   Let $\mathbf{x}$ be in $\mathbf{R}^n$. Since $AB = I_n$, we have

$$AB(A\mathbf{x}) = I_n(A\mathbf{x}) = A\mathbf{x} \quad \Longrightarrow \quad A(BA\mathbf{x}) = A\mathbf{x} \quad \Longrightarrow \quad A(BA\mathbf{x} - \mathbf{x}) = \mathbf{0}$$

Since A is invertible, $A\mathbf{y} = \mathbf{0}$ has only the trivial solution because the corresponding linear transformation is one-to-one. Hence $A(BA\mathbf{x} - \mathbf{x}) = \mathbf{0}$ implies that

$$BA\mathbf{x} - \mathbf{x} = \mathbf{0} \quad \Longrightarrow \quad BA\mathbf{x} = \mathbf{x}$$

Since $BA\mathbf{x} = \mathbf{x}$ for all $\mathbf{x}$ in $\mathbf{R}^n$, we may conclude that $BA = I_n$.

To show that B is unique, suppose that there is another $n \times n$ matrix C such that $AB = AC = I_n$. Then

$$B(AB) = B(AC) \quad \Longrightarrow \quad (BA)B = (BA)C \quad \Longrightarrow \quad B = C$$

because $BA = I_n$. Since $B = C$, it follows that B is unique.   ■

As a quick application of Theorem 3.21, for the matrices A and B in Example 1 we have

$$BA = \begin{bmatrix} 2 & -2 & 3 \\ -1 & 1 & -1 \\ -1 & 0 & 1 \end{bmatrix} \begin{bmatrix} 1 & 2 & -1 \\ 2 & 5 & -1 \\ 1 & 2 & 0 \end{bmatrix} = \begin{bmatrix} 1 & 0 & 0 \\ 0 & 1 & 0 \\ 0 & 0 & 1 \end{bmatrix} = I_3$$

Since for an invertible matrix A there is exactly one matrix B such that $AB = BA = I_n$, the next definition makes sense.

DEFINITION 3.22

Definition Inverse Matrix

If an $n \times n$ matrix A is invertible, then A^{-1} is called the **inverse** of A and denotes the unique $n \times n$ matrix such that $AA^{-1} = A^{-1}A = I_n$.

Definition Nonsingular, Singular

A square matrix A that is invertible is also called **nonsingular**. If A does not have an inverse, it is **singular**. Definition 3.22 is symmetric in that if A^{-1} is the inverse of A, then A is the inverse of A^{-1}.

Our next theorem gives several important properties of invertible matrices. Note the distinction between (c) and (d) of Theorem 3.23 and those given in Theorem 3.14 in Section 3.2.

THEOREM 3.23

Let A and B be invertible $n \times n$ matrices and C and D be $n \times m$ matrices. Then

(a) A^{-1} is invertible, with $\left(A^{-1}\right)^{-1} = A$.

(b) AB is invertible, with $(AB)^{-1} = B^{-1}A^{-1}$.

(c) If $AC = AD$ then $C = D$.

(d) If $AC = 0_{nm}$, then $C = 0_{nm}$.

Proof We prove (a) and (b) here and leave (c) and (d) as an exercise. For (a), since A is invertible we know that $A^{-1}A = I_n$. This implies that A is the inverse of A^{-1} — that is, $A = \left(A^{-1}\right)^{-1}$.

For (b), note that

$$(AB)(B^{-1}A^{-1}) = A(BB^{-1})A^{-1} = AI_nA^{-1} = AA^{-1} = I_n$$

Hence AB is invertible and $(AB)^{-1} = B^{-1}A^{-1}$. ∎

Finding A^{-1}

We now develop a method for computing A^{-1}. Let's start by supposing that A and B are $n \times n$ matrices with $AB = I_n$, where

$$A = \begin{bmatrix} \mathbf{a}_1 & \cdots & \mathbf{a}_n \end{bmatrix}, \quad B = \begin{bmatrix} \mathbf{b}_1 & \cdots & \mathbf{b}_n \end{bmatrix}, \quad \text{and} \quad I_n = \begin{bmatrix} \mathbf{e}_1 & \cdots & \mathbf{e}_n \end{bmatrix}$$

(Later we will set $B = A^{-1}$, but for now using B simplifies notation.) If $AB = I_n$, then, taking multiplication one column at a time, we have

$$A\mathbf{b}_1 = \mathbf{e}_1, \quad A\mathbf{b}_2 = \mathbf{e}_2, \quad \ldots, \quad A\mathbf{b}_n = \mathbf{e}_n$$

Thus $\mathbf{b}_1$ is a solution to the linear system $A\mathbf{x} = \mathbf{e}_1$, $\mathbf{b}_2$ is a solution to the linear system $A\mathbf{x} = \mathbf{e}_2$, and so on. We could solve these systems one at a time by transforming each of the augmented matrices

$$\begin{bmatrix} \mathbf{a}_1 & \cdots & \mathbf{a}_n & \mathbf{e}_1 \end{bmatrix}, \quad \begin{bmatrix} \mathbf{a}_1 & \cdots & \mathbf{a}_n & \mathbf{e}_2 \end{bmatrix}, \quad \ldots \quad \begin{bmatrix} \mathbf{a}_1 & \cdots & \mathbf{a}_n & \mathbf{e}_n \end{bmatrix}$$

to reduced row echelon form. However, since we do the same row operations for each matrix, we can save ourselves some work by setting up one large augmented matrix

$$\begin{bmatrix} \mathbf{a}_1 & \cdots & \mathbf{a}_n & \mathbf{e}_1 & \cdots & \mathbf{e}_n \end{bmatrix}$$

and go through the row operations once to transform the left half from A to I_n. If this can be done, then the right half will be transformed from I_n to $B = A^{-1}$. In brief, we want

$$\begin{bmatrix} A & I_n \end{bmatrix} \quad \text{transformed to} \quad \begin{bmatrix} I_n & A^{-1} \end{bmatrix}$$

Let's look at an example.

EXAMPLE 2   Find the inverse of $A = \begin{bmatrix} 1 & 3 \\ 2 & 5 \end{bmatrix}$.

Solution We begin by setting up the augmented matrix

$$\begin{bmatrix} A & I_2 \end{bmatrix} = \begin{bmatrix} 1 & 3 & 1 & 0 \\ 2 & 5 & 0 & 1 \end{bmatrix}$$

Now we use our usual row operations to transform the left half to I_2,

$$\begin{bmatrix} 1 & 3 & 1 & 0 \\ 2 & 5 & 0 & 1 \end{bmatrix} \overset{-2R_1+R_2 \Rightarrow R_2}{\sim} \begin{bmatrix} 1 & 3 & 1 & 0 \\ 0 & -1 & -2 & 1 \end{bmatrix}$$

$$\overset{-R_2 \Rightarrow R_2}{\sim} \begin{bmatrix} 1 & 3 & 1 & 0 \\ 0 & 1 & 2 & -1 \end{bmatrix}$$

$$\overset{-3R_2+R_1 \Rightarrow R_1}{\sim} \begin{bmatrix} 1 & 0 & -5 & 3 \\ 0 & 1 & 2 & -1 \end{bmatrix}$$

Thus we find that $A^{-1} = \begin{bmatrix} -5 & 3 \\ 2 & -1 \end{bmatrix}$. Let's test it out:

$$A A^{-1} = \begin{bmatrix} 1 & 3 \\ 2 & 5 \end{bmatrix} \begin{bmatrix} -5 & 3 \\ 2 & -1 \end{bmatrix} = \begin{bmatrix} 1 & 0 \\ 0 & 1 \end{bmatrix}$$

and

$$A^{-1} A = \begin{bmatrix} -5 & 3 \\ 2 & -1 \end{bmatrix} \begin{bmatrix} 1 & 3 \\ 2 & 5 \end{bmatrix} = \begin{bmatrix} 1 & 0 \\ 0 & 1 \end{bmatrix}$$

∎

For a given $n \times n$ matrix A, if A is invertible, then this procedure will find A^{-1}. If A is not invertible, then the reduced row echelon form of $\begin{bmatrix} A & I_n \end{bmatrix}$ will not have I_n on the left side.

EXAMPLE 3   Let $T(\mathbf{x}) = A\mathbf{x}$, where $A = \begin{bmatrix} 1 & -2 & 1 \\ -3 & 7 & -6 \\ 2 & -3 & 0 \end{bmatrix}$. Find T^{-1}, if it exists.

Solution T^{-1} exists if and only if A^{-1} exists. To find A^{-1}, we set up the augmented matrix $\begin{bmatrix} A & I_3 \end{bmatrix}$ and then use row operations to transform to $\begin{bmatrix} I_3 & A^{-1} \end{bmatrix}$:

$$\begin{bmatrix} 1 & -2 & 1 & 1 & 0 & 0 \\ -3 & 7 & -6 & 0 & 1 & 0 \\ 2 & -3 & 0 & 0 & 0 & 1 \end{bmatrix} \overset{\substack{3R_1+R_2 \Rightarrow R_2 \\ -2R_1+R_3 \Rightarrow R_3}}{\sim} \begin{bmatrix} 1 & -2 & 1 & 1 & 0 & 0 \\ 0 & 1 & -3 & 3 & 1 & 0 \\ 0 & 1 & -2 & -2 & 0 & 1 \end{bmatrix}$$

$$\overset{-R_2+R_3 \Rightarrow R_3}{\sim} \begin{bmatrix} 1 & -2 & 1 & 1 & 0 & 0 \\ 0 & 1 & -3 & 3 & 1 & 0 \\ 0 & 0 & 1 & -5 & -1 & 1 \end{bmatrix}$$

$$\overset{\substack{3R_3+R_2 \Rightarrow R_2 \\ -R_3+R_1 \Rightarrow R_1}}{\sim} \begin{bmatrix} 1 & -2 & 0 & 6 & 1 & -1 \\ 0 & 1 & 0 & -12 & -2 & 3 \\ 0 & 0 & 1 & -5 & -1 & 1 \end{bmatrix}$$

$$\overset{2R_2+R_1 \Rightarrow R_1}{\sim} \begin{bmatrix} 1 & 0 & 0 & -18 & -3 & 5 \\ 0 & 1 & 0 & -12 & -2 & 3 \\ 0 & 0 & 1 & -5 & -1 & 1 \end{bmatrix}$$

Thus we have $T^{-1}(\mathbf{x}) = A^{-1}\mathbf{x}$, where

$$A^{-1} = \begin{bmatrix} -18 & -3 & 5 \\ -12 & -2 & 3 \\ -5 & -1 & 1 \end{bmatrix}$$

The next example shows what happens when there is no inverse.

EXAMPLE 4 Find the inverse, if it exists, of $A = \begin{bmatrix} 1 & 1 & -2 \\ 2 & 1 & -3 \\ -3 & -1 & 4 \end{bmatrix}$.

Solution We set up the augmented matrix $\begin{bmatrix} A & I_3 \end{bmatrix}$, and using row operations we get

$$\begin{bmatrix} 1 & 1 & -2 & 1 & 0 & 0 \\ 2 & 1 & -3 & 0 & 1 & 0 \\ -3 & -1 & 4 & 0 & 0 & 1 \end{bmatrix} \overset{-2R_1+R_2\Rightarrow R_2}{\underset{3R_1+R_3\Rightarrow R_3}{\sim}} \begin{bmatrix} 1 & 1 & -2 & 1 & 0 & 0 \\ 0 & -1 & 1 & -2 & 1 & 0 \\ 0 & 2 & -2 & 3 & 0 & 1 \end{bmatrix}$$

$$\overset{2R_2+R_3\Rightarrow R_3}{\sim} \begin{bmatrix} 1 & 1 & -2 & 1 & 0 & 0 \\ 0 & -1 & 1 & -2 & 1 & 0 \\ 0 & 0 & 0 & -1 & 2 & 1 \end{bmatrix}$$

We can stop right there. The left half of the bottom row consists entirely of zeros, so there is no way to use row operations to transform the left half of the augmented matrix to I_3. Thus A has no inverse. ∎

The next theorem highlights how invertibility is related to the solutions of linear systems.

THEOREM 3.24 Let A be an $n \times n$ matrix. Then the following are equivalent:

(a) A is invertible.

(b) $A\mathbf{x} = \mathbf{b}$ has a unique solution for all $\mathbf{b}$, given by $\mathbf{x} = A^{-1}\mathbf{b}$.

(c) $A\mathbf{x} = \mathbf{0}$ has only the trivial solution.

Proof A is invertible if and only if $T(\mathbf{x}) = A\mathbf{x}$ is an invertible linear transformation, which in turn is true if and only if T is one-to-one and onto. This implies that (a) and (b) are equivalent. By setting $\mathbf{b} = \mathbf{0}$, we see that (b) implies (c), so all that remains to complete the proof is to show that (c) implies (a).

If $A\mathbf{x} = \mathbf{0}$ has only the trivial solution, then by Theorem 3.5 we know that T is one-to-one. Therefore, since A is $n \times n$, by Theorem 3.9 in Section 3.1, T also must be onto. Since T is one-to-one and onto, we can conclude that T is invertible, and so A is invertible. Therefore (c) implies (a).

Finally, we note that if A is invertible, then we have

$$A\mathbf{x} = \mathbf{b} \implies A^{-1}A\mathbf{x} = A^{-1}\mathbf{b} \implies I_n\mathbf{x} = A^{-1}\mathbf{b} \implies \mathbf{x} = A^{-1}\mathbf{b}$$

∎

EXAMPLE 5 Find the unique solution to the linear system

$$x_1 + 3x_2 = 4$$
$$2x_1 + 5x_2 = -3$$

Solution Start by setting

$$A = \begin{bmatrix} 1 & 3 \\ 2 & 5 \end{bmatrix} \quad \text{and} \quad \mathbf{b} = \begin{bmatrix} 4 \\ -3 \end{bmatrix}$$

Then our system is equivalent to $A\mathbf{x} = \mathbf{b}$, so that by Theorem 3.24, if A is invertible, then the solution is given by $\mathbf{x} = A^{-1}\mathbf{b}$. Happily, in Example 2 we found A^{-1}, which we use here to give us

$$\mathbf{x} = A^{-1}\mathbf{b} = \begin{bmatrix} -5 & 3 \\ 2 & -1 \end{bmatrix} \begin{bmatrix} 4 \\ -3 \end{bmatrix} = \begin{bmatrix} -29 \\ 11 \end{bmatrix}$$ ∎

A Quick Formula

Typically, we use the row reduction method to compute inverses for matrices. However, in the case of a 2×2 matrix

$$A = \begin{bmatrix} a & b \\ c & d \end{bmatrix}$$

there exists a quick formula. It can be shown that A has an inverse exactly when $ad - bc \neq 0$, and that the inverse is

$$A^{-1} = \frac{1}{ad - bc} \begin{bmatrix} d & -b \\ -c & a \end{bmatrix}$$

Instead of using an augmented matrix and row operations to verify the formula, we check it by multiplying,

$$A^{-1} A = \frac{1}{ad - bc} \begin{bmatrix} d & -b \\ -c & a \end{bmatrix} \begin{bmatrix} a & b \\ c & d \end{bmatrix} = \frac{1}{ad - bc} \begin{bmatrix} ad - bc & 0 \\ 0 & ad - bc \end{bmatrix} = \begin{bmatrix} 1 & 0 \\ 0 & 1 \end{bmatrix} = I_2$$

Since $A^{-1} A = I_2$ and the inverse is unique, it must be that the formula for A^{-1} is correct.

EXAMPLE 6   Use the Quick Formula to find A^{-1} for

$$A = \begin{bmatrix} 2 & 7 \\ 1 & 5 \end{bmatrix}$$

Solution From the Quick Formula, we have

$$A^{-1} = \frac{1}{(2)(5) - (7)(1)} \begin{bmatrix} 5 & -7 \\ -1 & 2 \end{bmatrix} = \begin{bmatrix} 5/3 & -7/3 \\ -1/3 & 2/3 \end{bmatrix}$$ ∎

The Big Theorem, Version 3

▶ This updates the Big Theorem, Version 2, from Section 3.1.

Summarizing the results we have proved about invertible matrices, we add one more important condition to the Big Theorem.

THEOREM 3.25

(**THE BIG THEOREM, VERSION 3**)   Let $\mathcal{A} = \{\mathbf{a}_1, \ldots, \mathbf{a}_n\}$ be a set of n vectors in $\mathbf{R}^n$, let $A = \begin{bmatrix} \mathbf{a}_1 & \cdots & \mathbf{a}_n \end{bmatrix}$, and let $T : \mathbf{R}^n \to \mathbf{R}^n$ be given by $T(\mathbf{x}) = A\mathbf{x}$. Then the following are equivalent:

(a) $\mathcal{A}$ spans $\mathbf{R}^n$.

(b) $\mathcal{A}$ is linearly independent.

(c) $A\mathbf{x} = \mathbf{b}$ has a unique solution for all $\mathbf{b}$ in $\mathbf{R}^n$.

(d) T is onto.

(e) T is one-to-one.

(f) A is invertible.

Proof From the Big Theorem, Version 2, we know that (a) through (e) are equivalent. Moreover, from Theorem 3.24, we know that (c) and (f) are equivalent. Thus we can conclude that all six conditions are equivalent. ∎

► The material on partitioned matrices is optional.

Partitioned Matrices

For matrices in certain special forms, it is possible to use matrix partitions to efficiently compute the inverse. For instance, suppose that

$$A = \begin{bmatrix} A_{11} & 0_{12} \\ 0_{21} & A_{22} \end{bmatrix}$$

where A_{11} is $n_1 \times n_1$, A_{22} is $n_2 \times n_2$, both A_{11} and A_{22} are invertible, and 0_{12} and 0_{21} represent matrices with all zero entries of dimension $n_1 \times n_2$ and $n_2 \times n_1$, respectively. A is an example of a **block diagonal** matrix.

To find the inverse of A, let

$$B = \begin{bmatrix} B_{11} & B_{12} \\ B_{21} & B_{22} \end{bmatrix}$$

where the blocks of B have the appropriate dimensions so that AB is defined and can be computed using block multiplication. Now we assume that $AB = I$, and from this we determine the form required of B that will give us a formula for A^{-1}.

To start, write

$$I = \begin{bmatrix} I_{n_1} & 0_{12} \\ 0_{21} & I_{n_2} \end{bmatrix}$$

where I_{n_1} and I_{n_2} are the $n_1 \times n_1$ and $n_2 \times n_2$ identity matrices, respectively, and 0_{12} and 0_{21} are defined as above. If $AB = I$, then from our block multiplication formulas we must have

$$A_{11} B_{11} + 0_{12} B_{21} = I_{n_1}$$
$$A_{11} B_{12} + 0_{12} B_{22} = 0_{12}$$
$$0_{21} B_{11} + A_{22} B_{21} = 0_{21}$$
$$0_{21} B_{12} + A_{22} B_{22} = I_{n_2}$$

Focusing on the first and fourth equations, we see that these imply that $B_{11} = A_{11}^{-1}$ and $B_{22} = A_{22}^{-1}$. Next, the second equation reduces to $A_{11} B_{12} = 0_{12}$, and since A_{11} is invertible it follows that $B_{12} = 0_{12}$. A similar argument can be used to show that $B_{21} = 0_{21}$, and so we can conclude that

$$A^{-1} = \begin{bmatrix} A_{11}^{-1} & 0_{12} \\ 0_{21} & A_{22}^{-1} \end{bmatrix} \tag{4}$$

We can stretch this method further to find an inverse if A has the form

$$A = \begin{bmatrix} A_{11} & 0_{12} \\ A_{21} & A_{22} \end{bmatrix}$$

which is a **block lower triangular** matrix. Defining B as we did previously (with the assumptions required for A so that the inverse exists) and computing the product AB

using block multiplication, we arrive at the equations

$$A_{11}B_{11} + 0_{12}B_{21} = I_{n_1}$$
$$A_{11}B_{12} + 0_{12}B_{22} = 0_{12}$$
$$A_{21}B_{11} + A_{22}B_{21} = 0_{21}$$
$$A_{21}B_{12} + A_{22}B_{22} = I_{n_2}$$

Using the same line of reasoning as above, we find that $B_{11} = A_{11}^{-1}$, $B_{12} = 0_{12}$, $B_{21} = -A_{22}^{-1}A_{21}A_{11}^{-1}$, and $B_{22} = A_{22}^{-1}$. That is,

$$A^{-1} = \begin{bmatrix} A_{11}^{-1} & 0_{12} \\ -A_{22}^{-1}A_{21}A_{11}^{-1} & A_{22}^{-1} \end{bmatrix} \tag{5}$$

EXAMPLE 7   Find the inverse for $A = \begin{bmatrix} 2 & 3 & 0 & 0 & 0 \\ 3 & 4 & 0 & 0 & 0 \\ 1 & -1 & -2 & 0 & 0 \\ 2 & -4 & 0 & 9 & 4 \\ 0 & 3 & 0 & 7 & 3 \end{bmatrix}$.

Solution   All of the zeros in the upper right of A suggest that we partition A into the blocks

$$A_{11} = \begin{bmatrix} 2 & 3 \\ 3 & 4 \end{bmatrix} \qquad A_{12} = \begin{bmatrix} 0 & 0 & 0 \\ 0 & 0 & 0 \end{bmatrix}$$

$$A_{21} = \begin{bmatrix} 1 & -1 \\ 2 & -4 \\ 0 & 3 \end{bmatrix} \qquad A_{22} = \begin{bmatrix} -2 & 0 & 0 \\ 0 & 9 & 4 \\ 0 & 7 & 3 \end{bmatrix}$$

Since A is block lower triangular, we can apply the formula given in (5). To do so, we need A_{11}^{-1} and A_{22}^{-1}. As A_{11} is 2×2, we can apply our quick inverse formula to find

$$A_{11}^{-1} = \frac{1}{8-9} \begin{bmatrix} 4 & -3 \\ -3 & 2 \end{bmatrix} = \begin{bmatrix} -4 & 3 \\ 3 & -2 \end{bmatrix}$$

The block A_{22} is itself block diagonal, so we can apply 4 to find A_{22}^{-1}. The upper-left entry has inverse $-1/2$, and the quick inverse formula can be used to find the inverse for the lower-right sub-block. Combining these gives us

$$A_{22}^{-1} = \begin{bmatrix} -1/2 & 0 & 0 \\ 0 & -3 & 4 \\ 0 & 7 & -9 \end{bmatrix}$$

Finally, we compute

$$-A_{22}^{-1}A_{21}A_{11}^{-1} = -\begin{bmatrix} -1/2 & 0 & 0 \\ 0 & -3 & 4 \\ 0 & 7 & -9 \end{bmatrix}\begin{bmatrix} 1 & -1 \\ 2 & -4 \\ 0 & 3 \end{bmatrix}\begin{bmatrix} -4 & 3 \\ 3 & -2 \end{bmatrix} = \begin{bmatrix} -7/2 & 5/2 \\ -96 & 66 \\ 221 & -152 \end{bmatrix}$$

Combining all these ingredients together as specified in 5, we arrive at

$$A^{-1} = \begin{bmatrix} -4 & 3 & 0 & 0 & 0 \\ 3 & -2 & 0 & 0 & 0 \\ -7/2 & 5/2 & -1/2 & 0 & 0 \\ -96 & 66 & 0 & -3 & 4 \\ 221 & -152 & 0 & 7 & -9 \end{bmatrix}$$

■

The formulas illustrated here can readily be extended to other, more complex block matrices.

| EXERCISES |

In Exercises 1–4, use the Quick Formula to find the inverse of the given matrix, if it exists.

1. $\begin{bmatrix} 7 & 3 \\ 2 & 1 \end{bmatrix}$

2. $\begin{bmatrix} 5 & -2 \\ -4 & 3 \end{bmatrix}$

3. $\begin{bmatrix} 2 & -5 \\ -4 & 10 \end{bmatrix}$

4. $\begin{bmatrix} -6 & 2 \\ 5 & -1 \end{bmatrix}$

In Exercises 5–16, use an augmented matrix and row operations to find the inverse of the given matrix, if it exists.

5. $\begin{bmatrix} 1 & 4 \\ 2 & 9 \end{bmatrix}$

6. $\begin{bmatrix} 4 & 13 \\ 1 & 3 \end{bmatrix}$

7. $\begin{bmatrix} 1 & 0 & 1 \\ 0 & 1 & 0 \\ 1 & 1 & 1 \end{bmatrix}$

8. $\begin{bmatrix} 0 & 1 & 1 \\ 0 & 1 & 0 \\ 1 & 0 & 1 \end{bmatrix}$

9. $\begin{bmatrix} 1 & 2 & -1 \\ 0 & 1 & 3 \\ 0 & 0 & 1 \end{bmatrix}$

10. $\begin{bmatrix} 1 & 2 & -1 \\ -4 & -7 & 7 \\ -1 & -1 & 5 \end{bmatrix}$

11. $\begin{bmatrix} 1 & -3 & 1 \\ 2 & -5 & 4 \\ -2 & 3 & -8 \end{bmatrix}$

12. $\begin{bmatrix} 3 & -1 & 9 \\ 1 & -1 & 4 \\ 2 & -2 & 10 \end{bmatrix}$

13. $\begin{bmatrix} 0 & 0 & 1 & 0 \\ 1 & 0 & 0 & 0 \\ 0 & 0 & 0 & 1 \\ 0 & 1 & 0 & 0 \end{bmatrix}$

14. $\begin{bmatrix} 1 & 0 & 0 & -2 \\ 0 & 1 & -1 & 0 \\ 0 & -2 & 3 & 0 \\ 2 & 0 & 0 & -3 \end{bmatrix}$

15. $\begin{bmatrix} 1 & 3 & 1 & -4 \\ 0 & 1 & -2 & 2 \\ 0 & 0 & 1 & 1 \\ 0 & 0 & 0 & 1 \end{bmatrix}$

16. $\begin{bmatrix} 1 & -3 & 1 & -2 \\ 2 & -5 & 4 & -2 \\ -3 & 9 & -2 & 5 \\ 4 & -12 & 4 & -7 \end{bmatrix}$

17. Use the answer to Exercise 6 to find the solutions to the linear system

$$4x_1 + 13x_2 = -3$$
$$x_1 + 3x_2 = 2$$

18. Use the answer to Exercise 10 to find the solutions to the linear system

$$x_1 + 2x_2 - x_3 = -2$$
$$-4x_1 - 7x_2 + 7x_3 = 1$$
$$-x_1 - x_2 + 5x_3 = -1$$

19. Use the answer to Exercise 12 to find the solutions to the linear system

$$3x_1 - x_2 + 9x_3 = 4$$
$$x_1 - x_2 + 4x_3 = -1$$
$$2x_1 - 2x_2 + 10x_3 = 3$$

20. Use the answer to Exercise 14 to find the solutions to the linear system

$$x_1 \qquad\quad - 2x_4 = -1$$
$$x_2 - x_3 \qquad = -2$$
$$-2x_2 + 3x_3 \qquad = 2$$
$$2x_1 \qquad\quad - 3x_4 = -1$$

In Exercises 21–26, determine if the given linear transformation T is invertible, and if so, find T^{-1}. (HINT: Start by finding the matrix A such that $T(\mathbf{x}) = A\mathbf{x}$.)

21. $T\left(\begin{bmatrix} x_1 \\ x_2 \end{bmatrix}\right) = \begin{bmatrix} 4x_1 + 3x_2 \\ 3x_1 + 2x_2 \end{bmatrix}$

22. $T\left(\begin{bmatrix} x_1 \\ x_2 \end{bmatrix}\right) = \begin{bmatrix} 2x_1 - 5x_2 \\ -x_1 + 4x_2 \\ x_1 + x_2 \end{bmatrix}$

23. $T\left(\begin{bmatrix} x_1 \\ x_2 \end{bmatrix}\right) = \begin{bmatrix} x_1 - 5x_2 \\ -2x_1 + 10x_2 \end{bmatrix}$

24. $T\left(\begin{bmatrix} x_1 \\ x_2 \\ x_3 \end{bmatrix}\right) = \begin{bmatrix} x_1 + x_3 \\ x_2 - x_3 \\ x_1 - x_2 + x_3 \end{bmatrix}$

25. $T\left(\begin{bmatrix} x_1 \\ x_2 \\ x_3 \end{bmatrix}\right) = \begin{bmatrix} x_1 + 2x_2 - x_3 \\ x_1 + x_2 - x_3 \end{bmatrix}$

26. $T\left(\begin{bmatrix} x_1 \\ x_2 \\ x_3 \end{bmatrix}\right) = \begin{bmatrix} x_1 + x_2 - x_3 \\ x_2 - x_3 \\ x_1 - x_2 + x_3 \end{bmatrix}$

27. Let T_1 and T_2 be linear transformations given by

$$T_1\left(\begin{bmatrix} x_1 \\ x_2 \end{bmatrix}\right) = \begin{bmatrix} 2x_1 + x_2 \\ x_1 + x_2 \end{bmatrix}$$

$$T_2\left(\begin{bmatrix} x_1 \\ x_2 \end{bmatrix}\right) = \begin{bmatrix} 3x_1 + 2x_2 \\ x_1 + x_2 \end{bmatrix}$$

Find the matrix A such that

(a) $T_1^{-1}(T_2(\mathbf{x})) = A\mathbf{x}$

(b) $T_1(T_2^{-1}(\mathbf{x})) = A\mathbf{x}$

(c) $T_2^{-1}(T_1(\mathbf{x})) = A\mathbf{x}$

(d) $T_2(T_1^{-1}(\mathbf{x})) = A\mathbf{x}$

28. Let T_1 and T_2 be linear transformations given by

$$T_1\left(\begin{bmatrix} x_1 \\ x_2 \end{bmatrix}\right) = \begin{bmatrix} 3x_1 + 5x_2 \\ 4x_1 + 7x_2 \end{bmatrix}$$

$$T_2\left(\begin{bmatrix} x_1 \\ x_2 \end{bmatrix}\right) = \begin{bmatrix} 2x_1 + 9x_2 \\ x_1 + 5x_2 \end{bmatrix}$$

Find the matrix A such that

(a) $T_1^{-1}(T_2(\mathbf{x})) = A\mathbf{x}$

(b) $T_1(T_2^{-1}(\mathbf{x})) = A\mathbf{x}$

(c) $T_2^{-1}(T_1(\mathbf{x})) = A\mathbf{x}$

(d) $T_2(T_1^{-1}(\mathbf{x})) = A\mathbf{x}$

In Exercises 29–34, use an appropriate partitioning of the matrix A to find A^{-1}.

29. $A = \begin{bmatrix} 1 & 0 & 0 \\ 0 & 2 & 7 \\ 0 & 1 & 4 \end{bmatrix}$

30. $A = \begin{bmatrix} 5 & 2 & 0 \\ 2 & 1 & 0 \\ 0 & 0 & 1 \end{bmatrix}$

31. $A = \begin{bmatrix} 2 & 5 & 0 & 0 \\ 3 & 8 & 0 & 0 \\ 0 & 0 & 1 & 4 \\ 0 & 0 & 1 & 3 \end{bmatrix}$

32. $A = \begin{bmatrix} 1 & 3 & 0 & 0 \\ 3 & 8 & 0 & 0 \\ -1 & 2 & 2 & 5 \\ 4 & 3 & 1 & 3 \end{bmatrix}$

33. $A = \begin{bmatrix} 1 & 3 & 0 & 0 & 0 \\ 3 & 8 & 0 & 0 & 0 \\ -1 & 2 & 1 & 2 & -2 \\ 4 & 3 & 0 & 1 & 0 \\ 1 & -2 & 0 & 0 & 1 \end{bmatrix}$

34. $A = \begin{bmatrix} 7 & 2 & 0 & 0 & 0 \\ 4 & 1 & 0 & 0 & 0 \\ 1 & 3 & 1 & 0 & 0 \\ -2 & 3 & 0 & 1 & -2 \\ 5 & -2 & 0 & 3 & -5 \end{bmatrix}$

FIND AN EXAMPLE For Exercises 35–40, find an example that meets the given specifications.

35. A diagonal 3×3 invertible matrix A.

36. A singular 3×3 matrix A that has no zero entries.

37. 2×2 matrices A and B such that $AB = 3I_2$.

38. 3×3 matrices A and B such that $BA = -2I_2$.

39. A 2×3 matrix A and a 3×2 matrix B such that $AB = I_2$ but $BA \neq I_3$.

40. A 3×4 matrix A and a 4×3 matrix B such that $AB = I_3$ but $BA \neq I_4$.

TRUE OR FALSE For Exercises 41–50, determine if the statement is true or false, and justify your answer.

41. If A is an invertible $n \times n$ matrix, then the number of solutions to $A\mathbf{x} = \mathbf{b}$ depends on the vector $\mathbf{b}$ in $\mathbf{R}^n$.

42. A must be a square matrix to be invertible.

43. If an $n \times n$ matrix A is equivalent to I_n, then A^{-1} is also equivalent to I_n.

44. If an $n \times n$ matrix A is singular, then the columns of A must be linearly independent.

45. The Caesar cipher encoding system is an example of a linear transformation.

46. If the columns of an $n \times n$ matrix A span $\mathbf{R}^n$, then A is singular.

47. If A and B are invertible $n \times n$ matrices, then the inverse of AB is $B^{-1}A^{-1}$.

48. If A and B are invertible $n \times n$ matrices, then the inverse of $A + B$ is $A^{-1} + B^{-1}$.

49. If A is invertible, then $\left(A^{-1}\right)^{-1} = A$.

50. If $AB = 2I_3$, then $BA = 2I_3$.

In Exercises 51–54, solve for the matrix X. Assume that all matrices are $n \times n$ and invertible as needed.

51. $AX = B$

52. $BX = A + CX$

53. $B(X + A)^{-1} = C$

54. $AX(D + BX)^{-1} = C$

55. Find all 2×2 matrices A such that $A^{-1} = A$.

56. Suppose that A is a square matrix with two equal rows. Is A invertible? Justify your answer.

57. Suppose that A is a square matrix with two equal columns. Is A invertible? Justify your answer.

58. Let $A = \begin{bmatrix} a & 0 \\ 0 & d \end{bmatrix}$ be a 2×2 diagonal matrix. For what values of a and d will A be invertible?

59. For what values of c will $A = \begin{bmatrix} 1 & 1 \\ c & c^2 \end{bmatrix}$ be invertible?

60. Suppose that $A^{-1} = \begin{bmatrix} 4 & -6 \\ 2 & 14 \end{bmatrix}$. Find $(2A)^{-1}$.

61. Let A be an $n \times n$ matrix and $\mathbf{b}$ be in $\mathbf{R}^n$. If $A\mathbf{x} = \mathbf{b}$ has a unique solution, show that A must be invertible.

62. Suppose that $A = PDP^{-1}$, where all matrices are square. Find an expression for each of A^2 and A^3, and then give a general formula for A^n.

63. Suppose that A, B, and C are $n \times n$ invertible matrices. Solve $AC = CB$ for B.

64. Suppose that A is an invertible $n \times n$ matrix and that X and B are $n \times m$ matrices. If $AX = B$, prove that $X = A^{-1}B$.

65. Suppose that A is an invertible $m \times m$ matrix and that B and C are $n \times m$ matrices. If $(B - C)A = 0_{nm}$, prove that $B = C$.

66. Let A and B be $n \times n$ matrices, and suppose that B and AB are both invertible. Prove that A is also invertible.

67. Let A and B be $n \times n$ matrices. Prove that if B is singular, then so is AB.

68. Let A and B be $n \times n$ matrices. Prove that if A is singular, then so is AB.

69. Complete the proof of Theorem 3.19: If T is an invertible linear transformation, then $T^{-1}(r\mathbf{x}) = r T^{-1}(\mathbf{x})$.

70. Complete the proof of Theorem 3.23: Let A be an invertible $n \times n$ matrix and C and D be $n \times m$ matrices.

(a) If $AC = AD$, then $C = D$.

(b) If $AC = 0_{nm}$, then $C = 0_{nm}$.

Ⓒ For Exercises 71–74, refer to the MP3 scenario given at the beginning of the Section 3.1. There a linear transformation $T(\mathbf{x})$ is defined that gives the costs associated with a production vector $\mathbf{x}$. Determine T^{-1}, and use it to find the production level for each type of player that will result in the given costs. If the given costs are not possible, explain why.

71. Labor = \$1070, Materials = \$1002, and Overhead = \$1368.

72. Labor = \$1148, Materials = \$1076, and Overhead = \$1452.

73. Labor = \$2045, Materials = \$1965, and Overhead = \$2615.

74. Labor = \$2348, Materials = \$2200, and Overhead = \$2983.

Ⓒ Suppose that you are in the garden supply business. Naturally, one of the things that you sell is fertilizer. You have three brands available: Vigoro and Parker's as introduced in Section 2.1, and a third brand, Bleyer's SuperRich. The amount of nitrogen, phosphoric acid, and potash per 100 pounds for each brand is given by the nutrient vectors

$$\mathbf{v} = \begin{bmatrix} 29 \\ 3 \\ 4 \end{bmatrix} \qquad \mathbf{p} = \begin{bmatrix} 18 \\ 25 \\ 6 \end{bmatrix} \qquad \mathbf{b} = \begin{bmatrix} 50 \\ 19 \\ 9 \end{bmatrix}$$

Vigoro   Parker's   Bleyer's

In Exercises 75–78, determine the linear transformation $T : \mathbf{R}^3 \rightarrow \mathbf{R}^3$ that takes a vector of brand amounts (in hundreds of pounds) as input and gives the nutrient vector as output. Then find a formula

for T^{-1}, and use it to determine the amount of Vigoro, Parker's, and Bleyer's required to produce the specified nutrient mix.

75. 409 pounds of nitrogen, 204 pounds of phosphoric acid, and 81 pounds of potash.

76. 439 pounds of nitrogen, 147 pounds of phosphoric acid, and 76 pounds of potash.

77. 1092 pounds of nitrogen, 589 pounds of phosphoric acid, and 223 pounds of potash.

78. 744 pounds of nitrogen, 428 pounds of phosphoric acid, and 156 pounds of potash.

Ⓒ In Exercises 79–82, the given set of vectors are the encoded version of a short message. Decode the message, given that the encoding matrix is

$$A = \begin{bmatrix} 1 & -3 & 2 \\ 2 & -7 & 9 \\ -4 & 14 & -17 \end{bmatrix}$$

79. $\left\{ \begin{bmatrix} 41 \\ 161 \\ -306 \end{bmatrix}, \begin{bmatrix} 7 \\ 79 \\ -142 \end{bmatrix} \right\}$

80. $\left\{ \begin{bmatrix} -30 \\ -45 \\ 96 \end{bmatrix}, \begin{bmatrix} 1 \\ 22 \\ -39 \end{bmatrix} \right\}$

81. $\left\{ \begin{bmatrix} 7 \\ 75 \\ -136 \end{bmatrix}, \begin{bmatrix} -25 \\ -37 \\ 79 \end{bmatrix}, \begin{bmatrix} 47 \\ 158 \\ -303 \end{bmatrix} \right\}$

82. $\left\{ \begin{bmatrix} 44 \\ 157 \\ -300 \end{bmatrix}, \begin{bmatrix} -46 \\ -105 \\ 211 \end{bmatrix}, \begin{bmatrix} 23 \\ 88 \\ -167 \end{bmatrix}, \begin{bmatrix} 40 \\ 152 \\ -289 \end{bmatrix} \right\}$

Ⓒ In Exercises 83–86, find A^{-1} if A is invertible.

83. $A = \begin{bmatrix} 3 & 1 & -2 & 0 \\ 2 & 2 & 5 & 1 \\ -3 & 0 & -2 & 2 \\ 4 & 1 & 2 & 3 \end{bmatrix}$

84. $A = \begin{bmatrix} 5 & 2 & -1 & 0 \\ 2 & -3 & 1 & 4 \\ 2 & 1 & -3 & 2 \\ 3 & 5 & -2 & -4 \end{bmatrix}$

85. $A = \begin{bmatrix} 5 & 1 & 2 & 1 & 2 \\ -3 & 2 & 2 & 1 & 0 \\ 2 & 3 & 1 & 0 & 1 \\ 5 & -1 & -1 & -1 & 3 \\ 0 & 0 & 3 & 2 & 1 \end{bmatrix}$

86. $A = \begin{bmatrix} 2 & 2 & 9 & 0 & 4 \\ 9 & 5 & 5 & 2 & 1 \\ 2 & 3 & 0 & 0 & 5 \\ 9 & 5 & 0 & 7 & 0 \\ 4 & 9 & 9 & 5 & 1 \end{bmatrix}$

▶ This section is optional. However, elementary matrices (introduced in this section) are used in optional Section 5.2 and LU factorizations are revisited in optional Section 11.2.

3.4 LU Factorization

In this section we revisit the problem of finding solutions to a system of linear equations, but we develop a new approach using matrices that can be more efficient in certain situations.

Figure 1 shows a diagram of three one-way streets that intersect. The arrows indicate the direction of traffic flow, and the numbers shown give the number of cars per minute passing along that stretch of road at a particular time. Our goal is to use this information to find x_1, x_2, and x_3.

The number of cars entering and exiting an intersection must be equal, so that for our three intersections a, b, and c we have the following equations:

$$\begin{array}{ll} \text{a:} & 20 + x_3 = x_1 \\ \text{b:} & 70 + x_1 = 30 + x_2 \\ \text{c:} & x_2 = 60 + x_3 \end{array}$$

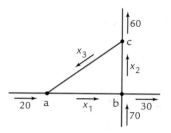

Figure 1 Traffic flow rates. (See Section 1.4 for additional discussion about this type of problem.)

In addition, there is a safety metering system in place that constrains $x_1 + x_2 = 160$. Combining these four equations into a linear system gives us

$$\begin{array}{rl} x_1 + x_2 & = 160 \\ x_1 \quad - x_3 & = 20 \\ x_1 - x_2 & = -40 \\ x_2 - x_3 & = 60 \end{array} \tag{1}$$

EXAMPLE 1   Find the solution to the linear system 1.

Solution This system can be expressed in matrix form $A\mathbf{x} = \mathbf{b}$, where

$$A = \begin{bmatrix} 1 & 1 & 0 \\ 1 & 0 & -1 \\ 1 & -1 & 0 \\ 0 & 1 & -1 \end{bmatrix} \quad \text{and} \quad \mathbf{b} = \begin{bmatrix} 160 \\ 20 \\ -40 \\ 60 \end{bmatrix}$$

Here we use a new approach to find the solutions to this system. It can be shown (we will see how later) that

$$A = \begin{bmatrix} 1 & 0 & 0 & 0 \\ 1 & 1 & 0 & 0 \\ 1 & 2 & 1 & 0 \\ 0 & -1 & -1 & 1 \end{bmatrix} \begin{bmatrix} 1 & 1 & 0 \\ 0 & -1 & -1 \\ 0 & 0 & 2 \\ 0 & 0 & 0 \end{bmatrix}$$

Denoting the left matrix L (for Lower triangular) and the right by U (for Upper triangular, which it nearly is), we have $A = LU$. Thus the system can be written

$$LU\mathbf{x} = \mathbf{b}$$

which looks more complicated but actually allows us to break the problem into two systems that can both be solved with back substitution.

The first step is to set $\mathbf{y} = U\mathbf{x}$, so that our system reduces to $L\mathbf{y} = \mathbf{b}$, or

$$\begin{bmatrix} 1 & 0 & 0 & 0 \\ 1 & 1 & 0 & 0 \\ 1 & 2 & 1 & 0 \\ 0 & -1 & -1 & 1 \end{bmatrix} \begin{bmatrix} y_1 \\ y_2 \\ y_3 \\ y_4 \end{bmatrix} = \begin{bmatrix} 160 \\ 20 \\ -40 \\ 60 \end{bmatrix}$$

This is equivalent to the system

$$
\begin{aligned}
y_1 &&&&&= 160 \\
y_1 + y_2 &&&&&= 20 \\
y_1 + 2y_2 + y_3 &&&&&= -40 \\
-y_2 - y_3 + y_4 &&&&&= 60
\end{aligned}
$$

To find the solution, we modify our usual back substitution method by starting at the top (the details are left to the reader) to find

$$
\mathbf{y} = \begin{bmatrix} 160 \\ -140 \\ 80 \\ 0 \end{bmatrix}
$$

Now that we know $\mathbf{y}$, we can return to the system $\mathbf{y} = U\mathbf{x}$, or

$$
\begin{bmatrix} 1 & 1 & 0 \\ 0 & -1 & -1 \\ 0 & 0 & 2 \\ 0 & 0 & 0 \end{bmatrix} \begin{bmatrix} x_1 \\ x_2 \\ x_3 \end{bmatrix} = \begin{bmatrix} 160 \\ -140 \\ 80 \\ 0 \end{bmatrix}
$$

This is equivalent to

$$
\begin{aligned}
x_1 + x_2 && &= 160 \\
-x_2 - &x_3 &= -140 \\
2x_3 && &= 80
\end{aligned}
$$

This system can be solved using standard back substitution, which gives us $x_1 = 60$, $x_2 = 100$, and $x_3 = 40$. ∎

In the previous example, having L and U made it easier to find the solution, but usually L and U are not known. Below we present the method for finding L and U, but before we get to that, we pause to discuss whether doing so is worth the trouble.

The number of computations required to find L and U from A, then perform the two back substitutions, is about the same as simply reducing A to echelon form and back substituting, so for a single system $A\mathbf{x} = \mathbf{b}$ there is no benefit to finding L and U first. However, some applications require the solutions to many systems that all have the same coefficient matrix,

$$
A\mathbf{x} = \mathbf{b}_1, \qquad A\mathbf{x} = \mathbf{b}_2, \qquad A\mathbf{x} = \mathbf{b}_3, \qquad \ldots
$$

For instance, the traffic rates in our example are likely to change from minute to minute, each time generating new systems with the same coefficient matrix. In such situations, once we have L and U we can use them over and over. Thus, from the second system on, all we have to do are the two back substitutions, which is much faster than solving each individual system from scratch.

Finding L and U

Definition LU factorization

If $A = LU$, where U is in echelon form and L is lower triangular with 1's on the diagonal (making L *unit lower triangular*), then the product is called an **LU factorization** of A. In the following examples, we show how to find an LU factorization.

EXAMPLE 2 Find an LU factorization for

$$A = \begin{bmatrix} -3 & 1 & 2 \\ 6 & 2 & -5 \\ 9 & 5 & -6 \end{bmatrix}$$

Solution We obtain U by reducing A to echelon form, and build up L one column at a time as we transform A.

Step 1a: Take the first column of A, divide each entry by the pivot (-3), and use the resulting values to form the first column of L.

▶ The • symbol represents a matrix entry that has not yet been determined.

$$A = \begin{bmatrix} \mathbf{-3} & 1 & 2 \\ \mathbf{6} & 2 & -5 \\ \mathbf{9} & 5 & -6 \end{bmatrix} \implies L = \begin{bmatrix} 1 & • & • \\ -2 & • & • \\ -3 & • & • \end{bmatrix}$$

Step 1b: Perform row operations on A as usual to introduce zeros down the first column of A.

Note: Do not scale rows by constants (i.e., dividing the first row by -3). Doing so will give incorrect results.

▶ It is possible to modify this procedure to allow row scaling, but it makes the algorithm more complicated. Since scaling is not necessary to reduce a matrix to echelon form, we leave it out.

$$A = \begin{bmatrix} -3 & 1 & 2 \\ 6 & 2 & -5 \\ 9 & 5 & -6 \end{bmatrix} \begin{smallmatrix} 2R_1+R_2\Rightarrow R_2 \\ 3R_1+R_3\Rightarrow R_3 \\ \sim \end{smallmatrix} \begin{bmatrix} -3 & 1 & 2 \\ 0 & 4 & -1 \\ 0 & 8 & 0 \end{bmatrix} = A_1$$

Step 2a: Take the second column of A_1, starting from the pivot entry (4) down (the entries are shown in boldface), and divide each entry by the pivot. Use the resulting values to form the lower portion of the second column of L.

$$A_1 = \begin{bmatrix} -3 & 1 & 2 \\ 0 & \mathbf{4} & -1 \\ 0 & \mathbf{8} & 0 \end{bmatrix} \implies L = \begin{bmatrix} 1 & • & • \\ -2 & 1 & • \\ -3 & 2 & • \end{bmatrix}$$

Step 2b: Perform row operations on A_1 to introduce zeros down the second column of A_1.

$$A_1 = \begin{bmatrix} -3 & 1 & 2 \\ 0 & 4 & -1 \\ 0 & 8 & 0 \end{bmatrix} \begin{smallmatrix} -2R_2+R_3\Rightarrow R_3 \\ \sim \end{smallmatrix} \begin{bmatrix} -3 & 1 & 2 \\ 0 & 4 & -1 \\ 0 & 0 & 2 \end{bmatrix} = A_2$$

Step 3: Now that the original matrix is in echelon form (indeed, it is upper triangular), two small items remain. First, set U equal to A_2, our echelon form of A. Second, finish filling in L. Since L must be unit lower triangular, the only choice is to add a 1 in the lower right corner and fill in the remain entries with zeros. Thus we end up with

$$L = \begin{bmatrix} 1 & 0 & 0 \\ -2 & 1 & 0 \\ -3 & 2 & 1 \end{bmatrix} \quad \text{and} \quad U = \begin{bmatrix} -3 & 1 & 2 \\ 0 & 4 & -1 \\ 0 & 0 & 2 \end{bmatrix}$$

as in the previous example. Standard matrix multiplication can be used to verify that $A = LU$. ∎

Not all matrices have an LU factorization. The following theorem gives a condition that will insure such a factorization exists.

THEOREM 3.26

Let A be an $n \times m$ matrix. If A can be transformed to echelon form without interchanging rows, then there exists an $n \times n$ unit lower triangular matrix L and an $n \times m$ echelon matrix U such that $A = LU$.

Note that if U is a square matrix, then since it is an echelon matrix it is also upper triangular. Later in this section, we explain why the algorithm for finding an LU factorization works and justify Theorem 3.26. For now let's consider additional examples.

EXAMPLE 3 Find an LU factorization for

$$A = \begin{bmatrix} 2 & 1 & 0 & 4 \\ -4 & -3 & 5 & -10 \\ 6 & 4 & -8 & 17 \\ 2 & -3 & 29 & -9 \end{bmatrix}$$

Solution We proceed in the same manner as in the previous example.

Step 1a: Divide by the pivot to form the first column of L.

$$A = \begin{bmatrix} \mathbf{2} & 1 & 0 & 4 \\ \mathbf{-4} & -3 & 5 & -10 \\ \mathbf{6} & 4 & -8 & 17 \\ \mathbf{2} & -3 & 29 & -9 \end{bmatrix} \implies L = \begin{bmatrix} 1 & \bullet & \bullet & \bullet \\ -2 & \bullet & \bullet & \bullet \\ 3 & \bullet & \bullet & \bullet \\ 1 & \bullet & \bullet & \bullet \end{bmatrix}$$

Step 1b: Perform row operations on A to introduce zeros down the first column.

$$A = \begin{bmatrix} 2 & 1 & 0 & 4 \\ -4 & -3 & 5 & -10 \\ 6 & 4 & -8 & 17 \\ 2 & -3 & 29 & -9 \end{bmatrix} \begin{matrix} 2R_1+R_2 \Rightarrow R_2 \\ -3R_1+R_3 \Rightarrow R_3 \\ -R_1+R_4 \Rightarrow R_4 \\ \sim \end{matrix} \begin{bmatrix} 2 & 1 & 0 & 4 \\ 0 & -1 & 5 & -2 \\ 0 & 1 & -8 & 5 \\ 0 & -4 & 29 & -13 \end{bmatrix} = A_1$$

Step 2a: Divide by the pivot to form the second column of L.

$$A_1 = \begin{bmatrix} 2 & 1 & 0 & 4 \\ 0 & \mathbf{-1} & 5 & -2 \\ 0 & \mathbf{1} & -8 & 5 \\ 0 & \mathbf{-4} & 29 & -13 \end{bmatrix} \implies L = \begin{bmatrix} 1 & \bullet & \bullet & \bullet \\ -2 & 1 & \bullet & \bullet \\ 3 & -1 & \bullet & \bullet \\ 1 & 4 & \bullet & \bullet \end{bmatrix}$$

Step 2b: Perform row operations on A_1 to introduce zeros down the second column.

$$A_1 = \begin{bmatrix} 2 & 1 & 0 & 4 \\ 0 & -1 & 5 & -2 \\ 0 & 1 & -8 & 5 \\ 0 & -4 & 29 & -13 \end{bmatrix} \begin{matrix} R_2+R_3 \Rightarrow R_3 \\ -4R_2+R_4 \Rightarrow R_4 \\ \sim \end{matrix} \begin{bmatrix} 2 & 1 & 0 & 4 \\ 0 & -1 & 5 & -2 \\ 0 & 0 & -3 & 3 \\ 0 & 0 & 9 & -5 \end{bmatrix} = A_2$$

Step 3a: Divide by the pivot to form the third column of L.

$$A_2 = \begin{bmatrix} 2 & 1 & 0 & 4 \\ 0 & -1 & 5 & -2 \\ 0 & 0 & \mathbf{-3} & 3 \\ 0 & 0 & \mathbf{9} & -5 \end{bmatrix} \implies L = \begin{bmatrix} 1 & \bullet & \bullet & \bullet \\ -2 & 1 & \bullet & \bullet \\ 3 & -1 & 1 & \bullet \\ 1 & 4 & -3 & \bullet \end{bmatrix}$$

Step 3b: Perform row operations on A_2 to introduce zeros down the third column.

$$A_2 = \begin{bmatrix} 2 & 1 & 0 & 4 \\ 0 & -1 & 5 & -2 \\ 0 & 0 & -3 & 3 \\ 0 & 0 & 9 & -5 \end{bmatrix} \begin{matrix} 3R_3+R_4 \Rightarrow R_4 \\ \sim \end{matrix} \begin{bmatrix} 2 & 1 & 0 & 4 \\ 0 & -1 & 5 & -2 \\ 0 & 0 & -3 & 3 \\ 0 & 0 & 0 & 4 \end{bmatrix} = A_3$$

Step 4: Finish up by placing in L a 1 in the bottom right entry and zeros elsewhere and setting $U = A_3$.

$$L = \begin{bmatrix} 1 & 0 & 0 & 0 \\ -2 & 1 & 0 & 0 \\ 3 & -1 & 1 & 0 \\ 1 & 4 & -3 & 1 \end{bmatrix} \quad \text{and} \quad U = \begin{bmatrix} 2 & 1 & 0 & 4 \\ 0 & -1 & 5 & -2 \\ 0 & 0 & -3 & 3 \\ 0 & 0 & 0 & 4 \end{bmatrix}$$

Thus L is unit lower triangular, U is in echelon form (and upper triangular), and it is easily verified that $A = LU$. ∎

EXAMPLE 4   Find an LU factorization for

$$A = \begin{bmatrix} 2 & 1 & -1 \\ -4 & -2 & 5 \\ 6 & 2 & 11 \end{bmatrix}$$

Solution Since A has three rows, L will be a 3×3 matrix.

Step 1a: Divide by the pivot to form the first column of L:

$$A = \begin{bmatrix} \mathbf{2} & 1 & -1 \\ \mathbf{-4} & -2 & 5 \\ \mathbf{6} & 2 & 11 \end{bmatrix} \quad \Longrightarrow \quad L = \begin{bmatrix} 1 & \bullet & \bullet \\ -2 & \bullet & \bullet \\ 3 & \bullet & \bullet \end{bmatrix}$$

Step 1b: Perform row operations on A to introduce zeros down the first column:

$$A = \begin{bmatrix} 2 & 1 & -1 \\ -4 & -2 & 5 \\ 6 & 2 & 11 \end{bmatrix} \begin{matrix} 2R_1 + R_2 \Rightarrow R_2 \\ -3R_1 + R_3 \Rightarrow R_3 \\ \sim \end{matrix} \begin{bmatrix} 2 & 1 & -1 \\ 0 & 0 & 3 \\ 0 & -1 & 14 \end{bmatrix} = A_1$$

At this point, we can see from the second column of A_1 that there is no way to transform A to echelon form without interchanging the second and third rows. Our algorithm will not work, so we stop. There is no LU factorization. ∎

The matrix A need not be square to have an LU factorization. In general, if A is $n \times m$, then L will be $n \times n$ and U will be $n \times m$. The next two examples consider nonsquare matrices.

EXAMPLE 5   Find an LU factorization for

$$A = \begin{bmatrix} 4 & -3 & -1 & 5 & 2 \\ -16 & 12 & 2 & -17 & -7 \\ 8 & -6 & -12 & 22 & 10 \end{bmatrix}$$

Solution Since A has three rows, L will be a 3×3 matrix. The solution method is similar to previous examples.

Step 1a: Divide by the pivot to form the first column of L.

$$A = \begin{bmatrix} \mathbf{4} & -3 & -1 & 5 & 2 \\ \mathbf{-16} & 12 & 2 & -17 & -7 \\ \mathbf{8} & -6 & -12 & 22 & 10 \end{bmatrix} \quad \Longrightarrow \quad L = \begin{bmatrix} 1 & \bullet & \bullet \\ -4 & \bullet & \bullet \\ 2 & \bullet & \bullet \end{bmatrix}$$

Step 1b: Perform row operations on A to introduce zeros down the first column.

$$A = \begin{bmatrix} 4 & -3 & -1 & 5 & 2 \\ -16 & 12 & 2 & -17 & -7 \\ 8 & -6 & -12 & 22 & 10 \end{bmatrix} \xrightarrow[\substack{4R_1+R_2 \Rightarrow R_2 \\ -2R_1+R_3 \Rightarrow R_3}]{} \begin{bmatrix} 4 & -3 & -1 & 5 & 2 \\ 0 & 0 & -2 & 3 & 1 \\ 0 & 0 & -10 & 12 & 6 \end{bmatrix} = A_1$$

Step 2a: Since the second and third entries in the second column of A_1 are both zero, we bypass it and move to the third column. Divide by the pivot of the third column to form the next column of L.

$$A_1 = \begin{bmatrix} 4 & -3 & -1 & 5 & 2 \\ 0 & 0 & -2 & 3 & 1 \\ 0 & 0 & -10 & 12 & 6 \end{bmatrix} \implies L = \begin{bmatrix} 1 & \bullet & \bullet \\ -4 & 1 & \bullet \\ 2 & 5 & \bullet \end{bmatrix}$$

Step 2b: Perform row operations on A_1 to introduce zeros down the third column.

$$A_1 = \begin{bmatrix} 4 & -3 & -1 & 5 & 2 \\ 0 & 0 & -2 & 3 & 1 \\ 0 & 0 & -10 & 12 & 6 \end{bmatrix} \xrightarrow[]{-5R_2+R_3 \Rightarrow R_3} \begin{bmatrix} 4 & -3 & -1 & 5 & 2 \\ 0 & 0 & -2 & 3 & 1 \\ 0 & 0 & 0 & -3 & 1 \end{bmatrix} = A_2$$

Step 3: Our matrix is now in echelon form. We place a one in the lower-right position of L and zeros elsewhere and set $U = A_2$, giving us

$$L = \begin{bmatrix} 1 & 0 & 0 \\ -4 & 1 & 0 \\ 2 & 5 & 1 \end{bmatrix}, \quad U = \begin{bmatrix} 4 & -3 & -1 & 5 & 2 \\ 0 & 0 & -2 & 3 & 1 \\ 0 & 0 & 0 & -3 & 1 \end{bmatrix}$$

■

EXAMPLE 6   Find an LU factorization for

$$A = \begin{bmatrix} 3 & -1 & 4 \\ 9 & -5 & 15 \\ 15 & -1 & 10 \\ -6 & 2 & -4 \\ -3 & -3 & 10 \end{bmatrix}$$

Solution Since A has five rows, L will be a 5×5 matrix.

Step 1a: Divide by the pivot to form the first column of L.

$$A = \begin{bmatrix} 3 & -1 & 4 \\ 9 & -5 & 15 \\ 15 & -1 & 10 \\ -6 & 2 & -4 \\ -3 & -3 & 10 \end{bmatrix} \implies L = \begin{bmatrix} 1 & \bullet & \bullet & \bullet & \bullet \\ 3 & \bullet & \bullet & \bullet & \bullet \\ 5 & \bullet & \bullet & \bullet & \bullet \\ -2 & \bullet & \bullet & \bullet & \bullet \\ -1 & \bullet & \bullet & \bullet & \bullet \end{bmatrix}$$

Step 1b: Perform row operations on A to introduce zeros down the first column.

$$A = \begin{bmatrix} 3 & -1 & 4 \\ 9 & -5 & 15 \\ 15 & -1 & 10 \\ -6 & 2 & -4 \\ -3 & -3 & 10 \end{bmatrix} \xrightarrow[\substack{-3R_1+R_2 \Rightarrow R_2 \\ -5R_1+R_3 \Rightarrow R_3 \\ 2R_1+R_4 \Rightarrow R_4 \\ R_1+R_5 \Rightarrow R_5}]{} \begin{bmatrix} 3 & -1 & 4 \\ 0 & -2 & 3 \\ 0 & 4 & -10 \\ 0 & 0 & 4 \\ 0 & -4 & 14 \end{bmatrix} = A_1$$

Step 2a: Divide by the pivot to form the second column of L.

$$A_1 = \begin{bmatrix} 3 & -1 & 4 \\ 0 & -2 & 3 \\ 0 & 4 & -10 \\ 0 & 0 & 4 \\ 0 & -4 & 14 \end{bmatrix} \implies L = \begin{bmatrix} 1 & \bullet & \bullet & \bullet & \bullet \\ 3 & 1 & \bullet & \bullet & \bullet \\ 5 & -2 & \bullet & \bullet & \bullet \\ -2 & 0 & \bullet & \bullet & \bullet \\ -1 & 2 & \bullet & \bullet & \bullet \end{bmatrix}$$

Step 2b: Perform row operations on A_1 to introduce zeros down the second column.

$$A_1 = \begin{bmatrix} 3 & -1 & 4 \\ 0 & -2 & 3 \\ 0 & 4 & -10 \\ 0 & 0 & 4 \\ 0 & -4 & 14 \end{bmatrix} \begin{array}{c} 2R_2+R_3 \Rightarrow R_3 \\ -2R_2+R_5 \Rightarrow R_5 \\ \sim \end{array} \begin{bmatrix} 3 & -1 & 4 \\ 0 & -2 & 3 \\ 0 & 0 & -4 \\ 0 & 0 & 4 \\ 0 & 0 & 8 \end{bmatrix} = A_2$$

Step 3a: Divide by the pivot to form the third column of L.

$$A_2 = \begin{bmatrix} 3 & -1 & 4 \\ 0 & -2 & 3 \\ 0 & 0 & -4 \\ 0 & 0 & 4 \\ 0 & 0 & 8 \end{bmatrix} \implies L = \begin{bmatrix} 1 & \bullet & \bullet & \bullet & \bullet \\ 3 & 1 & \bullet & \bullet & \bullet \\ 5 & -2 & 1 & \bullet & \bullet \\ -2 & 0 & -1 & \bullet & \bullet \\ -1 & 2 & -2 & \bullet & \bullet \end{bmatrix}$$

Step 3b: Perform row operations on A_2 to introduce zeros down the third column.

$$A_2 = \begin{bmatrix} 3 & -1 & 4 \\ 0 & -2 & 3 \\ 0 & 0 & -4 \\ 0 & 0 & 4 \\ 0 & 0 & 8 \end{bmatrix} \begin{array}{c} R_3+R_4 \Rightarrow R_4 \\ 2R_3+R_5 \Rightarrow R_5 \\ \sim \end{array} \begin{bmatrix} 3 & -1 & 4 \\ 0 & -2 & 3 \\ 0 & 0 & -4 \\ 0 & 0 & 0 \\ 0 & 0 & 0 \end{bmatrix} = A_3$$

Step 4: Since A_3 is in echelon form, we can set $U = A_3$. However, we still need the last two columns of L. Since the bottom two rows of U are all zeros, a moment's thought reveals that the product LU will be the same no matter which entries are in the last two columns of L. (See Exercise 70.) Since we need L to be unit lower triangular and in applications we do back substitution, the smart way to fill out the last two columns of L is with the last two columns of I_5. Thus we end up with

$$L = \begin{bmatrix} 1 & 0 & 0 & 0 & 0 \\ 3 & 1 & 0 & 0 & 0 \\ 5 & -2 & 1 & 0 & 0 \\ -2 & 0 & -1 & 1 & 0 \\ -1 & 2 & -2 & 0 & 1 \end{bmatrix}, \quad U = \begin{bmatrix} 3 & -1 & 4 \\ 0 & -2 & 3 \\ 0 & 0 & -4 \\ 0 & 0 & 0 \\ 0 & 0 & 0 \end{bmatrix} \quad \blacksquare$$

Regarding the LU factorization algorithm:

- The number of zero rows in U at the end of the algorithm is equal to the number of columns from I (taken from the right side) required to fill out the remainder of L (see Example 6).

- There can be more than one LU factorization for a given matrix.

- As seen in Example 4, if at any point in the algorithm a row interchange is required, stop. The matrix does not have an LU factorization.

LDU Factorization

A variant of LU factorization is called *LDU factorization*. Starting with a matrix A, now the goal is to write $A = LDU$, where L and U are as in LU factorization (except that U has 1's in the pivot positions) and D is a diagonal matrix with the same dimensions as L. To find the LDU factorization, we follow the same procedure as in finding the LU factorization and then at the end form D and modify U.

EXAMPLE 7 Find an LDU factorization for

$$A = \begin{bmatrix} 2 & 8 & 0 \\ 4 & 18 & -4 \\ -2 & -2 & -13 \end{bmatrix}$$

Solution We start by finding an LU factorization. Since A has three rows, L will be a 3×3 matrix.

Step 1a: Take the first column of A, divide each entry by the pivot, and use the resulting values to form the first column of L.

$$A = \begin{bmatrix} \mathbf{2} & 8 & 0 \\ \mathbf{4} & 18 & -4 \\ \mathbf{-2} & -2 & -13 \end{bmatrix} \implies L = \begin{bmatrix} 1 & \bullet & \bullet \\ 2 & \bullet & \bullet \\ -1 & \bullet & \bullet \end{bmatrix}$$

Step 1b: Perform row operations on A as usual to introduce zeros down the first column.

$$A = \begin{bmatrix} 2 & 8 & 0 \\ 4 & 18 & -4 \\ -2 & -2 & -13 \end{bmatrix} \overset{-2R_1+R_2 \Rightarrow R_2}{\underset{\sim}{R_1+R_3 \Rightarrow R_3}} \begin{bmatrix} 2 & 8 & 0 \\ 0 & 2 & -4 \\ 0 & 6 & -13 \end{bmatrix} = A_1$$

Step 2a: Take the second column of A_1, starting from the pivot entry (2), and divide each entry by the pivot. Use the resulting values to form the second column of L.

$$A_1 = \begin{bmatrix} 2 & 8 & 0 \\ 0 & \mathbf{2} & -4 \\ 0 & \mathbf{6} & -13 \end{bmatrix} \implies L = \begin{bmatrix} 1 & \bullet & \bullet \\ 2 & 1 & \bullet \\ -1 & 3 & \bullet \end{bmatrix}$$

Step 2b: Perform row operations on A_1 to introduce zeros down the second column of A_1.

$$A_1 = \begin{bmatrix} 2 & 8 & 0 \\ 0 & 2 & -4 \\ 0 & 6 & -13 \end{bmatrix} \overset{-3R_2+R_3 \Rightarrow R_3}{\underset{\sim}{}} \begin{bmatrix} 2 & 8 & 0 \\ 0 & 2 & -4 \\ 0 & 0 & -1 \end{bmatrix} = A_2$$

Step 3: The matrix A_2 is in echelon form. We have

$$L = \begin{bmatrix} 1 & 0 & 0 \\ 2 & 1 & 0 \\ -1 & 3 & 1 \end{bmatrix} \quad \text{and} \quad U = \begin{bmatrix} 2 & 8 & 0 \\ 0 & 2 & -4 \\ 0 & 0 & -1 \end{bmatrix}$$

Step 4a: We now have $A = LU$. The diagonal matrix D has entries taken from the pivots of U.

$$U = \begin{bmatrix} \mathbf{2} & 8 & 0 \\ 0 & \mathbf{2} & -4 \\ 0 & 0 & \mathbf{-1} \end{bmatrix} \implies D = \begin{bmatrix} 2 & 0 & 0 \\ 0 & 2 & 0 \\ 0 & 0 & -1 \end{bmatrix}$$

Step 4b: We modify U by dividing each row by the row pivot, leaving 1's in the pivot positions.

$$(\text{old})\ U = \begin{bmatrix} \mathbf{2} & 8 & 0 \\ 0 & \mathbf{2} & -4 \\ 0 & 0 & \mathbf{-1} \end{bmatrix} \implies (\text{new})\ U = \begin{bmatrix} 1 & 4 & 0 \\ 0 & 1 & -2 \\ 0 & 0 & 1 \end{bmatrix}$$

It is not hard to check that

$$\begin{bmatrix} 2 & 8 & 0 \\ 0 & 2 & -4 \\ 0 & 0 & -1 \end{bmatrix} = \begin{bmatrix} 2 & 0 & 0 \\ 0 & 2 & 0 \\ 0 & 0 & -1 \end{bmatrix} \begin{bmatrix} 1 & 4 & 0 \\ 0 & 1 & -2 \\ 0 & 0 & 1 \end{bmatrix}$$

Setting $U = (\text{new})\ U$, we have $A = LDU$, where

$$L = \begin{bmatrix} 1 & 0 & 0 \\ 2 & 1 & 0 \\ -1 & 3 & 1 \end{bmatrix}, \quad D = \begin{bmatrix} 2 & 0 & 0 \\ 0 & 2 & 0 \\ 0 & 0 & -1 \end{bmatrix}, \quad U = \begin{bmatrix} 1 & 4 & 0 \\ 0 & 1 & -2 \\ 0 & 0 & 1 \end{bmatrix} \quad ■$$

EXAMPLE 8 Find an LDU factorization for the matrix in Example 5,

$$A = \begin{bmatrix} 4 & -3 & -1 & 5 & 2 \\ -16 & 12 & 2 & -17 & -7 \\ 8 & -6 & -12 & 22 & 10 \end{bmatrix}$$

Solution In Example 8 we found that $A = LU$, where

$$L = \begin{bmatrix} 1 & 0 & 0 \\ -4 & 1 & 0 \\ 2 & 5 & 1 \end{bmatrix}, \quad U = \begin{bmatrix} 4 & -3 & -1 & 5 & 2 \\ 0 & 0 & -2 & 3 & 1 \\ 0 & 0 & 0 & -3 & 1 \end{bmatrix}$$

All that is left is to form the 3×3 matrix D and modify U.

Step 1a: Form the 3×3 diagonal matrix D, with the diagonal terms coming from the values in the pivot positions of U (in boldface).

$$U = \begin{bmatrix} \mathbf{4} & -3 & -1 & 5 & 2 \\ 0 & 0 & \mathbf{-2} & 3 & 1 \\ 0 & 0 & 0 & \mathbf{-3} & 1 \end{bmatrix} \implies D = \begin{bmatrix} 4 & 0 & 0 \\ 0 & -2 & 0 \\ 0 & 0 & -3 \end{bmatrix}$$

Step 1b: Modify U by dividing each row by its pivot, leaving 1's in the pivot positions.

$$(\text{old})\ U = \begin{bmatrix} 4 & -3 & -1 & 5 & 2 \\ 0 & 0 & -2 & 3 & 1 \\ 0 & 0 & 0 & -3 & 1 \end{bmatrix} \implies U = \begin{bmatrix} 1 & -3/4 & -1/4 & 5/4 & 1/2 \\ 0 & 0 & 1 & -3/2 & -1/2 \\ 0 & 0 & 0 & 1 & -1/3 \end{bmatrix}$$

By forming D and modifying U in this manner, it follows that the product DU is equal to the matrix U found earlier. Therefore we have $A = LDU$, where

$$L = \begin{bmatrix} 1 & 0 & 0 \\ -4 & 1 & 0 \\ 2 & 5 & 1 \end{bmatrix}, \quad D = \begin{bmatrix} 4 & 0 & 0 \\ 0 & -2 & 0 \\ 0 & 0 & -3 \end{bmatrix}, \quad U = \begin{bmatrix} 1 & -3/4 & -1/4 & 5/4 & 1/2 \\ 0 & 0 & 1 & -3/2 & -1/2 \\ 0 & 0 & 0 & 1 & -1/3 \end{bmatrix} \quad ■$$

EXAMPLE 9 Find an LDU factorization for the matrix in Example 6,

$$A = \begin{bmatrix} 3 & -1 & 4 \\ 9 & -5 & 15 \\ 15 & -1 & 10 \\ -6 & 2 & -4 \\ -3 & -3 & 10 \end{bmatrix}$$

Solution In Example 6 we found that $A = LU$, where

$$L = \begin{bmatrix} 1 & 0 & 0 & 0 & 0 \\ 3 & 1 & 0 & 0 & 0 \\ 5 & -2 & 1 & 0 & 0 \\ -2 & 0 & -1 & 1 & 0 \\ -1 & 2 & -2 & 0 & 1 \end{bmatrix}, \quad U = \begin{bmatrix} 3 & -1 & 4 \\ 0 & -2 & 3 \\ 0 & 0 & -4 \\ 0 & 0 & 0 \\ 0 & 0 & 0 \end{bmatrix}$$

The last steps are to form the 5×5 matrix D and modify U.

Step 1a: To find the terms of D, we start as in Example 8, setting the diagonal terms in the first three columns equal to the pivots of U. As with L, we have $A = LDU$ regardless of which entries are in the last two diagonal positions of D. In applications it is convenient to have D invertible, so we put 1's in the last two diagonal positions.

$$D = \begin{bmatrix} 3 & 0 & 0 & 0 & 0 \\ 0 & -2 & 0 & 0 & 0 \\ 0 & 0 & -4 & 0 & 0 \\ 0 & 0 & 0 & 1 & 0 \\ 0 & 0 & 0 & 0 & 1 \end{bmatrix}$$

Step 1b: As in the previous example, we modify U by dividing each nonzero row by its pivot. Doing so, we have $A = LDU$, where

$$L = \begin{bmatrix} 1 & 0 & 0 & 0 & 0 \\ 3 & 1 & 0 & 0 & 0 \\ 5 & -2 & 1 & 0 & 0 \\ -2 & 0 & -1 & 1 & 0 \\ -1 & 2 & -2 & 0 & 1 \end{bmatrix}, \quad D = \begin{bmatrix} 3 & 0 & 0 & 0 & 0 \\ 0 & -2 & 0 & 0 & 0 \\ 0 & 0 & -4 & 0 & 0 \\ 0 & 0 & 0 & 1 & 0 \\ 0 & 0 & 0 & 0 & 1 \end{bmatrix}, \quad U = \begin{bmatrix} 1 & -1/3 & 4/3 \\ 0 & 1 & -3/2 \\ 0 & 0 & 1 \\ 0 & 0 & 0 \\ 0 & 0 & 0 \end{bmatrix}$$

■

Why Does the Algorithm Work?

As illustrated in our examples, a central part of the LU factorization algorithm is using elementary row operations to reduce the original matrix to echelon form. For example, for the following matrix A, the first step toward echelon form in the LU factorization algorithm is

$$A = \begin{bmatrix} 3 & 1 & -1 & 2 \\ 6 & -2 & 0 & 1 \\ 4 & 0 & 1 & -1 \end{bmatrix} \xrightarrow[\sim]{-2R_1 + R_2 \Rightarrow R_2} \begin{bmatrix} 3 & 1 & -1 & 2 \\ 0 & -4 & 2 & -3 \\ 4 & 0 & 1 & -1 \end{bmatrix} = A_1$$

Now suppose that we start with the 3×3 identity matrix, and perform the same row operation,

$$I_3 = \begin{bmatrix} 1 & 0 & 0 \\ 0 & 1 & 0 \\ 0 & 0 & 1 \end{bmatrix} \xrightarrow[\sim]{-2R_1 + R_2 \Rightarrow R_2} \begin{bmatrix} 1 & 0 & 0 \\ -2 & 1 & 0 \\ 0 & 0 & 1 \end{bmatrix} = E$$

Forming the product $E A$, we get

$$E A = \begin{bmatrix} 1 & 0 & 0 \\ -2 & 1 & 0 \\ 0 & 0 & 1 \end{bmatrix} \begin{bmatrix} 3 & 1 & -1 & 2 \\ 6 & -2 & 0 & 1 \\ 4 & 0 & 1 & -1 \end{bmatrix} = \begin{bmatrix} 3 & 1 & -1 & 2 \\ 0 & -4 & 2 & -3 \\ 4 & 0 & 1 & -1 \end{bmatrix} = A_1$$

It turns out that we can perform *any* row operation by using the same approach. Given an $n \times m$ matrix A and a desired row operation, all we do is start with I_n, perform the row operation on I_n to produce a new matrix E, and then compute $E A$. (See Exercise 71.) Such matrices E are called **elementary matrices**.

Definition **Elementary Matrix**

Some examples of row operations on a matrix A and the corresponding 3×3 elementary matrices:

Row Operation Performed on A	Multiply A by the Elementary Matrix
$-4 R_1 + R_3 \Rightarrow R_3$	$\begin{bmatrix} 1 & 0 & 0 \\ 0 & 1 & 0 \\ -4 & 0 & 1 \end{bmatrix}$
$5 R_2 \Rightarrow R_2$	$\begin{bmatrix} 1 & 0 & 0 \\ 0 & 5 & 0 \\ 0 & 0 & 1 \end{bmatrix}$
$R_1 \Leftrightarrow R_2$	$\begin{bmatrix} 0 & 1 & 0 \\ 1 & 0 & 0 \\ 0 & 0 & 1 \end{bmatrix}$

Now let's return to LU factorization. The algorithm involves a number of row operations to transform the original matrix A to echelon form, producing the matrix U. Let $E_1, E_2, \ldots, E_k$ denote the elementary matrices corresponding to the row operations, in the order performed. Then we have

$$(E_k \cdots E_2 E_1) A = U \tag{2}$$

In our algorithm, the only type of row operation that we perform involves adding a multiple of one row to a row below. The elementary matrix corresponding to such a row operation will be unit lower triangular. Thus each of $E_1, E_2, \ldots, E_k$ is unit lower triangular, and it follows that

$$(E_k \cdots E_2 E_1) \quad \text{and} \quad (E_k \cdots E_2 E_1)^{-1}$$

also are both unit lower triangular (see Exercise 72).

Returning to 2 and solving for A, we have

$$A = (E_k \cdots E_2 E_1)^{-1} U$$

Thus, if we let $L = (E_k \cdots E_2 E_1)^{-1}$, then $A = LU$ where L is unit lower triangular and U is in echelon form. This shows that an LU factorization is possible. To see why the construction of L in the algorithm works, note that

$$(E_k \cdots E_2 E_1) L = I$$

Comparing to 2, we see that the same sequence of row operations that transform A into U will also transform L into I. This provides the rationale for the construction method for L — we set the columns of L so that the row operations that transform A to U will also transform L to I.

Computational Comments

- For many systems, interchanging rows is necessary when reducing to echelon form, especially when partial pivoting is being used to minimize round-off error. It is possible to modify the LU factorization algorithm given here to accommodate row swaps, although one ends up with a matrix L that is *permuted lower triangular*, meaning that the rows can be reorganized to form a lower triangular matrix.

- When presented with the problem of solving many systems of the form

$$A\mathbf{x} = \mathbf{b}_1, \quad A\mathbf{x} = \mathbf{b}_2, \quad A\mathbf{x} = \mathbf{b}_3, \ldots$$

where A is a square invertible matrix, it is tempting to compute A^{-1} and then use this to find the solution to each system. However, for an $n \times n$ matrix A, it takes about $2n^3/3$ flops to find L and U as opposed to $2n^3$ flops to find A^{-1}. In addition, if A is a sparse matrix (that is, most of the entries are 0), then L and U typically will be sparse, while A^{-1} will not. In this instance, using the LU factorization usually will be more efficient than using A^{-1}.

EXERCISES

In Exercises 1–4, a few terms are missing from the given LU factorization for A. Find them.

1. $\begin{bmatrix} 1 & 0 \\ -7 & 1 \end{bmatrix} \begin{bmatrix} 2 & 2 & -3 \\ 0 & 1 & -4 \end{bmatrix} = \begin{bmatrix} 2 & a & -3 \\ b & -13 & 17 \end{bmatrix}$

2. $\begin{bmatrix} 1 & 0 & 0 \\ -4 & 1 & 0 \\ -2 & -1 & 1 \end{bmatrix} \begin{bmatrix} -2 & 0 & 3 \\ 0 & 1 & 1 \\ 0 & 0 & 2 \end{bmatrix} = \begin{bmatrix} -2 & a & 3 \\ 8 & 1 & b \\ c & -1 & -5 \end{bmatrix}$

3. $\begin{bmatrix} 1 & 0 & 0 \\ 3 & 1 & 0 \\ a & 2 & 1 \end{bmatrix} \begin{bmatrix} 5 & 2 \\ 0 & b \\ 0 & 0 \end{bmatrix} = \begin{bmatrix} 5 & c \\ 15 & 9 \\ 20 & 14 \end{bmatrix}$

4. $\begin{bmatrix} 1 & 0 \\ a & 1 \end{bmatrix} \begin{bmatrix} 4 & 2 & 3 & b \\ 0 & 2 & 3 & 1 \end{bmatrix} = \begin{bmatrix} 4 & c & 3 & 1 \\ 8 & 6 & d & 3 \end{bmatrix}$

In Exercises 5–12, use the given LU factorization to find all solutions to $A\mathbf{x} = \mathbf{b}$.

5. $A = \begin{bmatrix} 1 & 0 \\ -2 & 1 \end{bmatrix} \begin{bmatrix} 2 & -2 \\ 0 & 3 \end{bmatrix}, \quad \mathbf{b} = \begin{bmatrix} 2 \\ 2 \end{bmatrix}$

6. $A = \begin{bmatrix} 1 & 0 \\ 3 & 1 \end{bmatrix} \begin{bmatrix} 1 & 4 \\ 0 & -2 \end{bmatrix}, \quad \mathbf{b} = \begin{bmatrix} -7 \\ -17 \end{bmatrix}$

7. $\mathbf{A} = \begin{bmatrix} 1 & 0 & 0 \\ -1 & 1 & 0 \\ 2 & -2 & 1 \end{bmatrix} \begin{bmatrix} 2 & -1 & 3 \\ 0 & 1 & 2 \\ 0 & 0 & -2 \end{bmatrix}, \quad \mathbf{b} = \begin{bmatrix} 4 \\ 0 \\ -4 \end{bmatrix}$

8. $A = \begin{bmatrix} 1 & 0 & 0 \\ -2 & 1 & 0 \\ 1 & 3 & 1 \end{bmatrix} \begin{bmatrix} 1 & -2 & 0 \\ 0 & 3 & -1 \\ 0 & 0 & -2 \end{bmatrix}, \quad \mathbf{b} = \begin{bmatrix} -4 \\ 11 \\ 5 \end{bmatrix}$

9. $A = \begin{bmatrix} 1 & 0 & 0 \\ 2 & 1 & 0 \\ -3 & 4 & 1 \end{bmatrix} \begin{bmatrix} 1 & -2 \\ 0 & 1 \\ 0 & 0 \end{bmatrix}, \quad \mathbf{b} = \begin{bmatrix} 0 \\ 1 \\ 4 \end{bmatrix}$

10. $A = \begin{bmatrix} 1 & 0 \\ 3 & 1 \end{bmatrix} \begin{bmatrix} 1 & -1 & 2 \\ 0 & -2 & -1 \end{bmatrix}, \quad \mathbf{b} = \begin{bmatrix} 2 \\ 13 \end{bmatrix}$

11. $A = \begin{bmatrix} 1 & 0 & 0 & 0 \\ -2 & 1 & 0 & 0 \\ 0 & 3 & 1 & 0 \\ 2 & -1 & 0 & 1 \end{bmatrix} \begin{bmatrix} 1 & -2 & 0 & -1 \\ 0 & 1 & 1 & 3 \\ 0 & 0 & 1 & -1 \\ 0 & 0 & 0 & 1 \end{bmatrix}$,

$\mathbf{b} = \begin{bmatrix} 0 \\ 0 \\ -1 \\ 0 \end{bmatrix}$

12. $A = \begin{bmatrix} 1 & 0 & 0 & 0 \\ 2 & 1 & 0 & 0 \\ -1 & 3 & 1 & 0 \\ -3 & 0 & 0 & 1 \end{bmatrix} \begin{bmatrix} 1 & 3 & 1 \\ 0 & 1 & 2 \\ 0 & 0 & 1 \\ 0 & 0 & 0 \end{bmatrix}, \quad \mathbf{b} = \begin{bmatrix} -1 \\ -3 \\ -2 \\ 3 \end{bmatrix}$

In Exercises 13–24, find an LU factorization for A.

13. $A = \begin{bmatrix} 1 & -4 \\ -2 & 9 \end{bmatrix}$

14. $A = \begin{bmatrix} 2 & 3 \\ 6 & 10 \end{bmatrix}$

15. $A = \begin{bmatrix} -2 & -1 & 1 \\ -6 & 0 & 4 \\ 2 & -2 & -1 \end{bmatrix}$

16. $A = \begin{bmatrix} -3 & 2 & 1 \\ -6 & 2 & 3 \\ 0 & -8 & 6 \end{bmatrix}$

17. $A = \begin{bmatrix} -1 & 0 & -1 & 2 \\ 1 & 3 & 2 & -2 \\ -2 & -9 & -3 & 3 \\ -1 & 9 & -2 & 5 \end{bmatrix}$

18. $A = \begin{bmatrix} -3 & 2 & 1 & 4 \\ 0 & 2 & 0 & 3 \\ 6 & -6 & -1 & -6 \\ -6 & 2 & -1 & -9 \end{bmatrix}$

19. $A = \begin{bmatrix} -1 & 2 & 1 & 3 \\ 4 & -7 & -7 & -17 \\ -2 & 6 & -3 & -2 \end{bmatrix}$

20. $A = \begin{bmatrix} -2 & 0 & -1 & 1 & 3 \\ 4 & 1 & -1 & 2 & -5 \\ -2 & 3 & -8 & 14 & 5 \end{bmatrix}$

21. $A = \begin{bmatrix} 1 & 1 & 0 \\ 1 & 0 & -1 \\ 1 & -1 & 0 \\ 0 & 1 & -1 \end{bmatrix}$

22. $A = \begin{bmatrix} -1 & 2 & 4 \\ 4 & -6 & -17 \\ -3 & 2 & 15 \\ 2 & -4 & -9 \end{bmatrix}$

23. $A = \begin{bmatrix} -2 & 1 & 3 \\ 2 & 0 & 8 \\ -4 & 1 & 12 \\ 2 & 0 & -10 \\ -4 & 2 & 7 \end{bmatrix}$

24. $A = \begin{bmatrix} 1 & 3 & 2 \\ -1 & -5 & -1 \\ 0 & -6 & -3 \\ 1 & 5 & 7 \\ -3 & 9 & 3 \end{bmatrix}$

In Exercises 25–30, find an LDU factorization for A given in the referenced exercise.

25. Exercise 5.

26. Exercise 8.

27. Exercise 10.

28. Exercise 13.

29. Exercise 15.

30. Exercise 19.

In Exercises 31–36, assume that A is a matrix with three rows. Find the elementary matrix E such that $E A$ gives the matrix resulting from A after the given row operation is performed.

31. $4 R_1 \Rightarrow R_1$

32. $-3 R_2 \Rightarrow R_2$

33. $R_2 \Leftrightarrow R_1$

34. $R_3 \Leftrightarrow R_2$

35. $2 R_1 + R_3 \Rightarrow R_3$

36. $-4 R_3 + R_2 \Rightarrow R_2$

In Exercises 37–42, assume that A is a matrix with three rows. Find the matrix B such that $B A$ gives the matrix resulting from A after the given row operations are performed.

37. $\quad -2 R_1 + R_2 \Rightarrow R_2,$
$\qquad\qquad 5 R_3 \Rightarrow R_3$

38. $\quad -6 R_2 + R_3 \Rightarrow R_3,$
$\qquad\qquad R_1 \Leftrightarrow R_3$

39. $\qquad R_2 \Leftrightarrow R_1,$
$\qquad 3 R_1 + R_2 \Rightarrow R_2$

40. $\qquad -2 R_1 \Rightarrow R_1,$
$\qquad 7 R_2 + R_3 \Rightarrow R_3$

41. $\qquad -3 R_1 \Rightarrow R_1,$
$\qquad\quad R_1 \Leftrightarrow R_2,$
$\qquad 4 R_1 + R_2 \Rightarrow R_2$

42. $\quad -3 R_1 + R_2 \Rightarrow R_2,$
$\qquad 2 R_1 + R_3 \Rightarrow R_3,$
$\qquad\; - R_2 + R_3 \Rightarrow R_3$

In Exercises 43–48, assume that A is a matrix with four rows. Find the elementary matrix E such that $E A$ gives the matrix resulting from A after the given row operation is performed. Then find E^{-1} and give the elementary row operation corresponding to E^{-1}.

43. $-6 R_2 \Rightarrow R_2$

44. $2 R_4 \Rightarrow R_4$

45. $R_3 \Leftrightarrow R_4$

46. $R_1 \Leftrightarrow R_3$

47. $-5 R_1 + R_2 \Rightarrow R_2$

48. $2 R_3 + R_1 \Rightarrow R_1$

If L and U are invertible, then $(LU)^{-1} = U^{-1} L^{-1}$. In Exercises 49–54, find A^{-1} from the LU factorization for A given in the referenced exercise.

49. Exercise 5.

50. Exercise 8.

51. Exercise 11.

52. Exercise 13.

53. Exercise 16.

54. Exercise 17.

FIND AN EXAMPLE For Exercises 55–60, find an example that meets the given specifications.

55. A matrix A that has an LU factorization where L is 4×4 and U is 4×3.

56. A matrix A that has an LU factorization where L is 3×3 and U is 3×6.

57. A 2×2 matrix A that has an LU factorization where U is diagonal.

58. A 3×2 matrix A that has an LU factorization where L is diagonal.

59. A 4×4 matrix A that has an LU factorization where L and U are both diagonal.

60. A 3×4 matrix A that has an LDU factorization where $D = I_3$.

TRUE OR FALSE For Exercises 61–68, determine if the statement is true or false, and justify your answer.

61. The dimensions of L and U are the same in an LU factorization.

62. A matrix has an LU factorization provided that it can be transformed to echelon form without the use of row interchanges.

63. If an $n \times n$ matrix A is lower triangular, then A has an LU factorization.

64. If A is a nonsingular $n \times n$ matrix, then A has an LU factorization.

65. The LU factorization for a given matrix is unique.

66. If E is an $n \times n$ elementary matrix corresponding to swapping two rows, then $E^2 = I_n$.

67. Suppose that E_1 and E_2 are two elementary matrices. Then $E_1 E_2 = E_2 E_1$.

68. If A is an $n \times n$ upper triangular matrix, we can have $L = I_n$ in an LU factorization of A.

69. Let U be an $n \times m$ matrix that is in echelon form, and let D be an $n \times n$ diagonal matrix. Prove that the product DU is equal to the matrix U with each row multiplied by the corresponding diagonal entry of D. (That is, the first row of U multiplied by d_{11}, the second row of U multiplied by d_{22}, and so on.)

70. Let U be an $n \times m$ matrix in echelon form, with the bottom k rows of U having all zero entries. Suppose that L_1 and L_2 are both $n \times n$ matrices and that the leftmost $n - k$ columns of both are identical. Prove that $L_1 U = L_2 U$.

71. Let A be an $n \times m$ matrix and I_n the $n \times n$ identity matrix. Prove that the matrix obtained after performing a given row operation on A is the same as the matrix obtained when computing EA, where E is the matrix obtained by performing the same row operation on I_n.

72. Let $L_1, L_2, \ldots, L_k$ be unit lower triangular matrices.

(a) Prove that $L_2 L_1$ is unit lower triangular.

(b) Use induction to prove that $L_k \cdots L_2 L_1$ is unit lower triangular.

(c) Prove that L_i^{-1} is unit lower triangular for $i = 1, \ldots, k$.

(d) Prove that $(L_k \cdots L_2 L_1)^{-1}$ is unit lower triangular.

73. Ⓒ In graph theory, an *adjacency matrix* A has an entry of 1 at a_{ij} if there is an edge connecting node i with node j, and a zero otherwise. (Such matrices come up in network analysis.) Figure 2 shows a *cyclic graph* with six nodes.

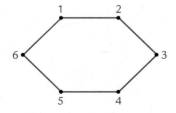

Figure 2 A cyclic graph with six nodes.

The adjacency matrix for this graph is

$$A = \begin{bmatrix} 0 & 1 & 0 & 0 & 0 & 1 \\ 1 & 0 & 1 & 0 & 0 & 0 \\ 0 & 1 & 0 & 1 & 0 & 0 \\ 0 & 0 & 1 & 0 & 1 & 0 \\ 0 & 0 & 0 & 1 & 0 & 1 \\ 1 & 0 & 0 & 0 & 1 & 0 \end{bmatrix}$$

If possible find an LU factorization for the matrix A.

74. Ⓒ *Band matrices* arise in numerous applications, such as finite difference problems in engineering. Such matrices have their nonzero entries clustered along the diagonal, as with

$$A = \begin{bmatrix} 1 & 2 & & & & & \\ 3 & 1 & 1 & & & & \\ & 2 & 2 & 1 & & & \\ & & 1 & 2 & 1 & & \\ & & & 2 & 1 & 3 & \\ & & & & 1 & 4 & 1 \\ & & & & & 3 & 2 \end{bmatrix}$$

The missing entries are zero. Find an LU factorization for the matrix A.

Ⓒ In Exercises 75–76, find an LU factorization for A if one exists. Note that some computer algorithms do not compute LU factorizations in the same manner as presented here, so use caution.

75. $A = \begin{bmatrix} 10 & 2 & 0 & -4 & 2 \\ 5 & 1 & -14 & 5 & 22 \\ 10 & 2 & 0 & -9 & 2 \\ 15 & 3 & -4 & -3 & -7 \end{bmatrix}$

76. $A = \begin{bmatrix} -15 & -3 & 21 \\ -10 & 3 & 29 \\ 5 & -1 & 3 \\ 5 & 4 & -26 \end{bmatrix}$

3.5 Markov Chains

Sequences of vectors arise naturally in certain applications. In those considered here, each vector in the sequence is found by multiplying a special matrix times the preceding vector. To illustrate this, we begin by recalling the following scenario, first introduced in Example 7, Section 3.2.

▶ This section is optional and can be omitted without loss of continuity.

EXAMPLE 1 In a small town there are 10,000 homes. When it comes to television viewing, the residents have three choices: they can subscribe to cable, they can pay for satellite service, or they watch no TV. (The town is sufficiently remote so that an antenna does not work.) In a given year, 80% of the cable customers stick with cable, 10% switch to satellite, and 10% get totally disgusted and quit watching TV. Over the same time period, 90% of satellite viewers continue with satellite service, 5% switch to cable, and 5% quit watching TV. And of those people who start the year not watching TV, 85% continue not watching, 5% subscribe to cable, and 10% get satellite service (see Figure 1). If the current distribution is 6000 homes with cable, 2500 with satellite service, and 1500 with no TV, how many of each type will there be a year from now? How about two years from now? Three years from now? ■

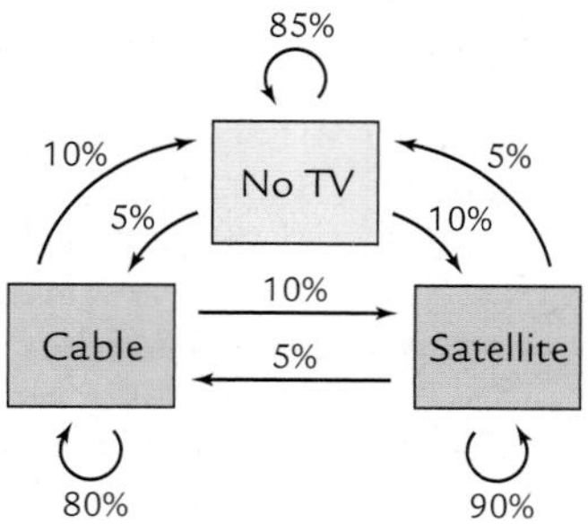

Figure 1 Customer transition percentages between cable, satellite, and no television.

To answer the questions posed, we started by forming the matrix

$$A = \begin{bmatrix} 0.80 & 0.05 & 0.05 \\ 0.10 & 0.90 & 0.10 \\ 0.10 & 0.05 & 0.85 \end{bmatrix}$$

Definition Stochastic Matrix, Doubly Stochastic Matrix

A square matrix like A that has nonnegative entries and columns that each add to 1 is called a **stochastic matrix**. If a stochastic matrix also has rows that add to 1, then it is a **doubly stochastic matrix**.

Previously, we formed the vector **x** giving the initial number of homes with each of satellite, cable, and no television. This time, we start with the vector that gives the initial *proportion* of homes of each type,

$$\mathbf{x}_0 = \begin{bmatrix} 0.60 \\ 0.25 \\ 0.15 \end{bmatrix} \qquad \begin{array}{l} 60\% \text{ of homes have cable} \\ 25\% \text{ of homes have satellite} \\ 15\% \text{ of homes have no television} \end{array}$$

Definition Initial State Vector

Here $\mathbf{x}_0$ denotes the **initial state vector**. Each entry in $\mathbf{x}_0$ can be thought of as representing the probability that a household falls into one of the television-watching groups. In general, any vector that has nonnegative entries that add up to 1 is called a **probability vector**. Thus a stochastic matrix consists of columns that are probability vectors.

Definition Probability Vector

After one year has passed, the distribution of households changes. The new distribution is given by

▶ A is called a **transition matrix** because it is used to make the transition from one state vector to the next.

$$\mathbf{x}_1 = A\mathbf{x}_0 = \begin{bmatrix} 0.80 & 0.05 & 0.05 \\ 0.10 & 0.90 & 0.10 \\ 0.10 & 0.05 & 0.85 \end{bmatrix} \begin{bmatrix} 0.60 \\ 0.25 \\ 0.15 \end{bmatrix} = \begin{bmatrix} 0.50 \\ 0.30 \\ 0.20 \end{bmatrix}$$

That is, after one year, 50% have cable, 30% have satellite, and 20% have no television. Two and three years later we have, respectively,

$$\mathbf{x}_2 = A\mathbf{x}_1 = \begin{bmatrix} 0.425 \\ 0.340 \\ 0.235 \end{bmatrix} \quad \text{and} \quad \mathbf{x}_3 = A\mathbf{x}_2 = \begin{bmatrix} 0.3688 \\ 0.3720 \\ 0.2592 \end{bmatrix}$$

Definition State Vector

Each vector in a sequence generated in this manner is called a **state vector** and is found by multiplying the stochastic matrix by the previous state vector. This process can be continued indefinitely — additional vectors are given below. (Some intermediate vectors are not shown to save space.)

▶ One can also compute state vectors using the formula $\mathbf{x}_i = A^i\mathbf{x}_0$ (as in Example 7, Section 3.2). However, calculating matrix powers $A^2, A^3, \ldots$ is more computationally intensive than using the recursive definition $\mathbf{x}_{i+1} = A\mathbf{x}_i$.

$$\mathbf{x}_4 = \begin{bmatrix} 0.3266 \\ 0.3976 \\ 0.2758 \end{bmatrix}, \mathbf{x}_5 = \begin{bmatrix} 0.2949 \\ 0.4181 \\ 0.2770 \end{bmatrix}, \mathbf{x}_6 = \begin{bmatrix} 0.2712 \\ 0.4345 \\ 0.2943 \end{bmatrix}, \mathbf{x}_7 = \begin{bmatrix} 0.2534 \\ 0.4476 \\ 0.2990 \end{bmatrix}, \mathbf{x}_8 = \begin{bmatrix} 0.2400 \\ 0.4581 \\ 0.3019 \end{bmatrix}$$

$$\mathbf{x}_{10} = \begin{bmatrix} 0.2225 \\ 0.4732 \\ 0.3043 \end{bmatrix}, \mathbf{x}_{12} = \begin{bmatrix} 0.2127 \\ 0.4828 \\ 0.3045 \end{bmatrix}, \mathbf{x}_{14} = \begin{bmatrix} 0.2071 \\ 0.4890 \\ 0.3039 \end{bmatrix}, \mathbf{x}_{16} = \begin{bmatrix} 0.2040 \\ 0.4930 \\ 0.3030 \end{bmatrix}, \mathbf{x}_{20} = \begin{bmatrix} 0.2013 \\ 0.4971 \\ 0.3016 \end{bmatrix}$$

$$\mathbf{x}_{24} = \begin{bmatrix} 0.2004 \\ 0.4988 \\ 0.3008 \end{bmatrix}, \mathbf{x}_{28} = \begin{bmatrix} 0.2001 \\ 0.4995 \\ 0.3004 \end{bmatrix}, \mathbf{x}_{32} = \begin{bmatrix} 0.2000 \\ 0.4998 \\ 0.3002 \end{bmatrix}, \mathbf{x}_{36} = \begin{bmatrix} 0.2000 \\ 0.4999 \\ 0.3001 \end{bmatrix}, \mathbf{x}_{40} = \begin{bmatrix} 0.2000 \\ 0.5000 \\ 0.3000 \end{bmatrix}$$

Definition Markov Chain

In general, a sequence of vectors $\mathbf{x}_0, \mathbf{x}_1, \ldots$ generated in this way is called a **Markov Chain.**

▶ Informally, by *converging* we mean that the entries in the sequence of vectors are getting closer and closer to fixed values.

Examining the state vectors $\mathbf{x}_0, \mathbf{x}_1, \ldots, \mathbf{x}_{40}$ above, we see that our sequence appears to be converging toward the vector

$$\mathbf{x} = \begin{bmatrix} 0.2 \\ 0.5 \\ 0.3 \end{bmatrix} \qquad (1)$$

Since $\mathbf{x}_{i+1} = A\mathbf{x}_i$, if for large i we have $\mathbf{x}_{i+1} \approx \mathbf{x}_i$, then this implies that $A\mathbf{x}_i \approx \mathbf{x}_i$. Hence the vector $\mathbf{x}$ that we are looking for should satisfy $A\mathbf{x} = \mathbf{x}$. Let's try this out for our A and $\mathbf{x}$ in 1.

$$A\mathbf{x} = \begin{bmatrix} 0.80 & 0.05 & 0.05 \\ 0.10 & 0.90 & 0.10 \\ 0.10 & 0.05 & 0.85 \end{bmatrix} \begin{bmatrix} 0.2 \\ 0.5 \\ 0.3 \end{bmatrix} = \begin{bmatrix} 0.2 \\ 0.5 \\ 0.3 \end{bmatrix} = \mathbf{x}$$

Definition Steady-State Vector

The vector $\mathbf{x}$ is called a **steady-state vector** for A.

EXAMPLE 2 The computer support group at a large corporation maintains thousands of machines using a well-known operating system. As the computers age, some tend to become less reliable. Based on past records, if a given computer crashes this week, then there is a 92% chance that it will crash again next week. On the other hand, if a computer did not crash this week, then there is a 60% chance that it will not crash next week (see Figure 2). Suppose that 70% of computers crashed this week. What percentage will crash two weeks from now? Is there a steady-state vector? If so, what is it?

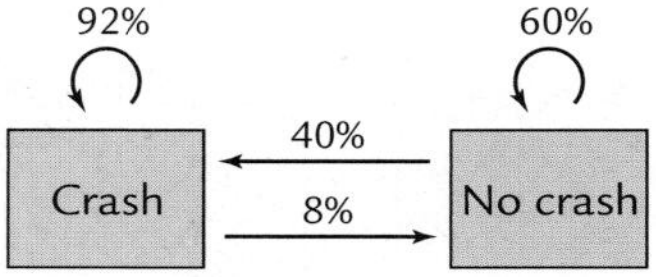

Figure 2 Transition percentages between crashing and noncrashing computers.

Solution The information given is summarized in the table and corresponding transition matrix A below.

		This Week	
		Crash	**No Crash**
Next Week	**Crash**	92%	40%
	No Crash	8%	60%

$$\Longrightarrow \quad A = \begin{bmatrix} 0.92 & 0.40 \\ 0.08 & 0.60 \end{bmatrix}$$

This week 70% of computers crashed and 30% did not, giving an initial state vector of

$$\mathbf{x}_0 = \begin{bmatrix} 0.70 \\ 0.30 \end{bmatrix}$$

The next two vectors in the sequence are

$$\mathbf{x}_1 = A\mathbf{x}_0 = \begin{bmatrix} 0.92 & 0.40 \\ 0.08 & 0.60 \end{bmatrix} \begin{bmatrix} 0.70 \\ 0.30 \end{bmatrix} = \begin{bmatrix} 0.764 \\ 0.236 \end{bmatrix}$$

and

$$\mathbf{x}_2 = A\mathbf{x}_1 = \begin{bmatrix} 0.92 & 0.40 \\ 0.08 & 0.60 \end{bmatrix} \begin{bmatrix} 0.764 \\ 0.236 \end{bmatrix} = \begin{bmatrix} 0.7973 \\ 0.2027 \end{bmatrix}$$

From $\mathbf{x}_2$ we see that two weeks from now we can expect 79.73% of computers to crash. Calculating more state vectors in the sequence gives us

$$\mathbf{x}_3 = \begin{bmatrix} 0.8146 \\ 0.1854 \end{bmatrix}, \quad \mathbf{x}_4 = \begin{bmatrix} 0.8236 \\ 0.1764 \end{bmatrix}, \quad \mathbf{x}_5 = \begin{bmatrix} 0.8283 \\ 0.1717 \end{bmatrix}, \quad \mathbf{x}_6 = \begin{bmatrix} 0.8307 \\ 0.1693 \end{bmatrix}$$

$$\mathbf{x}_7 = \begin{bmatrix} 0.8320 \\ 0.1680 \end{bmatrix}, \quad \mathbf{x}_8 = \begin{bmatrix} 0.8326 \\ 0.1674 \end{bmatrix}, \quad \mathbf{x}_9 = \begin{bmatrix} 0.8330 \\ 0.1670 \end{bmatrix}, \quad \mathbf{x}_{10} = \begin{bmatrix} 0.8331 \\ 0.1669 \end{bmatrix}$$

$$\mathbf{x}_{11} = \begin{bmatrix} 0.8332 \\ 0.1668 \end{bmatrix}, \quad \mathbf{x}_{12} = \begin{bmatrix} 0.8333 \\ 0.1667 \end{bmatrix}, \quad \mathbf{x}_{13} = \begin{bmatrix} 0.8333 \\ 0.1667 \end{bmatrix}, \quad \mathbf{x}_{14} = \begin{bmatrix} 0.8333 \\ 0.1667 \end{bmatrix}$$

This suggests a steady-state vector $\mathbf{x} = \begin{bmatrix} \frac{5}{6} \\ \frac{1}{6} \end{bmatrix}$. Let's test it out:

$$A\mathbf{x} = \begin{bmatrix} 0.92 & 0.40 \\ 0.08 & 0.60 \end{bmatrix} \begin{bmatrix} \frac{5}{6} \\ \frac{1}{6} \end{bmatrix} = \begin{bmatrix} \frac{5}{6} \\ \frac{1}{6} \end{bmatrix}$$

This confirms our observation. ∎

In practice, it may take many terms in the sequence for a steady-state vector to emerge, so computing lots of state vectors is usually not a practical way to find a steady-state vector. Fortunately, there is a direct algebraic method that we can use.

Finding Steady-State Vectors

We know that a steady-state vector $\mathbf{x}$ satisfies $A\mathbf{x} = \mathbf{x}$. Since $\mathbf{x} = I\mathbf{x}$, we have

$$A\mathbf{x} = \mathbf{x} \quad \Longrightarrow \quad A\mathbf{x} = I\mathbf{x} \quad \Longrightarrow \quad A\mathbf{x} - I\mathbf{x} = \mathbf{0} \quad \Longrightarrow \quad (A - I)\mathbf{x} = \mathbf{0}$$

Thus a steady-state vector for A will satisfy the homogeneous system with coefficient matrix $A - I$. For the matrix A in Example 2, we have

$$A - I = \begin{bmatrix} -0.08 & 0.40 \\ 0.08 & -0.40 \end{bmatrix}$$

The homogeneous system $(A - I)\mathbf{x} = \mathbf{0}$ has augmented matrix

$$\begin{bmatrix} -0.08 & 0.40 & 0 \\ 0.08 & -0.40 & 0 \end{bmatrix} \sim \begin{bmatrix} -0.08 & 0.40 & 0 \\ 0 & 0 & 0 \end{bmatrix}$$

This leaves the single equation $-0.08x_1 + 0.40x_2 = 0$. Setting $x_2 = s$ and back-substituting gives the general solution

$$\mathbf{x} = s \begin{bmatrix} 5 \\ 1 \end{bmatrix}$$

Since $\mathbf{x}$ is to be a probability vector, the entries need to add to 1. Setting $s = \frac{1}{5+1} = \frac{1}{6}$ gives

$$\mathbf{x} = \frac{1}{6} \begin{bmatrix} 5 \\ 1 \end{bmatrix} = \begin{bmatrix} \frac{5}{6} \\ \frac{1}{6} \end{bmatrix}$$

which is the steady-state vector found earlier.

EXAMPLE 3   Find a steady-state vector for the matrix A given in Example 1.

$$A = \begin{bmatrix} 0.80 & 0.05 & 0.05 \\ 0.10 & 0.90 & 0.10 \\ 0.10 & 0.05 & 0.85 \end{bmatrix}$$

Solution The augmented matrix for the system $(A - I)\mathbf{x} = \mathbf{0}$ and corresponding echelon form are

$$\begin{bmatrix} -0.20 & 0.05 & 0.05 & 0 \\ 0.10 & -0.10 & 0.10 & 0 \\ 0.10 & 0.05 & -0.15 & 0 \end{bmatrix} \sim \begin{bmatrix} -0.20 & 0.05 & 0.05 & 0 \\ 0 & -0.75 & -0.125 & 0 \\ 0 & 0 & 0 & 0 \end{bmatrix}$$

Back substitution yields the general solution

$$\mathbf{x} = s \begin{bmatrix} 0.2 \\ 0.5 \\ 0.3 \end{bmatrix}$$

Setting $s = 1$ gives the steady-state vector and matches the vector in (1) found earlier. ∎

Properties of Stochastic Matrices

The next theorem summarizes some properties of stochastic matrices. Proofs are left as exercises.

THEOREM 3.27

Let A be an $n \times n$ stochastic matrix and $\mathbf{x}_0$ a probability vector. Then

(a) If $\mathbf{x}_{i+1} = A\mathbf{x}_i$ for $i = 0, 1, 2, \ldots$, then each of $\mathbf{x}_1, \mathbf{x}_2, \ldots$ in the Markov chain is a probability vector.

(b) If B is another $n \times n$ stochastic matrix, then the product AB is also an $n \times n$ stochastic matrix.

(c) For each of $i = 2, 3, 4, \ldots$, A^i is an $n \times n$ stochastic matrix.

In the examples considered thus far, we could always find a steady-state vector for a given initial state vector. However, not all stochastic matrices have a steady-state vector for each initial state vector. For instance, if

$$A = \begin{bmatrix} 0 & 1 \\ 1 & 0 \end{bmatrix} \quad \text{and} \quad \mathbf{x}_0 = \begin{bmatrix} 1 \\ 0 \end{bmatrix}$$

then it is easy to verify that

$$\mathbf{x}_1 = \begin{bmatrix} 0 \\ 1 \end{bmatrix}, \quad \mathbf{x}_2 = \begin{bmatrix} 1 \\ 0 \end{bmatrix}, \quad \mathbf{x}_3 = \begin{bmatrix} 0 \\ 1 \end{bmatrix}, \quad \mathbf{x}_4 = \begin{bmatrix} 1 \\ 0 \end{bmatrix}, \quad \cdots$$

so that we will not reach a steady-state vector for this choice of $\mathbf{x}_0$. On the other hand, if we start with $\mathbf{x}_0 = \begin{bmatrix} 0.5 \\ 0.5 \end{bmatrix}$, then we have

$$\mathbf{x}_1 = \begin{bmatrix} 0.5 \\ 0.5 \end{bmatrix}, \quad \mathbf{x}_2 = \begin{bmatrix} 0.5 \\ 0.5 \end{bmatrix}, \quad \mathbf{x}_3 = \begin{bmatrix} 0.5 \\ 0.5 \end{bmatrix}, \quad \cdots$$

Thus this choice for $\mathbf{x}_0$ yields a steady-state vector. In fact, for the matrix A, this choice for $\mathbf{x}_0$ is the *only* initial state vector that leads to a steady-state vector. (See Exercise 42.)

A stochastic matrix may have many different steady-state vectors, depending on the initial state vector. For instance, suppose

$$A = \begin{bmatrix} 1 & 0 & 1/3 \\ 0 & 1 & 1/3 \\ 0 & 0 & 1/3 \end{bmatrix}$$

Then it can be verified (see Exercise 52) that

$$\text{Initial State: } \mathbf{x}_0 = \begin{bmatrix} 0.5 \\ 0.25 \\ 0.25 \end{bmatrix} \implies \text{steady-state vector: } \mathbf{x} = \begin{bmatrix} 0.625 \\ 0.375 \\ 0 \end{bmatrix}$$

and

$$\text{Initial State: } \mathbf{x}_0 = \begin{bmatrix} 0.2 \\ 0.6 \\ 0.2 \end{bmatrix} \implies \text{steady-state vector: } \mathbf{x} = \begin{bmatrix} 0.3 \\ 0.7 \\ 0 \end{bmatrix}.$$

In general for this case, solving the system $(A - I)\mathbf{x} = \mathbf{0}$ yields the general solution

$$\mathbf{x} = s_1 \begin{bmatrix} 1 \\ 0 \\ 0 \end{bmatrix} + s_2 \begin{bmatrix} 0 \\ 1 \\ 0 \end{bmatrix} = \begin{bmatrix} s_1 \\ s_2 \\ 0 \end{bmatrix}$$

Thus being a stochastic matrix is not enough to ensure that there will be a unique steady-state vector. It turns out that we need one additional condition.

DEFINITION 3.28

Let A be a stochastic matrix. Then A is **regular** if for some integer $k \geq 1$ the matrix A^k has all strictly positive entries.

Definition Regular Matrix

Note that both $A = \begin{bmatrix} 0 & 1 \\ 1 & 0 \end{bmatrix}$ and $A = \begin{bmatrix} 1 & 0 & 1/3 \\ 0 & 1 & 1/3 \\ 0 & 0 & 1/3 \end{bmatrix}$ are not regular. On the other hand, even though

$$B = \begin{bmatrix} 1/2 & 1/2 & 1/2 \\ 1/2 & 0 & 1/2 \\ 0 & 1/2 & 0 \end{bmatrix} \quad \text{and} \quad B^2 = \begin{bmatrix} 1/2 & 1/2 & 1/2 \\ 1/4 & 1/2 & 1/4 \\ 1/4 & 0 & 1/4 \end{bmatrix}$$

both have zero entries, B is regular because B^3 has all positive entries,

$$B^3 = \begin{bmatrix} 1/2 & 1/2 & 1/2 \\ 3/8 & 1/4 & 3/8 \\ 1/8 & 1/4 & 1/8 \end{bmatrix}$$

THEOREM 3.29

Let A be a regular stochastic matrix. Then

(a) For any initial state vector $\mathbf{x}_0$, the Markov chain $\mathbf{x}_0, \mathbf{x}_1, \mathbf{x}_2, \ldots$ converges to a unique steady-state vector $\mathbf{x}$.

(b) The sequence $A, A^2, A^3, \ldots$ converges to the matrix $\begin{bmatrix} \mathbf{x} & \mathbf{x} & \cdots & \mathbf{x} \end{bmatrix}$, where $\mathbf{x}$ is the unique steady-state vector given in part (a).

Proof   The proof of part (a) is beyond the scope of this book, but it can be found in texts on Markov chains. To prove part (b), we begin by noting that

$$\begin{aligned} A^n &= A^{n-1} A \\ &= A^{n-1} \begin{bmatrix} \mathbf{a}_1 & \mathbf{a}_2 & \cdots & \mathbf{a}_n \end{bmatrix} \\ &= \begin{bmatrix} A^{n-1}\mathbf{a}_1 & A^{n-1}\mathbf{a}_2 & \cdots & A^{n-1}\mathbf{a}_n \end{bmatrix} \end{aligned}$$

From part (a) we know that as n grows, $A^{n-1}\mathbf{a}_j$ converges to $\mathbf{x}$ for each of $j = 1, 2, \ldots, n$. Thus it follows that

$$A^n \to \begin{bmatrix} \mathbf{x} & \mathbf{x} & \cdots & \mathbf{x} \end{bmatrix}$$

completing the proof.   ∎

Theorem 3.29 shows that a regular stochastic matrix will have a unique steady-state vector that is independent of the initial state vector, which explains why we had no trouble solving our earlier examples.

| EXERCISES |

In Exercises 1–4, determine if A is a stochastic matrix.

1. $A = \begin{bmatrix} 0.2 & 0.6 \\ 0.8 & 0.4 \end{bmatrix}$

2. $A = \begin{bmatrix} 1.5 & 0.15 \\ -0.5 & 0.85 \end{bmatrix}$

3. $A = \begin{bmatrix} \frac{1}{5} & 1 & 0 \\ \frac{2}{5} & 0 & \frac{1}{2} \\ \frac{2}{5} & 0 & \frac{1}{2} \end{bmatrix}$

4. $A = \begin{bmatrix} \frac{3}{14} & \frac{3}{8} & \frac{1}{2} \\ \frac{1}{2} & \frac{3}{8} & \frac{1}{2} \\ \frac{3}{14} & \frac{1}{4} & \frac{1}{2} \end{bmatrix}$

In Exercises 5–8, fill in the missing values to make A a stochastic matrix.

5. $A = \begin{bmatrix} a & 0.45 \\ 0.65 & b \end{bmatrix}$

6. $A = \begin{bmatrix} a & 0.7 & 0.2 \\ 0.35 & b & 0.4 \\ 0.2 & 0.25 & c \end{bmatrix}$

7. $A = \begin{bmatrix} \frac{2}{13} & \frac{3}{7} & c \\ a & \frac{3}{7} & \frac{1}{5} \\ \frac{3}{13} & b & \frac{7}{10} \end{bmatrix}$

8. $A = \begin{bmatrix} a & 0.5 & 0.2 & 0.05 \\ 0.45 & 0.15 & 0.4 & d \\ 0.1 & b & c & 0.25 \\ 0 & 0.2 & 0.30 & 0.15 \end{bmatrix}$

In Exercises 9–12, if possible, fill in the missing values to make A a doubly stochastic matrix.

9. $A = \begin{bmatrix} a & 0.3 \\ 0.3 & b \end{bmatrix}$

10. $A = \begin{bmatrix} 0.4 & 0.6 \\ a & b \end{bmatrix}$

11. $A = \begin{bmatrix} 0.2 & a & 0.3 \\ b & 0.1 & c \\ d & 0.4 & 0.2 \end{bmatrix}$

12. $A = \begin{bmatrix} a & 0.5 & b \\ 0.2 & c & 0.2 \\ 0.5 & 0.1 & d \end{bmatrix}$

In Exercises 13–16, find the state vector $\mathbf{x}_3$ for the given stochastic matrix and initial state vector.

13. $A = \begin{bmatrix} 0.2 & 0.6 \\ 0.8 & 0.4 \end{bmatrix}$, $\mathbf{x}_0 = \begin{bmatrix} 0.2 \\ 0.8 \end{bmatrix}$

14. $A = \begin{bmatrix} 0.5 & 0.3 \\ 0.5 & 0.7 \end{bmatrix}$, $\mathbf{x}_0 = \begin{bmatrix} 0.3 \\ 0.7 \end{bmatrix}$

15. $A = \begin{bmatrix} \frac{1}{3} & \frac{2}{5} \\ \frac{2}{3} & \frac{3}{5} \end{bmatrix}$, $\mathbf{x}_0 = \begin{bmatrix} \frac{1}{2} \\ \frac{1}{2} \end{bmatrix}$

16. $A = \begin{bmatrix} \frac{1}{4} & \frac{3}{7} \\ \frac{3}{4} & \frac{4}{7} \end{bmatrix}$, $\mathbf{x}_0 = \begin{bmatrix} \frac{1}{3} \\ \frac{2}{3} \end{bmatrix}$

In Exercises 17–20, find all steady-state vectors for the given stochastic matrix.

17. $A = \begin{bmatrix} 0.8 & 0.5 \\ 0.2 & 0.5 \end{bmatrix}$

18. $A = \begin{bmatrix} 0.3 & 0.6 \\ 0.7 & 0.4 \end{bmatrix}$

19. $A = \begin{bmatrix} 0.4 & 0.5 & 0.3 \\ 0.2 & 0.3 & 0.4 \\ 0.4 & 0.2 & 0.3 \end{bmatrix}$

20. $A = \begin{bmatrix} 0.3 & 0 & 0 \\ 0.2 & 1 & 0 \\ 0.5 & 0 & 1 \end{bmatrix}$

In Exercises 21–24, determine if the given stochastic matrix is regular.

21. $A = \begin{bmatrix} 1 & 0.4 \\ 0 & 0.6 \end{bmatrix}$

22. $A = \begin{bmatrix} 0.3 & 0 \\ 0.7 & 1 \end{bmatrix}$

23. $A = \begin{bmatrix} 0.7 & 0.2 & 0.1 \\ 0.3 & 0.8 & 0.4 \\ 0 & 0 & 0.5 \end{bmatrix}$

24. $A = \begin{bmatrix} 0 & 0.2 & 0.5 \\ 0.9 & 0 & 0.5 \\ 0.1 & 0.8 & 0 \end{bmatrix}$

FIND AN EXAMPLE For Exercises 25–30, find an example that meets the given specifications.

25. A 4×4 stochastic matrix.

26. A 4×4 doubly stochastic matrix.

27. A 2×2 stochastic matrix A that has $\begin{bmatrix} 2/3 \\ 1/3 \end{bmatrix}$ for a steady-state vector.

28. A 2×3 stochastic matrix A that has $\begin{bmatrix} 0.5 \\ 0.25 \\ 0.25 \end{bmatrix}$ for a steady-state vector.

29. A 3×3 stochastic matrix A and initial state vector $\mathbf{x}_0$ such that the Markov chain $A\mathbf{x}_0$, $A^2\mathbf{x}_0$, ... does not converge to a steady-state vector.

30. A 3×3 stochastic matrix A that has exactly one initial state vector $\mathbf{x}_0$ that will generate a Markov chain with a steady-state vector.

TRUE OR FALSE For Exercises 31–36, determine if the statement is true or false, and justify your answer.

31. If A is a stochastic matrix, then so is A^T.

32. If A is an $n \times n$ stochastic matrix and

$$\mathbf{x} = \begin{bmatrix} 1/n \\ \vdots \\ 1/n \end{bmatrix}, \quad \text{then} \quad A\mathbf{x} = \begin{bmatrix} 1 \\ \vdots \\ 1 \end{bmatrix}$$

33. If A and B are stochastic $n \times n$ matrices, then AB^T is also stochastic.

34. If A is a symmetric stochastic matrix, then A is doubly stochastic.

35. All Markov chains converge to a steady-state vector.

36. Every 2×2 stochastic matrix has at least one steady-state vector.

37. Prove Theorem 3.27(a): Show that each state vector is a probability vector.

38. Prove Theorem 3.27(b): Show that the product of two stochastic matrices is a stochastic matrix. (HINT: Appeal to Theorem 3.27(a).)

39. Prove Theorem 3.27(c): Show that if A is a stochastic matrix, then so is A^2, A^3, ... (HINT: Use induction and appeal to Theorem 3.27(b).)

40. Suppose that A is a regular stochastic matrix. Show that A^2 is also a regular stochastic matrix.

41. Let $A = \begin{bmatrix} a & b \\ c & d \end{bmatrix}$ be a doubly stochastic matrix. Prove that $a = d$ and $b = c$.

42. If $A = \begin{bmatrix} 0 & 1 \\ 1 & 0 \end{bmatrix}$, prove that $\mathbf{x}_0 = \begin{bmatrix} 0.5 \\ 0.5 \end{bmatrix}$ is the only initial state vector that will lead to a steady-state vector.

43. Let A be a regular stochastic matrix, and suppose that k is the smallest integer such that A^k has all strictly positive entries. Show that each of A^{k+1}, A^{k+2}, ... will have strictly positive entries.

44. Let A be an upper or lower triangular stochastic matrix. Show that A is not regular.

45. Suppose that $A = \begin{bmatrix} \alpha & 0 \\ (1-\alpha) & 1 \end{bmatrix}$, where $0 < \alpha < 1$.

(a) Explain why A is a stochastic matrix.

(b) Find a formula for A^k, and use it to show that A is not regular.

(c) As $k \to \infty$, what matrix does A^k converge to?

(d) Find the one steady-state vector for A. (Note that this shows the converse of Theorem 3.29 does not hold: A stochastic matrix that is not regular can still have a unique steady-state vector.)

46. Ⓒ In an office complex of 1000 employees, on any given day some are at work and the rest are absent. It is known that if an employee is at work today, there is an 85% chance that she will be at work tomorrow, and if the employee is absent today, there is a 60% chance that she will be absent tomorrow. Suppose that today there are 760 employees at work.

(a) Find the transition matrix for this scenario.

(b) Predict the number that will be at work five days from now.

(c) Find the steady-state vector.

47. Ⓒ The star quarterback of a university football team has decided to return for one more season. He tells one person, who in turn tells someone else, and so on, with each person talking to someone who has not heard the news. At each step in this chain, if the message heard is "yes" (he is returning), then there is a 10% chance it will be changed to "no," and if the message heard is "no," then there is a 15% chance that it will be changed to "yes."

(a) Find the transition matrix for this scenario.

(b) Determine the probability that the sixth person in the chain hears the wrong news.

(c) Find the steady-state vector.

48. Ⓒ It has been claimed that the best predictor of today's weather is yesterday's weather. Suppose that in the town of Springfield, if it rained yesterday, then there is a 60% chance of rain today, and if it did not rain yesterday, then there is an 85% chance of no rain today.

(a) Find the transition matrix describing the rain probabilities.

(b) If it rained Tuesday, what is the probability of rain Thursday?

(c) If it did not rain Friday, what is the probability of rain Monday?

(d) If the probability of rain today is 30%, what is the probability of rain tomorrow?

(e) Find the steady-state vector.

49. Ⓒ Consumers in Shelbyville have a choice of one of two fast food restaurants, Krusty's and McDonald's. Both have trouble keeping customers. Of those who last went to Krusty's, 65% will go to McDonald's next time, and of those who last went to McDonald's, 80% will go to Krusty's next time.

(a) Find the transition matrix describing this situation.

(b) A customer goes out for fast food every Sunday, and just went to Krusty's.

i. What is the probability that two Sundays from now she will go to McDonald's?

ii. What is the probability that three Sundays from now she will go to McDonald's?

(c) Suppose a consumer has just moved to Shelbyville, and there is a 40% chance that he will go to Krusty's for his first fast food outing. What is the probability that his third fast food experience will be at Krusty's?

(d) Find the steady-state vector.

50. Ⓒ An assembly line turns out two types of pastries, Chocolate Zots and Rainbow Wahoos. The pastries come out one at a time; 40% of the time, a Wahoo follows a Zot, and 25% of the time, a Zot follows a Wahoo.

(a) If a Zot has just emerged from the line, what is the probability that a Wahoo will come two pastries later?

(b) If a Zot has just emerged from the line, what is the probability that a Zot will come three pastries later?

(c) What is the long-term probability that a randomly emerging pastry will be a Wahoo?

51. Ⓒ A medium-size town has three public library branches, designated A, B, and C. Patrons checking out books can return them to any of the three branches, where the books stay until checked out again. History shows that books borrowed from each branch are returned to a given location based on the following probabilities:

		Borrowed		
		A	B	C
	A	0.4	0.1	0.2
Returned	B	0.3	0.7	0.7
	C	0.3	0.2	0.1

(a) If a book is borrowed from A, what is the probability that it ends up at C after two more circulations?

(b) If a book is borrowed from B, what is the probability that it ends up at B after three more circulations?

(c) What is the steady-state vector?

52. Ⓒ Let $A = \begin{bmatrix} 1 & 0 & 1/3 \\ 0 & 1 & 1/3 \\ 0 & 0 & 1/3 \end{bmatrix}$.

Numerically verify that each initial state vector x_0 has the given steady-state vector x.

(a) $x_0 = \begin{bmatrix} 0.5 \\ 0.25 \\ 0.25 \end{bmatrix} \Longrightarrow x = \begin{bmatrix} 0.625 \\ 0.375 \\ 0 \end{bmatrix}$

(b) $x_0 = \begin{bmatrix} 0.2 \\ 0.6 \\ 0.2 \end{bmatrix} \Longrightarrow x = \begin{bmatrix} 0.3 \\ 0.7 \\ 0 \end{bmatrix}$

Ⓒ For Exercises 53–54, determine to six decimal places the steady-state vector corresponding to the given initial state vector. Also find the smallest integer k such that $x_k = x_{k+1}$ to 6 decimal places for all entries. (NOTE: Even if it is less computationally efficient, it may be easier to compute state vectors using powers of A instead of the recursive formula.)

53. $A = \begin{bmatrix} 0.2 & 0.3 & 0.1 & 0.4 \\ 0.3 & 0.5 & 0.6 & 0.2 \\ 0.1 & 0.1 & 0.2 & 0.2 \\ 0.4 & 0.1 & 0.1 & 0.2 \end{bmatrix}$, $\mathbf{x}_0 = \begin{bmatrix} 0.25 \\ 0.25 \\ 0.25 \\ 0.25 \end{bmatrix}$

54. $A = \begin{bmatrix} 0 & 1 & 0.2 & 0.5 \\ 0.2 & 0 & 0.3 & 0 \\ 0.5 & 0 & 0.4 & 0.5 \\ 0.3 & 0 & 0.1 & 0 \end{bmatrix}$, $\mathbf{x}_0 = \begin{bmatrix} 0.1 \\ 0.2 \\ 0.3 \\ 0.4 \end{bmatrix}$

Ⓒ For Exercises 55–56, use computational experimentation to find two initial state vectors that lead to different steady-state vectors. (NOTE: Even if it is less computationally efficient, it may be easier to compute state vectors using powers of A instead of the recursive formula.)

55. $A = \begin{bmatrix} 0.5 & 0 & 0 & 0.5 \\ 0 & 1 & 0 & 0 \\ 0 & 0 & 1 & 0.5 \\ 0.5 & 0 & 0 & 0 \end{bmatrix}$

56. $A = \begin{bmatrix} 1 & 0 & 0 & 0 \\ 0 & 0.5 & 0 & 0.5 \\ 0 & 0 & 1 & 0.5 \\ 0 & 0.5 & 0 & 0 \end{bmatrix}$

Subspaces

The Oresund Bridge connects Denmark and Sweden in a combination of an above ground, cable-stayed bridge and underground tunnel. The bridge allows vehicular traffic to reach either country from the other in less than 10 minutes. At 1,611 feet, its cable-stayed main span is the longest in the world. The span joins the Drogden tunnel on the artificial island of Peberholm, compiled of rock and soil excavated during construction. The undersea tube tunnel, which is made from prefabricated, interlocking concrete segments, carries both roads and railroad tracks.

Bridge suggested by Alain D'Amour, Southern Connecticut State University (Pierre Mens/Øresundsbro)

Subspaces are a special type of subset of Euclidean space $\mathbf{R}^n$. They arise naturally in connection with spanning sets, linear transformations, and systems of linear equations. Section 4.1 provides an introduction to, and includes examples of, subspaces and a general procedure for determining if a subset is a subspace. Section 4.2 introduces an important type of vector set and a means for measuring (roughly) the size of a subspace. Section 4.3 connects the concept of subspaces to matrices.

4.1 Introduction to Subspaces

In Section 2.2, we introduced the hypothetical VecMobile II in $\mathbf{R}^3$, the vehicle that can move only in the direction of vectors

$$\mathbf{u}_1 = \begin{bmatrix} 2 \\ 1 \\ 1 \end{bmatrix} \quad \text{and} \quad \mathbf{u}_2 = \begin{bmatrix} 1 \\ 2 \\ 3 \end{bmatrix}$$

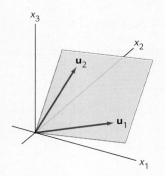

Figure 1 Subspace traversed by the VecMobile II.

Recall that this model of the VecMobile II can travel to any location in span$\{\mathbf{u}_1, \mathbf{u}_2\}$, the set of all linear combinations of $\mathbf{u}_1$ and $\mathbf{u}_2$, which forms a plane in $\mathbf{R}^3$ (Figure 1). This subset of $\mathbf{R}^3$ is an example of a *subspace*. In many ways, a subspace of $\mathbf{R}^n$ resembles $\mathbf{R}^m$ for some $m \leq n$.

DEFINITION 4.1

Definition Subspace

A subset S of $\mathbf{R}^n$ is a **subspace** if S satisfies the following three conditions:

   (a) S contains $\mathbf{0}$, the zero vector.

   (b) If $\mathbf{u}$ and $\mathbf{v}$ are in S, then $\mathbf{u} + \mathbf{v}$ is also in S.

   (c) If r is a real number and $\mathbf{u}$ is in S, then $r\mathbf{u}$ is also in S.

Definition Closed Under Addition, Closed Under Scalar Multiplication

A subset of $\mathbf{R}^n$ that satisfies condition (b) above is said to be **closed under addition**, and if it satisfies condition (c), then it is **closed under scalar multiplication**. Closure under addition and scalar multiplication ensures that arithmetic performed on vectors in a subspace produce other vectors in the subspace.

▶ Geometrically, condition (a) says that the graph of a subspace must pass through the origin.

Is S a Subspace?

To determine if a given subset S is a subspace, an easy place to start is with condition (a) of Definition 4.1, which states that every subspace must contain $\mathbf{0}$. A moment's thought reveals that this is equivalent to the statement

    If $\mathbf{0}$ is not in a subset S, then S is not a subspace.

Note that the converse is not true. Just because $\mathbf{0}$ is in S does not guarantee that S is a subspace, because conditions (b) and (c) must also be satisfied. For example, despite containing $\mathbf{0}$, the subset of $\mathbf{R}^2$ consisting of the x-axis and y-axis (Figure 2) is not a subspace of $\mathbf{R}^2$, because the set is not closed under addition. (However, it is closed under scalar multiplication.)

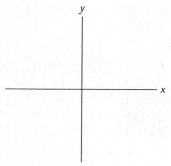

Figure 2 The coordinate axes are not a subspace of $\mathbf{R}^2$.

| **EXAMPLE 1** | Let S consist of all solutions $\mathbf{x} = (x_1, x_2)$ to the linear system

$$-3x_1 + 2x_2 = 17$$
$$x_1 - 5x_2 = -1$$

Is S a subspace of $\mathbf{R}^2$?

Solution We know that S is a subset of $\mathbf{R}^2$. However, note that $\mathbf{x} = (0, 0)$ is *not* a solution to the given system. Hence $\mathbf{0}$ is not in S, and so S cannot be a subspace of $\mathbf{R}^2$. ■

This section opened with the statement that the span of two vectors forms a subspace of $\mathbf{R}^3$. This claim generalizes to the span of any finite set of vectors in $\mathbf{R}^n$ and provides a useful way to determine if a set of vectors is a subspace.

THEOREM 4.2

▶ Theorem 4.2 also holds for the span of an infinite set of vectors. However, we do not require this case and including it would introduce additional technical issues, so we leave it out.

Let $S = \text{span}\{\mathbf{u}_1, \mathbf{u}_2, \ldots, \mathbf{u}_m\}$ be a subset of $\mathbf{R}^n$. Then S is a subspace of $\mathbf{R}^n$.

Proof To show that a subset is a subspace, we need to verify that the three conditions given in the definition are satisfied.

(a) Since $\mathbf{0} = 0\mathbf{u}_1 + \cdots + 0\mathbf{u}_m$, it follows that S contains $\mathbf{0}$.

(b) Suppose that $\mathbf{v}$ and $\mathbf{w}$ are in S. Then there exist scalars $r_1, r_2, \ldots r_m$ and $s_1, s_2, \ldots s_m$ such that

$$\mathbf{v} = r_1\mathbf{u}_1 + r_2\mathbf{u}_2 + \cdots r_m\mathbf{u}_m$$
$$\mathbf{w} = s_1\mathbf{u}_1 + s_2\mathbf{u}_2 + \cdots s_m\mathbf{u}_m$$

Then

$$\mathbf{v} + \mathbf{w} = (r_1\mathbf{u}_1 + r_2\mathbf{u}_2 + \cdots r_m\mathbf{u}_m) + (s_1\mathbf{u}_1 + s_2\mathbf{u}_2 + \cdots s_m\mathbf{u}_m)$$
$$= (r_1 + s_1)\mathbf{u}_1 + (r_2 + s_2)\mathbf{u}_2 + \cdots (r_m + s_m)\mathbf{u}_m$$

so $\mathbf{v} + \mathbf{w}$ is in $\text{span}\{\mathbf{u}_1, \mathbf{u}_2, \ldots, \mathbf{u}_m\} = S$.

(c) If t is a real number, then taking $\mathbf{v}$ as in part (b), we have

$$t\mathbf{v} = t(r_1\mathbf{u}_1 + r_2\mathbf{u}_2 + \cdots r_m\mathbf{u}_m)$$
$$= tr_1\mathbf{u}_1 + tr_2\mathbf{u}_2 + \cdots tr_m\mathbf{u}_m$$

so that $t\mathbf{v}$ is in S.

Since parts (a)–(c) of the definition hold, S is a subspace. ■

Definition Subspace Spanned, Subspace Generated

If $S = \mathrm{span}\{\mathbf{u}_1, \mathbf{u}_2, \ldots, \mathbf{u}_m\}$, then it is common to say that S is the **subspace spanned** (or **subspace generated**) by $\{\mathbf{u}_1, \mathbf{u}_2, \ldots, \mathbf{u}_m\}$.

DETERMINING IF S IS A SUBSPACE   To determine if a subset S is a subspace, apply the following steps.

Step 1. Check if $\mathbf{0}$ is in S. If not, then S is not a subspace.

Step 2. If you can show that S is generated by a set of vectors, then by Theorem 4.2 S is a subspace.

Step 3. Try to verify that conditions (b) and (c) of the definition are met. If so, then S is a subspace. If you cannot show that they hold, then you are likely to uncover a counterexample showing that they do not hold, which demonstrates that S is not a subspace.

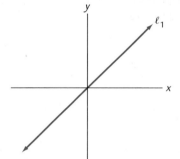

Figure 3 ℓ_1 is a subspace.

Let's try this out on some examples.

EXAMPLE 2   Determine if $S = \{\mathbf{0}\}$ and $S = \mathbf{R}^n$ are subspaces of $\mathbf{R}^n$.

Solution Since $\mathbf{0}$ is in both $S = \{\mathbf{0}\}$ and $S = \mathbf{R}^n$, Step 1 is no help, so we move to Step 2. Since $\{\mathbf{0}\} = \mathrm{span}\{\mathbf{0}\}$, by Theorem 4.2 the set $S = \{\mathbf{0}\}$ is a subspace. We also have $\mathbf{R}^n = \mathrm{span}\{\mathbf{e}_1, \mathbf{e}_2, \ldots, \mathbf{e}_n\}$, where

$$\mathbf{e}_1 = \begin{bmatrix} 1 \\ 0 \\ \vdots \\ 0 \end{bmatrix}, \quad \mathbf{e}_2 = \begin{bmatrix} 0 \\ 1 \\ \vdots \\ 0 \end{bmatrix}, \quad \cdots, \quad \mathbf{e}_n = \begin{bmatrix} 0 \\ 0 \\ \vdots \\ 1 \end{bmatrix} \tag{1}$$

Thus $\mathbf{R}^n$ is a subspace of itself. (These are sometimes called the **trivial subspaces** of $\mathbf{R}^n$.) ■

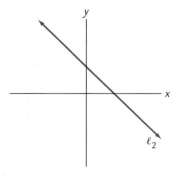

Figure 4 ℓ_2 does not pass through the origin, so is not a subspace.

EXAMPLE 3   Let ℓ_1 denote a line through the origin in $\mathbf{R}^2$ (Figure 3), and let ℓ_2 denote a line that does not pass through the origin in $\mathbf{R}^2$ (Figure 4). Do the points on ℓ_1 form a subspace? Do the points on ℓ_2 form a subspace?

Solution Since ℓ_1 passes through the origin, $\mathbf{0}$ is on ℓ_1, so Step 1 is not helpful. Moving to Step 2, suppose that we pick any nonzero vector $\mathbf{u}$ on ℓ_1. Then all points on ℓ_1 have the form $r\mathbf{u}$ for some scalar r. Thus $\ell_1 = \mathrm{span}\{\mathbf{u}\}$, so ℓ_1 is a subspace.

Focusing now on the line ℓ_2, we note that it does not contain $\mathbf{0}$, so ℓ_2 is not a subspace. ■

EXAMPLE 4 Let S be the subset of $\mathbf{R}^3$ consisting of all vectors of the form

$$\mathbf{v} = \begin{bmatrix} v_1 \\ v_2 \\ v_3 \end{bmatrix}$$

such that $v_1 + v_2 + v_3 = 0$. Is S a subspace of $\mathbf{R}^3$?

Solution Starting with Step 1, we see that setting $v_1 = v_2 = v_3 = 0$ implies $\mathbf{0}$ is in S, so we still cannot conclude anything from this. A spanning set for S is not immediately presenting itself (although we could find one), so let's skip to Step 3 and determine if conditions (b) and (c) of the definition are satisfied.

(b) Let $\mathbf{u} = \begin{bmatrix} u_1 \\ u_2 \\ u_3 \end{bmatrix}$ and $\mathbf{v} = \begin{bmatrix} v_1 \\ v_2 \\ v_3 \end{bmatrix}$ be in S. Then

$$\mathbf{u} + \mathbf{v} = \begin{bmatrix} u_1 + v_1 \\ u_2 + v_2 \\ u_3 + v_3 \end{bmatrix}$$

and since

$$(u_1 + v_1) + (u_2 + v_2) + (u_3 + v_3) = (u_1 + u_2 + u_3) + (v_1 + v_2 + v_3)$$
$$= 0 + 0 = 0$$

it follows that $\mathbf{u} + \mathbf{v}$ is in S.

(c) With $\mathbf{v} = \begin{bmatrix} v_1 \\ v_2 \\ v_3 \end{bmatrix}$ in S as above, for any scalar r we have $r\mathbf{v} = \begin{bmatrix} rv_1 \\ rv_2 \\ rv_3 \end{bmatrix}$. Since

$$rv_1 + rv_2 + rv_3 = r(v_1 + v_2 + v_3) = 0$$

$r\mathbf{v}$ is also in S.

Since all conditions of the definition are satisfied, we conclude that S is a subspace of $\mathbf{R}^3$. ∎

It is not hard to extend the result in Example 4 to $\mathbf{R}^n$. Let S be the set of all vectors of the form

$$\mathbf{v} = \begin{bmatrix} v_1 \\ \vdots \\ v_n \end{bmatrix}$$

such that $v_1 + \cdots + v_n = 0$. Then S is a subspace of $\mathbf{R}^n$ (see Exercise 15).

Homogeneous Systems and Null Spaces

The set of solutions to a homogeneous linear system forms a subspace. For instance, let A be the 3×4 matrix

$$A = \begin{bmatrix} 3 & -1 & 7 & -6 \\ 4 & -1 & 9 & -7 \\ -2 & 1 & -5 & 5 \end{bmatrix}$$

Using our usual algorithm, we find that all solutions to the homogeneous linear system $A\mathbf{x} = \mathbf{0}$ have the form

$$\mathbf{x} = s_1 \begin{bmatrix} -2 \\ 1 \\ 1 \\ 0 \end{bmatrix} + s_2 \begin{bmatrix} 1 \\ -3 \\ 0 \\ 1 \end{bmatrix}$$

where s_1 and s_2 can be any real numbers. Thus the set of solutions to $A\mathbf{x} = \mathbf{0}$ is equal to

$$\text{span} \left\{ \begin{bmatrix} -2 \\ 1 \\ 1 \\ 0 \end{bmatrix}, \begin{bmatrix} 1 \\ -3 \\ 0 \\ 1 \end{bmatrix} \right\}$$

and so the set of solutions is a subspace of $\mathbf{R}^4$. This result generalizes to the set of solutions of any homogeneous linear system.

THEOREM 4.3

If A is an $n \times m$ matrix, then the set of solutions to the homogeneous linear system $A\mathbf{x} = \mathbf{0}$ forms a subspace of $\mathbf{R}^m$.

Proof We verify the three conditions from Definition 4.1 to show that the set forms a subspace.

(a) Since $\mathbf{x} = \mathbf{0}$ is a solution to $A\mathbf{x} = \mathbf{0}$, the zero vector $\mathbf{0}$ is in the set of solutions.

(b) Suppose that $\mathbf{u}$ and $\mathbf{v}$ are both solutions to $A\mathbf{x} = \mathbf{0}$. Then

$$A(\mathbf{u} + \mathbf{v}) = A\mathbf{u} + A\mathbf{v} = \mathbf{0} + \mathbf{0} = \mathbf{0}$$

so that $\mathbf{u} + \mathbf{v}$ is in the set of solutions.

(c) Let $\mathbf{u}$ be a solution to $A\mathbf{x} = \mathbf{0}$, and let r be a scalar. Then

$$A(r\mathbf{u}) = r(A\mathbf{u}) = r\mathbf{0} = \mathbf{0}$$

and so $r\mathbf{u}$ is also in the set of solutions.

Since all three conditions of the definition are met, the set of solutions to $A\mathbf{x} = \mathbf{0}$ is a subspace of $\mathbf{R}^n$. ∎

A subspace given by the set of solutions to a homogeneous linear system goes by a special name.

DEFINITION 4.4

Definition Null Space

If A is an $n \times m$ matrix, then the set of solutions to $A\mathbf{x} = \mathbf{0}$ is called the **null space** of A and is denoted by null(A).

From Theorem 4.3 it follows that a null space is a subspace.

EXAMPLE 5 Ethane burns in oxygen to produce carbon dioxide and steam. The chemical reaction is described using the notation

$$x_1 C_2H_6 + x_2 O_2 \longrightarrow x_3 CO_2 + x_4 H_2O$$

where the subscripts on the elements indicate the number of atoms in each molecule. Describe the subspace of values that will balance this equation.

▶ Ethane is a gas similar to propane. Its primary use in the chemical industry is to make polyethylene, a common form of plastic.

▶ Balancing chemical equations is discussed in detail in Section 1.4.

Solution To balance the equation, we need to find values for x_1, x_2, x_3, and x_4 so that the number of atoms for each element is the same on both sides of the equation. Doing so yields the linear system

$$
\begin{array}{rcll}
2x_1 \quad - \quad x_3 \quad\quad &= 0 & \text{(carbon atoms)} \\
6x_1 \quad\quad\quad - 2x_4 &= 0 & \text{(hydrogen atoms)} \\
2x_2 - 2x_3 - \quad x_4 &= 0 & \text{(oxygen atoms)}
\end{array}
$$

Applying our usual methods, we find that the general solution to this system is

$$
\begin{array}{ll}
\begin{array}{l}
x_1 = 2s \\
x_2 = 7s \\
x_3 = 4s \\
x_4 = 6s
\end{array}
&
\text{or} \quad \mathbf{x} = s \begin{bmatrix} 2 \\ 7 \\ 4 \\ 6 \end{bmatrix}
\end{array}
$$

where s can be any real number. Put another way, the set of solutions is equal to

$$
\text{span} \left\{ \begin{bmatrix} 2 \\ 7 \\ 4 \\ 6 \end{bmatrix} \right\}
$$

which makes it clear that the set is a subspace of $\mathbf{R}^4$. ∎

Kernel and Range of a Linear Transformation

There are two sets associated with any linear transformation T that are subspaces. Recall that the range of T is the set of all vectors $\mathbf{y}$ such that $T(\mathbf{x}) = \mathbf{y}$ for some $\mathbf{x}$ and is denoted by range(T). The **kernel** of T is the set of vectors $\mathbf{x}$ such that $T(\mathbf{x}) = \mathbf{0}$. The kernel of T is denoted by ker(T) (see Figure 5). Theorem 4.5 shows that the range and kernel are subspaces.

Definition Kernel

THEOREM 4.5

Let $T : \mathbf{R}^m \to \mathbf{R}^n$ be a linear transformation. Then the kernel of T is a subspace of the domain $\mathbf{R}^m$ and the range of T is a subspace of the codomain $\mathbf{R}^n$.

Proof Because $T : \mathbf{R}^m \to \mathbf{R}^n$ is a linear transformation, it follows (Theorem 3.8, Section 3.1) that there exists an $n \times m$ matrix $A = \begin{bmatrix} \mathbf{a}_1 & \cdots & \mathbf{a}_m \end{bmatrix}$ such that $T(\mathbf{x}) = A\mathbf{x}$. Thus $T(\mathbf{x}) = \mathbf{0}$ if and only if $A\mathbf{x} = \mathbf{0}$. This implies that

$$\ker(T) = \text{null}(A)$$

and therefore by Theorem 4.3 the kernel of T is a subspace of the domain $\mathbf{R}^m$.

Now consider the range of T. By Theorem 3.3(b), we have

$$\text{range}(T) = \text{span}\{\mathbf{a}_1, \ldots, \mathbf{a}_m\}$$

Since range(T) is equal to the span of a set of vectors, by Theorem 4.2 the range of T is a subspace of the codomain $\mathbf{R}^n$. ∎

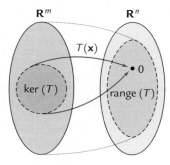

Figure 5 The kernel and range of T.

EXAMPLE 6 Suppose that $T : \mathbf{R}^2 \longrightarrow \mathbf{R}^3$ is defined by

$$
T\left(\begin{bmatrix} x_1 \\ x_2 \end{bmatrix} \right) = \begin{bmatrix} x_1 - 2x_2 \\ -3x_1 + 6x_2 \\ 2x_1 - 4x_2 \end{bmatrix}
$$

Find ker(T) and range(T).

Solution We have $T(\mathbf{x}) = A\mathbf{x}$ for

$$A = \begin{bmatrix} 1 & -2 \\ -3 & 6 \\ 2 & -4 \end{bmatrix}$$

To find the null space of A, we solve the homogeneous linear system $A\mathbf{x} = \mathbf{0}$. We have

$$\begin{bmatrix} 1 & -2 & 0 \\ -3 & 6 & 0 \\ 2 & -4 & 0 \end{bmatrix} \sim \begin{bmatrix} 1 & -2 & 0 \\ 0 & 0 & 0 \\ 0 & 0 & 0 \end{bmatrix}$$

which is equivalent to the single equation $x_1 - 2x_2 = 0$. Since $\ker(T) = \text{null}(A)$, it follows that if we let $x_2 = s$, then $x_1 = 2s$ and thus

▶ Remember that the kernel is a subspace of the *domain*, while the range is a subspace of the *codomain*.

$$\ker(T) = s \begin{bmatrix} 2 \\ 1 \end{bmatrix} \ (s \text{ real}) \quad \text{or} \quad \ker(T) = \text{span}\left\{ \begin{bmatrix} 2 \\ 1 \end{bmatrix} \right\}$$

Because the range of T is equal to the span of the columns of A, we have

$$\text{range}(T) = \text{span}\{\mathbf{a}_1, \mathbf{a}_2\} = \text{span}\left\{ \begin{bmatrix} 1 \\ -3 \\ 2 \end{bmatrix}, \begin{bmatrix} -2 \\ 6 \\ -4 \end{bmatrix} \right\} = \text{span}\left\{ \begin{bmatrix} 1 \\ -3 \\ 2 \end{bmatrix} \right\}$$

because $\mathbf{a}_1 = -2\mathbf{a}_2$. ∎

In Theorem 3.5 in Section 3.1 we showed that a linear transformation T is one-to-one if and only if $T(\mathbf{x}) = \mathbf{0}$ has only the trivial solution. The next theorem formulates this result in terms of $\ker(T)$.

THEOREM 4.6

Let $T : \mathbf{R}^m \to \mathbf{R}^n$ be a linear transformation. Then T is one-to-one if and only if $\ker(T) = \{\mathbf{0}\}$.

The proof is covered in Exercise 72. As a quick application, in Example 6 we saw that $\ker(T) \neq \{\mathbf{0}\}$, so we can conclude from Theorem 4.6 that T is not one-to-one.

▶ This updates the Big Theorem, Version 3, from Section 3.3.

The Big Theorem, Version 4

Theorem 4.6 gives us another condition to add to the Big Theorem.

THEOREM 4.7

(THE BIG THEOREM, VERSION 4) Let $\mathcal{A} = \{\mathbf{a}_1, \ldots, \mathbf{a}_n\}$ be a set of n vectors in $\mathbf{R}^n$, let $A = \begin{bmatrix} \mathbf{a}_1 & \cdots & \mathbf{a}_n \end{bmatrix}$, and let $T : \mathbf{R}^n \to \mathbf{R}^n$ be given by $T(\mathbf{x}) = A\mathbf{x}$. Then the following are equivalent:

(a) $\mathcal{A}$ spans $\mathbf{R}^n$.

(b) $\mathcal{A}$ is linearly independent.

(c) $A\mathbf{x} = \mathbf{b}$ has a unique solution for all $\mathbf{b}$ in $\mathbf{R}^n$.

(d) T is onto.

(e) T is one-to-one.

(f) A is invertible.

(g) $\ker(T) = \{\mathbf{0}\}$.

Proof From The Big Theorem, Version 3, we know that (a) through (f) are equivalent. From Theorem 4.6 we know that T is one-to-one if and only if $\ker(T) = \{\mathbf{0}\}$, so (e) and (g) are equivalent. Thus (a)–(g) are all equivalent. ■

EXERCISES

In Exercises 1–16, determine if the described set is a subspace. If so, give a proof. If not, explain why not. Unless stated otherwise, a, b, and c are real numbers.

1. The subset of $\mathbf{R}^3$ consisting of vectors of the form $\begin{bmatrix} a \\ 0 \\ b \end{bmatrix}$.

2. The subset of $\mathbf{R}^3$ consisting of vectors of the form $\begin{bmatrix} a \\ a \\ 0 \end{bmatrix}$.

3. The subset of $\mathbf{R}^2$ consisting of vectors of the form $\begin{bmatrix} a \\ b \end{bmatrix}$, where $a + b = 1$.

4. The subset of $\mathbf{R}^3$ consisting of vectors of the form $\begin{bmatrix} a \\ b \\ c \end{bmatrix}$, where $a = b = c$.

5. The subset of $\mathbf{R}^4$ consisting of vectors of the form $\begin{bmatrix} a \\ 1 \\ 0 \\ b \end{bmatrix}$.

6. The subset of $\mathbf{R}^4$ consisting of vectors of the form $\begin{bmatrix} a \\ a+b \\ 2a-b \\ 3b \end{bmatrix}$.

7. The subset of $\mathbf{R}^2$ consisting of vectors of the form $\begin{bmatrix} a \\ b \end{bmatrix}$, where a and b are integers.

8. The subset of $\mathbf{R}^3$ consisting of vectors of the form $\begin{bmatrix} a \\ b \\ c \end{bmatrix}$, where $c = b - a$.

9. The subset of $\mathbf{R}^3$ consisting of vectors of the form $\begin{bmatrix} a \\ b \\ c \end{bmatrix}$, where $abc = 0$.

10. The subset of $\mathbf{R}^2$ consisting of vectors of the form $\begin{bmatrix} a \\ b \end{bmatrix}$, where $a^2 + b^2 \le 1$.

11. The subset of $\mathbf{R}^3$ consisting of vectors of the form $\begin{bmatrix} a \\ b \\ c \end{bmatrix}$, where $a \ge 0$, $b \ge 0$, and $c \ge 0$.

12. The subset of $\mathbf{R}^3$ consisting of vectors of the form $\begin{bmatrix} a \\ b \\ c \end{bmatrix}$, where at most one of a, b, and c is nonzero.

13. The subset of $\mathbf{R}^2$ consisting of vectors of the form $\begin{bmatrix} a \\ b \end{bmatrix}$, where $a \le b$.

14. The subset of $\mathbf{R}^2$ consisting of vectors of the form $\begin{bmatrix} a \\ b \end{bmatrix}$, where $|a| = |b|$.

15. The subset of $\mathbf{R}^n$ consisting of vectors of the form

$$\mathbf{v} = \begin{bmatrix} v_1 \\ \vdots \\ v_n \end{bmatrix}$$

such that $v_1 + \cdots + v_n = 0$.

16. The subset of $\mathbf{R}^n$ (n even) consisting of vectors of the form

$$\mathbf{v} = \begin{bmatrix} v_1 \\ \vdots \\ v_n \end{bmatrix}$$

such that $v_1 - v_2 + v_3 - v_4 + v_5 - \cdots - v_n = 0$.

In Exercises 17–20, the shaded region is not a subspace of $\mathbf{R}^2$. Explain why.

17.

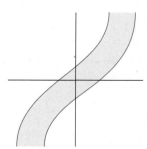

18.

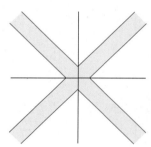

19.

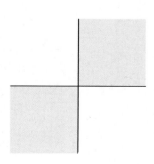

20.

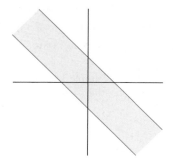

In Exercises 21–32, find the null space for A.

21. $A = \begin{bmatrix} 1 & -3 \\ 0 & 1 \end{bmatrix}$

22. $A = \begin{bmatrix} 3 & 5 \\ 6 & 4 \end{bmatrix}$

23. $A = \begin{bmatrix} 1 & 0 & -5 \\ 0 & 1 & 2 \end{bmatrix}$

24. $A = \begin{bmatrix} 1 & 2 & -2 \\ 0 & 1 & 4 \end{bmatrix}$

25. $A = \begin{bmatrix} 1 & -2 & 2 \\ -2 & 5 & -7 \end{bmatrix}$

26. $A = \begin{bmatrix} 3 & 0 & -4 \\ -1 & 6 & 2 \end{bmatrix}$

27. $A = \begin{bmatrix} 1 & 3 \\ -2 & 1 \\ 3 & 2 \end{bmatrix}$

28. $A = \begin{bmatrix} 2 & -10 \\ -3 & 15 \\ 1 & -5 \end{bmatrix}$

29. $A = \begin{bmatrix} 1 & -1 & 1 \\ 0 & 1 & 3 \\ 0 & 0 & 3 \end{bmatrix}$

30. $A = \begin{bmatrix} 1 & 2 & 0 \\ -3 & -4 & -1 \\ 2 & -2 & 3 \end{bmatrix}$

31. $A = \begin{bmatrix} 1 & 1 & -2 & 1 \\ 0 & 1 & 1 & -1 \\ 0 & 0 & 0 & 2 \end{bmatrix}$

32. $A = \begin{bmatrix} 1 & 0 & 0 & 1 \\ 0 & 2 & 1 & 0 \\ 0 & 0 & 1 & 0 \\ 1 & 0 & 1 & 1 \end{bmatrix}$

In Exercises 33–36, let $T(\mathbf{x}) = A\mathbf{x}$ for the matrix A. Determine if the vector $\mathbf{b}$ is in the kernel of T and if the vector $\mathbf{c}$ is in the range of T.

33. $A = \begin{bmatrix} 1 & -2 \\ -3 & -1 \end{bmatrix}$, $\mathbf{b} = \begin{bmatrix} 2 \\ 1 \end{bmatrix}$, $\mathbf{c} = \begin{bmatrix} 4 \\ -7 \end{bmatrix}$

34. $A = \begin{bmatrix} 2 & -3 & 0 \\ 1 & 4 & -2 \end{bmatrix}$, $\mathbf{b} = \begin{bmatrix} 6 \\ 4 \\ 11 \end{bmatrix}$, $\mathbf{c} = \begin{bmatrix} 4 \\ 13 \end{bmatrix}$

35. $A = \begin{bmatrix} 4 & -2 \\ 1 & 3 \\ 2 & 7 \end{bmatrix}$, $\mathbf{b} = \begin{bmatrix} -5 \\ 2 \end{bmatrix}$, $\mathbf{c} = \begin{bmatrix} 1 \\ 3 \end{bmatrix}$

36. $A = \begin{bmatrix} 1 & 2 & 3 \\ 4 & 5 & 6 \\ 7 & 8 & 9 \end{bmatrix}$, $\mathbf{b} = \begin{bmatrix} 1 \\ -2 \\ 1 \end{bmatrix}$, $\mathbf{c} = \begin{bmatrix} 2 \\ 5 \\ 8 \end{bmatrix}$

FIND AN EXAMPLE For Exercises 37–44, find an example that meets the given specifications.

37. An infinite subset of $\mathbf{R}^2$ that is not a subspace of $\mathbf{R}^2$.

38. Two subspaces S_1 and S_2 of $\mathbf{R}^3$ such that $S_1 \cup S_2$ is not a subspace of $\mathbf{R}^3$.

39. Two nonsubspace subsets S_1 and S_2 of $\mathbf{R}^3$ such that $S_1 \cup S_2$ is a subspace of $\mathbf{R}^3$.

40. Two nonsubspace subsets S_1 and S_2 of $\mathbf{R}^3$ such that $S_1 \cap S_2$ is a subspace of $\mathbf{R}^3$.

41. A linear transformation $T : \mathbf{R}^2 \to \mathbf{R}^2$ such that range$(T) =$ span $\left\{ \begin{bmatrix} 1 \\ 1 \end{bmatrix} \right\}$.

42. A linear transformation $T : \mathbf{R}^2 \to \mathbf{R}^3$ such that range$(T) =$ span $\left\{ \begin{bmatrix} 1 \\ -1 \\ 2 \end{bmatrix} \right\}$.

43. A linear transformation $T : \mathbf{R}^3 \to \mathbf{R}^3$ such that range$(T) = \mathbf{R}^3$.

44. A linear transformation $T : \mathbf{R}^3 \to \mathbf{R}^3$ such that range$(T) =$ span $\left\{ \begin{bmatrix} 3 \\ 1 \\ 4 \end{bmatrix}, \begin{bmatrix} 1 \\ 2 \\ -2 \end{bmatrix} \right\}$.

TRUE OR FALSE For Exercises 45–60, determine if the statement is true or false, and justify your answer.

45. If A is an $n \times n$ matrix and $\mathbf{b} \neq \mathbf{0}$ is in $\mathbf{R}^n$, then the solutions to $A\mathbf{x} = \mathbf{b}$ do not form a subspace.

46. If A is a 5×3 matrix, then null(A) forms a subspace of $\mathbf{R}^5$.

47. If A is a 4×7 matrix, then null(A) forms a subspace of $\mathbf{R}^7$.

48. Let $T : \mathbf{R}^6 \to \mathbf{R}^3$ be a linear transformation. Then ker(T) is a subspace of $\mathbf{R}^6$.

49. Let $T : \mathbf{R}^5 \to \mathbf{R}^8$ be a linear transformation. Then ker(T) is a subspace of $\mathbf{R}^8$.

50. Let $T : \mathbf{R}^2 \to \mathbf{R}^7$ be a linear transformation. Then range(T) is a subspace of $\mathbf{R}^2$.

51. Let $T : \mathbf{R}^3 \to \mathbf{R}^9$ be a linear transformation. Then range(T) is a subspace of $\mathbf{R}^9$.

52. The union of two subspaces of $\mathbf{R}^n$ forms another subspace of $\mathbf{R}^n$.

53. The intersection of two subspaces of $\mathbf{R}^n$ forms another subspace of $\mathbf{R}^n$.

54. Let S_1 and S_2 be subspaces of $\mathbf{R}^n$, and define S to be the set of all vectors of the form $\mathbf{s}_1 + \mathbf{s}_2$, where $\mathbf{s}_1$ is in S_1 and $\mathbf{s}_2$ is in S_2. Then S is a subspace of $\mathbf{R}^n$.

55. Let S_1 and S_2 be subspaces of $\mathbf{R}^n$, and define S to be the set of all vectors of the form $\mathbf{s}_1 - \mathbf{s}_2$, where $\mathbf{s}_1$ is in S_1 and $\mathbf{s}_2$ is in S_2. Then S is a subspace of $\mathbf{R}^n$.

56. The set of integers forms a subspace of $\mathbf{R}$.

57. A subspace $S \neq \{\mathbf{0}\}$ can have a finite number of vectors.

58. If $\mathbf{u}$ and $\mathbf{v}$ are in a subspace S, then every point on the line connecting $\mathbf{u}$ and $\mathbf{v}$ is also in S.

59. If S_1 and S_2 are subsets of $\mathbf{R}^n$ but *not* subspaces, then the union of S_1 and S_2 cannot be a subspace of $\mathbf{R}^n$.

60. If S_1 and S_2 are subsets of $\mathbf{R}^n$ but *not* subspaces, then the intersection of S_1 and S_2 cannot be a subspace of $\mathbf{R}^n$.

61. Show that every subspace of $\mathbf{R}$ is either $\{\mathbf{0}\}$ or $\mathbf{R}$.

62. Suppose that S is a subspace of $\mathbf{R}^n$ and c is a scalar. Let cS denote the set of vectors $c\mathbf{s}$ where $\mathbf{s}$ is in S. Prove that cS is also a subspace of $\mathbf{R}^n$.

63. Prove that if $\mathbf{b} \neq \mathbf{0}$, then the set of solutions to $A\mathbf{x} = \mathbf{b}$ is not a subspace.

64. Describe the geometric form of all subspaces of $\mathbf{R}^2$.

65. Describe the geometric form of all subspaces of $\mathbf{R}^3$.

66. Some texts use just conditions (b) and (c) in Definition 4.1, along with S nonempty, as the definition of a subspace. Explain why this is equivalent to our definition.

67. Let A be an $n \times m$ matrix, and suppose that $\mathbf{y} \neq \mathbf{0}$ is in $\mathbf{R}^n$. Show that the set of all vectors $\mathbf{x}$ in $\mathbf{R}^m$ such that $A\mathbf{x} = \mathbf{y}$ is *not* a subspace of $\mathbf{R}^m$.

68. Let $A = \begin{bmatrix} \mathbf{a}_1 & \mathbf{a}_2 & \mathbf{a}_3 & \mathbf{a}_4 \end{bmatrix}$, and suppose that $\mathbf{x} = (2, -5, 4, 1)$ is in $\mathbf{null}(A)$. Write $\mathbf{a}_4$ as a linear combination of the other three vectors.

69. Let A be a matrix and $T(\mathbf{x}) = A\mathbf{x}$ a linear transformation. Show that $\ker(T) = \{\mathbf{0}\}$ if and only if the columns of A are linearly independent.

70. If T is a linear transformation, show that $\mathbf{0}$ is always in $\ker(T)$.

71. Prove that if $\mathbf{u}$ and $\mathbf{v}$ are in a subspace S, then so is $\mathbf{u} - \mathbf{v}$.

72. Prove Theorem 4.6: If T is a linear transformation, then T is one-to-one if and only if $\ker(T) = \{\mathbf{0}\}$.

[C] In Exercises 73–76, use Example 5 as a guide to find the subspace of values that balances the given chemical equation.

73. Glucose ferments to form ethyl alcohol and carbon dioxide.

$$x_1 C_6H_{12}O_6 \longrightarrow x_2 C_2H_5OH + x_3 CO_2$$

74. Methane burns in oxygen to form carbon dioxide and steam.

$$x_1 CH_4 + x_2 O_2 \longrightarrow x_3 CO_2 + x_4 H_2O$$

75. An antacid (calcium hydroxide) neutralizes stomach acid (hydrochloric acid) to form calcium chloride and water.

$$x_1 Ca(OH)_2 + x_2 HCl \longrightarrow x_3 CaCl_2 + x_4 H_2O$$

76. Ethyl alcohol reacts with oxygen to form vinegar and water.

$$x_1 C_2H_5OH + x_2 O_2 \longrightarrow x_3 HC_2H_3O_2 + x_4 H_2O$$

[C] In Exercises 77–80, find the null space for the given matrix.

77. $A = \begin{bmatrix} 1 & 7 & -2 & 14 & 0 \\ 3 & 0 & 1 & -2 & 3 \\ 6 & 1 & -1 & 0 & 4 \end{bmatrix}$

78. $A = \begin{bmatrix} -1 & 0 & 0 & 4 & 5 & 2 \\ 6 & 2 & 1 & 2 & 4 & 0 \\ 3 & 2 & -5 & -1 & 0 & 2 \end{bmatrix}$

79. $A = \begin{bmatrix} 3 & 1 & 2 & 4 \\ 5 & 0 & 2 & -1 \\ 2 & 2 & 2 & 2 \\ -1 & 0 & 3 & 1 \\ 0 & 2 & 0 & 4 \end{bmatrix}$

80. $A = \begin{bmatrix} 2 & 0 & 5 \\ -1 & 6 & 2 \\ 4 & 4 & -1 \\ 5 & 1 & 0 \\ 4 & 1 & 1 \end{bmatrix}$

4.2 Basis and Dimension

In this section we combine the concepts of linearly independent sets and spanning sets to learn more about subspaces. Let $S = \text{span}\{\mathbf{u}_1, \mathbf{u}_2, \ldots, \mathbf{u}_m\}$ be a subspace of $\mathbf{R}^n$. Then every element $\mathbf{s}$ of S can be written as a linear combination

$$\mathbf{s} = r_1\mathbf{u}_1 + r_2\mathbf{u}_2 + \cdots + r_m\mathbf{u}_m$$

If $\mathbf{u}_1, \ldots, \mathbf{u}_m$ is a linearly dependent set, then by Theorem 2.14 we know that one of the vectors in the set – say, $\mathbf{u}_1$ – is in the span of the remaining vectors. Thus it follows that every element of S can be written as a linear combination of $\mathbf{u}_2, \ldots, \mathbf{u}_m$, so that

$$S = \text{span}\{\mathbf{u}_2, \ldots, \mathbf{u}_m\}$$

If after eliminating $\mathbf{u}_1$ the remaining set of vectors is still linearly dependent, then we can repeat this process to eliminate another dependent vector. We can do this over and over, and since we started with a finite number of vectors the process must eventually lead us to a set that both spans S and is linearly independent. Such a set is particularly important and goes by a special name.

DEFINITION 4.8

Definition Basis

A set $\mathcal{B} = \{\mathbf{u}_1, \ldots, \mathbf{u}_m\}$ is a **basis** for a subspace S if

(a) $\mathcal{B}$ spans S.

(b) $\mathcal{B}$ is linearly independent.

Figure 1 and Figure 2 show basis vectors for $\mathbf{R}^2$ and $\mathbf{R}^3$, respectively. Note that there is one subspace for which the above procedure will not work: $S = \{\mathbf{0}\} = \mathrm{span}\{\mathbf{0}\}$, the zero subspace. The set $\{\mathbf{0}\}$ is not linearly independent, and there are no vectors that can be removed. The zero subspace is the only subspace of $\mathbf{R}^n$ that does not have a basis.

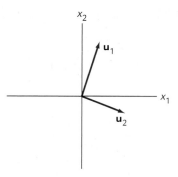

Figure 1 Any two nonzero vectors that do not lie on the same line forms a basis for $\mathbf{R}^2$.

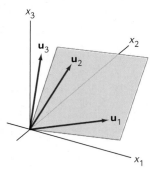

Figure 2 Any three nonzero vectors that do not lie in the same plane forms a basis for $\mathbf{R}^3$.

Each basis has the following important property.

THEOREM 4.9

Let $\mathcal{B} = \{\mathbf{u}_1, \ldots, \mathbf{u}_m\}$ be a basis for a subspace S. Then every vector $\mathbf{s}$ in S can be written as a linear combination

$$\mathbf{s} = s_1\mathbf{u}_1 + \cdots + s_m\mathbf{u}_m$$

in exactly one way.

Proof   Because $\mathcal{B}$ is a basis for S, the vectors in $\mathcal{B}$ span S, so that every vector $\mathbf{s}$ can be written as a linear combination of vectors in $\mathcal{B}$ in *at least* one way. To show that there can only be one way to write $\mathbf{s}$, let's suppose that there are two, say,

$$\mathbf{s} = r_1\mathbf{u}_1 + \cdots + r_m\mathbf{u}_m \quad \text{and} \quad \mathbf{s} = t_1\mathbf{u}_1 + \cdots + t_m\mathbf{u}_m$$

Then $r_1\mathbf{u}_1 + \cdots + r_m\mathbf{u}_m = t_1\mathbf{u}_1 + \cdots + t_m\mathbf{u}_m$, so that after reorganizing we have

$$(r_1 - t_1)\mathbf{u}_1 + \cdots + (r_m - t_m)\mathbf{u}_m = \mathbf{0}$$

Since $\mathcal{B}$ is a basis, it is also a linearly independent set, and therefore it must be that $r_1 - t_1 = 0, \ldots, r_m - t_m = 0$. Hence $r_1 = t_1, \ldots, r_m = t_m$, so that there is just one way to express $\mathbf{s}$ as a linear combination of the vectors in $\mathcal{B}$. ■

This is important enough to repeat: Theorem 4.9 tells us that every vector in a subspace S can be expressed in *exactly* one way as a linear combination of vectors in a basis $\mathcal{B}$.

Finding a Basis

Frequently, a subspace S is described as the span of a set of vectors; that is, $S = \text{span}\{\mathbf{u}_1, \mathbf{u}_2, \ldots, \mathbf{u}_m\}$. Example 1 demonstrates a way to find a basis in this situation. Before getting to the example, we pause to give a theorem that we will need soon. The proof is left as an exercise.

THEOREM 4.10

▶ Recall that two matrices A and B are equivalent if A can be transformed into B through a sequence of elementary row operations.

Let A and B be equivalent matrices. Then the subspace spanned by the rows of A is the same as the subspace spanned by the rows of B.

The next example shows one way to find a basis from a spanning set.

EXAMPLE 1 Let S be the subspace of $\mathbf{R}^4$ spanned by the vectors

$$\mathbf{u}_1 = \begin{bmatrix} 2 \\ -1 \\ 3 \\ 1 \end{bmatrix}, \quad \mathbf{u}_2 = \begin{bmatrix} 7 \\ -6 \\ 5 \\ 2 \end{bmatrix}, \quad \mathbf{u}_3 = \begin{bmatrix} -3 \\ 4 \\ 1 \\ 0 \end{bmatrix}$$

Find a basis for S.

Solution Start by using the vectors $\mathbf{u}_1$, $\mathbf{u}_2$, $\mathbf{u}_3$ to form the *rows* of a matrix.

$$A = \begin{bmatrix} \mathbf{u}_1 \\ \mathbf{u}_2 \\ \mathbf{u}_3 \end{bmatrix} = \begin{bmatrix} 2 & -1 & 3 & 1 \\ 7 & -6 & 5 & 2 \\ -3 & 4 & 1 & 0 \end{bmatrix}$$

Next, use row operations to transform A into the equivalent matrix B that is in echelon form.

$$A = \begin{bmatrix} 2 & -1 & 3 & 1 \\ 7 & -6 & 5 & 2 \\ -3 & 4 & 1 & 0 \end{bmatrix} \sim \begin{bmatrix} 5 & 0 & 13 & 4 \\ 0 & 5 & 11 & 3 \\ 0 & 0 & 0 & 0 \end{bmatrix} = B$$

By Theorem 4.10, we know that the subspace spanned by the rows of B is the same as the subspace spanned by the rows of A, so the rows of B span S. Moreover, since B is in echelon form, the nonzero rows are linearly independent (see Exercise 62, Section 2.3). Thus the set

$$\left\{ \begin{bmatrix} 5 \\ 0 \\ 13 \\ 4 \end{bmatrix}, \begin{bmatrix} 0 \\ 5 \\ 11 \\ 3 \end{bmatrix} \right\}$$

forms a basis for S. ■

Summarizing this solution method: To find a basis for $S = \text{span}\{\mathbf{u}_1, \ldots, \mathbf{u}_m\}$,

1. Use the vectors $\mathbf{u}_1, \ldots, \mathbf{u}_m$ to form the rows of a matrix A.

2. Transform A to echelon form B.

3. The nonzero rows of B give a basis for S.

Before proceeding, we pause to state the following useful result that will be used to show a second method for finding a basis for a subspace S. (The proof is left as an exercise.)

THEOREM 4.11

Suppose that $U = \begin{bmatrix} \mathbf{u}_1 & \cdots & \mathbf{u}_m \end{bmatrix}$ and $V = \begin{bmatrix} \mathbf{v}_1 & \cdots & \mathbf{v}_m \end{bmatrix}$ are two equivalent matrices. Then any linear dependence that exists among the vectors $\mathbf{u}_1, \ldots, \mathbf{u}_m$ also exists among the vectors $\mathbf{v}_1, \ldots, \mathbf{v}_m$.

For example, Theorem 4.11 tells us that

$$\text{if} \quad 3\mathbf{v}_1 - 2\mathbf{v}_4 + \mathbf{v}_6 = 5\mathbf{v}_2 \quad \text{then} \quad 3\mathbf{u}_1 - 2\mathbf{u}_4 + \mathbf{u}_6 = 5\mathbf{u}_2$$

Theorem 4.11 gives us another way to find a basis from a spanning set.

EXAMPLE 2   Let S be the subspace of $\mathbf{R}^4$ spanned by the vectors $\mathbf{u}_1$, $\mathbf{u}_2$, and $\mathbf{u}_3$ given in Example 1. Find a basis for S.

Solution This time we start by using the vectors $\mathbf{u}_1$, $\mathbf{u}_2$, $\mathbf{u}_3$ to form the *columns* of a matrix

$$A = \begin{bmatrix} \mathbf{u}_1 & \mathbf{u}_2 & \mathbf{u}_3 \end{bmatrix} = \begin{bmatrix} 2 & 7 & -3 \\ -1 & -6 & 4 \\ 3 & 5 & 1 \\ 1 & 2 & 0 \end{bmatrix}$$

Using row operations to transform A to echelon form, we have

$$B = \begin{bmatrix} \mathbf{v}_1 & \mathbf{v}_2 & \mathbf{v}_3 \end{bmatrix} = \begin{bmatrix} 1 & 6 & -4 \\ 0 & 1 & -1 \\ 0 & 0 & 0 \\ 0 & 0 & 0 \end{bmatrix}$$

The nice thing about the matrix B is that it is not hard to find the dependence relationship among the columns. For instance, we can readily verify that

$$2\mathbf{v}_1 - \mathbf{v}_2 = \mathbf{v}_3$$

Now we apply Theorem 4.11. Since $2\mathbf{v}_1 - \mathbf{v}_2 = \mathbf{v}_3$, then we also have $2\mathbf{u}_1 - \mathbf{u}_2 = \mathbf{u}_3$. Therefore

$$2\begin{bmatrix} 2 \\ -1 \\ 3 \\ 1 \end{bmatrix} - \begin{bmatrix} 7 \\ -6 \\ 5 \\ 2 \end{bmatrix} = \begin{bmatrix} -3 \\ 4 \\ 1 \\ 0 \end{bmatrix}$$

For B we have

$$\text{span}\{\mathbf{v}_1, \mathbf{v}_2, \mathbf{v}_3\} = \text{span}\{\mathbf{v}_1, \mathbf{v}_2\}$$

and $\mathbf{v}_1$ and $\mathbf{v}_2$ are linearly independent. Hence it follows that for A,

$$S = \text{span}\{\mathbf{u}_1, \mathbf{u}_2, \mathbf{u}_3\} = \text{span}\{\mathbf{u}_1, \mathbf{u}_2\}$$

and that $\mathbf{u}_1$ and $\mathbf{u}_2$ are linearly independent. Thus the set

$$\left\{ \begin{bmatrix} 2 \\ -1 \\ 3 \\ 1 \end{bmatrix}, \begin{bmatrix} 7 \\ -6 \\ 5 \\ 2 \end{bmatrix} \right\}$$

forms a basis for S.   ∎

Summarizing this solution method: To find a basis for $S = \text{span}\{\mathbf{u}_1, \ldots, \mathbf{u}_m\}$,

1. Use the vectors $\mathbf{u}_1, \ldots, \mathbf{u}_m$ to form the columns of a matrix A.

2. Transform A to echelon form B. The pivot columns of B will be linearly independent, and the other columns will be linearly dependent on the pivot columns.

3. The columns of A corresponding to the pivot columns of B form a basis for S.

The solution method in Example 1 will usually produce a subspace basis that is relatively "simple" in that the basis vectors will contain some zeros. The solution method in Example 2 produces a basis from a subset of the original spanning vectors, which is sometimes desirable. In general, each method will produce a different basis, so that a basis need not be unique.

Dimension

Example 1 and Example 2 show that a subspace can have more than one basis. However, note that each basis has two vectors. Although a given nonzero subspace will have more than one basis, the next theorem shows that a nonzero subspace has a fixed number of basis vectors.

THEOREM 4.12   If S is a subspace of $\mathbf{R}^n$, then every basis of S has the same number of vectors.

A proof of this theorem is given at the end of the section.

Since every basis for a subspace S has the same number of vectors, the following definition makes sense.

DEFINITION 4.13

Definition Dimension

Let S be a subspace of $\mathbf{R}^n$. Then the **dimension** of S is the number of vectors in any basis of S.

The zero subspace $S = \{\mathbf{0}\}$ has no basis and is defined to have dimension 0. At the other extreme, $\mathbf{R}^n$ is a subspace of itself, and in Example 2, Section 4.1, we showed that $\{\mathbf{e}_1, \ldots, \mathbf{e}_n\}$ spans $\mathbf{R}^n$. It is also clear that these vectors are linearly independent, so that $\{\mathbf{e}_1, \ldots, \mathbf{e}_n\}$ forms a basis—called the **standard basis**—of $\mathbf{R}^n$ (see Figure 3). Thus the dimension of $\mathbf{R}^n$ is n. It can be shown that $\mathbf{R}^n$ is the only subspace of $\mathbf{R}^n$ of dimension n (see Exercise 57).

Definition Standard Basis

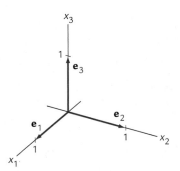

Figure 3 The standard basis for R^3.

EXAMPLE 3   Suppose that S is the subspace of $\mathbf{R}^5$ given by

$$S = \text{span}\left\{ \begin{bmatrix} -1 \\ 2 \\ 5 \\ -1 \\ -4 \end{bmatrix}, \begin{bmatrix} 3 \\ -6 \\ -15 \\ 3 \\ 12 \end{bmatrix}, \begin{bmatrix} -3 \\ 8 \\ 19 \\ -5 \\ -18 \end{bmatrix}, \begin{bmatrix} 5 \\ -3 \\ -11 \\ -2 \\ -1 \end{bmatrix} \right\}$$

Find the dimension of S.

Solution   Since our set has four vectors, we know that the dimension of S will be 4 or less. To find the dimension, we need to find a basis for S. It makes no difference how we do this, so let's use the solution method given in Example 2. Our vectors form the columns of the matrix on the left, with an echelon form given on the right.

$$\begin{bmatrix} -1 & 3 & -3 & 5 \\ 2 & -6 & 8 & -3 \\ 5 & -15 & 19 & -11 \\ -1 & 3 & -5 & -2 \\ -4 & 12 & -18 & -1 \end{bmatrix} \sim \begin{bmatrix} -1 & 3 & -3 & 5 \\ 0 & 0 & 2 & 7 \\ 0 & 0 & 0 & 0 \\ 0 & 0 & 0 & 0 \\ 0 & 0 & 0 & 0 \end{bmatrix}$$

Since the first and third columns of the echelon matrix are the pivot columns, we conclude that the first and third vectors from the original set

$$\left\{ \begin{bmatrix} -1 \\ 2 \\ 5 \\ -1 \\ -4 \end{bmatrix}, \begin{bmatrix} -3 \\ 8 \\ 19 \\ -5 \\ -18 \end{bmatrix} \right\}$$

form a basis for S. Hence the dimension of S is 2. ∎

In many instances it is handy to be able to modify a given set of vectors to serve as a basis. The following theorem gives two cases when this is possible.

THEOREM 4.14

Let $\mathcal{U} = \{\mathbf{u}_1, \ldots, \mathbf{u}_m\}$ be a set of vectors in a subspace $S \neq \{\mathbf{0}\}$ of $\mathbf{R}^n$.

(a) If $\mathcal{U}$ is linearly independent, then either $\mathcal{U}$ is a basis for S or additional vectors can be added to $\mathcal{U}$ to form a basis for S.

(b) If $\mathcal{U}$ spans S, then either $\mathcal{U}$ is a basis for S or vectors can be removed from $\mathcal{U}$ to form a basis for S.

Proof Taking part (a) first, if $\mathcal{U}$ also spans S, then we are done. If not, then select a vector $\mathbf{s}_1$ from S that is not in the span of $\mathcal{U}$ and form a new set

$$\mathcal{U}_1 = \{\mathbf{u}_1, \ldots, \mathbf{u}_m, \mathbf{s}_1\}$$

Then $\mathcal{U}_1$ must also be linearly independent, for if not then $\mathbf{s}_1$ would be in the span of $\mathcal{U}$. If $\mathcal{U}_1$ spans S, then we are done. If not, select a vector $\mathbf{s}_2$ that is not in the span of $\mathcal{U}_1$ and let

$$\mathcal{U}_2 = \{\mathbf{u}_1, \ldots, \mathbf{u}_m, \mathbf{s}_1, \mathbf{s}_2\}$$

▶ The process cannot go on indefinitely. Since all of the vectors are in $\mathbf{R}^n$, no set can have more than n linearly independent vectors (see Theorem 2.13, Section 2.3).

As before, $\mathcal{U}_2$ must be linearly independent. If $\mathcal{U}_2$ spans S, then we are done. If not, repeat this procedure again and again, until we finally have a linearly independent set that also spans S, giving a basis.

For part (b), we start with a spanning set. All we need to do is employ the solution method from Example 2, which will give a subset of $\mathcal{U}$ that forms a basis for S. (Or we can use the method described at the beginning of the section, removing one vector at a time until reaching a basis.) ∎

EXAMPLE 4 Expand the set $\mathcal{U} = \left\{ \begin{bmatrix} 1 \\ 1 \\ -2 \end{bmatrix}, \begin{bmatrix} 3 \\ 2 \\ -4 \end{bmatrix} \right\}$ to a basis for $\mathbf{R}^3$.

Solution Since $\mathbf{R}^3$ has dimension 3, we know that $\mathcal{U}$ does not have enough vectors to be a basis. We can see that the two vectors in $\mathcal{U}$ are linearly independent, so by

Theorem 4.14(a) we can expand $\mathcal{U}$ to a basis of $\mathbf{R}^3$. We know that the standard basis $\{\mathbf{e}_1, \mathbf{e}_2, \mathbf{e}_3\}$ forms a basis for $\mathbf{R}^3$, so that

$$\mathbf{R}^3 = \text{span}\left\{\begin{bmatrix} 1 \\ 1 \\ -2 \end{bmatrix}, \begin{bmatrix} 3 \\ 2 \\ -4 \end{bmatrix}, \begin{bmatrix} 1 \\ 0 \\ 0 \end{bmatrix}, \begin{bmatrix} 0 \\ 1 \\ 0 \end{bmatrix}, \begin{bmatrix} 0 \\ 0 \\ 1 \end{bmatrix}\right\}$$

Now we form the matrix

$$A = \begin{bmatrix} 1 & 3 & 1 & 0 & 0 \\ 1 & 2 & 0 & 1 & 0 \\ -2 & -4 & 0 & 0 & 1 \end{bmatrix}$$

and then apply the solution method from Example 2, which will give us a basis for $\mathbf{R}^3$. Since we placed the vectors that we want to include in the left columns, we are assured that they will end up among the basis vectors. Employing our usual row operations, we find an echelon form equivalent to A is

$$A = \begin{bmatrix} 1 & 3 & 1 & 0 & 0 \\ 1 & 2 & 0 & 1 & 0 \\ -2 & -4 & 0 & 0 & 1 \end{bmatrix} \sim \begin{bmatrix} 2 & 0 & -4 & 0 & -3 \\ 0 & 2 & 2 & 0 & 1 \\ 0 & 0 & 0 & 2 & 1 \end{bmatrix} = B$$

Since the pivots are in the 1st, 2nd, and 4th columns of B, referring back to A we see that the vectors

$$\left\{\begin{bmatrix} 1 \\ 1 \\ -2 \end{bmatrix}, \begin{bmatrix} 3 \\ 2 \\ -4 \end{bmatrix}, \begin{bmatrix} 0 \\ 1 \\ 0 \end{bmatrix}\right\}$$

must be linearly independent and span $\mathbf{R}^3$, so the set forms a basis for $\mathbf{R}^3$. ■

EXAMPLE 5 The vector $\mathbf{x}_1$ is in the null space of A,

$$\mathbf{x}_1 = \begin{bmatrix} 7 \\ 3 \\ -6 \\ 4 \end{bmatrix}, \quad A = \begin{bmatrix} -3 & 3 & -6 & -6 \\ 2 & 6 & 0 & -8 \\ 0 & -8 & 4 & 12 \\ -3 & -7 & -1 & 9 \\ 2 & 10 & -2 & -14 \end{bmatrix}$$

Find a basis for the null space that includes $\mathbf{x}_1$.

Solution In Example 4, we were able to exploit the fact that we knew a basis for $\mathbf{R}^3$. Here we do not know a basis for the null space, so we start by determining the vector form of the general solution to $A\mathbf{x} = \mathbf{0}$ and use the vectors to form the initial basis. We skip the details and just report the news that

$$\text{null}(A) = \text{span}\left\{\begin{bmatrix} -1 \\ 3 \\ 0 \\ 2 \end{bmatrix}, \begin{bmatrix} -3 \\ 1 \\ 2 \\ 0 \end{bmatrix}\right\} \tag{1}$$

From this point we follow the procedure in Example 4, by forming the matrix with our given vector $\mathbf{x}_1$ and the two basis vectors in (1), and then finding an echelon form.

$$\begin{bmatrix} 7 & -1 & -3 \\ 3 & 3 & 1 \\ -6 & 0 & 2 \\ 4 & 2 & 0 \end{bmatrix} \sim \begin{bmatrix} 3 & 0 & -1 \\ 0 & 3 & 2 \\ 0 & 0 & 0 \\ 0 & 0 & 0 \end{bmatrix}$$

Since the pivots are in the first two columns, it follows that

$$\left\{ \begin{bmatrix} 7 \\ 3 \\ -6 \\ 4 \end{bmatrix}, \begin{bmatrix} -1 \\ 3 \\ 0 \\ 2 \end{bmatrix} \right\} \tag{2}$$

forms a basis for the null space of A that contains $\mathbf{x}_1$.

Note that (2) is not the only basis containing $\mathbf{x}_1$. For instance, if we reverse the order of the two vectors in (1) and follow the same procedure, we end up with the basis

$$\left\{ \begin{bmatrix} 7 \\ 3 \\ -6 \\ 4 \end{bmatrix}, \begin{bmatrix} -3 \\ 1 \\ 2 \\ 0 \end{bmatrix} \right\} \quad \blacksquare$$

Definition Nullity   The **nullity** of a matrix A is the dimension of the null space of A and is denoted by nullity(A). Thus in Example 5 we have nullity(A) = 2.

If we happen to know the dimension of a subspace S, then the following theorem makes it easier to determine if a given set forms a basis.

THEOREM 4.15

Let $\mathcal{U} = \{\mathbf{u}_1, \ldots, \mathbf{u}_m\}$ be a set of m vectors in a subspace S of dimension m. If $\mathcal{U}$ is either linearly independent or spans S, then $\mathcal{U}$ is a basis for S.

Proof First, suppose that $\mathcal{U}$ is linearly independent. If $\mathcal{U}$ does not span S, then by Theorem 4.14 we can add additional vectors to $\mathcal{U}$ to form a basis for S. But this gives a basis with more than m vectors, contradicting the assumption that the dimension of S equals m. Hence $\mathcal{U}$ also must span S and so is a basis.

A similar argument can be used to show that if $\mathcal{U}$ spans S then $\mathcal{U}$ is a basis. The details are left as an exercise. $\blacksquare$

EXAMPLE 6   Suppose that S is a subspace of $\mathbf{R}^3$ of dimension 2 containing the vectors in the set

$$\mathcal{U} = \left\{ \begin{bmatrix} -1 \\ 2 \\ 0 \end{bmatrix}, \begin{bmatrix} 3 \\ 7 \\ 1 \end{bmatrix} \right\}$$

Show that $\mathcal{U}$ is a basis for S.

Solution Since S has dimension 2 and $\mathcal{U}$ has two vectors, by Theorem 4.15 all we need to do to show that $\mathcal{U}$ is a basis for S is verify that $\mathcal{U}$ is linearly independent or spans S. We do not know enough about S to show that $\mathcal{U}$ spans S, but since the two vectors are not multiples of each other, $\mathcal{U}$ is a linearly independent set. Hence we can conclude that $\mathcal{U}$ is a basis for S. $\blacksquare$

Theorems 4.16 and 4.17 present more properties of the dimension of a subspace that are useful in certain situations. The proofs are left as exercises.

THEOREM 4.16

Suppose that S_1 and S_2 are both subspaces of $\mathbf{R}^n$ and that S_1 is a subset of S_2. Then $\dim(S_1) \leq \dim(S_2)$, and $\dim(S_1) = \dim(S_2)$ only if $S_1 = S_2$.

THEOREM 4.17

Let $\mathcal{U} = \{\mathbf{u}_1, \ldots, \mathbf{u}_m\}$ be a set of vectors in a subspace S of dimension k.

(a) If $m < k$, then $\mathcal{U}$ does not span S.

(b) If $m > k$, then $\mathcal{U}$ is not linearly independent.

The Big Theorem, Version 5

The results of this section give us another condition for the Big Theorem.

THEOREM 4.18

▶ This updates the Big Theorem, Version 4, given in Section 4.1.

(THE BIG THEOREM, VERSION 5) Let $\mathcal{A} = \{\mathbf{a}_1, \ldots, \mathbf{a}_n\}$ be a set of n vectors in $\mathbf{R}^n$, let $A = \begin{bmatrix} \mathbf{a}_1 & \cdots & \mathbf{a}_n \end{bmatrix}$, and let $T : \mathbf{R}^n \to \mathbf{R}^n$ be given by $T(\mathbf{x}) = A\mathbf{x}$. Then the following are equivalent:

(a) $\mathcal{A}$ spans $\mathbf{R}^n$.

(b) $\mathcal{A}$ is linearly independent.

(c) $A\mathbf{x} = \mathbf{b}$ has a unique solution for all $\mathbf{b}$ in $\mathbf{R}^n$.

(d) T is onto.

(e) T is one-to-one.

(f) A is invertible.

(g) $\ker(T) = \{\mathbf{0}\}$.

(h) $\mathcal{A}$ is a basis for $\mathbf{R}^n$.

Proof From The Big Theorem, Version 4, we know that (a) through (g) are equivalent. By Definition 4.8, (a) and (b) are equivalent to (h), completing the proof. ■

EXAMPLE 7 Let $x_1, x_2, \ldots, x_n$ be real numbers. The *Vandermonde matrix*, which arises in signal processing and coding theory, is given by

$$V = \begin{bmatrix} 1 & x_1 & x_1^2 & \cdots & x_1^{n-1} \\ 1 & x_2 & x_2^2 & \cdots & x_2^{n-1} \\ \vdots & \vdots & \vdots & \ddots & \vdots \\ 1 & x_n & x_n^2 & \cdots & x_n^{n-1} \end{bmatrix}$$

Show that if $x_1, x_2, \ldots, x_n$ are distinct, then the columns of V form a basis for $\mathbf{R}^n$.

Solution By The Big Theorem, Version 5, we can show that the columns of V form a basis for $\mathbf{R}^n$ by showing that the columns are linearly independent. Given real numbers $a_0, a_1, \ldots, a_{n-1}$, we have

$$a_0 \begin{bmatrix} 1 \\ 1 \\ \vdots \\ 1 \end{bmatrix} + a_1 \begin{bmatrix} x_1 \\ x_2 \\ \vdots \\ x_n \end{bmatrix} + \cdots + a_{n-1} \begin{bmatrix} x_1^{n-1} \\ x_2^{n-1} \\ \vdots \\ x_n^{n-1} \end{bmatrix} = \begin{bmatrix} a_0 + a_1 x_1 + \cdots + a_{n-1} x_1^{n-1} \\ a_0 + a_1 x_2 + \cdots + a_{n-1} x_2^{n-1} \\ \vdots \\ a_0 + a_1 x_n + \cdots + a_{n-1} x_n^{n-1} \end{bmatrix} \tag{3}$$

If the polynomial $f(x) = a_0 + a_1 x + a_2 x^2 + \cdots + a_{n-1} x^{n-1}$, then the right side of (3) is equal to

$$\begin{bmatrix} f(x_1) \\ f(x_2) \\ \vdots \\ f(x_n) \end{bmatrix}$$

▶ One consequence of the *Fundamental Theorem of Algebra* (proved by Gauss) is that a polynomial of degree m can have at most m distinct roots.

This is the zero vector only if each of $x_1, x_2, \ldots, x_n$ are roots of the polynomial f. But since the roots are distinct and f has degree at most $n-1$, the only way this can happen is if $f(x) = 0$, the identically zero polynomial. Hence $a_0 = \cdots = a_{n-1} = 0$, and so the columns of V are linearly independent. Therefore the columns of V form a basis for $\mathbf{R}^n$. ■

Proof of Theorem 4.12

We state the theorem again:

THEOREM 4.12

If S is a subspace of $\mathbf{R}^n$, then every basis of S has the same number of vectors.

Proof Suppose that we have a subspace S with two bases of different sizes. The argument that follows can be generalized (this is left as an exercise), but to simplify notation we assume that S has bases

$$\mathcal{U} = \{\mathbf{u}_1, \mathbf{u}_2\} \quad \text{and} \quad \mathcal{V} = \{\mathbf{v}_1, \mathbf{v}_2, \mathbf{v}_3\}$$

Since $\mathcal{U}$ spans S, it follows that $\mathbf{v}_1$, $\mathbf{v}_2$, and $\mathbf{v}_3$ can each be expressed as linear combinations of $\mathbf{u}_1$ and $\mathbf{u}_2$,

$$\begin{aligned} \mathbf{v}_1 &= c_{11}\mathbf{u}_1 + c_{12}\mathbf{u}_2 \\ \mathbf{v}_2 &= c_{21}\mathbf{u}_1 + c_{22}\mathbf{u}_2 \\ \mathbf{v}_3 &= c_{31}\mathbf{u}_1 + c_{32}\mathbf{u}_2 \end{aligned} \tag{4}$$

Now consider the equation

$$a_1\mathbf{v}_1 + a_2\mathbf{v}_2 + a_3\mathbf{v}_3 = \mathbf{0} \tag{5}$$

Substituting into (5) from (4) for $\mathbf{v}_1$, $\mathbf{v}_2$, and $\mathbf{v}_3$ gives

$$\begin{aligned} \mathbf{0} &= a_1(c_{11}\mathbf{u}_1 + c_{12}\mathbf{u}_2) + a_2(c_{21}\mathbf{u}_1 + c_{22}\mathbf{u}_2) + a_3(c_{31}\mathbf{u}_1 + c_{32}\mathbf{u}_2) \\ &= (a_1 c_{11} + a_2 c_{21} + a_3 c_{31})\mathbf{u}_1 + (a_1 c_{12} + a_2 c_{22} + a_3 c_{32})\mathbf{u}_2 \end{aligned}$$

Since $\mathcal{U}$ is linearly independent, we must have

$$\begin{aligned} a_1 c_{11} + a_2 c_{21} + a_3 c_{31} &= 0 \\ a_1 c_{12} + a_2 c_{22} + a_3 c_{32} &= 0 \end{aligned}$$

Now view a_1, a_2, and a_3 as variables in this homogeneous system. Since there are more variables than equations, the system must have infinitely many solutions. But this means that there are nontrivial solutions to (5), which implies that $\mathcal{V}$ is linearly dependent, a contradiction. (Remember that $\mathcal{V}$ is a basis.) Hence our assumption that there can be bases of two different sizes is incorrect, so all bases for a subspace must have the same number of vectors. ■

| EXERCISES |

In Exercises 1–4, determine if the vectors shown form a basis for **R**². Justify your answer.

1.

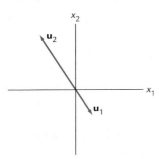

2.

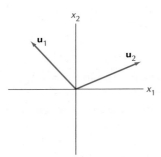

3.

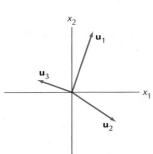

4.

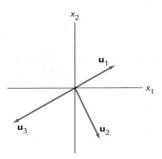

In Exercises 5–10, use the solution method from Example 1 to find a basis for the given subspace and give the dimension.

5. $S = \text{span}\left\{ \begin{bmatrix} 1 \\ -4 \end{bmatrix}, \begin{bmatrix} -5 \\ 20 \end{bmatrix} \right\}$

6. $S = \text{span}\left\{ \begin{bmatrix} 3 \\ 5 \end{bmatrix}, \begin{bmatrix} 9 \\ -2 \end{bmatrix} \right\}$

7. $S = \text{span}\left\{ \begin{bmatrix} 1 \\ 3 \\ -2 \end{bmatrix}, \begin{bmatrix} 2 \\ 4 \\ 1 \end{bmatrix}, \begin{bmatrix} -1 \\ 1 \\ -8 \end{bmatrix} \right\}$

8. $S = \text{span}\left\{ \begin{bmatrix} 2 \\ -1 \\ 3 \end{bmatrix}, \begin{bmatrix} 4 \\ -1 \\ 2 \end{bmatrix}, \begin{bmatrix} 2 \\ 1 \\ -5 \end{bmatrix} \right\}$

9. $S = \text{span}\left\{ \begin{bmatrix} 1 \\ -2 \\ 3 \\ -2 \end{bmatrix}, \begin{bmatrix} 0 \\ 2 \\ -5 \\ 1 \end{bmatrix}, \begin{bmatrix} 2 \\ -2 \\ 1 \\ -3 \end{bmatrix} \right\}$

10. $S = \text{span}\left\{ \begin{bmatrix} 1 \\ 0 \\ -1 \\ 1 \end{bmatrix}, \begin{bmatrix} 2 \\ 1 \\ 0 \\ 2 \end{bmatrix}, \begin{bmatrix} 0 \\ 1 \\ 2 \\ 0 \end{bmatrix}, \begin{bmatrix} 3 \\ 1 \\ -1 \\ 3 \end{bmatrix} \right\}$

In Exercises 11–16, use the solution method from Example 2 to find a basis for the given subspace and give the dimension.

11. $S = \text{span}\left\{ \begin{bmatrix} 1 \\ 3 \end{bmatrix}, \begin{bmatrix} 4 \\ -12 \end{bmatrix} \right\}$

12. $S = \text{span}\left\{ \begin{bmatrix} 2 \\ -6 \end{bmatrix}, \begin{bmatrix} -5 \\ 15 \end{bmatrix} \right\}$

13. $S = \text{span}\left\{ \begin{bmatrix} 1 \\ 2 \\ 4 \end{bmatrix}, \begin{bmatrix} 0 \\ 1 \\ -3 \end{bmatrix}, \begin{bmatrix} 3 \\ -2 \\ -1 \end{bmatrix} \right\}$

14. $S = \text{span}\left\{ \begin{bmatrix} 1 \\ 2 \\ 3 \end{bmatrix}, \begin{bmatrix} 3 \\ 7 \\ 5 \end{bmatrix}, \begin{bmatrix} -1 \\ -3 \\ 1 \end{bmatrix} \right\}$

15. $S = \text{span}\left\{ \begin{bmatrix} 1 \\ -1 \\ 0 \\ 2 \end{bmatrix}, \begin{bmatrix} 2 \\ -5 \\ 9 \\ 7 \end{bmatrix}, \begin{bmatrix} 0 \\ 1 \\ -3 \\ -1 \end{bmatrix} \right\}$

16. $S = \text{span}\left\{ \begin{bmatrix} 1 \\ 0 \\ 3 \\ 1 \end{bmatrix}, \begin{bmatrix} 4 \\ 2 \\ 13 \\ 4 \end{bmatrix}, \begin{bmatrix} 2 \\ 1 \\ 6 \\ 3 \end{bmatrix}, \begin{bmatrix} -1 \\ 1 \\ -2 \\ -2 \end{bmatrix} \right\}$

In Exercises 17–22, find a basis for the given subspace by deleting linearly dependent vectors, and give the dimension. Very little computation should be required.

17. $S = \text{span}\left\{ \begin{bmatrix} 2 \\ -6 \end{bmatrix}, \begin{bmatrix} -3 \\ 9 \end{bmatrix} \right\}$

18. $S = \text{span}\left\{ \begin{bmatrix} 12 \\ -3 \end{bmatrix}, \begin{bmatrix} -18 \\ 6 \end{bmatrix} \right\}$

19. $S = \text{span}\left\{ \begin{bmatrix} 1 \\ 1 \\ 1 \end{bmatrix}, \begin{bmatrix} 2 \\ 2 \\ 2 \end{bmatrix}, \begin{bmatrix} 3 \\ 3 \\ 3 \end{bmatrix} \right\}$

20. $S = \text{span}\left\{ \begin{bmatrix} 1 \\ -1 \\ 1 \end{bmatrix}, \begin{bmatrix} -5 \\ 5 \\ -5 \end{bmatrix}, \begin{bmatrix} 4 \\ 3 \\ 2 \end{bmatrix} \right\}$

21. $S = \text{span} \left\{ \begin{bmatrix} 0 \\ 0 \\ 0 \end{bmatrix}, \begin{bmatrix} 3 \\ 0 \\ 0 \end{bmatrix}, \begin{bmatrix} 2 \\ 1 \\ 0 \end{bmatrix}, \begin{bmatrix} 1 \\ 2 \\ 3 \end{bmatrix} \right\}$

22. $S = \text{span} \left\{ \begin{bmatrix} 1 \\ 2 \\ 3 \\ 4 \end{bmatrix}, \begin{bmatrix} 5 \\ 5 \\ 5 \\ 5 \end{bmatrix}, \begin{bmatrix} 4 \\ 2 \\ 2 \\ 1 \end{bmatrix}, \begin{bmatrix} 6 \\ 7 \\ 8 \\ 9 \end{bmatrix} \right\}$

In Exercises 23–24, expand the given set to form a basis for $\mathbf{R}^2$.

23. $\left\{ \begin{bmatrix} 1 \\ -3 \end{bmatrix} \right\}$

24. $\left\{ \begin{bmatrix} 0 \\ 4 \end{bmatrix} \right\}$

In Exercises 25–28, expand the given set to form a basis for $\mathbf{R}^3$.

25. $\left\{ \begin{bmatrix} -1 \\ 2 \\ 1 \end{bmatrix} \right\}$

26. $\left\{ \begin{bmatrix} 1 \\ 0 \\ 5 \end{bmatrix} \right\}$

27. $\left\{ \begin{bmatrix} 1 \\ 3 \\ -2 \end{bmatrix}, \begin{bmatrix} 2 \\ -1 \\ 0 \end{bmatrix} \right\}$

28. $\left\{ \begin{bmatrix} 2 \\ 1 \\ 3 \end{bmatrix}, \begin{bmatrix} 3 \\ 2 \\ 6 \end{bmatrix} \right\}$

In Exercises 29–32, find a basis for the null space of the given matrix and give nullity(A).

29. $A = \begin{bmatrix} -2 & -5 \\ 1 & 3 \end{bmatrix}$

30. $A = \begin{bmatrix} 2 & 1 & 0 \\ 1 & 1 & 1 \end{bmatrix}$

31. $A = \begin{bmatrix} 1 & 1 & 2 & 1 \\ 0 & 0 & 1 & -3 \end{bmatrix}$

32. $A = \begin{bmatrix} 1 & 0 & -2 & 1 & -1 \\ 0 & 1 & 0 & 2 & 0 \\ 0 & 0 & 0 & 1 & 4 \end{bmatrix}$

FIND AN EXAMPLE For Exercises 33–40, find an example that meets the given specifications.

33. A set of four vectors in $\mathbf{R}^2$ such that, when two are removed, the remaining two are a basis for $\mathbf{R}^2$.

34. A set of three vectors in $\mathbf{R}^4$ such that, when one is removed and then two more are added, the new set is a basis for $\mathbf{R}^4$.

35. A subspace S of $\mathbf{R}^n$ with dim(S) $= m$, where $0 < m < n$.

36. Two subspaces S_1 and S_2 of $\mathbf{R}^5$ such that $S_1 \subset S_2$ and dim(S_1) $+ 2 = $ dim(S_2).

37. Two two-dimensional subspaces S_1 and S_2 of $\mathbf{R}^4$ such that $S_1 \cap S_2 = \{\mathbf{0}\}$.

38. Two three-dimensional subspaces S_1 and S_2 of $\mathbf{R}^5$ such that dim($S_1 \cap S_2$) $= 1$.

39. Two vectors $\mathbf{u}_1$ and $\mathbf{u}_2$ in $\mathbf{R}^3$ that produce the same set of vectors when the methods of Example 1 and Example 2 are applied.

40. Three vectors $\mathbf{u}_1$ and $\mathbf{u}_2$ in $\mathbf{R}^3$ that produce the same set of vectors when the methods of Example 1 and Example 2 are applied.

TRUE OR FALSE For Exercises 41–54, determine if the statement is true or false, and justify your answer.

41. If S_1 and S_2 are subspaces of $\mathbf{R}^n$ of the same dimension, then $S_1 = S_2$.

42. If $S = \text{span}\{\mathbf{u}_1, \mathbf{u}_2, \mathbf{u}_3\}$, then dim($S$) $= 3$.

43. If a set of vectors $\mathcal{U}$ spans a subspace S, then vectors can be added to $\mathcal{U}$ to create a basis for S.

44. If a set of vectors $\mathcal{U}$ is linearly independent in a subspace S, then vectors can be added to $\mathcal{U}$ to create a basis for S.

45. If a set of vectors $\mathcal{U}$ spans a subspace S, then vectors can be removed from $\mathcal{U}$ to create a basis for S.

46. If a set of vectors $\mathcal{U}$ is linearly independent in a subspace S, then vectors can be removed from $\mathcal{U}$ to create a basis for S.

47. Three nonzero vectors that lie in a plane in $\mathbf{R}^3$ might form a basis for $\mathbf{R}^3$.

48. If S_1 is a subspace of dimension 3 in $\mathbf{R}^4$, then there cannot exist a subspace S_2 of $\mathbf{R}^4$ such that $S_1 \subset S_2 \subset \mathbf{R}^4$ but $S_1 \neq S_2$ and $S_2 \neq \mathbf{R}^4$.

49. The set $\{\mathbf{0}\}$ forms a basis for the zero subspace.

50. $\mathbf{R}^n$ has exactly one subspace of dimension m for each of $m = 0, 1, 2, \ldots, n$.

51. Let $m > n$. Then $\mathcal{U} = \{\mathbf{u}_1, \mathbf{u}_2, \ldots, \mathbf{u}_m\}$ in $\mathbf{R}^n$ can form a basis for $\mathbf{R}^n$ if the correct $m - n$ vectors are removed from $\mathcal{U}$.

52. Let $m < n$. Then $\mathcal{U} = \{\mathbf{u}_1, \mathbf{u}_2, \ldots, \mathbf{u}_m\}$ in $\mathbf{R}^n$ can form a basis for $\mathbf{R}^n$ if the correct $n - m$ vectors are added to $\mathcal{U}$.

53. If $\{\mathbf{u}_1, \mathbf{u}_2, \mathbf{u}_3\}$ is a basis for $\mathbf{R}^3$, then span$\{\mathbf{u}_1, \mathbf{u}_2\}$ is a plane.

54. The nullity of a matrix A is the same as the dimension of the subspace spanned by the columns of A.

55. Suppose that S_1 and S_2 are nonzero subspaces, with S_1 contained inside S_2. Suppose that dim(S_2) $= 3$.

(a) What are the possible dimensions of S_1?

(b) If $S_1 \neq S_2$, then what are the possible dimensions of S_1?

56. Suppose that S_1 and S_2 are nonzero subspaces, with S_1 contained inside S_2. Suppose that dim(S_2) $= 4$.

(a) What are the possible dimensions of S_1?

(b) If $S_1 \neq S_2$, then what are the possible dimensions of S_1?

57. Show that the only subspace of $\mathbf{R}^n$ that has dimension n is $\mathbf{R}^n$.

58. Explain why $\mathbf{R}^n$ ($n > 1$) has infinitely many subspaces of dimension 1.

59. Prove the converse of Theorem 4.9: If every vector $\mathbf{s}$ of a subspace S can be written uniquely as a linear combination of the vectors $\mathbf{s}_1, \ldots, \mathbf{s}_m$ (all in S), then the vectors form a basis for S.

60. Complete the proof of Theorem 4.15: Let $\mathcal{U} = \{\mathbf{u}_1, \ldots, \mathbf{u}_m\}$ be a set of m vectors in a subspace S of dimension m. Show that if $\mathcal{U}$ spans S, then $\mathcal{U}$ is a basis for S.

61. Prove Theorem 4.16: Suppose that S_1 and S_2 are both subspaces of $\mathbf{R}^n$, with S_1 a subset of S_2. Then $\dim(S_1) \leq \dim(S_2)$, and $\dim(S_1) = \dim(S_2)$ only if $S_1 = S_2$.

62. Prove Theorem 4.17: Let $\mathcal{U} = \{\mathbf{u}_1, \ldots, \mathbf{u}_m\}$ be a set of vectors in a subspace S of dimension k.

(a) If $m < k$, show that $\mathcal{U}$ does not span S.

(b) If $m > k$, show that $\mathcal{U}$ is not linearly independent.

63. Suppose that a matrix A is in echelon form. Prove that the nonzero rows of A are linearly independent.

64. If the set $\{\mathbf{u}_1, \mathbf{u}_2, \mathbf{u}_3\}$ spans $\mathbf{R}^3$ and

$$A = \begin{bmatrix} \mathbf{u}_1 & \mathbf{u}_2 & \mathbf{u}_3 \end{bmatrix},$$

what is nullity(A)?

65. Suppose that S_1 and S_2 are subspaces of $\mathbf{R}^n$, with $\dim(S_1) = m_1$ and $\dim(S_2) = m_2$. If S_1 and S_2 have only $\mathbf{0}$ in common, then what is the maximum value of $m_1 + m_2$?

66. Prove Theorem 4.10: Let A and B be equivalent matrices. Then the subspace spanned by the rows of A is the same as the subspace spanned by the rows of B.

67. Prove Theorem 4.11: Suppose that $U = \begin{bmatrix} \mathbf{u}_1 & \cdots & \mathbf{u}_m \end{bmatrix}$ and $V = \begin{bmatrix} \mathbf{v}_1 & \cdots & \mathbf{v}_m \end{bmatrix}$ are two equivalent matrices. Then any linear dependence that exists among the vectors $\mathbf{u}_1, \ldots, \mathbf{u}_m$ also exists among the vectors $\mathbf{v}_1, \ldots, \mathbf{v}_m$.

68. Give a general proof of Theorem 4.12: If S is a subspace of $\mathbf{R}^n$, then every basis of S has the same number of vectors.

Ⓒ In Exercises 69–70, determine if the given set of vectors is a basis of $\mathbf{R}^3$. If not, then determine the dimension of the subspace spanned by the vectors.

69. $\left\{ \begin{bmatrix} 2 \\ -1 \\ 5 \end{bmatrix}, \begin{bmatrix} -3 \\ 4 \\ -2 \end{bmatrix}, \begin{bmatrix} -5 \\ 10 \\ 4 \end{bmatrix} \right\}$

70. $\left\{ \begin{bmatrix} 4 \\ 2 \\ -7 \end{bmatrix}, \begin{bmatrix} -1 \\ 5 \\ -3 \end{bmatrix}, \begin{bmatrix} 3 \\ 7 \\ -9 \end{bmatrix} \right\}$

Ⓒ In Exercises 71–72, determine if the given set of vectors is a basis of $\mathbf{R}^4$. If not, then determine the dimension of the subspace spanned by the vectors.

71. $\left\{ \begin{bmatrix} 3 \\ 0 \\ 1 \\ -2 \end{bmatrix}, \begin{bmatrix} 2 \\ -4 \\ 5 \\ 0 \end{bmatrix}, \begin{bmatrix} -2 \\ 7 \\ 0 \\ 4 \end{bmatrix}, \begin{bmatrix} -2 \\ 5 \\ -5 \\ 4 \end{bmatrix} \right\}$

72. $\left\{ \begin{bmatrix} 6 \\ 0 \\ -5 \\ 2 \end{bmatrix}, \begin{bmatrix} 5 \\ -1 \\ 1 \\ 3 \end{bmatrix}, \begin{bmatrix} -3 \\ 4 \\ 1 \\ -5 \end{bmatrix}, \begin{bmatrix} 7 \\ -2 \\ 6 \\ 8 \end{bmatrix} \right\}$

Ⓒ In Exercises 73–74, determine if the given set of vectors is a basis of $\mathbf{R}^5$. If not, then determine the dimension of the subspace spanned by the vectors.

73. $\left\{ \begin{bmatrix} 1 \\ 1 \\ -1 \\ 1 \\ 1 \end{bmatrix}, \begin{bmatrix} -1 \\ 0 \\ 1 \\ 2 \\ -1 \end{bmatrix}, \begin{bmatrix} 2 \\ 1 \\ -2 \\ 1 \\ 2 \end{bmatrix}, \begin{bmatrix} -2 \\ 1 \\ 2 \\ 1 \\ -2 \end{bmatrix}, \begin{bmatrix} 1 \\ 2 \\ -1 \\ 0 \\ 1 \end{bmatrix} \right\}$

74. $\left\{ \begin{bmatrix} 1 \\ 2 \\ 3 \\ 4 \\ 5 \end{bmatrix}, \begin{bmatrix} 2 \\ 3 \\ 4 \\ 5 \\ 1 \end{bmatrix}, \begin{bmatrix} 3 \\ 4 \\ 5 \\ 1 \\ 2 \end{bmatrix}, \begin{bmatrix} 4 \\ 5 \\ 1 \\ 2 \\ 3 \end{bmatrix}, \begin{bmatrix} 5 \\ 1 \\ 2 \\ 3 \\ 4 \end{bmatrix} \right\}$

4.3 Row and Column Spaces

In Example 7 of Section 4.2, it was shown that if $x_1, \ldots, x_n$ are distinct real numbers, then the columns of the Vandermonde matrix

$$V = \begin{bmatrix} 1 & x_1 & x_1^2 & \cdots & x_1^{n-1} \\ 1 & x_2 & x_2^2 & \cdots & x_2^{n-1} \\ \vdots & \vdots & \vdots & \ddots & \vdots \\ 1 & x_n & x_n^2 & \cdots & x_n^{n-1} \end{bmatrix}$$

form a basis for $\mathbf{R}^n$. But suppose that the x_i's are not distinct. Can we tell if the columns are linearly independent or linearly dependent? One result that we shall develop will make this question easy to answer.

In this section we round out our knowledge of subspaces of $\mathbf{R}^n$. As we have seen, subspaces arise naturally in the context of a matrix. For instance, suppose that

$$A = \begin{bmatrix} 1 & -2 & 7 & 5 \\ -2 & -1 & -9 & -7 \\ 1 & 13 & -8 & -4 \end{bmatrix}$$

Definition Row Vectors

The **row vectors** of A come from viewing the rows of A as vectors. For the matrix A, the set of row vectors is

$$\left\{ \begin{bmatrix} 1 \\ -2 \\ 7 \\ 5 \end{bmatrix}, \begin{bmatrix} -2 \\ -1 \\ -9 \\ -7 \end{bmatrix}, \begin{bmatrix} 1 \\ 13 \\ -8 \\ -4 \end{bmatrix} \right\}$$

Definition Column Vectors

Similarly, the **column vectors** of A come from viewing the columns of A as vectors, so in this case we have

$$\left\{ \begin{bmatrix} 1 \\ -2 \\ 1 \end{bmatrix}, \begin{bmatrix} -2 \\ -1 \\ 13 \end{bmatrix}, \begin{bmatrix} 7 \\ -9 \\ -8 \end{bmatrix}, \begin{bmatrix} 5 \\ -7 \\ -4 \end{bmatrix} \right\}$$

Taking the span of the row or column vectors yields the subspaces defined below.

DEFINITION 4.19

Definition Row Space

Definition Column Space

Let A be an $n \times m$ matrix.

(a) The **row space** of A is the subspace of $\mathbf{R}^m$ spanned by the row vectors of A and is denoted by row(A).

(b) The **column space** of A is the subspace of $\mathbf{R}^n$ spanned by the column vectors of A and is denoted by col(A).

In Section 4.2 we proved Theorem 4.10 and Theorem 4.11, which concern the rows and columns of matrices and can be used to find a basis for a subspace. Theorem 4.20 is a reformulation of those theorems, stated in terms of row and column spaces.

THEOREM 4.20

Let A be a matrix and B an echelon form of A.

(a) The nonzero rows of B form a basis for row(A).

(b) The columns of A corresponding to the pivot columns of B form a basis for col(A).

EXAMPLE 1   Find a basis and the dimension for the row space and the column space of A.

$$A = \begin{bmatrix} 1 & -2 & 7 & 5 \\ -2 & -1 & -9 & -7 \\ 1 & 13 & -8 & -4 \end{bmatrix}$$

Solution To use Theorem 4.20, we start by finding an echelon form of A, which is given by

$$A = \begin{bmatrix} 1 & -2 & 7 & 5 \\ -2 & -1 & -9 & -7 \\ 1 & 13 & -8 & -4 \end{bmatrix} \sim \begin{bmatrix} 1 & -2 & 7 & 5 \\ 0 & -5 & 5 & 3 \\ 0 & 0 & 0 & 0 \end{bmatrix} = B$$

By Theorem 4.20(a), we know that a basis for the row space of A is given by the nonzero rows of B,

$$\left\{ \begin{bmatrix} 1 \\ -2 \\ 7 \\ 5 \end{bmatrix}, \begin{bmatrix} 0 \\ -5 \\ 5 \\ 3 \end{bmatrix} \right\}$$

By Theorem 4.20(b), we know that a basis for the column space of A is given by the columns of A corresponding to the pivot columns of B, which in this case are the first and second columns. Thus a basis for $\text{col}(A)$ is

$$\left\{ \begin{bmatrix} 1 \\ -2 \\ 1 \end{bmatrix}, \begin{bmatrix} -2 \\ -1 \\ 13 \end{bmatrix} \right\}$$

Since both $\text{row}(A)$ and $\text{col}(A)$ have two basis vectors, the dimension of both subspaces is 2. ■

In Example 1, the row space and the column space of A have the same dimension. This is not a coincidence.

THEOREM 4.21

For any matrix A, the dimension of the row space equals the dimension of the column space.

Proof Given a matrix A, use the usual row operations to find an equivalent echelon form matrix B. From Theorem 4.20(a), we know that the dimension of the row space of A is equal to the number of nonzero rows of B. Next note that each nonzero row of B has exactly one pivot, and that different rows have pivots in different columns. Thus the number of pivot columns equals the number of nonzero rows. But by Theorem 4.20(b), the number of pivot columns of B equals the number of vectors in a basis for the column space of A. Thus the dimension of the column space is equal to the number of nonzero rows of B, and so the dimensions of the row space and column space are the same. ■

Now let's return to the question about the Vandermonde matrix from the start of the section.

EXAMPLE 2 Suppose that two or more of $x_1, \ldots, x_n$ are the same. Are the columns of

$$V = \begin{bmatrix} 1 & x_1 & x_1^2 & \cdots & x_1^{n-1} \\ 1 & x_2 & x_2^2 & \cdots & x_2^{n-1} \\ \vdots & \vdots & \vdots & \ddots & \vdots \\ 1 & x_n & x_n^2 & \cdots & x_n^{n-1} \end{bmatrix}$$

linearly independent or linearly dependent?

Solution If two or more of $x_1, \ldots, x_n$ are the same, then two or more of the rows of V are the same. Hence the rows of V are linearly dependent, so by The Big Theorem (applied to the rows of V) the rows of V do not span $\mathbf{R}^n$. Therefore the dimension of $\text{row}(V)$ is less than n, and thus by Theorem 4.21 the dimension of $\text{col}(V)$ is less than n. Finally, again by The Big Theorem (applied to the columns of V), we conclude that the columns are linearly dependent. ■

Because the dimensions of the row and column spaces for a given matrix A are the same, the following definition makes sense.

DEFINITION 4.22

Definition Rank of a Matrix

The **rank** of a matrix A is the dimension of the row (or column) space of A, and is denoted by $\text{rank}(A)$.

EXAMPLE 3   Find the rank and the nullity for the matrix

$$A = \begin{bmatrix} 1 & -2 & 3 & 0 & -1 \\ 2 & -4 & 7 & -3 & 3 \\ 3 & -6 & 8 & 3 & -8 \end{bmatrix}$$

▶ Recall that the nullity is the dimension of the null space.

Solution   Applying the standard row operation procedure to A yields the echelon form

$$A = \begin{bmatrix} 1 & -2 & 3 & 0 & -1 \\ 2 & -4 & 7 & -3 & 3 \\ 3 & -6 & 8 & 3 & -8 \end{bmatrix} \sim \begin{bmatrix} 1 & -2 & 3 & 0 & -1 \\ 0 & 0 & 1 & -3 & 5 \\ 0 & 0 & 0 & 0 & 0 \end{bmatrix} = B$$

Since B has two nonzero rows, the rank of A is 2. To find the nullity of A, we need to determine the dimension of the subspace of solutions to $A\mathbf{x} = \mathbf{0}$. Adding a column of zeros to A and B gives the augmented matrix for $A\mathbf{x} = \mathbf{0}$ and the corresponding echelon form. (Why?)

$$\begin{bmatrix} 1 & -2 & 3 & 0 & -1 & 0 \\ 2 & -4 & 7 & -3 & 3 & 0 \\ 3 & -6 & 8 & 3 & -8 & 0 \end{bmatrix} \sim \begin{bmatrix} 1 & -2 & 3 & 0 & -1 & 0 \\ 0 & 0 & 1 & -3 & 5 & 0 \\ 0 & 0 & 0 & 0 & 0 & 0 \end{bmatrix}$$

The matrix on the right corresponds to the system

$$\begin{aligned} x_1 - 2x_2 + 3x_3 \quad\quad - \; x_5 &= 0 \\ x_3 - 3x_4 + 5x_5 &= 0 \end{aligned} \tag{1}$$

For this system, x_2, x_4, and x_5 are free variables, so we assign the parameters $x_2 = s_1$, $x_4 = s_2$, and $x_5 = s_3$. Back substitution gives us

$$\begin{aligned} x_3 &= 3x_4 - 5x_5 = 3s_2 - 5s_3 \\ x_1 &= 2x_2 - 3x_3 + x_5 = 2s_1 - 3(3s_2 - 5s_3) + s_3 = 2s_1 - 9s_2 + 16s_3 \end{aligned}$$

▶ Once we know that the system $A\mathbf{x} = \mathbf{0}$ has three free variables, we can conclude that nullity(A) = 3. For the sake of completeness, we continue to the vector form of the solution.

In vector form, the general solution is

$$\mathbf{x} = s_1 \begin{bmatrix} 2 \\ 1 \\ 0 \\ 0 \\ 0 \end{bmatrix} + s_2 \begin{bmatrix} -9 \\ 0 \\ 3 \\ 1 \\ 0 \end{bmatrix} + s_3 \begin{bmatrix} 16 \\ 0 \\ -5 \\ 0 \\ 1 \end{bmatrix}$$

The three vectors in the general solution form a basis for the null space, which shows that nullity(A) = 3. ∎

Let's look at another example and see if a pattern emerges.

EXAMPLE 4   Determine the rank and nullity for the matrix A given in Example 5 of Section 4.2.

Solution   In Example 5, Section 4.2, we showed that a basis for the null space of A is given by

$$\left\{ \begin{bmatrix} -1 \\ 3 \\ 0 \\ 2 \end{bmatrix}, \begin{bmatrix} -3 \\ 1 \\ 2 \\ 0 \end{bmatrix} \right\}$$

so that nullity(A) = 2. Since we did not show an echelon form for A earlier, we report it at right.

$$A = \begin{bmatrix} -3 & 3 & -6 & -6 \\ 2 & 6 & 0 & -8 \\ 0 & -8 & 4 & 12 \\ -3 & -7 & -1 & 9 \\ 2 & 10 & -2 & -14 \end{bmatrix} \sim \begin{bmatrix} 1 & 1 & 1 & -1 \\ 0 & 2 & -1 & -3 \\ 0 & 0 & 0 & 0 \\ 0 & 0 & 0 & 0 \\ 0 & 0 & 0 & 0 \end{bmatrix}$$

From the echelon form, we see that the rank of A is 2. ∎

Let's review what we have seen:

- Example 3: rank(A) = 2, nullity(A) = 3, total number of columns is 5.
- Example 4: rank(A) = 2, nullity(A) = 2, total number of columns is 4.

In both cases, rank(A) + nullity(A) equals the number of columns of A. This is not a coincidence.

THEOREM 4.23

(RANK – NULLITY THEOREM) Let A be an $n \times m$ matrix. Then

$$\text{rank}(A) + \text{nullity}(A) = m.$$

Proof Transform A to echelon form B.

- The rank of A is equal to the number of nonzero rows of B. Each nonzero row has a pivot, and each pivot appears in a different column. Hence the number of pivot columns equals rank(A).

- Every nonpivot column corresponds to a free variable in the system $A\mathbf{x} = \mathbf{0}$. Each free variable becomes a parameter, and each parameter is multiplied times a basis vector of null(A). (This is shown in detail in Example 3.) Therefore the number of nonpivot columns equals nullity(A).

Since the number of pivot columns plus the number of nonpivot columns must equal the total number of columns m, we have

$$\text{rank}(A) + \text{nullity}(A) = m \quad ∎$$

EXAMPLE 5 Suppose that A is a 5×13 matrix and that $T(\mathbf{x}) = A\mathbf{x}$. If the dimension of the kernel of T is 9, what is the dimension of the range of T?

Solution Since Theorem 4.23 is expressed in terms of the properties of a matrix A, we first convert the given information into equivalent statements about A. We are told that the dimension of ker(T) equals 9. Since ker(T) = null(A), then nullity(A) = 9. By Theorem 4.23, $m - $ nullity(A) = rank(A), so rank(A) = 4 because A has 13 columns. Recall that range(T) is equal to the span of the columns of A (Theorem 3.3), which is the same as col(A). Therefore the dimension of range(T) is 4. ∎

EXAMPLE 6 Find a linear transformation T that has kernel equal to span $\{\mathbf{x}_1, \mathbf{x}_2\}$, where

$$x_1 = \begin{bmatrix} 1 \\ 0 \\ -2 \\ 1 \end{bmatrix}, \quad x_2 = \begin{bmatrix} 0 \\ 1 \\ 3 \\ 2 \end{bmatrix}$$

Solution Since T is a linear transformation, we know that there exists a matrix A such that $T(\mathbf{x}) = A\mathbf{x}$. Since the kernel of T equals the null space of A, another way to state our problem is that we need a matrix A such that $\text{null}(A) = \text{span}\{\mathbf{x}_1, \mathbf{x}_2\}$.

To get us started, since $\mathbf{x}_1$ and $\mathbf{x}_2$ are linearly independent (why?), they form a basis for $\text{null}(A)$, and so $\text{nullity}(A) = 2$. Moreover, A must have four columns because $\mathbf{x}_1$ and $\mathbf{x}_2$ are in $\mathbf{R}^4$. Thus $\text{rank}(A) = 4 - 2 = 2$ by the Rank–Nullity Theorem. This tells us that A must have at least two rows, so let's assume that A has the form and see if we can solve the problem.

$$A = \begin{bmatrix} a & b & c & d \\ e & f & g & h \end{bmatrix}$$

In order for $\mathbf{x}_1$ and $\mathbf{x}_2$ to be in $\text{null}(A)$, we must have $A\mathbf{x}_1 = \mathbf{0}$ and $A\mathbf{x}_2 = \mathbf{0}$. Computing the first entry of $A\mathbf{x}_1$ and $A\mathbf{x}_2$ and setting each equal to zero produces the linear system

$$\begin{aligned} a \quad - 2c + \ d &= 0 \\ b + 3c + 2d &= 0 \end{aligned}$$

The system is in echelon form, and after back substituting we find that the general solution is given by

$$\begin{bmatrix} a \\ b \\ c \\ d \end{bmatrix} = s_1 \begin{bmatrix} 2 \\ -3 \\ 1 \\ 0 \end{bmatrix} + s_2 \begin{bmatrix} -1 \\ -2 \\ 0 \\ 1 \end{bmatrix} \tag{2}$$

There are many choices for s_1 and s_2, but let's make it easy on ourselves by setting $s_1 = 1$ and $s_2 = 0$, so that $a = 2$, $b = -3$, $c = 1$, and $d = 0$. This gives us half of A,

$$A = \begin{bmatrix} 2 & -3 & 1 & 0 \\ e & f & g & h \end{bmatrix}$$

In order to find e, f, g, and h, we could repeat the same analysis. However, we will just get the same answers, with e, f, g, and h replacing a, b, c, and d. So we can set $s_1 = 0$ and $s_2 = 1$ and use the second vector in (2) as the second row of A,

$$A = \begin{bmatrix} 2 & -3 & 1 & 0 \\ -1 & -2 & 0 & 1 \end{bmatrix}$$

Since the two rows of A are linearly independent, we know that $\text{rank}(A) = 2$. This ensures that $\text{nullity}(A) = 2$, so that $\text{null}(A) = \text{span}\{\mathbf{x}_1, \mathbf{x}_2\}$. ∎

We wrap up this subsection with a theorem that relates row and column spaces to other topics that we previously encountered. The proofs of both parts are left as exercises.

THEOREM 4.24

Let A be an $n \times m$ matrix and $\mathbf{b}$ a vector in $\mathbf{R}^n$.

(a) The system $A\mathbf{x} = \mathbf{b}$ is consistent if and only if $\mathbf{b}$ is in the column space of A.

(b) The system $A\mathbf{x} = \mathbf{b}$ has a unique solution if and only if $\mathbf{b}$ is in the column space of A and the columns of A are linearly independent.

▶ This updates the Big Theorem, Version 5, given in Section 4.2.

The Big Theorem, Version 6

We can add three more conditions to the Big Theorem based on our work in this section. The Big Theorem is starting to get really big. This theorem provides great flexibility—do not hesitate to use it.

THEOREM 4.25

(THE BIG THEOREM, VERSION 6) Let $\mathcal{A} = \{\mathbf{a}_1, \ldots, \mathbf{a}_n\}$ be a set of n vectors in $\mathbf{R}^n$, let $A = \begin{bmatrix} \mathbf{a}_1 & \cdots & \mathbf{a}_n \end{bmatrix}$, and let $T : \mathbf{R}^n \to \mathbf{R}^n$ be given by $T(\mathbf{x}) = A\mathbf{x}$. Then the following are equivalent:

(a) $\mathcal{A}$ spans $\mathbf{R}^n$.

(b) $\mathcal{A}$ is linearly independent.

(c) $A\mathbf{x} = \mathbf{b}$ has a unique solution for all $\mathbf{b}$ in $\mathbf{R}^n$.

(d) T is onto.

(e) T is one-to-one.

(f) A is invertible.

(g) $\ker(T) = \{\mathbf{0}\}$.

(h) $\mathcal{A}$ is a basis for $\mathbf{R}^n$.

(i) $\text{col}(A) = \mathbf{R}^n$.

(j) $\text{row}(A) = \mathbf{R}^n$.

(k) $\text{rank}(A) = n$.

Proof From the Big Theorem, Version 5, we know that (a) through (h) are equivalent. Theorem 4.21 and Definition 4.22 imply that (i), (j), and (k) are equivalent, and by definition (a) and (i) are equivalent. Hence the 11 conditions are all one big equivalent family. ■

EXERCISES

In Exercises 1–4, find bases for the column space of A, the row space of A, and the null space of A. Verify that the Rank–Nullity Theorem holds. (To make your job easier, an equivalent echelon form is given for each matrix.)

1. $A = \begin{bmatrix} 1 & -3 & 2 \\ -2 & 5 & 0 \\ -3 & 8 & -2 \end{bmatrix} \sim \begin{bmatrix} 1 & 0 & -10 \\ 0 & 1 & -4 \\ 0 & 0 & 0 \end{bmatrix}$

2. $A = \begin{bmatrix} 1 & 0 & -4 & -3 \\ -2 & 1 & 13 & 5 \\ 0 & 1 & 5 & -1 \end{bmatrix} \sim \begin{bmatrix} 1 & 0 & -4 & -3 \\ 0 & 1 & 5 & -1 \\ 0 & 0 & 0 & 0 \end{bmatrix}$

3. $A = \begin{bmatrix} 1 & 0 & -4 & -3 \\ -2 & 1 & 13 & 5 \\ 0 & 1 & 5 & -1 \end{bmatrix} \sim \begin{bmatrix} 1 & 0 & -4 & -3 \\ 0 & 1 & 5 & -1 \\ 0 & 0 & 0 & 0 \end{bmatrix}$

4. $A = \begin{bmatrix} 1 & -2 & 5 \\ 2 & 4 & 1 \\ -4 & 0 & 2 \\ 1 & -2 & 0 \\ 3 & 1 & 1 \end{bmatrix} \sim \begin{bmatrix} 1 & 0 & 0 \\ 0 & 1 & 0 \\ 0 & 0 & 1 \\ 0 & 0 & 0 \\ 0 & 0 & 0 \end{bmatrix}$

In Exercises 5–8, find bases for the column space of A, the row space of A, and the null space of A. Verify that the Rank–Nullity Theorem holds.

5. $A = \begin{bmatrix} 1 & -2 & 2 \\ 2 & -2 & 3 \\ -1 & -2 & 0 \end{bmatrix}$

6. $A = \begin{bmatrix} 1 & 2 & -1 & 1 \\ 2 & 1 & -1 & 4 \\ 1 & -4 & 1 & 5 \end{bmatrix}$

7. $A = \begin{bmatrix} 1 & 3 & 2 & 0 \\ 3 & 11 & 7 & 1 \\ 1 & 1 & 4 & 0 \end{bmatrix}$

8. $A = \begin{bmatrix} 1 & 4 & -1 & 1 \\ 3 & 11 & -1 & 4 \\ 1 & 5 & 2 & 3 \\ 2 & 8 & -2 & 2 \end{bmatrix}$

In Exercises 9–12, find all values of x so that $\text{rank}(A) = 2$.

9. $A = \begin{bmatrix} 1 & -4 \\ -2 & x \end{bmatrix}$

10. $A = \begin{bmatrix} 2 & 3 & 4 \\ -1 & x & 0 \end{bmatrix}$

11. $A = \begin{bmatrix} -1 & 2 & 1 \\ 3 & 1 & 11 \\ 4 & 3 & x \end{bmatrix}$

12. $A = \begin{bmatrix} -2 & 1 & 0 & 7 \\ 0 & 1 & x & 9 \\ 1 & 0 & -3 & 1 \end{bmatrix}$

13. Suppose that A is a 6×8 matrix. If the dimension of the row space of A is 5, what is the dimension of the column space of A?

14. Suppose that A is a 9×7 matrix. If the dimension of col(A) is 5, what is the dimension of row(A)?

15. Suppose that A is a 4×7 matrix that has an echelon form with one zero row. Find the dimension of the row space of A, the column space of A, and the null space of A.

16. Suppose that A is a 6×11 matrix that has an echelon form with two zero rows. Find the dimension of the row space of A, the column space of A, and the null space of A.

17. A 8×5 matrix A has a null space of dimension 3. What is the rank of A?

18. A 5×13 matrix A has a null space of dimension 10. What is the rank of A?

19. A 7×11 matrix A has rank 4. What is the dimension of the null space of A?

20. A 14×9 matrix A has rank 7. What is the dimension of the null space of A?

21. Suppose that A is a 6×11 matrix and that $T(\mathbf{x}) = A\mathbf{x}$. If nullity($A$) = 7, what is the dimension of the range of T?

22. Suppose that A is a 17×12 matrix and that $T(\mathbf{x}) = A\mathbf{x}$. If rank($A$) = 8, what is the dimension of the kernel of T?

23. Suppose that A is a 13×5 matrix and that $T(\mathbf{x}) = A\mathbf{x}$. If T is one-to-one, then what is the dimension of the null space of A?

24. Suppose that A is a 5×13 matrix and that $T(\mathbf{x}) = A\mathbf{x}$. If T is onto, then what is the dimension of the null space of A?

25. Suppose that A is a 5×13 matrix. What is the maximum possible value for the rank of A, and what is the minimum possible value for the nullity of A?

26. Suppose that A is a 12×7 matrix. What is the minimum possible value for the rank of A, and what is the maximum possible value for the nullity of A?

In Exercises 27–32, suppose that A is a 9×5 matrix and that B is an equivalent matrix in echelon form.

27. If B has three nonzero rows, what is rank(A)?

28. If B has two pivot columns, what is rank(A)?

29. If B has three nonzero rows, what is nullity(A)?

30. If B has one pivot column, what is nullity(A)?

31. If rank(A) = 3, how many nonzero rows does B have?

32. If rank(A) = 1, how many pivot columns does B have?

33. Suppose that A is an $n \times m$ matrix, that col(A) is a subspace of $\mathbf{R}^7$, and that row(A) is a subspace of $\mathbf{R}^5$. What are the dimensions of A?

34. Suppose that A is an $n \times m$ matrix, with rank(A) = 4, nullity(A) = 3, and col(A) a subspace of $\mathbf{R}^5$. What are the dimensions of A?

FIND AN EXAMPLE For Exercises 35–42, find an example that meets the given specifications.

35. A 2×3 matrix A with nullity(A) = 1.

36. A 4×3 matrix A with nullity(A) = 0.

37. A 9×4 matrix A with rank(A) = 3.

38. A 5×7 matrix A with rank(A) = 4.

39. A matrix A with rank(A) = 3 and nullity(A) = 1.

40. A matrix A with rank(A) = 2 and nullity(A) = 2.

41. A 2×2 matrix A such that row(A) = col(A).

42. A 2×2 matrix A such that row(A) = col(A).

TRUE OR FALSE For Exercises 43–54, determine if the statement is true or false, and justify your answer.

43. If A is a matrix, then the dimension of the row space of A is equal to the dimension of the column space of A.

44. If A is a square matrix, then row(A) = col(A).

45. The rank of a matrix A cannot exceed the number of rows of A.

46. If A and B are equivalent matrices, then row(A) = row(B).

47. If A and B are equivalent matrices, then col(A) = col(B).

48. If $A\mathbf{x} = \mathbf{b}$ is a consistent linear system, then $\mathbf{b}$ is in row(A).

49. If $\mathbf{x}_0$ is a solution to $A\mathbf{x} = \mathbf{b}$, then $\mathbf{x}_0$ is in row(A).

50. If A is a 4×13 matrix, then the nullity of A could be equal to 5.

51. Suppose that A is a 9×5 matrix and that $T(\mathbf{x}) = A\mathbf{x}$ is a linear transformation. Then T can be onto.

52. Suppose that A is a 9×5 matrix and that $T(\mathbf{x}) = A\mathbf{x}$ is a linear transformation. Then T can be one-to-one.

53. Suppose that A is a 4×13 matrix and that $T(\mathbf{x}) = A\mathbf{x}$ is a linear transformation. Then T can be onto.

54. Suppose that A is a 4×13 matrix and that $T(\mathbf{x}) = A\mathbf{x}$ is a linear transformation. Then T can be one-to-one.

55. Prove that if A is an $n \times m$ matrix then rank(A) = rank(A^T).

56. Prove that if A is an $n \times m$ matrix and $c \neq 0$ is a scalar, then rank(A) = rank(cA).

57. Prove that if A is an $n \times m$ matrix and rank(A) < m, then $A\mathbf{x} = \mathbf{0}$ has nontrivial solutions.

58. Prove that if A is an $n \times m$ matrix and rank(A) < n, then the reduced row echelon form of A has a row of zeros.

59. Suppose that A is an $n \times m$ matrix with $n \neq m$. Prove that either nullity(A) > 0 or nullity(A^T) > 0 (or both).

60. Prove Theorem 4.24: Let A be an $n \times m$ matrix and $\mathbf{b}$ a vector in $\mathbf{R}^n$.

(a) Show that the system $A\mathbf{x} = \mathbf{b}$ is consistent if and only if $\mathbf{b}$ is in the column space of A.

(b) Show that the system $A\mathbf{x} = \mathbf{b}$ has a unique solution if and only if $\mathbf{b}$ is in the column space of A and the columns of A are linearly independent.

Ⓒ In Exercises 61–64, determine the rank and nullity of the given matrix.

61. $A = \begin{bmatrix} 1 & 3 & 2 & 4 & -1 \\ 1 & 5 & -3 & 3 & -4 \\ 2 & 8 & -1 & 7 & -5 \end{bmatrix}$

62. $A = \begin{bmatrix} 2 & -1 & 0 & 1 \\ 5 & 2 & 1 & -4 \\ -1 & -4 & -1 & 6 \\ -8 & -5 & -2 & 9 \end{bmatrix}$

63. $A = \begin{bmatrix} 4 & 8 & 2 \\ 3 & 5 & 1 \\ 9 & 19 & 5 \\ 7 & 13 & 3 \\ 5 & 11 & 3 \end{bmatrix}$

64. $A = \begin{bmatrix} 4 & 3 & 2 & 1 \\ 5 & -1 & 3 & 2 \\ 2 & 1 & 3 & 6 \\ 7 & 10 & 3 & 1 \\ 6 & 4 & 5 & 7 \\ -2 & -2 & 1 & 5 \end{bmatrix}$

Determinants

Perhaps the most famous, and undeniably the most photographed, bridge in the world, San Francisco's Golden Gate Bridge has been declared one of the modern Wonders of the World by the American Society of Civil Engineers. In 1987, the bridge was closed to vehicular traffic for its 50th anniversary to allow pedestrians to walk across the bridge. Estimations before the celebration suggested 75,000 people would be on the bridge, but more than 300,000 ended up on the bridge at the same time. As a result, the suspension bridge lost its characteristic arch (left photo) until the pedestrian traffic cleared. For the 75th anniversary in 2012, the pedestrian bridge walk was not repeated.

Bridge suggested by Jeff Holt, Author, University of Virginia (Courtesy of Maurice Bizzarri; Stefano Politi Markovina/Alamy)

The *determinant* is a function that takes a matrix as input and produces a real number as output. Determinants have a rich history and a variety of useful interpretations. In Section 5.1 we define the determinant and find formulas for the determinant for certain special types of matrices. The properties of the determinant are further developed in Section 5.2, including how row operations and matrix arithmetic influence the determinant. In Section 5.3 we see how to use the determinant to find the solution to a linear system and a matrix inverse, and how determinants give us information about the behavior of linear transformations.

5.1 The Determinant Function

The determinant of a square matrix A combines the entries of A to produce a single real number. There are several different interpretations and characterizations of the determinant, and the formulas are generally complicated. The development that we give here is guided by an important property of the determinant, namely, that a square matrix is invertible exactly when it has a nonzero determinant.

Let's start with the easiest case. Let $A = \begin{bmatrix} a_{11} \end{bmatrix}$ be a 1×1 matrix. Then A is invertible exactly when $a_{11} \neq 0$. This motivates our first definition.

DEFINITION 5.1

Definition Determinant of a
1 × 1 Matrix

Let $A = [a_{11}]$ be a 1 × 1 matrix. Then the **determinant** of A is given by

$$\det(A) = a_{11}$$

Thus a 1 × 1 matrix A is invertible if and only if $\det(A) \neq 0$.
    Next, suppose that

$$A = \begin{bmatrix} a_{11} & a_{12} \\ a_{21} & a_{22} \end{bmatrix} \tag{1}$$

In Section 3.3 we developed the "Quick Formula" for the inverse of A, which says that

$$A^{-1} = \frac{1}{a_{11}a_{22} - a_{12}a_{21}} \begin{bmatrix} a_{22} & -a_{12} \\ -a_{21} & a_{11} \end{bmatrix} \tag{2}$$

provided that $a_{11}a_{22} - a_{12}a_{21} \neq 0$. If $a_{11}a_{22} - a_{12}a_{21} = 0$, then (2) is undefined and A has no inverse. Hence $a_{11}a_{22} - a_{12}a_{21}$ is nonzero exactly when A is invertible, so we take this as the definition of the determinant for a 2 × 2 matrix A.

DEFINITION 5.2

Definition Determinant of a
2 × 2 Matrix

Let A be the 2 × 2 matrix in (1). Then the **determinant** of A is given by

$$\det(A) = a_{11}a_{22} - a_{12}a_{21} \tag{3}$$

Summarizing, we have that A is invertible if and only if $\det(A) \neq 0$. This gives us an easy-to-apply formula for determining when a 2 × 2 matrix is invertible.

EXAMPLE 1 Let $A = \begin{bmatrix} 3 & 5 \\ -1 & 4 \end{bmatrix}$. Find $\det(A)$ and determine if A is invertible.

Solution Applying Definition 5.2, we have

$$\det(A) = (3)(4) - (5)(-1) = 17$$

Since $\det(A) \neq 0$, we can conclude that A is invertible. ∎

The definition of the determinant is more complicated for larger matrices. Let's consider the 3 × 3 case. Suppose

$$A = \begin{bmatrix} a_{11} & a_{12} & a_{13} \\ a_{21} & a_{22} & a_{23} \\ a_{31} & a_{32} & a_{33} \end{bmatrix} \tag{4}$$

▶The row operations are (in order performed):
$$a_{11}R_2 \Rightarrow R_2$$
$$a_{11}R_3 \Rightarrow R_3$$
$$-a_{21}R_1 + R_2 \Rightarrow R_2$$
$$-a_{31}R_1 + R_3 \Rightarrow R_3$$

In order for A to be invertible, at least one of a_{11}, a_{21}, or a_{31} must be nonzero. For the moment, assume that $a_{11} \neq 0$. Applying the row operations listed in the margin, we have

$$A \sim \begin{bmatrix} a_{11} & a_{12} & a_{13} \\ 0 & (a_{11}a_{22} - a_{12}a_{21}) & (a_{11}a_{23} - a_{13}a_{21}) \\ 0 & (a_{11}a_{32} - a_{12}a_{31}) & (a_{11}a_{33} - a_{13}a_{31}) \end{bmatrix}$$

Since $a_{11} \neq 0$, the matrix on the right is invertible exactly when

$$\begin{bmatrix} (a_{11}a_{22} - a_{12}a_{21}) & (a_{11}a_{23} - a_{13}a_{21}) \\ (a_{11}a_{32} - a_{12}a_{31}) & (a_{11}a_{33} - a_{13}a_{31}) \end{bmatrix}$$

is invertible. The determinant of this 2×2 matrix is

$$(a_{11}a_{22} - a_{12}a_{21})(a_{11}a_{33} - a_{13}a_{31}) - (a_{11}a_{23} - a_{13}a_{21})(a_{11}a_{32} - a_{12}a_{31})$$
$$= a_{11}\left[a_{11}a_{22}a_{33} + a_{12}a_{23}a_{31} + a_{13}a_{21}a_{32} - a_{11}a_{23}a_{32} - a_{12}a_{21}a_{33} - a_{13}a_{22}a_{31}\right]$$

Since $a_{11} \neq 0$, A is invertible if and only if the expression in brackets is nonzero. Other than a sign change, the bracketed expression is the same if we start with a_{21} or a_{31}. Furthermore, this expression is zero if $a_{11} = a_{21} = a_{31} = 0$. Hence the term in brackets is nonzero exactly when A is invertible, so we use it for the determinant.

DEFINITION 5.3

Definition Determinant of a 3×3 Matrix

Let A be the 3×3 matrix in (4). Then the **determinant** of A is given by

$$\det(A) = a_{11}a_{22}a_{33} + a_{12}a_{23}a_{31} + a_{13}a_{21}a_{32} - a_{11}a_{23}a_{32} \tag{5}$$
$$-a_{12}a_{21}a_{33} - a_{13}a_{22}a_{31}$$

EXAMPLE 2   Find $\det(A)$ for

$$A = \begin{bmatrix} -3 & 1 & 2 \\ 5 & 5 & -8 \\ 4 & 2 & -5 \end{bmatrix}$$

Solution   From (5) we have

$$\det(A) = (-3)(5)(-5) + (1)(-8)(4) + (2)(5)(2)$$
$$-(-3)(-8)(2) - (1)(5)(-5) - (2)(5)(4)$$
$$= 75 - 32 + 20 - 48 + 25 - 40 = 0$$

Note that this implies that A is not invertible. ■

Our formulas for determinants may appear unconnected, but in fact the 3×3 determinant is related to the 2×2 determinant. To see how, we start by reorganizing (5) and factoring out common terms,

$$\det(A) = a_{11}(a_{22}a_{33} - a_{23}a_{32}) - a_{12}(a_{21}a_{33} - a_{23}a_{31}) + a_{13}(a_{21}a_{32} - a_{22}a_{31}) \tag{6}$$

Each expression in parentheses can be viewed as the determinant of a 2×2 matrix. For instance,

▶ We replace $[\ ]$ with $|\ |$ around a matrix to indicate the determinant. For example,

$$\begin{vmatrix} 4 & 2 \\ 3 & 7 \end{vmatrix} = (4)(7) - (2)(3) = 22$$

$$a_{22}a_{33} - a_{23}a_{32} = \begin{vmatrix} a_{22} & a_{23} \\ a_{32} & a_{33} \end{vmatrix}$$

Combining this observation with (6) gives us

$$\det(A) = a_{11}\begin{vmatrix} a_{22} & a_{23} \\ a_{32} & a_{33} \end{vmatrix} - a_{12}\begin{vmatrix} a_{21} & a_{23} \\ a_{31} & a_{33} \end{vmatrix} + a_{13}\begin{vmatrix} a_{21} & a_{22} \\ a_{31} & a_{32} \end{vmatrix} \tag{7}$$

Each of these 2×2 matrices can be found within A by crossing out the row and column containing a_{11}, a_{12}, and a_{13}, respectively, then forming 2×2 matrices from the entries that remain.

$$a_{11} \Rightarrow \begin{bmatrix} a_{11} & a_{12} & a_{13} \\ a_{21} & a_{22} & a_{23} \\ a_{31} & a_{32} & a_{33} \end{bmatrix} \qquad a_{12} \Rightarrow \begin{bmatrix} a_{11} & a_{12} & a_{13} \\ a_{21} & a_{22} & a_{23} \\ a_{31} & a_{32} & a_{33} \end{bmatrix} \qquad a_{13} \Rightarrow \begin{bmatrix} a_{11} & a_{12} & a_{13} \\ a_{21} & a_{22} & a_{23} \\ a_{31} & a_{32} & a_{33} \end{bmatrix}$$

Definition Minor

In general, if A is an $n \times n$ matrix, then M_{ij} denotes the $(n-1) \times (n-1)$ matrix that we get from A after deleting the row and column containing a_{ij}. The determinant $\det(M_{ij})$ is called the **minor** of a_{ij}.

EXAMPLE 3 Suppose

$$A = \begin{bmatrix} 4 & -1 & 1 & 0 \\ 1 & 7 & 3 & 5 \\ 0 & -3 & -2 & 1 \\ 2 & 4 & 8 & -1 \end{bmatrix}$$

Find M_{23} and M_{42}.

Solution The term a_{23} is located in row 2 and column 3 of A. To find M_{23}, we cross out row 2 and column 3,

$$\begin{bmatrix} 4 & -1 & 1 & 0 \\ 1 & 7 & 3 & 5 \\ 0 & -3 & -2 & 1 \\ 2 & 4 & 8 & -1 \end{bmatrix} \implies M_{23} = \begin{bmatrix} 4 & -1 & 0 \\ 0 & -3 & 1 \\ 2 & 4 & -1 \end{bmatrix}$$

Similarly, M_{42} is found by deleting row 4 and column 2,

$$\begin{bmatrix} 4 & -1 & 1 & 0 \\ 1 & 7 & 3 & 5 \\ 0 & -3 & -2 & 1 \\ 2 & 4 & 8 & -1 \end{bmatrix} \implies M_{42} = \begin{bmatrix} 4 & 1 & 0 \\ 1 & 3 & 5 \\ 0 & -2 & 1 \end{bmatrix}$$ ∎

Referring to (7), we see that our formula for the determinant of a 3×3 matrix can be expressed in terms of minors as

$$\det(A) = a_{11} \det(M_{11}) - a_{12} \det(M_{12}) + a_{13} \det(M_{13})$$

Definition **Cofactor**

This formula can be further simplified with the introduction of C_{ij}, the **cofactor** of a_{ij}, given by

$$C_{ij} = (-1)^{i+j} \det(M_{ij})$$

Thus we have

$$\det(A) = a_{11} C_{11} + a_{12} C_{12} + a_{13} C_{13} \tag{8}$$

The formula (8) provides us with a model for the general definition of the determinant.

DEFINITION 5.4

Let A be the $n \times n$ matrix

$$A = \begin{bmatrix} a_{11} & a_{12} & \cdots & a_{1n} \\ \vdots & \vdots & \ddots & \vdots \\ a_{n1} & a_{n2} & \cdots & a_{nn} \end{bmatrix} \tag{9}$$

Definition **Determinant**

Then the **determinant** of A is

$$\det(A) = a_{11} C_{11} + a_{12} C_{12} + \cdots + a_{1n} C_{1n} \tag{10}$$

where $C_{11}, \ldots, C_{1n}$ are the cofactors of $a_{11}, \ldots, a_{1n}$, respectively. When $n = 1$, $A = [a_{11}]$ and $\det(A) = a_{11}$.

Definition 5.4 is an example of a *recursive* definition, because the determinant of an $n \times n$ matrix is defined in terms of the determinants of $(n - 1) \times (n - 1)$ matrices.

EXAMPLE 4 Use Definition 5.4 and cofactors to find $\det(A)$ for

$$A = \begin{bmatrix} 2 & -1 & 3 \\ 1 & 4 & 0 \\ 3 & 1 & 2 \end{bmatrix}$$

Solution The first step is to find the cofactors C_{11}, C_{12}, and C_{13}:

▶ The determinants for the cofactors were calculated using the formula (3).

$$C_{11} = (-1)^{1+1} \begin{vmatrix} 4 & 0 \\ 1 & 2 \end{vmatrix} = (1)\big((4)(2) - (0)(1)\big) = 8$$

$$C_{12} = (-1)^{1+2} \begin{vmatrix} 1 & 0 \\ 3 & 2 \end{vmatrix} = (-1)\big((1)(2) - (0)(3)\big) = -2$$

$$C_{13} = (-1)^{1+3} \begin{vmatrix} 1 & 4 \\ 3 & 1 \end{vmatrix} = (1)\big((1)(1) - (4)(3)\big) = -11$$

By Definition 5.4 we have

$$\det(A) = 2(8) + (-1)(-2) + 3(-11) = -15 \quad \blacksquare$$

We can use induction and Definition 5.4 to prove that $\det(I_n) = 1$.

THEOREM 5.5

For $n \geq 1$ we have $\det(I_n) = 1$.

▶ This chapter contains theorems and exercises requiring proof by induction. If you are unfamiliar with this method of proof or are just a bit rusty, consult the appendix "Reading and Writing Proofs."

Proof We shall carry out this proof by induction on n, the number of rows. First suppose that $n = 1$. Then

$$I_1 = \begin{bmatrix} 1 \end{bmatrix} \implies \det(I_1) = 1$$

so that the theorem is true in this case. Next suppose that $n \geq 2$ and that the theorem is true for I_{n-1}. (This is the *induction hypothesis*.) Since the top row of I_n has a single 1 followed by zeros, by Definition 5.4 we have

$$\det(I_n) = (1)C_{11} + (0)C_{12} + \cdots + (0)C_{1n} = C_{11} = (-1)^2 \det(M_{11}) = \det(M_{11})$$

Since M_{11} is the matrix we get from deleting the first row and column of I_n, we have

$$M_{11} = \begin{bmatrix} 1 & 0 & 0 & \cdots & 0 \\ 0 & 1 & 0 & \cdots & 0 \\ 0 & 0 & 1 & \cdots & 0 \\ \vdots & \vdots & \vdots & \ddots & \vdots \\ 0 & 0 & 0 & \cdots & 1 \end{bmatrix} = I_{n-1}$$

By the induction hypothesis we know $\det(I_{n-1}) = 1$. Therefore

$$\det(I_n) = \det(M_{11}) = \det(I_{n-1}) = 1$$

The two parts of the induction proof are verified, so the proof is complete. $\quad \blacksquare$

A remarkable fact about the general definition of the determinant given in Definition 5.4 is that it has the same property as the determinant of 2×2 and 3×3 matrices, namely, that for any $n \times n$ matrix A, $\det(A)$ is nonzero exactly when A is invertible.

THEOREM 5.6

Let A be an $n \times n$ matrix. Then A is invertible if and only if $\det(A) \neq 0$.

We have already see that Theorem 5.6 is true when $n = 1, 2$, and 3. The proof of Theorem 5.6 for larger matrices is given in Section 5.2.

Theorem 5.6 allows us to add a condition involving the determinant to the Big Theorem.

THEOREM 5.7

▶ This updates the Big Theorem, Version 6, given in Section 4.3.

(THE BIG THEOREM, VERSION 7) Let $\mathcal{A} = \{\mathbf{a}_1, \ldots, \mathbf{a}_n\}$ be a set of n vectors in $\mathbf{R}^n$, let $A = \begin{bmatrix} \mathbf{a}_1 & \cdots & \mathbf{a}_n \end{bmatrix}$, and let $T : \mathbf{R}^n \to \mathbf{R}^n$ be given by $T(\mathbf{x}) = A\mathbf{x}$. Then the following are equivalent:

(a) $\mathcal{A}$ spans $\mathbf{R}^n$.

(b) $\mathcal{A}$ is linearly independent.

(c) $A\mathbf{x} = \mathbf{b}$ has a unique solution for all $\mathbf{b}$ in $\mathbf{R}^n$.

(d) T is onto.

(e) T is one-to-one.

(f) A is invertible.

(g) $\ker(T) = \{\mathbf{0}\}$.

(h) $\mathcal{A}$ is a basis for $\mathbf{R}^n$.

(i) $\text{col}(A) = \mathbf{R}^n$.

(j) $\text{row}(A) = \mathbf{R}^n$.

(k) $\text{rank}(A) = n$.

(l) $\det(A) \neq 0$.

Proof   From the Big Theorem, Version 6 in Section 4.3, we know that (a) through (k) are equivalent. Theorem 5.6 tells us that (f) and (l) are equivalent, and so we conclude that all 12 conditions are equivalent. ∎

The Shortcut Method

For 2×2 and 3×3 matrices there exist nice visual aids for computing determinants that we refer to as *the Shortcut Method*. For the 2×2 case, start by drawing diagonal arrows through the terms of the matrix, labeled with $+$ and $-$ as shown below. Multiply the terms of each arrow, and then add or subtract as indicated by the $+$ or $-$.

$$\begin{vmatrix} a_{11} & a_{12} \\ a_{21} & a_{22} \end{vmatrix} = a_{11}a_{22} - a_{12}a_{21}$$

Note that this matches the formula given in (3).

For a 3×3 matrix, we write down the matrix, copy the left two columns to the right, and then draw six diagonal arrows with labels as shown.

▶ **Warning!** The Shortcut Method does not work for 4×4 or larger matrices.

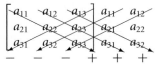

As in the 2×2 case, for each arrow we multiply terms and then add or subtract based on the labels. This yields

$$a_{11}a_{22}a_{33} + a_{12}a_{23}a_{31} + a_{13}a_{21}a_{32} - a_{13}a_{22}a_{31} - a_{11}a_{23}a_{32} - a_{12}a_{21}a_{33}$$

which matches the formula given in (5).

EXAMPLE 5 Find $\det(A)$ from Example 4 using the Shortcut Method.

Solution Adding on the two extra columns and drawing the diagonals, we have

$$\begin{vmatrix} 2 & -1 & 3 \\ 1 & 4 & 0 \\ 3 & 1 & 2 \end{vmatrix} \begin{matrix} 2 & -1 \\ 1 & 4 \\ 3 & 1 \end{matrix} \implies \det(A) = (16 + 0 + 3) - (36 + 0 - 2) = -15$$

which matches what we found earlier. ∎

EXAMPLE 6 Show that the set

$$\mathcal{A} = \left\{ \begin{bmatrix} 3 \\ -1 \\ 5 \end{bmatrix}, \begin{bmatrix} -2 \\ 0 \\ 7 \end{bmatrix}, \begin{bmatrix} 4 \\ 3 \\ 1 \end{bmatrix} \right\}$$

forms a basis for $\mathbf{R}^3$.

Solution Let

$$A = \begin{bmatrix} 3 & -2 & 4 \\ -1 & 0 & 3 \\ 5 & 7 & 1 \end{bmatrix}$$

By the Big Theorem, Version 7, $\mathcal{A}$ is a basis for $\mathbf{R}^3$ if and only if $\det(A) \neq 0$. Applying the Shortcut Method, we find that

$$\begin{vmatrix} 3 & -2 & 4 \\ -1 & 0 & 3 \\ 5 & 7 & 1 \end{vmatrix} \begin{matrix} 3 & -2 \\ -1 & 0 \\ 5 & 7 \end{matrix} \implies \det(A) = (0 - 30 - 28) - (0 + 63 + 2) = -123$$

Since $\det(A) \neq 0$, $\mathcal{A}$ is a basis for $\mathbf{R}^3$. ∎

Cofactor Expansion

In our definition of the determinant, we use cofactors for the entries along the top row of the matrix. The next theorem allows us to generalize to entries in other rows or columns. The proof is omitted.

THEOREM 5.8

Let A be the $n \times n$ matrix in (9). Then

   (a) $\det(A) = a_{i1}C_{i1} + a_{i2}C_{i2} + \cdots + a_{in}C_{in}$ (Expand across row i)

   (b) $\det(A) = a_{1j}C_{1j} + a_{2j}C_{2j} + \cdots + a_{nj}C_{nj}$ (Expand down column j)

where C_{ij} denotes the cofactor of a_{ij}. These formulas are referred to collectively as the **cofactor expansions**.

Definition Cofactor Expansions

Theorem 5.8 tells us that we can compute the determinant by taking cofactors along *any* row or column of the matrix. This theorem is handy in cases where a matrix has a row or column containing several zeros, because we can save ourselves some work.

EXAMPLE 7 Find $\det(A)$ for

$$A = \begin{bmatrix} -2 & 1 & 4 & -1 \\ 1 & 0 & -1 & 2 \\ 5 & -1 & 2 & 1 \\ 0 & 0 & 3 & -1 \end{bmatrix}$$

Solution The cofactor expansion down the 2nd column is

$$\det(A) = a_{12}C_{12} + a_{22}C_{22} + a_{32}C_{32} + a_{42}C_{42}$$

Since $a_{22} = 0$ and $a_{42} = 0$, there is no need to calculate C_{22} and C_{42}. The cofactors C_{12} and C_{32} are given by

$$C_{12} = (-1)^{1+2}\begin{vmatrix} 1 & -1 & 2 \\ 5 & 2 & 1 \\ 0 & 3 & -1 \end{vmatrix} = (-1)(20) = -20$$

$$C_{32} = (-1)^{3+2}\begin{vmatrix} -2 & 4 & -1 \\ 1 & -1 & 2 \\ 0 & 3 & -1 \end{vmatrix} = (-1)(11) = -11$$

▶ The 3 × 3 determinants in Example 7 were found using the Shortcut Method.

Pulling everything together, we have

$$\det(A) = 1 \cdot C_{12} + (-1) \cdot C_{32} = 1(-20) + (-1)(-11) = -9$$

Other than the amount of work involved, it makes no difference which row or column we choose. Expanding along the 4th row of A, we have

$$\det(A) = a_{41}C_{41} + a_{42}C_{42} + a_{43}C_{43} + a_{44}C_{44}$$

Since $a_{41} = a_{42} = 0$, we need only compute C_{43} and C_{44}, which are

$$C_{43} = (-1)^{4+3}\begin{vmatrix} -2 & 1 & -1 \\ 1 & 0 & 2 \\ 5 & -1 & 1 \end{vmatrix} = (-1)(6) = -6$$

$$C_{44} = (-1)^{4+4}\begin{vmatrix} -2 & 1 & 4 \\ 1 & 0 & -1 \\ 5 & -1 & 2 \end{vmatrix} = (1)(-9) = -9$$

Therefore

$$\det(A) = 3 \cdot C_{43} + (-1) \cdot C_{44} = 3(-6) + (-1)(-9) = -9$$

as before. ■

Matrices that have certain special forms or characteristics have determinants that are easy to compute.

EXAMPLE 8 Find $\det(A)$ for

$$A = \begin{bmatrix} -1 & 2 & 7 & -5 & 8 \\ 0 & 3 & 4 & 1 & -9 \\ 0 & 0 & -2 & 4 & 11 \\ 0 & 0 & 0 & -4 & 5 \\ 0 & 0 & 0 & 0 & 1 \end{bmatrix}$$

Solution Note that A is upper triangular. Let's take advantage of all the zeros in the first column. Since the first entry in that column is the only one that is nonzero, the cofactor expansion down the first column is

$$\det(A) = (-1)C_{11} = (-1)(-1)^2 \det(M_{11}) = (-1)\begin{vmatrix} 3 & 4 & 1 & -9 \\ 0 & -2 & 4 & 11 \\ 0 & 0 & -4 & 5 \\ 0 & 0 & 0 & 1 \end{vmatrix}$$

For the remaining determinant $\det(M_{11})$, we again use cofactor expansion down the first column, giving us

$$\det(A) = (-1)(3)C_{11} = (-1)(3) \begin{vmatrix} -2 & 4 & 11 \\ 0 & -4 & 5 \\ 0 & 0 & 1 \end{vmatrix}$$

The remaining 3×3 determinant can be computed using the Shortcut Method, which produces only one nonzero term, the product of the diagonal $(-2)(-4)(1)$. Therefore

$$\det(A) = (-1)(3)(-2)(-4)(1) = -24 \quad \blacksquare$$

In Example 8, A is a square triangular matrix and $\det(A)$ is equal to the product of the diagonal terms. This suggests the next theorem.

THEOREM 5.9

If A is a triangular $n \times n$ matrix, then $\det(A)$ is the product of the terms along the diagonal.

The proof of Theorem 5.9 is left as an exercise.

Recall that if we interchange the rows and columns of a matrix A, we get A^T, the transpose of A. Interestingly, taking the transpose has no effect on the determinant.

THEOREM 5.10

Let A be a square matrix. Then $\det(A^T) = \det(A)$.

▶ For

$$A = \begin{bmatrix} a_{11} & a_{12} & a_{13} & \cdots & a_{1n} \\ a_{21} & a_{22} & a_{23} & \cdots & a_{2n} \\ a_{31} & a_{32} & a_{33} & \cdots & a_{3n} \\ \vdots & \vdots & \vdots & \ddots & \vdots \\ a_{n1} & a_{n2} & a_{n3} & \cdots & a_{nn} \end{bmatrix}$$

the transpose is

$$A^T = \begin{bmatrix} a_{11} & a_{21} & a_{31} & \cdots & a_{n1} \\ a_{12} & a_{22} & a_{32} & \cdots & a_{n2} \\ a_{13} & a_{23} & a_{33} & \cdots & a_{n3} \\ \vdots & \vdots & \vdots & \ddots & \vdots \\ a_{1n} & a_{2n} & a_{3n} & \cdots & a_{nn} \end{bmatrix}$$

Proof We use induction. First note that if $A = \begin{bmatrix} a_{11} \end{bmatrix}$ is a 1×1 matrix, then $A = A^T$ so that $\det(A) = \det(A^T)$.

Now for the induction hypothesis: Suppose that $n \geq 2$ and that the theorem holds for all $(n-1) \times (n-1)$ matrices. If A is an $n \times n$ matrix, then the cofactor expansion along the top row of A gives us

$$\det(A) = a_{11}\det(M_{11}) - a_{12}\det(M_{12}) + \cdots + (-1)^{n+1}a_{1n}\det(M_{1n})$$

Since $M_{11}, \ldots, M_{1n}$ are all $(n-1) \times (n-1)$ matrices, by the induction hypothesis we have $\det(M_{11}) = \det(M_{11}^T), \ldots, \det(M_{1n}) = \det(M_{1n}^T)$. Hence

$$\det(A) = a_{11}\det(M_{11}^T) - a_{12}\det(M_{12}^T) + \cdots + (-1)^{n+1}a_{1n}\det(M_{1n}^T) \quad (11)$$

Next note that the first *column* of A^T has entries $a_{11}, \ldots, a_{1n}$. Thus the right side of (11) also gives the cofactor expansion down the first column of A^T, and so it follows that $\det(A^T) = \det(A)$. $\blacksquare$

EXAMPLE 9 Show that $\det(A) = \det(A^T)$ for

$$A = \begin{bmatrix} 2 & 3 & -1 \\ 4 & 2 & 0 \\ 5 & -2 & -4 \end{bmatrix}$$

Solution Applying the Shortcut Method twice, we find that

$$\det(A) = \begin{vmatrix} 2 & 3 & -1 \\ 4 & 2 & 0 \\ 5 & -2 & -4 \end{vmatrix} = (-16 + 0 + 8) - (-10 + 0 - 48) = 50$$

$$\det(A^T) = \begin{vmatrix} 2 & 4 & 5 \\ 3 & 2 & -2 \\ -1 & 0 & -4 \end{vmatrix} = (-16 + 8 + 0) - (-10 + 0 - 48) = 50$$

Hence $\det(A) = \det(A^T)$. ■

THEOREM 5.11

Let A be a square matrix.

(a) If A has a row or column of zeros, then $\det(A) = 0$.

(b) If A has two identical rows or columns, then $\det(A) = 0$.

Proof   The proof of both parts of this theorem are left as exercises. ■

EXAMPLE 10   Show that $\det(A) = 0$ and $\det(B) = 0$, where

$$A = \begin{bmatrix} 3 & 0 & 0 & 2 \\ 0 & 4 & 0 & 5 \\ 9 & 0 & 0 & 7 \\ 1 & 1 & 0 & 0 \end{bmatrix} \quad \text{and} \quad B = \begin{bmatrix} -1 & 4 & 2 & 0 \\ 6 & 2 & 5 & -4 \\ -1 & 4 & 2 & 0 \\ 0 & 8 & 4 & 7 \end{bmatrix}$$

Solution   Since the third column of A consists of zeros, by Theorem 5.11(a) we have $\det(A) = 0$. Since rows 1 and 3 of B are identical, by Theorem 5.11(b) we have $\det(B) = 0$. ■

Multiplying matrices and computing determinants are both processes requiring numerous arithmetic operations. However, the relationship between the determinant of the product of two matrices and the product of the individual determinants is remarkably simple, as shown in Theorem 5.12.

THEOREM 5.12

If A and B are both $n \times n$ matrices, then

$$\det(AB) = \det(A)\det(B)$$

A proof of Theorem 5.12, and further discussion of the determinant of products of matrices, is given in Section 5.2.

EXAMPLE 11   Show that $\det(AB) = \det(A)\det(B)$ for the matrices

$$A = \begin{bmatrix} 2 & -4 \\ -1 & 1 \end{bmatrix} \quad \text{and} \quad B = \begin{bmatrix} 3 & -1 \\ 2 & 1 \end{bmatrix}$$

Solution   Starting with A and B, we have

$$\det(A) = (2)(1) - (-4)(-1) = -2 \quad \text{and} \quad \det(B) = (3)(1) - (-1)(2) = 5$$

Hence $\det(A)\det(B) = -10$. Computing AB, we have

$$AB = \begin{bmatrix} 2 & -4 \\ -1 & 1 \end{bmatrix} \begin{bmatrix} 3 & -1 \\ 2 & 1 \end{bmatrix} = \begin{bmatrix} -2 & -6 \\ -1 & 2 \end{bmatrix}$$

Therefore $\det(AB) = (-2)(2) - (-6)(-1) = -10 = \det(A)\det(B)$. ∎

Computational Comment

Except when a matrix has mostly zero entries, computing determinants using cofactor expansion is slow for even a modest-sized matrix. Working recursively eventually generates a lot of 3×3 determinants that all require evaluation. For an $n \times n$ matrix, the number of multiplications needed is about $n!$. Thus, for example, a 20×20 matrix will require about $20! = 2,432,902,008,176,640,000$ multiplications, far more than is remotely practical. In the next section, we see how to use row operations to speed things up.

EXERCISES

In Exercises 1–6, find M_{23} and M_{31} for the given matrix A.

1. $A = \begin{bmatrix} 7 & 0 & -4 \\ 3 & 6 & 2 \\ 5 & 1 & 5 \end{bmatrix}$

2. $A = \begin{bmatrix} -5 & 3 & 1 \\ 6 & 2 & 2 \\ 4 & 4 & 0 \end{bmatrix}$

3. $A = \begin{bmatrix} 6 & 1 & -1 & 5 \\ 0 & 2 & 3 & 0 \\ 7 & 1 & 1 & 1 \\ 4 & 3 & 1 & 2 \end{bmatrix}$

4. $A = \begin{bmatrix} 0 & -2 & 4 & 0 \\ 5 & 1 & -1 & 0 \\ 0 & 2 & -1 & 0 \\ 3 & 6 & 1 & 6 \end{bmatrix}$

5. $A = \begin{bmatrix} 4 & 3 & 2 & 1 & 0 \\ 6 & 1 & 2 & 0 & 5 \\ 3 & 2 & 2 & 4 & 4 \\ 5 & 1 & 0 & 0 & 3 \\ 2 & 2 & 4 & 1 & 0 \end{bmatrix}$

6. $A = \begin{bmatrix} 1 & 0 & 1 & 0 & 1 \\ 3 & 7 & 9 & 4 & 5 \\ 8 & 1 & 0 & 0 & 2 \\ 2 & 5 & 3 & 4 & 1 \\ 6 & 1 & 2 & 3 & 3 \end{bmatrix}$

In Exercises 7–10, find C_{13} and C_{22} for the given matrix A.

7. $A = \begin{bmatrix} 2 & 1 & 3 \\ 0 & -1 & 4 \\ 4 & 0 & 1 \end{bmatrix}$

8. $A = \begin{bmatrix} 6 & 1 & 0 \\ 2 & -1 & -3 \\ 3 & 4 & 1 \end{bmatrix}$

9. $A = \begin{bmatrix} 6 & 1 & 2 \\ 4 & 3 & 0 \\ 1 & 1 & 1 \end{bmatrix}$

10. $A = \begin{bmatrix} 0 & -1 & -1 \\ 3 & 2 & 1 \\ 4 & 0 & 2 \end{bmatrix}$

In Exercises 11–18, find the determinant for the given matrix A in two ways, by using cofactor expansion (a) along the row of your choosing, and (b) along the column of your choosing. Use the determinant to decide if $T(\mathbf{x}) = A\mathbf{x}$ is invertible.

11. $A = \begin{bmatrix} 1 & 2 & -3 \\ 0 & 4 & 0 \\ 5 & -1 & 0 \end{bmatrix}$

12. $A = \begin{bmatrix} -2 & 0 & 0 \\ 4 & 5 & 0 \\ 3 & 0 & 6 \end{bmatrix}$

13. $A = \begin{bmatrix} -1 & 1 & -1 & 2 \\ 0 & 3 & 2 & 0 \\ 1 & 4 & 0 & 1 \\ 0 & -1 & 3 & -1 \end{bmatrix}$

14. $A = \begin{bmatrix} 2 & 1 & 3 & 0 \\ 1 & 2 & 0 & 1 \\ 4 & 2 & 0 & 1 \\ 0 & 1 & 2 & 0 \end{bmatrix}$

15. $A = \begin{bmatrix} -1 & 1 & 0 & 0 \\ 2 & 3 & 3 & 2 \\ 0 & -1 & 0 & 5 \\ 3 & 1 & 4 & -1 \end{bmatrix}$

16. $A = \begin{bmatrix} 0 & 2 & 5 & 4 \\ 3 & 0 & -1 & 0 \\ 1 & 1 & -2 & 1 \\ -2 & 0 & 3 & 0 \end{bmatrix}$

17. $A = \begin{bmatrix} 4 & 2 & 1 & 0 & 1 \\ 0 & 3 & -1 & 1 & 2 \\ 0 & -1 & 0 & 0 & 1 \\ 0 & 1 & 2 & 0 & 3 \\ 0 & 0 & 1 & 0 & 1 \end{bmatrix}$

18. $A = \begin{bmatrix} 1 & -1 & 0 & 1 & 0 \\ 3 & 0 & 0 & -1 & 1 \\ -1 & 1 & 0 & -2 & 0 \\ 0 & 0 & 1 & 2 & 0 \\ 2 & 0 & -2 & 0 & 1 \end{bmatrix}$

In Exercises 19–26, when possible use the Shortcut Method to compute det(A). If the Shortcut Method is not applicable, explain why.

19. $A = \begin{bmatrix} 4 & 6 \\ -1 & 2 \end{bmatrix}$

20. $A = \begin{bmatrix} 5 & 1 \\ -3 & -2 \end{bmatrix}$

21. $A = \begin{bmatrix} 1 & 2 & -1 \\ 3 & 1 & 0 \end{bmatrix}$

22. $A = \begin{bmatrix} 6 & -1 \\ 1 & 0 \\ 2 & 2 \end{bmatrix}$

23. $A = \begin{bmatrix} 3 & 1 & -1 \\ 2 & 0 & 4 \\ 1 & 6 & 1 \end{bmatrix}$

24. $A = \begin{bmatrix} 2 & 2 & 3 \\ -1 & 4 & 1 \\ 3 & 1 & -2 \end{bmatrix}$

25. $A = \begin{bmatrix} 6 & 1 & 2 & 1 \\ 3 & 1 & 0 & 0 \\ 0 & 2 & 2 & 1 \\ 1 & 2 & 3 & -1 \end{bmatrix}$

26. $A = \begin{bmatrix} 2 & 1 & -1 & 2 \\ 3 & 1 & 1 & 0 \\ 5 & 1 & 2 & 1 \\ 4 & 3 & -3 & 2 \end{bmatrix}$

In Exercises 27–34, find all values of a such that the given matrix is not invertible. (HINT: Think determinants, not row operations.)

27. $A = \begin{bmatrix} 2 & 3 \\ 6 & a \end{bmatrix}$

28. $A = \begin{bmatrix} 12 & a \\ a & 3 \end{bmatrix}$

29. $A = \begin{bmatrix} a & a \\ 3 & -1 \end{bmatrix}$

30. $A = \begin{bmatrix} a & -3 \\ 2 & a \end{bmatrix}$

31. $A = \begin{bmatrix} 1 & -1 & 3 \\ 0 & a & -2 \\ 2 & 4 & 3 \end{bmatrix}$

32. $A = \begin{bmatrix} -1 & 2 & a \\ 0 & 1 & 1 \\ 3 & 0 & -1 \end{bmatrix}$

33. $A = \begin{bmatrix} 1 & a & -2 \\ -1 & 0 & 1 \\ a & 3 & -4 \end{bmatrix}$

34. $A = \begin{bmatrix} 0 & 4 & a \\ a & 1 & 3 \\ 0 & a & 1 \end{bmatrix}$

In Exercises 35–40, find det(A). No cofactor expansions are required, but you should explain your answer.

35. $A = \begin{bmatrix} 2 & 0 & 0 & 0 \\ 5 & -1 & 0 & 0 \\ 7 & 2 & 1 & 0 \\ 13 & 37 & 11 & 4 \end{bmatrix}$

36. $A = \begin{bmatrix} 3 & 4 & 5 & 7 \\ 0 & -2 & 5 & -9 \\ 0 & 0 & 1 & 6 \\ 0 & 0 & 0 & 5 \end{bmatrix}$

37. $A = \begin{bmatrix} 6 & 1 & 0 & 4 \\ 0 & 2 & 0 & 3 \\ 1 & 0 & 0 & 6 \\ 6 & 1 & 0 & -7 \end{bmatrix}$

38. $A = \begin{bmatrix} 2 & 0 & 0 & 1 \\ 0 & 0 & 0 & 0 \\ 4 & 1 & 0 & 2 \\ 0 & 1 & 2 & 3 \end{bmatrix}$

39. $A = \begin{bmatrix} 2 & 1 & 6 & 2 \\ 3 & -2 & 4 & 1 \\ 2 & 1 & 6 & 2 \\ 3 & 5 & 2 & 4 \end{bmatrix}$

40. $A = \begin{bmatrix} 1 & 4 & 3 & 1 & 4 \\ 3 & 2 & 4 & 3 & 2 \\ 0 & 1 & 6 & 0 & -1 \\ 2 & -1 & 1 & 2 & 1 \\ 1 & 2 & 0 & 1 & -1 \end{bmatrix}$

In Exercises 41–44, verify that det(A) = det(A^T).

41. $A = \begin{bmatrix} 3 & -2 \\ 4 & 1 \end{bmatrix}$

42. $A = \begin{bmatrix} 6 & 1 \\ 2 & 3 \end{bmatrix}$

43. $A = \begin{bmatrix} 0 & 7 & 1 \\ 2 & 3 & 1 \\ 4 & -1 & -1 \end{bmatrix}$

44. $A = \begin{bmatrix} -1 & 2 & 1 \\ 2 & -1 & 0 \\ 1 & 0 & 4 \end{bmatrix}$

In Exercises 45–48, determine all real values of λ such that det($A - \lambda I_2$) = 0 for the given matrix A.

45. $A = \begin{bmatrix} 2 & 4 \\ 5 & 3 \end{bmatrix}$

46. $A = \begin{bmatrix} 0 & 3 \\ 5 & 2 \end{bmatrix}$

47. $A = \begin{bmatrix} 1 & 0 \\ -5 & 1 \end{bmatrix}$

48. $A = \begin{bmatrix} 3 & -6 \\ 2 & -1 \end{bmatrix}$

In Exercises 49–52, determine all real values of λ such that $\det(A - \lambda I_3) = 0$ for the given matrix A.

49. $A = \begin{bmatrix} 1 & 0 & 0 \\ 5 & 3 & 0 \\ -4 & 7 & -2 \end{bmatrix}$

50. $A = \begin{bmatrix} -5 & -2 & -3 \\ 0 & 0 & 6 \\ 0 & 0 & 4 \end{bmatrix}$

51. $A = \begin{bmatrix} 0 & 2 & 0 \\ 1 & 0 & 2 \\ 2 & -1 & 0 \end{bmatrix}$

52. $A = \begin{bmatrix} 0 & 1 & 2 \\ 1 & 1 & 1 \\ 2 & -1 & 0 \end{bmatrix}$

In Exercises 53–56, for each given matrix A, first compute $\det(A)$. Then interchange two rows of your choosing and compute the determinant of the resulting matrix. Form a conjecture about the effect of row interchanges on determinants.

53. (a) $A = \begin{bmatrix} 3 & 5 \\ -2 & 4 \end{bmatrix}$ **(b)** $A = \begin{bmatrix} 1 & 2 & -1 \\ 3 & 0 & 2 \\ 0 & 1 & -1 \end{bmatrix}$

54. (a) $A = \begin{bmatrix} 1 & 0 \\ 2 & 1 \end{bmatrix}$ **(b)** $A = \begin{bmatrix} 2 & 2 & 1 \\ 1 & -1 & 2 \\ 1 & 0 & 0 \end{bmatrix}$

55. (a) $A = \begin{bmatrix} 3 & 1 & 0 \\ 1 & 2 & 3 \\ 0 & 2 & 1 \end{bmatrix}$ **(b)** $A = \begin{bmatrix} 4 & -1 & 0 \\ 0 & 2 & 1 \\ 1 & 1 & 1 \end{bmatrix}$

56. (a) $A = \begin{bmatrix} 0 & -1 & 1 \\ -1 & 2 & 1 \\ 4 & 0 & 3 \end{bmatrix}$ **(b)** $A = \begin{bmatrix} 3 & 2 & 1 \\ 0 & 1 & 1 \\ 0 & 0 & -2 \end{bmatrix}$

In Exercises 57–60, for each given matrix A, first compute $\det(A)$. Then multiply a row of your choosing by 3 and compute the determinant of the resulting matrix. Form a conjecture about the effect on determinants of multiplying a row times a scalar.

57. (a) $A = \begin{bmatrix} 3 & 5 \\ -2 & 4 \end{bmatrix}$ **(b)** $A = \begin{bmatrix} 1 & 2 & -1 \\ 3 & 0 & 2 \\ 0 & 1 & -1 \end{bmatrix}$

58. (a) $A = \begin{bmatrix} 1 & 0 \\ 2 & 1 \end{bmatrix}$ **(b)** $A = \begin{bmatrix} 2 & 2 & 1 \\ 1 & -1 & 2 \\ 1 & 0 & 0 \end{bmatrix}$

59. (a) $A = \begin{bmatrix} 3 & 1 & 0 \\ 1 & 2 & 3 \\ 0 & 2 & 1 \end{bmatrix}$ **(b)** $A = \begin{bmatrix} 4 & -1 & 0 \\ 0 & 2 & 1 \\ 1 & 1 & 1 \end{bmatrix}$

60. (a) $A = \begin{bmatrix} 0 & -1 & 1 \\ -1 & 2 & 1 \\ 4 & 0 & 3 \end{bmatrix}$ **(b)** $A = \begin{bmatrix} 3 & 2 & 1 \\ 0 & 1 & 1 \\ 0 & 0 & -2 \end{bmatrix}$

FIND AN EXAMPLE For Exercises 61–68, find an example that meets the given specifications.

61. A 2×2 matrix A with $\det(A) = 12$.

62. A 3×3 matrix A with $\det(A) = 21$.

63. A 2×2 matrix A with nonzero entries and $\det(A) = -3$.

64. A 3×3 matrix A with nonzero entries and $\det(A) = 5$.

65. A 3×3 matrix A with

$$M_{11} = \begin{bmatrix} 0 & 4 \\ 6 & -3 \end{bmatrix}, \quad M_{23} = \begin{bmatrix} 5 & -1 \\ 2 & 6 \end{bmatrix}$$

66. A 4×4 matrix A with

$$M_{14} = \begin{bmatrix} 3 & 1 & 4 \\ 4 & 5 & 0 \\ -1 & 7 & 3 \end{bmatrix}, \quad M_{33} = \begin{bmatrix} 1 & -2 & 1 \\ 3 & 1 & 0 \\ -1 & 7 & 6 \end{bmatrix}$$

67. A 3×3 matrix A with cofactors $C_{12} = -8$ and $C_{31} = -5$.

68. A 3×3 matrix A with cofactors $C_{22} = -2$, $C_{11} = 6$, and $C_{32} = -2$.

TRUE OR FALSE For Exercises 69–76, determine if the statement is true or false, and justify your answer.

69. Every matrix A has a determinant.

70. If A is an $n \times n$ matrix, then each cofactor of A is an $(n-1) \times (n-1)$ matrix.

71. If A is an $n \times n$ matrix with all positive entries, then $\det(A) > 0$.

72. If A is an $n \times n$ matrix such that $C_{i1} = \cdots = C_{in} = 0$ for some i, then $\det(A) = 0$.

73. If A is an upper triangular $n \times n$ matrix, then $\det(A) \neq 0$.

74. If A is a diagonal matrix, then M_{ij} is also diagonal for all i and j.

75. If the cofactors of an $n \times n$ matrix A are all nonzero, then $\det(A) \neq 0$.

76. If A and B are 2×2 matrices, then $\det(A - B) = \det(A) - \det(B)$.

77. Let (x_1, y_1) and (x_2, y_2) be two distinct points in the plane. Prove that

$$\begin{vmatrix} x & y & 1 \\ x_1 & y_1 & 1 \\ x_2 & y_2 & 1 \end{vmatrix} = 0$$

gives an equation for the line passing through (x_1, y_1) and (x_2, y_2).

78. Find a general formula for the determinant of

$$A = \begin{bmatrix} 0 & \cdots & 0 & 0 & a_{1n} \\ 0 & \cdots & 0 & a_{2(n-1)} & a_{2n} \\ 0 & \cdots & a_{3(n-2)} & a_{3(n-1)} & a_{3n} \\ \vdots & \cdots & \vdots & \vdots & \vdots \\ a_{n1} & \cdots & a_{n(n-2)} & a_{n(n-1)} & a_{nn} \end{bmatrix}$$

79. Let

$$A = \begin{bmatrix} a_{11} & a_{12} & \cdots & a_{1n} \\ a_{21} & a_{22} & \cdots & a_{2n} \\ \vdots & \vdots & \ddots & \vdots \\ a_{n1} & a_{n2} & \cdots & a_{nn} \end{bmatrix}$$

Let $C_{j1}, \ldots, C_{jn}$ be the cofactors of A along row j. For $i \neq j$ prove that

$$a_{i1}C_{j1} + a_{i2}C_{j2} + \cdots + a_{in}C_{jn} = 0$$

80. Use induction to complete the proof of Theorem 5.9: If A is an $n \times n$ lower triangular matrix, then $\det(A)$ is the product of the terms along the diagonal of A.

81. Prove Theorem 5.11(a): Let A be a square matrix. If A has a row or column of zeros, then $\det(A) = 0$.

82. Prove Theorem 5.11(b): Let A be a square matrix. If A has two identical rows or columns, then $\det(A) = 0$. (HINT: Use induction.)

© In Exercises 83–86, find $\det(A)$.

83. $A = \begin{bmatrix} 3 & -4 & 0 & 5 \\ 2 & 1 & -7 & 1 \\ 0 & -3 & 2 & 2 \\ 5 & 8 & -2 & -1 \end{bmatrix}$

84. $A = \begin{bmatrix} 0 & 3 & 7 & 9 \\ -1 & 4 & -1 & 0 \\ 2 & 9 & -4 & 3 \\ 2 & 3 & -3 & -2 \end{bmatrix}$

85. $A = \begin{bmatrix} 3 & 5 & 0 & 0 & 2 \\ 0 & 1 & -2 & -3 & -2 \\ 7 & -2 & -1 & 0 & 0 \\ 4 & 1 & 1 & 1 & 4 \\ -5 & -1 & 0 & 5 & 3 \end{bmatrix}$

86. $A = \begin{bmatrix} 3 & 2 & 1 & 2 & 3 \\ 7 & 8 & 9 & 1 & 3 \\ -1 & -2 & 3 & -2 & -1 \\ 3 & 6 & 9 & 6 & 2 \\ 4 & 2 & 1 & 9 & 4 \end{bmatrix}$

5.2 Properties of the Determinant

▶ This section is optional and can be omitted without loss of continuity.

▶ We already noted that it takes about $n!$ multiplications to compute the determinant of an $n \times n$ matrix using cofactor expansion. By contrast, using row operations requires roughly n^3 multiplications to compute the determinant. The difference is modest for small matrices but highly significant for larger matrices.

At the end of Section 5.1, we noted that computing determinants using cofactor expansion is too slow for use with even modest-sized matrices. In this section we show how to use row operations to make computing determinants more efficient. We will also develop additional properties of the determinant.

Instead of using cofactor expansion to compute the determinant, it is typically faster to first convert the matrix to echelon form using row operations and then multiply the terms on the diagonal. Example 1 examines the influence of row operations on the determinant.

EXAMPLE 1 Suppose that $A = \begin{bmatrix} 2 & -1 & 4 \\ -6 & 3 & -3 \\ 1 & 5 & 0 \end{bmatrix}$. Compare $\det(A)$ with $\det(B)$, where B is the matrix we get from A after performing the given row operation.

(a) Interchange Row 1 and Row 3 ($R_1 \Leftrightarrow R_3$).

(b) Multiply Row 2 by $\frac{1}{3}$ ($\frac{1}{3} R_2 \Rightarrow R_2$).

(c) Add -2 times Row 1 to Row 3 ($-2R_1 + R_3 \Rightarrow R_3$).

Solution We use the Shortcut Method to compute the determinants in this example, starting with

$$\det(A) = (0 + 3 - 120) - (12 + 0 - 30) = -99$$

For part (a), we interchange Row 1 with Row 3 and then compute the determinant.

$$B = \begin{bmatrix} 1 & 5 & 0 \\ -6 & 3 & -3 \\ 2 & -1 & 4 \end{bmatrix} \implies \det(B) = (12 - 30 + 0) - (0 - 120 + 3) = 99$$

Hence $\det(A) = -\det(B)$, so interchanging two rows changed the sign of the determinant. For part (b), multiplying the second row of A by $\frac{1}{3}$ and then computing the determinant gives us

$$B = \begin{bmatrix} 2 & -1 & 4 \\ -2 & 1 & -1 \\ 1 & 5 & 0 \end{bmatrix} \implies \det(B) = (0 + 1 - 40) - (4 + 0 - 10) = -33$$

Thus $\det(A) = 3\det(B)$. For part (c), we add -2 times Row 1 to Row 3 and then compute the determinant.

$$B = \begin{bmatrix} 2 & -1 & 4 \\ -6 & 3 & -3 \\ -3 & 7 & -8 \end{bmatrix} \implies \det(B) = (-48 - 9 - 168) - (-36 - 48 - 42) = -99$$

This time $\det(A) = \det(B)$, so adding a multiple of one row to another did not change the determinant. ■

Theorem 5.13 summarizes the influence of row operations on determinants.

THEOREM 5.13

▶ Theorem 5.13 is also true if the rows are replaced with columns.

Let A be a square matrix.

(a) Suppose that B is produced by interchanging two rows of A. Then $\det(A) = -\det(B)$.

(b) Suppose that B is produced by multiplying a row of A by c. Then $\det(A) = \frac{1}{c} \cdot \det(B)$.

(c) Suppose that B is produced by adding a multiple of one row of A to another. Then $\det(A) = \det(B)$.

A proof of Theorem 5.13 is given at the end of the section.

EXAMPLE 2 Use row operations together with Theorem 5.13 to find $\det(A)$ for

$$A = \begin{bmatrix} -2 & 1 & 4 & -1 \\ 1 & 0 & -1 & 2 \\ 5 & -1 & 2 & 1 \\ 0 & 0 & 3 & -1 \end{bmatrix}$$

Solution In Example 7 of Section 5.1, we used cofactor expansion to show that $\det(A) = -9$. Here we use row operations to transform A to triangular form while applying Theorem 5.13 to track the effect of the row operations on $\det(A)$.

To reduce A to triangular form, we start with the first column. It is handy to have a 1 in the pivot position, so we start by interchanging the first two rows.

$$A = \begin{bmatrix} -2 & 1 & 4 & -1 \\ 1 & 0 & -1 & 2 \\ 5 & -1 & 2 & 1 \\ 0 & 0 & 3 & -1 \end{bmatrix} \overset{R_1 \Leftrightarrow R_2}{\sim} \begin{bmatrix} 1 & 0 & -1 & 2 \\ -2 & 1 & 4 & -1 \\ 5 & -1 & 2 & 1 \\ 0 & 0 & 3 & -1 \end{bmatrix} = A_1$$

By Theorem 5.13(a) we have $\det(A) = -\det(A_1)$. Next, we introduce zeros down the first and second columns (the row operations are combined for brevity) with

$$A_1 = \begin{bmatrix} 1 & 0 & -1 & 2 \\ -2 & 1 & 4 & -1 \\ 5 & -1 & 2 & 1 \\ 0 & 0 & 3 & -1 \end{bmatrix} \overset{\underset{R_2 + R_3 \Rightarrow R_3}{\overset{2R_1 + R_2 \Rightarrow R_2}{-5R_1 + R_3 \Rightarrow R_3}}}{\sim} \begin{bmatrix} 1 & 0 & -1 & 2 \\ 0 & 1 & 2 & 3 \\ 0 & 0 & 9 & -6 \\ 0 & 0 & 3 & -1 \end{bmatrix} = A_2$$

By Theorem 5.13(c), none of these row operations changes the determinant, so that $\det(A_1) = \det(A_2)$ and hence $\det(A) = -\det(A_1) = -\det(A_2)$.

► This step could be combined with the one that follows.

Next we multiply the third row of A_2 by $(-1/3)$ to introduce a -3 in the pivot position of the third column.

$$A_2 = \begin{bmatrix} 1 & 0 & -1 & 2 \\ 0 & 1 & 2 & 3 \\ 0 & 0 & 9 & -6 \\ 0 & 0 & 3 & -1 \end{bmatrix} \overset{-\frac{1}{3}R_3 \Rightarrow R_3}{\sim} \begin{bmatrix} 1 & 0 & -1 & 2 \\ 0 & 1 & 2 & 3 \\ 0 & 0 & -3 & 2 \\ 0 & 0 & 3 & -1 \end{bmatrix} = A_3$$

By Theorem 5.13(b) we have $\det(A_2) = 1/(-1/3)\det(A_3) = -3\det(A_3)$, so that

$$\det(A) = -\det(A_2) = -(-3)\det(A_3) = 3\det(A_3)$$

The last step is to introduce a zero at the bottom of the third column of A_3.

$$A_3 = \begin{bmatrix} 1 & 0 & -1 & 2 \\ 0 & 1 & 2 & 3 \\ 0 & 0 & -3 & 2 \\ 0 & 0 & 3 & -1 \end{bmatrix} \overset{R_3+R_4 \Rightarrow R_4}{\sim} \begin{bmatrix} 1 & 0 & -1 & 2 \\ 0 & 1 & 2 & 3 \\ 0 & 0 & -3 & 2 \\ 0 & 0 & 0 & 1 \end{bmatrix} = A_4$$

Since this row operation has no effect on the determinant, we have $\det(A_3) = \det(A_4)$ and so $\det(A) = 3\det(A_3) = 3\det(A_4)$. Since A_4 is a triangular matrix, by Theorem 5.9

$$\det(A_4) = (1)(1)(-3)(1) = -3$$

and hence $\det(A) = 3\det(A_4) = -9$, matching the answer we obtained using cofactor expansion. ■

EXAMPLE 3 Use row operations and Theorem 5.13 to find $\det(A)$ for

$$A = \begin{bmatrix} 1 & -2 & 3 & -1 \\ 3 & -6 & 11 & 1 \\ -2 & 4 & -9 & 4 \\ 2 & -4 & 8 & 1 \end{bmatrix}$$

Solution We proceed just as in Example 2. Starting with the first column, we introduce zeros with the row operations

$$\begin{bmatrix} 1 & -2 & 3 & -1 \\ 3 & -6 & 11 & 1 \\ -2 & 4 & -9 & 4 \\ 2 & -4 & 8 & 1 \end{bmatrix} \overset{\substack{-3R_1+R_2 \Rightarrow R_2 \\ 2R_1+R_3 \Rightarrow R_3 \\ -2R_1+R_4 \Rightarrow R_4}}{\sim} \begin{bmatrix} 1 & -2 & 3 & -1 \\ 0 & 0 & 2 & 4 \\ 0 & 0 & -3 & 2 \\ 0 & 0 & 2 & 3 \end{bmatrix}$$

By Theorem 5.13(c), none of these row operations changes the determinant, so that

$$\det(A) = \begin{vmatrix} 1 & -2 & 3 & -1 \\ 0 & 0 & 2 & 4 \\ 0 & 0 & -3 & 2 \\ 0 & 0 & 2 & 3 \end{vmatrix}$$

We can continue row operations to triangular form, but the zero in the a_{22} position will remain. Hence the product of the diagonal terms of any eventual triangular matrix will be zero. Since $\det(A)$ is a multiple of this product, it must be that $\det(A) = 0$. ■

Theorem 5.13 gives us the tools to prove Theorem 5.6 from Section 5.1. The theorem is stated again below.

THEOREM 5.6 Let A be an $n \times n$ matrix. Then A is invertible if and only if $\det(A) \neq 0$.

Proof We can transform any square matrix A to an upper triangular matrix B using two types of row operations:

- Row interchanges, which only change the sign of the determinant.

- Adding a multiple of one row to another, which does not change the determinant.

Therefore $\det(A) = \pm \det(B)$. Since B is triangular, $\det(B)$ is equal to the product of the diagonal terms of B. Thus $\det(A) \neq 0$ exactly when all the diagonal terms of B are nonzero.

When the diagonal terms of B are all nonzero, all of the pivots are also nonzero. Hence transforming B to reduced echelon form will yield the identity matrix I_n, which implies A is invertible. On the other hand, if there is a zero among the diagonal terms of B, then $\det(A) = 0$. In this case, the zero pivot means that the reduced echelon form of A cannot be equal to I_n, so that A is not invertible. Hence A is invertible if and only if $\det(A) \neq 0$. ∎

EXAMPLE 4 Suppose that $T(\mathbf{x}) = A\mathbf{x}$ is a linear transformation, with

$$A = \begin{bmatrix} 2 & 0 & 0 & 8 \\ 1 & -7 & -5 & 0 \\ 3 & 8 & 6 & 0 \\ 0 & 7 & 5 & 4 \end{bmatrix}$$

Determine if T is invertible.

Solution The linear transformation T is invertible if and only if A is invertible. To apply Theorem 5.6, we need the determinant of A. There are enough zeros in A to make cofactor expansion attractive, so we use that approach. Expanding along the top row of A, we have

▶ The 3×3 determinants were computed using the Shortcut Method.

$$\det(A) = 2 \begin{vmatrix} -7 & -5 & 0 \\ 8 & 6 & 0 \\ 7 & 5 & 4 \end{vmatrix} - 8 \begin{vmatrix} 1 & -7 & -5 \\ 3 & 8 & 6 \\ 0 & 7 & 5 \end{vmatrix}$$

$$= 2\big[(-168 + 0 + 0) - (0 + 0 - 160)\big] - 8\big[(40 + 0 - 105) - (0 - 105 + 42)\big]$$
$$= 2(-8) - 8(-2) = 0$$

Since $\det(A) = 0$, we conclude that A is not invertible, so T is not invertible. ∎

Determinants of Products

We now return to Theorem 5.12 from Section 5.1, which says that the determinant of the product of two matrices is equal to the product of the individual determinants. The theorem is stated again below.

THEOREM 5.12 | If A and B are both $n \times n$ matrices, then $\det(AB) = \det(A)\det(B)$.

We will get to the proof of Theorem 5.12 shortly. First, note that an interesting consequence of Theorem 5.12 is that while generally $AB \neq BA$, it is true that $\det(AB) = \det(BA)$, because

$$\det(AB) = \det(A)\det(B) = \det(B)\det(A) = \det(BA)$$

Here we used the fact that $\det(A)\det(B) = \det(B)\det(A)$ because multiplication of real numbers is commutative. Let's look at a specific example illustrating Theorem 5.12.

EXAMPLE 5 Show that $\det(AB) = \det(A)\det(B)$ for the matrices

$$A = \begin{bmatrix} 1 & 3 \\ 2 & 5 \end{bmatrix} \quad \text{and} \quad B = \begin{bmatrix} 4 & -5 \\ -3 & 2 \end{bmatrix}$$

Then show that $\det(A + B) \neq \det(A) + \det(B)$.

Solution Starting with A and B, we have

$$\det(A) = (1)(5) - (3)(2) = -1 \quad \text{and} \quad \det(B) = (4)(2) - (-5)(-3) = -7$$

Hence $\det(A)\det(B) = 7$. Multiplying the matrices, we find that

$$AB = \begin{bmatrix} 1 & 3 \\ 2 & 5 \end{bmatrix} \begin{bmatrix} 4 & -5 \\ -3 & 2 \end{bmatrix} = \begin{bmatrix} -5 & 1 \\ -7 & 0 \end{bmatrix}$$

Therefore $\det(AB) = (-5)(0) - (-7)(1) = 7 = \det(A)\det(B)$. Turning to $A + B$, we have

$$A + B = \begin{bmatrix} 5 & -2 \\ -1 & 7 \end{bmatrix}$$

▶ For $n \times n$ matrices A and B, in general

$$\det(A + B) \neq \det(A) + \det(B)$$

Thus $\det(A + B) = (5)(7) - (-2)(-1) = 33$ and $\det(A) + \det(B) = -8$, so $\det(A + B) \neq \det(A) + \det(B)$. ∎

To prove Theorem 5.12, we start with a special case involving elementary matrices. Recall that elementary matrices, introduced in Section 3.4, are square matrices E such that the product EA performs an elementary row operation on A.

THEOREM 5.14

If E and B are both $n \times n$ matrices and E is an elementary matrix, then $\det(EB) = \det(E)\det(B)$.

Proof Suppose that E is an elementary matrix corresponding to interchanging two rows. Then E is equal to I_n after the same two rows have been interchanged. Since $\det(I_n) = 1$, it follows from Theorem 5.13(a) that $\det(E) = -1$. As the product EB is the same as B with two rows interchanged, we have (again by Theorem 5.13(a)) $\det(EB) = -\det(B)$. Therefore

$$\det(EB) = -\det(B) = \det(E)\det(B)$$

completing the proof for this type of elementary matrix. The proofs for the other two types of elementary matrices are similar and are left as exercises. ∎

We now use Theorem 5.14 to prove Theorem 5.12.

Proof of Theorem 5.12 First, if A is singular, then so is AB (see Exercise 68, Section 3.3), so that $\det(A)\det(B) = 0$ and $\det(AB) = 0$, proving the theorem in this case. Now suppose that A is nonsingular and hence has an inverse. Then there exists a sequence of row operations that will transform A into I_n. Let $E_1, E_2, \ldots E_k$ denote the corresponding elementary matrices, with E_1 inducing the first row operation, E_2 the second, and so on. Then $E_k \cdots E_2 E_1 A = I_n$, so that

$$A = (E_k \cdots E_2 E_1)^{-1} = E_1^{-1} E_2^{-1} \cdots E_k^{-1}$$

It is not hard to verify that the inverse of an elementary matrix is another elementary matrix (see Exercise 68), so that A is the product of elementary matrices. By repeatedly applying Theorem 5.14, we have

$$
\begin{aligned}
\det(AB) &= \det(E_1^{-1} E_2^{-1} \cdots E_k^{-1} B) \\
&= \det(E_1^{-1}) \det(E_2^{-1} \cdots E_k^{-1} B) \\
&\quad\vdots \\
&= \det(E_1^{-1}) \det(E_2^{-1}) \cdots \det(E_k^{-1}) \det(B) \\
&= \det(E_1^{-1} E_2^{-1} \cdots E_k^{-1}) \det(B) = \det(A) \det(B)
\end{aligned}
$$

so that $\det(AB) = \det(A) \det(B)$.   ■

Theorem 5.15 is an immediate consequence of Theorem 5.12.

THEOREM 5.15

Let A be an $n \times n$ invertible matrix. Then

$$
\det(A^{-1}) = \frac{1}{\det(A)}
$$

Proof   Since A is invertible, A^{-1} exists and $AA^{-1} = I_n$. Therefore

$$
1 = \det(I_n) = \det(AA^{-1}) = \det(A) \det(A^{-1})
$$

with the last equality holding by Theorem 5.12. Since $\det(A) \det(A^{-1}) = 1$, we have

$$
\det(A^{-1}) = \frac{1}{\det(A)}
$$
■

Determinants of Partitioned Matrices

▶ Partitioned matrices were introduced in Section 3.2. Recall that 0_{nm} denotes an $n \times m$ matrix with all entries equal to zero.

Suppose that we have the 5×5 partitioned matrix $P = \begin{bmatrix} A & 0_{23} \\ 0_{32} & D \end{bmatrix}$ with blocks

$$
A = \begin{bmatrix} a_{11} & a_{12} \\ a_{21} & a_{22} \end{bmatrix} \quad \text{and} \quad D = \begin{bmatrix} d_{11} & d_{12} & d_{13} \\ d_{21} & d_{22} & d_{23} \\ d_{31} & d_{32} & d_{33} \end{bmatrix}
$$

Our goal is to find a formula for $\det(P)$ in terms of $\det(A)$ and $\det(D)$. Thinking about 2×2 matrices, we have

$$
\begin{vmatrix} a & 0 \\ 0 & d \end{vmatrix} = ad
$$

which suggests the possibility that $\det(P) = \det(A) \det(D)$. To see if this is true, we employ cofactor expansion across the top row of P, which produces

$$
\det(P) = a_{11} \begin{vmatrix} a_{22} & 0 & 0 & 0 \\ 0 & d_{11} & d_{12} & d_{13} \\ 0 & d_{21} & d_{22} & d_{23} \\ 0 & d_{31} & d_{32} & d_{33} \end{vmatrix} - a_{12} \begin{vmatrix} a_{21} & 0 & 0 & 0 \\ 0 & d_{11} & d_{12} & d_{13} \\ 0 & d_{21} & d_{22} & d_{23} \\ 0 & d_{31} & d_{32} & d_{33} \end{vmatrix}
$$

Applying cofactor expansion across the top rows of both determinants, we have

$$
\det(P) = a_{11} a_{22} \begin{vmatrix} d_{11} & d_{12} & d_{13} \\ d_{21} & d_{22} & d_{23} \\ d_{31} & d_{32} & d_{33} \end{vmatrix} - a_{12} a_{21} \begin{vmatrix} d_{11} & d_{12} & d_{13} \\ d_{21} & d_{22} & d_{23} \\ d_{31} & d_{32} & d_{33} \end{vmatrix}
$$

$$
= (a_{11} a_{22} - a_{12} a_{21}) \det(D) = \det(A) \det(D)
$$

Hence it is true that $\det(P) = \det(A)\det(D)$. Looking back over our computations, we see that we would have arrived at the same formula even if

$$P = \begin{bmatrix} A & 0_{23} \\ C & D \end{bmatrix}$$

for any 3×2 matrix C, because the entries of C would not contribute to the cofactor expansions. This is again consistent with

$$\begin{vmatrix} a & 0 \\ c & d \end{vmatrix} = ad$$

A similar argument using cofactor expansions down columns can be used to show that

$$\begin{vmatrix} A & B \\ 0_{32} & D \end{vmatrix} = \det(A)\det(D)$$

for any 2×3 matrix B (see Exercise 70). These observations generalize to partitioned square matrices of higher dimension.

THEOREM 5.16

Let P be a partitioned $n \times n$ matrix of the form

$$P = \begin{bmatrix} A & B \\ 0 & D \end{bmatrix} \quad \text{or} \quad P = \begin{bmatrix} A & 0 \\ C & D \end{bmatrix}$$

where A and D are square block submatrices. Then $\det(P) = \det(A)\det(D)$.

A proof can be formulated using induction on the dimension of P. This is left as an exercise.

It is tempting to speculate that

$$\begin{vmatrix} A & B \\ C & D \end{vmatrix} = \det(A)\det(D) - \det(B)\det(C)$$

when A, B, C, and D are square submatrices, but this turns out not to be true in general. See Exercises 33–34 for counterexamples.

Proof of Theorem 5.13

Proof of Theorem 5.13 We take each part of the theorem in turn.

(a) Suppose that the matrix B results from A by interchanging two adjacent rows R_i and R_{i+1}. Using cofactor expansion along R_i of A gives

$$\det(A) = a_{i1}(-1)^{i+1}\det(M_{i1}) + \cdots + a_{in}(-1)^{i+n}\det(M_{in})$$

Now suppose that we compute $\det(B)$ by using cofactor expansion along R_{i+1} of B. Since we interchanged R_i and R_{i+1} to get B, the entries of R_{i+1} of B are the same as those of R_i of A, as are the matrices M_{ij} corresponding to these entries. Hence

$$\det(B) = a_{i1}(-1)^{i+2}\det(M_{i1}) + \cdots + a_{in}(-1)^{i+1+n}\det(M_{in})$$
$$= (-1)\{a_{i1}(-1)^{i+1}\det(M_{i1}) + \cdots + a_{in}(-1)^{i+n}\det(M_{in})\} = -\det(A)$$

Keeping this in mind, let's consider the general case. Suppose that B results from A by the operation $R_i \Leftrightarrow R_j$ (interchanging rows R_i and R_j), where for convenience we assume that $i < j$. This operation can be accomplished by two sequences of interchanges of adjacent rows. Start with the sequence $R_i \Leftrightarrow R_{i+1}$, $R_{i+1} \Leftrightarrow R_{i+2}$,

$$\begin{bmatrix} \vdots \\ r_{i-1} \\ r_i \\ \vdots \\ r_{j-1} \\ r_j \\ \vdots \end{bmatrix} \sim \begin{bmatrix} \vdots \\ r_{i-1} \\ r_{i+1} \\ \vdots \\ r_j \\ r_i \\ \vdots \end{bmatrix} \begin{matrix} \\ \\ \leftarrow \text{Row } i \\ \\ \\ \leftarrow \text{Row } j \\ \\ \end{matrix}$$

Figure 1 The result of the first $j-i$ adjacent row interchanges.

$$\begin{bmatrix} \vdots \\ r_{i-1} \\ r_{i+1} \\ \vdots \\ r_j \\ r_i \\ \vdots \end{bmatrix} \sim \begin{bmatrix} \vdots \\ r_{i-1} \\ r_j \\ \vdots \\ r_{j-1} \\ r_i \\ \vdots \end{bmatrix} \begin{matrix} \\ \\ \leftarrow \text{Row } i \\ \\ \\ \leftarrow \text{Row } j \\ \\ \end{matrix}$$

Figure 2 The result of the second $j-i-1$ adjacent row interchanges.

$\ldots, R_{j-1} \Leftrightarrow R_j$. When these $j-i$ interchanges are complete, the elements of rows $i+1$ through j are shifted up one row, and the elements in row i are moved to row j (see Figure 1).

We shift the elements originally in row j up to row i with the sequence of $j-i-1$ interchanges of adjacent rows $R_{j-2} \Leftrightarrow R_{j-1}$, $R_{j-3} \Leftrightarrow R_{j-2}, \ldots, R_i \Leftrightarrow R_{i+1}$ (see Figure 2). At this point we have $R_i \Leftrightarrow R_j$, and all the other rows are back where they started.

Returning to the relationship between $\det(A)$ and $\det(B)$, by our earlier observation each interchange of adjacent rows multiplied the determinant by -1. Since there are a total of $2(j-i)-1$ such interchanges, we have

$$\det(B) = (-1)^{2(j-i)-1}\det(A) = -\det(A)$$

as stated in part (a) of the theorem.

(b) Suppose that B is produced by multiplying row i of A by a scalar c. Using cofactor expansion along row i of B, we find that

$$\begin{aligned}\det(B) &= ca_{i1}C_{i1} + \cdots + ca_{in}C_{in} \\ &= c\,(a_{i1}C_{i1} + \cdots + a_{in}C_{in}) = c\det(A)\end{aligned}$$

so that $\det(A) = \frac{1}{c}\det(B)$.

(c) Suppose that B results from applying to A the row operation $cR_i + R_j \Rightarrow R_j$. Using cofactor expansion along row j of B yields

$$\begin{aligned}\det(B) &= (ca_{i1} + a_{j1})C_{j1} + \cdots + (ca_{in} + a_{jn})C_{jn} \\ &= \left(a_{j1}C_{j1} + \cdots + a_{jn}C_{jn}\right) + c\left(a_{i1}C_{j1} + \cdots + a_{in}C_{jn}\right)\end{aligned}$$

The term in the left parentheses is the cofactor expansion along row j of A and so is equal to $\det(A)$. The term in the right parentheses is equal to zero (see Exercise 79 in Section 5.1). Hence

$$\det(B) = \det(A) + c(0) = \det(A)$$

completing the proof of part (c) and the theorem. ■

EXERCISES

In Exercises 1–6, compute the determinant of the given matrix A by using row operations to reduce to echelon form, as illustrated in Example 2.

1. $A = \begin{bmatrix} 2 & 8 \\ -1 & -3 \end{bmatrix}$

2. $A = \begin{bmatrix} 9 & -1 \\ 3 & 2 \end{bmatrix}$

3. $A = \begin{bmatrix} 1 & -1 & -3 \\ -2 & 2 & 6 \\ -3 & -3 & 10 \end{bmatrix}$

4. $A = \begin{bmatrix} -4 & 2 & -2 \\ 1 & 0 & 1 \\ 3 & -1 & 1 \end{bmatrix}$

5. $A = \begin{bmatrix} 0 & 1 & 0 & 0 \\ 1 & 0 & 0 & 0 \\ 0 & 0 & 0 & 1 \\ 0 & 0 & 1 & 0 \end{bmatrix}$

6. $A = \begin{bmatrix} -3 & 2 & -1 & 6 \\ 7 & -3 & 2 & -11 \\ 0 & 0 & 0 & -2 \\ 1 & 0 & 0 & -1 \end{bmatrix}$

In Exercises 7–14, the given row operations, when performed on a matrix A, result in the given matrix B. Find the determinant of A, and decide if A is invertible.

7. $\begin{matrix} R_1 \Leftrightarrow R_2 \\ 2R_1 + R_2 \Rightarrow R_2 \end{matrix} \implies B = \begin{bmatrix} 1 & -5 \\ 0 & -4 \end{bmatrix}$

8. $\begin{matrix} -3R_1 \Rightarrow R_1 \\ -R_1 + R_2 \Rightarrow R_2 \end{matrix} \implies B = \begin{bmatrix} 1 & 7 \\ 0 & 3 \end{bmatrix}$

9. $\begin{matrix} R_3 \Leftrightarrow R_1 \\ -2R_1 + R_2 \Rightarrow R_2 \\ 5R_2 + R_3 \Rightarrow R_3 \end{matrix} \implies B = \begin{bmatrix} 1 & -4 & 9 \\ 0 & -3 & 2 \\ 0 & 0 & 7 \end{bmatrix}$

10. $\begin{matrix} (1/2)R_1 \Rightarrow R_1 \\ 5R_1 + R_2 \Rightarrow R_2 \\ R_2 \Leftrightarrow R_3 \\ -4R_2 + R_3 \Rightarrow R_3 \end{matrix} \implies B = \begin{bmatrix} 2 & 0 & -7 \\ 0 & -1 & 1 \\ 0 & 0 & 6 \end{bmatrix}$

11.
$$\begin{matrix} -7R_2 \Rightarrow R_2 \\ -3R_1 + R_3 \Rightarrow R_3 \\ 5R_2 + R_3 \Rightarrow R_3 \end{matrix} \implies B = \begin{bmatrix} -6 & 2 & -1 \\ 0 & 0 & 1 \\ 0 & 0 & -4 \end{bmatrix}$$

12.
$$\begin{matrix} R_1 \Leftrightarrow R_3 \\ 2R_1 + R_3 \Rightarrow R_3 \\ R_2 \Leftrightarrow R_3 \\ -3R_3 \Rightarrow R_3 \end{matrix} \implies B = \begin{bmatrix} 1 & -2 & 0 \\ 0 & 4 & 3 \\ 0 & 0 & -2 \end{bmatrix}$$

13.
$$\begin{matrix} R_1 \Leftrightarrow R_4 \\ -5R_1 + R_3 \Rightarrow R_3 \\ -2R_1 + R_4 \Rightarrow R_4 \\ R_2 \Leftrightarrow R_4 \end{matrix} \implies B = \begin{bmatrix} 1 & 0 & 5 & 2 \\ 0 & 1 & 2 & 0 \\ 0 & 0 & 2 & 1 \\ 0 & 0 & 0 & 4 \end{bmatrix}$$

14.
$$\begin{matrix} R_2 \Leftrightarrow R_3 \\ R_2 + R_3 \Rightarrow R_3 \\ -7R_2 + R_4 \Rightarrow R_4 \\ R_3 \Leftrightarrow R_4 \\ -5R_4 \Rightarrow R_4 \end{matrix} \implies B = \begin{bmatrix} 1 & 2 & 0 & 0 \\ 0 & 3 & 1 & 1 \\ 0 & 0 & 1 & 6 \\ 0 & 0 & 0 & 5 \end{bmatrix}$$

In Exercises 15–18, suppose that

$$\det(A) = \begin{vmatrix} a & b & c \\ d & e & f \\ g & h & i \end{vmatrix} = 3$$

and find the determinant of the given matrix.

15. $\begin{bmatrix} d & e & f \\ a & b & c \\ g & h & i \end{bmatrix}$

16. $\begin{bmatrix} d & e & f \\ g & h & i \\ a & b & c \end{bmatrix}$

17. $\begin{bmatrix} a & b & c \\ -2d & -2e & -2f \\ a+g & b+h & c+i \end{bmatrix}$

18. $\begin{bmatrix} a & b & c \\ 5g & 5h & 5i \\ d & e & f \end{bmatrix}$

In Exercises 19–22, verify that $\det(A)\det(B) = \det(AB)$ and that $\det(A + B) \neq \det(A) + \det(B)$.

19. $A = \begin{bmatrix} 2 & 3 \\ 1 & -4 \end{bmatrix}$, $B = \begin{bmatrix} 0 & -1 \\ 3 & 7 \end{bmatrix}$

20. $A = \begin{bmatrix} -1 & 5 \\ 2 & 4 \end{bmatrix}$, $B = \begin{bmatrix} 7 & 3 \\ 2 & 1 \end{bmatrix}$

21. $A = \begin{bmatrix} 2 & 0 & -1 \\ 1 & 1 & 0 \\ 0 & 1 & 1 \end{bmatrix}$, $B = \begin{bmatrix} 2 & 2 & 1 \\ 2 & 0 & 4 \\ -1 & 3 & 1 \end{bmatrix}$

22. $A = \begin{bmatrix} 0 & 0 & -2 \\ 2 & 0 & 3 \\ 1 & 1 & 1 \end{bmatrix}$, $B = \begin{bmatrix} 1 & -2 & 0 \\ 1 & -1 & 3 \\ 1 & 4 & -1 \end{bmatrix}$

23. Suppose that A is a square matrix with $\det(A) = 3$. Find each of the following:

(a) $\det(A^2)$ **(c)** $\det(A^2 A^T)$

(b) $\det(A^4)$ **(d)** $\det(A^{-1})$

24. Suppose that B is a square matrix with $\det(B) = -2$. Find each of the following:

(a) $\det(B^3)$ **(c)** $\det(BB^T)$

(b) $\det(B^5)$ **(d)** $\det((B^{-1})^3)$

25. Suppose that A and B are $n \times n$ matrices with $\det(A) = 3$ and $\det(B) = -2$. Find each of the following:

(a) $\det(A^2 B^3)$ **(c)** $\det(B^3 A^T)$

(b) $\det(AB^{-1})$ **(d)** $\det(A^2 B^3 B^T)$

26. Suppose that A and B are $n \times n$ matrices with $\det(A) = 4$ and $\det(B) = -3$. Find each of the following:

(a) $\det(B^3 A^2)$ **(c)** $\det(A^2 B^T)$

(b) $\det(A^{-2} B^4)$ **(d)** $\det(B^{-2} A^3 A^T)$

In Exercises 27–32, partition the matrix A in order to compute $\det(A)$.

27. $A = \begin{bmatrix} 1 & -3 & 0 & 0 \\ 2 & 5 & 0 & 0 \\ 0 & 0 & 5 & -1 \\ 0 & 0 & 3 & 3 \end{bmatrix}$

28. $A = \begin{bmatrix} -2 & 0 & 2 & 0 & 0 \\ 1 & 3 & 1 & 0 & 0 \\ 2 & 0 & 4 & 0 & 0 \\ 0 & 0 & 0 & -3 & 2 \\ 0 & 0 & 0 & 1 & 2 \end{bmatrix}$

29. $A = \begin{bmatrix} 4 & 2 & 1 & 1 \\ 3 & 1 & 5 & 9 \\ 0 & 0 & 3 & 2 \\ 0 & 0 & 1 & 0 \end{bmatrix}$

30. $A = \begin{bmatrix} 1 & 2 & 6 & 7 & 9 \\ 2 & 3 & 8 & 2 & 5 \\ 0 & 0 & 1 & 2 & 1 \\ 0 & 0 & 2 & 1 & 3 \\ 0 & 0 & 3 & 2 & 1 \end{bmatrix}$

31. $A = \begin{bmatrix} 9 & 3 & 0 & 0 \\ 3 & 1 & 0 & 0 \\ 3 & 9 & 2 & 2 \\ 8 & 1 & 3 & 1 \end{bmatrix}$

32. $A = \begin{bmatrix} 2 & 3 & 0 & 0 & 0 \\ -1 & 5 & 0 & 0 & 0 \\ 6 & 7 & 1 & 3 & 2 \\ -4 & 1 & 0 & 3 & 1 \\ 0 & 5 & 2 & 1 & 0 \end{bmatrix}$

In Exercises 33–34, for the given matrices A, B, C, and D verify that $\begin{vmatrix} A & B \\ C & D \end{vmatrix} \neq \det(A)\det(D) - \det(B)\det(C)$.

33. $A = \begin{bmatrix} 1 & 2 \\ 0 & 4 \end{bmatrix}$, $B = \begin{bmatrix} -1 & 0 \\ 5 & 1 \end{bmatrix}$,

$C = \begin{bmatrix} 0 & -2 \\ -1 & 3 \end{bmatrix}$, $D = \begin{bmatrix} 2 & 1 \\ 4 & 0 \end{bmatrix}$

34. $A = \begin{bmatrix} -1 & 0 \\ 2 & 4 \end{bmatrix}$, $B = \begin{bmatrix} 0 & 1 \\ -3 & 2 \end{bmatrix}$,

$C = \begin{bmatrix} 3 & 2 \\ 0 & -3 \end{bmatrix}$, $D = \begin{bmatrix} 1 & -1 \\ 5 & 1 \end{bmatrix}$

In Exercises 35–40, determine if a unique solution exists for the given linear system.

35. $\begin{aligned} 6x_1 - 5x_2 &= 12 \\ -2x_1 + 7x_2 &= 0 \end{aligned}$

36. $\begin{aligned} 10x_1 - 5x_2 &= 5 \\ -4x_1 + 2x_2 &= -3 \end{aligned}$

37. $\begin{aligned} 3x_1 + 2x_2 + 7x_3 &= 0 \\ - 3x_3 &= -3 \\ - x_2 - 4x_3 &= 13 \end{aligned}$

38. $\begin{aligned} -2x_1 + 5x_2 - 10x_3 &= 4 \\ x_1 - 2x_2 + 3x_3 &= -1 \\ 7x_1 - 17x_2 + 34x_3 &= -16 \end{aligned}$

39. $\begin{aligned} x_1 + x_2 - 2x_3 &= -3 \\ 3x_1 - 2x_2 + 2x_3 &= 9 \\ 6x_1 - 7x_2 - x_3 &= 4 \end{aligned}$

40. $\begin{aligned} x_1 - 3x_2 + 2x_3 &= 4 \\ -2x_1 + 7x_2 - 2x_3 &= -7 \\ 4x_1 - 13x_2 + 7x_3 &= 12 \end{aligned}$

FIND AN EXAMPLE For Exercises 41–46, find an example that meets the given specifications.

41. A nonzero 2×2 matrix A such that $3 \det(A) = \det(3A)$.

42. A nonzero 3×3 matrix A such that $-2 \det(A) = \det(-2A)$.

43. Find 2×2 matrices A and B, both nonzero, such that $\det(A + B) = \det(A) + \det(B)$. (NOTE: This identity is not generally true, but there are examples where it holds.)

44. Find 3×3 matrices A and B, both nonzero, such that $\det(A + B) = \det(A) + \det(B)$. (NOTE: This identity is not generally true, but there are examples where it holds.)

45. Find a 3×3 matrix A such that $\det(A) = 1$ and all entries of A are nonzero. (HINT: Start with an upper triangular matrix that has the specified determinant, then use row operations to obtain A.)

46. Find a 4×4 matrix A such that $\det(A) = 1$ and all entries of A are nonzero. (HINT: Start with an upper triangular matrix that has the specified determinant, then use row operations to obtain A.)

TRUE OR FALSE For Exercises 47–56, determine if the statement is true or false, and justify your answer.

47. Interchanging the rows of a matrix has no effect on its determinant.

48. If $\det(A) \neq 0$, then the columns of A are linearly independent.

49. If E is an elementary matrix, then $\det(E) = 1$.

50. If A and B are $n \times n$ matrices, then $\det(A + B) = \det(A) + \det(B)$.

51. If A is a 3×3 matrix and $\det(A) = 0$, then $\text{rank}(A) = 0$.

52. If A is a 4×4 matrix and $\det(A) = 4$, then $\text{nullity}(A) = 0$.

53. Suppose A, B, and S are $n \times n$ matrices, and that S is invertible. If $B = S^{-1}AS$, then $\det(A) = \det(B)$.

54. If A is an $n \times n$ matrix with all entries equal to 1, then $\det(A) = n$.

55. Suppose that A is a 4×4 matrix and that B is the matrix obtained by multiplying the third column of A by 2. Then $\det(B) = 2 \det(A)$.

56. If A is an invertible matrix, then at least one of the submatrices M_{ij} of A is also invertible.

In Exercises 57–66, assume that A is an $n \times n$ matrix.

57. Prove that if A has two identical rows, then $\det(A) = 0$.

58. Prove that if A has two identical columns, then $\det(A) = 0$. (HINT: Apply Theorem 5.10 and Exercise 57.)

59. Prove that $\det(A^T A) \geq 0$.

60. Suppose that $\det(A) = 2$. Prove that A^{-1} cannot have all integer entries.

61. Prove that $\det(-A) = (-1)^n \det(A)$.

62. Prove that $\det(cA) = c^n \det(A)$.

63. Suppose that A is *idempotent*, which means $A = A^2$. What are the possible values of $\det(A)$?

64. Suppose that A is *skew symmetric*, which means $A = -A^T$. Show that if n is odd, then $\det(A) = 0$.

65. Prove that if n is odd, then $A^2 \neq -I_n$. (HINT: Compare determinants of A^2 and $-I_n$.)

66. Show that if the entries of each row of A add to zero, then $\det(A) = 0$. (HINT: Think linear independence and The Big Theorem.)

67. Prove that a square matrix A has an echelon form B such that $\det(A) = \pm \det(B)$.

68. Suppose that E is an elementary matrix. Show that E^{-1} is also an elementary matrix.

69. This exercise completes the proof of Theorem 5.14. Let B be an $n \times n$ matrix and E be an $n \times n$ elementary matrix.

(a) Suppose that E corresponds to multiplying a row by a scalar c. Show that $\det(EB) = \det(E) \det(B) = c \det(B)$.

(b) Suppose that E corresponds to adding a multiple of one row to another. Show that $\det(EB) = \det(E) \det(B) = \det(B)$.

70. Prove that

$$\begin{vmatrix} A & B \\ 0_{32} & D \end{vmatrix} = \det(A) \det(D)$$

where A is a 2×2 matrix, B is a 2×3 matrix, D is a 3×3 matrix, and 0_{32} is a 3×2 matrix with all entries equal to zero.

71. Prove Theorem 5.16: Let M be a partitioned $n \times n$ matrix of either of the forms

$$M = \begin{bmatrix} A & B \\ 0 & D \end{bmatrix} \quad \text{or} \quad M = \begin{bmatrix} A & 0 \\ C & D \end{bmatrix}$$

where A and D are square block submatrices. Then $\det(M) = \det(A)\det(D)$. (HINT: Show the formula holds for the first form of M using induction on the number of rows of M, and then take the transpose to show that the formula holds for the second form.)

72. The *Vandermonde matrix* is given by

$$V = \begin{bmatrix} 1 & x_1 & x_1^2 & \cdots & x_1^{n-1} \\ 1 & x_2 & x_2^2 & \cdots & x_2^{n-1} \\ \vdots & \vdots & \vdots & \ddots & \vdots \\ 1 & x_n & x_n^2 & \cdots & x_n^{n-1} \end{bmatrix}$$

(a) For the Vandermonde matrix with $n = 3$, show that

$$\begin{vmatrix} 1 & x_1 & x_1^2 \\ 1 & x_2 & x_2^2 \\ 1 & x_3 & x_3^2 \end{vmatrix} = (x_3 - x_2)(x_3 - x_1)(x_2 - x_1)$$

(b) For any $n > 1$, prove that

$$\det(V) = \prod_{1 \le i < j \le n} (x_j - x_i)$$

C In Exercises 73–74, verify *Sylvester's determinant theorem*, which states that

$$\det(I_m + AB) = \det(I_n + BA)$$

for any $m \times n$ matrix A and $n \times m$ matrix B.

73. $A = \begin{bmatrix} 3 & -2 & 6 \\ 4 & 0 & 5 \\ 2 & -9 & 1 \\ 5 & -1 & -4 \end{bmatrix}$

$B = \begin{bmatrix} 0 & -3 & 2 & 6 \\ 1 & 4 & 4 & 2 \\ -8 & 3 & 0 & 5 \end{bmatrix}$

74. $A = \begin{bmatrix} 6 & 2 & 3 & -1 & 4 \\ 7 & 0 & -2 & 4 & 5 \\ -8 & 2 & 4 & 9 & 0 \end{bmatrix}$

$B = \begin{bmatrix} 2 & 4 & -6 \\ 0 & 5 & 2 \\ -3 & 7 & 7 \\ 0 & 2 & 8 \\ -9 & 3 & 5 \end{bmatrix}$

5.3 Applications of the Determinant

▶ This section is optional and can be omitted without loss of continuity.

In this section we consider a few applications of the determinant, beginning with a method for using determinants to find the solution to the linear systems $A\mathbf{x} = \mathbf{b}$ when A is an invertible square matrix. From the Big Theorem, Version 7, we know that in this case there will be a unique solution.

Before stating the theorem, we need to introduce some notation. If $A = \begin{bmatrix} \mathbf{a}_1 & \mathbf{a}_2 & \cdots & \mathbf{a}_n \end{bmatrix}$ is an $n \times n$ matrix and $\mathbf{b}$ is in $\mathbf{R}^n$, then let A_i denote the matrix A after replacing $\mathbf{a}_i$ with $\mathbf{b}$. That is,

$$A_i = \begin{bmatrix} \mathbf{a}_1 & \cdots & \mathbf{a}_{i-1} & \mathbf{b} & \mathbf{a}_{i+1} & \cdots & \mathbf{a}_n \end{bmatrix}$$

For instance, if

$$A = \begin{bmatrix} 4 & -1 & 2 & 0 \\ 3 & 7 & 5 & -2 \\ -5 & 2 & 0 & 4 \\ 0 & 6 & 1 & 1 \end{bmatrix} \quad \text{and} \quad \mathbf{b} = \begin{bmatrix} 8 \\ 9 \\ 2 \\ -3 \end{bmatrix}$$

then

$$A_1 = \begin{bmatrix} 8 & -1 & 2 & 0 \\ 9 & 7 & 5 & -2 \\ 2 & 2 & 0 & 4 \\ -3 & 6 & 1 & 1 \end{bmatrix} \quad \text{and} \quad A_3 = \begin{bmatrix} 4 & -1 & 8 & 0 \\ 3 & 7 & 9 & -2 \\ -5 & 2 & 2 & 4 \\ 0 & 6 & -3 & 1 \end{bmatrix}$$

Matrices of this type are used in the next theorem.

THEOREM 5.17

(CRAMER'S RULE) Let A be an invertible $n \times n$ matrix. Then the components of the unique solution $\mathbf{x}$ to $A\mathbf{x} = \mathbf{b}$ are given by

$$x_i = \frac{\det(A_i)}{\det(A)} \quad \text{for } i = 1, 2, \ldots, n$$

The proof of Theorem 5.17 is given at the end of this section. For now, let's look at an example.

EXAMPLE 1   Use Cramer's Rule to find the solution to the system

$$\begin{array}{rcrcrcr} 3x_1 & + & x_2 & & & = & 5 \\ -x_1 & + & 2x_2 & + & x_3 & = & -2 \\ & & -x_2 & + & 2x_3 & = & -1 \end{array}$$

Solution   The system is equivalent to $A\mathbf{x} = \mathbf{b}$, where

$$A = \begin{bmatrix} 3 & 1 & 0 \\ -1 & 2 & 1 \\ 0 & -1 & 2 \end{bmatrix} \quad \text{and} \quad \mathbf{b} = \begin{bmatrix} 5 \\ -2 \\ -1 \end{bmatrix}$$

We have

$$A_1 = \begin{bmatrix} 5 & 1 & 0 \\ -2 & 2 & 1 \\ -1 & -1 & 2 \end{bmatrix}, \quad A_2 = \begin{bmatrix} 3 & 5 & 0 \\ -1 & -2 & 1 \\ 0 & -1 & 2 \end{bmatrix}, \quad A_3 = \begin{bmatrix} 3 & 1 & 5 \\ -1 & 2 & -2 \\ 0 & -1 & -1 \end{bmatrix}$$

▶ The determinants in Example 1 were computed using the Shortcut Method.

Computing determinants gives us $\det(A) = 17$, $\det(A_1) = 28$, $\det(A_2) = 1$, and $\det(A_3) = -8$. Therefore, by Cramer's Rule, the solution to $A\mathbf{x} = \mathbf{b}$ is

$$x_1 = \frac{\det(A_1)}{\det(A)} = \frac{28}{17}, \quad x_2 = \frac{\det(A_2)}{\det(A)} = \frac{1}{17}, \quad x_3 = \frac{\det(A_3)}{\det(A)} = \frac{-8}{17} \quad \blacksquare$$

Cramer's Rule is easy to implement when the coefficient matrix A is 2×2 or 3×3. Unfortunately, for larger systems all the required determinants generally make this method computationally impractical. (See the "Computational Comment" at the end of Section 5.1.)

Inverses from Determinants

In Chapter 3 we showed how to adapt our row operation algorithm for finding the solutions to a linear system to determine the inverse of a square matrix. Here we do something similar, using Cramer's Rule to develop a formula for finding inverses. We start with a statement of the formula and then explain why it works.

Definition Cofactor Matrix

For an $n \times n$ matrix A, the **cofactor matrix** is given by

$$C = \begin{bmatrix} C_{11} & C_{12} & \cdots & C_{1n} \\ C_{21} & C_{22} & \cdots & C_{2n} \\ \vdots & \vdots & & \vdots \\ C_{n1} & C_{n2} & \cdots & C_{nn} \end{bmatrix}$$

Definition Adjoint Matrix

where the cofactors C_{ij} are as defined in Section 5.1. Now we define the **adjoint** of A, denoted adj(A), by

$$\text{adj}(A) = C^T = \begin{bmatrix} C_{11} & C_{21} & \cdots & C_{n1} \\ C_{12} & C_{22} & \cdots & C_{n2} \\ \vdots & \vdots & \ddots & \vdots \\ C_{1n} & C_{2n} & \cdots & C_{nn} \end{bmatrix}$$

THEOREM 5.18

If A is an invertible matrix, then

$$A^{-1} = \frac{1}{\det(A)} \text{adj}(A) \tag{1}$$

Proof We prove this theorem by showing that $A\left(\frac{1}{\det(A)}\text{adj}(A)\right) = I_n$. Start by forming the product

$$A(\text{adj}(A)) = \begin{bmatrix} a_{11} & a_{12} & \cdots & a_{1n} \\ a_{21} & a_{22} & \cdots & a_{2n} \\ \vdots & \vdots & \ddots & \vdots \\ a_{n1} & a_{n2} & \cdots & a_{nn} \end{bmatrix} \begin{bmatrix} C_{11} & C_{21} & \cdots & C_{n1} \\ C_{12} & C_{22} & \cdots & C_{n2} \\ \vdots & \vdots & \ddots & \vdots \\ C_{1n} & C_{2n} & \cdots & C_{nn} \end{bmatrix}$$

The entry in row i and column j of $A(\text{adj}(A))$ is

$$a_{i1}C_{j1} + a_{i2}C_{j2} + \cdots + a_{in}C_{in} \tag{2}$$

When $i = j$, (2) is the cofactor expansion across row i of A and is equal to $\det(A)$ by Theorem 5.8, Section 5.1. If $i \neq j$, then (2) is equal to zero (see Exercise 79 in Section 5.1). Hence we have

$$A(\text{adj}(A)) = \det(A)I_n \implies A\left(\frac{1}{\det(A)}\text{adj}(A)\right) = I_n$$

and so $A^{-1} = \frac{1}{\det(A)}\text{adj}(A)$. ∎

EXAMPLE 2 Use Theorem 5.18 to find the inverse of the matrix

$$A = \begin{bmatrix} 3 & 1 & 0 \\ -1 & 2 & 1 \\ 0 & -1 & 2 \end{bmatrix}$$

Solution From Example 1 we know that $\det(A) = 17$. Since A is 3×3, A has nine cofactors. Four of them are

$$C_{11} = (-1)^2 \begin{vmatrix} 2 & 1 \\ -1 & 2 \end{vmatrix} = 5 \quad C_{21} = (-1)^3 \begin{vmatrix} 1 & 0 \\ -1 & 2 \end{vmatrix} = -2$$

$$C_{31} = (-1)^4 \begin{vmatrix} 1 & 0 \\ 2 & 1 \end{vmatrix} = 1 \quad C_{12} = (-1)^3 \begin{vmatrix} -1 & 1 \\ 0 & 2 \end{vmatrix} = 2$$

The remaining five are computed similarly, yielding $C_{22} = 6$, $C_{32} = -3$, $C_{13} = 1$, $C_{23} = 3$, and $C_{33} = 7$. Filling out the adjoint of A gives the inverse

$$A^{-1} = \frac{1}{\det(A)}\text{adj}(A) = \frac{1}{17}\begin{bmatrix} 5 & -2 & 1 \\ 2 & 6 & -3 \\ 1 & 3 & 7 \end{bmatrix} = \begin{bmatrix} \frac{5}{17} & -\frac{2}{17} & \frac{1}{17} \\ \frac{2}{17} & \frac{6}{17} & -\frac{3}{17} \\ \frac{1}{17} & \frac{3}{17} & \frac{7}{17} \end{bmatrix} \quad ∎$$

In Section 3.3, we encountered the "Quick Formula" for computing the inverse of a 2×2 matrix. This is revisited in the next example.

EXAMPLE 3   Use Theorem 5.18 to find the inverse of the matrix

$$A = \begin{bmatrix} a & b \\ c & d \end{bmatrix}$$

Solution   We have $\det(A) = ad - bc$, which we know must be nonzero for an inverse to exist. The cofactors of A are

$$C_{11} = d, \quad C_{21} = -b, \quad C_{12} = -c, \quad C_{22} = a$$

Therefore

$$A^{-1} = \frac{1}{\det(A)}\operatorname{adj}(A) = \frac{1}{ad - bc}\begin{bmatrix} d & -b \\ -c & a \end{bmatrix}$$

as given in Section 3.3.   ∎

Area and Determinants

There is a striking relationship between area, determinants, and linear transformations. Although we focus on $\mathbf{R}^2$ here, the results developed are also true in higher dimensions.

Let $\mathcal{S}$ denote the unit square in the first quadrant of $\mathbf{R}^2$, and suppose that $T : \mathbf{R}^2 \to \mathbf{R}^2$ is a linear transformation. Let $A = \begin{bmatrix} \mathbf{a}_1 & \mathbf{a}_2 \end{bmatrix}$ be the 2×2 matrix such that $T(\mathbf{x}) = A\mathbf{x}$. If $\mathcal{P} = T(\mathcal{S})$ denotes the image of $\mathcal{S}$ under T and A is invertible, then $\mathcal{P}$ is a parallelogram in $\mathbf{R}^2$ (see Exercise 65 in Section 3.1). An example is shown in Figure 1.

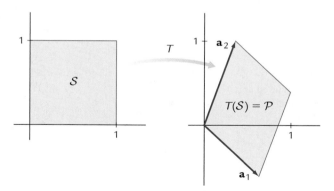

Figure 1 The square $\mathcal{S}$ and $\mathcal{P} = T(\mathcal{S})$, where $T(\mathbf{x}) = A\mathbf{x}$ and $A = [\mathbf{a}_1 \ \mathbf{a}_2]$.

Theorem 5.19 shows how the area of $\mathcal{P}$ is related to the determinant of A.

THEOREM 5.19

Let $\mathcal{S}$ be the unit square in the first quadrant of $\mathbf{R}^2$, and let $T : \mathbf{R}^2 \to \mathbf{R}^2$ with $T(\mathbf{x}) = A\mathbf{x}$. If $\mathcal{P} = T(\mathcal{S})$ is the image of $\mathcal{S}$ under T, then

$$\operatorname{area}(\mathcal{P}) = |\det(A)| \tag{3}$$

where area$(\mathcal{P})$ denotes the area of $\mathcal{P}$.

Proof   Let $A = \begin{bmatrix} \mathbf{a}_1 & \mathbf{a}_2 \end{bmatrix}$. First suppose that the columns of A are linearly dependent. Then $T(\mathcal{S})$ is a line segment (see Exercise 65, Section 3.1) so that area$(\mathcal{P}) = 0$. We also have $\det(A) = 0$ by the Big Theorem, Version 7, and so (3) is true in this case.

Next suppose that the columns of A are linearly independent, so that $\mathcal{P}$ is a parallelogram. Rotate $\mathcal{P}$ about the origin through the angle θ so that the rotated image of $\mathbf{a}_1$ ends up on the x-axis and the rotated parallelogram $\mathcal{P}^*$ is above the x-axis (see Figure 2).

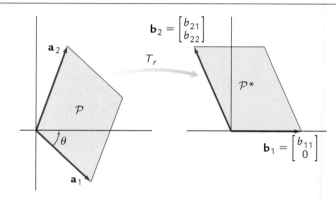

Figure 2 Rotating $\mathcal{P}$ about the origin by an angle θ yields a new region $\mathcal{P}^*$ of equal area.

Rotation preserves area, so that

$$\text{area}(\mathcal{P}) = \text{area}(\mathcal{P}^*) \tag{4}$$

The rotation of $\mathcal{P}$ to $\mathcal{P}^*$ is achieved with the linear transformation $T_r(\mathbf{x}) = C\mathbf{x}$, where (see Section 3.1)

$$C = \begin{bmatrix} \cos\theta & -\sin\theta \\ \sin\theta & \cos\theta \end{bmatrix}$$

If $B = \begin{bmatrix} \mathbf{b}_1 & \mathbf{b}_2 \end{bmatrix}$, where $\mathbf{b}_1$ and $\mathbf{b}_2$ are as in Figure 2, then formulas from geometry tell us that

▶ It is possible that $b_{11} < 0$. Absolute values are included so that (5) is true in general.

$$\text{area}(\mathcal{P}^*) = |b_{11}b_{22}| = |\det(B)| \tag{5}$$

If $T_1(\mathbf{x}) = B\mathbf{x}$, then

$$T_1(\mathbf{e}_1) = \begin{bmatrix} \mathbf{b}_1 & \mathbf{b}_2 \end{bmatrix} \begin{bmatrix} 1 \\ 0 \end{bmatrix} = \mathbf{b}_1 = T_r(\mathbf{a}_1) = T_r\big(T(\mathbf{e}_1)\big)$$

$$T_1(\mathbf{e}_2) = \begin{bmatrix} \mathbf{b}_1 & \mathbf{b}_2 \end{bmatrix} \begin{bmatrix} 0 \\ 1 \end{bmatrix} = \mathbf{b}_2 = T_r(\mathbf{a}_2) = T_r\big(T(\mathbf{e}_2)\big)$$

Since $\{\mathbf{e}_1, \mathbf{e}_2\}$ is a basis for $\mathbf{R}^2$, it must be that $T_1(\mathbf{x})$ and $T_r\big(T(\mathbf{x})\big)$ are the same linear transformation. Since $T_1(\mathbf{x}) = B\mathbf{x}$ and $T_r\big(T(\mathbf{x})\big) = CA\mathbf{x}$, we have $B = CA$. Therefore

▶ $\det(C) = \cos^2\theta + \sin^2\theta = 1$

$$\det(B) = \det(CA) = \det(C)\det(A) = \det(A) \tag{6}$$

Combining (4), (5), and (6), we conclude that $\text{area}(\mathcal{P}) = |\det(A)|$. ■

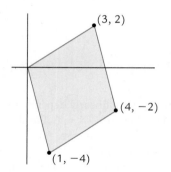

Figure 3 A parallelogram in $\mathbf{R}^2$.

EXAMPLE 4 Use Theorem 5.19 to find the area of the parallelogram in Figure 3.

Solution Let $A = \begin{bmatrix} 3 & 1 \\ 2 & -4 \end{bmatrix}$. Then $T(\mathbf{x}) = A\mathbf{x}$ will map $\mathcal{S}$, the unit square in the first quadrant, onto $\mathcal{P}$. Hence, by Theorem 5.19,

$$\text{area}(\mathcal{P}) = |\det(A)| = |-14| = 14 \quad ■$$

We now generalize Theorem 5.19 to arbitrary regions of finite area.

THEOREM 5.20

Let $\mathcal{D}$ be a region of finite area in $\mathbf{R}^2$, and suppose that $T : \mathbf{R}^2 \to \mathbf{R}^2$ with $T(\mathbf{x}) = A\mathbf{x}$. If $T(\mathcal{D})$ denotes the image of $\mathcal{D}$ under T, then

$$\text{area}\big(T(\mathcal{D})\big) = |\det(A)| \cdot \text{area}(\mathcal{D}) \tag{7}$$

Proof We give a complete proof for rectangular regions $\mathcal{R}$, and after that we sketch the method of proof for general regions $\mathcal{D}$.

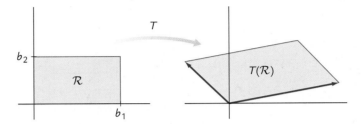

Figure 4 A rectangular region $\mathcal{R}$ and its image $T(\mathcal{R})$.

Our proof strategy is to find the linear transformation T_1 such that $T_1(\mathcal{S}) = T(\mathcal{R})$ ($\mathcal{S}$ denotes the unit square) and then apply Theorem 5.19 (Figure 4). If $T_B(\mathbf{x}) = B\mathbf{x}$ for

$$B = \begin{bmatrix} b_1 & 0 \\ 0 & b_2 \end{bmatrix}$$

then $T_B(\mathcal{S}) = \mathcal{R}$ (see Figure 5 and Exercise 51).

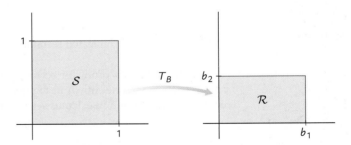

Figure 5 The image of $\mathcal{S}$ under T_B is the rectangular region $\mathcal{R}$.

Since $T(\mathcal{R})$ is the image of $\mathcal{R}$ under T, then $T(\mathcal{R})$ is also the image of $\mathcal{S}$ under the composition

$$T_1(\mathbf{x}) = T\big(T_B(\mathbf{x})\big)$$

That is, $T_1(\mathcal{S}) = T(\mathcal{R})$. Since $T(\mathbf{x}) = A\mathbf{x}$, we have $T_1(\mathbf{x}) = T\big(T_B(\mathbf{x})\big) = AB\mathbf{x}$. Hence, by Theorem 5.19, we have

$$\begin{aligned}
\text{area}\big(T(\mathcal{R})\big) &= |\det(AB)| \\
&= |\det(A) \cdot \det(B)| \\
&= |b_1 b_2 \cdot \det(A)| \\
&= |\det(A)| \cdot \text{area}(\mathcal{R})
\end{aligned}$$

▶ The area of $\mathcal{R}$ is $b_1 b_2$.

Thus (7) is true in this case.

An arbitrary rectangular region $\mathcal{R}^*$ with sides parallel to the coordinate axes (see Figure 6) has the form $\mathbf{v} + \mathcal{R}$, where $\mathbf{v}$ is a fixed vector and $\mathcal{R}$ is a rectangular region of the type in Figure 4.

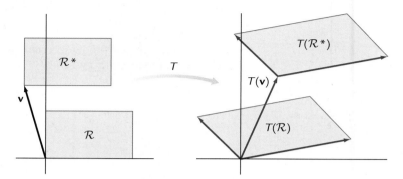

Figure 6 A rectangular region $\mathcal{R}^*$ and its image $T(\mathcal{R}^*)$.

Since each vector $\mathbf{r}^*$ in $\mathcal{R}^*$ has the form $\mathbf{v} + \mathbf{r}$ for some $\mathbf{r}$ in $\mathcal{R}$, we have

$$T(\mathbf{r}^*) = T(\mathbf{v} + \mathbf{r}) = T(\mathbf{v}) + T(\mathbf{r})$$

Hence $T(\mathcal{R}^*)$ is a translation of $T(\mathcal{R})$, with $T(\mathcal{R}^*) = T(\mathbf{v}) + T(\mathcal{R})$. As translation does not change area, we have area$(\mathcal{R}^*) = $ area$(\mathcal{R})$ and area$\big(T(\mathcal{R}^*)\big) = $ area$\big(T(\mathcal{R})\big)$. Therefore

$$\begin{aligned} \text{area}\big(T(\mathcal{R}^*)\big) &= \text{area}\big(T(\mathcal{R})\big) \\ &= |\det(A)| \cdot \text{area}(\mathcal{R}) \quad \text{(by the previous case)} \\ &= |\det(A)| \cdot \text{area}(\mathcal{R}^*) \end{aligned}$$

so (7) also holds for translated rectangles. ∎

With the formal proof for rectangles complete, now suppose that $\mathcal{D}$ is a general region such as the one shown in Figure 7. Then (7) also holds for this case. The proof uses techniques found in calculus—we give the basic elements of the argument here.

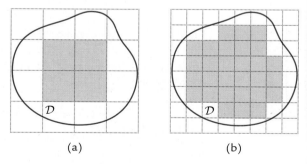

(a)   (b)

Figure 7 (a) A grid of squares over $\mathcal{D}$. (b) In general, a finer grid of squares over $\mathcal{D}$ gives a more accurate area estimate.

We can approximate the area of $\mathcal{D}$ by superimposing a grid of squares and then counting the number of squares that are inside $\mathcal{D}$ (see Figure 7). Multiplying this count by the area of a single square—which we can easily calculate—gives the area approximation.

We can make the approximation as accurate as we like by making the grid of squares sufficiently fine.

When we apply a linear transformation T to $\mathcal{D}$, each grid square is mapped to a corresponding parallelogram (see Figure 8).

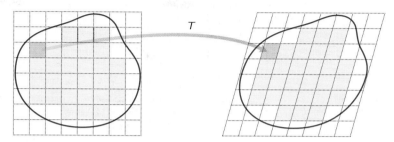

Figure 8 A grid of squares over $\mathcal{D}$ and the corresponding parallelograms over $T(\mathcal{D})$.

There is a one-to-one correspondence between grid squares within $\mathcal{D}$ and parallelograms within $T(\mathcal{D})$. Furthermore, from (7) we have

$$(\text{area of parallelogram}) = |\det(A)| \cdot (\text{area of grid square})$$

Since the grid squares provide an approximation to the area of $\mathcal{D}$ and the parallelograms provide an approximation to the area of $T(\mathcal{D})$, and there are the same number of each, then

$$\text{area}\big(T(\mathcal{D})\big) \approx |\det(A)| \cdot \text{area}(\mathcal{D})$$

In general, the approximation gets better as the grid of squares becomes finer, so that as the size of the grid squares shrinks, in the limit we get (7).

Theorem 5.20 has a higher-dimensional analog, stated below without proof. See Exercises 33–36 for a brief discussion of this theorem in $\mathbf{R}^3$.

THEOREM 5.21

Let $\mathcal{D}$ be a region of finite volume in $\mathbf{R}^n$, and suppose that $T : \mathbf{R}^n \rightarrow \mathbf{R}^n$ with $T(\mathbf{x}) = A\mathbf{x}$. If $T(\mathcal{D})$ denotes the image of $\mathcal{D}$ under T, then

$$\text{volume}\big(T(\mathcal{D})\big) = |\det(A)| \cdot \text{volume}(\mathcal{D}) \tag{8}$$

EXAMPLE 5 Use Theorem 5.20 and the known area of the unit circle $\mathcal{U}$ to find the area of the ellipse $\mathcal{E}$ in Figure 9.

Solution The regions inside the unit circle $\mathcal{U}$ and the ellipse $\mathcal{E}$ are given by the set of all (x_1, x_2) such that

$$x_1^2 + x_2^2 \leq 1 \qquad\qquad \left(\frac{x_1}{a}\right)^2 + \left(\frac{x_2}{b}\right)^2 \leq 1$$

$$\text{Unit circle } \mathcal{U} \qquad\qquad\qquad \text{Ellipse } \mathcal{E}$$

The linear transformation

$$T\left(\begin{bmatrix} x_1 \\ x_2 \end{bmatrix}\right) = \begin{bmatrix} ax_1 \\ bx_2 \end{bmatrix} = \underbrace{\begin{bmatrix} a & 0 \\ 0 & b \end{bmatrix}}_{A} \begin{bmatrix} x_1 \\ x_2 \end{bmatrix}$$

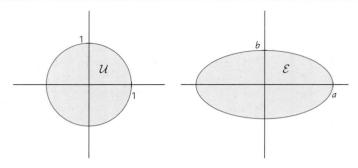

Figure 9 The region inside the unit circle $\mathcal{U}$ is related by a linear transformation to the region inside the ellipse $\mathcal{E}$.

gives a one-to-one correspondence from $\mathcal{U}$ to $\mathcal{E}$ (see Exercise 52). Since $T(\mathbf{x}) = A\mathbf{x}$ and the area of $\mathcal{U}$ is π, by Theorem 5.20 we have

$$\text{area}(\mathcal{E}) = |\det(A)| \cdot \text{area}(\mathcal{U}) = ab\pi \quad \blacksquare$$

Proof of Theorem 5.17 (Cramer's Rule)

Proof of Theorem 5.17   Since A is invertible, the system $A\mathbf{x} = \mathbf{b}$ has a unique solution. If $I_{n,i}$ denotes the $n \times n$ identity matrix with the ith column replaced by $\mathbf{x}$ (see margin), then we have

$$\begin{aligned} AI_{n,i} &= A\begin{bmatrix} \mathbf{e}_1 & \cdots & \mathbf{x} & \cdots & \mathbf{e}_n \end{bmatrix} \\ &= \begin{bmatrix} A\mathbf{e}_1 & \cdots & A\mathbf{x} & \cdots & A\mathbf{e}_n \end{bmatrix} \\ &= \begin{bmatrix} \mathbf{a}_1 & \cdots & \mathbf{b} & \cdots & \mathbf{a}_n \end{bmatrix} = A_i \end{aligned}$$

where we replace $A\mathbf{x}$ with $\mathbf{b}$. Since $AI_{n,i} = A_i$, it follows that $\det(AI_{n,i}) = \det(A_i)$. Applying Theorem 5.12, we have $\det(AI_{n,i}) = \det(A)\det(I_{n,i})$, so that $\det(I_{n,i}) = \det(A)/\det(A_i)$. On the other hand, cofactor expansion along the ith row of $I_{n,i}$ yields $\det(I_{n,i}) = x_i \det(I_{n-1}) = x_i$. Therefore

$$x_i = \frac{\det(A_i)}{\det(A)}$$

as stated in the theorem. $\blacksquare$

$\blacktriangleright\ I_{n,i} =$

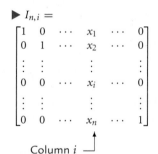

$$\begin{bmatrix} 1 & 0 & \cdots & x_1 & \cdots & 0 \\ 0 & 1 & \cdots & x_2 & \cdots & 0 \\ \vdots & \vdots & & \vdots & & \vdots \\ 0 & 0 & \cdots & x_i & \cdots & 0 \\ \vdots & \vdots & & \vdots & & \vdots \\ 0 & 0 & \cdots & x_n & \cdots & 1 \end{bmatrix}$$

Column i

EXERCISES

In Exercises 1–6, determine if Cramer's Rule can be applied to find the solution for the given linear system, and if so, then find the solution.

1. $\begin{aligned} 6x_1 - 5x_2 &= 12 \\ -2x_1 + 7x_2 &= 0 \end{aligned}$

2. $\begin{aligned} 10x_1 - 5x_2 &= 5 \\ -4x_1 + 2x_2 &= -3 \end{aligned}$

3. $\begin{aligned} 3x_1 + 2x_2 + 7x_3 &= 0 \\ -3x_3 &= -3 \\ -x_2 - 4x_3 &= 13 \end{aligned}$

4. $\begin{aligned} -2x_1 + 5x_2 - 10x_3 &= 4 \\ x_1 - 2x_2 + 3x_3 &= -1 \\ 7x_1 - 17x_2 + 34x_3 &= -16 \end{aligned}$

5. $\begin{aligned} x_1 + x_2 - 2x_3 &= -3 \\ 3x_1 - 2x_2 + 2x_3 &= 9 \\ 6x_1 - 7x_2 - x_3 &= 4 \end{aligned}$

6. $\begin{aligned} x_1 - 3x_2 + 2x_3 &= 4 \\ -2x_1 + 7x_2 - 2x_3 &= -7 \\ 4x_1 - 13x_2 + 7x_3 &= 12 \end{aligned}$

In Exercises 7–12, find the value of x_2 in the unique solution of the given linear system.

7. $\begin{aligned} -2x_1 + 3x_2 &= 3 \\ -3x_1 - 7x_2 &= -1 \end{aligned}$

8. $\begin{aligned} x_1 - 4x_2 &= 11 \\ -3x_1 + x_2 &= -2 \end{aligned}$

9.
$$\begin{aligned}
3x_1 \quad\quad + 2x_3 &= 1 \\
3x_2 + 2x_3 &= 3 \\
2x_1 + 3x_2 + x_3 &= -4
\end{aligned}$$

10.
$$\begin{aligned}
x_1 - x_2 \quad\quad &= 0 \\
3x_1 - x_2 - 3x_3 &= 2 \\
x_1 - 3x_2 - 2x_3 &= -3
\end{aligned}$$

11.
$$\begin{aligned}
3x_2 \quad\quad - 3x_4 &= 1 \\
2x_1 - x_2 + 3x_3 - 3x_4 &= -2 \\
-2x_1 + 3x_2 + 2x_3 + 2x_4 &= 0 \\
2x_1 \quad\quad + 2x_3 + x_4 &= -1
\end{aligned}$$

12.
$$\begin{aligned}
3x_1 - 3x_2 \quad\quad - 3x_4 &= 5 \\
-x_1 \quad\quad + 2x_3 + x_4 &= 0 \\
x_1 \quad\quad + 3x_3 \quad\quad &= 3 \\
-2x_2 + 3x_3 + 3x_4 &= 1
\end{aligned}$$

In Exercises 13–18, for the given matrix A, find adj(A) and then use it to compute A^{-1}.

13. $A = \begin{bmatrix} 2 & 5 \\ 3 & 7 \end{bmatrix}$

14. $A = \begin{bmatrix} 1 & 7 \\ 1 & 6 \end{bmatrix}$

15. $A = \begin{bmatrix} 0 & 1 & 0 \\ 0 & 0 & 1 \\ 1 & 0 & 0 \end{bmatrix}$

16. $A = \begin{bmatrix} 2 & 0 & 1 \\ 0 & 0 & 2 \\ 1 & 1 & 0 \end{bmatrix}$

17. $A = \begin{bmatrix} 1 & 2 & 1 \\ 0 & 1 & 2 \\ 0 & 0 & 1 \end{bmatrix}$

18. $A = \begin{bmatrix} 3 & 0 & 0 \\ 1 & 2 & 0 \\ 1 & 1 & 1 \end{bmatrix}$

In Exercises 19–22, sketch the parallelogram with the given vertices, then determine its area using determinants.

19. $(0, 0), (2, 3), (5, 1), (7, 4)$

20. $(0, 0), (2, 7), (4, -5), (6, 2)$

21. $(2, 3), (-1, 4), (5, 7), (2, 8)$

22. $(3, -1), (0, -2), (5, -6), (2, -7)$

In Exercises 23–28, find the area of $T(\mathcal{D})$ for $T(\mathbf{x}) = A\mathbf{x}$.

23. $\mathcal{D}$ is the rectangle with vertices $(2, 2), (7, 2), (7, 5), (2, 5)$, and $A = \begin{bmatrix} 3 & -1 \\ 5 & 2 \end{bmatrix}$.

24. $\mathcal{D}$ is the rectangle with vertices $(-3, 4), (5, 4), (5, 7), (-3, 7)$, and $A = \begin{bmatrix} -2 & 7 \\ 3 & 4 \end{bmatrix}$.

25. $\mathcal{D}$ is the parallelogram with vertices $(0, 0), (5, 1), (2, 4), (7, 5)$, and $A = \begin{bmatrix} 1 & 4 \\ 2 & 5 \end{bmatrix}$.

26. $\mathcal{D}$ is the parallelogram with vertices $(0, 0), (-2, 3), (3, 5), (1, 8)$, and $A = \begin{bmatrix} 5 & 2 \\ 9 & 1 \end{bmatrix}$.

27. $\mathcal{D}$ is the parallelogram with vertices $(1, 2), (6, 4), (2, 6), (7, 8)$, and $A = \begin{bmatrix} 1 & 4 \\ 2 & 5 \end{bmatrix}$.

28. $\mathcal{D}$ is the parallelogram with vertices $(-2, 1), (-4, 4), (1, 6), (-1, 9)$, and $A = \begin{bmatrix} 5 & 2 \\ 9 & 1 \end{bmatrix}$.

In Exercises 29–32, find a linear transformation T that gives a one-to-one correspondence between the unit circle and the given ellipse.

29.

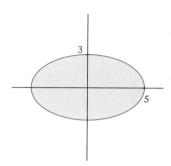

30.

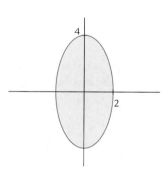

31.

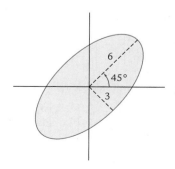

32.

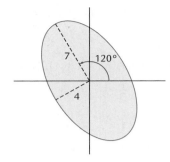

For Exercises 33–36: In three dimensions, Theorem 5.21 states that if $\mathcal{D}$ is a region of finite volume in $\mathbf{R}^3$, and $T : \mathbf{R}^3 \rightarrow \mathbf{R}^3$ with $T(\mathbf{x}) = A\mathbf{x}$, then

$$\text{volume}\big(T(\mathcal{D})\big) = |\det(A)| \cdot \text{volume}(\mathcal{D}) \qquad (9)$$

The image of the unit sphere (below left) under a linear transformation is an *ellipsoid* (example below right).

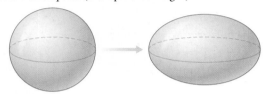

In Exercises 33–34, use (9) to determine the volume of the described ellipsoid.

33. An ellipsoid centered at the origin with axis intercepts $x = \pm 4$, $y = \pm 3$, and $z = \pm 5$.

34. An ellipsoid centered at the origin with axis intercepts $x = \pm 2$, $y = \pm 6$, and $z = \pm 4$.

The image of the unit cube (below left) under a linear transformation is a *parallelepiped* (example below right).

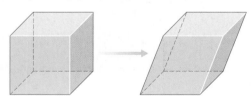

In Exercises 35–36, use (9) to determine the volume of the described parallelepiped.

35. The parallelepiped with sides described by the vectors

$$\begin{bmatrix} 3 \\ 5 \\ 2 \end{bmatrix}, \quad \begin{bmatrix} 6 \\ 1 \\ 3 \end{bmatrix}, \quad \begin{bmatrix} 2 \\ 0 \\ 4 \end{bmatrix}$$

36. The parallelepiped with sides described by the vectors

$$\begin{bmatrix} -1 \\ 3 \\ 5 \end{bmatrix}, \quad \begin{bmatrix} 0 \\ 4 \\ 2 \end{bmatrix}, \quad \begin{bmatrix} 6 \\ 1 \\ 1 \end{bmatrix}$$

FIND AN EXAMPLE For Exercises 37–42, find an example that meets the given specifications.

37. A linear system with two equations and two unknowns that is consistent but cannot be solved with Cramer's Rule.

38. A linear system with three equations and three unknowns that is consistent but cannot be solved with Cramer's Rule.

39. A parallelogram that has vertices with integer coordinates and area 5.

40. A parallelogram with area 7 that has vertices with integer coordinates that are not on the coordinate axes.

41. A 2×2 matrix A such that $\text{adj}(A) = \begin{bmatrix} 5 & -3 \\ -2 & 1 \end{bmatrix}$.

42. A 2×2 matrix A such that $\text{adj}(A) = \begin{bmatrix} 2 & 3 \\ 5 & 7 \end{bmatrix}$.

TRUE OR FALSE For Exercises 43–50, determine if the statement is true or false, and justify your answer.

43. Cramer's Rule can be used to find the solution to any system that has the same number of equations as unknowns.

44. If A is a square matrix with integer entries, then so is $\text{adj}(A)$.

45. If A is a 3×3 matrix, then $\text{adj}(2A) = 2\text{adj}(A)$.

46. If A is a square matrix that has all positive entries, then so does $\text{adj}(A)$.

47. If A is a 2×2 matrix, $\mathcal{S}$ is the unit square, and $T(\mathbf{x}) = A\mathbf{x}$, then $T(\mathcal{S})$ is a parallelogram of nonzero area.

48. If A is an $n \times n$ matrix with $\det(A) = 1$, then $A^{-1} = \text{adj}(A)$.

49. Suppose that A is an invertible $n \times n$ matrix with integer entries. If $\det(A) = 1$, then A^{-1} also has integer entries.

50. If A is a square matrix, then $\big(\text{adj}(A)\big)^T = \text{adj}(A^T)$.

51. Prove that the linear transformation $T(\mathbf{x}) = B\mathbf{x}$ with

$$B = \begin{bmatrix} b_1 & 0 \\ 0 & b_2 \end{bmatrix}$$

gives a one-to-one correspondence between the interior of the unit square $\mathcal{S}$ and the interior of the rectangle $\mathcal{R}$ shown in Figure 5.

52. Prove that the linear transformation $T(\mathbf{x}) = A\mathbf{x}$ with

$$A = \begin{bmatrix} a & 0 \\ 0 & b \end{bmatrix}$$

gives a one-to-one correspondence between the interior of the unit circle $\mathcal{U}$ and the interior of the ellipse $\mathcal{E}$ shown in Figure 9.

53. Suppose that A is a 3×3 matrix with $\det(A) = -2$. Show that $A \cdot \text{adj}(A)$ is a 3×3 diagonal matrix, and determine the diagonal terms.

54. Prove that if A is an $n \times n$ matrix with linearly independent columns, then so is $\text{adj}(A)$.

55. Show that if A is an $n \times n$ symmetric matrix, then $\text{adj}(A)$ is also symmetric.

56. Prove that if A is an $n \times n$ diagonal matrix, then so is $\text{adj}(A)$.

57. Suppose that A is an $n \times n$ matrix and c is a scalar. Prove that $\text{adj}(cA) = c^{n-1}\text{adj}(A)$.

58. Prove that if A is an invertible $n \times n$ matrix, then $\det(\text{adj}(A)) = \left(\det(A)\right)^{n-1}$.

59. Suppose that A is an invertible $n \times n$ matrix. Show that $\left(\text{adj}(A)\right)^{-1} = \text{adj}(A^{-1})$.

60. Suppose that A is an invertible $n \times n$ matrix and that both A and A^{-1} have integer entries. Show that $\det(A) = \pm 1$.

61. Prove that if A is a diagonal matrix, then so is $\text{adj}(A)$.

62. In this problem, we show that

$$\text{area}(\mathcal{T}) = \tfrac{1}{2}|\det(A)|,$$

where $\text{area}(\mathcal{T})$ is the area of the triangle $\mathcal{T}$ with vertices (x_1, y_1), (x_2, y_2), and (x_3, y_3) (Figure 10) and

$$A = \begin{bmatrix} x_1 & y_1 & 1 \\ x_2 & y_2 & 1 \\ x_3 & y_3 & 1 \end{bmatrix}.$$

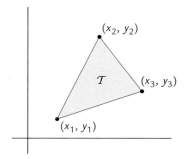

Figure 10 The triangle $\mathcal{T}$.

(a) Explain why $\text{area}(\mathcal{T}) = \text{area}(\mathcal{T}^*)$, where $\mathcal{T}^*$ is the triangle with vertices $(0, 0)$, $(x_2 - x_1, y_2 - y_1)$, and $(x_3 - x_1, y_3 - y_1)$ (Figure 11).

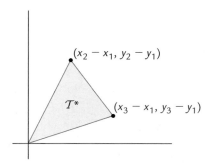

Figure 11 The triangle $\mathcal{T}^*$.

(b) Show that $\text{area}(\mathcal{T}^*) = \tfrac{1}{2}|\det(B)|$, where

$$B = \begin{bmatrix} (x_2 - x_1) & (y_2 - y_1) \\ (x_3 - x_1) & (y_3 - y_1) \end{bmatrix}$$

HINT: Apply Theorem 5.19 to compute the area of the parallelogram in Figure 12, and use the fact that $\det(B) = \det(B^T)$.

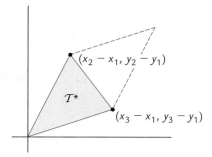

Figure 12 $\mathcal{T}^*$ and parallelogram.

(c) Show that $\det(B) = \det(C)$, where

$$C = \begin{bmatrix} x_1 & y_1 & 1 \\ (x_2 - x_1) & (y_2 - y_1) & 0 \\ (x_3 - x_1) & (y_3 - y_1) & 0 \end{bmatrix}$$

(d) Use row operations to show that $\det(C) = \det(A)$, and from this conclude that $\text{area}(\mathcal{T}) = \tfrac{1}{2}|\det(A)|$.

Ⓒ In Exercises 63–66, use Cramer's Rule to find the solution to the system.

63.
$$\begin{aligned}
-x_1 + 7x_2 + 5x_3 &= 13 \\
6x_1 - 2x_2 + x_3 &= 9 \\
3x_1 + 11x_2 - 9x_3 &= 4
\end{aligned}$$

64.
$$\begin{aligned}
8x_1 - 5x_2 \quad\quad &= 6 \\
-2x_1 - 4x_2 + 8x_3 &= -13 \\
5x_1 + 7x_2 - 11x_3 &= 17
\end{aligned}$$

65.
$$\begin{aligned}
3x_1 + 5x_2 - x_3 - 4x_4 &= -5 \\
-2x_1 - 4x_2 - 3x_3 + 7x_4 &= 0 \\
x_1 + 2x_2 + 4x_3 - 2x_4 &= 2 \\
4x_1 + x_2 - 5x_3 - x_4 &= 4
\end{aligned}$$

66.
$$\begin{aligned}
-5x_1 + 3x_2 + 2x_3 + x_4 &= 2 \\
x_1 - 7x_2 - 5x_3 + 7x_4 &= -3 \\
4x_1 + x_2 + x_3 + 2x_4 &= 0 \\
-4x_1 - 11x_2 \quad\quad - 5x_4 &= -9
\end{aligned}$$

Ⓒ In Exercises 67–70, for the given matrix A, find $\text{adj}(A)$ and then use it to compute A^{-1}.

67. $A = \begin{bmatrix} 4 & -2 & 5 \\ 8 & 3 & 0 \\ -1 & 7 & 9 \end{bmatrix}$

68. $A = \begin{bmatrix} 0 & 3 & 7 \\ -3 & 6 & 2 \\ 5 & 11 & -1 \end{bmatrix}$

69. $A = \begin{bmatrix} 4 & 2 & 5 & -1 \\ -2 & 3 & 0 & 6 \\ 5 & 7 & 2 & 11 \\ 3 & 0 & 1 & -5 \end{bmatrix}$

70. $A = \begin{bmatrix} 8 & -2 & 1 & 1 \\ -5 & 3 & 5 & 3 \\ 0 & 4 & 4 & -4 \\ 3 & 1 & 9 & 2 \end{bmatrix}$

Eigenvalues and Eigenvectors

The Mackinac Bridge crosses the Straits of Mackinac and connects Michigan's Upper and Lower Peninsulas. Five miles long, the bridge, also known as "Big Mac," is currently the third-longest suspension bridge in the U.S. The total length of wire in the cables is 42,000 miles. Though discussions of a bridge that crossed the Straits began in the 1880s, the bridge was not actually completed until 1957. The designer, David B. Steinman, studied the 1940 collapse of the Tacoma Narrows Bridge. In his design for Big Mac, Steinman responded to the Tacoma Narrows failure by incorporating stiffening trusses and an open-grid road deck to combat potential bridge instability caused by high winds.

Bridge suggested by Robert Fossum, University of Illinois Urbana-Champaign (Vito Palmisano/ Getty Images)

The focus of this chapter is *eigenvalues* and *eigenvectors*, which are characteristics of matrices and linear transformations. Eigenvalues and eigenvectors arise in a wide range of fields, including finance, quantum mechanics, image processing, and mechanical engineering.

In Section 6.1 we define eigenvalues and eigenvectors and develop an algebraic method for finding them. Algebraic methods typically are not practical for large matrices, so in Section 6.2 we develop a numerical approach similar to the one used to solve systems of linear equations (Section 1.3). In Section 6.3 and Section 6.4 we focus on change of basis transformations and using an eigenvector basis to diagonalize a matrix. Section 6.5 treats eigenvalues and eigenvectors involving complex numbers, and Section 6.6 focuses on solving systems of differential equations by using eigenvalues and eigenvectors.

6.1 Eigenvalues and Eigenvectors

Let $T : \mathbf{R}^2 \to \mathbf{R}^2$ be a linear transformation, with $T(\mathbf{x}) = A\mathbf{x}$ for a 2×2 matrix A. For a given vector $\mathbf{u}$ in $\mathbf{R}^2$, we can think of the multiplication $A\mathbf{u}$ as changing the direction and length of $\mathbf{u}$. Figure 1 shows $\mathbf{u}$ and $A\mathbf{u}$ for several vectors $\mathbf{u}$ and $A = \begin{bmatrix} 1 & 1 \\ 2 & 0 \end{bmatrix}$.

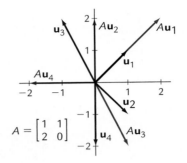

► Note that $\mathbf{u}_1$ and $A\mathbf{u}_1$ are parallel, which means that they point in the same (or opposite) direction. This is also true of $\mathbf{u}_3$ and $A\mathbf{u}_3$.

Figure 1 Plots of $\mathbf{u}$ and $A\mathbf{u}$ for the vectors $\mathbf{u}_1$, $\mathbf{u}_2$, $\mathbf{u}_3$, and $\mathbf{u}_4$.

Of particular importance in applications and in analyzing the behavior of a linear transformation are those vectors $\mathbf{u}$ such that $\mathbf{u}$ and $A\mathbf{u}$ are parallel. In Figure 1, $\mathbf{u}_1$ and $\mathbf{u}_3$ are such vectors. Algebraically, $\mathbf{u}$ and $A\mathbf{u}$ are parallel if there exists a scalar λ such that $A\mathbf{u} = \lambda\mathbf{u}$. Nonzero vectors that satisfy this equation are called *eigenvectors*.

DEFINITION 6.1

Definition Eigenvector, Eigenvalue

► Note that an eigenvalue λ can be equal to zero, but an eigenvector $\mathbf{u}$ must be a nonzero vector.

Let A be an $n \times n$ matrix. Then a nonzero vector $\mathbf{u}$ is an **eigenvector** of A if there exists a scalar λ such that

$$A\mathbf{u} = \lambda\mathbf{u} \tag{1}$$

The scalar λ is called an **eigenvalue** of A.

When λ and $\mathbf{u}$ are related as in equation (1), we say that λ is the eigenvalue associated with $\mathbf{u}$ and that $\mathbf{u}$ is an eigenvector associated with λ.

EXAMPLE 1 Let $A = \begin{bmatrix} 3 & 5 \\ 4 & 2 \end{bmatrix}$. Determine if each of

$$\mathbf{u}_1 = \begin{bmatrix} 5 \\ 4 \end{bmatrix}, \quad \mathbf{u}_2 = \begin{bmatrix} 4 \\ -1 \end{bmatrix}, \quad \text{and} \quad \mathbf{u}_3 = \begin{bmatrix} -1 \\ 1 \end{bmatrix}$$

is an eigenvector of A. For those that are, find the associated eigenvalue.

Solution Starting with $\mathbf{u}_1$, we have

$$A\mathbf{u}_1 = \begin{bmatrix} 3 & 5 \\ 4 & 2 \end{bmatrix} \begin{bmatrix} 5 \\ 4 \end{bmatrix} = \begin{bmatrix} 35 \\ 28 \end{bmatrix} = 7 \begin{bmatrix} 5 \\ 4 \end{bmatrix} = 7\mathbf{u}_1$$

Thus $A\mathbf{u}_1 = 7\mathbf{u}_1$, so $\mathbf{u}_1$ is an eigenvector of A with associated eigenvalue $\lambda = 7$.
Calculating $A\mathbf{u}_2$, we have

$$A\mathbf{u}_2 = \begin{bmatrix} 3 & 5 \\ 4 & 2 \end{bmatrix} \begin{bmatrix} 4 \\ -1 \end{bmatrix} = \begin{bmatrix} 7 \\ 14 \end{bmatrix}$$

Since $A\mathbf{u}_2$ is not a multiple of $\mathbf{u}_2$, this tells us that $\mathbf{u}_2$ is not an eigenvector of A. Finally,

$$A\mathbf{u}_3 = \begin{bmatrix} 3 & 5 \\ 4 & 2 \end{bmatrix} \begin{bmatrix} -1 \\ 1 \end{bmatrix} = \begin{bmatrix} 2 \\ -2 \end{bmatrix} = -2 \begin{bmatrix} -1 \\ 1 \end{bmatrix} = -2\mathbf{u}_3$$

Since $A\mathbf{u}_3 = -2\mathbf{u}_3$, it follows that $\mathbf{u}_3$ is an eigenvector of A with associated eigenvalue $\lambda = -2$. See Figure 2 for graphs of vectors. ∎

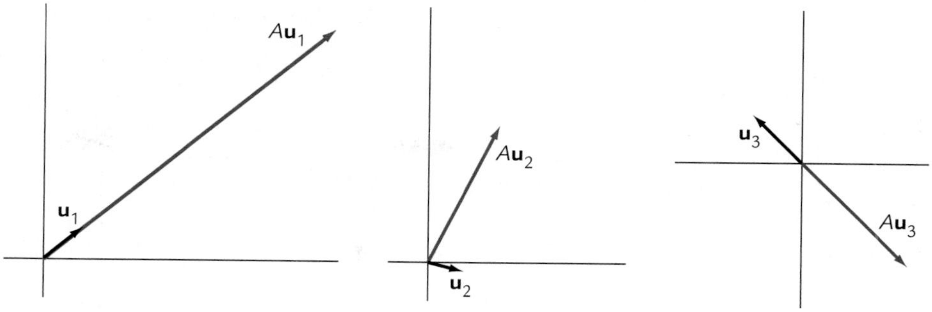

Figure 2 Graphs of vectors from Example 1. If **u** and $A\mathbf{u}$ are parallel vectors, then **u** is an eigenvector.

Referring back to $\mathbf{u}_1$ in Example 1, suppose $\mathbf{u}_4 = 3\mathbf{u}_1 = 3\begin{bmatrix} 5 \\ 4 \end{bmatrix} = \begin{bmatrix} 15 \\ 12 \end{bmatrix}$. Then we have

$$A\mathbf{u}_4 = \begin{bmatrix} 3 & 5 \\ 4 & 2 \end{bmatrix}\begin{bmatrix} 15 \\ 12 \end{bmatrix} = \begin{bmatrix} 105 \\ 84 \end{bmatrix} = 7\begin{bmatrix} 15 \\ 12 \end{bmatrix} = 7\mathbf{u}_4$$

Therefore $\mathbf{u}_4 = 3\mathbf{u}_1$ is also an eigenvector of A associated with $\lambda = 7$. We could have used any nonzero scalar in place of 3 and achieved the same result, so any nonzero multiple of $\mathbf{u}_1$ is also an eigenvector of A associated with $\lambda = 7$. Theorem 6.2 generalizes this observation.

THEOREM 6.2

Let A be a square matrix, and suppose that **u** is an eigenvector of A associated with eigenvalue λ. Then for any scalar $c \neq 0$, $c\mathbf{u}$ is also an eigenvector of A associated with λ.

▶ We require $c \neq 0$ because eigenvectors must be nonzero.

Proof   Let **u** be an eigenvector of A with associated eigenvalue λ. Then for any scalar $c \neq 0$, we have

$$A(c\mathbf{u}) = c(A\mathbf{u}) = c(\lambda\mathbf{u}) = \lambda(c\mathbf{u})$$

so that $c\mathbf{u}$ is also an eigenvector of A associated with eigenvalue λ. ■

Finding Eigenvectors

▶ We address the problem of finding the eigenvalues shortly.

Here we consider the problem of finding the eigenvectors associated with a known eigenvalue. Let's start with an example.

EXAMPLE 2   Take as known that $\lambda = 3$ and $\lambda = 2$ are eigenvalues for

$$A = \begin{bmatrix} 4 & 4 & -2 \\ 1 & 4 & -1 \\ 3 & 6 & -1 \end{bmatrix}$$

Find the eigenvectors associated with each eigenvalue.

Solution Starting with $\lambda = 3$, we need to find all nonzero vectors **u** such that $A\mathbf{u} = 3\mathbf{u}$. Since $3\mathbf{u} = 3I_3\mathbf{u}$, our equation becomes

$$A\mathbf{u} = 3I_3\mathbf{u} \quad \Longrightarrow \quad A\mathbf{u} - 3I_3\mathbf{u} = \mathbf{0} \quad \Longrightarrow \quad (A - 3I_3)\mathbf{u} = \mathbf{0}$$

Thus we need to find the solutions to the homogeneous system with coefficient matrix

$$A - 3I_3 = \begin{bmatrix} 4 & 4 & -2 \\ 1 & 4 & -1 \\ 3 & 6 & -1 \end{bmatrix} - \begin{bmatrix} 3 & 0 & 0 \\ 0 & 3 & 0 \\ 0 & 0 & 3 \end{bmatrix} = \begin{bmatrix} 1 & 4 & -2 \\ 1 & 1 & -1 \\ 3 & 6 & -4 \end{bmatrix}$$

Forming an augmented matrix and performing the indicated row operations, we have

$$\begin{bmatrix} 1 & 4 & -2 & 0 \\ 1 & 1 & -1 & 0 \\ 3 & 6 & -4 & 0 \end{bmatrix} \overset{\substack{-R_1+R_2 \Rightarrow R_2 \\ -3R_1+R_3 \Rightarrow R_3 \\ -2R_2+R_3 \Rightarrow R_3}}{\underset{\sim}{}} \begin{bmatrix} 1 & 4 & -2 & 0 \\ 0 & -3 & 1 & 0 \\ 0 & 0 & 0 & 0 \end{bmatrix}$$

After back substitution, we find that the system $(A - 3I_3)\mathbf{u} = \mathbf{0}$ has general solution

$$\mathbf{u} = s \begin{bmatrix} 2 \\ 1 \\ 3 \end{bmatrix}$$

We can verify that $\mathbf{u}$ is an eigenvector associated with $\lambda = 3$ by computing

$$A\mathbf{u} = \begin{bmatrix} 4 & 4 & -2 \\ 1 & 4 & -1 \\ 3 & 6 & -1 \end{bmatrix} \left(s \begin{bmatrix} 2 \\ 1 \\ 3 \end{bmatrix} \right) = s \begin{bmatrix} 4 & 4 & -2 \\ 1 & 4 & -1 \\ 3 & 6 & -1 \end{bmatrix} \begin{bmatrix} 2 \\ 1 \\ 3 \end{bmatrix} = s \begin{bmatrix} 6 \\ 3 \\ 9 \end{bmatrix} = 3\mathbf{u}$$

The procedure is the same for $\lambda = 2$. This time we need to find the general solution to the homogeneous system $(A - 2I_3)\mathbf{u} = \mathbf{0}$, where

$$A - 2I_3 = \begin{bmatrix} 4 & 4 & -2 \\ 1 & 4 & -1 \\ 3 & 6 & -1 \end{bmatrix} - \begin{bmatrix} 2 & 0 & 0 \\ 0 & 2 & 0 \\ 0 & 0 & 2 \end{bmatrix} = \begin{bmatrix} 2 & 4 & -2 \\ 1 & 2 & -1 \\ 3 & 6 & -3 \end{bmatrix}$$

The augmented matrix and corresponding echelon form are

$$\begin{bmatrix} 2 & 4 & -2 & 0 \\ 1 & 2 & -1 & 0 \\ 3 & 6 & -3 & 0 \end{bmatrix} \overset{\substack{R_1 \Leftrightarrow R_2 \\ -2R_1+R_2 \Rightarrow R_2 \\ -3R_1+R_3 \Rightarrow R_3}}{\underset{\sim}{}} \begin{bmatrix} 1 & 2 & -1 & 0 \\ 0 & 0 & 0 & 0 \\ 0 & 0 & 0 & 0 \end{bmatrix}$$

After back substitution, we find that this system has general solution

$$\mathbf{u} = s_1 \begin{bmatrix} -2 \\ 1 \\ 0 \end{bmatrix} + s_2 \begin{bmatrix} 1 \\ 0 \\ 1 \end{bmatrix} \tag{2}$$

As long as at least one of s_1 and s_2 is nonzero, then $\mathbf{u}$ will be an eigenvector associated with $\lambda = 2$. We can check our answer by computing

$$A\mathbf{u} = A \left(s_1 \begin{bmatrix} -2 \\ 1 \\ 0 \end{bmatrix} + s_2 \begin{bmatrix} 1 \\ 0 \\ 1 \end{bmatrix} \right) = s_1 A \begin{bmatrix} -2 \\ 1 \\ 0 \end{bmatrix} + s_2 A \begin{bmatrix} 1 \\ 0 \\ 1 \end{bmatrix}$$

$$= s_1 \begin{bmatrix} -4 \\ 2 \\ 0 \end{bmatrix} + s_2 \begin{bmatrix} 2 \\ 0 \\ 2 \end{bmatrix} = 2 \left(s_1 \begin{bmatrix} -2 \\ 1 \\ 0 \end{bmatrix} + s_2 \begin{bmatrix} 1 \\ 0 \\ 1 \end{bmatrix} \right) = 2\mathbf{u} \qquad ■$$

▶ Recall that a nonempty set of vectors is a subspace if the set is closed under addition and scalar multiplication.

If the zero vector is included, the eigenvectors in Example 2 associated with each eigenvalue form a subspace. This is always true of the set of eigenvectors associated with a given eigenvalue.

THEOREM 6.3

Let A be an $n \times n$ matrix with eigenvalue λ. Let S denote the set of all eigenvectors associated with λ, together with the zero vector $\mathbf{0}$. Then S is a subspace of $\mathbf{R}^n$.

Proof We show S is a subspace by verifying the three required conditions of Definition 4.1. First, by definition $\mathbf{0}$ is in S. Second, Theorem 6.2 tells us that if $\mathbf{u}$ is an eigenvector associated with λ, then so is $c\mathbf{u}$ for $c \neq 0$. If $c = 0$, then $c\mathbf{u} = \mathbf{0}$ is in S, so S is closed under scalar multiplication. Third, if $\mathbf{u}_1$ and $\mathbf{u}_2$ are both eigenvectors associated with λ, then

$$A(\mathbf{u}_1 + \mathbf{u}_2) = A\mathbf{u}_1 + A\mathbf{u}_2 = \lambda\mathbf{u}_1 + \lambda\mathbf{u}_2 = \lambda(\mathbf{u}_1 + \mathbf{u}_2)$$

so that $\mathbf{u}_1 + \mathbf{u}_2$ is also an eigenvector associated with λ. Therefore S is closed under addition, and so S is a subspace. ■

DEFINITION 6.4
Definition Eigenspace

Let A be a square matrix with eigenvalue λ. The subspace of all eigenvectors associated with λ, together with the zero vector, is called the **eigenspace** of λ.

▶ Each distinct eigenvalue has its own associated eigenspace.

For instance, in Example 2 we see from (2) that the set

$$\left\{ \begin{bmatrix} -2 \\ 1 \\ 0 \end{bmatrix}, \begin{bmatrix} 1 \\ 0 \\ 1 \end{bmatrix} \right\}$$

forms a basis for the eigenspace of $\lambda = 2$.

Finding Eigenvalues

Let's review what we have learned so far. If we know an eigenvalue λ for a given $n \times n$ matrix A, then we can find the associated eigenvectors by solving the linear system $A\mathbf{u} = \lambda\mathbf{u}$, or equivalently, the homogeneous system

$$(A - \lambda I_n)\mathbf{u} = \mathbf{0} \tag{3}$$

If we know the eigenvalue, then this is a problem that we know how to solve. Finding the eigenvalues is a different problem that we have not yet considered. The next theorem shows how to use determinants to find eigenvalues.

THEOREM 6.5

Let A be an $n \times n$ matrix. Then λ is an eigenvalue of A if and only if $\det(A - \lambda I_n) = 0$.

Proof λ is an eigenvalue of A if and only if there exists a nontrivial solution to $A\mathbf{u} = \lambda\mathbf{u}$. This is equivalent to the existence of a nontrivial solution to (3), which by the Big Theorem, Version 7 is true if and only if $\det(A - \lambda I_n) = 0$. ■

EXAMPLE 3 Find the eigenvalues for $A = \begin{bmatrix} 3 & 3 \\ 6 & -4 \end{bmatrix}$.

Solution Our strategy is to determine the values of λ that satisfy $\det(A - \lambda I_2) = 0$. We have

$$A - \lambda I_2 = \begin{bmatrix} 3 & 3 \\ 6 & -4 \end{bmatrix} - \begin{bmatrix} \lambda & 0 \\ 0 & \lambda \end{bmatrix} = \begin{bmatrix} (3 - \lambda) & 3 \\ 6 & (-4 - \lambda) \end{bmatrix}$$

Next, we compute the determinant,

$$\det(A - \lambda I_2) = (3 - \lambda)(-4 - \lambda) - 18 = \lambda^2 + \lambda - 30$$

Setting $\det(A - \lambda I) = 0$, we have

$$\lambda^2 + \lambda - 30 = 0 \implies (\lambda - 5)(\lambda + 6) = 0 \implies \lambda = 5 \quad \text{or} \quad \lambda = -6$$

Thus the eigenvalues for A are $\lambda = 5$ and $\lambda = -6$. ■

Definition Characteristic Polynomial, Characteristic Equation

The polynomial that we get from $\det(A - \lambda I)$ is called the **characteristic polynomial** of A, and the equation $\det(A - \lambda I) = 0$ is called the **characteristic equation**. The eigenvalues for a matrix A are given by the real roots of the characteristic equation. (It is also possible to consider the complex roots. They are covered in Section 6.5.)

EXAMPLE 4 Find the eigenvalues and a basis for each eigenspace for the matrix

$$A = \begin{bmatrix} 2 & -1 \\ -1 & 2 \end{bmatrix}$$

Solution We start by finding the eigenvalues of A by computing

$$\det(A - \lambda I_2) = \begin{vmatrix} (2 - \lambda) & -1 \\ -1 & (2 - \lambda) \end{vmatrix}$$
$$= (2 - \lambda)^2 - 1$$
$$= \lambda^2 - 4\lambda + 3 = (\lambda - 1)(\lambda - 3)$$

Therefore the eigenvalues are $\lambda_1 = 1$ and $\lambda_2 = 3$. Next, we find the eigenvectors, starting with those associated with $\lambda_1 = 1$. We find the associated eigenvectors by solving the homogeneous linear system $(A - 1 \cdot I_2)\mathbf{u} = (A - I_2)\mathbf{u} = \mathbf{0}$. Since

$$A - I_2 = \begin{bmatrix} 1 & -1 \\ -1 & 1 \end{bmatrix}$$

the augmented matrix and corresponding echelon form are

$$\begin{bmatrix} 1 & -1 & 0 \\ -1 & 1 & 0 \end{bmatrix} \overset{R_1 + R_2 \Rightarrow R_2}{\sim} \begin{bmatrix} 1 & -1 & 0 \\ 0 & 0 & 0 \end{bmatrix}$$

Back substitution gives us

General Solution: $\mathbf{u}_1 = s \begin{bmatrix} 1 \\ 1 \end{bmatrix} \implies$ Basis for Eigenspace of $\lambda_1 = 1$: $\left\{ \begin{bmatrix} 1 \\ 1 \end{bmatrix} \right\}$

The eigenvectors associated with $\lambda_2 = 3$ are found by solving the homogeneous linear system $(A - 3I_2)\mathbf{u} = \mathbf{0}$. We have

$$A - 3I_2 = \begin{bmatrix} -1 & -1 \\ -1 & -1 \end{bmatrix}$$

so the augmented matrix and corresponding echelon form are

$$\begin{bmatrix} -1 & -1 & 0 \\ -1 & -1 & 0 \end{bmatrix} \overset{-R_1 + R_2 \Rightarrow R_2}{\sim} \begin{bmatrix} -1 & -1 & 0 \\ 0 & 0 & 0 \end{bmatrix}$$

Back substitution gives us

General Solution: $\mathbf{u}_2 = s \begin{bmatrix} -1 \\ 1 \end{bmatrix} \implies$ Basis for Eigenspace of $\lambda_2 = 3$: $\left\{ \begin{bmatrix} -1 \\ 1 \end{bmatrix} \right\}$ ∎

EXAMPLE 5   Find the eigenvalues and a basis for each eigenspace of

$$A = \begin{bmatrix} 1 & -3 & 3 \\ 2 & -2 & 2 \\ 2 & 0 & 0 \end{bmatrix}$$

Solution   We determine the eigenvalues of A by calculating the characteristic polynomial,

$$\det(A - \lambda I_3) = \begin{vmatrix} (1 - \lambda) & -3 & 3 \\ 2 & (-2 - \lambda) & 2 \\ 2 & 0 & (0 - \lambda) \end{vmatrix} = -\lambda^3 - \lambda^2 + 2\lambda = -\lambda(\lambda + 2)(\lambda - 1)$$

Hence the eigenvalues are $\lambda_1 = -2$, $\lambda_2 = 0$, and $\lambda_3 = 1$. Taking them in order, the eigenvectors associated with $\lambda_1 = -2$ are found by solving the homogeneous system $(A + 2I_3)\mathbf{u} = \mathbf{0}$. The augmented matrix and echelon form are

$$\begin{bmatrix} 3 & -3 & 3 & 0 \\ 2 & 0 & 2 & 0 \\ 2 & 0 & 2 & 0 \end{bmatrix} \overset{\substack{R_1 \Leftrightarrow R_2 \\ -\frac{3}{2}R_1 + R_2 \Rightarrow R_2 \\ -R_1 + R_3 \Rightarrow R_3}}{\sim} \begin{bmatrix} 2 & 0 & 2 & 0 \\ 0 & -3 & 0 & 0 \\ 0 & 0 & 0 & 0 \end{bmatrix}$$

Back substitution gives us

General Solution: $\mathbf{u}_1 = s \begin{bmatrix} 1 \\ 0 \\ -1 \end{bmatrix} \implies$ Basis for Eigenspace of $\lambda_1 = -2$: $\left\{ \begin{bmatrix} 1 \\ 0 \\ -1 \end{bmatrix} \right\}$

For $\lambda_2 = 0$, the homogeneous system is $(A - 0 \cdot I_2)\mathbf{u} = A\mathbf{u} = \mathbf{0}$. The augmented matrix and echelon form are

$$\begin{bmatrix} 1 & -3 & 3 & 0 \\ 2 & -2 & 2 & 0 \\ 2 & 0 & 0 & 0 \end{bmatrix} \overset{\substack{-2R_1 + R_2 \Rightarrow R_2 \\ -2R_1 + R_3 \Rightarrow R_3 \\ -\frac{3}{2}R_2 + R_3 \Rightarrow R_3}}{\sim} \begin{bmatrix} 1 & -3 & 3 & 0 \\ 0 & 4 & -4 & 0 \\ 0 & 0 & 0 & 0 \end{bmatrix}$$

This time back substitution yields

General Solution: $\mathbf{u}_2 = s \begin{bmatrix} 0 \\ 1 \\ 1 \end{bmatrix} \implies$ Basis for Eigenspace of $\lambda_2 = 0$: $\left\{ \begin{bmatrix} 0 \\ 1 \\ 1 \end{bmatrix} \right\}$

The final eigenvalue is $\lambda_3 = 1$. The homogeneous system is $(A - I_3)\mathbf{u} = \mathbf{0}$, and the augmented matrix and echelon form are

$$\begin{bmatrix} 0 & -3 & 3 & 0 \\ 2 & -3 & 2 & 0 \\ 2 & 0 & -1 & 0 \end{bmatrix} \overset{\substack{R_1 \Leftrightarrow R_3 \\ -R_1 + R_2 \Rightarrow R_2 \\ -R_2 + R_3 \Rightarrow R_3 \\ \sim}}{} \begin{bmatrix} 2 & 0 & -1 & 0 \\ 0 & -3 & 3 & 0 \\ 0 & 0 & 0 & 0 \end{bmatrix}$$

With back substitution, we find

$$\text{General Solution: } \mathbf{u}_3 = s \begin{bmatrix} 1 \\ 2 \\ 2 \end{bmatrix} \implies \text{Basis for Eigenspace of } \lambda_3 = 1: \left\{ \begin{bmatrix} 1 \\ 2 \\ 2 \end{bmatrix} \right\}$$ ∎

EXAMPLE 6   Find the eigenvalues and a basis for each eigenspace of

$$A = \begin{bmatrix} 1 & -2 & 1 \\ -1 & 0 & 1 \\ -1 & -2 & 3 \end{bmatrix}$$

Solution We start out by finding the eigenvalues for A by computing

$$\det(A - \lambda I_3) = \begin{vmatrix} (1 - \lambda) & -2 & 1 \\ -1 & -\lambda & 1 \\ -1 & -2 & (3 - \lambda) \end{vmatrix} = -\lambda^3 + 4\lambda^2 - 4\lambda = -\lambda(\lambda - 2)^2$$

From the factored form we see that our matrix has two distinct eigenvalues, $\lambda_1 = 0$ and $\lambda_2 = 2$.

Now we find the eigenvectors. Starting with those associated with $\lambda_1 = 0$, we solve the homogeneous system $(A - 0 \cdot I_3)\mathbf{u} = A\mathbf{u} = \mathbf{0}$. The augmented matrix and echelon form are

$$\begin{bmatrix} 1 & -2 & 1 & 0 \\ -1 & 0 & 1 & 0 \\ -1 & -2 & 3 & 0 \end{bmatrix} \overset{\substack{R_1 + R_2 \Rightarrow R_2 \\ R_1 + R_3 \Rightarrow R_3 \\ -2R_2 + R_3 \Rightarrow R_3 \\ \sim}}{} \begin{bmatrix} 1 & -2 & 1 & 0 \\ 0 & -2 & 2 & 0 \\ 0 & 0 & 0 & 0 \end{bmatrix}$$

Back substitution produces

$$\text{General Solution: } \mathbf{u}_1 = s \begin{bmatrix} 1 \\ 1 \\ 1 \end{bmatrix} \implies \text{Basis for Eigenspace of } \lambda_1 = 0: \left\{ \begin{bmatrix} 1 \\ 1 \\ 1 \end{bmatrix} \right\}$$

Turning to the eigenvalue $\lambda_2 = 2$, we form the augmented matrix for the system $(A - 2I_3)\mathbf{u} = \mathbf{0}$ and reduce to echelon form,

$$\begin{bmatrix} -1 & -2 & 1 & 0 \\ -1 & -2 & 1 & 0 \\ -1 & -2 & 1 & 0 \end{bmatrix} \overset{\substack{-R_1 + R_2 \Rightarrow R_2 \\ -R_1 + R_3 \Rightarrow R_3 \\ \sim}}{} \begin{bmatrix} -1 & -2 & 1 & 0 \\ 0 & 0 & 0 & 0 \\ 0 & 0 & 0 & 0 \end{bmatrix}$$

This time back substitution produces

$$\text{General Solution: } \mathbf{u}_2 = s_1 \begin{bmatrix} 1 \\ 0 \\ 1 \end{bmatrix} + s_2 \begin{bmatrix} -2 \\ 1 \\ 0 \end{bmatrix}$$

$$\implies \text{Basis for Eigenspace of } \lambda_2 = 2: \left\{ \begin{bmatrix} 1 \\ 0 \\ 1 \end{bmatrix}, \begin{bmatrix} -2 \\ 1 \\ 0 \end{bmatrix} \right\}$$ ∎

Multiplicities

In Example 6 the factored form of the characteristic polynomial is

$$-\lambda(\lambda - 2)^2 = -(\lambda - 0)^1(\lambda - 2)^2$$

The *multiplicity* of an eigenvalue is equal to its factor's exponent. In this case, we say that $\lambda = 0$ has multiplicity 1 and $\lambda = 2$ has multiplicity 2.

Definition Multiplicity

▶ Informally, the multiplicity is the number of times a root is repeated.

In general, for a polynomial $P(x)$, a root α of $P(x) = 0$ has **multiplicity** r if $P(x) = (x - \alpha)^r Q(x)$ with $Q(\alpha) \neq 0$. When discussing eigenvalues, the phrase "λ has multiplicity r" means that λ is a root of the characteristic polynomial with multiplicity r.

Reviewing our previous examples, we see that the dimension of the eigenspaces matched the multiplicities of the associated eigenvalues. This happens most of the time, but not always.

EXAMPLE 7   Find the eigenvalues and a basis for each eigenspace of

$$A = \begin{bmatrix} 0 & 2 & -1 \\ 1 & -1 & 0 \\ 1 & -2 & 0 \end{bmatrix}$$

Solution The characteristic polynomial of A is

$$\det(A - \lambda I_3) = \begin{vmatrix} -\lambda & 2 & -1 \\ 1 & (-1 - \lambda) & 0 \\ 1 & -2 & -\lambda \end{vmatrix} = -\lambda^3 - \lambda^2 + \lambda + 1 = -(\lambda - 1)(\lambda + 1)^2$$

Thus A has two distinct eigenvalues, $\lambda_1 = -1$ (multiplicity 2) and $\lambda_2 = 1$ (multiplicity 1).

To find the eigenvectors associated with $\lambda_1 = -1$, we solve the homogeneous system $(A + I_3)\mathbf{u} = \mathbf{0}$. The augmented matrix and echelon form are

$$\begin{bmatrix} 1 & 2 & -1 & 0 \\ 1 & 0 & 0 & 0 \\ 1 & -2 & 1 & 0 \end{bmatrix} \begin{smallmatrix} -R_1+R_2 \Rightarrow R_2 \\ -R_1+R_3 \Rightarrow R_3 \\ -2R_2+R_3 \Rightarrow R_3 \\ \sim \end{smallmatrix} \begin{bmatrix} 1 & 2 & -1 & 0 \\ 0 & -2 & 1 & 0 \\ 0 & 0 & 0 & 0 \end{bmatrix}$$

Back substitution produces

$$\text{General Solution: } \mathbf{u}_1 = s \begin{bmatrix} 0 \\ 1 \\ 2 \end{bmatrix} \implies \text{Basis for Eigenspace of } \lambda_1 = -1: \left\{ \begin{bmatrix} 0 \\ 1 \\ 2 \end{bmatrix} \right\}$$

Note that although the eigenvalue $\lambda_1 = -1$ has multiplicity 2, the eigenspace has dimension 1.

For $\lambda_2 = 1$, the augmented matrix for the system $(A - I_3)\mathbf{u} = \mathbf{0}$ and echelon form are

$$\begin{bmatrix} -1 & 2 & -1 & 0 \\ 1 & -2 & 0 & 0 \\ 1 & -2 & -1 & 0 \end{bmatrix} \begin{smallmatrix} R_1+R_2 \Rightarrow R_2 \\ R_1+R_3 \Rightarrow R_3 \\ -2R_2+R_3 \Rightarrow R_3 \\ \sim \end{smallmatrix} \begin{bmatrix} -1 & 2 & -1 & 0 \\ 0 & 0 & -1 & 0 \\ 0 & 0 & 0 & 0 \end{bmatrix}$$

Back substitution gives us

$$\text{General Solution: } \mathbf{u}_2 = s \begin{bmatrix} 2 \\ 1 \\ 0 \end{bmatrix} \implies \text{Basis for Eigenspace of } \lambda_2 = 1: \left\{ \begin{bmatrix} 2 \\ 1 \\ 0 \end{bmatrix} \right\} \blacksquare$$

As demonstrated in Example 7, it is possible for the dimension of an eigenspace to be less than the multiplicity of the associated eigenvalue. However, the opposite cannot happen.

THEOREM 6.6

Let A be a square matrix with eigenvalue λ. Then the dimension of the associated eigenspace is less than or equal to the multiplicity of λ.

The proof is beyond the scope of this book and is omitted.

The methods developed in this section work well for small matrices, but they can be impractical for large complicated matrices. For instance, suppose

$$A = \begin{bmatrix} 245 & -254 & -252 & -46 & -224 \\ 161 & -168 & -174 & -32 & -148 \\ -39 & 40 & 45 & 7 & 38 \\ 27 & -28 & -32 & -6 & -26 \\ 110 & -113 & -110 & -21 & -101 \end{bmatrix} \tag{4}$$

Computing the determinant by hand to find the characteristic polynomial for A is not easy. Using computer software, we find that

$$\det(A - \lambda I_5) = -\lambda^5 + 15\lambda^4 - 3\lambda^3 - 287\lambda^2 - 192\lambda + 468$$

This polynomial is challenging to factor. In fact, there is no general algorithm for factoring polynomials of degree 5 or more. (Not even on a computer.) Thus we cannot find the eigenvalues and eigenvectors using our existing methods.

Numerous applications require eigenvalues and eigenvectors from large matrices. In principle, we could use numerical methods to find approximations to the roots of the characteristic polynomial (and hence the eigenvalues) and use these to find the eigenvectors. However, for various reasons this does not work well in practice. Instead, there are algorithms that lead directly to approximations to the eigenvectors, bypassing the need to first find the eigenvalues. These are described in Section 6.2, where we analyze the matrix A in (4).

The Big Theorem, Version 8

Although $\mathbf{u} = \mathbf{0}$ is not allowed as an eigenvector, it is fine to have $\lambda = 0$ as an eigenvalue. From Theorem 6.5, we know that $\lambda = 0$ is an eigenvalue of an $n \times n$ matrix A if and only if $\det(A - 0I_n) = \det(A) = 0$. Put another way, $\lambda = 0$ is *not* an eigenvalue of A if and only if $\det(A) \neq 0$. This observation provides us with another condition for the Big Theorem.

THEOREM 6.7

(**THE BIG THEOREM, VERSION 8**) Let $\mathcal{A} = \{\mathbf{a}_1, \ldots, \mathbf{a}_n\}$ be a set of n vectors in $\mathbf{R}^n$, let $A = [\mathbf{a}_1 \cdots \mathbf{a}_n]$, and let $T : \mathbf{R}^n \to \mathbf{R}^n$ be given by $T(\mathbf{x}) = A\mathbf{x}$. Then the following are equivalent:

(a) $\mathcal{A}$ spans $\mathbf{R}^n$.

(b) $\mathcal{A}$ is linearly independent.

(c) $A\mathbf{x} = \mathbf{b}$ has a unique solution for all $\mathbf{b}$ in $\mathbf{R}^n$.

(d) T is onto.

(e) T is one-to-one.

(f) A is invertible.

(g) $\ker(T) = \{\mathbf{0}\}$.

(h) $\mathcal{A}$ is a basis for $\mathbf{R}^n$.

(i) $\text{col}(A) = \mathbf{R}^n$.

(j) $\text{row}(A) = \mathbf{R}^n$.

(k) $\text{rank}(A) = n$.

(l) $\det(A) \neq 0$.

(m) $\lambda = 0$ is not an eigenvalue of A.

Proof From the Big Theorem, Version 7, we know that (a) through (l) are equivalent. The comments above show that (l) and (m) are equivalent, and so it follows that all 13 conditions are equivalent. ■

EXAMPLE 8 Show that $\lambda = 0$ is an eigenvalue for the matrix

$$A = \begin{bmatrix} 3 & -1 & 5 \\ 2 & 1 & 0 \\ 4 & 1 & 2 \end{bmatrix}$$

Solution From the Big Theorem, Version 8, $\lambda = 0$ is an eigenvalue of A if and only if $\det(A) = 0$. By the Shortcut Method, we have

$$\det(A) = (6 + 0 + 10) - (20 - 4 + 0) = 0$$

so $\lambda = 0$ is an eigenvalue of A. ■

EXERCISES

In Exercises 1–6, determine which of $\mathbf{x}_1$, $\mathbf{x}_2$, and $\mathbf{x}_3$ is an eigenvector for the matrix A. For those that are, determine the associated eigenvalue.

1. $A = \begin{bmatrix} 1 & 3 \\ 2 & 2 \end{bmatrix}$, $\mathbf{x}_1 = \begin{bmatrix} -3 \\ 2 \end{bmatrix}$,

$\mathbf{x}_2 = \begin{bmatrix} 1 \\ -1 \end{bmatrix}$, $\mathbf{x}_3 = \begin{bmatrix} -2 \\ -2 \end{bmatrix}$

2. $A = \begin{bmatrix} -1 & 2 \\ 0 & 3 \end{bmatrix}$, $\mathbf{x}_1 = \begin{bmatrix} 0 \\ 2 \end{bmatrix}$,

$\mathbf{x}_2 = \begin{bmatrix} 1 \\ 3 \end{bmatrix}$, $\mathbf{x}_3 = \begin{bmatrix} 1 \\ 2 \end{bmatrix}$

3. $A = \begin{bmatrix} 2 & 7 & 2 \\ 0 & -1 & 0 \\ 0 & -2 & 1 \end{bmatrix}$, $\mathbf{x}_1 = \begin{bmatrix} -3 \\ 1 \\ 1 \end{bmatrix}$,

$\mathbf{x}_2 = \begin{bmatrix} -2 \\ 0 \\ 1 \end{bmatrix}$, $\mathbf{x}_3 = \begin{bmatrix} 1 \\ 0 \\ 0 \end{bmatrix}$

4. $A = \begin{bmatrix} 3 & -1 & 0 \\ -1 & 3 & 0 \\ -1 & 1 & 2 \end{bmatrix}$, $\mathbf{x}_1 = \begin{bmatrix} 1 \\ 1 \\ 1 \end{bmatrix}$,

$\mathbf{x}_2 = \begin{bmatrix} 1 \\ 1 \\ 0 \end{bmatrix}$, $\mathbf{x}_3 = \begin{bmatrix} 1 \\ 2 \\ -1 \end{bmatrix}$

5. $A = \begin{bmatrix} 6 & -3 & 1 & 0 \\ 0 & 3 & 1 & 0 \\ -6 & 6 & 0 & 0 \\ -3 & 3 & -2 & 3 \end{bmatrix}$, $\mathbf{x}_1 = \begin{bmatrix} 1 \\ 1 \\ 0 \\ 0 \end{bmatrix}$,

$\mathbf{x}_2 = \begin{bmatrix} 1 \\ 2 \\ -1 \\ 0 \end{bmatrix}$, $\mathbf{x}_3 = \begin{bmatrix} 1 \\ 1 \\ -3 \\ -2 \end{bmatrix}$

6. $A = \begin{bmatrix} 5 & 5 & 1 & 8 \\ 8 & 2 & 1 & 8 \\ -6 & 6 & -9 & 0 \\ -7 & -1 & -2 & -10 \end{bmatrix}$, $\mathbf{x}_1 = \begin{bmatrix} 1 \\ 1 \\ 0 \\ -2 \end{bmatrix}$,

$\mathbf{x}_2 = \begin{bmatrix} 1 \\ 1 \\ 0 \\ 0 \end{bmatrix}$, $\mathbf{x}_3 = \begin{bmatrix} 1 \\ 2 \\ -2 \\ -1 \end{bmatrix}$

In Exercises 7–10, use the characteristic polynomial to determine if λ is an eigenvalue for the matrix A.

7. $A = \begin{bmatrix} 2 & 7 \\ -1 & 6 \end{bmatrix}$, $\lambda = 3$

8. $A = \begin{bmatrix} 1 & 5 \\ 4 & 2 \end{bmatrix}$, $\lambda = 6$

9. $A = \begin{bmatrix} 0 & 2 & 0 \\ 2 & 0 & 0 \\ 2 & 2 & -2 \end{bmatrix}$, $\quad \lambda = -2$

10. $A = \begin{bmatrix} 6 & 3 & -1 \\ -4 & -1 & 1 \\ 18 & 6 & -4 \end{bmatrix}$, $\quad \lambda = 1$

In Exercises 11–20, find a basis for the eigenspace of A associated with the given eigenvalue λ.

11. $A = \begin{bmatrix} 1 & -3 \\ 1 & 5 \end{bmatrix}$, $\quad \lambda = 4$

12. $A = \begin{bmatrix} -2 & 4 \\ 3 & -1 \end{bmatrix}$, $\quad \lambda = -5$

13. $A = \begin{bmatrix} 6 & -10 \\ 2 & -3 \end{bmatrix}$, $\quad \lambda = 2$

14. $A = \begin{bmatrix} -11 & 12 \\ -8 & 9 \end{bmatrix}$, $\quad \lambda = -3$

15. $A = \begin{bmatrix} 6 & -3 & 7 \\ 4 & 1 & 5 \\ 4 & -3 & 9 \end{bmatrix}$, $\quad \lambda = 4$

16. $A = \begin{bmatrix} -2 & 5 & -7 \\ -2 & 11 & -13 \\ -2 & 5 & -7 \end{bmatrix}$, $\quad \lambda = -4$

17. $A = \begin{bmatrix} 5 & -1 & 2 \\ 2 & 2 & 2 \\ 2 & -1 & 5 \end{bmatrix}$, $\quad \lambda = 6$

18. $A = \begin{bmatrix} -1 & 2 & -7 \\ -10 & 2 & 2 \\ -10 & 2 & 2 \end{bmatrix}$, $\quad \lambda = 9$

19. $A = \begin{bmatrix} 11 & -3 & -3 & 8 \\ 13 & -5 & -5 & 8 \\ 2 & -2 & -2 & 0 \\ -3 & 3 & 3 & 0 \end{bmatrix}$, $\quad \lambda = -4$

20. $A = \begin{bmatrix} 15 & -3 & -15 & 8 \\ 21 & -9 & -17 & 8 \\ 2 & -2 & -6 & 0 \\ -3 & 3 & 15 & 4 \end{bmatrix}$, $\quad \lambda = -8$

In Exercises 21–30, find the characteristic polynomial, the eigenvalues, and a basis for each eigenspace for the matrix A.

21. $A = \begin{bmatrix} 2 & 0 \\ 4 & -3 \end{bmatrix}$

22. $A = \begin{bmatrix} 2 & 6 \\ 1 & 1 \end{bmatrix}$

23. $A = \begin{bmatrix} 1 & -2 \\ 2 & -3 \end{bmatrix}$

24. $A = \begin{bmatrix} -2 & 8 \\ 1 & -4 \end{bmatrix}$

25. $A = \begin{bmatrix} 3 & 0 & 0 \\ 1 & 2 & 0 \\ -4 & 5 & -1 \end{bmatrix}$

26. $A = \begin{bmatrix} 0 & 0 & 1 \\ 1 & 0 & 0 \\ 0 & 1 & 0 \end{bmatrix}$

27. $A = \begin{bmatrix} 2 & 5 & 1 \\ 0 & -3 & -1 \\ 2 & 14 & 4 \end{bmatrix}$

28. $A = \begin{bmatrix} 0 & -3 & -1 \\ -1 & 2 & 1 \\ 3 & -9 & -4 \end{bmatrix}$

29. $A = \begin{bmatrix} -1 & 0 & 0 & 0 \\ 5 & -2 & 0 & 0 \\ 0 & 3 & 1 & 0 \\ 2 & 0 & 1 & 1 \end{bmatrix}$

30. $A = \begin{bmatrix} 0 & 0 & 1 & 0 \\ 0 & 1 & 0 & 0 \\ 1 & 0 & 0 & 0 \\ 0 & 0 & 0 & 1 \end{bmatrix}$

FIND AN EXAMPLE For Exercises 31–36, find an example that meets the given specifications.

31. A 2×2 matrix A with eigenvalues $\lambda = 1$ and $\lambda = 2$.

32. A 2×2 matrix A with eigenvalues $\lambda = -3$ and $\lambda = 0$.

33. A 3×3 matrix A with eigenvalues $\lambda = 1$, $\lambda = -2$, and $\lambda = 3$.

34. A 3×3 matrix A with eigenvalues $\lambda = -1$ (multiplicity 2) and $\lambda = 4$.

35. A 2×2 matrix that has no real eigenvalues.

36. A 4×4 matrix that has no real eigenvalues.

TRUE OR FALSE For Exercises 37–46, determine if the statement is true or false, and justify your answer.

37. An eigenvalue λ must be nonzero, but an eigenvector **u** can be equal to the zero vector.

38. The dimension of an eigenspace is always less than or equal to the multiplicity of the associated eigenvalue.

39. If **u** is a nonzero eigenvector of A, then **u** and A**u** point in the same direction.

40. If λ_1 and λ_2 are eigenvalues of a matrix, then so is $\lambda_1 + \lambda_2$.

41. If A is a diagonal matrix, then the eigenvalues of A lie along the diagonal.

42. If 0 is an eigenvalue of an $n \times n$ matrix A, then the columns of A span $\mathbf{R}^n$.

43. If 0 is an eigenvalue of A, then nullity(A) > 0.

44. Row operations do not change the eigenvalues of a matrix.

45. If 0 is the only eigenvalue of A, then A must be the zero matrix.

46. The product of the eigenvalues (counting multiplicities) of A is equal to the constant term of the characteristic polynomial of A.

47. Suppose that A is a square matrix with characteristic polynomial $(\lambda - 3)^3(\lambda - 2)^2(\lambda + 1)$.

(a) What are the dimensions of A?

(b) What are the eigenvalues of A?

(c) Is A invertible?

(d) What is the largest possible dimension for an eigenspace of A?

48. Suppose that A is a square matrix with characteristic polynomial $-\lambda(\lambda - 1)^3(\lambda + 2)^3$.

(a) What are the dimensions of A?

(b) What are the eigenvalues of A?

(c) Is A invertible?

(d) What is the largest possible dimension for an eigenspace of A?

49. Let $T : \mathbf{R}^n \to \mathbf{R}^n$ be given by $T(\mathbf{x}) = A\mathbf{x}$. Prove that if 0 is not an eigenvalue of A, then T is onto.

50. Prove that if λ is an eigenvalue of A, then 4λ is an eigenvalue of $4A$.

51. Prove that if $\lambda = 1$ is an eigenvalue of an $n \times n$ matrix A, then $A - I_n$ is singular.

52. Prove that if $\mathbf{u}$ is an eigenvector of A, then $\mathbf{u}$ is also an eigenvector of A^2.

53. Prove that $\mathbf{u}$ cannot be an eigenvector associated with two distinct eigenvalues λ_1 and λ_2 of A.

54. Prove that if 5 is an eigenvalue of A, then 25 is an eigenvalue of A^2.

55. If $\mathbf{u} = \mathbf{0}$ was allowed to be an eigenvector, then which values of λ would be associated eigenvalues? (This is one reason why eigenvectors are defined to be nonzero.)

56. Suppose that A is a square matrix that is either upper or lower triangular. Show that the eigenvalues of A are the diagonal terms of A.

57. Let A be an invertible matrix. Prove that if λ is an eigenvalue of A with associated eigenvector $\mathbf{u}$, then λ^{-1} is an eigenvalue of A^{-1} with associated eigenvector $\mathbf{u}$.

58. Let $A = \begin{bmatrix} a & b \\ c & d \end{bmatrix}$. Find a formula for the eigenvalues of A in terms of $a, b, c,$ and d. (HINT: The quadratic formula can be handy here.)

59. Suppose that A and B are both $n \times n$ matrices, and that $\mathbf{u}$ is an eigenvector for both A and B. Prove that $\mathbf{u}$ is an eigenvector for the product AB.

60. Suppose that A is an $n \times n$ matrix with eigenvalue λ and associated eigenvector $\mathbf{u}$. Show that for each positive integer k, the matrix A^k has eigenvalue λ^k and associated eigenvector $\mathbf{u}$.

61. Suppose that the entries of each row of a square matrix A add to zero. Prove that $\lambda = 0$ is an eigenvalue of A.

62. Suppose that $A = \begin{bmatrix} a & b \\ c & d \end{bmatrix}$, where $a, b, c,$ and d satisfy $a + b = c + d$. Show that $\lambda_1 = a + b$ and $\lambda_2 = a - c$ are both eigenvalues of A.

63. Suppose that A is a square matrix. Prove that if λ is an eigenvalue of A, then λ is also an eigenvalue of A^T. (HINT: Recall that $\det(A) = \det(A^T)$.)

64. Suppose that A is an $n \times n$ matrix and c is a scalar. Prove that if λ is an eigenvalue of A with associated eigenvector $\mathbf{u}$, then $\lambda - c$ is an eigenvalue of $A - cI_n$ with associated eigenvector $\mathbf{u}$.

65. Suppose that the entries of each row of a square matrix A add to c for some scalar c. Prove that $\lambda = c$ is an eigenvalue of A.

66. Let A be an $n \times n$ matrix.

(a) Prove that the characteristic polynomial of A has degree n.

(b) What is the coefficient on λ^n in the characteristic polynomial?

(c) Show that the constant term in the characteristic polynomial is equal to $\det(A)$.

(d) Suppose that A has eigenvalues $\lambda_1, \ldots, \lambda_n$ that are all real numbers. Prove that $\det(A) = \lambda_1\lambda_2\cdots\lambda_n$.

Ⓒ In Exercises 67–70, find the eigenvalues and bases for the eigenspaces of A.

67. $A = \begin{bmatrix} 0 & 0 & -2 & -1 \\ 1 & 1 & 6 & 5 \\ 2 & 0 & 4 & 1 \\ -2 & 0 & -2 & 1 \end{bmatrix}$

68. $A = \begin{bmatrix} -20 & -9 & 14 & 18 \\ 40 & 17 & -18 & -28 \\ 17 & 9 & -10 & -11 \\ -17 & -9 & 14 & 15 \end{bmatrix}$

69. $A = \begin{bmatrix} 10 & 0 & 1 & -3 & 3 \\ 23 & -1 & 6 & -3 & 2 \\ -24 & 0 & -1 & 9 & -9 \\ 14 & 0 & 1 & -5 & 5 \\ -10 & 0 & -1 & 3 & -3 \end{bmatrix}$

70. $A = \begin{bmatrix} 5 & 0 & 2 & 1 & -1 \\ 6 & 1 & 4 & 3 & -3 \\ -6 & 0 & -3 & -3 & 3 \\ 2 & 0 & 2 & 3 & -2 \\ -4 & 0 & -2 & -1 & 2 \end{bmatrix}$

6.2 Approximation Methods

▶ This section is optional and can be omitted without loss of continuity.

In Section 6.1 we saw how to use the characteristic polynomial to find the eigenvalues (and then the eigenvectors) for matrices. Such methods work fine for the small matrices considered there, but they are not practical for many larger matrices.

Since large matrices turn up in all kinds of applications, we need another way to find eigenvalues and eigenvectors. Several related approaches for dealing with large matrices are described in this section. All have their basis in the Power Method, so we start with that.

The Power Method

The Power Method is an iterative algorithm that gets its name from how it is implemented. Given a square matrix A, we start with a fixed vector $\mathbf{x}_0$ and compute the sequence $A\mathbf{x}_0$, $A^2\mathbf{x}_0$, $A^3\mathbf{x}_0, \ldots$. Remarkably, in many cases the resulting sequence of vectors will approach an eigenvector of A. We begin with an example and leave the discussion of when and why the Power Method works for later in this section.

Suppose that

$$A = \begin{bmatrix} 1 & 3 \\ 2 & 2 \end{bmatrix}, \quad \mathbf{x}_0 = \begin{bmatrix} 1 \\ 0 \end{bmatrix} \tag{1}$$

and let

$$\mathbf{x}_1 = A\mathbf{x}_0, \quad \mathbf{x}_2 = A\mathbf{x}_1 = A^2\mathbf{x}_0, \quad \mathbf{x}_3 = A\mathbf{x}_2 = A^3\mathbf{x}_0, \ldots$$

Table 1 gives $\mathbf{x}_0$ to $\mathbf{x}_7$ for A and $\mathbf{x}_0$ above.

k	0	1	2	3	4	5	6	7
$\mathbf{x}_k$	$\begin{bmatrix}1\\0\end{bmatrix}$	$\begin{bmatrix}1\\2\end{bmatrix}$	$\begin{bmatrix}7\\6\end{bmatrix}$	$\begin{bmatrix}25\\26\end{bmatrix}$	$\begin{bmatrix}103\\102\end{bmatrix}$	$\begin{bmatrix}409\\410\end{bmatrix}$	$\begin{bmatrix}1639\\1638\end{bmatrix}$	$\begin{bmatrix}6553\\6554\end{bmatrix}$

Table 1 $\mathbf{x}_k = A^k\mathbf{x}_0$ for $k = 0, 1, \ldots, 7$

Often the components of $\mathbf{x}_k$ grow with k when forming this type of sequence. Since scalar multiples of eigenvectors are still eigenvectors, we control the size of each vector in the sequence by scaling.

Definition Scaling Factor

THE POWER METHOD: For each $k \geq 0$,

(a) Let s_k denote the largest component (in absolute value) of $A\mathbf{x}_k$. (We call s_k a **scaling factor.**)

(b) Set $\mathbf{x}_{k+1} = \dfrac{1}{s_k} A\mathbf{x}_k$.

Repeat (a) and (b) to generate the sequence $\mathbf{x}_0, \mathbf{x}_1, \mathbf{x}_2, \ldots$.

Scaling in this way ensures that the largest component (in absolute value) of each vector is either 1 or -1. For example, starting with $\mathbf{x}_0$ and A in (1), we compute $\mathbf{x}_1$ by

$$A\mathbf{x}_0 = \begin{bmatrix} 1 & 3 \\ 2 & 2 \end{bmatrix} \begin{bmatrix} 1 \\ 0 \end{bmatrix} = \begin{bmatrix} 1 \\ 2 \end{bmatrix} \implies s_0 = 2 \implies \mathbf{x}_1 = \frac{1}{2} A\mathbf{x}_0 = \begin{bmatrix} 0.5 \\ 1 \end{bmatrix}$$

We then compute $\mathbf{x}_2$ by

$$A\mathbf{x}_1 = \begin{bmatrix} 1 & 3 \\ 2 & 2 \end{bmatrix} \begin{bmatrix} 0.5 \\ 1 \end{bmatrix} = \begin{bmatrix} 3.5 \\ 3 \end{bmatrix} \implies s_1 = 3.5 \implies \mathbf{x}_2 = \frac{1}{3.5} A\mathbf{x}_1 = \begin{bmatrix} 1 \\ 0.8571 \end{bmatrix}$$

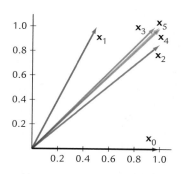

Figure 1 Vectors $\mathbf{x}_0, \ldots, \mathbf{x}_5$ from Table 2.

Table 2 shows $\mathbf{x}_0$ to $\mathbf{x}_7$, together with the scaling factors.

k	0	1	2	3	4	5	6	7
$\mathbf{x}_k$	$\begin{bmatrix} 1 \\ 0 \end{bmatrix}$	$\begin{bmatrix} 0.5 \\ 1 \end{bmatrix}$	$\begin{bmatrix} 1.000 \\ 0.8571 \end{bmatrix}$	$\begin{bmatrix} 0.9615 \\ 1.000 \end{bmatrix}$	$\begin{bmatrix} 1.000 \\ 0.9903 \end{bmatrix}$	$\begin{bmatrix} 0.9976 \\ 1.000 \end{bmatrix}$	$\begin{bmatrix} 1.000 \\ 0.9994 \end{bmatrix}$	$\begin{bmatrix} 0.9998 \\ 1.000 \end{bmatrix}$
s_k	2	3.5	3.714	3.962	3.981	3.998	3.999	4.000

Table 2 The Power Method applied to A and $\mathbf{x}_0$ in (1)

The entries in Table 2 and Figure 1 suggest that the sequence of vectors is getting closer and closer to $\mathbf{u} = \begin{bmatrix} 1 \\ 1 \end{bmatrix}$. Computing $A\mathbf{u}$, we find that

$$A\mathbf{u} = \begin{bmatrix} 1 & 3 \\ 2 & 2 \end{bmatrix} \begin{bmatrix} 1 \\ 1 \end{bmatrix} = \begin{bmatrix} 4 \\ 4 \end{bmatrix} = 4\mathbf{u} \tag{2}$$

Hence $\mathbf{u} = \begin{bmatrix} 1 \\ 1 \end{bmatrix}$ is an eigenvector of A with associated eigenvalue of $\lambda = 4$.

In Table 2, not only does the sequence of vectors converge to an eigenvector, but the sequence of scaling factors converges to the associated eigenvalue. This latter observation makes sense, because our scaling rule $\mathbf{x}_{k+1} = \frac{1}{s_k} A\mathbf{x}_k$ can be expressed as

$$A\mathbf{x}_k = s_k \mathbf{x}_{k+1}$$

Since the vectors are getting closer to an eigenvector (and each other), it follows that s_k must be getting closer to an eigenvalue.

Typically, any nonzero $\mathbf{x}_0$ can be used as an initial vector for the Power Method. (However, see the note in the Computational Comments at the end of the section.) If an approximate value of an eigenvector is known, then using it for $\mathbf{x}_0$ can speed convergence.

EXAMPLE 1 Find an eigenvector and associated eigenvalue for the matrix

▶ The matrix A is from (4) in Section 6.1.

$$A = \begin{bmatrix} 245 & -254 & -252 & -46 & -224 \\ 161 & -168 & -174 & -32 & -148 \\ -39 & 40 & 45 & 7 & 38 \\ 27 & -28 & -32 & -6 & -26 \\ 110 & -113 & -110 & -21 & -101 \end{bmatrix} \tag{3}$$

▶ To save space, in this section we often write vectors horizontally instead of vertically.

Solution We apply the Power Method starting with $\mathbf{x}_0 = (1, 1, 1, 1, 1)$. Table 3 gives the vectors $\mathbf{x}_1, \ldots \mathbf{x}_9$.

k	$\mathbf{x}_k$	s_k
1	$(1.0000, 0.6798, -0.1714, 0.1224, 0.4426)$	10.74
2	$(1.0000, 0.6693, -0.1717, 0.1124, 0.4431)$	13.83
3	$(1.0000, 0.6688, -0.1683, 0.1126, 0.4438)$	12.94
4	$(1.0000, 0.6674, -0.1676, 0.1116, 0.4442)$	13.07
5	$(1.0000, 0.6671, -0.1670, 0.1114, 0.4443)$	13.01
6	$(1.0000, 0.6668, -0.1668, 0.1112, 0.4444)$	13.01
7	$(1.0000, 0.6668, -0.1667, 0.1112, 0.4444)$	13.00
8	$(1.0000, 0.6667, -0.1667, 0.1111, 0.4444)$	13.00
9	$(1.0000, 0.6667, -0.1667, 0.1111, 0.4444)$	13.00

Table 3 The Power Method Applied to A in Example 1

The sequence settles to the vector $(1.0000, 0.6667, -0.1667, 0.1111, 0.4444)$, and the sequence of scaling factors to 13.00. The components of the vector are recognizable as decimal approximations to rationals. Changing the decimals to equivalent rationals and multiplying by 18 to eliminate the fractions gives us

▶ We express the entries as fractions and multiply by 18 to make it easier to check our answer. This would not be done in applications. Instead, typically we would compute enough iterations to achieve a desired accuracy and take the last vector in the sequence as the eigenvector.

$$
\mathbf{u} = \begin{bmatrix} 1.0000 \\ 0.6667 \\ -0.1667 \\ 0.1111 \\ 0.4444 \end{bmatrix} \approx \begin{bmatrix} 1 \\ 2/3 \\ -1/6 \\ 1/9 \\ 4/9 \end{bmatrix} \implies \mathbf{u} = \begin{bmatrix} 18 \\ 12 \\ -3 \\ 2 \\ 8 \end{bmatrix}
$$

To test our answer, we calculate

$$
A\mathbf{u} = \begin{bmatrix} 245 & -254 & -252 & -46 & -224 \\ 161 & -168 & -174 & -32 & -148 \\ -39 & 40 & 45 & 7 & 38 \\ 27 & -28 & -32 & -6 & -26 \\ 110 & -113 & -110 & -21 & -101 \end{bmatrix} \begin{bmatrix} 18 \\ 12 \\ -3 \\ 2 \\ 8 \end{bmatrix} = \begin{bmatrix} 234 \\ 156 \\ -39 \\ 26 \\ 104 \end{bmatrix} = 13 \begin{bmatrix} 18 \\ 12 \\ -3 \\ 2 \\ 8 \end{bmatrix}
$$

confirming that $\mathbf{u}$ is an eigenvector and that $\lambda = 13$ is the associated eigenvalue. ■

The Power Method will frequently find an eigenvalue and associated eigenvector. But which ones? To find out, first suppose that a matrix A has eigenvalues $\lambda_1, \lambda_2, \ldots, \lambda_n$ such that

$$
|\lambda_1| > |\lambda_2| \geq |\lambda_3| \geq \cdots \geq |\lambda_n|
$$

Definition Dominant Eigenvalue

In this case λ_1 is called the **dominant eigenvalue** of A.

The next theorem tells us that the Power Method will (usually) converge to an eigenvector associated with the dominant eigenvalue. The proof of this theorem is given at the end of the section.

THEOREM 6.8

Let A be an $n \times n$ matrix with linearly independent eigenvectors $\mathbf{u}_1, \mathbf{u}_2, \ldots, \mathbf{u}_n$ and associated eigenvalues $\lambda_1, \lambda_2, \ldots, \lambda_n$, where λ_1 is dominant. Suppose that

$$
\mathbf{x}_0 = c_1 \mathbf{u}_1 + \cdots + c_n \mathbf{u}_n
$$

where $c_1 \neq 0$. Then multiples of $A\mathbf{x}_0$, $A^2\mathbf{x}_0, \ldots$ converge to a scalar multiple of the eigenvector $\mathbf{u}_1$.

The Shifted Power Method

The Power Method can be used to find eigenvectors other than those associated with the dominant eigenvalue. Suppose that λ_1 is the dominant eigenvalue of A, and then let λ be the dominant eigenvalue of $A - \lambda_1 I_n$ with associated eigenvector $\mathbf{u}_2$. Then

$$
(A - \lambda_1 I_n)\mathbf{u}_2 = \lambda \mathbf{u}_2 \implies A\mathbf{u}_2 = (\lambda + \lambda_1)\mathbf{u}_2
$$

Definition Shifted Power Method

Thus, if $\lambda_2 = \lambda + \lambda_1$, then λ_2 is an eigenvalue of A. Furthermore, since λ is the dominant eigenvalue of $A - \lambda_1 I_n$ and $\lambda = \lambda_2 - \lambda_1$, it follows that λ_2 is the eigenvalue of A that is farthest from λ_1. (Why?) Hence applying the Power Method to $A - \lambda_1 I_n$ will produce another eigenvalue and eigenvector of A. This is called the **Shifted Power Method** and is illustrated in the next example.

EXAMPLE 2 For the matrix A given in Example 1, find the eigenvalue (and an associated eigenvector) farthest from $\lambda_1 = 13$.

Solution To find λ_2, the eigenvalue of A farthest from $\lambda_1 = 13$, we first apply the Power Method to the matrix

$$B = A - 13I_5 = \begin{bmatrix} 232 & -254 & -252 & -46 & -224 \\ 161 & -181 & -174 & -32 & -148 \\ -39 & 40 & 32 & 7 & 38 \\ 27 & -28 & -32 & -19 & -26 \\ 110 & -113 & -110 & -21 & -114 \end{bmatrix}$$

As in Example 1, we start with $\mathbf{x}_0 = (1, 1, 1, 1, 1)$. In this case, the convergence is slower, so we report only every 10th iteration in Table 4.

k	$\mathbf{x}_k$	s_k
10	$(1.0000, 0.7033, -0.1483, 0.1483, 0.4450)$	-15.76
20	$(1.0000, 0.7078, -0.1461, 0.1461, 0.4383)$	-15.86
30	$(1.0000, 0.7107, -0.1447, 0.1447, 0.4340)$	-15.92
40	$(1.0000, 0.7123, -0.1438, 0.1438, 0.4315)$	-15.96
50	$(1.0000, 0.7132, -0.1434, 0.1434, 0.4302)$	-15.98
60	$(1.0000, 0.7137, -0.1431, 0.1431, 0.4294)$	-15.99
70	$(1.0000, 0.7140, -0.1430, 0.1430, 0.4290)$	-15.99
80	$(1.0000, 0.7141, -0.1429, 0.1429, 0.4288)$	-16.00
90	$(1.0000, 0.7142, -0.1429, 0.1429, 0.4287)$	-16.00
100	$(1.0000, 0.7142, -0.1429, 0.1429, 0.4286)$	-16.00

Table 4 The Shifted Power Method Applied to $B = A - 13I_5$

We see that $s_k \to -16$, which implies that $\lambda = -16$ is an eigenvalue for B. Therefore $\lambda_2 = -16 + 13 = -3$ is the eigenvalue of A farthest from $\lambda_1 = 13$. Table 4 shows $\mathbf{u} = (1.000, 0.7142, -0.1429, 0.1429, 0.4286)$ is the associated eigenvector. We check this by computing

$$Au = \begin{bmatrix} 245 & -254 & -252 & -46 & -224 \\ 161 & -168 & -174 & -32 & -148 \\ -39 & 40 & 45 & 7 & 38 \\ 27 & -28 & -32 & -6 & -26 \\ 110 & -113 & -110 & -21 & -101 \end{bmatrix} \begin{bmatrix} 1.0000 \\ 0.7142 \\ -0.1429 \\ 0.1429 \\ 0.4286 \end{bmatrix}$$

$$\approx \begin{bmatrix} -2.9758 \\ -2.1266 \\ 0.4246 \\ -0.4258 \\ -1.2751 \end{bmatrix} \approx -3 \begin{bmatrix} 0.9919 \\ 0.7089 \\ -0.1415 \\ 0.1419 \\ 0.4250 \end{bmatrix} \approx -3\mathbf{u}$$

The approximations are a bit rough, but they can be refined by carrying more decimal places and computing additional iterations (see margin). ∎

▶ Carrying more decimal places in Example 2 gives (after 250 iterations)

$$\mathbf{u} = \begin{bmatrix} 1.00000000 \\ 0.71428571 \\ -0.14285714 \\ 0.14285714 \\ 0.42857143 \end{bmatrix}$$

Then

$$Au = \begin{bmatrix} -3.00000000 \\ -2.14285714 \\ 0.42857143 \\ -0.42857143 \\ -1.28571428 \end{bmatrix}$$

$$= -3 \begin{bmatrix} 1.00000000 \\ 0.71428571 \\ -0.14285714 \\ 0.14285714 \\ 0.42857143 \end{bmatrix} = -3\mathbf{u}$$

The Inverse Power Method

For the matrix A in Examples 1 and 2, we have found eigenvalues $\lambda_1 = 13$ and $\lambda_2 = -3$, and we also know that these are the positive and negative eigenvalues farthest from 0.

Here we develop a method for finding the eigenvalue λ_3 that is *closest* to 0. Exercise 57 in Section 6.1 showed that if A is an invertible matrix with eigenvalue λ and associated eigenvector $\mathbf{u}$, then λ^{-1} is an eigenvalue of A^{-1} with associated eigenvector $\mathbf{u}$. Thus, in particular, if λ is the *largest* (in absolute value) eigenvalue of A^{-1}, then λ^{-1} must be the *smallest* (in absolute value) eigenvalue of A. (Why?) Therefore, applying the Power Method to A^{-1} will yield the smallest eigenvalue (and corresponding eigenvector) of A.

EXAMPLE 3   For the matrix A in Example 1, find the smallest eigenvalue (in absolute value) and an associated eigenvector.

Solution Here we apply the Power Method to

$$A^{-1} = \frac{1}{78} \begin{bmatrix} -646 & 698 & -420 & -1430 & 620 \\ -461 & 487 & -306 & -988 & 448 \\ 99 & -112 & 70 & 221 & -86 \\ -27 & 40 & -64 & -143 & 14 \\ -290 & 329 & -178 & -663 & 264 \end{bmatrix}$$

generating a sequence of vectors of the form

$$\mathbf{x}_{k+1} = \frac{1}{s_k} A^{-1} \mathbf{x}_k \tag{4}$$

▶ For large matrices it can be difficult to compute A^{-1} accurately due to round-off error. This can be avoided by multiplying each side of (4) by A, yielding the sequence of linear systems

$$A\mathbf{x}_{k+1} = \frac{1}{s_k}\mathbf{x}_k$$

Applying LU-factorization (or a related method) to A can greatly improve computational efficiency when solving these systems.

For a change of pace (and another reason to be discussed later), let's take the initial vector to be $\mathbf{x}_0 = (1, 2, 3, 4, 5)$. Table 5 gives the results of every other iteration.

k	$\mathbf{x}_k$	s_k
2	$(1.0000, 0.6842, -0.1565, 0.1178, 0.4734)$	-1.253
4	$(1.0000, 0.6887, -0.1556, 0.0921, 0.4669)$	2.185
6	$(1.0000, 0.6912, -0.1544, 0.0811, 0.4632)$	1.168
8	$(1.0000, 0.6920, -0.1540, 0.0780, 0.4620)$	1.037
10	$(1.0000, 0.6922, -0.1539, 0.0772, 0.4616)$	1.009
12	$(1.0000, 0.6923, -0.1539, 0.0770, 0.4616)$	1.002
14	$(1.0000, 0.6923, -0.1538, 0.0769, 0.4615)$	1.001
16	$(1.0000, 0.6923, -0.1538, 0.0769, 0.4615)$	1.000

Table 5 The Inverse Power Method Applied to A^{-1}

From the output, we see that A^{-1} has eigenvalue $\lambda = 1$, so that $\lambda_3 = \lambda^{-1} = 1$ is an eigenvalue of A. We check the vector $\mathbf{u} = (1.0000, 0.6923, -0.1538, 0.0769, 0.4615)$ from Table 5 by computing

$$A\mathbf{u} = \begin{bmatrix} 245 & -254 & -252 & -46 & -224 \\ 161 & -168 & -174 & -32 & -148 \\ -39 & 40 & 45 & 7 & 38 \\ 27 & -28 & -32 & -6 & -26 \\ 110 & -113 & -110 & -21 & -101 \end{bmatrix} \begin{bmatrix} 1.0000 \\ 0.6923 \\ -0.1538 \\ 0.0769 \\ 0.4615 \end{bmatrix} \approx \begin{bmatrix} 1.0000 \\ 0.6920 \\ -0.1537 \\ 0.0768 \\ 0.4617 \end{bmatrix} \approx \mathbf{u}$$

Thus $\mathbf{u}$ is an eigenvector associated with eigenvalue $\lambda_3 = 1$. ∎

The Shifted Inverse Power Method

So far we have found three eigenvalues (and associated eigenvectors) for A,

- The largest (in absolute value) λ_1.
- The eigenvalue λ_2 that is farthest from λ_1.
- The eigenvalue λ_3 that is closest to the origin.

Another way to look for eigenvalues is to start with the matrix $B = A - cI_n$ for some scalar c. Applying the Inverse Power Method to B will find the eigenvalue λ of B that is closest to the origin. Since $B = A - cI_n$, $\lambda + c$ is the eigenvalue of A that is closest to c.

EXAMPLE 4   For the matrix A given in Example 1, find the eigenvalue that is closest to $c = 4$.

Solution We start by setting

$$
B = A - 4I_5 = \begin{bmatrix} 241 & -254 & -252 & -46 & -224 \\ 161 & -172 & -174 & -32 & -148 \\ -39 & 40 & 41 & 7 & 38 \\ 27 & -28 & -32 & -10 & -26 \\ 110 & -113 & -110 & -21 & -105 \end{bmatrix}
$$

Now we apply the Inverse Power Method to B, starting out with the vector $\mathbf{x}_0 = (5, 4, 3, 2, 1)$. Table 6 includes every fifth iteration.

k	$\mathbf{x}_k$	s_k
5	$(1.0000, 0.5363, -0.0292, 0.0141, 0.4931)$	0.9416
10	$(1.0000, 0.4936, 0.0051, -0.0026, 0.5013)$	0.4616
15	$(1.0000, 0.5008, -0.0006, 0.0003, 0.4998)$	0.5053
20	$(1.0000, 0.4999, 0.0001, 0.0000, 0.5000)$	0.4993
25	$(1.0000, 0.5000, 0.0000, 0.0000, 0.5000)$	0.5001
30	$(1.0000, 0.5000, 0.0000, 0.0000, 0.5000)$	0.5000
35	$(1.0000, 0.5000, 0.0000, 0.0000, 0.5000)$	0.5000

Table 6   The Shifted Inverse Power Method (Example 4)

We can see that $\lambda = 0.5$ is an eigenvalue for B^{-1}, so that $\lambda = 2$ is an eigenvalue for B. By shifting back, we find that $\lambda_4 = 2 + 4 = 6$ is an eigenvalue for A, with associated eigenvector $\mathbf{u} = (1, 0.5, 0, 0, 0.5)$. We check this by computing

$$
A\mathbf{u} = \begin{bmatrix} 245 & -254 & -252 & -46 & -224 \\ 161 & -168 & -174 & -32 & -148 \\ -39 & 40 & 45 & 7 & 38 \\ 27 & -28 & -32 & -6 & -26 \\ 110 & -113 & -110 & -21 & -101 \end{bmatrix} \begin{bmatrix} 1.0 \\ 0.5 \\ 0 \\ 0 \\ 0.5 \end{bmatrix} = \begin{bmatrix} 6 \\ 3 \\ 0 \\ 0 \\ 3 \end{bmatrix} = 6\mathbf{u}
$$

■

Applying the Shifted Inverse Power Method with different choices of c can yield other eigenvalues and eigenvectors. It is typically not efficient to do this for random values of c, but in some applications rough estimates of the eigenvalues are known. In these instances, the Shifted Inverse Power Method can turn estimates into accurate approximations.

Computational Comments

Some general remarks about the Power Method and related techniques discussed in this section:

- Similar to the approximation methods for finding solutions to linear systems given in Section 1.3, the Power Method is not overly sensitive to round-off error. In fact, each successive vector can be viewed as a starting point for the algorithm, so even if an error occurs, it typically will be corrected.

- The rate of convergence has differed in the examples we have considered. With the Power Method and its relatives, the larger the dominant eigenvalue is relative to the other eigenvalues, the faster the rate of convergence.

- The Power Method is guaranteed to work only on matrices whose eigenvectors span $\mathbf{R}^n$. However, in practice this method often also will work on other matrices, although convergence may be slower.

- Earlier we stated that the choice of starting vector $\mathbf{x}_0$ does not matter, and in virtually all cases this will be true. However it is possible to get misleading results from an unlucky choice of $\mathbf{x}_0$. For instance, in Example 3 we used $\mathbf{x}_0 = (1, 2, 3, 4, 5)$ in place of $(1, 1, 1, 1, 1)$, claiming at the time that this was done for "a change of pace." However, there was another reason. Table 7 gives the results when starting with $\mathbf{x}_0 = (1, 1, 1, 1, 1)$.

k	s_k	$\mathbf{x}_k$
5	−0.4559	(1.0000, 0.6758, −0.1624, 0.1624, 0.4866)
10	−0.4928	(1.0000, 0.6681, −0.1660, 0.1660, 0.4979)
15	−0.4990	(1.0000, 0.6669, −0.1666, 0.1666, 0.4997)
20	−0.4999	(1.0000, 0.6667, −0.1667, 0.1667, 0.5000)
25	−0.5000	(1.0000, 0.6667, −0.1667, 0.1667, 0.5000)
30	−0.5000	(1.0000, 0.6667, −0.1667, 0.1667, 0.5000)

Table 7 New Application of the Inverse Power Method

This gives us $\lambda = -2$, which is an eigenvalue of A but not the eigenvalue closest to 0. We did not find $\lambda = 1$ because this choice of $\mathbf{x}_0$ happens to be in the span of the eigenvectors *not* associated with $\lambda = 1$. If we look at the statement of Theorem 6.8 again, we see that one of the conditions is violated.

In most applications, the entries of a matrix A are decimals and are subject to some degree of rounding. The likelihood of such an unlucky choice of $\mathbf{x}_0$ happening in practice is small.

- In cases where the eigenvalues satisfy

$$|\lambda_1| = |\lambda_2| = \cdots = |\lambda_k| > |\lambda_{k+1}| \geq \cdots \geq |\lambda_n|$$

and $\lambda_1 = \lambda_2 = \cdots = \lambda_k$, the Power Method will still work fine. However, if (for example) instead $\lambda_1 = -\lambda_2$, then the Power Method can produce strange results (see Exercises 41–42).

Proof of Theorem 6.8

Proof of Theorem 6.8 To understand why the Power Method works, suppose that we have an $n \times n$ matrix A that has eigenvalues $\lambda_1, \cdots, \lambda_n$ such that

$$|\lambda_1| > |\lambda_2| \geq \cdots \geq |\lambda_n| \tag{5}$$

Assume that the associated eigenvectors $\mathbf{u}_1, \ldots, \mathbf{u}_n$ form a basis for $\mathbf{R}^n$. If $\mathbf{x}_0$ is an arbitrary vector, then there exist scalars $c_1, \ldots, c_n$ such that

$$\mathbf{x}_0 = c_1\mathbf{u}_1 + c_2\mathbf{u}_2 + \cdots + c_n\mathbf{u}_n$$

Since $A^k\mathbf{u}_j = \lambda_j^k\mathbf{u}_j$ for each eigenvector $\mathbf{u}_j$ and each positive integer k (see Exercise 60 in Section 6.1), if we form the product $A^k\mathbf{x}_0$, we get

$$\begin{aligned} A^k\mathbf{x}_0 &= A^k\left(c_1\mathbf{u}_1 + c_2\mathbf{u}_2 + \cdots + c_n\mathbf{u}_n\right) \\ &= c_1\lambda_1^k\mathbf{u}_1 + c_2\lambda_2^k\mathbf{u}_2 + \cdots + c_n\lambda_n^k\mathbf{u}_n \end{aligned}$$

Next, divide both sides by λ_1^k. (This is similar to our scaling in each step of the Power Method.) This gives us

$$\left(\frac{1}{\lambda_1^k}\right) A^k\mathbf{x}_0 = c_1\mathbf{u}_1 + \left(\frac{\lambda_2}{\lambda_1}\right)^k \mathbf{u}_2 + \cdots + \left(\frac{\lambda_n}{\lambda_1}\right)^k \mathbf{u}_n$$

By (5), as k gets large, each of $\left(\frac{\lambda_2}{\lambda_1}\right)^k, \ldots, \left(\frac{\lambda_n}{\lambda_1}\right)^k$ gets smaller. Hence as $k \to \infty$,

$$\left(\frac{1}{\lambda_1^k}\right) A^k\mathbf{x}_0 \to c_1\mathbf{u}_1$$

as claimed and observed in our examples. ∎

Note that the larger $|\lambda_1|$ is relative to $|\lambda_2|, \ldots, |\lambda_n|$, the faster $\left(\frac{\lambda_2}{\lambda_1}\right)^k, \ldots, \left(\frac{\lambda_n}{\lambda_1}\right)^k$ converge to 0. This is why the Power Method converges more rapidly when the dominant eigenvalue is much larger than the other eigenvalues.

| EXERCISES |

In Exercises 1–6, compute the first three iterations of the Power Method without scaling, starting with the given $\mathbf{x}_0$.

1. $A = \begin{bmatrix} 1 & -3 \\ 1 & 5 \end{bmatrix}$, $\quad \mathbf{x}_0 = \begin{bmatrix} 1 \\ 0 \end{bmatrix}$

2. $A = \begin{bmatrix} -2 & 4 \\ 3 & -1 \end{bmatrix}$, $\quad \mathbf{x}_0 = \begin{bmatrix} 0 \\ 1 \end{bmatrix}$

3. $A = \begin{bmatrix} 6 & -3 & 7 \\ 4 & 1 & 5 \\ 4 & -3 & 9 \end{bmatrix}$, $\quad \mathbf{x}_0 = \begin{bmatrix} 1 \\ 0 \\ 0 \end{bmatrix}$

4. $A = \begin{bmatrix} -2 & 5 & -7 \\ -2 & 11 & -13 \\ -2 & 5 & -7 \end{bmatrix}$, $\quad \mathbf{x}_0 = \begin{bmatrix} 1 \\ 1 \\ 0 \end{bmatrix}$

5. $A = \begin{bmatrix} 5 & -1 & 2 \\ 2 & 2 & 2 \\ 2 & -1 & 5 \end{bmatrix}$, $\quad \mathbf{x}_0 = \begin{bmatrix} 1 \\ 0 \\ -1 \end{bmatrix}$

6. $A = \begin{bmatrix} -1 & 2 & -7 \\ -10 & 2 & 2 \\ -10 & 2 & 2 \end{bmatrix}$, $\quad \mathbf{x}_0 = \begin{bmatrix} 0 \\ 0 \\ 1 \end{bmatrix}$

In Exercises 7–12, compute the first two iterations of the Power Method with scaling, starting with the given $\mathbf{x}_0$. Round any numerical values to two decimal places.

7. $A = \begin{bmatrix} 2 & -1 \\ 0 & 1 \end{bmatrix}$, $\quad \mathbf{x}_0 = \begin{bmatrix} 0 \\ 1 \end{bmatrix}$

8. $A = \begin{bmatrix} 3 & 1 \\ 2 & 0 \end{bmatrix}$, $\quad \mathbf{x}_0 = \begin{bmatrix} -1 \\ 1 \end{bmatrix}$

9. $A = \begin{bmatrix} -1 & 0 & 2 \\ 1 & 1 & 0 \\ 0 & -2 & 1 \end{bmatrix}$, $\quad \mathbf{x}_0 = \begin{bmatrix} 0 \\ 1 \\ 0 \end{bmatrix}$

10. $A = \begin{bmatrix} -2 & 1 & 1 \\ 0 & 3 & -2 \\ 2 & 0 & 0 \end{bmatrix}$, $\quad \mathbf{x}_0 = \begin{bmatrix} -1 \\ 1 \\ 0 \end{bmatrix}$

11. $A = \begin{bmatrix} 1 & 0 & 0 \\ -1 & 3 & 0 \\ 2 & -1 & 1 \end{bmatrix}$, $\quad \mathbf{x}_0 = \begin{bmatrix} 0 \\ 1 \\ -1 \end{bmatrix}$

12. $A = \begin{bmatrix} 0 & 2 & -1 \\ 0 & 2 & 1 \\ 2 & 0 & 0 \end{bmatrix}$, $\quad \mathbf{x}_0 = \begin{bmatrix} -1 \\ 1 \\ 1 \end{bmatrix}$

In Exercises 13–18, the eigenvalues of a 3×3 matrix A are given. Determine if it is assured that the Power Method will converge to an eigenvector and eigenvalue, and if so, identify the eigenvalue.

13. $\lambda_1 = 5, \lambda_2 = -2, \lambda_3 = 7$

14. $\lambda_1 = 3, \lambda_2 = -4, \lambda_3 = 0$

15. $\lambda_1 = -6, \lambda_2 = 2, \lambda_3 = 2$

16. $\lambda_1 = 5, \lambda_2 = 5, \lambda_3 = 4$

17. $\lambda_1 = -4, \lambda_2 = 4, \lambda_3 = 6$

18. $\lambda_1 = 3, \lambda_2 = -3, \lambda_3 = 2$

In Exercises 19–22, the given λ is the dominant eigenvalue for A. To which matrix B would you apply the Power Method in order to find the eigenvalue that is farthest from λ?

19. $A = \begin{bmatrix} 1 & 2 \\ 3 & 2 \end{bmatrix}, \quad \lambda = 4$

20. $A = \begin{bmatrix} -1 & 0 \\ 2 & 3 \end{bmatrix}, \quad \lambda = 3$

21. $A = \begin{bmatrix} -1 & 2 & -7 \\ -10 & 2 & 2 \\ -10 & 2 & 2 \end{bmatrix}, \quad \lambda = 9$

22. $A = \begin{bmatrix} 5 & -1 & 2 \\ 2 & 2 & 2 \\ 2 & -1 & 5 \end{bmatrix}, \quad \lambda = 6$

In Exercises 23–26, to which matrix B would you apply the Inverse Power Method in order to find the eigenvalue that is closest to c?

23. $A = \begin{bmatrix} -3 & 1 \\ 5 & 2 \end{bmatrix}, \quad c = 4$

24. $A = \begin{bmatrix} 1 & 2 \\ 3 & 4 \end{bmatrix}, \quad c = -5$

25. $A = \begin{bmatrix} 3 & 1 & 4 \\ 1 & 5 & 9 \\ 2 & 6 & 1 \end{bmatrix}, \quad c = -1$

26. $A = \begin{bmatrix} 2 & 7 & 1 \\ 8 & 2 & 8 \\ 1 & 8 & 2 \end{bmatrix}, \quad c = 3$

27. Below is the output resulting from applying the Inverse Power Method to a matrix A. Identify the eigenvalue and eigenvector.

k	$\mathbf{x}_k$	s_k
5	$(1.0000, 0.5363, -0.0292)$	0.4415
10	$(1.0000, 0.4837, -0.0026)$	0.3623
15	$(1.0000, 0.5091, -0.0006)$	0.2503
20	$(1.0000, 0.4997, -0.0001)$	0.2501
25	$(1.0000, 0.5000, 0.0000)$	0.2500

28. Below is the output resulting from applying the Shifted Inverse Power Method to a matrix A with $c = 3$. Identify the eigenvalue and eigenvector.

k	$\mathbf{x}_k$	s_k
2	$(0.3577, 0.0971, 1.0000)$	0.5102
4	$(0.4697, 0.1021, 1.0000)$	0.5063
6	$(0.4925, 0.1007, 1.0000)$	0.5021
8	$(0.4997, 0.1002, 1.0000)$	0.5003
10	$(0.5000, 0.1000, 1.0000)$	0.5000

FIND AN EXAMPLE For Exercises 29–34, find an example that meets the given specifications.

29. A 2×2 matrix A and an initial vector $\mathbf{x}_0$ such that the Power Method converges immediately. That is, $\mathbf{x}_0 = \mathbf{x}_1 = \cdots$.

30. A 3×3 matrix A and an initial vector $\mathbf{x}_0$ such that the Power Method converges immediately. That is, $\mathbf{x}_0 = \mathbf{x}_1 = \cdots$.

31. A 2×2 matrix A and an initial vector $\mathbf{x}_0$ such that the Power Method alternates between two different vectors. Thus $\mathbf{x}_0 = \mathbf{x}_2 = \cdots$ and $\mathbf{x}_1 = \mathbf{x}_3 = \cdots$, but $\mathbf{x}_0 \neq \mathbf{x}_1$.

32. A 3×3 matrix A and an initial vector $\mathbf{x}_0$ such that the Power Method alternates between two different vectors. Thus $\mathbf{x}_0 = \mathbf{x}_2 = \cdots$ and $\mathbf{x}_1 = \mathbf{x}_3 = \cdots$, but $\mathbf{x}_0 \neq \mathbf{x}_1$.

33. A 2×2 matrix A and an initial vector $\mathbf{x}_0$ such that the Power Method without scaling alternates between three different vectors. Thus $\mathbf{x}_0 = \mathbf{x}_3 = \cdots$, $\mathbf{x}_1 = \mathbf{x}_4 = \cdots$, and $\mathbf{x}_2 = \mathbf{x}_5 = \cdots$, with $\mathbf{x}_0$, $\mathbf{x}_1$, and $\mathbf{x}_2$ distinct.

34. A 3×3 matrix A and an initial vector $\mathbf{x}_0$ such that the Power Method alternates between three different vectors. Thus $\mathbf{x}_0 = \mathbf{x}_3 = \cdots$, $\mathbf{x}_1 = \mathbf{x}_4 = \cdots$, and $\mathbf{x}_2 = \mathbf{x}_5 = \cdots$, with $\mathbf{x}_0$, $\mathbf{x}_1$, and $\mathbf{x}_2$ distinct.

TRUE OR FALSE For Exercises 35–40, determine if the statement is true or false, and justify your answer.

35. If a square matrix A has a dominant eigenvalue, then the Power Method will converge.

36. The Power Method is generally sensitive to round-off error.

37. The Inverse Power Method can only be applied to invertible matrices.

38. If the Power Method converges, then it will converge to the same eigenvector for any initial vector $\mathbf{x}_0$.

39. Typically, the closer an initial vector $\mathbf{x}_0$ is to a dominant eigenvector $\mathbf{u}$, the faster the Power Method will converge.

40. If $\lambda_1 = \lambda_2$ are the two largest eigenvalues of a matrix A, then the Power Method will not converge.

41. For the matrix A and vector $\mathbf{x}_0$, compute the first four iterations of the Power Method, and then explain the behavior that you observe.

$$A = \begin{bmatrix} 1 & 1 \\ 0 & -1 \end{bmatrix}, \quad \mathbf{x}_0 = \begin{bmatrix} 0 \\ 1 \end{bmatrix}$$

42. For the matrix A and vector $\mathbf{x}_0$, compute the first four iterations of the Power Method with scaling, and then explain the behavior that you observe.

$$A = \begin{bmatrix} -2 & 0 \\ 1 & 2 \end{bmatrix}, \quad \mathbf{x}_0 = \begin{bmatrix} 1 \\ 0 \end{bmatrix}$$

43. For the matrix A, $\lambda = 2$ is the largest eigenvalue. Use the given value of $\mathbf{x}_0$ to generate enough iterations of the Power Method to estimate an eigenvalue. Explain the results that you get.

$$A = \begin{bmatrix} 1 & 0 \\ 2 & 2 \end{bmatrix}, \quad \mathbf{x}_0 = \begin{bmatrix} -1 \\ 2 \end{bmatrix}$$

44. For the matrix A, $\lambda = 3$ is the largest eigenvalue. Use the given value of x_0 to generate enough iterations of the Power Method to estimate an eigenvalue. Explain the results that you get.

$$A = \begin{bmatrix} 1 & 2 \\ 2 & 1 \end{bmatrix}, \quad x_0 = \begin{bmatrix} -1 \\ 1 \end{bmatrix}$$

© In Exercises 45–50, compute the first six iterations of the Power Method without scaling, starting with the given x_0.

45. $A = \begin{bmatrix} 1 & -3 \\ 1 & 5 \end{bmatrix}, \quad x_0 = \begin{bmatrix} 1 \\ 0 \end{bmatrix}$

46. $A = \begin{bmatrix} -2 & 4 \\ 3 & -1 \end{bmatrix}, \quad x_0 = \begin{bmatrix} 0 \\ 1 \end{bmatrix}$

47. $A = \begin{bmatrix} 6 & -3 & 7 \\ 4 & 1 & 5 \\ 4 & -3 & 9 \end{bmatrix}, \quad x_0 = \begin{bmatrix} 1 \\ 0 \\ 0 \end{bmatrix}$

48. $A = \begin{bmatrix} -2 & 5 & -7 \\ -2 & 11 & -13 \\ -2 & 5 & -7 \end{bmatrix}, \quad x_0 = \begin{bmatrix} 1 \\ 1 \\ 0 \end{bmatrix}$

49. $A = \begin{bmatrix} 5 & -1 & 2 \\ 2 & 2 & 2 \\ 2 & -1 & 5 \end{bmatrix}, \quad x_0 = \begin{bmatrix} 1 \\ 0 \\ -1 \end{bmatrix}$

50. $A = \begin{bmatrix} -1 & 2 & -7 \\ -10 & 2 & 2 \\ -10 & 2 & 2 \end{bmatrix}, \quad x_0 = \begin{bmatrix} 0 \\ 0 \\ 1 \end{bmatrix}$

© In Exercises 51–56, compute as many iterations of the Power Method with scaling as are needed to estimate an eigenvalue and eigenvector for A, starting with the given x_0.

51. $A = \begin{bmatrix} 2 & -1 \\ 0 & 1 \end{bmatrix}, \quad x_0 = \begin{bmatrix} 0 \\ 1 \end{bmatrix}$

52. $A = \begin{bmatrix} 3 & 1 \\ 2 & 0 \end{bmatrix}, \quad x_0 = \begin{bmatrix} -1 \\ 1 \end{bmatrix}$

53. $A = \begin{bmatrix} -1 & 1 & 2 \\ 1 & 3 & -2 \\ 2 & -2 & 1 \end{bmatrix}, \quad x_0 = \begin{bmatrix} 0 \\ 1 \\ 0 \end{bmatrix}$

54. $A = \begin{bmatrix} -2 & 1 & 1 \\ 0 & 3 & -2 \\ 2 & 0 & 0 \end{bmatrix}, \quad x_0 = \begin{bmatrix} -1 \\ 1 \\ 0 \end{bmatrix}$

55. $A = \begin{bmatrix} 1 & 0 & 0 \\ -1 & 3 & 0 \\ 2 & -1 & 1 \end{bmatrix}, \quad x_0 = \begin{bmatrix} 0 \\ 1 \\ -1 \end{bmatrix}$

56. $A = \begin{bmatrix} 0 & 2 & -1 \\ 0 & 2 & 1 \\ 2 & 0 & 0 \end{bmatrix}, \quad x_0 = \begin{bmatrix} -1 \\ 1 \\ 1 \end{bmatrix}$

6.3 Change of Basis

We have seen that there are numerous different bases for $\mathbf{R}^n$ (or a subspace of $\mathbf{R}^n$). In this section we develop a general procedure for changing from one basis to another.

One use for a change of basis is to analyze linear transformations. Suppose that $T : \mathbf{R}^2 \to \mathbf{R}^2$ is given by $T(\mathbf{x}) = A\mathbf{x}$, where A is a 2×2 matrix with linearly independent eigenvectors $\{\mathbf{u}_1, \mathbf{u}_2\}$ and corresponding eigenvalues $\{\lambda_1, \lambda_2\}$. Then for any $\mathbf{v}$ in $\mathbf{R}^2$ there exist scalars c_1 and c_2 such that

$$\mathbf{v} = c_1 \mathbf{u}_1 + c_2 \mathbf{u}_2$$

Because $\mathbf{u}_1$ and $\mathbf{u}_2$ are eigenvectors of A, expressing $\mathbf{v}$ in this form makes $T(\mathbf{v})$ easy to compute:

$$T(\mathbf{v}) = A\big(c_1 \mathbf{u}_1 + c_2 \mathbf{u}_2\big) = c_1 A\mathbf{u}_1 + c_2 A\mathbf{u}_2 = c_1 \lambda_1 \mathbf{u}_1 + c_2 \lambda_2 \mathbf{u}_2$$

Thus expressing $\mathbf{v}$ in terms of an eigenvector basis makes clearer the behavior of T. There is more on using different bases to analyze linear transformations in later sections.

To develop a systematic procedure for changing from one basis to another, we start with the standard basis for $\mathbf{R}^2$,

$$S = \{\mathbf{e}_1, \mathbf{e}_2\} = \left\{ \begin{bmatrix} 1 \\ 0 \end{bmatrix}, \begin{bmatrix} 0 \\ 1 \end{bmatrix} \right\}$$

Then, for example, if $\mathbf{x} = \begin{bmatrix} 3 \\ -2 \end{bmatrix}$, we have

$$\mathbf{x} = 3\mathbf{e}_1 - 2\mathbf{e}_2$$

We can view the entries in $\begin{bmatrix} 3 \\ -2 \end{bmatrix}$ as the coefficients needed to write $\mathbf{x}$ as a linear combination of $\{\mathbf{e}_1, \mathbf{e}_2\}$.

Now suppose that

$$\mathcal{B} = \{\mathbf{u}_1, \mathbf{u}_2\} = \left\{ \begin{bmatrix} 2 \\ 7 \end{bmatrix}, \begin{bmatrix} 1 \\ 4 \end{bmatrix} \right\}$$

Then $\mathcal{B}$ is another basis for $\mathbf{R}^2$. It is not difficult to verify that if $\mathbf{x} = \begin{bmatrix} 3 \\ -2 \end{bmatrix}$, then $\mathbf{x} = 14\mathbf{u}_1 - 25\mathbf{u}_2$. The compact notation that we use to express this relationship is

$$\mathbf{x}_{\mathcal{B}} = \begin{bmatrix} 14 \\ -25 \end{bmatrix}_{\mathcal{B}}$$

▶ Although we continue to use set notation for bases, in this section the order of the vectors in the basis matters.

More generally, suppose that $\mathcal{B} = \{\mathbf{u}_1, \ldots, \mathbf{u}_n\}$ forms a basis for $\mathbf{R}^n$. If $\mathbf{y} = y_1 \mathbf{u}_1 + \cdots + y_n \mathbf{u}_n$, then we write

$$\mathbf{y}_{\mathcal{B}} = \begin{bmatrix} y_1 \\ \vdots \\ y_n \end{bmatrix}_{\mathcal{B}}$$

Definition Coordinate Vector

for the **coordinate vector of y with respect to** $\mathcal{B}$. As above, the coordinate vector contains the coefficients required to express $\mathbf{y}$ as a linear combination of the vectors in basis $\mathcal{B}$.

Now define the $n \times n$ matrix $U = \begin{bmatrix} \mathbf{u}_1 & \cdots & \mathbf{u}_n \end{bmatrix}$. Multiplying as usual, we have

$$U \begin{bmatrix} y_1 \\ \vdots \\ y_n \end{bmatrix}_{\mathcal{B}} = y_1 \mathbf{u}_1 + \cdots + y_n \mathbf{u}_n$$

Hence multiplying by U transforms the coordinate vector with respect to $\mathcal{B}$ to the standard basis. Put symbolically, we have $\mathbf{y} = U\mathbf{y}_{\mathcal{B}}$.

EXAMPLE 1 Let

$$\mathcal{B} = \left\{ \begin{bmatrix} 1 \\ 3 \\ -2 \end{bmatrix}, \begin{bmatrix} 2 \\ 0 \\ 1 \end{bmatrix}, \begin{bmatrix} 4 \\ 5 \\ -1 \end{bmatrix} \right\} \quad \text{and} \quad \mathbf{x}_{\mathcal{B}} = \begin{bmatrix} -2 \\ 3 \\ 1 \end{bmatrix}_{\mathcal{B}}$$

Then $\mathcal{B}$ forms a basis for $\mathbf{R}^3$. Find $\mathbf{x}$ with respect to the standard basis $\mathcal{S}$.

Solution Start by setting $U = \begin{bmatrix} 1 & 2 & 4 \\ 3 & 0 & 5 \\ -2 & 1 & -1 \end{bmatrix}$. Then we have

$$\mathbf{x} = U\mathbf{x}_{\mathcal{B}} = \begin{bmatrix} 1 & 2 & 4 \\ 3 & 0 & 5 \\ -2 & 1 & -1 \end{bmatrix} \begin{bmatrix} -2 \\ 3 \\ 1 \end{bmatrix}_{\mathcal{B}} = \begin{bmatrix} 8 \\ -1 \\ 6 \end{bmatrix}_{\mathcal{S}}$$

Thus $\mathbf{x} = \begin{bmatrix} 8 \\ -1 \\ 6 \end{bmatrix}_{\mathcal{S}}$ with respect to the standard basis. ■

When working with coordinate vectors, there is potential for confusion about which basis is in use.

> ## NOTATION CONVENTION
> If no subscript is given on a vector, then it is expressed with respect to the standard basis. For any other basis, a subscript will be included.

Definition Change of Basis
Matrix

The matrix U in Example 1, called a **change of basis matrix**, allows us to switch from a basis $\mathcal{B}$ to the standard basis $\mathcal{S}$. Example 2 shows how to go the other direction, from the standard basis $\mathcal{S}$ to another basis $\mathcal{B}$.

EXAMPLE 2 Let

$$\mathbf{x} = \begin{bmatrix} 3 \\ -2 \end{bmatrix} \quad \text{and} \quad \mathcal{B} = \{\mathbf{u}_1, \mathbf{u}_2\} = \left\{ \begin{bmatrix} 2 \\ 7 \end{bmatrix}, \begin{bmatrix} 1 \\ 4 \end{bmatrix} \right\}$$

as before, and set $U = \begin{bmatrix} \mathbf{u}_1 & \mathbf{u}_2 \end{bmatrix}$. Find the change of basis matrix from $\mathcal{S}$ to $\mathcal{B}$.

Solution To write $\mathbf{x}$ in terms of $\mathcal{B}$, we need to find x_1 and x_2 such that

$$x_1 \mathbf{u}_1 + x_2 \mathbf{u}_2 = \begin{bmatrix} 3 \\ -2 \end{bmatrix} \quad \Longrightarrow \quad U \begin{bmatrix} x_1 \\ x_2 \end{bmatrix} = \begin{bmatrix} 3 \\ -2 \end{bmatrix}$$

Since the columns of U are linearly independent, U is invertible, and hence the solution is

$$\begin{bmatrix} x_1 \\ x_2 \end{bmatrix} = U^{-1} \begin{bmatrix} 3 \\ -2 \end{bmatrix}$$

▶ Recall the Quick Formula,

$$\begin{bmatrix} a & b \\ c & d \end{bmatrix}^{-1} = \frac{1}{ad - bc} \begin{bmatrix} d & -b \\ -c & a \end{bmatrix}$$

This shows that the change of basis matrix is U^{-1}. To test this, we compute (using the Quick Formula from Section 3.3)

$$U^{-1} = \begin{bmatrix} 4 & -1 \\ -7 & 2 \end{bmatrix}$$

Then

$$U^{-1} \begin{bmatrix} 3 \\ -2 \end{bmatrix} = \begin{bmatrix} 4 & -1 \\ -7 & 2 \end{bmatrix} \begin{bmatrix} 3 \\ -2 \end{bmatrix} = \begin{bmatrix} 14 \\ -25 \end{bmatrix}$$

which tells us that

$$\mathbf{x}_\mathcal{B} = \begin{bmatrix} 14 \\ -25 \end{bmatrix}_\mathcal{B}$$

as we saw previously. ■

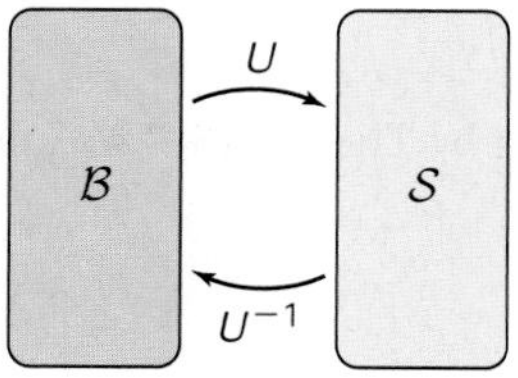

Figure 1 Change of basis between $\mathcal{S}$ and $\mathcal{B}$.

Example 2 illustrates a general fact: If U is the change of basis matrix from $\mathcal{B}$ to $\mathcal{S}$, then U^{-1} is the change of basis matrix from $\mathcal{S}$ to $\mathcal{B}$. This approach generalizes to $\mathbf{R}^n$ and is summarized in Theorem 6.9. The full proof is left as an exercise; a graphical depiction is given in Figure 1.

THEOREM 6.9

Let $\mathbf{x}$ be expressed with respect to the standard basis, and let $\mathcal{B} = \{\mathbf{u}_1, \ldots, \mathbf{u}_n\}$ be any basis for $\mathbf{R}^n$. If $U = \begin{bmatrix} \mathbf{u}_1 & \cdots & \mathbf{u}_n \end{bmatrix}$, then

(a) $\mathbf{x} = U\mathbf{x}_\mathcal{B}$

(b) $\mathbf{x}_\mathcal{B} = U^{-1}\mathbf{x}$

EXAMPLE 3   Let

$$\mathbf{x} = \begin{bmatrix} 3 \\ 4 \\ 4 \end{bmatrix} \quad \text{and} \quad \mathcal{B} = \left\{ \begin{bmatrix} 1 \\ 0 \\ 1 \end{bmatrix}, \begin{bmatrix} 1 \\ -3 \\ 0 \end{bmatrix}, \begin{bmatrix} 2 \\ 1 \\ 2 \end{bmatrix} \right\}$$

Find $\mathbf{x}_\mathcal{B}$, the coordinate vector of $\mathbf{x}$ with respect to the basis $\mathcal{B}$.

Solution   We start by letting U be the matrix with columns given by the vectors in $\mathcal{B}$,

$$U = \begin{bmatrix} 1 & 1 & 2 \\ 0 & -3 & 1 \\ 1 & 0 & 2 \end{bmatrix}$$

Then by Theorem 6.9, we have

$$\mathbf{x}_\mathcal{B} = U^{-1}\mathbf{x} = \begin{bmatrix} -6 & -2 & 7 \\ 1 & 0 & -1 \\ 3 & 1 & -3 \end{bmatrix} \begin{bmatrix} 3 \\ 4 \\ 4 \end{bmatrix} = \begin{bmatrix} 2 \\ -1 \\ 1 \end{bmatrix}_\mathcal{B}$$ ∎

Two Nonstandard Bases

Now suppose that we have nonstandard bases $\mathcal{B}_1 = \{\mathbf{u}_1, \ldots, \mathbf{u}_n\}$ and $\mathcal{B}_2 = \{\mathbf{v}_1, \ldots, \mathbf{v}_n\}$ for $\mathbf{R}^n$. How do we get from $\mathbf{x}_{\mathcal{B}_1}$ to $\mathbf{x}_{\mathcal{B}_2}$—that is, from $\mathbf{x}$ expressed with respect to $\mathcal{B}_1$ to $\mathbf{x}$ expressed with respect to $\mathcal{B}_2$?

The simple solution uses two steps. We apply Theorem 6.9 twice, first to go from $\mathbf{x}_{\mathcal{B}_1}$ to $\mathbf{x}_S$, and then to go from $\mathbf{x}_S$ to $\mathbf{x}_{\mathcal{B}_2}$. Matrix multiplication is used to combine the steps.

THEOREM 6.10

Let $\mathcal{B}_1 = \{\mathbf{u}_1, \ldots, \mathbf{u}_n\}$ and $\mathcal{B}_2 = \{\mathbf{v}_1, \ldots, \mathbf{v}_n\}$ be bases for $\mathbf{R}^n$. If $U = \begin{bmatrix} \mathbf{u}_1 & \cdots & \mathbf{u}_n \end{bmatrix}$ and $V = \begin{bmatrix} \mathbf{v}_1 & \cdots & \mathbf{v}_n \end{bmatrix}$, then

(a) $\mathbf{x}_{\mathcal{B}_2} = V^{-1}U\mathbf{x}_{\mathcal{B}_1}$

(b) $\mathbf{x}_{\mathcal{B}_1} = U^{-1}V\mathbf{x}_{\mathcal{B}_2}$

Proof   If $U = \begin{bmatrix} \mathbf{u}_1 & \cdots & \mathbf{u}_n \end{bmatrix}$ and $V = \begin{bmatrix} \mathbf{v}_1 & \cdots & \mathbf{v}_n \end{bmatrix}$, then by Theorem 6.9 we know that

$$U\mathbf{x}_{\mathcal{B}_1} = \mathbf{x}_S \quad \text{and} \quad V^{-1}\mathbf{x}_S = \mathbf{x}_{\mathcal{B}_2}$$

Combining these gives

$$\mathbf{x}_{\mathcal{B}_2} = V^{-1}\mathbf{x}_S = V^{-1}(U\mathbf{x}_{\mathcal{B}_1}) = V^{-1}U\mathbf{x}_{\mathcal{B}_1}$$

Thus the change of basis matrix from $\mathcal{B}_1$ to $\mathcal{B}_2$ is $V^{-1}U$. The change of basis matrix from $\mathcal{B}_2$ to $\mathcal{B}_1$ is the inverse,

$$(V^{-1}U)^{-1} = U^{-1}V$$

A graphical depiction is given in Figure 2.   ∎

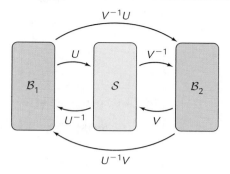

Figure 2 Change of basis between $\mathcal{B}_1$ and $\mathcal{B}_2$.

EXAMPLE 4 Suppose that

$$\mathcal{B}_1 = \left\{ \begin{bmatrix} 1 \\ 3 \end{bmatrix}, \begin{bmatrix} 2 \\ 7 \end{bmatrix} \right\} \quad \text{and} \quad \mathcal{B}_2 = \left\{ \begin{bmatrix} 3 \\ 5 \end{bmatrix}, \begin{bmatrix} 2 \\ 3 \end{bmatrix} \right\}$$

Find $\mathbf{x}_{\mathcal{B}_2}$ if $\mathbf{x}_{\mathcal{B}_1} = \begin{bmatrix} -1 \\ 4 \end{bmatrix}_{\mathcal{B}_1}$.

Solution We start by setting $U = \begin{bmatrix} 1 & 2 \\ 3 & 7 \end{bmatrix}$ and $V = \begin{bmatrix} 3 & 2 \\ 5 & 3 \end{bmatrix}$. Then by Theorem 6.10, the change of basis matrix from $\mathcal{B}_1$ to $\mathcal{B}_2$ is

$$V^{-1}U = \begin{bmatrix} -3 & 2 \\ 5 & -3 \end{bmatrix} \begin{bmatrix} 1 & 2 \\ 3 & 7 \end{bmatrix} = \begin{bmatrix} 3 & 8 \\ -4 & -11 \end{bmatrix}$$

Hence it follows that

$$\mathbf{x}_{\mathcal{B}_2} = V^{-1}U\mathbf{x}_{\mathcal{B}_1} = \begin{bmatrix} 3 & 8 \\ -4 & -11 \end{bmatrix} \begin{bmatrix} -1 \\ 4 \end{bmatrix}_{\mathcal{B}_1} = \begin{bmatrix} 29 \\ -40 \end{bmatrix}_{\mathcal{B}_2}$$

Therefore we have $\begin{bmatrix} -1 \\ 4 \end{bmatrix}_{\mathcal{B}_1} = \begin{bmatrix} 29 \\ -40 \end{bmatrix}_{\mathcal{B}_2}$. We can check our results by converting both $\begin{bmatrix} -1 \\ 4 \end{bmatrix}_{\mathcal{B}_1}$ and $\begin{bmatrix} 29 \\ -40 \end{bmatrix}_{\mathcal{B}_2}$ to the standard basis. We have

$$U\mathbf{x}_{\mathcal{B}_1} = \begin{bmatrix} 1 & 2 \\ 3 & 7 \end{bmatrix} \begin{bmatrix} -1 \\ 4 \end{bmatrix}_{\mathcal{B}_1} = \begin{bmatrix} 7 \\ 25 \end{bmatrix} \quad \text{and} \quad V\mathbf{x}_{\mathcal{B}_2} = \begin{bmatrix} 3 & 2 \\ 5 & 3 \end{bmatrix} \begin{bmatrix} 29 \\ -40 \end{bmatrix}_{\mathcal{B}_2} = \begin{bmatrix} 7 \\ 25 \end{bmatrix}$$

which also shows that $\begin{bmatrix} -1 \\ 4 \end{bmatrix}_{\mathcal{B}_1} = \begin{bmatrix} 29 \\ -40 \end{bmatrix}_{\mathcal{B}_2}$. ∎

EXAMPLE 5 Suppose that

$$\mathcal{B}_1 = \left\{ \begin{bmatrix} 1 \\ 1 \\ 3 \end{bmatrix}, \begin{bmatrix} 1 \\ 4 \\ 2 \end{bmatrix}, \begin{bmatrix} 2 \\ 1 \\ 6 \end{bmatrix} \right\} \quad \text{and} \quad \mathcal{B}_2 = \left\{ \begin{bmatrix} 1 \\ 0 \\ 1 \end{bmatrix}, \begin{bmatrix} 1 \\ -3 \\ 0 \end{bmatrix}, \begin{bmatrix} 2 \\ 1 \\ 2 \end{bmatrix} \right\}$$

Find $\mathbf{x}_{\mathcal{B}_1}$ if $\mathbf{x}_{\mathcal{B}_2} = \begin{bmatrix} 3 \\ 2 \\ 1 \end{bmatrix}_{\mathcal{B}_2}$.

Solution We start by setting

$$U = \begin{bmatrix} 1 & 1 & 2 \\ 1 & 4 & 1 \\ 3 & 2 & 6 \end{bmatrix} \quad \text{and} \quad V = \begin{bmatrix} 1 & 1 & 2 \\ 0 & -3 & 1 \\ 1 & 0 & 2 \end{bmatrix}$$

By Theorem 6.10 we have

$$\mathbf{x}_{\mathcal{B}_1} = U^{-1} V \mathbf{x}_{\mathcal{B}_2}$$
$$= \begin{bmatrix} -22 & 2 & 7 \\ 3 & 0 & -1 \\ 10 & -1 & -3 \end{bmatrix} \begin{bmatrix} 1 & 1 & 2 \\ 0 & -3 & 1 \\ 1 & 0 & 2 \end{bmatrix} \begin{bmatrix} 3 \\ 2 \\ 1 \end{bmatrix}_{\mathcal{B}_2} = \begin{bmatrix} -129 \\ 16 \\ 60 \end{bmatrix}_{\mathcal{B}_1} \quad \blacksquare$$

EXAMPLE 6 Suppose that $S = \text{span}\{\mathbf{v}_1, \mathbf{v}_2\}$ is a subspace of $\mathbf{R}^3$, where

$$\mathbf{v}_1 = \begin{bmatrix} 2 \\ -1 \\ 1 \end{bmatrix} \quad \text{and} \quad \mathbf{v}_2 = \begin{bmatrix} 1 \\ 0 \\ 1 \end{bmatrix}$$

Let $T : \mathbf{R}^3 \to \mathbf{R}^3$ be the linear transformation $T(\mathbf{x}) = A\mathbf{x}$, where

$$A = \begin{bmatrix} -22 & -24 & 18 \\ 17 & 19 & -13 \\ -10 & -10 & 10 \end{bmatrix}$$

Express $T(S)$, the image of S under T, in terms of a basis of eigenvectors of A.

Solution As noted at the section opening, it is easy to evaluate a linear transformation when the input value is given as a linear combination of eigenvectors—we just multiply each eigenvector by the associated eigenvalue. To exploit this, we shall use a change of basis to express $\mathbf{v}_1$ and $\mathbf{v}_2$ in terms of eigenvectors of A.

We need two bases to apply Theorem 6.10. For the first, we let $\mathbf{v}_3 = \mathbf{e}_3$ and then define

$$V = \begin{bmatrix} \mathbf{v}_1 & \mathbf{v}_2 & \mathbf{v}_3 \end{bmatrix} = \begin{bmatrix} 2 & 1 & 0 \\ -1 & 0 & 0 \\ 1 & 1 & 1 \end{bmatrix}$$

Since $\det(V) = 1$, the set $\mathcal{B}_2 = \{\mathbf{v}_1, \mathbf{v}_2, \mathbf{v}_3\}$ is a basis for $\mathbf{R}^3$ and V is a change of basis matrix from $\mathcal{B}_2$ to S. The basis $\mathcal{B}_2$ is attractive because the coordinate vectors of $\mathbf{v}_1$ and $\mathbf{v}_2$ are simple.

Our second basis $\mathcal{B}_1 = \{\mathbf{u}_1, \mathbf{u}_2, \mathbf{u}_3\}$ is made up of eigenvectors of A. We find these using the methods from Section 6.1, starting with the characteristic polynomial

$$\det(A - \lambda I_3) = -\lambda^3 + 7\lambda^2 - 10\lambda = -\lambda(\lambda - 2)(\lambda - 5)$$

The calculations to find the associated eigenvectors are similar to those in Section 6.1 and so are omitted. We have

$$\lambda_1 = 5 \Rightarrow \mathbf{u}_1 = \begin{bmatrix} 4 \\ -3 \\ 2 \end{bmatrix}, \qquad \lambda_2 = 2 \Rightarrow \mathbf{u}_2 = \begin{bmatrix} -1 \\ 1 \\ 0 \end{bmatrix}, \qquad \lambda_3 = 0 \Rightarrow \mathbf{u}_3 = \begin{bmatrix} 3 \\ -1 \\ 1 \end{bmatrix}$$

Setting

$$U = \begin{bmatrix} \mathbf{u}_1 & \mathbf{u}_2 & \mathbf{u}_3 \end{bmatrix} = \begin{bmatrix} 4 & -1 & 3 \\ -3 & 1 & -2 \\ 2 & 0 & 1 \end{bmatrix}$$

we have $\det(U) = -1$, so that $\mathcal{B}_1$ is also a basis and U is a change of basis matrix from $\mathcal{B}_1$ to $\mathcal{S}$. By Theorem 6.10 the change of basis matrix from $\mathcal{B}_2$ to $\mathcal{B}_1$ is

$$
U^{-1}V = \begin{bmatrix} -1 & -1 & 1 \\ 1 & 2 & 1 \\ 2 & 2 & -1 \end{bmatrix} \begin{bmatrix} 2 & 1 & 0 \\ -1 & 0 & 0 \\ 1 & 1 & 1 \end{bmatrix} = \begin{bmatrix} 0 & 0 & 1 \\ 1 & 2 & 1 \\ 1 & 1 & -1 \end{bmatrix}
$$

The next step is to use the change of basis matrix to express $\mathbf{v}_1$ and $\mathbf{v}_2$ in terms of the eigenvectors. Since $\mathbf{v}_1$ is the first vector in $\mathcal{B}_2$, we have

> ▶ The square brackets in terms such as $[\mathbf{v}_1]_{\mathcal{B}_2}$ are to separate the two subscripts.

$$
[\mathbf{v}_1]_{\mathcal{B}_2} = \begin{bmatrix} 1 \\ 0 \\ 0 \end{bmatrix}_{\mathcal{B}_2} \implies [\mathbf{v}_1]_{\mathcal{B}_1} = U^{-1}V[\mathbf{v}_1]_{\mathcal{B}_2} = \begin{bmatrix} 0 & 0 & 1 \\ 1 & 2 & 1 \\ 1 & 1 & -1 \end{bmatrix} \begin{bmatrix} 1 \\ 0 \\ 0 \end{bmatrix}_{\mathcal{B}_2}
$$

$$
= \begin{bmatrix} 0 \\ 1 \\ 1 \end{bmatrix}_{\mathcal{B}_1} = \mathbf{u}_2 + \mathbf{u}_3
$$

Similarly,

$$
[\mathbf{v}_2]_{\mathcal{B}_2} = \begin{bmatrix} 0 \\ 1 \\ 0 \end{bmatrix}_{\mathcal{B}_2} \implies [\mathbf{v}_2]_{\mathcal{B}_1} = U^{-1}V[\mathbf{v}_2]_{\mathcal{B}_2} = \begin{bmatrix} 0 & 0 & 1 \\ 1 & 2 & 1 \\ 1 & 1 & -1 \end{bmatrix} \begin{bmatrix} 0 \\ 1 \\ 0 \end{bmatrix}_{\mathcal{B}_2}
$$

$$
= \begin{bmatrix} 0 \\ 2 \\ 1 \end{bmatrix}_{\mathcal{B}_1} = 2\mathbf{u}_2 + \mathbf{u}_3
$$

Since $\mathbf{u}_2$ and $\mathbf{u}_3$ are eigenvectors of A associated with eigenvalues $\lambda_2 = 2$ and $\lambda_3 = 0$, we have

$$
T(\mathbf{u}_2) = A\mathbf{u}_2 = 2\mathbf{u}_2 \quad \text{and} \quad T(\mathbf{u}_3) = A\mathbf{u}_3 = 0\mathbf{u}_3 = \mathbf{0}
$$

If $\mathbf{v} = c_1\mathbf{v}_1 + c_2\mathbf{v}_2$ is a typical element of the subspace S, then

$$
\mathbf{v} = c_1(\mathbf{u}_2 + \mathbf{u}_3) + c_2(2\mathbf{u}_2 + \mathbf{u}_3) = (c_1 + 2c_2)\mathbf{u}_2 + (c_1 + c_2)\mathbf{u}_3
$$

Therefore

$$
T(\mathbf{v}) = T\big((c_1 + 2c_2)\mathbf{u}_1 + (c_1 + c_2)\mathbf{u}_2\big)
$$
$$
= (c_1 + 2c_2)T(\mathbf{u}_2) + (c_1 + c_2)T(\mathbf{u}_3) = 2(c_1 + 2c_2)\mathbf{u}_2
$$

Thus S is mapped by T to the span of a single eigenvector $\mathbf{u}_2$. ■

Change of Basis in Subspaces

Suppose that a subspace S of $\mathbf{R}^n$ has two bases $\mathcal{B}_1 = \{\mathbf{u}_1, \ldots, \mathbf{u}_k\}$ and $\mathcal{B}_2 = \{\mathbf{v}_1, \ldots, \mathbf{v}_k\}$. There exists a change of basis matrix, but it cannot be found using Theorem 6.10, because the matrices U and V are not square and so are not invertible. Instead, we can use Theorem 6.11.

THEOREM 6.11

Let S be a subspace of $\mathbf{R}^n$ with bases $\mathcal{B}_1 = \{\mathbf{u}_1, \ldots, \mathbf{u}_k\}$ and $\mathcal{B}_2 = \{\mathbf{v}_1, \ldots, \mathbf{v}_k\}$. If

$$
C = \begin{bmatrix} [\mathbf{u}_1]_{\mathcal{B}_2} & \cdots & [\mathbf{u}_k]_{\mathcal{B}_2} \end{bmatrix}
$$

then $\mathbf{x}_{\mathcal{B}_2} = C\mathbf{x}_{\mathcal{B}_1}$.

Proof For a vector $\mathbf{x}$ in S, there exist scalars $x_1, \ldots, x_k$ such that

$$\mathbf{x} = x_1\mathbf{u}_1 + \cdots + x_k\mathbf{u}_k = \begin{bmatrix} x_1 \\ \vdots \\ x_k \end{bmatrix}_{\mathcal{B}_1}$$

Therefore

$$
\begin{aligned}
C\mathbf{x}_{\mathcal{B}_1} &= \begin{bmatrix} [\mathbf{u}_1]_{\mathcal{B}_2} & \cdots & [\mathbf{u}_k]_{\mathcal{B}_2} \end{bmatrix} \begin{bmatrix} x_1 \\ \vdots \\ x_k \end{bmatrix}_{\mathcal{B}_1} \\
&= x_1[\mathbf{u}_1]_{\mathcal{B}_2} + \cdots + x_k[\mathbf{u}_k]_{\mathcal{B}_2} \\
&= [x_1\mathbf{u}_1 + \cdots + x_k\mathbf{u}_k]_{\mathcal{B}_2} \qquad \text{(See Exercise 48)} \\
&= \mathbf{x}_{\mathcal{B}_2}
\end{aligned}
$$

◼

When $S = \mathbf{R}^n$, the matrix C in Theorem 6.11 is equal to the matrix $V^{-1}U$ given in Theorem 6.10 (see Exercise 49).

EXAMPLE 7 Let

$$\mathcal{B}_1 = \left\{ \begin{bmatrix} 1 \\ -5 \\ 8 \end{bmatrix}, \begin{bmatrix} 3 \\ -8 \\ 3 \end{bmatrix} \right\} \quad \text{and} \quad \mathcal{B}_2 = \left\{ \begin{bmatrix} 1 \\ -3 \\ 2 \end{bmatrix}, \begin{bmatrix} -1 \\ 2 \\ 1 \end{bmatrix} \right\}$$

be two bases of a subspace S of $\mathbf{R}^3$. Find the change of basis matrix from $\mathcal{B}_1$ to $\mathcal{B}_2$, and find $\mathbf{x}_{\mathcal{B}_2}$ if $\mathbf{x}_{\mathcal{B}_1} = \begin{bmatrix} 3 \\ -1 \end{bmatrix}_{\mathcal{B}_1}$.

Solution To apply Theorem 6.11, we need to write each vector in $\mathcal{B}_1$ in terms of the vectors in $\mathcal{B}_2$. The system

$$\begin{bmatrix} 1 \\ -5 \\ 8 \end{bmatrix} = c_{11} \begin{bmatrix} 1 \\ -3 \\ 2 \end{bmatrix} + c_{21} \begin{bmatrix} -1 \\ 2 \\ 1 \end{bmatrix}$$

has solution $c_{11} = 3$ and $c_{21} = 2$, so that $\begin{bmatrix} 1 \\ -5 \\ 8 \end{bmatrix} = \begin{bmatrix} 3 \\ 2 \end{bmatrix}_{\mathcal{B}_2}$. Similarly, the system

$$\begin{bmatrix} 3 \\ -8 \\ 3 \end{bmatrix} = c_{12} \begin{bmatrix} 1 \\ -3 \\ 2 \end{bmatrix} + c_{22} \begin{bmatrix} -1 \\ 2 \\ 1 \end{bmatrix}$$

has solution $c_{12} = 2$ and $c_{22} = -1$, so that $\begin{bmatrix} 3 \\ -8 \\ 3 \end{bmatrix} = \begin{bmatrix} 2 \\ -1 \end{bmatrix}_{\mathcal{B}_2}$. Thus

$$C = \begin{bmatrix} 3 & 2 \\ 2 & -1 \end{bmatrix}$$

By Theorem 6.11,

$$\mathbf{x}_{\mathcal{B}_2} = C\mathbf{x}_{\mathcal{B}_1} = \begin{bmatrix} 3 & 2 \\ 2 & -1 \end{bmatrix} \begin{bmatrix} 3 \\ -1 \end{bmatrix}_{\mathcal{B}_1} = \begin{bmatrix} 7 \\ 7 \end{bmatrix}_{\mathcal{B}_2}$$

◼

| EXERCISES |

In Exercises 1–6, convert the coordinate vector $\mathbf{x}_\mathcal{B}$ from the given basis $\mathcal{B}$ to the standard basis.

1. $\mathcal{B} = \left\{ \begin{bmatrix} 3 \\ -2 \end{bmatrix}, \begin{bmatrix} 2 \\ 5 \end{bmatrix} \right\}, \quad \mathbf{x}_\mathcal{B} = \begin{bmatrix} 1 \\ -1 \end{bmatrix}_\mathcal{B}$

2. $\mathcal{B} = \left\{ \begin{bmatrix} -5 \\ 3 \end{bmatrix}, \begin{bmatrix} -2 \\ 1 \end{bmatrix} \right\}, \quad \mathbf{x}_\mathcal{B} = \begin{bmatrix} -2 \\ 3 \end{bmatrix}_\mathcal{B}$

3. $\mathcal{B} = \left\{ \begin{bmatrix} 4 \\ 3 \end{bmatrix}, \begin{bmatrix} 2 \\ 1 \end{bmatrix} \right\}, \quad \mathbf{x}_\mathcal{B} = \begin{bmatrix} 2 \\ -4 \end{bmatrix}_\mathcal{B}$

4. $\mathcal{B} = \left\{ \begin{bmatrix} -6 \\ 1 \end{bmatrix}, \begin{bmatrix} 5 \\ -3 \end{bmatrix} \right\}, \quad \mathbf{x}_\mathcal{B} = \begin{bmatrix} -3 \\ 2 \end{bmatrix}_\mathcal{B}$

5. $\mathcal{B} = \left\{ \begin{bmatrix} 1 \\ -2 \\ -1 \end{bmatrix}, \begin{bmatrix} -1 \\ 2 \\ 0 \end{bmatrix}, \begin{bmatrix} 2 \\ -1 \\ 3 \end{bmatrix} \right\}, \quad \mathbf{x}_\mathcal{B} = \begin{bmatrix} 1 \\ 2 \\ 1 \end{bmatrix}_\mathcal{B}$

6. $\mathcal{B} = \left\{ \begin{bmatrix} 0 \\ 3 \\ 1 \end{bmatrix}, \begin{bmatrix} 1 \\ 2 \\ 3 \end{bmatrix}, \begin{bmatrix} 0 \\ -1 \\ 2 \end{bmatrix} \right\}, \quad \mathbf{x}_\mathcal{B} = \begin{bmatrix} 1 \\ 1 \\ -2 \end{bmatrix}_\mathcal{B}$

In Exercises 7–12, find the change of basis matrix from the standard basis $\mathcal{S}$ to $\mathcal{B}$, and then convert $\mathbf{x}$ to the coordinate vector with respect to $\mathcal{B}$.

7. $\mathcal{B} = \left\{ \begin{bmatrix} 1 \\ 2 \end{bmatrix}, \begin{bmatrix} 1 \\ 3 \end{bmatrix} \right\}, \quad \mathbf{x} = \begin{bmatrix} 3 \\ -1 \end{bmatrix}$

8. $\mathcal{B} = \left\{ \begin{bmatrix} 5 \\ 4 \end{bmatrix}, \begin{bmatrix} 1 \\ 1 \end{bmatrix} \right\}, \quad \mathbf{x} = \begin{bmatrix} 1 \\ 2 \end{bmatrix}$

9. $\mathcal{B} = \left\{ \begin{bmatrix} -2 \\ 1 \end{bmatrix}, \begin{bmatrix} 5 \\ -3 \end{bmatrix} \right\}, \quad \mathbf{x} = \begin{bmatrix} 1 \\ -1 \end{bmatrix}$

10. $\mathcal{B} = \left\{ \begin{bmatrix} 7 \\ 5 \end{bmatrix}, \begin{bmatrix} 4 \\ 3 \end{bmatrix} \right\}, \quad \mathbf{x} = \begin{bmatrix} 4 \\ -3 \end{bmatrix}$

11. $\mathcal{B} = \left\{ \begin{bmatrix} 1 \\ 0 \\ 0 \end{bmatrix}, \begin{bmatrix} -1 \\ 2 \\ 1 \end{bmatrix}, \begin{bmatrix} 1 \\ -1 \\ 0 \end{bmatrix} \right\}, \quad \mathbf{x} = \begin{bmatrix} 1 \\ 2 \\ -1 \end{bmatrix}$

12. $\mathcal{B} = \left\{ \begin{bmatrix} 2 \\ 2 \\ -1 \end{bmatrix}, \begin{bmatrix} -1 \\ -2 \\ 1 \end{bmatrix}, \begin{bmatrix} 1 \\ -1 \\ 1 \end{bmatrix} \right\}, \quad \mathbf{x} = \begin{bmatrix} -2 \\ 1 \\ 2 \end{bmatrix}$

In Exercises 13–18, find the change of basis matrix from $\mathcal{B}_1$ to $\mathcal{B}_2$.

13. $\mathcal{B}_1 = \left\{ \begin{bmatrix} 2 \\ 1 \end{bmatrix}, \begin{bmatrix} 1 \\ 1 \end{bmatrix} \right\}, \quad \mathcal{B}_2 = \left\{ \begin{bmatrix} 5 \\ 7 \end{bmatrix}, \begin{bmatrix} 3 \\ 4 \end{bmatrix} \right\}$

14. $\mathcal{B}_1 = \left\{ \begin{bmatrix} 2 \\ 1 \end{bmatrix}, \begin{bmatrix} 3 \\ 2 \end{bmatrix} \right\}, \quad \mathcal{B}_2 = \left\{ \begin{bmatrix} 3 \\ 1 \end{bmatrix}, \begin{bmatrix} 7 \\ 2 \end{bmatrix} \right\}$

15. $\mathcal{B}_1 = \left\{ \begin{bmatrix} -1 \\ 0 \\ 4 \end{bmatrix}, \begin{bmatrix} 2 \\ 3 \\ 3 \end{bmatrix}, \begin{bmatrix} 1 \\ -1 \\ -2 \end{bmatrix} \right\}$

$\mathcal{B}_2 = \left\{ \begin{bmatrix} 1 \\ 3 \\ -1 \end{bmatrix}, \begin{bmatrix} 0 \\ 1 \\ -1 \end{bmatrix}, \begin{bmatrix} 3 \\ 7 \\ 0 \end{bmatrix} \right\}$

16. $\mathcal{B}_1 = \left\{ \begin{bmatrix} 2 \\ 3 \\ 0 \end{bmatrix}, \begin{bmatrix} 1 \\ -1 \\ 2 \end{bmatrix}, \begin{bmatrix} -4 \\ 1 \\ 5 \end{bmatrix} \right\}$

$\mathcal{B}_2 = \left\{ \begin{bmatrix} 1 \\ 3 \\ 2 \end{bmatrix}, \begin{bmatrix} -2 \\ -5 \\ -4 \end{bmatrix}, \begin{bmatrix} 1 \\ 2 \\ 3 \end{bmatrix} \right\}$

17. $\mathcal{B}_1 = \left\{ \begin{bmatrix} 3 \\ -5 \\ 5 \end{bmatrix}, \begin{bmatrix} -1 \\ 4 \\ -3 \end{bmatrix} \right\} \quad \mathcal{B}_2 = \left\{ \begin{bmatrix} 2 \\ -1 \\ 2 \end{bmatrix}, \begin{bmatrix} 1 \\ 3 \\ -1 \end{bmatrix} \right\}$

18. $\mathcal{B}_1 = \left\{ \begin{bmatrix} 5 \\ -6 \\ 3 \end{bmatrix}, \begin{bmatrix} -1 \\ 4 \\ 3 \end{bmatrix} \right\} \quad \mathcal{B}_2 = \left\{ \begin{bmatrix} 2 \\ -1 \\ 3 \end{bmatrix}, \begin{bmatrix} -3 \\ 5 \\ 0 \end{bmatrix} \right\}$

In Exercises 19–24, find the change of basis matrix from $\mathcal{B}_2$ to $\mathcal{B}_1$.

19. $\mathcal{B}_1 = \left\{ \begin{bmatrix} 2 \\ 1 \end{bmatrix}, \begin{bmatrix} 1 \\ 1 \end{bmatrix} \right\}, \quad \mathcal{B}_2 = \left\{ \begin{bmatrix} 5 \\ 7 \end{bmatrix}, \begin{bmatrix} 3 \\ 4 \end{bmatrix} \right\}$

20. $\mathcal{B}_1 = \left\{ \begin{bmatrix} 2 \\ 1 \end{bmatrix}, \begin{bmatrix} 3 \\ 2 \end{bmatrix} \right\}, \quad \mathcal{B}_2 = \left\{ \begin{bmatrix} 3 \\ 1 \end{bmatrix}, \begin{bmatrix} 7 \\ 2 \end{bmatrix} \right\}$

21. $\mathcal{B}_1 = \left\{ \begin{bmatrix} -1 \\ 1 \\ -1 \end{bmatrix}, \begin{bmatrix} 1 \\ 0 \\ 2 \end{bmatrix}, \begin{bmatrix} -2 \\ 5 \\ 0 \end{bmatrix} \right\}$

$\mathcal{B}_2 = \left\{ \begin{bmatrix} -2 \\ 1 \\ 3 \end{bmatrix}, \begin{bmatrix} 2 \\ 0 \\ 1 \end{bmatrix}, \begin{bmatrix} 4 \\ 1 \\ -1 \end{bmatrix} \right\}$

22. $\mathcal{B}_1 = \left\{ \begin{bmatrix} -1 \\ -3 \\ 1 \end{bmatrix}, \begin{bmatrix} 1 \\ 4 \\ -2 \end{bmatrix}, \begin{bmatrix} -2 \\ -3 \\ -2 \end{bmatrix} \right\}$

$\mathcal{B}_2 = \left\{ \begin{bmatrix} 1 \\ 4 \\ 1 \end{bmatrix}, \begin{bmatrix} 4 \\ 2 \\ 0 \end{bmatrix}, \begin{bmatrix} 3 \\ 1 \\ -2 \end{bmatrix} \right\}$

23. $\mathcal{B}_1 = \left\{ \begin{bmatrix} -1 \\ 4 \\ 2 \end{bmatrix}, \begin{bmatrix} 2 \\ 3 \\ 1 \end{bmatrix} \right\} \quad \mathcal{B}_2 = \left\{ \begin{bmatrix} 1 \\ 7 \\ 3 \end{bmatrix}, \begin{bmatrix} -4 \\ 5 \\ 3 \end{bmatrix} \right\}$

24. $\mathcal{B}_1 = \left\{ \begin{bmatrix} -3 \\ 1 \\ 1 \end{bmatrix}, \begin{bmatrix} 2 \\ -5 \\ 4 \end{bmatrix} \right\} \quad \mathcal{B}_2 = \left\{ \begin{bmatrix} -5 \\ 6 \\ -3 \end{bmatrix}, \begin{bmatrix} 1 \\ -9 \\ 9 \end{bmatrix} \right\}$

25. For $\mathcal{B}_1$ and $\mathcal{B}_2$ in Exercise 13, find $\mathbf{x}_{\mathcal{B}_2}$ if $\mathbf{x}_{\mathcal{B}_1} = \begin{bmatrix} 2 \\ -1 \end{bmatrix}_{\mathcal{B}_1}$.

26. For $\mathcal{B}_1$ and $\mathcal{B}_2$ in Exercise 16, find $\mathbf{x}_{\mathcal{B}_2}$ if $\mathbf{x}_{\mathcal{B}_1} = \begin{bmatrix} 1 \\ 3 \\ -2 \end{bmatrix}_{\mathcal{B}_1}$.

27. For $\mathcal{B}_1$ and $\mathcal{B}_2$ in Exercise 17, find $\mathbf{x}_{\mathcal{B}_2}$ if $\mathbf{x}_{\mathcal{B}_1} = \begin{bmatrix} 2 \\ 5 \end{bmatrix}_{\mathcal{B}_1}$.

28. For $\mathcal{B}_1$ and $\mathcal{B}_2$ in Exercise 19, find $\mathbf{x}_{\mathcal{B}_1}$ if $\mathbf{x}_{\mathcal{B}_2} = \begin{bmatrix} 1 \\ 2 \end{bmatrix}_{\mathcal{B}_2}$.

29. For $\mathcal{B}_1$ and $\mathcal{B}_2$ in Exercise 21, find $\mathbf{x}_{\mathcal{B}_1}$ if $\mathbf{x}_{\mathcal{B}_2} = \begin{bmatrix} -1 \\ 1 \\ 3 \end{bmatrix}_{\mathcal{B}_2}$.

30. For $\mathcal{B}_1$ and $\mathcal{B}_2$ in Exercise 24, find $\mathbf{x}_{\mathcal{B}_1}$ if $\mathbf{x}_{\mathcal{B}_2} = \begin{bmatrix} -2 \\ 3 \end{bmatrix}_{\mathcal{B}_2}$.

31. Suppose that $\mathcal{B}_1 = \{\mathbf{u}_1, \mathbf{u}_2\}$ and $\mathcal{B}_2 = \{\mathbf{u}_2, \mathbf{u}_1\}$ are bases of $\mathbf{R}^2$. Find $\mathbf{x}_{\mathcal{B}_2}$ if $\mathbf{x}_{\mathcal{B}_1} = \begin{bmatrix} a \\ b \end{bmatrix}_{\mathcal{B}_1}$.

32. Suppose that $\mathcal{B}_1 = \{\mathbf{u}_1, \mathbf{u}_2, \mathbf{u}_3\}$ and $\mathcal{B}_2 = \{\mathbf{u}_2, \mathbf{u}_3, \mathbf{u}_1\}$ are bases of $\mathbf{R}^3$. Find $\mathbf{x}_{\mathcal{B}_1}$ if $\mathbf{x}_{\mathcal{B}_2} = \begin{bmatrix} a \\ b \\ c \end{bmatrix}_{\mathcal{B}_2}$.

FIND AN EXAMPLE For Exercises 33–38, find an example that meets the given specifications.

33. A basis $\mathcal{B}$ of $\mathbf{R}^2$ such that $\begin{bmatrix} 1 \\ 3 \end{bmatrix}_{\mathcal{B}} = \begin{bmatrix} -2 \\ 1 \end{bmatrix}$.

34. A basis $\mathcal{B}$ of $\mathbf{R}^3$ such that $\begin{bmatrix} 3 \\ 1 \\ -2 \end{bmatrix}_{\mathcal{B}} = \begin{bmatrix} 1 \\ 2 \\ 5 \end{bmatrix}$.

35. Bases $\mathcal{B}_1$ and $\mathcal{B}_2$ of $\mathbf{R}^2$ such that $\begin{bmatrix} 2 \\ -2 \end{bmatrix}_{\mathcal{B}_1} = \begin{bmatrix} 4 \\ 1 \end{bmatrix}_{\mathcal{B}_2}$.

36. Bases $\mathcal{B}_1$ and $\mathcal{B}_2$ of $\mathbf{R}^3$ such that $\begin{bmatrix} 3 \\ 0 \\ -1 \end{bmatrix}_{\mathcal{B}_1} = \begin{bmatrix} 2 \\ -4 \\ 1 \end{bmatrix}_{\mathcal{B}_2}$.

37. Bases $\mathcal{B}_1$ and $\mathcal{B}_2$ of $\mathbf{R}^2$ with change of basis matrix $C = \begin{bmatrix} 1 & 3 \\ 2 & 7 \end{bmatrix}$ from $\mathcal{B}_1$ to $\mathcal{B}_2$.

38. Bases $\mathcal{B}_1$ and $\mathcal{B}_2$ of $\mathbf{R}^3$ with change of basis matrix $C = \begin{bmatrix} 1 & -1 & 2 \\ -2 & 3 & 0 \\ 1 & 4 & 1 \end{bmatrix}$ from $\mathcal{B}_2$ to $\mathcal{B}_1$.

TRUE OR FALSE For Exercises 39–44, determine if the statement is true or false, and justify your answer.

39. If U is a change of basis matrix between bases $\mathcal{B}_1$ and $\mathcal{B}_2$ of $\mathbf{R}^n$, then U must be an $n \times n$ matrix.

40. A change of basis matrix from one basis of $\mathbf{R}^n$ to another basis of $\mathbf{R}^n$ is unique.

41. If U is a change of basis matrix between bases $\mathcal{B}_1$ and $\mathcal{B}_2$ of $\mathbf{R}^n$, then U must be invertible.

42. Any change of basis matrix must have linearly independent columns.

43. Let $\mathcal{B}_1 = \{\mathbf{u}_1, \mathbf{u}_2\}$ and $\mathcal{B}_2 = \{\mathbf{v}_1, \mathbf{v}_2\}$ be two bases of $\mathbf{R}^2$, and suppose that $\mathbf{u}_1 = a\mathbf{v}_1 + b\mathbf{v}_2$ and $\mathbf{u}_2 = c\mathbf{v}_1 + d\mathbf{v}_2$. Then the change of basis matrix from $\mathcal{B}_1$ to $\mathcal{B}_2$ is $\begin{bmatrix} a & b \\ c & d \end{bmatrix}$.

44. If C_1 is the change of basis matrix from $\mathcal{B}_1$ to $\mathcal{B}_2$ and C_2 is the change of basis matrix from $\mathcal{B}_2$ to $\mathcal{B}_3$, then $C_1 C_2$ is the change of basis matrix from $\mathcal{B}_1$ to $\mathcal{B}_3$.

45. Let $\mathcal{B}$ be a basis. Prove that $[\mathbf{u} + \mathbf{v}]_{\mathcal{B}} = \mathbf{u}_{\mathcal{B}} + \mathbf{v}_{\mathcal{B}}$ for vectors $\mathbf{u}$ and $\mathbf{v}$.

46. Let $\mathcal{B}$ be a basis. Prove that $[c\mathbf{u}]_{\mathcal{B}} = c\mathbf{u}_{\mathcal{B}}$, where $\mathbf{u}$ is a vector and c a scalar.

47. Let $T : \mathbf{R}^n \to \mathbf{R}^n$ be given by $T(\mathbf{x}) = \mathbf{x}_{\mathcal{B}}$ for a basis $\mathcal{B}$. Prove that T is a linear transformation.

48. Let $\mathcal{B}$ be a basis, $c_1, \ldots, c_k$ be scalars, and $\mathbf{u}_1, \ldots, \mathbf{u}_k$ vectors. Prove that

$$c_1[\mathbf{u}_1]_{\mathcal{B}} + \cdots + c_k[\mathbf{u}_k]_{\mathcal{B}} = [c_1\mathbf{u}_1 + \cdots c_k\mathbf{u}_k]_{\mathcal{B}}$$

49. Prove that if $S = \mathbf{R}^n$, then the matrix C in Theorem 6.11 is equal to the matrix $V^{-1}U$ given in Theorem 6.10.

50. Let $\mathcal{B}_1 = \{\mathbf{u}_1, \ldots, \mathbf{u}_n\}$ and $\mathcal{B}_2 = \{\mathbf{v}_1, \ldots, \mathbf{v}_n\}$ be bases for $\mathbf{R}^n$, and set $U = \begin{bmatrix} \mathbf{u}_1 & \cdots & \mathbf{u}_n \end{bmatrix}$ and $V = \begin{bmatrix} \mathbf{v}_1 & \cdots & \mathbf{v}_n \end{bmatrix}$. Show that the change of basis matrix from $\mathcal{B}_1$ to $\mathcal{B}_2$ can be found by extracting the right half of the row reduced echelon form of $\begin{bmatrix} V & U \end{bmatrix}$.

Ⓒ In Exercises 51–56, find the change of basis matrix from $\mathcal{B}_1$ to $\mathcal{B}_2$.

51. $\mathcal{B}_1 = \left\{ \begin{bmatrix} 3 \\ 5 \\ 9 \end{bmatrix}, \begin{bmatrix} 4 \\ -2 \\ 7 \end{bmatrix}, \begin{bmatrix} 2 \\ 11 \\ -6 \end{bmatrix} \right\}$

$\mathcal{B}_2 = \left\{ \begin{bmatrix} 3 \\ 2 \\ -7 \end{bmatrix}, \begin{bmatrix} 2 \\ 5 \\ 8 \end{bmatrix}, \begin{bmatrix} -4 \\ -6 \\ 1 \end{bmatrix} \right\}$

52. $\mathcal{B}_1 = \left\{ \begin{bmatrix} 6 \\ 4 \\ 2 \end{bmatrix}, \begin{bmatrix} 5 \\ 0 \\ -5 \end{bmatrix}, \begin{bmatrix} -3 \\ 3 \\ 2 \end{bmatrix} \right\}$

$\mathcal{B}_2 = \left\{ \begin{bmatrix} 7 \\ -2 \\ -1 \end{bmatrix}, \begin{bmatrix} 4 \\ 3 \\ 9 \end{bmatrix}, \begin{bmatrix} -8 \\ 1 \\ 5 \end{bmatrix} \right\}$

53. $\mathcal{B}_1 = \left\{ \begin{bmatrix} 1 \\ 0 \\ 1 \\ -1 \end{bmatrix}, \begin{bmatrix} 2 \\ -2 \\ -5 \\ 3 \end{bmatrix}, \begin{bmatrix} 5 \\ 0 \\ 0 \\ -2 \end{bmatrix}, \begin{bmatrix} 3 \\ 7 \\ -9 \\ -2 \end{bmatrix} \right\}$

$\mathcal{B}_2 = \left\{ \begin{bmatrix} 3 \\ 4 \\ -2 \\ 1 \end{bmatrix}, \begin{bmatrix} 7 \\ -2 \\ -1 \\ 0 \end{bmatrix}, \begin{bmatrix} 5 \\ -3 \\ 2 \\ 3 \end{bmatrix}, \begin{bmatrix} 4 \\ 0 \\ 3 \\ -1 \end{bmatrix} \right\}$

54. $\mathcal{B}_1 = \left\{ \begin{bmatrix} 7 \\ -2 \\ -2 \\ 3 \end{bmatrix}, \begin{bmatrix} -3 \\ 0 \\ 1 \\ 1 \end{bmatrix}, \begin{bmatrix} 0 \\ 4 \\ -4 \\ 7 \end{bmatrix}, \begin{bmatrix} 5 \\ 4 \\ 3 \\ 1 \end{bmatrix} \right\}$

$\mathcal{B}_2 = \left\{ \begin{bmatrix} 3 \\ 5 \\ 7 \\ -8 \end{bmatrix}, \begin{bmatrix} -3 \\ -4 \\ -5 \\ -6 \end{bmatrix}, \begin{bmatrix} 2 \\ 0 \\ 1 \\ 4 \end{bmatrix}, \begin{bmatrix} 1 \\ 7 \\ 4 \\ 2 \end{bmatrix} \right\}$

55. $\mathcal{B}_1 = \left\{ \begin{bmatrix} -10 \\ 4 \\ 2 \\ -5 \end{bmatrix}, \begin{bmatrix} -5 \\ 5 \\ 0 \\ -4 \end{bmatrix}, \begin{bmatrix} -8 \\ 5 \\ 2 \\ -6 \end{bmatrix} \right\}$

$\mathcal{B}_2 = \left\{ \begin{bmatrix} 2 \\ 1 \\ 0 \\ -1 \end{bmatrix}, \begin{bmatrix} -3 \\ 0 \\ 2 \\ -2 \end{bmatrix}, \begin{bmatrix} 6 \\ -3 \\ 2 \\ 0 \end{bmatrix} \right\}$

56. $\mathcal{B}_1 = \left\{ \begin{bmatrix} 7 \\ -7 \\ -4 \\ 5 \end{bmatrix}, \begin{bmatrix} -5 \\ 3 \\ 4 \\ -5 \end{bmatrix}, \begin{bmatrix} -12 \\ 11 \\ 7 \\ -9 \end{bmatrix} \right\}$

$\mathcal{B}_2 = \left\{ \begin{bmatrix} 3 \\ -2 \\ -1 \\ 2 \end{bmatrix}, \begin{bmatrix} 4 \\ -1 \\ -1 \\ 3 \end{bmatrix}, \begin{bmatrix} -5 \\ 4 \\ 3 \\ -4 \end{bmatrix} \right\}$

6.4 Diagonalization

If D is a diagonal matrix, then it is relatively easy to analyze the behavior of the linear transformation $T(\mathbf{x}) = D\mathbf{x}$ because for $\mathbf{x}$ in $\mathbf{R}^n$,

$$D = \begin{bmatrix} d_{11} & 0 & \cdots & 0 \\ 0 & d_{22} & \cdots & 0 \\ \vdots & \vdots & \ddots & \vdots \\ 0 & 0 & \cdots & d_{nn} \end{bmatrix} \implies D\mathbf{x} = \begin{bmatrix} d_{11}x_1 \\ d_{22}x_2 \\ \vdots \\ d_{nn}x_n \end{bmatrix}$$

In this section we develop a procedure for expressing a square matrix A as the product of three matrices. The process is called *diagonalizing* A, because the middle matrix in the product is diagonal.

DEFINITION 6.12

Definition Diagonalizable Matrix

An $n \times n$ matrix A is **diagonalizable** if there exist $n \times n$ matrices D and P, with D diagonal and P invertible, such that

$$A = PDP^{-1}$$

Because D is a diagonal matrix, expressing $A = PDP^{-1}$ makes it easier to analyze the linear transformation $T(\mathbf{x}) = A\mathbf{x}$. Diagonalizing A also allows for more efficient computation of matrix powers A^2, A^3, ..., which arise in modeling systems that evolve over time. Matrix powers are discussed at the end of the section.

EXAMPLE 1 Let $A = \begin{bmatrix} -2 & 2 \\ -6 & 5 \end{bmatrix}$. Show that if

$$P = \begin{bmatrix} 1 & 2 \\ 2 & 3 \end{bmatrix} \quad \text{and} \quad D = \begin{bmatrix} 2 & 0 \\ 0 & 1 \end{bmatrix}$$

then $A = PDP^{-1}$ and hence A is diagonalizable.

Solution Applying the Quick Formula for the inverse of a 2×2 matrix (Section 3.3), we find that $P^{-1} = \begin{bmatrix} -3 & 2 \\ 2 & -1 \end{bmatrix}$. Thus we have

$$PDP^{-1} = \begin{bmatrix} 1 & 2 \\ 2 & 3 \end{bmatrix} \begin{bmatrix} 2 & 0 \\ 0 & 1 \end{bmatrix} \begin{bmatrix} -3 & 2 \\ 2 & -1 \end{bmatrix} = \begin{bmatrix} -2 & 2 \\ -6 & 5 \end{bmatrix} = A$$

Therefore A is diagonalizable. ∎

We now turn to the problem of how to find matrices P and D to diagonalize a matrix A. The key is to view the problem in terms of changing bases. Suppose that A is an $n \times n$ matrix with linearly independent eigenvectors $\mathbf{u}_1, \ldots, \mathbf{u}_n$, and let

$$\mathcal{B} = \{\mathbf{u}_1, \ldots, \mathbf{u}_n\}$$

Since $\mathcal{B}$ forms a basis for $\mathbf{R}^n$, we know that for every vector $\mathbf{x}$ in $\mathbf{R}^n$ there exists a unique set of scalars $c_1, \ldots, c_n$ such that

▶ Recall that the subscript $\mathcal{B}$ indicates the coordinate vector with respect to the basis $\mathcal{B}$.

$$\mathbf{x} = c_1\mathbf{u}_1 + \cdots + c_n\mathbf{u}_n = \begin{bmatrix} c_1 \\ \vdots \\ c_n \end{bmatrix}_{\mathcal{B}}$$

Since $\mathbf{u}_1, \ldots, \mathbf{u}_n$ are eigenvectors of A, we have

$$\begin{aligned} A\mathbf{x} &= A(c_1\mathbf{u}_1 + \cdots + c_n\mathbf{u}_n) \\ &= c_1 A\mathbf{u}_1 + \cdots + c_n A\mathbf{u}_n \\ &= c_1\lambda_1\mathbf{u}_1 + \cdots + c_n\lambda_n\mathbf{u}_n \\ &= \begin{bmatrix} c_1\lambda_1 \\ \vdots \\ c_n\lambda_n \end{bmatrix}_{\mathcal{B}} \end{aligned}$$

where $\lambda_1, \ldots, \lambda_n$ are the eigenvalues associated with $\mathbf{u}_1, \ldots, \mathbf{u}_n$, respectively. Thus the product $A\mathbf{x}$ is the entries $c_1, \ldots, c_n$ of $\mathbf{x}_{\mathcal{B}}$ multiplied by $\lambda_1, \ldots, \lambda_n$. We can produce the same product using the diagonal matrix

$$D = \begin{bmatrix} \lambda_1 & 0 & \cdots & 0 \\ 0 & \lambda_2 & & \vdots \\ \vdots & & \ddots & 0 \\ 0 & \cdots & 0 & \lambda_n \end{bmatrix}$$

We have

$$D\mathbf{x}_{\mathcal{B}} = \begin{bmatrix} \lambda_1 & 0 & \cdots & 0 \\ 0 & \lambda_2 & & \vdots \\ \vdots & & \ddots & 0 \\ 0 & \cdots & 0 & \lambda_n \end{bmatrix} \begin{bmatrix} c_1 \\ \vdots \\ c_n \end{bmatrix}_{\mathcal{B}} = \begin{bmatrix} c_1\lambda_1 \\ \vdots \\ c_n\lambda_n \end{bmatrix}_{\mathcal{B}} = A\mathbf{x}$$

This shows that we can compute $A\mathbf{x}$ by:

1. Converting $\mathbf{x}$ from the standard basis to $\mathbf{x}_{\mathcal{B}}$, where $\mathcal{B}$ is an eigenvector basis.

2. Computing $D\mathbf{x}_{\mathcal{B}}$, where D is diagonal with diagonal entries $\lambda_1, \ldots, \lambda_n$.

3. Converting $D\mathbf{x}_{\mathcal{B}}$ from the eigenvector basis back to the standard basis.

To implement these steps, we start by forming the matrix

$$P = \begin{bmatrix} \mathbf{u}_1 & \cdots & \mathbf{u}_n \end{bmatrix}$$

where $\mathbf{u}_1, \ldots, \mathbf{u}_n$ are eigenvectors of A that form a basis for $\mathbf{R}^n$. From Section 6.3 we know that to convert from the standard basis to the basis $\mathcal{B}$, we compute

$$\mathbf{x}_{\mathcal{B}} = P^{-1}\mathbf{x}$$

Next, we compute

$$D\mathbf{x}_{\mathcal{B}} = D(P^{-1}\mathbf{x}) = DP^{-1}\mathbf{x}$$

Finally, to convert back to the standard basis, we multiply by P. Hence

$$A\mathbf{x} = P\left(DP^{-1}\mathbf{x}\right) = PDP^{-1}\mathbf{x}$$

Since this holds for all vectors $\mathbf{x}$, it follows that $A = PDP^{-1}$, which diagonalizes A. To sum up, if $\{\mathbf{u}_1, \ldots \mathbf{u}_n\}$ are linearly independent eigenvectors of A with associated eigenvalues $\{\lambda_1, \ldots, \lambda_n\}$, and

$$P = \begin{bmatrix} \mathbf{u}_1 & \cdots & \mathbf{u}_n \end{bmatrix} \quad \text{and} \quad D = \begin{bmatrix} \lambda_1 & 0 & \cdots & 0 \\ 0 & \lambda_2 & & \vdots \\ \vdots & & \ddots & 0 \\ 0 & \cdots & 0 & \lambda_n \end{bmatrix}$$

then $A = PDP^{-1}$.

EXAMPLE 2   Find matrices P and D to diagonalize $A = \begin{bmatrix} 3 & 1 \\ -2 & 0 \end{bmatrix}$.

Solution   To diagonalize A, we need to find the eigenvalues and eigenvectors. Starting with the eigenvalues, we have

$$\det(A - \lambda I_2) = \begin{vmatrix} 3 - \lambda & 1 \\ -2 & -\lambda \end{vmatrix}$$

$$= (3 - \lambda)(-\lambda) - (-2) = \lambda^2 - 3\lambda + 2 = (\lambda - 2)(\lambda - 1)$$

Thus we have eigenvalues $\lambda_1 = 1$ and $\lambda_2 = 2$. Starting with $\lambda_1 = 1$, the homogeneous system $(A - I_2)\mathbf{u} = \mathbf{0}$ has augmented matrix and echelon form

$$\begin{bmatrix} 2 & 1 & 0 \\ -2 & -1 & 0 \end{bmatrix} \overset{R_1 + R_2 \Rightarrow R_2}{\sim} \begin{bmatrix} 2 & 1 & 0 \\ 0 & 0 & 0 \end{bmatrix}$$

Back substitution gives us the associated eigenvector

$$\mathbf{u}_1 = \begin{bmatrix} 1 \\ -2 \end{bmatrix}$$

For $\lambda_2 = 2$, the system $(A - 2I_2)\mathbf{u} = \mathbf{0}$ has augmented matrix and echelon form

$$\begin{bmatrix} 1 & 1 & 0 \\ -2 & -2 & 0 \end{bmatrix} \overset{2R_1 + R_2 \Rightarrow R_2}{\sim} \begin{bmatrix} 1 & 1 & 0 \\ 0 & 0 & 0 \end{bmatrix}$$

This time back substitution gives us the eigenvector

$$\mathbf{u}_2 = \begin{bmatrix} 1 \\ -1 \end{bmatrix}$$

Now we define P and D, with

$$P = \begin{bmatrix} \mathbf{u}_1 & \mathbf{u}_2 \end{bmatrix} = \begin{bmatrix} 1 & 1 \\ -2 & -1 \end{bmatrix} \quad \text{and} \quad D = \begin{bmatrix} \lambda_1 & 0 \\ 0 & \lambda_2 \end{bmatrix} = \begin{bmatrix} 1 & 0 \\ 0 & 2 \end{bmatrix}$$

▶ To check diagonalization computations, we can save the trouble of computing P^{-1} by instead verifying that $PD = AP$.

To check our work, we compute P^{-1} and then the product

$$PDP^{-1} = \begin{bmatrix} 1 & 1 \\ -2 & -1 \end{bmatrix} \begin{bmatrix} 1 & 0 \\ 0 & 2 \end{bmatrix} \begin{bmatrix} -1 & -1 \\ 2 & 1 \end{bmatrix} = \begin{bmatrix} 3 & 1 \\ -2 & 0 \end{bmatrix} = A \qquad \blacksquare$$

We can use the diagonalization formula to construct a matrix A that has specified eigenvalues and eigenvectors.

EXAMPLE 3   Find a 2×2 matrix A that has eigenvalues $\lambda_1 = -1$ and $\lambda_2 = 2$ and corresponding eigenvectors $\mathbf{u}_1 = \begin{bmatrix} 5 \\ 3 \end{bmatrix}$ and $\mathbf{u}_2 = \begin{bmatrix} 3 \\ 2 \end{bmatrix}$.

Solution   We start by defining the diagonalization matrices P and D and multiply to find A. Let

$$P = \begin{bmatrix} \mathbf{u}_1 & \mathbf{u}_2 \end{bmatrix} = \begin{bmatrix} 5 & 3 \\ 3 & 2 \end{bmatrix} \quad \text{and} \quad D = \begin{bmatrix} \lambda_1 & 0 \\ 0 & \lambda_2 \end{bmatrix} = \begin{bmatrix} -1 & 0 \\ 0 & 2 \end{bmatrix}$$

We have $P^{-1} = \begin{bmatrix} 2 & -3 \\ -3 & 5 \end{bmatrix}$ from the Quick Formula for 2×2 matrix inverses, so that

$$A = PDP^{-1} = \begin{bmatrix} 5 & 3 \\ 3 & 2 \end{bmatrix} \begin{bmatrix} -1 & 0 \\ 0 & 2 \end{bmatrix} \begin{bmatrix} 2 & -3 \\ -3 & 5 \end{bmatrix} = \begin{bmatrix} -28 & 45 \\ -18 & 29 \end{bmatrix}$$

To verify that A has the required specifications, we compute

$$A\mathbf{u}_1 = \begin{bmatrix} -28 & 45 \\ -18 & 29 \end{bmatrix} \begin{bmatrix} 5 \\ 3 \end{bmatrix} = \begin{bmatrix} -5 \\ -3 \end{bmatrix} = (-1) \begin{bmatrix} 5 \\ 3 \end{bmatrix} = \lambda_1 \mathbf{u}_1$$

$$A\mathbf{u}_2 = \begin{bmatrix} -28 & 45 \\ -18 & 29 \end{bmatrix} \begin{bmatrix} 3 \\ 2 \end{bmatrix} = \begin{bmatrix} 6 \\ 4 \end{bmatrix} = (2) \begin{bmatrix} 3 \\ 2 \end{bmatrix} = \lambda_2 \mathbf{u}_2$$

Hence $\{\lambda_1, \lambda_2\}$ are eigenvalues of A associated with eigenvectors $\{\mathbf{u}_1, \mathbf{u}_2\}$. ∎

Most square matrices are diagonalizable, but not all. The next theorem tells us exactly when a matrix is diagonalizable.

THEOREM 6.13

An $n \times n$ matrix A is diagonalizable if and only if A has eigenvectors that form a basis for $\mathbf{R}^n$.

Proof   We have seen how to diagonalize A if A has eigenvectors that form a basis for $\mathbf{R}^n$, so half of the proof is done. For the other half, suppose that A is diagonalizable, with

$$A = PDP^{-1} \tag{1}$$

where $\mathbf{p}_1, \ldots, \mathbf{p}_n$ are the columns of P and $d_{11}, \ldots, d_{nn}$ are the diagonal entries of D. Since P is invertible, the columns $\mathbf{p}_1, \ldots, \mathbf{p}_n$ of P are nonzero and linearly independent. Multiplying by P on the right of both sides of (1), we have $AP = PD$. Since column i of AP is equal to $A\mathbf{p}_i$ and column i of PD is equal to $d_{ii}\mathbf{p}_i$, we have

$$A\mathbf{p}_i = d_{ii}\mathbf{p}_i$$

Therefore $\mathbf{p}_i$ is an eigenvector of A with associated eigenvalue d_{ii}. Since P is invertible, A has eigenvectors that form a basis for $\mathbf{R}^n$. ∎

The proof tells us both when A is diagonalizable and how A is diagonalized. The diagonal elements of D must be the eigenvalues and the columns of P must be associated eigenvectors. Of course, there are many possibilities for the associated eigenvectors, so the diagonalization is not unique. However, we now know the only path for finding a diagonalization of an $n \times n$ matrix A.

DIAGONALIZING AN $n \times n$ **MATRIX** A Find the eigenvalues and the associated linearly independent eigenvectors.

- If A has n linearly independent eigenvectors $\mathbf{u}_1, \ldots, \mathbf{u}_n$, then A is diagonalizable, with $P = \begin{bmatrix} \mathbf{u}_1 & \cdots & \mathbf{u}_n \end{bmatrix}$ and the diagonal entries of D given by the corresponding eigenvalues.

- If there are not n linearly independent eigenvectors, then A is not diagonalizable.

Note that the order of the eigenvalues in D does not matter, as long as it matches the order of the corresponding eigenvectors in P.

The next theorem tells us that eigenvectors associated with distinct eigenvalues must be linearly independent. This theorem comes in handy when trying to diagonalize a matrix.

THEOREM 6.14

If $\{\lambda_1, \ldots, \lambda_k\}$ are distinct eigenvalues of a matrix A, then a set of associated eigenvectors $\{\mathbf{u}_1, \ldots, \mathbf{u}_k\}$ is linearly independent.

Proof Suppose that the set of eigenvectors $\{\mathbf{u}_1, \ldots, \mathbf{u}_k\}$ is linearly dependent. Since all eigenvectors are nonzero, it follows from Theorem 2.14 and Exercise 64 of Section 2.3 that one of the eigenvectors can be written as a linear combination of a linearly independent subset of the remaining eigenvectors, with the coefficients nonzero and unique for the given subset. Thus, without loss of generality, let $c_2, \ldots, c_j$ be nonzero scalars such that

$$\mathbf{u}_1 = c_2 \mathbf{u}_2 + \cdots + c_j \mathbf{u}_j \tag{2}$$

Then

$$\begin{aligned} \lambda_1 \mathbf{u}_1 = A\mathbf{u}_1 &= A\left(c_2 \mathbf{u}_2 + \cdots + c_j \mathbf{u}_j\right) \\ &= c_2 A\mathbf{u}_2 + \cdots + c_j A\mathbf{u}_j \\ &= c_2 \lambda_2 \mathbf{u}_2 + \cdots + c_j \lambda_j \mathbf{u}_j \end{aligned}$$

If $\lambda_1 \neq 0$, then

$$\mathbf{u}_1 = c_2 \left(\frac{\lambda_2}{\lambda_1}\right) \mathbf{u}_2 + \cdots + c_j \left(\frac{\lambda_j}{\lambda_1}\right) \mathbf{u}_j$$

which is a different linear combination equal to $\mathbf{u}_1$, contradicting the uniqueness of (2). If $\lambda_1 = 0$, then

$$c_2 \lambda_2 \mathbf{u}_2 + \cdots + c_j \lambda_j \mathbf{u}_j = \mathbf{0}$$

Since $c_2, \ldots, c_j$ and $\lambda_2, \ldots, \lambda_j$ are nonzero, this contradicts the linear independence of $\{\mathbf{u}_2, \ldots, \mathbf{u}_j\}$. Hence either way we reach a contradiction, so it must be that the set of eigenvectors $\{\mathbf{u}_1, \ldots, \mathbf{u}_k\}$ is linearly independent. ∎

EXAMPLE 4 If possible, diagonalize the matrix $A = \begin{bmatrix} 1 & 1 & 1 \\ -2 & -2 & -1 \\ 0 & 0 & -1 \end{bmatrix}$.

Solution We start by finding the eigenvalues by factoring the characteristic polynomial of A,

$$\det(A - \lambda I_3) = -\lambda^3 - 2\lambda^2 - \lambda = -\lambda(\lambda + 1)^2$$

Thus we have eigenvalues $\lambda_1 = 0$ and $\lambda_2 = -1$. Starting with $\lambda_1 = 0$, the augmented matrix for the system $(A - 0I_3)\mathbf{u} = \mathbf{0}$ and the corresponding echelon form are

$$\begin{bmatrix} 1 & 1 & 1 & 0 \\ -2 & -2 & -1 & 0 \\ 0 & 0 & -1 & 0 \end{bmatrix} \overset{\substack{2R_1+R_2 \Rightarrow R_2 \\ R_2+R_3 \Rightarrow R_3}}{\sim} \begin{bmatrix} 1 & 1 & 1 & 0 \\ 0 & 0 & 1 & 0 \\ 0 & 0 & 0 & 0 \end{bmatrix}$$

Back substitution can be used to show that a basis for the eigenspace associated with $\lambda_1 = 0$ is $\left\{ \begin{bmatrix} 1 \\ -1 \\ 0 \end{bmatrix} \right\}$.

▶ Recall that an *eigenspace* is the subspace of eigenvectors associated with a particular eigenvalue.

For $\lambda_2 = -1$, the augmented matrix and echelon form are

$$\begin{bmatrix} 2 & 1 & 1 & 0 \\ -2 & -1 & -1 & 0 \\ 0 & 0 & 0 & 0 \end{bmatrix} \overset{R_1+R_2 \Rightarrow R_2}{\sim} \begin{bmatrix} 2 & 1 & 1 & 0 \\ 0 & 0 & 0 & 0 \\ 0 & 0 & 0 & 0 \end{bmatrix}$$

Back substituting shows that the eigenspace of $\lambda_2 = -1$ has dimension 2 and basis $\left\{ \begin{bmatrix} 1 \\ -2 \\ 0 \end{bmatrix}, \begin{bmatrix} 1 \\ 0 \\ -2 \end{bmatrix} \right\}$. By Theorem 6.14 we know that eigenvectors associated with distinct eigenvalues are linearly independent. Hence the set

$$\left\{ \begin{bmatrix} 1 \\ -1 \\ 0 \end{bmatrix}, \begin{bmatrix} 1 \\ -2 \\ 0 \end{bmatrix}, \begin{bmatrix} 1 \\ 0 \\ -2 \end{bmatrix} \right\}$$

is linearly independent and thus forms a basis for $\mathbf{R}^3$. Since there are two linearly independent eigenvectors associated with $\lambda_2 = -1$, this eigenvalue appears twice in D. We have

$$D = \begin{bmatrix} 0 & 0 & 0 \\ 0 & -1 & 0 \\ 0 & 0 & -1 \end{bmatrix} \quad \text{and} \quad P = \begin{bmatrix} 1 & 1 & 1 \\ -1 & -2 & 0 \\ 0 & 0 & -2 \end{bmatrix}$$

Note that the eigenvalues along the diagonal of D are in columns corresponding to the columns of P containing the associated eigenvectors. We check that D and P are correct by computing

$$PD = \begin{bmatrix} 1 & 1 & 1 \\ -1 & -2 & 0 \\ 0 & 0 & -2 \end{bmatrix} \begin{bmatrix} 0 & 0 & 0 \\ 0 & -1 & 0 \\ 0 & 0 & -1 \end{bmatrix} = \begin{bmatrix} 0 & -1 & -1 \\ 0 & 2 & 0 \\ 0 & 0 & 2 \end{bmatrix}$$

and

$$AP = \begin{bmatrix} 1 & 1 & 1 \\ -2 & -2 & -1 \\ 0 & 0 & -1 \end{bmatrix} \begin{bmatrix} 1 & 1 & 1 \\ -1 & -2 & 0 \\ 0 & 0 & -2 \end{bmatrix} = \begin{bmatrix} 0 & -1 & -1 \\ 0 & 2 & 0 \\ 0 & 0 & 2 \end{bmatrix}$$

Since $AP = PD$ and P is invertible, we have $A = PDP^{-1}$. ∎

The next theorem provides a set of conditions required for a matrix to be diagonalizable.

THEOREM 6.15

▶ See Section 6.1 for the definition of the multiplicity of an eigenvector.

Suppose that an $n \times n$ matrix A has only real eigenvalues. Then A is diagonalizable if and only if the dimension of each eigenspace is equal to the multiplicity of the corresponding eigenvalue.

Proof This theorem follows from things that we already know:

- An $n \times n$ matrix A is diagonalizable if and only if A has n linearly independent eigenvectors. (This is from Theorem 6.13.)

- Each eigenspace has a dimension no greater than the multiplicity of its associated eigenvector. (This is from Theorem 6.6 in Section 6.1.)

- If A is an $n \times n$ matrix, then the multiplicities of the eigenvalues sum to n. (This is because the degree of the characteristic polynomial is equal to n, and the multiplicities must add up to the degree.)

- Vectors from distinct eigenspaces are linearly independent. (This is from Theorem 6.14.)

Pulling these together, we see that A is diagonalizable when the dimension of each eigenspace is as large as possible. Otherwise, there will not be enough linearly independent eigenvectors to form a basis. If the dimension of each eigenspace is as large as possible, then since vectors from distinct eigenspaces are linearly independent, we are assured that there will be enough linearly independent eigenvectors to form a basis for $\mathbf{R}^n$. ∎

EXAMPLE 5 If possible, diagonalize the matrix $A = \begin{bmatrix} 3 & 6 & 5 \\ 3 & 2 & 3 \\ -5 & -6 & -7 \end{bmatrix}$.

Solution The characteristic polynomial for A is

$$\det(A - \lambda I_3) = -(\lambda + 2)^2(\lambda - 2)$$

giving us eigenvalues $\lambda_1 = -2$ and $\lambda_2 = 2$. For $\lambda_1 = -2$, the augmented matrix for the system $(A + 2I_3)\mathbf{u} = \mathbf{0}$ and the corresponding echelon form are

$$\begin{bmatrix} 5 & 6 & 5 & 0 \\ 3 & 4 & 3 & 0 \\ -5 & -6 & -5 & 0 \end{bmatrix} \overset{\substack{-\frac{3}{5}R_1 + R_2 \Rightarrow R_2 \\ R_1 + R_3 \Rightarrow R_3}}{\sim} \begin{bmatrix} 5 & 6 & 5 & 0 \\ 0 & \frac{2}{5} & 0 & 0 \\ 0 & 0 & 0 & 0 \end{bmatrix}$$

Back substitution shows that the eigenspace associated with $\lambda_1 = -2$ has basis $\left\{ \begin{bmatrix} 1 \\ 1 \\ 0 \end{bmatrix} \right\}$.

Clearly this eigenspace has dimension 1, which is less than the multiplicity of $\lambda_1 = -2$. By Theorem 6.15, we know immediately that A is not diagonalizable. ∎

In general, we cannot tell simply by looking at a matrix if it will be diagonalizable. Usually we must determine the eigenvalues and eigenvectors, comparing the multiplicity of the eigenvalues with the dimensions of the eigenspaces. However, there is one special case where we are guaranteed a matrix will be diagonalizable.

THEOREM 6.16

If A is an $n \times n$ matrix with n distinct real eigenvalues, then A is diagonalizable.

Proof Every eigenvalue has an eigenvector, ensuring that the associated eigenspace has dimension at least 1. On the other hand, if the eigenvalues are distinct, then each has multiplicity 1, so that each eigenspace must have dimension 1—there is no other option. Thus, by Theorem 6.15, A is diagonalizable. ∎

EXAMPLE 6 If possible, diagonalize the matrix $A = \begin{bmatrix} 5 & 0 & 0 \\ -4 & 3 & 0 \\ 1 & -3 & -2 \end{bmatrix}$.

Solution Since A is lower triangular, the eigenvalues lie along the main diagonal (Exercise 56, Section 6.1), with $\lambda_1 = 5$, $\lambda_2 = 3$, and $\lambda_3 = -2$. These are distinct, so we know from Theorem 6.16 that A is diagonalizable. Bases for the associated eigenspaces are (computations omitted)

$$\lambda_1 = 5 \Rightarrow \left\{ \begin{bmatrix} 1 \\ -2 \\ 1 \end{bmatrix} \right\}, \qquad \lambda_2 = 3 \Rightarrow \left\{ \begin{bmatrix} 0 \\ -5 \\ 3 \end{bmatrix} \right\}, \qquad \lambda_3 = -2 \Rightarrow \left\{ \begin{bmatrix} 0 \\ 0 \\ 1 \end{bmatrix} \right\}$$

Therefore $A = PDP^{-1}$, with

$$D = \begin{bmatrix} 5 & 0 & 0 \\ 0 & 3 & 0 \\ 0 & 0 & -2 \end{bmatrix} \quad \text{and} \quad P = \begin{bmatrix} 1 & 0 & 0 \\ -2 & -5 & 0 \\ 1 & 3 & 1 \end{bmatrix}$$

∎

Matrix Powers

Suppose that A is diagonalizable, with $A = PDP^{-1}$. Then

$$A^2 = \left(PDP^{-1}\right)\left(PDP^{-1}\right) = PD\left(P^{-1}P\right)DP^{-1} = PD^2 P^{-1}$$

$$A^3 = A\left(A^2\right) = \left(PDP^{-1}\right)\left(PD^2 P^{-1}\right) = PD\left(P^{-1}P\right)D^2 P^{-1} = PD^3 P^{-1}$$

and in general

$$A^k = PD^k P^{-1}$$

Next, note that

$$D = \begin{bmatrix} d_{11} & 0 & \cdots & 0 \\ 0 & d_{22} & & \vdots \\ \vdots & & \ddots & 0 \\ 0 & \cdots & 0 & d_{nn} \end{bmatrix} \implies D^k = \begin{bmatrix} d_{11}^k & 0 & \cdots & 0 \\ 0 & d_{22}^k & & \vdots \\ \vdots & & \ddots & 0 \\ 0 & \cdots & 0 & d_{nn}^k \end{bmatrix}$$

For example, if $D = \begin{bmatrix} -3 & 0 \\ 0 & 2 \end{bmatrix}$, then

$$D^2 = \begin{bmatrix} -3 & 0 \\ 0 & 2 \end{bmatrix}\begin{bmatrix} -3 & 0 \\ 0 & 2 \end{bmatrix} = \begin{bmatrix} 9 & 0 \\ 0 & 4 \end{bmatrix} = \begin{bmatrix} (-3)^2 & 0 \\ 0 & 2^2 \end{bmatrix}$$

$$D^3 = \begin{bmatrix} -3 & 0 \\ 0 & 2 \end{bmatrix}\begin{bmatrix} (-3)^2 & 0 \\ 0 & 2^2 \end{bmatrix} = \begin{bmatrix} (-3)^3 & 0 \\ 0 & 2^3 \end{bmatrix}$$

and so on. Hence we see that while directly calculating A^k can take many computations, calculating D^k is relatively easy.

EXAMPLE 7   Suppose that $A = \begin{bmatrix} \frac{1}{3} & \frac{2}{9} \\ \frac{2}{3} & \frac{7}{9} \end{bmatrix}$. Find P and D so that $A = PDP^{-1}$, and then use this to give a formula for A^k.

Solution   Leaving out the computational details, the eigenvalues and associated eigenvectors of A are

$$\lambda_1 = 1 \Rightarrow \mathbf{u}_1 = \begin{bmatrix} 1 \\ 3 \end{bmatrix}, \qquad \lambda_2 = \frac{1}{9} \Rightarrow \begin{bmatrix} -1 \\ 1 \end{bmatrix}$$

Therefore if

$$P = \begin{bmatrix} 1 & -1 \\ 3 & 1 \end{bmatrix} \quad \text{and} \quad D = \begin{bmatrix} 1 & 0 \\ 0 & \frac{1}{9} \end{bmatrix}$$

then $A = PDP^{-1}$. To compute A^k, we use

$$A^k = PD^k P^{-1} = \begin{bmatrix} 1 & -1 \\ 3 & 1 \end{bmatrix} \begin{bmatrix} 1^k & 0 \\ 0 & \left(\frac{1}{9}\right)^k \end{bmatrix} \begin{bmatrix} \frac{1}{4} & \frac{1}{4} \\ -\frac{3}{4} & \frac{1}{4} \end{bmatrix}$$

$$= \frac{1}{4} \begin{bmatrix} 1 + 3\left(\frac{1}{9}\right)^k & 1 - \left(\frac{1}{9}\right)^k \\ 3 - 3\left(\frac{1}{9}\right)^k & 3 + \left(\frac{1}{9}\right)^k \end{bmatrix}$$

▶ A is an example of a *probability matrix*, because the entries are nonnegative and each column adds to 1. Probability matrices are discussed in Section 3.5.

Since $\left(\frac{1}{9}\right)^k \to 0$ as $k \to \infty$, it follows that $A^k \to \frac{1}{4} \begin{bmatrix} 1 & 1 \\ 3 & 3 \end{bmatrix}$ as $k \to \infty$. ■

| EXERCISES |

In Exercises 1–4, compute A^5 if $A = PDP^{-1}$.

1. $P = \begin{bmatrix} 4 & 3 \\ 1 & 1 \end{bmatrix}$, $D = \begin{bmatrix} 2 & 0 \\ 0 & -1 \end{bmatrix}$

2. $P = \begin{bmatrix} 2 & 1 \\ 7 & 3 \end{bmatrix}$, $D = \begin{bmatrix} 1 & 0 \\ 0 & -3 \end{bmatrix}$

3. $P = \begin{bmatrix} 1 & 3 & 1 \\ 0 & -1 & 2 \\ 0 & 0 & -1 \end{bmatrix}$, $D = \begin{bmatrix} 1 & 0 & 0 \\ 0 & 2 & 0 \\ 0 & 0 & -1 \end{bmatrix}$

4. $P = \begin{bmatrix} 1 & 1 & -1 \\ 1 & 0 & 1 \\ 1 & 0 & 2 \end{bmatrix}$, $D = \begin{bmatrix} 3 & 0 & 0 \\ 0 & 1 & 0 \\ 0 & 0 & 1 \end{bmatrix}$

In Exercises 5–8, find the matrix A that has the given eigenvalues and corresponding eigenvectors.

5. $\lambda_1 = 1 \Rightarrow \left\{ \begin{bmatrix} 2 \\ 3 \end{bmatrix} \right\}$, $\lambda_2 = -1 \Rightarrow \left\{ \begin{bmatrix} 3 \\ 5 \end{bmatrix} \right\}$

6. $\lambda_1 = 3 \Rightarrow \left\{ \begin{bmatrix} 4 \\ 7 \end{bmatrix} \right\}$, $\lambda_2 = 1 \Rightarrow \left\{ \begin{bmatrix} 1 \\ 2 \end{bmatrix} \right\}$

7. $\lambda_1 = -1 \Rightarrow \left\{ \begin{bmatrix} 1 \\ 1 \\ 0 \end{bmatrix} \right\}$, $\lambda_2 = 0 \Rightarrow \left\{ \begin{bmatrix} 1 \\ 2 \\ 1 \end{bmatrix} \right\}$,

$\lambda_3 = 1 \Rightarrow \left\{ \begin{bmatrix} -1 \\ 1 \\ 1 \end{bmatrix} \right\}$

8. $\lambda_1 = 2 \Rightarrow \left\{ \begin{bmatrix} 1 \\ 3 \\ 1 \end{bmatrix} \right\}$, $\lambda_2 = 1 \Rightarrow \left\{ \begin{bmatrix} 2 \\ 1 \\ -1 \end{bmatrix}, \begin{bmatrix} 0 \\ 2 \\ 1 \end{bmatrix} \right\}$

In Exercises 9–18, diagonalize the given matrix, if possible.

9. $\begin{bmatrix} 1 & -2 \\ 0 & 1 \end{bmatrix}$

10. $\begin{bmatrix} -2 & 2 \\ 0 & 0 \end{bmatrix}$

11. $\begin{bmatrix} 7 & -8 \\ 4 & -5 \end{bmatrix}$

12. $\begin{bmatrix} 7 & -10 \\ 2 & -2 \end{bmatrix}$

13. $\begin{bmatrix} 1 & 2 & 1 \\ 0 & -3 & -2 \\ 2 & 4 & 2 \end{bmatrix}$

14. $\begin{bmatrix} 4 & -1 & -2 \\ -6 & 3 & 4 \\ 8 & -2 & -4 \end{bmatrix}$

15. $\begin{bmatrix} 0 & 1 & -1 \\ 1 & 0 & 1 \\ 1 & -1 & 2 \end{bmatrix}$

16. $\begin{bmatrix} 3 & 5 & 3 \\ -5 & -7 & -3 \\ 3 & 3 & 1 \end{bmatrix}$

17. $\begin{bmatrix} 1 & 0 & 0 & 1 \\ 0 & 2 & 0 & 0 \\ 0 & 0 & 3 & 0 \\ 0 & 0 & 0 & 4 \end{bmatrix}$

18. $\begin{bmatrix} 1 & 0 & 0 & 0 \\ 1 & 1 & 0 & 0 \\ 1 & 0 & 0 & 0 \\ -1 & 0 & 1 & -1 \end{bmatrix}$

In Exercises 19–22, compute A^{1000} for the given matrix A.

19. $A = \begin{bmatrix} -3 & 4 \\ -2 & 3 \end{bmatrix}$

20. $A = \begin{bmatrix} 5 & -4 \\ 2 & -1 \end{bmatrix}$

21. $A = \begin{bmatrix} 7 & -8 \\ 4 & -5 \end{bmatrix}$

22. $A = \begin{bmatrix} 7 & -10 \\ 2 & -2 \end{bmatrix}$

23. Suppose that a 4×4 diagonalizable matrix has two distinct eigenvalues, one with an eigenspace of dimension 2. What is the dimension of the other eigenspace?

24. Suppose that a 7×7 diagonalizable matrix has three distinct eigenvalues, one with an eigenspace of dimension 1 and another with an eigenspace of dimension 2. What is the dimension of the third eigenspace?

FIND AN EXAMPLE For Exercises 25–30, find an example that meets the given specifications.

25. A 2×2 matrix that is diagonalizable but not invertible.

26. A 3×3 matrix that is diagonalizable but not invertible.

27. A 2×2 matrix that is invertible but is not diagonalizable.

28. A 3×3 matrix that is invertible but is not diagonalizable.

29. A 3×3 diagonalizable (but not diagonal) matrix that has three distinct eigenvalues.

30. A 3×3 diagonalizable (but not diagonal) matrix that has two distinct eigenvalues.

TRUE OR FALSE For Exercises 31–38, determine if the statement is true or false, and justify your answer.

31. Suppose a square matrix A has only real eigenvalues. If each eigenspace of A has dimension equal to the multiplicity of the associated eigenvalue, then A is diagonalizable.

32. If an $n \times n$ matrix A has n distinct eigenvectors, then A is diagonalizable.

33. If A is not invertible, then A is not diagonalizable.

34. If A is diagonalizable, then so is A^T.

35. If A is a diagonalizable $n \times n$ matrix, then rank$(A) = n$.

36. If A and B are diagonalizable $n \times n$ matrices, then so is AB.

37. If A and B are diagonalizable $n \times n$ matrices, then so is $A + B$.

38. If A is a diagonalizable $n \times n$ matrix, then there exist eigenvectors of A that form a basis for $\mathbf{R}^n$.

39. Suppose that $\lambda_1 \neq \lambda_2$ are eigenvalues of a 2×2 matrix A with associated eigenvectors $\mathbf{u}_1$ and $\mathbf{u}_2$. Prove that $\det(U) \neq 0$ for $U = \begin{bmatrix} \mathbf{u}_1 & \mathbf{u}_2 \end{bmatrix}$.

40. Prove that if A is diagonalizable and $c \neq 0$ is a scalar, then cA is also diagonalizable.

41. Prove that if A is diagonalizable, then there are infinitely many distinct matrices P and D such that $A = PDP^{-1}$.

42. Suppose that A is an $n \times n$ matrix with eigenvectors that form a basis for $\mathbf{R}^n$. Prove that there exists an invertible matrix Q such that QAQ^{-1} is a diagonal matrix.

43. Prove that if A is diagonalizable, then so is A^T.

44. Suppose that A is a matrix that can be diagonalized using matrices P and D as in Definition 6.12. Prove that $\det(A) = \det(D)$.

45. Suppose that A and B are $n \times n$ matrices that can both be diagonalized using the same matrix P. Prove that $AB = BA$.

46. Suppose that A is a diagonalizable matrix with distinct nonzero eigenvalues. Prove that A^2 has positive eigenvalues.

© In Exercises 47–50, diagonalize the given matrix, if possible.

47. $\begin{bmatrix} 3 & 0 & -2 & -1 \\ -1 & 2 & 5 & 4 \\ 6 & 0 & -5 & -3 \\ -6 & 0 & 4 & 2 \end{bmatrix}$

48. $\begin{bmatrix} 0 & -1 & -1 & -1 \\ 10 & 3 & 2 & -4 \\ -8 & -2 & -1 & -5 \\ -4 & 2 & 2 & 0 \end{bmatrix}$

49. $\begin{bmatrix} 2 & -1 & 0 & 1 & 0 \\ -4 & 3 & 0 & -5 & -4 \\ 5 & -2 & -1 & 1 & -1 \\ -6 & 2 & 0 & -4 & -2 \\ 2 & -1 & 0 & 1 & 0 \end{bmatrix}$

50. $\begin{bmatrix} 3 & -5 & 3 & 2 & -3 \\ 5 & -7 & 3 & 1 & 1 \\ 6 & -6 & 2 & -1 & 5 \\ 0 & 0 & 0 & 1 & 3 \\ 0 & 0 & 0 & 0 & 2 \end{bmatrix}$

6.5 Complex Eigenvalues

▶ This section is optional. However, complex numbers and eigenvalues are revisited in optional Section 11.4 and optional Section 11.5.

Up until now we have only considered eigenvalues that are real numbers. However, some characteristic polynomials have roots that are not real numbers. For instance, the matrix

$$A = \begin{bmatrix} 3 & -4 \\ 1 & 3 \end{bmatrix}$$

has characteristic polynomial

$$|A - \lambda I_2| = (3 - \lambda)^2 + 4$$

This characteristic polynomial has no real roots. If we are willing to expand our horizons, we can consider complex roots. Although previously we avoided them, complex eigenvalues and eigenvectors are useful. We will get to some applications later in this section and in the next. We start here with a brief review of the properties of complex numbers.

Complex Numbers

We are not going to fully develop the complex numbers, but instead just focus on the aspects that are needed later. If a and b are real numbers, then a typical complex number has the form

$$z = a + ib$$

Definition Real Part, Imaginary Part

where i satisfies $i^2 = -1$, making i the square root of -1. Here a is called the **real part** of z, denoted by $\text{Re}(z)$, and b is the **imaginary part**, denoted by $\text{Im}(z)$. (Note that both the real and imaginary parts are real numbers.) The set of all complex numbers is denoted by **C**.

To add complex numbers $z_1 = a_1 + ib_1$ and $z_2 = a_2 + ib_2$, we just add the real parts and the imaginary parts separately,

$$z_1 + z_2 = (a_1 + ib_1) + (a_2 + ib_2) = (a_1 + a_2) + i(b_1 + b_2)$$

For example, if $z_1 = 3 + i$ and $z_2 = 2 + 3i$, then

$$z_1 + z_2 = (3 + i) + (2 + 3i) = (3 + 2) + i(1 + 3) = 5 + 4i$$

Adding complex numbers is similar to adding vectors in $\mathbf{R}^2$, with each component added separately.

We can also represent complex numbers geometrically just as we do vectors in $\mathbf{R}^2$, with $\text{Re}(z)$ on the x-axis and $\text{Im}(z)$ on the y-axis. For example, z_1 and z_2 from above are shown in Figure 1.

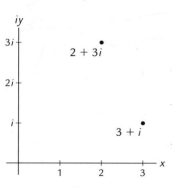

Figure 1 $z_1 = 3 + i$ and $z_2 = 2 + 3i$.

The product of complex numbers is found by multiplying term by term and then simplifying using the identity $i^2 = -1$,

$$z_1 z_2 = (a_1 + ib_1)(a_2 + ib_2)$$
$$= a_1 a_2 + ia_1 b_2 + ia_2 b_1 + i^2 b_1 b_2 = (a_1 a_2 - b_1 b_2) + i(a_1 b_2 + a_2 b_1)$$

For our two complex numbers $z_1 = 3 + i$ and $z_2 = 2 + 3i$, we have

$$z_1 z_2 = (3 + i)(2 + 3i) = \big((3)(2) - (1)(3)\big) + i\big((3)(3) + (2)(1)\big) = 3 + 11i$$

An alternate way to represent a complex number is to use polar coordinates. Define the **modulus** of a complex number $z = a + ib$ by

Definition Modulus

$$|z| = \sqrt{a^2 + b^2}.$$

The modulus generalizes the absolute value to complex numbers and gives the distance from z to the origin. The **argument** of z, denoted by $\arg(z)$, is the angle θ (in radians) in the counter clockwise direction from the positive x-axis to the ray from the origin to z (see Figure 2). Note that the argument is not unique, because we can always add or subtract multiples of 2π.

Definition Argument

If $r = |z|$ and $\theta = \arg(z)$, then we can express z in polar form as

$$z = r\big(\cos(\theta) + i\sin(\theta)\big)$$

For instance, if $r = 5$ and $\theta = \pi/3$, then we can convert to rectangular coordinates by evaluating, with

$$z = 5\big(\cos(\pi/3) + i\sin(\pi/3)\big) = \frac{5}{2} + i\frac{5\sqrt{3}}{2}$$

Converting from rectangular to polar coordinates is depicted in Figure 3. For $z = a + ib$, we set $r = |z| = \sqrt{a^2 + b^2}$ and have

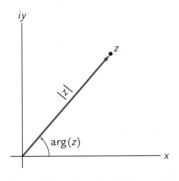

Figure 2 $|z|$ and $\arg(z)$.

$$\tan(\theta) = \frac{b}{a} \quad \Longrightarrow \quad \theta = \tan^{-1}\left(\frac{b}{a}\right)$$

For example, in the case of $z_1 = 3 + i$, we have

$$r = \sqrt{3^2 + 1^2} = \sqrt{10} \quad \text{and} \quad \theta = \tan^{-1}\left(\frac{1}{3}\right) \approx 0.3218 \text{ radians}$$

An interesting formula arises when multiplying complex numbers written in polar form. If

$$z_1 = r_1\big(\cos(\theta_1) + i\sin(\theta_1)\big) \quad \text{and} \quad z_2 = r_2\big(\cos(\theta_2) + i\sin(\theta_2)\big)$$

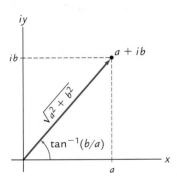

Figure 3 Converting from rectangular to polar coordinates.

then we have

$$\begin{aligned} z_1 z_2 &= r_1\big(\cos(\theta_1) + i\sin(\theta_1)\big) \cdot r_2\big(\cos(\theta_2) + i\sin(\theta_2)\big) \\ &= r_1 r_2 \big\{ \big(\cos(\theta_1)\cos(\theta_2) - \sin(\theta_1)\sin(\theta_2)\big) + i\big(\cos(\theta_1)\sin(\theta_2) + \cos(\theta_2)\sin(\theta_1)\big) \big\} \\ &= r_1 r_2 \big(\cos(\theta_1 + \theta_2) + i\sin(\theta_1 + \theta_2)\big) \end{aligned}$$

with the last line following from trigonometric identities. This formula tells us that

$$|z_1 z_2| = |z_1||z_2| \quad \text{and} \quad \arg(z_1 z_2) = \arg(z_1) + \arg(z_2)$$

Applying this repeatedly (by induction) to $z = r\big(\cos(\theta) + i\sin(\theta)\big)$, we find that for each positive integer k,

$$z^k = r^k\big(\cos(k\theta) + i\sin(k\theta)\big)$$

In the special case where $z = \cos(\theta) + i\sin(\theta)$ (that is, $r = 1$), this yields *DeMoivre's Formula*,

$$\big(\cos(\theta) + i\sin(\theta)\big)^k = \cos(k\theta) + i\sin(k\theta)$$

For $z = a + ib$, the exponential function e^x extends to the complex numbers by the definition

$$e^{a+ib} = e^a \big(\cos(b) + i\sin(b)\big)$$

Note that if $z = a$ is real, then $e^z = e^a$ reduces to the usual exponential function. On the other hand, if $z = ib$ is purely imaginary, then

$$e^{ib} = \cos(b) + i\sin(b)$$

and $|e^{ib}| = 1$.

Definition Complex Conjugate   The **complex conjugate** of $z = a + ib$ is given by $\bar{z} = a - ib$. Complex conjugation distributes across addition and multiplication, so that if z and w are complex numbers, then

$$\overline{z + w} = \bar{z} + \bar{w} \quad \text{and} \quad \overline{zw} = \bar{z} \cdot \bar{w}$$

One interesting consequence of these properties is that for polynomials with real coefficients, complex roots come in conjugate pairs. That is, if

$$f(z) = a_n z^n + \cdots + a_1 z + a_0$$

has real coefficients and $f(z_0) = 0$, then $f(\bar{z}_0) = 0$. To see why, note that $\bar{x} = x$ for any real number x. If $f(z_0) = 0$, then

$$
\begin{aligned}
0 = \bar{0} = \overline{f(z_0)} &= \overline{a_n z_0^n + \cdots + a_1 z_0 + a_0} \\
&= \bar{a}_n \bar{z}_0^n + \cdots + \bar{a}_1 \bar{z}_0 + \bar{a}_0 \\
&= a_n (\bar{z}_0)^n + \cdots + a_1 \bar{z}_0 + a_0 = f(\bar{z}_0)
\end{aligned}
$$

which tells us that $f(\bar{z}_0) = 0$ as well. For example, if $f(z) = z^2 + 2z + 4$, then the solutions to $f(z) = 0$ can be found by applying the quadratic formula,

$$z = \frac{-2 \pm \sqrt{2^2 - 4(1)(4)}}{2(1)} = \frac{-2 \pm \sqrt{-12}}{2} = -1 \pm i\sqrt{3}$$

Therefore we have two solutions, $z = -1 + i\sqrt{3}$ and the conjugate $\bar{z} = -1 - i\sqrt{3}$.

One benefit of expanding from the real numbers to the complex numbers is that we can completely factor polynomials. Specifically, any polynomial $f(z)$ of degree n with real or complex coefficients can be factored completely to

$$f(z) = c(z - z_1)(z - z_2) \cdots (z - z_n)$$

where $c, z_1, \ldots z_n$ are complex numbers.

Given an $n \times n$ matrix A, we know that the characteristic polynomial will have degree n. When working in the complex numbers, the characteristic polynomial has exactly n roots (counting multiplicities), so that A must have exactly n eigenvalues (again, counting multiplicities).

Complex Eigenvalues and Eigenvectors

Now that we have refreshed our knowledge of complex numbers, let's return to our opening problem.

EXAMPLE 1   Find the eigenvalues and associated eigenvectors for the matrix

$$A = \begin{bmatrix} 3 & -4 \\ 1 & 3 \end{bmatrix}$$

Solution   We know that $\det(A - \lambda I_2) = (3 - \lambda)^2 + 4$. Setting this equal to 0 and solving for λ, we have

$$(3 - \lambda)^2 = -4 \quad \Longrightarrow \quad 3 - \lambda = \pm\sqrt{-4} = \pm 2i \quad \Longrightarrow \quad \lambda = 3 \pm 2i$$

In general, it is algebraically messy to find complex eigenvectors by hand. However, it is manageable for 2×2 matrices. For the eigenvector $\lambda_1 = 3 - 2i$, we have

$$A - \lambda_1 I_2 = A - (3 - 2i) I_2 = \begin{bmatrix} 2i & -4 \\ 1 & 2i \end{bmatrix}$$

The augmented matrix of $(A - \lambda_1 I_2)\mathbf{u} = \mathbf{0}$ and the corresponding echelon form are

$$\begin{bmatrix} 2i & -4 & 0 \\ 1 & 2i & 0 \end{bmatrix} \begin{smallmatrix} \frac{1}{2}i R_1 + R_2 \Rightarrow R_2 \\ \sim \end{smallmatrix} \begin{bmatrix} 2i & -4 & 0 \\ 0 & 0 & 0 \end{bmatrix}$$

The echelon form is equivalent to the equation $2ix_1 - 4x_2 = 0$. A nontrivial solution is $x_1 = 2$ and $x_2 = i$, which gives us the eigenvector $\mathbf{u}_1 = \begin{bmatrix} 2 \\ i \end{bmatrix}$. Similar calculations applied to $\lambda_2 = 3 + 2i$ can be used to produce the associated eigenvector $\mathbf{u}_2 = \begin{bmatrix} 2 \\ -i \end{bmatrix}$. ∎

▶ For a vector $\mathbf{z}$, the complex conjugate $\bar{\mathbf{z}}$ means that we take the complex conjugate for each entry of $\mathbf{z}$. An analogous definition holds for $\overline{A}$, the complex conjugate of the matrix A.

The eigenvalues in Example 1 are a conjugate pair, with $\lambda_2 = \overline{\lambda}_1$. The corresponding eigenvectors are similarly related, with $\mathbf{u}_2 = \overline{\mathbf{u}}_1$. This is true for any square matrix with real entries.

THEOREM 6.17

Suppose that A is a real matrix with eigenvalue λ and associated eigenvector $\mathbf{u}$. Then $\overline{\lambda}$ is also an eigenvalue of A, with associated eigenvector $\overline{\mathbf{u}}$.

Proof   If A is a real matrix, then the characteristic polynomial has real coefficients. Previously we showed that complex roots of polynomials with real coefficients come in conjugate pairs. Thus, if λ is a complex eigenvalue of a matrix A, then so is $\overline{\lambda}$.

Next suppose that $\mathbf{u}$ is an eigenvector of A associated with λ. Since A has real entries, we have $A = \overline{A}$, so that

$$A\overline{\mathbf{u}} = \overline{A}\,\overline{\mathbf{u}} = \overline{A\mathbf{u}} = \overline{\lambda\mathbf{u}} = \overline{\lambda}\,\overline{\mathbf{u}}$$

Hence $\overline{\mathbf{u}}$ is an eigenvector of A associated with eigenvalue $\overline{\lambda}$. ∎

EXAMPLE 2   Find the eigenvalues and associated eigenvectors for the matrix

$$A = \begin{bmatrix} -1 & 3 & -4 \\ -2 & 3 & -4 \\ 1 & 1 & 3 \end{bmatrix}$$

Solution   Starting with the characteristic polynomial, we have

$$\det(A - \lambda I_3) = -\lambda^3 + 5\lambda^2 - 17\lambda + 13 = -(\lambda - 1)(\lambda^2 - 4\lambda + 13)$$

Thus one eigenvalue is $\lambda_1 = 1$. Applying the quadratic formula to the quadratic term shows the other eigenvalues are $\lambda_2 = 2 + 3i$ and $\lambda_3 = \overline{\lambda}_2 = 2 - 3i$.

For the eigenvectors associated with $\lambda_1 = 1$, the augmented matrix and corresponding echelon form are

$$
\begin{bmatrix} -2 & 3 & -4 & 0 \\ -2 & 2 & -4 & 0 \\ 1 & 1 & 2 & 0 \end{bmatrix}
\overset{\substack{-R_1+R_2 \Rightarrow R_2 \\ \frac{1}{2}R_1+R_3 \Rightarrow R_3 \\ \frac{5}{2}R_2+R_3 \Rightarrow R_3}}{\sim}
\begin{bmatrix} -2 & 3 & -4 & 0 \\ 0 & -1 & 0 & 0 \\ 0 & 0 & 0 & 0 \end{bmatrix}
$$

Back substitution yields the eigenvector $\mathbf{u}_1 = \begin{bmatrix} 2 \\ 0 \\ -1 \end{bmatrix}$. For $\lambda_2 = 2 + 3i$, we have

▶ The row operations are

$$-\frac{2}{3+3i}R_1 + R_2 \Rightarrow R_2,$$

$$\frac{1}{3+3i}R_1 + R_3 \Rightarrow R_3,$$

$$-\frac{1+3i}{4}R_2 + R_3 \Rightarrow R_3$$

$$
\begin{bmatrix} (-3-3i) & 3 & -4 & 0 \\ -2 & (1-3i) & -4 & 0 \\ 1 & 1 & (1-3i) & 0 \end{bmatrix}
\sim
\begin{bmatrix} (-3-3i) & 3 & -4 & 0 \\ 0 & -2i & -(8+4i)/3 & 0 \\ 0 & 0 & 0 & 0 \end{bmatrix}
$$

After back substitution and scaling, we find that $\mathbf{u}_2 = \begin{bmatrix} -1+5i \\ -2+4i \\ 3 \end{bmatrix}$.

There is no need for row operations to find $\mathbf{u}_3$. Since $\lambda_3 = \overline{\lambda}_2$, we know from Theorem 6.17 that $\mathbf{u}_3 = \overline{\mathbf{u}}_2 = \begin{bmatrix} -1-5i \\ -2-4i \\ 3 \end{bmatrix}$. ∎

Rotation–Dilation Matrices

One application of complex eigenvalues and eigenvectors is in analyzing the behavior of a special class of 2×2 matrices. Suppose that $\mathbf{x}_1, \ldots, \mathbf{x}_8$ are the eight vectors distributed evenly around the unit circle as shown in Figure 4(a). Now define the matrix

$$A = \begin{bmatrix} 1 & -2 \\ 2 & 1 \end{bmatrix}$$

Figure 4(b) shows the vectors $A\mathbf{x}_1, \ldots, A\mathbf{x}_8$.

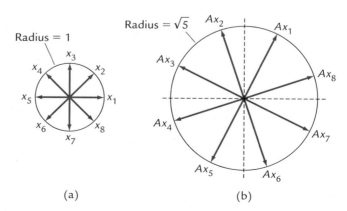

(a)   (b)

Figure 4 (a) The vectors $\mathbf{x}_1, \ldots, \mathbf{x}_8$ are evenly spaced around the unit circle. (b) The vectors $\mathbf{x}_1, \ldots, \mathbf{x}_8$ after multiplication by A, which rotates each by $63.43°$ and dilates each by $\sqrt{5}$. (NOTE: Figures are not drawn to scale.)

Comparing $\mathbf{x}_i$ with $A\mathbf{x}_i$, we see that multiplication by A causes each $\mathbf{x}_i$ to be rotated and dilated by the same amount. It turns out that the same thing will happen for any vector $\mathbf{x}$ when multiplied by A.

This rotation–dilation behavior happens with any real matrix of the form

$$A = \begin{bmatrix} a & -b \\ b & a \end{bmatrix}$$

To see why, first note that if $\mathbf{x} = \begin{bmatrix} x_1 \\ x_2 \end{bmatrix}$, then

$$A\mathbf{x} = \begin{bmatrix} a & -b \\ b & a \end{bmatrix} \begin{bmatrix} x_1 \\ x_2 \end{bmatrix} = \begin{bmatrix} ax_1 - bx_2 \\ bx_1 + ax_2 \end{bmatrix}$$

It can be shown that $\lambda = a + ib$ is an eigenvalue of A (see Exercise 47). Next note that if we let $x = x_1 + ix_2$, then

$$\lambda x = (a + ib)(x_1 + ix_2) = (ax_1 - bx_2) + i(bx_1 + ax_2)$$

Thus the components of $A\mathbf{x}$ match the real and imaginary parts of λx, so the two products can be viewed as equivalent. From the properties of the products of complex numbers, we know that

$$\arg(\lambda x) = \arg(\lambda) + \arg(x) \qquad \text{(Rotation of } x \text{ in } \mathbf{C} \text{ by the angle } \arg(\lambda))$$

$$|\lambda x| = |\lambda||x| = \sqrt{a^2 + b^2}|x| \qquad \text{(Dilation of } x \text{ in } \mathbf{C} \text{ by the multiple } |\lambda|)$$

Therefore the eigenvalue tells us the amount of rotation and dilation induced by A. Returning to our matrix

$$A = \begin{bmatrix} 1 & -2 \\ 2 & 1 \end{bmatrix}$$

we have $\lambda = 1 + 2i$, so that $A\mathbf{x}$ will produce

$$\text{Rotation by } \arg(\lambda) = \tan^{-1}(2/1) \approx 1.107 \text{ radians}$$
$$\text{Dilation by } |\lambda| = \sqrt{1^2 + 2^2} = \sqrt{5}$$

Note that this is consistent with Figure 4.

EXAMPLE 3   Determine the rotation and dilation that result from multiplying $\mathbf{x}$ in $\mathbf{R}^2$ by

$$A = \begin{bmatrix} 7 & -4 \\ 4 & 7 \end{bmatrix}$$

Solution An eigenvalue of A is $\lambda = 7 + 4i$, so that we have

$$\text{Rotation by } \tan^{-1}(4/7) \approx 0.5191 \text{ radians}$$
$$\text{Dilation by } \sqrt{7^2 + 4^2} = \sqrt{65} \qquad \blacksquare$$

The Hidden Rotation–Dilation Matrix

As it happens, *any* 2×2 real matrix with complex eigenvalues has a rotation–dilation hidden within it. Finding this rotation–dilation requires a procedure reminiscent of diagonalization. We start by illustrating using the matrix from the beginning of this section,

$$A = \begin{bmatrix} 3 & -4 \\ 1 & 3 \end{bmatrix}$$

We have previously shown that $\mathbf{u} = \begin{bmatrix} 2 \\ i \end{bmatrix}$ is an eigenvector of A. Now form the matrix

$$P = \begin{bmatrix} \text{Re}(\mathbf{u}) & \text{Im}(\mathbf{u}) \end{bmatrix} = \begin{bmatrix} 2 & 0 \\ 0 & 1 \end{bmatrix}$$

where $\text{Re}(\mathbf{u})$ and $\text{Im}(\mathbf{u})$ denote the vectors formed by taking the real and imaginary parts of each component of $\mathbf{u}$. Then we compute

$$P^{-1}AP = \tfrac{1}{2}\begin{bmatrix} 1 & 0 \\ 0 & 2 \end{bmatrix}\begin{bmatrix} 3 & -4 \\ 1 & 3 \end{bmatrix}\begin{bmatrix} 2 & 0 \\ 0 & 1 \end{bmatrix} = \begin{bmatrix} 3 & -2 \\ 2 & 3 \end{bmatrix} = B$$

Thus $A = PBP^{-1}$, where B is a rotation–dilation matrix. Note that the first row of B corresponds to the real and imaginary parts of the eigenvalue $\lambda = 3 - 2i$ of A associated with $\mathbf{u}$. This example is generalized in the next theorem.

THEOREM 6.18

Let A be a nonzero real 2×2 matrix with complex eigenvalue $\lambda = a - ib$ and associated eigenvector $\mathbf{u}$. If $P = \begin{bmatrix} \text{Re}(\mathbf{u}) & \text{Im}(\mathbf{u}) \end{bmatrix}$, then

$$A = PBP^{-1}$$

where $B = \begin{bmatrix} a & -b \\ b & a \end{bmatrix}$ is a rotation–dilation matrix.

We do not give a proof here, but most of the pieces required are covered in Exercise 49. The form PBP^{-1} suggests viewing the transformation $A\mathbf{x}$ as the composition of transformations. The first is a change to the basis $\{\text{Re}(\mathbf{u}), \text{Im}(\mathbf{u})\}$, the second is a rotation–dilation, and the third is a change back to the standard basis. Note that the rotation–dilation is not applied to $\mathbf{x}$, but to the coordinate vector of $\mathbf{x}$ relative to the basis $\{\text{Re}(\mathbf{u}), \text{Im}(\mathbf{u})\}$.

EXAMPLE 4 Find the hidden rotation–dilation matrix within

$$A = \begin{bmatrix} 1 & 5 \\ -2 & 3 \end{bmatrix}$$

Solution The characteristic polynomial is $\det(A - \lambda I_2) = \lambda^2 - 4\lambda + 13$. A quick application of the quadratic formula reveals that one of the eigenvalues is $\lambda = 2 - 3i$. (What is the other?) The usual procedure yields an associated eigenvector $\mathbf{u} = \begin{bmatrix} 1 + 3i \\ 2 \end{bmatrix} = \begin{bmatrix} 1 \\ 2 \end{bmatrix} + i \begin{bmatrix} 3 \\ 0 \end{bmatrix}$. Applying Theorem 6.18 with $\lambda = 2 - 3i$, we have

$$P = \begin{bmatrix} 1 & 3 \\ 2 & 0 \end{bmatrix} \quad \text{and} \quad B = \begin{bmatrix} 2 & -3 \\ 3 & 2 \end{bmatrix}$$

Clearly B has the form of a rotation–dilation matrix. We can check our calculations with the computation

$$PBP^{-1} = \begin{bmatrix} 1 & 3 \\ 2 & 0 \end{bmatrix}\begin{bmatrix} 2 & -3 \\ 3 & 2 \end{bmatrix}\left(-\tfrac{1}{6}\begin{bmatrix} 0 & -3 \\ -2 & 1 \end{bmatrix}\right) = \begin{bmatrix} 1 & 5 \\ -2 & 3 \end{bmatrix} = A \quad \blacksquare$$

| EXERCISES |

In Exercises 1–2, suppose that $z_1 = 5 - 2i$ and $z_2 = 3 + 4i$ are both complex numbers.

1. Compute each of the following:

(a) $z_1 + z_2$

(b) $2z_1 + 3z_2$

(c) $z_1 - z_2$

(d) $z_1 z_2$

2. Compute each of the following:

(a) $-4z_1$

(b) $3z_1 + 2z_2$

(c) $z_2 - z_1$

(d) $(-5z_1)(3z_2)$

In Exercises 3–8, find the eigenvalues and a basis for each eigenspace for the given matrix.

3. $\begin{bmatrix} 3 & 1 \\ -5 & 1 \end{bmatrix}$

4. $\begin{bmatrix} 1 & -1 \\ 2 & 3 \end{bmatrix}$

5. $\begin{bmatrix} 1 & -2 \\ 1 & 3 \end{bmatrix}$

6. $\begin{bmatrix} 1 & 3 \\ -3 & 1 \end{bmatrix}$

7. $\begin{bmatrix} 4 & 2 \\ -1 & 2 \end{bmatrix}$

8. $\begin{bmatrix} 5 & 2 \\ -5 & -1 \end{bmatrix}$

In Exercises 9–14, determine the rotation and dilation for the given matrix.

9. $\begin{bmatrix} 2 & -1 \\ 1 & 2 \end{bmatrix}$

10. $\begin{bmatrix} 3 & -2 \\ 2 & 3 \end{bmatrix}$

11. $\begin{bmatrix} 1 & -1 \\ 1 & 1 \end{bmatrix}$

12. $\begin{bmatrix} 3 & -4 \\ 4 & 3 \end{bmatrix}$

13. $\begin{bmatrix} 4 & 3 \\ -3 & 4 \end{bmatrix}$

14. $\begin{bmatrix} 5 & 2 \\ -2 & 5 \end{bmatrix}$

In Exercises 15–20, find the rotation–dilation matrix within the given matrix.

15. $\begin{bmatrix} 3 & 1 \\ -5 & 1 \end{bmatrix}$

16. $\begin{bmatrix} 1 & -1 \\ 2 & 3 \end{bmatrix}$

17. $\begin{bmatrix} 1 & -2 \\ 1 & 3 \end{bmatrix}$

18. $\begin{bmatrix} 1 & 3 \\ -3 & 1 \end{bmatrix}$

19. $\begin{bmatrix} 4 & 2 \\ -1 & 2 \end{bmatrix}$

20. $\begin{bmatrix} 5 & 2 \\ -5 & -1 \end{bmatrix}$

21. Suppose that some of the roots of a degree 5 polynomial with real coefficients are 2, $1 + 2i$, and $3 - i$. What are the other roots, and what are the multiplicities of all roots?

22. Suppose that some of the roots of a degree 7 polynomial with real coefficients are -3, $-2 + i$, $5 + i$, and i. What are the other roots, and what are the multiplicities of all roots?

FIND AN EXAMPLE For Exercises 23–30, find an example that meets the given specifications.

23. A complex number z such that $|z| = 3$ and $\mathrm{Re}(z) = 2\mathrm{Im}(z)$.

24. A complex number z such that $|z| = 5$ and $2\mathrm{Re}(z) = -\mathrm{Im}(z)$.

25. A rotation–dilation matrix A that rotates vectors by $90°$ and dilates vectors by 2.

26. A rotation–dilation matrix A that rotates vectors by $-45°$ and dilates vectors by $\frac{1}{2}$.

27. A 2×2 matrix A that is not itself a rotation–dilation matrix but does have the rotation–dilation matrix $B = \begin{bmatrix} 1 & -2 \\ 2 & 1 \end{bmatrix}$ hidden within it.

28. A 2×2 matrix A that is not itself a rotation–dilation matrix but does have the rotation–dilation matrix $B = \begin{bmatrix} 3 & -1 \\ 1 & 3 \end{bmatrix}$ hidden within it.

29. A 2×2 matrix A that has complex entries but real eigenvalues.

30. A 4×4 matrix with real entries and only complex eigenvalues.

TRUE OR FALSE For Exercises 31–40, determine if the statement is true or false, and justify your answer.

31. If z and w are complex numbers, then $|zw| = |z||w|$.

32. A 2×2 matrix A with real eigenvalues has a rotation–dilation matrix hidden within it.

33. If A is a square matrix with real entries and complex eigenvectors, then A has complex eigenvalues.

34. If z is a complex number, then $|z| = |\bar{z}|$.

35. If z and w are complex numbers, then $|z + w| = |z| + |w|$.

36. The amount of dilation imparted by a rotation–dilation matrix A is equal to $|\lambda|$, where λ is an eigenvalue of A.

37. If z is a complex number, then $z = \bar{\bar{z}}$.

38. If A is a rotation–dilation matrix, then the angle between a vector $\mathbf{x}_0$ and $A\mathbf{x}_0$ is the same for all nonzero $\mathbf{x}_0$ in $\mathbf{R}^2$.

39. If $A = \begin{bmatrix} a & -b \\ b & a \end{bmatrix}$ is a rotation–dilation matrix with eigenvalues λ_1 and λ_2, then $|\lambda_1| = |\lambda_2|$.

40. If A is a 2×2 matrix with complex entries, then the eigenvalues of A cannot be complex conjugates.

41. If z and w are complex numbers, prove that

(a) $\overline{z + w} = \bar{z} + \bar{w}$

(b) $\overline{zw} = \bar{z} \cdot \bar{w}$

42. Suppose z is a complex number.

(a) Prove that $z\bar{z} = |z|^2$.

(b) If w is also complex, use (a) to show that $\dfrac{w}{z} = \dfrac{w\bar{z}}{|z|^2}$ for $z \neq 0$.

(c) Use part (b) to simplify $\dfrac{2 + i}{4 - 3i}$ to the form $a + ib$, where a and b are real.

43. If c is a complex scalar and $\mathbf{v}$ is a vector with complex entries, prove that $\bar{c} \cdot \bar{\mathbf{v}} = \overline{c\mathbf{v}}$.

44. If A is a matrix with complex entries and $\mathbf{v}$ is a vector with complex entries, prove that $\bar{A}\bar{\mathbf{v}} = \overline{A\mathbf{v}}$.

45. Suppose that λ is complex and t is real. Prove that $\overline{e^{\lambda t}} = e^{\overline{\lambda} t}$.

46. If z is complex, prove that

(a) $\frac{1}{2}(z + \bar{z}) = \operatorname{Re}(z)$

(b) $\frac{1}{2i}(z - \bar{z}) = \operatorname{Im}(z)$

47. Prove that $\lambda = a + ib$ is an eigenvalue of the matrix $A = \begin{bmatrix} a & -b \\ b & a \end{bmatrix}$.

48. Prove that if n is odd and A is a real $n \times n$ matrix, then there exists a nonzero vector $\mathbf{u}$ such that $A\mathbf{u} = c\mathbf{u}$, where c is a real number.

49. In this exercise we prove that $AP = PC$ for the matrices in Theorem 6.18. (This combined with Exercise 50 proves the theorem.)

(a) Show that $A(\operatorname{Re}(\mathbf{u})) = \operatorname{Re}(A\mathbf{u})$ and $A(\operatorname{Im}(\mathbf{u})) = \operatorname{Im}(A\mathbf{u})$. (HINT: Recall that A is a real matrix.)

(b) Use (a) and the identity $\mathbf{u} = \operatorname{Re}(\mathbf{u}) + i\operatorname{Im}(\mathbf{u})$ to show that

$$A(\operatorname{Re}(\mathbf{u})) = a\operatorname{Re}(\mathbf{u}) + b\operatorname{Im}(\mathbf{u}),$$
$$A(\operatorname{Im}(\mathbf{u})) = -b\operatorname{Re}(\mathbf{u}) + a\operatorname{Im}(\mathbf{u})$$

(HINT: Recall that $\mathbf{u}$ is an eigenvector of A with eigenvalue $\lambda = a - ib$.)

(c) Apply (b) to show that the columns of AP are the same as the columns of PC, and conclude $AP = PC$.

50. In this exercise we prove that the columns of P in Theorem 6.18 are linearly independent and therefore P is invertible. The proof is by contradiction: Suppose that $\operatorname{Re}(\mathbf{u})$ and $\operatorname{Im}(\mathbf{u})$ are linearly dependent. Then there exists a real scalar c such that $\operatorname{Re}(\mathbf{u}) = c\operatorname{Im}(\mathbf{u})$.

(a) Prove that

$$\mathbf{u} = \operatorname{Re}(\mathbf{u}) + i\operatorname{Im}(\mathbf{u}) = (c + i)\operatorname{Im}(\mathbf{u})$$

and from this show $\lambda \mathbf{u} = \lambda(c + i)\operatorname{Im}(\mathbf{u})$.

(b) Show that $\operatorname{Re}(\lambda\mathbf{u}) = c\operatorname{Im}(\lambda\mathbf{u})$ by evaluating $A(\operatorname{Re}(\mathbf{u}))$ and $A(c\operatorname{Im}(\mathbf{u}))$ and setting the results equal to each other. (HINT: Use (a) from Exercise 49 and that $\mathbf{u}$ is an eigenvector with eigenvalue λ.)

(c) Show that $\lambda\mathbf{u} = \operatorname{Re}(\lambda\mathbf{u}) + i\operatorname{Im}(\lambda\mathbf{u})$, and combine this with the result from (b) to prove that $\lambda\mathbf{u} = (c + i)\operatorname{Im}(\lambda\mathbf{u})$.

(d) Prove that $\lambda(c + i)\operatorname{Im}(\mathbf{u}) = (c + i)\operatorname{Im}(\lambda\mathbf{u})$. Show that $\lambda\operatorname{Im}(\mathbf{u}) = \operatorname{Im}(\lambda\mathbf{u})$, and explain why this implies λ is a real number.

(e) Explain why λ being a real number is a contradiction, and from this complete the proof.

Ⓒ In Exercises 51–54, find the complex eigenvalues and a basis for each associated eigenspace for the given matrix.

51. $A = \begin{bmatrix} 1 & 3 & 2 \\ 4 & 2 & 1 \\ 0 & 5 & -2 \end{bmatrix}$

52. $A = \begin{bmatrix} 4 & -3 & 1 \\ 2 & 2 & 7 \\ 1 & -4 & 2 \end{bmatrix}$

53. $A = \begin{bmatrix} 0 & 5 & 3 & -1 \\ 2 & 2 & -1 & 2 \\ 4 & 0 & 2 & 4 \\ 3 & 9 & 7 & 9 \end{bmatrix}$

54. $A = \begin{bmatrix} 2 & -5 & 3 & 1 \\ 7 & 0 & -3 & -4 \\ 5 & 2 & 1 & 1 \\ 6 & 2 & 0 & 1 \end{bmatrix}$

6.6 Systems of Differential Equations

▶ This section is optional and can be omitted without loss of continuity.

In a variety of applications, systems of equations arise involving one or more functions and the derivatives of those functions. One example can be found in a simplified model of the concentration of insulin and glucose in an individual. Insulin is a hormone that reduces glucose concentrations.

Suppose that $y_1(t)$ and $y_2(t)$ give the deviation from normal of insulin and glucose concentrations, respectively. Then the rates of change $y_1'(t)$ and $y_2'(t)$ of insulin and glucose concentrations are related by

$$y_1'(t) = ay_1(t) + by_2(t)$$
$$y_2'(t) = cy_1(t) + dy_2(t)$$

where a, b, c, and d are constants. This is an example of a system of linear differential equations. We will return to this example shortly, after developing a method for finding the solutions to such systems.

▶ Here we assume a basic familiarity with differential equations, and so only provide a brief background.

If $y = y(t)$, one of the simplest *differential equations* is

$$y' = ay \tag{1}$$

where a is a constant. Setting $y = ce^{at}$ for any constant c, we have $y' = ace^{at}$, so that our function y satisfies $y' = ay$. Our function is called a *solution* to the differential equation, and in fact, the only solutions to this differential equation have this form.

In this section we describe how to find the solutions to a system of linear first-order differential equations, which has the form

$$
\begin{aligned}
y_1' &= a_{11}y_1 + a_{12}y_2 + a_{13}y_3 + \cdots + a_{1n}y_n \\
y_2' &= a_{21}y_1 + a_{22}y_2 + a_{23}y_3 + \cdots + a_{2n}y_n \\
y_3' &= a_{31}y_1 + a_{32}y_2 + a_{33}y_3 + \cdots + a_{3n}y_n \\
&\vdots \\
y_n' &= a_{n1}y_1 + a_{n2}y_2 + a_{n3}y_3 + \cdots + a_{nn}y_n
\end{aligned}
\tag{2}
$$

Here we assume that $y_1 = y_1(t), \ldots, y_n = y_n(t)$ are each differentiable functions. The system is linear because the functions are linearly related, and it is first-order because only the first derivative appears. If we denote

$$
\mathbf{y} = \begin{bmatrix} y_1 \\ \vdots \\ y_n \end{bmatrix}, \quad
\mathbf{y}' = \begin{bmatrix} y_1' \\ \vdots \\ y_n' \end{bmatrix}, \quad \text{and} \quad
A = \begin{bmatrix} a_{11} & a_{12} & \cdots & a_{1n} \\ a_{21} & a_{22} & \cdots & a_{2n} \\ \vdots & \vdots & \ddots & \vdots \\ a_{n1} & a_{n2} & \cdots & a_{nn} \end{bmatrix}
$$

then the system (2) can be expressed compactly as $\mathbf{y}' = A\mathbf{y}$. This matrix equation resembles the differential equation (1), which suggests that a solution to our system might have the form

$$
\mathbf{y} = \begin{bmatrix} u_1 e^{\lambda t} \\ u_2 e^{\lambda t} \\ \vdots \\ u_n e^{\lambda t} \end{bmatrix} = e^{\lambda t}\mathbf{u}
$$

If $\mathbf{y}$ is so defined, then $\mathbf{y}' = \lambda e^{\lambda t}\mathbf{u} = e^{\lambda t}(\lambda\mathbf{u})$ and $A\mathbf{y} = A(e^{\lambda t}\mathbf{u}) = e^{\lambda t}(A\mathbf{u})$. Since $e^{\lambda t} \neq 0$, this tells us that $\mathbf{y}$ is a solution to $\mathbf{y}' = A\mathbf{y}$ exactly when

$$A\mathbf{u} = \lambda\mathbf{u}$$

That is, $\mathbf{y} = e^{\lambda t}\mathbf{u}$ is a solution when λ is an eigenvalue of A with associated eigenvector $\mathbf{u}$.

In many cases, an $n \times n$ matrix A will have n linearly independent eigenvectors $\mathbf{u}_1, \ldots, \mathbf{u}_n$ with associated eigenvalues $\lambda_1, \ldots, \lambda_n$. If we form the linear combination

$$\mathbf{y} = c_1 e^{\lambda_1 t} \mathbf{u}_1 + \cdots + c_n e^{\lambda_n t} \mathbf{u}_n$$

where $c_1, \ldots, c_n$ are constants, then $\mathbf{y}' = c_1 \lambda_1 e^{\lambda_1 t} \mathbf{u}_1 + \cdots + c_n \lambda_n e^{\lambda_n t} \mathbf{u}_n$ and

$$\begin{aligned}
A\mathbf{y} &= A\left(c_1 e^{\lambda_1 t} \mathbf{u}_1 + \cdots + c_n e^{\lambda_n t} \mathbf{u}_n\right) \\
&= c_1 e^{\lambda_1 t} A\mathbf{u}_1 + \cdots + c_n e^{\lambda_n t} A\mathbf{u}_n \\
&= c_1 \lambda_1 e^{\lambda_1 t} \mathbf{u}_1 + \cdots + c_n \lambda_n e^{\lambda_n t} \mathbf{u}_n = \mathbf{y}'
\end{aligned}$$

Thus $\mathbf{y}$ is a solution to $\mathbf{y}' = A\mathbf{y}$. It turns out that all solutions to this type of system of differential equations will have this form. The set of all solutions is called the **general solution** for the system. This is summarized in the next theorem.

THEOREM 6.19

Suppose that $\mathbf{y}' = A\mathbf{y}$ is a first-order linear system of differential equations. If A is an $n \times n$ diagonalizable matrix, then the general solution to the system is given by

$$\mathbf{y} = c_1 e^{\lambda_1 t} \mathbf{u}_1 + \cdots + c_n e^{\lambda_n t} \mathbf{u}_n$$

where $\mathbf{u}_1, \ldots, \mathbf{u}_n$ are n linearly independent eigenvectors with associated eigenvalues $\lambda_1, \ldots, \lambda_n$, and $c_1, \ldots, c_n$ are constants.

Note that if A is diagonalizable, then there must be n linearly independent eigenvectors. Also, the eigenvalues may be repeated to reflect multiplicities.

EXAMPLE 1   The concentrations of insulin and glucose in an individual interact with each other and vary over time. A mathematical model for the concentrations of insulin and glucose in an individual is given by the system

$$\begin{aligned}
y_1' &= -0.05 y_1 + 0.225 y_2 \\
y_2' &= -0.3 y_1 - 0.65 y_2
\end{aligned}$$

Find the general solution for this system.

Solution The coefficient matrix is $A = \begin{bmatrix} -0.05 & 0.225 \\ -0.3 & -0.65 \end{bmatrix}$. Applying our standard methods, we find that the eigenvalues and eigenvectors are

$$\lambda_1 = -0.5 \;\Rightarrow\; \mathbf{u}_1 = \begin{bmatrix} -1 \\ 2 \end{bmatrix}, \qquad \lambda_2 = -0.2 \;\Rightarrow\; \mathbf{u}_2 = \begin{bmatrix} 3 \\ -2 \end{bmatrix}$$

By Theorem 6.19, the general solution is

$$\mathbf{y} = c_1 e^{-0.5t} \mathbf{u}_1 + c_2 e^{-0.2t} \mathbf{u}_2 = c_1 e^{-0.5t} \begin{bmatrix} -1 \\ 2 \end{bmatrix} + c_2 e^{-0.2t} \begin{bmatrix} 3 \\ -2 \end{bmatrix}$$

The functions giving the insulin and glucose concentrations are

$$\begin{aligned}
y_1 &= -c_1 e^{-0.5t} + 3c_2 e^{-0.2t} \\
y_2 &= 2c_1 e^{-0.5t} - 2c_2 e^{-0.2t}
\end{aligned}$$

■

Arms Races

After the end of World War I, Lewis F. Richardson (who pioneered the use of mathematics in meteorology) proposed a model to describe the evolution of an arms race between two countries. Here we consider a simplified version of this model. Let $y_1 = y_1(t)$ and $y_2 = y_2(t)$ denote the quantity of arms held by two different nations. The derivatives y_1' and y_2' represent the rate of change in the size of each nation's arsenal. Each nation is concerned about security against the other and acquires arms in proportion to those held by its opponent. There is also a cost of acquiring arms, which tends to reduce the rate of additions to each nation's arsenal in proportion to arsenal size. These factors are incorporated into the system of differential equations

$$\begin{aligned} y_1' &= -dy_1 + ey_2 \\ y_2' &= fy_1 - gy_2 \end{aligned} \tag{3}$$

The constants d, e, f, and g are all positive and depend on the particular situation, with e and f dictated by the degree of fear that each country has of the other and d and g the level of aversion to additional spending on arms in each country.

EXAMPLE 2   Suppose that we have two countries in an arms race modeled by the system of differential equations

$$\begin{aligned} y_1' &= -3y_1 + 2y_2 \\ y_2' &= 3y_1 - 2y_2 \end{aligned}$$

Find the general solution for this system. Then find the formula for $y_1(t)$ and $y_2(t)$ if $y_1(0) = 5$ and $y_2(0) = 15$.

Solution   The coefficient matrix for this system is $A = \begin{bmatrix} -3 & 2 \\ 3 & -2 \end{bmatrix}$. Applying our usual methods for finding the eigenvalues and eigenvectors, we have

$$\lambda_1 = -5 \ \Rightarrow \ \mathbf{u}_1 = \begin{bmatrix} -1 \\ 1 \end{bmatrix}, \qquad \lambda_2 = 0 \ \Rightarrow \ \mathbf{u}_2 = \begin{bmatrix} 2 \\ 3 \end{bmatrix}$$

Therefore, by Theorem 6.19, the general solution for this system is

$$\mathbf{y} = c_1 e^{-5t}\mathbf{u}_1 + c_2 e^0 \mathbf{u}_2 = c_1 e^{-5t} \begin{bmatrix} -1 \\ 1 \end{bmatrix} + c_2 \begin{bmatrix} 2 \\ 3 \end{bmatrix}$$

Extracting the two functions y_1 and y_2, we have

$$\begin{aligned} y_1 &= -c_1 e^{-5t} + 2c_2 \\ y_2 &= c_1 e^{-5t} + 3c_2 \end{aligned}$$

Evaluating these functions at $t = 0$ and using the equations $y_1(0) = 5$ and $y_2(0) = 15$ yield the system

$$\begin{aligned} -c_1 + 2c_2 &= 5 \\ c_1 + 3c_2 &= 15 \end{aligned}$$

This linear system has unique solution $c_1 = 3$ and $c_2 = 4$. Hence

$$\begin{aligned} y_1 &= -3e^{-5t} + 8 \\ y_2 &= 3e^{-5t} + 12 \end{aligned} \quad \blacksquare$$

The next example shows what we do if the coefficient matrix for a first-order linear system has repeated real eigenvalues.

EXAMPLE 3   Find the general solution for the system

$$\begin{aligned}
y_1' &= 2y_1 + y_2 - y_3 \\
y_2' &= 2y_1 + 3y_2 - 2y_3 \\
y_3' &= -3y_1 - 3y_2 + 4y_3
\end{aligned}$$

Solution   Here the coefficient matrix is given by

$$A = \begin{bmatrix} 2 & 1 & -1 \\ 2 & 3 & -2 \\ -3 & -3 & 4 \end{bmatrix}$$

The characteristic polynomial for A is $-(\lambda - 7)(\lambda - 1)^2$, and bases for the eigenspaces are

$$\lambda_1 = 7 \implies \text{Basis: } \left\{ \begin{bmatrix} 1 \\ 2 \\ -3 \end{bmatrix} \right\}, \qquad \lambda_2 = 1 \implies \text{Basis: } \left\{ \begin{bmatrix} 1 \\ 0 \\ 1 \end{bmatrix}, \begin{bmatrix} 1 \\ -1 \\ 0 \end{bmatrix} \right\}$$

Although $\lambda_2 = 1$ has multiplicity 2, since the eigenspace also has dimension 2, Theorem 6.19 still applies. (The case where an eigenspace has less than maximal dimension is more complicated and not included here.) We just repeat the term corresponding to $\lambda_2 = 1$ twice, once for each of the linearly independent eigenvectors. The general solution is

$$\mathbf{y} = c_1 e^{7t} \begin{bmatrix} 1 \\ 2 \\ -3 \end{bmatrix} + c_2 e^t \begin{bmatrix} 1 \\ 0 \\ 1 \end{bmatrix} + c_3 e^t \begin{bmatrix} 1 \\ -1 \\ 0 \end{bmatrix}$$

Writing the individual functions, we have

$$\begin{aligned}
y_1 &= c_1 e^{7t} + (c_2 + c_3)e^t \\
y_2 &= 2c_1 e^{7t} - c_3 e^t \\
y_3 &= -3c_1 e^{7t} + c_2 e^t
\end{aligned}$$ ∎

Complex Eigenvalues

Suppose that our system of linear differential equations has a real coefficient matrix A with complex eigenvalues. We can express the general solution just as we did with real eigenvalues. However, when λ and associated eigenvector $\mathbf{u}$ are both complex, the product $e^{\lambda t}\mathbf{u}$ typically is as well. It is generally preferable to have solutions free of complex terms, and with a bit of extra thought we can.

To get us started, we recall a few properties of complex numbers. (All are given in Section 6.5.)

- $e^{a+ib} = e^a\big(\cos(b) + i\sin(b)\big)$ • $e^{\bar{\lambda}t} = \overline{e^{\lambda t}}$
- $\frac{1}{2}(\bar{z} + z) = \operatorname{Re}(z)$ • $\frac{1}{2}i(\bar{z} - z) = \operatorname{Im}(z)$

Now let A be a real matrix with complex eigenvalue $\lambda = a + ib$ and associated eigenvector $\mathbf{u}$. Then $\bar{\lambda} = a - ib$ is also an eigenvalue and has associated eigenvector $\bar{\mathbf{u}}$. Instead of taking linear combinations of $e^{\lambda t}\mathbf{u}$ and $e^{\bar{\lambda}t}\bar{\mathbf{u}}$ for the general solution, we take linear combinations of

$$\mathbf{y}_1 = \frac{1}{2}\big(e^{\bar{\lambda}t}\bar{\mathbf{u}} + e^{\lambda t}\mathbf{u}\big) = \operatorname{Re}\big(e^{\lambda t}\mathbf{u}\big)$$

$$\mathbf{y}_2 = \frac{1}{2}i\big(e^{\bar{\lambda}t}\bar{\mathbf{u}} - e^{\lambda t}\mathbf{u}\big) = \operatorname{Im}\big(e^{\lambda t}\mathbf{u}\big)$$

Note that both $\text{Re}(e^{\lambda t}\mathbf{u})$ and $\text{Im}(e^{\lambda t}\mathbf{u})$ are real-valued, so that the general solution will be made up of real-valued functions. To find $\text{Re}(e^{\lambda t}\mathbf{u})$ and $\text{Im}(e^{\lambda t}\mathbf{u})$, we compute

$$
\begin{aligned}
e^{\lambda t}\mathbf{u} &= e^{at+ibt}\mathbf{u} \\
&= e^{at}(\cos(bt) + i\sin(bt))(\text{Re}(\mathbf{u}) + i\text{Im}(\mathbf{u})) \\
&= e^{at}(\cos(bt)\text{Re}(\mathbf{u}) - \sin(bt)\text{Im}(\mathbf{u})) + ie^{at}(\sin(bt)\text{Re}(\mathbf{u}) + \cos(bt)\text{Im}(\mathbf{u}))
\end{aligned}
$$

Separating the real and imaginary parts, we have

$$
\begin{aligned}
\mathbf{y}_1 &= e^{at}(\cos(bt)\text{Re}(\mathbf{u}) - \sin(bt)\text{Im}(\mathbf{u})) \\
\mathbf{y}_2 &= e^{at}(\sin(bt)\text{Re}(\mathbf{u}) + \cos(bt)\text{Im}(\mathbf{u}))
\end{aligned}
\tag{4}
$$

EXAMPLE 4   Find the general solution for the system

$$
\begin{aligned}
y_1' &= 6y_1 - 5y_2 \\
y_2' &= 5y_1 - 2y_2
\end{aligned}
$$

Solution Here the coefficient matrix is

$$
A = \begin{bmatrix} 6 & -5 \\ 5 & -2 \end{bmatrix}
$$

The characteristic polynomial for this matrix is $\det(A - \lambda I_2) = \lambda^2 - 4\lambda + 13$. Applying the quadratic formula and the usual matrix manipulations yields the eigenvalues and eigenvectors

$$
\lambda_1 = 2 + 3i \;\Rightarrow\; \mathbf{u}_1 = \begin{bmatrix} 4 + 3i \\ 5 \end{bmatrix}, \qquad \lambda_2 = 2 - 3i \;\Rightarrow\; \mathbf{u}_2 = \begin{bmatrix} 4 - 3i \\ 5 \end{bmatrix}
$$

The two eigenvalues are a complex conjugate pair, so we use the formulas for $\mathbf{y}_1$ and $\mathbf{y}_2$ given in (4) to find

$$
\mathbf{y}_1 = e^{2t}(\cos(3t)\text{Re}(\mathbf{u}_1) - \sin(3t)\text{Im}(\mathbf{u}_1)) = e^{2t}\left(\cos(3t)\begin{bmatrix}4\\5\end{bmatrix} - \sin(3t)\begin{bmatrix}3\\0\end{bmatrix} \right)
$$

$$
\mathbf{y}_2 = e^{2t}(\sin(3t)\text{Re}(\mathbf{u}_1) + \cos(3t)\text{Im}(\mathbf{u}_1)) = e^{2t}\left(\sin(3t)\begin{bmatrix}4\\5\end{bmatrix} + \cos(3t)\begin{bmatrix}3\\0\end{bmatrix} \right)
$$

Hence the general solution is

$$
\begin{aligned}
\mathbf{y} &= c_1\mathbf{y}_1 + c_2\mathbf{y}_2 \\
&= c_1 e^{2t}\left(\cos(3t)\begin{bmatrix}4\\5\end{bmatrix} - \sin(3t)\begin{bmatrix}3\\0\end{bmatrix} \right) + c_2 e^{2t}\left(\sin(3t)\begin{bmatrix}4\\5\end{bmatrix} + \cos(3t)\begin{bmatrix}3\\0\end{bmatrix} \right)
\end{aligned}
$$

The individual functions are

$$
\begin{aligned}
y_1 &= \big((4c_1 + 3c_2)\cos(3t) + (4c_2 - 3c_1)\sin(3t)\big)e^{2t} \\
y_2 &= \big(5c_1\cos(3t) + 5c_2\sin(3t)\big)e^{2t}
\end{aligned}
$$

■

The next example features a system with a combination of real and complex eigenvalues.

EXAMPLE 5 Find the general solution for the system

$$y_1' = -10y_1 + 6y_2 - 3y_3$$
$$y_2' = -12y_1 + 6y_2 - 5y_3$$
$$y_3' = 8y_1 - 4y_2 + 3y_3$$

Solution The coefficient matrix is

$$A = \begin{bmatrix} -10 & 6 & -3 \\ -12 & 6 & -5 \\ 8 & -4 & 3 \end{bmatrix}$$

and the characteristic polynomial is

$$\det(A - \lambda I_3) = -\lambda^3 - \lambda^2 - 4\lambda - 4 = -\lambda^2(\lambda + 1) - 4(\lambda + 1) = -(\lambda^2 + 4)(\lambda + 1)$$

giving us eigenvalues $\lambda = -1, \ \pm 2i$. The associated eigenvectors are

$$\lambda_1 = -1 \Rightarrow \mathbf{u}_1 = \begin{bmatrix} -1 \\ -1 \\ 1 \end{bmatrix}, \ \lambda_2 = 2i \Rightarrow \mathbf{u}_2 = \begin{bmatrix} 9 - 3i \\ 12 - 2i \\ -8 \end{bmatrix}, \ \lambda_3 = -2i \Rightarrow \mathbf{u}_3 = \begin{bmatrix} 9 + 3i \\ 12 + 2i \\ -8 \end{bmatrix}$$

We treat the real and complex eigenvalues separately and form a linear combination of the components at the end. For $\lambda_1 = -1$, we have

$$\mathbf{y}_1 = e^{-t}\mathbf{u}_1 = e^{-t} \begin{bmatrix} -1 \\ -1 \\ 1 \end{bmatrix}$$

For the complex conjugate pair $\pm 2i$, we have

$$\mathbf{y}_2 = e^0 \big(\cos(2t)\text{Re}(\mathbf{u}_2) - \sin(2t)\text{Im}(\mathbf{u}_2)\big) = \cos(2t) \begin{bmatrix} 9 \\ 12 \\ -8 \end{bmatrix} - \sin(2t) \begin{bmatrix} -3 \\ -2 \\ 0 \end{bmatrix}$$

$$\mathbf{y}_3 = e^0 \big(\sin(2t)\text{Re}(\mathbf{u}_2) + \cos(2t)\text{Im}(\mathbf{u}_2)\big) = \sin(2t) \begin{bmatrix} 9 \\ 12 \\ -8 \end{bmatrix} + \cos(2t) \begin{bmatrix} -3 \\ -2 \\ 0 \end{bmatrix}$$

The general solution is then

$$\mathbf{y} = c_1\mathbf{y}_1 + c_2\mathbf{y}_2 + c_3\mathbf{y}_3$$

for $\mathbf{y}_1, \mathbf{y}_2$, and $\mathbf{y}_3$ above. ∎

| EXERCISES |

In Exercises 1–10, the coefficient matrix for a system of linear differential equations of the form $\mathbf{y}' = A\mathbf{y}$ has the given eigenvalues and eigenspace bases. Find the general solution for the system.

1. $\lambda_1 = -1 \Rightarrow \left\{ \begin{bmatrix} 1 \\ 1 \end{bmatrix} \right\}, \ \lambda_2 = 2 \Rightarrow \left\{ \begin{bmatrix} 1 \\ -1 \end{bmatrix} \right\}$

2. $\lambda_1 = 1 \Rightarrow \left\{ \begin{bmatrix} 2 \\ -1 \end{bmatrix} \right\}, \ \lambda_2 = 3 \Rightarrow \left\{ \begin{bmatrix} 3 \\ 1 \end{bmatrix} \right\}$

3. $\lambda_1 = 2 \Rightarrow \left\{ \begin{bmatrix} 4 \\ 3 \\ 1 \end{bmatrix} \right\}, \ \lambda_2 = -2 \Rightarrow \left\{ \begin{bmatrix} 1 \\ 2 \\ 0 \end{bmatrix}, \begin{bmatrix} 2 \\ 3 \\ 1 \end{bmatrix} \right\}$

4. $\lambda_1 = 3 \Rightarrow \left\{ \begin{bmatrix} 1 \\ 1 \\ 0 \end{bmatrix} \right\}, \ \lambda_2 = 0 \Rightarrow \left\{ \begin{bmatrix} 1 \\ 5 \\ 1 \end{bmatrix}, \begin{bmatrix} 2 \\ 1 \\ 4 \end{bmatrix} \right\}$

5. $\lambda_1 = 2i \Rightarrow \left\{ \begin{bmatrix} 1+i \\ 2-i \end{bmatrix} \right\}$, $\lambda_2 = -2i \Rightarrow \left\{ \begin{bmatrix} 1-i \\ 2+i \end{bmatrix} \right\}$

6. $\lambda_1 = 3 + i \Rightarrow \left\{ \begin{bmatrix} 2i \\ i \end{bmatrix} \right\}$, $\lambda_2 = 3 - i \Rightarrow \left\{ \begin{bmatrix} -2i \\ -i \end{bmatrix} \right\}$

7. $\lambda_1 = 4 \Rightarrow \left\{ \begin{bmatrix} 3 \\ 1 \\ 5 \end{bmatrix} \right\}$, $\lambda_2 = 1 + i \Rightarrow \left\{ \begin{bmatrix} 4+i \\ -2i \\ 3+i \end{bmatrix} \right\}$,

$\lambda_3 = 1 - i \Rightarrow \left\{ \begin{bmatrix} 4-i \\ 2i \\ 3-i \end{bmatrix} \right\}$

8. $\lambda_1 = -1 \Rightarrow \left\{ \begin{bmatrix} 1 \\ 0 \\ 3 \end{bmatrix} \right\}$, $\lambda_2 = 3i \Rightarrow \left\{ \begin{bmatrix} 2-i \\ 1+i \\ 7i \end{bmatrix} \right\}$,

$\lambda_3 = -3i \Rightarrow \left\{ \begin{bmatrix} 2+i \\ 1-i \\ -7i \end{bmatrix} \right\}$

9. $\lambda_1 = 1 \Rightarrow \left\{ \begin{bmatrix} 6 \\ 2 \\ 5 \\ 0 \end{bmatrix} \right\}$, $\lambda_2 = 1 + i \Rightarrow \left\{ \begin{bmatrix} 3+2i \\ 6 \\ 2-3i \\ -5i \end{bmatrix} \right\}$,

$\lambda_3 = 4 \Rightarrow \left\{ \begin{bmatrix} 1 \\ 2 \\ 3 \\ 2 \end{bmatrix} \right\}$, $\lambda_4 = 1 - i \Rightarrow \left\{ \begin{bmatrix} 3-2i \\ 6 \\ 2+3i \\ 5i \end{bmatrix} \right\}$

10. $\lambda_1 = -2 \Rightarrow \left\{ \begin{bmatrix} 3 \\ 0 \\ 1 \\ 2 \end{bmatrix} \right\}$, $\lambda_2 = 4i \Rightarrow \left\{ \begin{bmatrix} -5 \\ 4+3i \\ 3i \\ 1-i \end{bmatrix} \right\}$,

$\lambda_3 = 0 \Rightarrow \left\{ \begin{bmatrix} 4 \\ 2 \\ 1 \\ 1 \end{bmatrix} \right\}$, $\lambda_4 = -4i \Rightarrow \left\{ \begin{bmatrix} -5 \\ 4-3i \\ -3i \\ 1+i \end{bmatrix} \right\}$

In Exercises 11–18, find the general solution for the system.

11. $y_1' = y_1 + 4y_2$
$y_2' = y_1 + y_2$

12. $y_1' = 4y_1 + 2y_2$
$y_2' = 6y_1 + 3y_2$

13. $y_1' = 7y_1 - 8y_2$
$y_2' = 4y_1 - 5y_2$

14. $y_1' = 7y_1 - 10y_2$
$y_2' = 2y_1 - 2y_2$

15. $y_1' = 3y_1 + y_2$
$y_2' = -5y_1 + y_2$

16. $y_1' = y_1 - y_2$
$y_2' = 2y_1 + 3y_2$

17. $y_1' = 7y_1 + 2y_2 - 8y_3$
$y_2' = -3y_1 + 3y_3$
$y_3' = 6y_1 + 2y_2 - 7y_3$

18. $y_1' = y_2 - y_3$
$y_2' = y_1 + y_3$
$y_3' = y_1 - y_2 + 2y_3$

In Exercises 19–24, find the solution for the system that satisfies the conditions at $t = 0$.

19. $y_1' = 8y_1 - 10y_2$, $\quad y_1(0) = 4$
$y_2' = 5y_1 - 7y_2$, $\quad y_2(0) = 1$

20. $y_1' = -4y_1 + 10y_2$, $\quad y_1(0) = 1$
$y_2' = -3y_1 + 7y_2$, $\quad y_2(0) = 1$

21. $y_1' = y_1 + 3y_2$, $\quad y_1(0) = 2$
$y_2' = -3y_1 + y_2$, $\quad y_2(0) = -1$

22. $y_1' = 2y_1 + 4y_2$, $\quad y_1(0) = -1$
$y_2' = -2y_1 - 2y_2$, $\quad y_2(0) = 3$

23. $y_1' = 2y_1 - y_2 - y_3$, $\quad y_1(0) = -1$
$y_2' = 6y_1 + 3y_2 - 2y_3$, $\quad y_2(0) = 0$
$y_3' = 6y_1 - 2y_2 - 3y_3$, $\quad y_3(0) = -4$

24. $y_1' = 3y_1 - 2y_2$, $\quad y_1(0) = 1$
$y_2' = 5y_1 - 3y_2$, $\quad y_2(0) = -1$
$y_3' = y_1 + 1y_2 - 1y_3$, $\quad y_3(0) = 2$

In Exercises 25–26, the given system of linear differential equations models the concentrations of insulin and glucose in an individual, as described earlier in this section. Find the general solution for the system.

25. $y_1' = -0.1y_1 + 0.2y_2$
$y_2' = -0.3y_1 - 0.6y_2$

26. $y_1' = -0.44y_1 + 0.12y_2$
$y_2' = -0.08y_1 - 0.16y_2$

27. For the system of differential equations given in Exercise 25, suppose that it is known that at time $t = 0$ the concentrations of insulin and glucose, respectively, are

$$y_1(0) = 10, \quad y_2(0) = 20$$

Find a formula for $y_1(t)$ and $y_2(t)$.

28. For the system of differential equations given in Exercise 26, suppose that it is known that at time $t = 0$ the concentrations of insulin and glucose, respectively, are

$$y_1(0) = 15, \quad y_2(0) = 50$$

Find a formula for $y_1(t)$ and $y_2(t)$.

In Exercises 29–30, the given system of linear differential equations models a two-country arms race, as described in this section. Find the general solution for the system, and provide a brief interpretation of the results.

29. $y_1' = -3y_1 + 5y_2$
$y_2' = 4y_1 - 4y_2$

30. $y_1' = y_1 + y_2$
$y_2' = 4y_1 + y_2$

31. For the system of differential equations given in Exercise 29, suppose that it is known that the initial quantity of arms in each country's arsenal is

$$y_1(0) = 1, \quad y_2(0) = 2$$

Find a formula for $y_1(t)$ and $y_2(t)$.

32. For the system of differential equations given in Exercise 30, suppose that it is known that the initial quantity of arms in each country's arsenal is

$$y_1(0) = 4, \quad y_2(0) = 1$$

Find a formula for $y_1(t)$ and $y_2(t)$.

FIND AN EXAMPLE For Exercises 33–38, find an example that meets the given specifications.

33. A system of two first-order linear differential equations that has general solution $y_1 = c_1 e^{-3t}$ and $y_2 = c_2 e^{2t}$.

34. A system of two first-order linear differential equations that has general solution $y_1 = c_1 e^t$ and $y_2 = c_2 e^{-2t}$.

35. A system of two first-order linear differential equations that has general solution $y_1 = 2c_1 e^t - c_2 e^{-2t}$ and $y_2 = -3c_1 e^t + 2c_2 e^{-2t}$. (HINT: Section 6.4 contains an example showing how to construct a matrix with specific eigenvalues and eigenvectors.)

36. A system of two first-order linear differential equations that has general solution $y_1 = 3c_1 e^{-t} + 7c_2 e^{4t}$ and $y_2 = c_1 e^{-t} + 2c_2 e^{4t}$. (HINT: Section 6.4 contains an example showing how to construct a matrix with specific eigenvalues and eigenvectors.)

37. A system of three first-order linear differential equations with general solution that is made up of linear combinations of e^t, e^{-2t}, and e^{5t}.

38. A system of two first-order linear differential equations with general solution that is made up of linear combinations of only trigonometric functions.

TRUE OR FALSE For Exercises 39–42, determine if the statement is true or false, and justify your answer.

39. The general solution to $y' = ky$ is $y = ce^{kt}$.

40. Every system of linear differential equations $\mathbf{y}' = A\mathbf{y}$ can be solved using the methods presented in this section.

41. The solution to any system of linear differential equations always includes a real exponential function e^{ct} for some nonzero constant c.

42. If $\mathbf{y}_a$ and $\mathbf{y}_b$ are solutions to the system of linear differential equations $\mathbf{y}' = A\mathbf{y}$, then so is a linear combination $c_a\mathbf{y}_a + c_b\mathbf{y}_b$.

© For Exercises 43–46, find the general solution for the system.

43. $y_1' = 2y_1 - 3y_2 + 6y_3$
$y_2' = -3y_1 - 5y_2 - 7y_3$
$y_3' = 4y_1 + 4y_2 + 4y_3$

44. $y_1' = 7y_1 + 2y_2 + y_3$
$y_2' = y_1 + 6y_2 - y_3$
$y_3' = 2y_1 + 3y_2 + 4y_3$

45. $y_1' = 3y_1 - y_2 + 5y_4$
$y_2' = -2y_1 + 3y_2 + 7y_3$
$y_3' = 4y_1 - 3y_2 - y_3 - y_4$
$y_4' = 5y_1 + 2y_2 + y_3 + y_4$

46. $y_1' = -y_1 - 5y_2 + 4y_3 + y_4$
$y_2' = y_1 + y_2 - 3y_3 - 2y_4$
$y_3' = -3y_1 + y_2 - 6y_3 - 7y_4$
$y_4' = 2y_1 - y_2 - y_3 - 5y_4$

© For Exercises 47–50, find the solution for the system that satisfies the conditions at $t = 0$.

47. $y_1' = 3y_1 - y_2 + 4y_3$, $y_1(0) = -1$
$y_2' = -2y_1 - 6y_2 + y_3$, $y_2(0) = -4$
$y_3' = 4y_1 + 5y_2 + 5y_3$, $y_3(0) = 3$

48. $y_1' = -4y_1 + 7y_2 + 2y_3$, $y_1(0) = 1$
$y_2' = 2y_1 + 4y_2 + 3y_3$, $y_2(0) = -5$
$y_3' = 3y_1 - 2y_2 + 4y_3$, $y_3(0) = 2$

49. $y_1' = -2y_1 + 3y_2 - 4y_3 + 5y_4$, $y_1(0) = 7$
$y_2' = -y_1 + 2y_2 - 3y_3 + 4y_4$, $y_2(0) = 2$
$y_3' = 4y_1 - 3y_2 - 2y_3 + y_4$, $y_3(0) = -2$
$y_4' = 5y_1 + 6y_2 + 7y_3 + 8y_4$, $y_4(0) = -5$

50. $y_1' = 7y_1 - 3y_2 + 5y_3 - y_4$, $y_1(0) = 2$
$y_2' = 4y_1 + 2y_2 - 3y_3 - 6y_4$, $y_2(0) = -9$
$y_3' = -y_1 + 4y_2 - 4y_4$, $y_3(0) = -4$
$y_4' = 3y_2 - 4y_3 - 2y_4$, $y_4(0) = 3$

Vector Spaces

The Chesapeake Bay Bridge-Tunnel (CBBT) provides direct vehicular passage between Virginia's Eastern Shore and south Hampton Roads, Virginia, where once only ferry service was available. Although elevated bridges were originally considered, the U.S. Navy objected, citing that either accidental or deliberate collapse would block access to the Atlantic Ocean from Norfolk Navy Base. The CBBT includes 12 miles of low-level trestle, two mile-long tunnels, two bridges, two miles of causeway, and four manmade islands. Shortly after opening in 1964, the bridge-tunnel was designated "One of Seven Engineering Wonders of the Modern World" by the American Society of Civil Engineers.

Bridge suggested by Eddie Boyd, Jr., University of Maryland – Eastern Shore (Hyunsoo Leo Kim)

▶ The order of Chapter 7 and Chapter 8 can be reversed if needed.

O ver the first six chapters we have focused our attention on understanding the structure of vectors in Euclidean space. In this chapter we adapt these results to a more general setting, where we adopt a broader notion of vectors and the spaces containing them.

In Section 7.1 we generalize the definition of *vector* and define a *vector space*. Section 7.2 describes how the concepts of span and linear independence carry over from Euclidean space to vector spaces, and in Section 7.3 we revisit the topics of basis and dimension.

This chapter is relatively brief because all of the material has analogs among the concepts that we developed for Euclidean space $\mathbf{R}^n$. References to comparable earlier definitions and theorems are provided to reinforce connections. As you read this chapter, think about how the concepts presented match up with those from Euclidean space.

7.1 Vector Spaces and Subspaces

In this section we describe how to expand the concept of Euclidean space and subspaces from $\mathbf{R}^n$ to a more general setting. To get us started, let $\mathbf{P}^2$ be the set of all polynomials with real coefficients that have degree 2 or less. A typical element of $\mathbf{P}^2$ has the form $p(x) = a_2 x^2 + a_1 x + a_0$, where a_0, a_1, and a_2 are real numbers. Let's compare $\mathbf{R}^3$ and $\mathbf{P}^2$.

- If $p(x) = a_2 x^2 + a_1 x + a_0$ and $q(x) = b_2 x^2 + b_1 x + b_0$ are two polynomials in $\mathbf{P}^2$, then the sum is

$$p(x) + q(x) = (a_2 + b_2)x^2 + (a_1 + b_1)x + (a_0 + b_0)$$

When adding polynomials, we add together the coefficients of like terms. This is similar to the componentwise addition of elements in $\mathbf{R}^3$.

- If c is a real number and $p(x)$ is as above, then

$$cp(x) = (ca_2)x^2 + (ca_1)x + (ca_0)$$

Scalar multiplication of polynomials distributes across terms, similar to how scalar multiples distribute across components in $\mathbf{R}^3$.

- Just as $\mathbf{R}^3$ is closed under addition and scalar multiplication, so is $\mathbf{P}^2$. The sum of two polynomials in $\mathbf{P}^2$ has degree no greater than 2, as is the degree of the scalar multiple of a polynomial in $\mathbf{P}^2$. (Note that these operations might *decrease* the degree but cannot *increase* it.)

- The zero polynomial $z(x) = 0$ satisfies $p(x) + z(x) = p(x)$ for every polynomial in $\mathbf{P}^2$, so that $z(x)$ plays the same role as $\mathbf{0} = \begin{bmatrix} 0 \\ 0 \\ 0 \end{bmatrix}$ does in $\mathbf{R}^3$.

- For every polynomial $p(x)$ in $\mathbf{P}^2$, there is another polynomial

$$-p(x) = -a_2 x^2 - a_1 x - a_0$$

such that $p(x) + (-p(x)) = 0 = z(x)$. This is also true in $\mathbf{R}^3$, where for each $\begin{bmatrix} a \\ b \\ c \end{bmatrix}$ there exists $\begin{bmatrix} -a \\ -b \\ -c \end{bmatrix}$ such that $\begin{bmatrix} a \\ b \\ c \end{bmatrix} + \begin{bmatrix} -a \\ -b \\ -c \end{bmatrix} = \begin{bmatrix} 0 \\ 0 \\ 0 \end{bmatrix} = \mathbf{0}$.

- The definitions of addition and scalar multiplication on $\mathbf{P}^2$ satisfy the same distributive and associative laws as those of $\mathbf{R}^3$ given in Theorem 2.3 in Section 2.1.

Although they look different, $\mathbf{R}^3$ and $\mathbf{P}^2$ have many similar features. Theorem 2.3 in Section 2.1 lists the algebraic properties of elements in $\mathbf{R}^n$. The preceding discussion shows that the polynomials in $\mathbf{P}^2$ have similar properties. Other sets of mathematical objects, such as matrices and continuous functions, also possess these properties. Theorem 2.3 serves as a guide for our broader definition of a vector space.

DEFINITION 7.1

Definition Vector Space, Vector

▶ Technically, Definition 7.1 is the definition of a *real* vector space, and we could replace the real scalars with complex numbers with minimal changes. But for our purposes real scalars will suffice, so from here on we assume that "vector space" refers to a real vector space.

A **vector space** consists of a set V of **vectors** together with operations of addition and scalar multiplication on the vectors that satisfy each of the following:

(1) If $\mathbf{v}_1$ and $\mathbf{v}_2$ are in V, then so is $\mathbf{v}_1 + \mathbf{v}_2$. Hence V is closed under addition.

(2) If c is a real scalar and $\mathbf{v}$ is in V, then so is $c\mathbf{v}$. Hence V is closed under scalar multiplication.

(3) There exists a **zero vector** $\mathbf{0}$ in V such that $\mathbf{0} + \mathbf{v} = \mathbf{v}$ for all $\mathbf{v}$ in V.

(4) For each $\mathbf{v}$ in V there exists an **additive inverse** (or **opposite**) vector $-\mathbf{v}$ in V such that $\mathbf{v} + (-\mathbf{v}) = \mathbf{0}$ for all $\mathbf{v}$ in V.

(5) For all $\mathbf{v}_1$, $\mathbf{v}_2$, and $\mathbf{v}_3$ in V and real scalars c_1 and c_2, we have

(a) $\mathbf{v}_1 + \mathbf{v}_2 = \mathbf{v}_2 + \mathbf{v}_1$

(b) $(\mathbf{v}_1 + \mathbf{v}_2) + \mathbf{v}_3 = \mathbf{v}_1 + (\mathbf{v}_2 + \mathbf{v}_3)$

(c) $c_1(\mathbf{v}_1 + \mathbf{v}_2) = c_1\mathbf{v}_1 + c_1\mathbf{v}_2$

(d) $(c_1 + c_2)\mathbf{v}_1 = c_1\mathbf{v}_1 + c_2\mathbf{v}_1$

(e) $(c_1 c_2)\mathbf{v}_1 = c_1(c_2\mathbf{v}_1)$

(f) $1 \cdot \mathbf{v}_1 = \mathbf{v}_1$

Note that Euclidean space $\mathbf{R}^n$ is a vector space. $\mathbf{P}^2$, together with the usual operations of addition and scalar multiplication, also forms a vector space.

Three important points:

- To describe a vector space, we need to specify both the set of vectors and the arithmetic operations (addition and scalar multiplication) that are performed on them. The set alone is not enough.

- **Vectors are not always columns of numbers!** (For instance, in $\mathbf{P}^2$ the polynomials are the vectors.) This takes getting used to but is crucial, so say it to yourself every night as you fall asleep until it sinks in. It is fine to think of a column of numbers as an *example* of a vector, as long as you do not assume that a vector is always a column of numbers.

- The phrase "vector space" refers to any set satisfying Definition 7.1. The phrase "Euclidean space" will be used for the specific vector space $\mathbf{R}^n$ with the standard definition of addition and scalar multiplication.

EXAMPLE 1 Let $V = \mathbf{R}^{2 \times 2}$ denote the set of real 2×2 matrices, together with the usual definition of matrix addition and multiplication by a constant scalar. Show that $\mathbf{R}^{2 \times 2}$ is a vector space.

Solution We have a clearly defined set of vectors (the real 2×2 matrices) and definitions for addition and scalar multiplication. It remains to verify that the five conditions of Definition 7.1 hold:

(1) If A and B are real 2×2 matrices, then we already know that $A + B$ is also a real 2×2 matrix. Hence $\mathbf{R}^{2 \times 2}$ is closed under addition.

(2) If c is a real scalar and A is in $\mathbf{R}^{2 \times 2}$, then cA is also a real 2×2 matrix. Thus $\mathbf{R}^{2 \times 2}$ is closed under scalar multiplication.

(3) If $0_{22} = \begin{bmatrix} 0 & 0 \\ 0 & 0 \end{bmatrix}$ and $A = \begin{bmatrix} a_{11} & a_{12} \\ a_{21} & a_{22} \end{bmatrix}$, then

$$0_{22} + A = \begin{bmatrix} 0 & 0 \\ 0 & 0 \end{bmatrix} + \begin{bmatrix} a_{11} & a_{12} \\ a_{21} & a_{22} \end{bmatrix} = \begin{bmatrix} (0 + a_{11}) & (0 + a_{12}) \\ (0 + a_{21}) & (0 + a_{22}) \end{bmatrix} = \begin{bmatrix} a_{11} & a_{12} \\ a_{21} & a_{22} \end{bmatrix} = A$$

Hence $0_{22} + A = A$ for all real 2×2 matrices, so that 0_{22} is the zero vector in $\mathbf{R}^{2 \times 2}$.

(4) If $A = \begin{bmatrix} a_{11} & a_{12} \\ a_{21} & a_{22} \end{bmatrix}$, then $-A = \begin{bmatrix} -a_{11} & -a_{12} \\ -a_{21} & -a_{22} \end{bmatrix}$ satisfies

$$A + (-A) = \begin{bmatrix} a_{11} & a_{12} \\ a_{21} & a_{22} \end{bmatrix} + \begin{bmatrix} -a_{11} & -a_{12} \\ -a_{21} & -a_{22} \end{bmatrix} = \begin{bmatrix} 0 & 0 \\ 0 & 0 \end{bmatrix} = 0_{22}$$

so that each vector in $\mathbf{R}^{2 \times 2}$ has an additive inverse.

(5) The six conditions (a)–(f) all follow directly from properties of the real numbers. Verification is left as an exercise.

Since all the required conditions hold, the set $\mathbf{R}^{2 \times 2}$ together with the given arithmetic operations form a vector space. ■

▶ $\mathbf{Q}^2$ consists of polynomials of degree 2, while $\mathbf{P}^2$ is the polynomials of degree 2 or *less*.

EXAMPLE 2 Let $\mathbf{Q}^2$ denote the set of polynomials with real coefficients that have degree equal to 2, together with the usual definition of addition and scalar multiplication for polynomials. Is $\mathbf{Q}^2$ a vector space?

Solution The set $\mathbf{Q}^2$ satisfies some of the conditions of Definition 7.1, but it falls short on others. For instance, if $q_1(x) = x^2$ and $q_2(x) = 5 - x^2$, then q_1 and q_2 are both in $\mathbf{Q}^2$, but $(q_1 + q_2)(x) = q_1(x) + q_2(x) = 5$ is not, so $\mathbf{Q}^2$ is not closed under addition. Hence $\mathbf{Q}^2$ is not a vector space. ■

Before moving on to the next example, we pause to report several properties of vector spaces that are consequences of Definition 7.1.

THEOREM 7.2

Let V be a vector space and suppose that $\mathbf{v}$ is in V. Then:

(a) If $\mathbf{0}$ is a zero vector of V, then $\mathbf{v} + \mathbf{0} = \mathbf{v}$.

(b) If $-\mathbf{v}$ is an additive inverse of $\mathbf{v}$, then $-\mathbf{v} + \mathbf{v} = \mathbf{0}$.

(c) $\mathbf{v}$ has a unique additive inverse $-\mathbf{v}$.

(d) The zero vector $\mathbf{0}$ is unique.

(e) $0 \cdot \mathbf{v} = \mathbf{0}$.

(f) $(-1) \cdot \mathbf{v} = -\mathbf{v}$.

Proof These may seem obvious based upon past experience with real numbers, but remember, all we can assume about a vector space are the properties given in Definition 7.1. We give a proof of (a) and (c), and leave the rest as Exercise 49.

For (a), we know from (3) of Definition 7.1 that there exists a zero vector $\mathbf{0}$ such that

$$\mathbf{0} + \mathbf{v} = \mathbf{v}$$

Property (5a) of Definition 7.1 (the Commutative Law) lets us interchange the order of addition. Doing so on the left above gives us the equation

$$\mathbf{v} + \mathbf{0} = \mathbf{v}$$

proving part (a) of our theorem.

For part (c), note that (4) of Definition 7.1 states that every vector in V has at least one additive inverse. To prove that the additive inverse is unique, we suppose to the contrary that there exists a vector $\mathbf{v}$ in V with two additive inverses $\mathbf{v}_1$ and $\mathbf{v}_2$. Then

$$\mathbf{v} + \mathbf{v}_1 = \mathbf{0} \quad \text{and} \quad \mathbf{v} + \mathbf{v}_2 = \mathbf{0}$$

so that

$$\mathbf{v} + \mathbf{v}_1 = \mathbf{v} + \mathbf{v}_2$$

Property (5a) of Definition 7.1 (the Commutative Law) lets us interchange the order of addition, so we also have

$$\mathbf{v}_1 + \mathbf{v} = \mathbf{v}_2 + \mathbf{v}$$

To cancel out $\mathbf{v}$, we start by adding $\mathbf{v}_1$ to both sides of the equation.

$$(\mathbf{v}_1 + \mathbf{v}) + \mathbf{v}_1 = (\mathbf{v}_2 + \mathbf{v}) + \mathbf{v}_1 \quad \Longrightarrow \quad \mathbf{v}_1 + (\mathbf{v} + \mathbf{v}_1) = \mathbf{v}_2 + (\mathbf{v} + \mathbf{v}_1)$$

The grouping on the right is justified by (5b) of Definition 7.1 (the Associative Law). Since $\mathbf{v} + \mathbf{v}_1 = \mathbf{v} + \mathbf{v}_2 = \mathbf{0}$, our equation simplifies to

$$\mathbf{v}_1 + \mathbf{0} = \mathbf{v}_2 + \mathbf{0} \quad \Longrightarrow \quad \mathbf{v}_1 = \mathbf{v}_2$$

with the cancellation of the zero vector $\mathbf{0}$ justified by part (a) of this theorem. Since $\mathbf{v}_1 = \mathbf{v}_2$, it follows that the additive inverse is unique. ■

EXAMPLE 3   Let $\mathbf{R}^\infty$ denote the set of all infinite sequences of real numbers $\mathbf{v} = (v_1, v_2, \ldots)$, so the elements in $\mathbf{R}^\infty$ have an infinite number of components. Addition and scalar multiplication are defined componentwise, by

$$(v_1, v_2, \ldots) + (u_1, u_2, \ldots) = (v_1 + u_1, v_2 + u_2, \ldots)$$

and

$$c(v_1, v_2, \ldots) = (cv_1, cv_2, \ldots)$$

Show that $\mathbf{R}^\infty$ is a vector space.

Solution Here $\mathbf{R}^\infty$ looks somewhat like $\mathbf{R}^n$ but with an infinite number of components. It is not hard to see that $\mathbf{R}^\infty$ is closed under addition and scalar multiplication, and that the zero vector is given by

$$\mathbf{0} = (0, 0, 0, \ldots)$$

Also, if $\mathbf{v} = (v_1, v_2, \ldots)$, then $-\mathbf{v} = (-v_1, -v_2, \ldots)$ satisfies $\mathbf{v} + (-\mathbf{v}) = \mathbf{0}$, so each vector in $\mathbf{R}^\infty$ has an additive inverse. Finally, the six conditions given in (5) of Definition 7.1 are inherited from the real numbers. Hence $\mathbf{R}^\infty$ is a vector space. ■

EXAMPLE 4   Suppose that $a < b$ are real numbers, and let $C[a, b]$ denote the set of all real-valued continuous functions on $[a, b]$. For f and g in $C[a, b]$ and a real scalar c, we define $(f + g)(x) = f(x) + g(x)$ and $(cf)(x) = c \cdot f(x)$. (These are the usual pointwise definitions of addition and scalar multiplication of functions.) Show that $C[a, b]$ is a vector space.

Solution The set $C[a, b]$ looks different than the other vector spaces we have seen, where the vectors had entries reminiscent of components of vectors in $\mathbf{R}^n$. Here, we have a set consisting of functions such as $f(x) = \sin(x)$ and $g(x) = e^{-x}$.

To determine if $C[a, b]$ is a vector space, we avoid being distracted by superficial appearances by focusing on the definition. First, the set $C[a, b]$ together with addition and scalar multiplication are clearly defined. Next we need to determine if the five conditions of Definition 7.1 are all met. We verify the first three here and leave the remaining conditions as Exercise 3.

(1) Is $C[a, b]$ closed under addition?

If f and g are both in $C[a, b]$, then both are continuous on the interval $[a, b]$. The sum of two continuous functions is also a continuous function, so that $f + g$ is in $C[a, b]$. Hence $C[a, b]$ is closed under addition.

(2) Is $C[a, b]$ closed under scalar multiplication?

If f is in $C[a, b]$ and c is a scalar, then $c \cdot f(x)$ is continuous on $[a, b]$, so that cf is also in $C[a, b]$. Thus $C[a, b]$ is also closed under scalar multiplication.

(3) Is there a zero function?

If $z(x) = 0$, the identically zero function, then $z(x)$ is continuous on $[a, b]$ and so is in $C[a, b]$. Thus if f is in $C[a, b]$, then $z(x) + f(x) = f(x)$ for all x in $[a, b]$. Hence $z + f = f$ for all f in $C[a, b]$, so that z is the zero function.

The remaining questions that need to be answered are:

(4) Does each function have an additive inverse?

(5) Do the six commutative, distributive, and associative laws hold?

Verifying that the answer to each is yes is left as an exercise. Therefore $C[a, b]$ is a vector space. ∎

Below is a brief list of vector spaces. Some have already been verified as vector spaces, and others are left as exercises.

▶ This list is far from exhaustive. There are many other vector spaces not included.

- Euclidean space $\mathbf{R}^n$ ($n > 0$ an integer), together with the standard addition and scalar multiplication of vectors.

- $\mathbf{P}^n$ ($n \geq 1$ an integer), the set of polynomials with real coefficients and degree no greater then n, together with the usual addition and scalar multiplication of polynomials.

- $\mathbf{R}^{m \times n}$, the set of real $m \times n$ matrices together with the usual definition of matrix addition and scalar multiplication.

- $\mathbf{P}$, the set of polynomials with real coefficients and any degree, together with the usual addition and scalar multiplication of polynomials.

- $\mathbf{R}^\infty$, together with addition and scalar multiplication of vectors described in Example 3.

- $C[a, b]$, the set of real-valued continuous functions on the interval $[a, b]$, together with the usual addition and scalar multiplication of functions.

- $C(\mathbf{R})$, the set of real-valued continuous functions on the real numbers $\mathbf{R}$, together with the usual addition and scalar multiplication of functions.

- $T(m, n)$, the set of linear transformations $T : \mathbf{R}^m \to \mathbf{R}^n$, together with the usual addition and scalar multiplication of functions.

Subspaces

Just as in Euclidean space, a subspace of a vector space can be thought of as a vector space contained within another vector space. Since a subspace inherits the properties of the parent vector space, all that is required to certify its subspace status is that it be closed under addition and scalar multiplication. The formal definition is essentially identical to the one given for Euclidean space.

DEFINITION 7.3

Definition Subspace

▶ Definition 7.3 generalizes Definition 4.1 in Section 4.1.

A subset S of a vector space V is a **subspace** if S satisfies the following three conditions:

(a) S contains $\mathbf{0}$, the zero vector.

(b) If $\mathbf{u}$ and $\mathbf{v}$ are in S, then $\mathbf{u} + \mathbf{v}$ is also in S.

(c) If c is a scalar and $\mathbf{v}$ is in S, then $c\mathbf{v}$ is also in S.

EXAMPLE 5 Let S denote the set of all polynomials p with real coefficients such that $p(0) = 0$. Is S a subspace of $\mathbf{P}$, the set of all polynomials with real coefficients?

Solution First, we note that S is a subset of $\mathbf{P}$. Next, we do the same thing that we did to show that a subset of $\mathbf{R}^n$ is a subspace: We check if the three conditions of the definition hold.

(a) The zero polynomial $z(x) = 0$ (the identically zero function) satisfies $z(0) = 0$ and hence is in S, so S contains the zero vector.

(b) If p and q are both in S, then $p(0) = 0$ and $q(0) = 0$. Therefore $(p + q)(0) = p(0) + q(0) = 0$, so that $p + q$ is also in S.

(c) If p is in S and c is a real scalar, then $(cp)(0) = c(p(0)) = 0$. Hence cp is also in S.

Since the three conditions of Definition 7.3 hold, S is a subspace of $\mathbf{P}$. ∎

EXAMPLE 6 Suppose that $m < n$ are both integers. Is $\mathbf{P}^m$ a subspace of $\mathbf{P}^n$?

Solution First, we note that $\mathbf{P}^m$ is a subset of $\mathbf{P}^n$. Moreover, since $\mathbf{P}^m$ is itself a vector space, we know that it contains the zero vector, is closed under addition, and is closed under scalar multiplication. Thus the three subspace conditions are met, so that $\mathbf{P}^m$ is a subspace of $\mathbf{P}^n$. ∎

EXAMPLE 7 Suppose that $m < n$ are both integers. Is $\mathbf{R}^m$ a subspace of $\mathbf{R}^n$?

Solution While it is true that $\mathbf{R}^m$ is itself a vector space, it is not a subset of $\mathbf{R}^n$ and so cannot be a subspace of $\mathbf{R}^n$. ∎

EXAMPLE 8 The **trace** of a square matrix is the sum of the diagonal terms. Suppose that S is the subset of $\mathbf{R}^{2\times 2}$ consisting of matrices with trace equal to 0. Is S a subspace of $\mathbf{R}^{2\times 2}$?

Solution By definition S is a subset of $\mathbf{R}^{2\times 2}$. Let's check the three conditions of Definition 7.3.

(a) The zero matrix of $\mathbf{R}^{2\times 2}$ is $0_{22} = \begin{bmatrix} 0 & 0 \\ 0 & 0 \end{bmatrix}$, which has trace 0. Hence 0_{22} is in S.

(b) Suppose that $A = \begin{bmatrix} a_{11} & a_{12} \\ a_{21} & a_{22} \end{bmatrix}$ and $B = \begin{bmatrix} b_{11} & b_{12} \\ b_{21} & b_{22} \end{bmatrix}$ are both in S. Then $a_{11} + a_{22} = 0$ and $b_{11} + b_{22} = 0$. Since

$$A + B = \begin{bmatrix} (a_{11} + b_{11}) & (a_{12} + b_{12}) \\ (a_{21} + b_{21}) & (a_{22} + b_{22}) \end{bmatrix}$$

the trace of $A + B$ is equal to

$$(a_{11} + b_{11}) + (a_{22} + b_{22}) = (a_{11} + a_{22}) + (b_{11} + b_{22}) = 0$$

Hence $A + B$ is also in S.

(c) For a scalar c and A in S as in (b), we have

$$cA = \begin{bmatrix} ca_{11} & ca_{12} \\ ca_{21} & ca_{22} \end{bmatrix}$$

Therefore the trace of cA is

$$ca_{11} + ca_{22} = c(a_{11} + a_{22}) = 0$$

which implies that cA is also in S.

Since all three conditions for a subspace are satisfied, S is a subspace. ∎

The trivial subspaces carry over from Euclidean space. Proof of the next theorem is left as an exercise.

THEOREM 7.4

▶ Theorem 7.4 generalizes the result found in Example 2 of Section 4.1.

Suppose that V is a vector space. Then $S = \{0\}$ and $S = V$ are both subspaces of V, sometimes called the trivial subspaces.

The last two examples require some knowledge of calculus.

EXAMPLE 9 Let $C(a, b)$ be the set of functions continuous on (a, b), and suppose S is the subset of $C(a, b)$ that consists of the set of differentiable functions on (a, b). Is S a subspace of $C(a, b)$?

Solution By definition, S is a subset of $C(a, b)$. Since $z(x) = 0$ is differentiable ($z'(x) = 0$), the zero vector is in S. Also, from calculus we know that sums and constant multiples of differentiable functions are differentiable, so that S is closed under addition and scalar multiplication. Hence S is a subspace of $C(a, b)$. ∎

EXAMPLE 10 Let $C(\mathbf{R})$ denote the set of functions that are continuous on all of $\mathbf{R}$. Let S denote the subset of $C(\mathbf{R})$ consisting of functions $y(t)$ that satisfy the differential equation

$$y''(t) + y(t) = 0 \tag{1}$$

Show that S is a subspace of $C(\mathbf{R})$.

Solution Differential equations of this type arise in the modeling of simple harmonic motion, such as a mass moving up and down while suspended by a spring (see Figure 1).

To show that S is a subspace, we start by noting that if $y(t) = 0$ is the zero function, then $y''(t) = 0$ and this function satisfies 1.

Next, suppose that y_1 and y_2 both satisfy (1). Then

$$\begin{aligned}\left(y_1(t) + y_2(t)\right)'' + \left(y_1(t) + y_2(t)\right) &= y_1''(t) + y_2''(t) + y_1(t) + y_2(t) \\ &= \left(y_1''(t) + y_1(t)\right) + \left(y_2''(t) + y_2(t)\right) = 0\end{aligned}$$

so $y_1 + y_2$ is also in S.

Finally, given a scalar c and a solution y_1 of (1), we have

$$\left(cy_1(t)\right)'' + cy_1(t) = c\left(y_1''(t) + y_1(t)\right) = 0$$

Thus cy_1 is also in S, and therefore S is a subspace. ∎

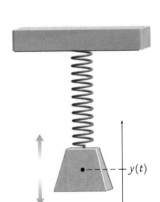

Figure 1 A mass–spring system. $y(t)$ gives the vertical displacement at time t.

We close this section with a reminder. As mentioned earlier, often it is difficult to transition away from the notion that elements of $\mathbf{R}^n$ are the only type of "vector." As we have seen, viewed from a more general perspective, we can think of polynomials, matrices, and continuous functions as vectors. So remember: Vectors are not always columns of numbers!

| EXERCISES |

1. Complete Example 1: Verify that $\mathbf{R}^{2\times2}$ with the usual definition of matrix addition and scalar multiplication satisfies the six conditions of (5) given in Definition 7.1.

2. Determine which properties of Definition 7.1 are not met by the set $\mathbf{Q}^2$ given in Example 2.

In Exercises 3–8, prove that V is a vector space.

3. $V = C[a, b]$, the set of continuous functions defined on the interval $[a, b]$, together with the standard pointwise definition of addition and scalar multiplication of functions. (Portions of this exercise are completed in Example 4.)

4. $V = \mathbf{R}^{m\times n}$, the set of real $m \times n$ matrices together with the usual definition of matrix addition and scalar multiplication.

5. $V = \mathbf{P}^n$, the set of polynomials with real coefficients and degree no greater than n, together with the usual definition of polynomial addition and scalar multiplication.

6. $V = \mathbf{P}$, the set of polynomials with real coefficients and any degree, together with the usual addition and scalar multiplication of polynomials.

7. $V = C(\mathbf{R})$, the set of real-valued continuous functions defined on $\mathbf{R}$, together with the usual pointwise addition and scalar multiplication of functions.

8. $V = T(m, n)$, the set of linear transformations $T : \mathbf{R}^m \to \mathbf{R}^n$, together with the usual addition and scalar multiplication of functions.

In Exercises 9–12, a set V is given, together with definitions of addition and scalar multiplication. Determine if V is a vector space, and if so, prove it. If not, identify a condition of Definition 7.1 that is not satisfied.

9. V is the set of polynomials with real coefficients and degree 2 or less. Addition is defined by

$$(a_2 x^2 + a_1 x + a_0) + (b_2 x^2 + b_1 x + b_0)$$
$$= (a_0 + b_0)x^2 + (a_1 + b_1)x + (a_2 + b_2)$$

and scalar multiplication by

$$c(a_2 x^2 + a_1 x + a_0) = ca_0 x^2 + ca_1 x + ca_2$$

10. V is the set of lines in the plane through the origin, excluding the y-axis. Addition of lines is defined by adding slopes, and scalar multiplication by the scalar multiple of the slope.

11. V is the set of vectors in $\mathbf{R}^2$ with the following definitions of addition and scalar multiplication:

Addition: $\begin{bmatrix} a_1 \\ b_1 \end{bmatrix} + \begin{bmatrix} a_2 \\ b_2 \end{bmatrix} = \begin{bmatrix} 0 \\ b_1 + b_2 \end{bmatrix}$

Scalar multiplication: $c \begin{bmatrix} a_1 \\ b_1 \end{bmatrix} = \begin{bmatrix} ca_1 \\ cb_1 \end{bmatrix}$

12. V is the set of vectors in $\mathbf{R}^2$ with the following definitions of addition and scalar multiplication:

Addition: $\begin{bmatrix} a_1 \\ b_1 \end{bmatrix} + \begin{bmatrix} a_2 \\ b_2 \end{bmatrix} = \begin{bmatrix} 0 \\ b_1 + b_2 \end{bmatrix}$

Scalar multiplication: $c \begin{bmatrix} a_1 \\ b_1 \end{bmatrix} = \begin{bmatrix} 0 \\ cb_1 \end{bmatrix}$

In Exercises 13–18, prove that the set S is a subspace of the vector space V.

13. $V = \mathbf{R}^{3 \times 3}$ and S is the set of upper triangular 3×3 matrices.

14. $V = \mathbf{P}^5$ and S is the set of polynomials of the form $p(x) = a_4 x^4 + a_2 x^2 + a_0$.

15. $V = C(\mathbf{R})$ and S is the subset of functions f in V such that $f(4) = 0$.

16. $V = \mathbf{P}$ and S is the set of all polynomials with terms of only even degree. (Thus $1, x^2, x^4, \ldots$ are allowed, but not $x, x^3, \ldots$.)

17. $V = T(2, 2)$ and S is the set of linear transformations T such that

$$T\left(\begin{bmatrix} x_1 \\ x_2 \end{bmatrix} \right) = \begin{bmatrix} a_1 x_1 \\ a_2 x_2 \end{bmatrix}$$

where a_1 and a_2 are scalars.

18. $V = \mathbf{R}^\infty$ and S consists of those vectors with a finite number of nonzero components.

In Exercises 19–28, a vector space V and a subset S are given. Determine if S is a subspace of V, and if so, prove it. If not, give an example showing one of the conditions of Definition 7.3 is not satisfied.

19. $V = C[-2, 2]$ and $S = \mathbf{P}$.

20. $V = C[-4, 7]$ and S consists of functions of the form ae^{bx} (a, b are real constants).

21. $V = \mathbf{R}^\infty$ and S is the subset consisting of vectors where the second component is equal to zero—that is, vectors $\mathbf{v} = (v_1, 0, v_3, v_4, \ldots)$.

22. $V = \mathbf{R}^\infty$ and S is the subset consisting of vectors where the components are all integers.

23. $V = \mathbf{R}^\infty$ and S is the subset of $\mathbf{R}^\infty$ consisting of vectors where all but a finite number of components are not equal to zero.

24. $V = C[-3, 3]$ and S is the set of real-valued functions f such that $f(-1) + f(1) = 0$.

25. $V = \mathbf{P}^4$ and S is the set of real-valued functions g in $\mathbf{P}^3$ such that $g(2) + g(3) = 0$.

26. $V = T(4, 5)$ and S is the set of linear transformations that are one-to-one.

27. $V = T(3, 3)$ and S is the set of invertible linear transformations.

28. $V = C[-3, 3]$ and S is the set of real-valued functions h in $C[-3, 3]$ such that $h(0) = 1$.

Calculus Required In Exercises 29–32, a vector space V and a subset S are given. Determine if S is a subspace of V, and if so, prove it. If not, explain which conditions of Definition 7.3 are not satisfied. (Assume $a < b$ as needed.)

29. $V = C(a, b)$ and $S = C^n(a, b)$, the set of functions on (a, b) with n continuous derivatives.

30. $V = C(\mathbf{R})$ and S is the set of all solutions to the differential equation $y'(t) - 4y(t) = 0$.

31. $V = C[a, b]$ and S is the set of functions g such that

$$\int_a^b g(x)\, dx = (b - a)$$

32. $V = C(\mathbf{R})$ and S is the set of functions h such that

$$\int_{-\infty}^{\infty} e^{-x^2} h(x)\, dx = 0$$

FIND AN EXAMPLE For Exercises 33–38, find an example that meets the given specifications.

33. A failed vector space—that is, a set of vectors and definitions of addition and scalar multiplication that meet some but not all of the conditions of Definition 7.1.

34. A vector space V not given in this section.

35. A vector space V and a subset S that is *almost* a subspace: S contains $\mathbf{0}$ and is closed under addition, but S is not closed under scalar multiplication.

36. A vector space V and a subset S that is *almost* a subspace: S contains $\mathbf{0}$ and is closed under scalar multiplication, but S is not closed under addition.

37. A single set of vectors and two different definitions of addition and scalar multiplication to produce two different vector spaces V_1 and V_2.

38. A vector space V and an infinite sequence of subspaces $S_1 \subset S_2 \subset S_3 \subset \cdots$, with each subset proper.

TRUE OR FALSE For Exercises 39–46, determine if the statement is true or false, and justify your answer.

39. A vector space V consists of a set of vectors together with definitions of addition and scalar multiplication of the vectors.

40. No two vector spaces can share the same set of vectors.

41. If $\mathbf{v}_1$ and $\mathbf{v}_2$ are in a vector space V, then so is $\mathbf{v}_1 - \mathbf{v}_2$.

42. If $\mathbf{v}_1 + \mathbf{v}_2$ is in a subspace S, then $\mathbf{v}_1$ and $\mathbf{v}_2$ must be in S.

43. If S_1 and S_2 are subspaces of a vector space V, then the intersection $S_1 \cap S_2$ is also a subspace of V.

44. If S_1 and S_2 are subspaces of a vector space V, then the union $S_1 \cup S_2$ is also a subspace of V.

45. Suppose S_1 and S_2 are subspaces of a vector space V, and define $S_1 + S_2$ to be the set of all vectors of the form $\mathbf{s}_1 + \mathbf{s}_2$, where $\mathbf{s}_1$ is in S_1 and $\mathbf{s}_2$ is in S_2. Then $S_1 + S_2$ is a subspace of V.

46. A vector space V must have an infinite number of distinct elements.

47. Prove that Definition 7.3 is unchanged if condition (a) is replaced with the condition "S is nonempty."

48. Prove Theorem 7.4: Suppose that V is a vector space. Prove that $S = \{\mathbf{0}\}$ and $S = V$ are subspaces of V.

49. Complete the proof of Theorem 7.2: Let V be a vector space and suppose that $\mathbf{v}$ is in V.

(a) If $-\mathbf{v}$ is an additive inverse of $\mathbf{v}$, then $-\mathbf{v} + \mathbf{v} = \mathbf{0}$.

(b) The zero vector $\mathbf{0}$ is unique.

(c) $0 \cdot \mathbf{v} = \mathbf{0}$.

(d) $(-1) \cdot \mathbf{v} = -\mathbf{v}$.

7.2 Span and Linear Independence

In this section we extend the concepts of span and linear independence from Euclidean space to vector spaces. The definitions here are similar to those for Euclidean space, so lean on your previous knowledge.

Span

Definition **Linear Combination**

Just as in $\mathbf{R}^n$, a **linear combination** of a set of vectors $\{\mathbf{v}_1, \mathbf{v}_2, \ldots, \mathbf{v}_m\}$ is a sum of the form

$$c_1\mathbf{v}_1 + c_2\mathbf{v}_2 + \cdots + c_m\mathbf{v}_m \tag{1}$$

where $c_1, c_2, \ldots, c_m$ are real numbers. For instance, for the vectors $f(x) = \cos(x)$ and $g(x) = \log(x)$ in $C[1, 5]$, one possible linear combination is

$$7\cos(x) - \pi \log(x)$$

As in Euclidean space, the span of a set is defined in terms of linear combinations.

DEFINITION 7.5

Definition **Span**

▶ Definition 7.5 generalizes Definition 2.5 in Section 2.2.

Let $\mathcal{V} = \{\mathbf{v}_1, \mathbf{v}_2, \ldots, \mathbf{v}_m\}$ be a nonempty set of vectors in a vector space V. The **span** of this set is denoted $\text{span}\{\mathbf{v}_1, \mathbf{v}_2, \ldots, \mathbf{v}_m\}$ and is defined to be the set of all linear combinations of the form

$$c_1\mathbf{v}_1 + c_2\mathbf{v}_2 + \cdots + c_m\mathbf{v}_m$$

where $c_1, c_2, \ldots, c_m$ can be any real numbers.

If $\mathcal{V}$ consists of infinitely many vectors, then we define $\text{span}(\mathcal{V})$ to be the set of all linear combinations of *finite* subsets of $\mathcal{V}$.

EXAMPLE 1   Let $S = \text{span}\{x^2 - 2x + 3, \ -2x^2 + 3x + 1\}$ be a subset of $\mathbf{P}^2$. Determine if $p(x) = 10x^2 - 17x + 9$ is in S.

Solution The vector $p(x) = 10x^2 - 17x + 9$ is in S if there exist real numbers c_1 and c_2 such that

$$c_1(x^2 - 2x + 3) + c_2(-2x^2 + 3x + 1) = 10x^2 - 17x + 9$$

Reorganizing the left side to collect common factors, we have

$$(c_1 - 2c_2)x^2 + (-2c_1 + 3c_2)x + (3c_1 + c_2) = 10x^2 - 17x + 9 \qquad (2)$$

The only way that two polynomials are equal is if coefficients of like terms are equal. Thus 2 is true only if there exist c_1 and c_2 that satisfy the system

$$
\begin{aligned}
c_1 - 2c_2 &= 10 \\
-2c_1 + 3c_2 &= -17 \\
3c_1 + c_2 &= 9
\end{aligned}
$$

Applying our usual solution methods, we can determine that the system has the unique solution $c_1 = 4$ and $c_2 = -3$. Hence $p(x)$ is in S.   ∎

EXAMPLE 2   Let $S = \text{span}\{1, \cos(x), \cos(2x)\}$ be a subset of $C[0, \pi]$. Determine if $f(x) = \sin^2(x)$ is in S.

Solution At first glance it may not appear that $f(x)$ is in S. However, recall from trigonometry the identity

$$\sin^2(x) = \frac{1 - \cos(2x)}{2}$$

Hence it follows that $\sin^2(x)$ is a linear combination of two vectors in S, with

$$\sin^2(x) = \tfrac{1}{2}(1) - \tfrac{1}{2}\cos(2x)$$

Therefore $f(x) = \sin^2(x)$ is in S.   ∎

The preceding examples both involved spanning subsets S of a vector space. In Euclidean space $\mathbf{R}^n$, such sets are subspaces. The same is true in vector spaces.

THEOREM 7.6

Suppose that $\mathcal{V}$ is a (possibly infinite) subset of a vector space V, and let $S = \text{span}(\mathcal{V})$. Then S is a subspace of V.

▶ Theorem 7.6 generalizes Theorem 4.2 in Section 4.1.

The proof follows from verifying that the three conditions required of a subspace (Definition 7.3 in Section 7.1) are met. It is left as an exercise.

EXAMPLE 3   Let $S = \text{span}\left\{\begin{bmatrix} 1 & 0 \\ 2 & 1 \end{bmatrix}, \begin{bmatrix} 0 & -1 \\ 1 & 3 \end{bmatrix}, \begin{bmatrix} 4 & 1 \\ -2 & 1 \end{bmatrix}\right\}$ be a subspace of the vector space $\mathbf{R}^{2\times 2}$. Is $\mathbf{v} = \begin{bmatrix} 2 & 5 \\ -3 & 4 \end{bmatrix}$ in S?

Solution In order for $\mathbf{v}$ to be in S, there must exist scalars c_1, c_2, and c_3 such that

$$c_1\begin{bmatrix} 1 & 0 \\ 2 & 1 \end{bmatrix} + c_2\begin{bmatrix} 0 & -1 \\ 1 & 3 \end{bmatrix} + c_3\begin{bmatrix} 4 & 1 \\ -2 & 1 \end{bmatrix} = \begin{bmatrix} 2 & 5 \\ -3 & 4 \end{bmatrix}$$

$$\implies \begin{bmatrix} (c_1 + 4c_3) & (-c_2 + c_3) \\ (2c_1 + c_2 - 2c_3) & (c_1 + 3c_2 + c_3) \end{bmatrix} = \begin{bmatrix} 2 & 5 \\ -3 & 4 \end{bmatrix}$$

For the components to be equal, c_1, c_2, and c_3 must satisfy the linear system

$$
\begin{array}{rcrcrcr}
c_1 & & & + & 4c_3 & = & 2 \\
& - & c_2 & + & c_3 & = & 5 \\
2c_1 & + & c_2 & - & 2c_3 & = & -3 \\
c_1 & + & 3c_2 & + & c_3 & = & 4
\end{array}
$$

Our standard solution methods can be used to show that this system has no solutions, so $\mathbf{v}$ is not in S. ∎

As we can see from our examples, even though vector spaces can be made up of polynomials, matrices, or other objects, answering questions about spanning sets often boils down to something that we have done again and again: solving a system of linear equations. But this is not always the case.

EXAMPLE 4 Can a finite set of vectors span **P**, the set of polynomials with real coefficients?

Solution Suppose that $\{ f_1(x),\ f_2(x), \ldots,\ f_m(x) \}$ is a set of polynomials in **P**, sorted by degree with

$$
\deg(f_1) \geq \deg(f_2) \geq \cdots \geq \deg(f_m)
$$

Then any linear combination

$$
c_1 f_1(x) + c_2 f_2(x) + \cdots + c_m f_m(x)
$$

has degree at most that of $f_1(x)$ (and possibly less, if there is cancellation). Thus, if $n = \deg(f_1) + 1$, then $g(x) = x^n$ has degree greater than any linear combination of our set and hence cannot be in the span of the set. As there is nothing special about our set, this argument shows that no finite set of vectors can possibly span all of **P**. ∎

The argument in Example 4 that shows that a finite set cannot span **P** does not apply to the infinite set $\{1, x, x^2, x^3, \ldots\}$. This infinite set does span **P**. Verification is left as an exercise.

Linear Independence

We now turn to the definition of linear independence. The definition is a near duplicate of the definition given in Section 2.3 for vectors in Euclidean space.

DEFINITION 7.7

Definition Linearly Independent, linearly Dependent

▶ Definition 7.7 generalizes Definition 2.11 in Section 2.3.

Let $\mathcal{V} = \{\mathbf{v}_1, \mathbf{v}_2, \ldots, \mathbf{v}_m\}$ be a set of vectors in a vector space V. Then $\mathcal{V}$ is **linearly independent** if the equation

$$
c_1\mathbf{v}_1 + c_2\mathbf{v}_2 + \cdots + c_m\mathbf{v}_m = \mathbf{0}
$$

has only the trivial solution $c_1 = \cdots = c_m = 0$. The set $\mathcal{V}$ is **linearly dependent** if the equation has any nontrivial solutions.

As with span, the definition of linear independence extends to infinite sets. If $\mathcal{V}$ consists of infinitely many vectors, then $\mathcal{V}$ is linearly independent if *every finite* subset of $\mathcal{V}$ is linearly independent.

EXAMPLE 5   Determine if

$$\{x^3 - 2x + 4, \ x^3 + 2x^2 - 2, \ 2x^3 + x^2 - 3x + 5\}$$

is a linearly independent subset of $\mathbf{P}^3$.

Solution Just as in Euclidean space, we start by setting

$$c_1(x^3 - 2x + 4) + c_2(x^3 + 2x^2 - 2) + c_3(2x^3 + x^2 - 3x + 5) = 0$$

and then determine the values of c_1, c_2, and c_3 that satisfy the equation. Collecting common factors gives us the equivalent equation

$$(c_1 + c_2 + 2c_3)x^3 + (2c_2 + c_3)x^2 + (-2c_1 - 3c_3)x + (4c_1 - 2c_2 + 5c_3) = 0$$

Our polynomial is identically zero if and only if the coefficients are all zero. This will be true for any solution to the homogeneous system

$$
\begin{aligned}
c_1 + \ c_2 + 2c_3 &= 0 \\
2c_2 + \ c_3 &= 0 \\
-2c_1 \qquad\quad - 3c_3 &= 0 \\
4c_1 - 2c_2 + 5c_3 &= 0
\end{aligned}
$$

Applying our standard solution methods shows that the system has infinitely many solutions, among them $c_1 = 3$, $c_2 = 1$, and $c_3 = -2$. Since a nontrivial linear combination of our vectors equals the zero vector, our set is linearly dependent. ■

EXAMPLE 6   Determine if the subset

$$\left\{ \begin{bmatrix} 1 & 0 & 2 \\ -1 & 1 & 0 \end{bmatrix}, \begin{bmatrix} 3 & 1 & 0 \\ 2 & 2 & 2 \end{bmatrix} \right\}$$

of $\mathbf{R}^{2 \times 3}$ is linearly independent.

Solution We proceed just as we did in the preceding example, by setting up the equation

$$c_1 \begin{bmatrix} 1 & 0 & 2 \\ -1 & 1 & 0 \end{bmatrix} + c_2 \begin{bmatrix} 3 & 1 & 0 \\ 2 & 2 & 2 \end{bmatrix} = \begin{bmatrix} 0 & 0 & 0 \\ 0 & 0 & 0 \end{bmatrix}$$

Comparing the components on each side gives the linear system

$$
\begin{aligned}
c_1 + 3c_2 &= 0 \\
c_2 &= 0 \\
2c_1 \qquad &= 0 \\
-c_1 + 2c_2 &= 0 \\
c_1 + 2c_2 &= 0 \\
2c_2 &= 0
\end{aligned}
$$

We can see that this system has only the trivial solution $c_1 = c_2 = 0$, so the set is linearly independent. ■

A few observations about linear independence carry over from Euclidean space (proofs left as exercises):

- The set $\{\mathbf{0}, \mathbf{v}_1, \cdots, \mathbf{v}_m\}$ is linearly dependent for all vectors $\mathbf{v}_1, \cdots, \mathbf{v}_m$.

▶ This generalizes Theorem 2.12 in Section 2.3.

- A set of two nonzero vectors $\{\mathbf{v}_1, \mathbf{v}_2\}$ is linearly dependent if and only if one is a scalar multiple of the other. (We could solve Example 6 using this fact.)

- A set with just one vector $\{\mathbf{v}_1\}$ is linearly dependent if and only if $\mathbf{v}_1 = \mathbf{0}$.

EXAMPLE 7   Determine if $\{1, x, x^2, x^3, \ldots\}$ is a linearly independent subset of **P**, the set of all polynomials with real coefficients.

Solution Here we are working with an infinite set, so we must show that *every* finite subset is linearly independent. Let's take a typical finite subset

$$\{x^{a_1}, x^{a_2}, \ldots, x^{a_m}\}$$

where $a_1, a_2, \ldots, a_m$ are distinct nonnegative integers. Suppose that

$$c_1 x^{a_1} + c_2 x^{a_2} + \cdots + c_m x^{a_m} = 0$$

We have the trivial solution $c_1 = c_2 = \cdots = c_m = 0$, but are there others? The answer is no, because a polynomial is identically zero (that is, zero for all x) only if all the coefficients are zero. Since we have only the trivial solution, our subset is linearly independent. Since our subset is arbitrary, it follows that all finite subsets are linearly independent and therefore our original infinite set is linearly independent. ∎

EXAMPLE 8   Determine if $\{1, x, e^x\}$ is a linearly independent subset of $C[0, 10]$.

Solution Just as in the other examples, we ask ourselves if there exist nontrivial scalars c_1, c_2, and c_3 such that

$$c_1(1) + c_2 x + c_3 e^x = 0 \tag{3}$$

An equivalent formulation is

$$c_1 + c_2 x = -c_3 e^x$$

Note that regardless of choice for c_1 and c_2, the left side is a linear equation. Since e^x is an exponential function, the only way the right side is linear is if $c_3 = 0$. This in turn forces $c_1 = c_2 = 0$ as well, which shows that the only solution to our original equation is the trivial one. Therefore our set is linearly independent. ∎

In Example 8 we used the fact that e^x is not a linear function. But suppose we did not know that? Let's look at this example again, this time using a different approach.

EXAMPLE 9   Determine if $\{1, x, e^x\}$ is a linearly independent subset of $C[0, 10]$.

Solution Suppose that c_1, c_2, and c_3 satisfy (3). Then the equation is satisfied for every value of x, and hence in particular for each of $x = 0$, $x = 1$, and $x = 2$. Setting x equal to each of these values and plugging into (3) yields the homogeneous linear system

$$
\begin{aligned}
c_1 \qquad\; + \;\; c_3 &= 0 \\
c_1 + \;\; c_2 + \;\; e c_3 &= 0 \\
c_1 + 2c_2 + e^2 c_3 &= 0
\end{aligned}
\tag{4}
$$

By applying our standard methods, we can verify that the system (4) has only the trivial solution $c_1 = c_2 = c_3 = 0$. Hence the only possible solution to (3) is also the trivial one, because any nontrivial solution to (3) would also satisfy (4). Therefore the set $\{1, x, e^x\}$ is linearly independent. ∎

A note of warning: The method in Example 9 cannot be used to show that a set of functions is linearly dependent. To see why, suppose that $f(x) = -x^3 + 3x^2 - 2x$ and

Figure 1 $f(x) = -x^3 + 3x^2 - 2x$ and $g(x) = \sin(2\pi x)$.

$g(x) = \sin(2\pi x)$. Then

$$f(0) = g(0), \quad f(1) = g(1), \quad f(2) = g(2)$$

So if we apply the approach in Example 9, we get a linear system

$$c_1 + c_2 = 0$$
$$c_1 + c_2 = 0$$
$$c_1 + c_2 = 0$$

Clearly this system has nontrivial solutions. However, the functions $f(x)$ and $g(x)$ are not multiples of each other (see Figure 1) and so are linearly independent. Functions as vectors are linearly dependent only if the dependence relation holds for all domain values, not just a few isolated values.

Span and Linear Independence

Sometimes it is difficult to keep straight the concepts of span and linear independence. The last two theorems in this section highlight the distinction between the two concepts.

THEOREM 7.8

▶ Theorem 7.8 generalizes Theorem 2.14 in Section 2.3.

The set $\mathcal{V} = \{\mathbf{v}_1, \mathbf{v}_2, \ldots, \mathbf{v}_m\}$ of nonzero vectors is linearly dependent if and only if one vector in the set is in the span of the others.

The proof of Theorem 7.8 is left as an exercise.

EXAMPLE 10   Determine if the set of vectors

$$\{1, \cos(x), \cos(2x), \sin^2(x)\} \tag{5}$$

is a linearly independent subset of $C[0, 10]$.

Solution In Example 2 we showed that $\sin^2(x)$ is in span$\{1, \cos(x), \cos(2x)\}$. Hence by Theorem 7.8 the set of vectors in (5) is linearly dependent. ∎

THEOREM 7.9

▶ Theorem 7.9 generalizes Theorem 2.10 in Section 2.2 and Theorem 2.18 in Section 2.3.

Let $\mathcal{V} = \{\mathbf{v}_1, \mathbf{v}_2, \ldots, \mathbf{v}_m\}$ be a subset of a vector space V. Then:

(a) The set $\mathcal{V}$ is *linearly independent* if and only if the equation $c_1\mathbf{v}_1 + \cdots + c_m\mathbf{v}_m = \mathbf{v}$ has *at most* one solution for each $\mathbf{v}$ in V.

(b) The set $\mathcal{V}$ *spans* V if and only if the equation $c_1\mathbf{v}_1 + \cdots + c_m\mathbf{v}_m = \mathbf{v}$ has *at least* one solution for each $\mathbf{v}$ in V.

The proof is left as an exercise. Note that Theorem 7.9 applies to subspaces as well as vector spaces.

EXAMPLE 11 Let S be the subspace of $T(2, 2)$ of linear transformations of the form

$$T\left(\begin{bmatrix} x_1 \\ x_2 \end{bmatrix}\right) = \begin{bmatrix} a_1 x_1 \\ a_2 x_2 \end{bmatrix} \tag{6}$$

where a_1 and a_2 are scalars (see Exercise 17 of Section 7.1). Suppose that

$$T_1\left(\begin{bmatrix} x_1 \\ x_2 \end{bmatrix}\right) = \begin{bmatrix} x_1 \\ 0 \end{bmatrix} \quad \text{and} \quad T_2\left(\begin{bmatrix} x_1 \\ x_2 \end{bmatrix}\right) = \begin{bmatrix} 0 \\ x_2 \end{bmatrix}$$

Show that the set $\{T_1, T_2\}$ is linearly independent and spans S.

Solution Suppose that T is a linear transformation in S and hence has the form in (6). To apply Theorem 7.9, we need to determine the number of solutions to the equation $c_1 T_1(\mathbf{x}) + c_2 T_2(\mathbf{x}) = T(\mathbf{x})$. This equation is equivalent to

$$c_1 \begin{bmatrix} x_1 \\ 0 \end{bmatrix} + c_2 \begin{bmatrix} 0 \\ x_2 \end{bmatrix} = \begin{bmatrix} a_1 x_1 \\ a_2 x_2 \end{bmatrix}$$

For this to hold for all x_1 and x_2, we must have $c_1 = a_1$ and $c_2 = a_2$. Thus the equation $c_1 T_1(\mathbf{x}) + c_2 T_2(\mathbf{x}) = T(\mathbf{x})$ has exactly one solution for any a_1 and a_2. Therefore by Theorem 7.9(a) the set $\{T_1, T_2\}$ is linearly independent, and by Theorem 7.9(b) the set $\{T_1, T_2\}$ spans S. ■

| EXERCISES |

In Exercises 1–4, determine if the vector is in the subspace of $\mathbf{P}^2$ given by

$$\text{span}\{3x^2 + x - 1, \ x^2 - 3x + 2\}$$

1. $\mathbf{v} = 3x^2 + 11x - 8$

2. $\mathbf{v} = 2x^2 - 9x + 7$

3. $\mathbf{v} = 10x - 7$

4. $\mathbf{v} = 7x^2 - x$

In Exercises 5–8, determine if the vector is in the subspace of $\mathbf{P}^3$ given by

$$\text{span}\{x^3 + x - 2, \ x^2 + 2x + 1, \ x^3 - x^2 + x\}$$

5. $\mathbf{v} = x^3 + 2x^2 - 3x$

6. $\mathbf{v} = 3x^2 + 4x$

7. $\mathbf{v} = x^2 + 4x + 4$

8. $\mathbf{v} = x^2 + 2x - 1$

In Exercises 9–12, determine if the vector is in the subspace of $\mathbf{R}^{2\times3}$ given by

$$\text{span}\left\{\begin{bmatrix} 1 & 2 & 1 \\ 0 & 1 & 3 \end{bmatrix}, \begin{bmatrix} 0 & 3 & 1 \\ -1 & 1 & 0 \end{bmatrix}\right\}$$

9. $\mathbf{v} = \begin{bmatrix} -1 & 4 & 1 \\ -2 & 1 & -3 \end{bmatrix}$

10. $\mathbf{v} = \begin{bmatrix} 2 & 1 & 1 \\ 1 & 1 & 5 \end{bmatrix}$

11. $\mathbf{v} = \begin{bmatrix} 2 & -5 & -1 \\ 3 & -1 & 6 \end{bmatrix}$

12. $\mathbf{v} = \begin{bmatrix} 3 & 3 & 2 \\ 1 & 2 & 9 \end{bmatrix}$

In Exercises 13–16, determine if the vector is in the subspace of $\mathbf{R}^{2\times2}$ given by

$$\text{span}\left\{\begin{bmatrix} -1 & 3 \\ 4 & 1 \end{bmatrix}, \begin{bmatrix} 0 & 2 \\ 5 & -3 \end{bmatrix}, \begin{bmatrix} 1 & 4 \\ 2 & 1 \end{bmatrix}\right\}$$

13. $\mathbf{v} = \begin{bmatrix} -4 & 3 \\ 5 & 5 \end{bmatrix}$

14. $\mathbf{v} = \begin{bmatrix} 2 & 3 \\ 3 & -3 \end{bmatrix}$

15. $\mathbf{v} = \begin{bmatrix} -2 & -1 \\ 2 & 0 \end{bmatrix}$

16. $\mathbf{v} = \begin{bmatrix} 1 & 2 \\ -3 & 4 \end{bmatrix}$

In Exercises 17–26, determine if the subset is linearly independent in the given vector space.

17. $\{x^2 - 3, \ 3x^2 + 1\}$ in $\mathbf{P}^2$

18. $\{2x^3 - x + 3,\ -4x^3 + 2x - 6\}$ in $\mathbf{P}^3$

19. $\{x^3 + 2x + 4,\ x^2 - x - 1,\ x^3 + 2x^2 + 2\}$ in $\mathbf{P}^3$

20. $\{x^2 + 3x + 2,\ x^3 - 2x^2,\ x^3 + x^2 - x - 1\}$ in $\mathbf{P}^3$

21. $\left\{\begin{bmatrix} 2 & -1 \\ 1 & 3 \end{bmatrix}, \begin{bmatrix} -4 & 2 \\ -2 & -6 \end{bmatrix}\right\}$ in $\mathbf{R}^{2\times 2}$

22. $\left\{\begin{bmatrix} -1 & 3 \\ 4 & 1 \end{bmatrix}, \begin{bmatrix} 3 & 5 \\ 0 & 1 \end{bmatrix}, \begin{bmatrix} 1 & 4 \\ 2 & 1 \end{bmatrix}\right\}$ in $\mathbf{R}^{2\times 2}$

23. $\left\{\begin{bmatrix} 1 & 0 & 1 \\ 2 & 1 & 4 \end{bmatrix}, \begin{bmatrix} 3 & 1 & 2 \\ 0 & 3 & 3 \end{bmatrix}\right\}$ in $\mathbf{R}^{2\times 3}$

24. $\left\{\begin{bmatrix} 1 & 2 \\ 0 & 1 \\ 1 & 4 \end{bmatrix}, \begin{bmatrix} 5 & 4 \\ 1 & 5 \\ 4 & 10 \end{bmatrix}, \begin{bmatrix} 3 & 0 \\ 1 & 3 \\ 2 & 3 \end{bmatrix}\right\}$ in $\mathbf{R}^{3\times 2}$

25. $\{\sin^2(x),\ \cos^2(x),\ 1\}$ in $C[0, \pi]$

26. $\{\sin(2x),\ \cos(2x),\ \sin(x)\cos(x)\}$ in $C[0, \pi]$

FIND AN EXAMPLE For Exercises 27–32, find an example that meets the given specifications.

27. A subset of $\mathbf{R}^{2\times 2}$ that spans $\mathbf{R}^{2\times 2}$ but is not linearly independent.

28. An infinite subset of $\mathbf{P}$ that is linearly independent but does not span $\mathbf{P}$.

29. A vector space V and an infinite linearly independent subset $\mathcal{V}$.

30. A set of nonzero vectors that is linearly dependent and yet has a vector in the set that is *not* a linear combination of the other vectors. Explain why this does not contradict Theorem 7.8.

31. Two infinite linearly independent subsets $\mathcal{V}_1$ and $\mathcal{V}_2$ of $\mathbf{R}^\infty$ such that $\text{span}(\mathcal{V}_1) \cap \text{span}(\mathcal{V}_2) = \{\mathbf{0}\}$.

32. A subset of $T(2, 2)$ that is linearly independent and spans $T(2, 2)$.

TRUE OR FALSE For Exercises 33–42, determine if the statement is true or false, and justify your answer.

33. Vectors must be columns of numbers.

34. A linearly independent set cannot have an infinite number of vectors.

35. A set of vectors $\mathcal{V}$ in a vector space V can be linearly independent or can span V, but cannot do both.

36. Suppose that f and g are linearly dependent functions in $C[1, 4]$. If $f(1) = -3g(1)$, then it must be that $f(4) = -3g(4)$.

37. Suppose that $\mathcal{V}_1 \subset \mathcal{V}_2$ are sets in a vector space V. If $\mathcal{V}_1$ is linearly independent, then so is $\mathcal{V}_2$.

38. Let $\{\mathbf{v}_1, \ldots, \mathbf{v}_k\}$ be a linearly independent subset of a vector space V. If $c \neq 0$ is a scalar, then $\{c\mathbf{v}_1, \ldots, c\mathbf{v}_k\}$ is also linearly independent.

39. Suppose that $\mathcal{V}_1 \subset \mathcal{V}_2$ are sets in a vector space V. If $\mathcal{V}_2$ spans V, then so does $\mathcal{V}_1$.

40. Let $\{\mathbf{v}_1, \ldots, \mathbf{v}_k\}$ be a linearly independent subset of a vector space V. For any $\mathbf{v} \neq \mathbf{0}$ in V, the set $\{\mathbf{v} + \mathbf{v}_1, \ldots, \mathbf{v} + \mathbf{v}_k\}$ is also linearly independent.

41. If $\{\mathbf{v}_1, \mathbf{v}_2, \mathbf{v}_3\}$ is a linearly independent set, then so is $\{\mathbf{v}_1, \mathbf{v}_2 - \mathbf{v}_1, \mathbf{v}_3 - \mathbf{v}_2 + \mathbf{v}_1\}$.

42. If $\mathcal{V}_1$ and $\mathcal{V}_2$ are linearly independent subsets of a vector space V and $\mathcal{V}_1 \cap \mathcal{V}_2$ is nonempty, then $\mathcal{V}_1 \cap \mathcal{V}_2$ is also linearly independent.

43. In Example 7 it is shown that $\{1, x, x^2, \ldots\}$ is linearly independent. Prove $\{1, x, x^2, \ldots\}$ also spans $\mathbf{P}$.

44. Prove Theorem 7.8: A set $\{\mathbf{v}_1, \mathbf{v}_2, \ldots, \mathbf{v}_m\}$ of nonzero vectors is linearly dependent if and only if one vector in the set is in the span of the others. (HINT: This is similar to Theorem 2.14 in Section 2.3.)

45. Prove that the subset $\{\mathbf{0}, \mathbf{v}_1, \ldots, \mathbf{v}_m\}$ of a vector space must be linearly dependent. (HINT: This is similar to Theorem 2.12 in Section 2.3.)

46. Prove that any set of two nonzero vectors $\{\mathbf{v}_1, \mathbf{v}_2\}$ is linearly dependent if and only if one is a scalar multiple of the other.

47. Prove that a set consisting of one vector $\{\mathbf{v}_1\}$ is linearly dependent if and only if $\mathbf{v}_1 = \mathbf{0}$.

48. Let $\mathcal{V} = \{\mathbf{v}_1, \mathbf{v}_2, \ldots, \mathbf{v}_m\}$ be a subset of a vector space V. Prove Theorem 7.9 by proving each of the following:

(a) The set $\mathcal{V}$ is linearly independent if and only if the equation $c_1\mathbf{v}_1 + \cdots + c_m\mathbf{v}_m = \mathbf{v}$ has *at most* one solution for each $\mathbf{v}$ in V.

(b) The set $\mathcal{V}$ spans V if and only if the equation $c_1\mathbf{v}_1 + \cdots + c_m\mathbf{v}_m = \mathbf{v}$ has *at least* one solution for each $\mathbf{v}$ in V.

49. Prove that no finite subset of $\mathbf{R}^\infty$ can span $\mathbf{R}^\infty$.

50. Suppose that $\mathcal{V}_1 \subset \mathcal{V}_2$ are sets in a vector space V. Prove that if $\mathcal{V}_1$ spans V, then so does $\mathcal{V}_2$.

51. Suppose that $\mathcal{V}_1 \subset \mathcal{V}_2$ are nonempty sets in a vector space V. Prove that if $\mathcal{V}_2$ is linearly independent, then so is $\mathcal{V}_1$.

52. Let $\mathbf{v}_1, \ldots, \mathbf{v}_m$ and $\mathbf{v}$ be vectors in a vector space V. If $\mathbf{v}$ is in the span of the set $\{\mathbf{v}_1, \ldots, \mathbf{v}_m\}$, prove that

$$\text{span}\{\mathbf{v}, \mathbf{v}_1, \ldots, \mathbf{v}_m\} = \text{span}\{\mathbf{v}_1, \ldots, \mathbf{v}_m\}$$

53. Suppose that $\mathcal{V} = \{\mathbf{v}_1, \ldots, \mathbf{v}_m\}$ spans a vector space V, and suppose that $\mathbf{v}$ is in V but not in $\mathcal{V}$. Prove that $\{\mathbf{v}, \mathbf{v}_1, \ldots, \mathbf{v}_m\}$ is linearly dependent.

54. Suppose that $\mathcal{V} = \{\mathbf{v}_1, \mathbf{v}_2, \ldots, \mathbf{v}_m\}$ is a linearly independent subset of a vector space V. Prove that $\{\mathbf{v}_2, \ldots, \mathbf{v}_m\}$ does not span V.

Ⓒ In Exercises 55–58, if possible use the method demonstrated in Example 9 to determine if the given subset is linearly independent in the given vector space. If this is not possible, explain why.

55. $\{x,\ \sin(\pi x/2),\ e^x\}$ in $C[0, \pi]$

56. $\{x,\ \sin(x),\ \cos(x)\}$ in $C[0, \pi]$

57. $\{e^x,\ \cos^2(x),\ \cos(2x),\ 1\}$ in $C[0, \pi]$

58. $\{\cos(2x),\ \sin(2x),\ \cos^2(x),\ \sin^2(x)\}$ in $C[0, \pi]$

7.3 Basis and Dimension

Now that we have developed the concepts of span and linear independence in the context of a vector space, we are ready to consider the notion of basis and dimension in the same setting. As with much of this chapter, the definitions of basis and dimension are essentially the same as in Euclidean space. Let's start with basis.

DEFINITION 7.10

Definition Basis

Let V be a subset of a vector space V. Then V is a **basis** of V if V is linearly independent and spans V.

▶ Definition 7.10 generalizes Definition 4.8 in Section 4.2.

EXAMPLE 1 Is the set $\{x^2 + 4x - 3,\ x^2 + 1,\ x - 2\}$ a basis for $\mathbf{P}^2$?

Solution We need to determine if the given set is linearly independent and spans $\mathbf{P}^2$. Of the two, span is generally more difficult to verify, so let's tackle that first. (We will also find out about linear independence along the way.) A typical vector in $\mathbf{P}^2$ has the form $a_2 x^2 + a_1 x + a_0$, and for each such vector we need to know if there exist corresponding scalars c_1, c_2, and c_3 such that

$$c_1(x^2 + 4x - 3) + c_2(x^2 + 1) + c_3(x - 2) = a_2 x^2 + a_1 x + a_0$$

Reorganizing the left side to collect common terms, we have

$$(c_1 + c_2)x^2 + (4c_1 + c_3)x + (-3c_1 + c_2 - 2c_3) = a_2 x^2 + a_1 x + a_0$$

Comparing coefficients gives us the linear system

$$\begin{aligned} c_1 + c_2 \quad\quad &= a_2 \\ 4c_1 \quad\ + c_3 &= a_1 \\ -3c_1 + c_2 - 2c_3 &= a_0 \end{aligned} \tag{1}$$

Using our standard solution methods, we find that for each choice of a_0, a_1, and a_2, the system has unique solution

$$c_1 = \frac{-a_2 + 2a_1 + a_0}{4}, \quad c_2 = \frac{5a_2 - 2a_1 - a_0}{4}, \quad c_3 = a_2 - a_1 - a_0$$

Applying both parts of Theorem 7.9 in Section 7.2, we can conclude that our set is both linearly independent and spans $\mathbf{P}^2$. Hence the set is a basis of $\mathbf{P}^2$. ∎

The method of solution in Example 1 suggests the following general theorem.

THEOREM 7.11

The set $V = \{\mathbf{v}_1, \ldots, \mathbf{v}_m\}$ is a basis for a vector space V if and only if the equation

$$c_1 \mathbf{v}_1 + \cdots + c_m \mathbf{v}_m = \mathbf{v} \tag{2}$$

has a unique solution $c_1, \ldots, c_m$ for every $\mathbf{v}$ in V.

▶ Theorem 7.11 generalizes Theorem 4.9 in Section 4.2.

The proof follows from applying Theorem 7.9 in Section 7.2 and is left as an exercise.

EXAMPLE 2 Verify that the set $\{1, x, x^2, \ldots, x^n\}$ is a basis for $\mathbf{P}^n$. (This is called the **standard basis** for $\mathbf{P}^n$.)

Solution Suppose that $a_n x^n + \cdots + a_1 x + a_0$ is a typical vector in $\mathbf{P}^n$. Forming a linear combination of our set and setting it equal to our typical vector produces

$$c_1(1) + c_2(x) + \cdots + c_{n+1}(x^n) = a_n x^n + \cdots + a_1 x + a_0$$

Comparing coefficients instantly shows that the unique linear combination is given by $c_1 = a_0, c_2 = a_1, \ldots, c_{n+1} = a_n$. Since this works for every vector in $\mathbf{P}^n$, by Theorem 7.11 our set forms a basis for $\mathbf{P}^n$. ∎

Setting $n = 2$ in Example 2 gives another basis for $\mathbf{P}^2$. Note that this basis has the same number of elements as the basis in Example 1, illustrating the following theorem, which carries over from Euclidean space.

THEOREM 7.12

Suppose that $\mathcal{V}_1$ and $\mathcal{V}_2$ are both bases of a vector space V. Then $\mathcal{V}_1$ and $\mathcal{V}_2$ have the same number of elements.

▶ Theorem 7.12 generalizes Theorem 4.12 in Section 4.2.

The proof is left as an exercise. The theorem also applies to vector spaces with bases that have infinitely many vectors: If one basis has infinitely many vectors, then they all do.

EXAMPLE 3 Show that every basis for $\mathbf{P}$ has infinitely many elements.

Solution In Example 7 of Section 7.2, it is shown that the set $\{1, x, x^2, \ldots\}$ is linearly independent, and in Exercise 43 of Section 7.2 it is shown that $\{1, x, x^2, \ldots\}$ spans $\mathbf{P}$. Therefore the set $\{1, x, x^2, \ldots\}$ is a basis for $\mathbf{P}$, and thus by Theorem 7.12 every basis for $\mathbf{P}$ must have infinitely many elements. ∎

Since every basis for a vector space has the same number of vectors, the following definition makes sense.

DEFINITION 7.13

Definition Dimension

The **dimension** of a vector space V is equal to the number of vectors in any basis of V. If a basis of V has infinitely many vectors, then we say that the dimension is infinite.

▶ Definition 7.13 generalizes Definition 4.13 in Section 4.2.

For example, we have seen that $\mathbf{P}^n$ has dimension $n+1$ and $\mathbf{P}$ has infinite dimension. Put briefly, $\dim(\mathbf{P}^n) = n + 1$ and $\dim(\mathbf{P}) = \infty$.

Note that a trivial vector space $V = \{\mathbf{0}\}$ consisting only of the zero vector has no basis because it has no linearly independent subsets. We define $\dim(V) = 0$ when V is a trivial vector space.

EXAMPLE 4 Find the dimension of $\mathbf{R}^{2\times2}$, the vector space of real 2×2 matrices.

Solution All we need to do is find any basis for $\mathbf{R}^{2\times2}$ and then count the vectors. The standard basis is given by

$$\left\{ \begin{bmatrix} 1 & 0 \\ 0 & 0 \end{bmatrix}, \begin{bmatrix} 0 & 1 \\ 0 & 0 \end{bmatrix}, \begin{bmatrix} 0 & 0 \\ 1 & 0 \end{bmatrix}, \begin{bmatrix} 0 & 0 \\ 0 & 1 \end{bmatrix} \right\}$$

Verification that this is a basis for $\mathbf{R}^{2\times2}$ is left as an exercise. Hence it follows that $\dim(\mathbf{R}^{2\times2}) = 4$. ∎

A similar argument to the one in Example 4 can be used to show that $\dim(\mathbf{R}^{n\times m}) = nm$ (see Exercise 50).

Since subspaces are essentially vector spaces within vector spaces, they also have dimensions.

EXAMPLE 5 Find the dimension of the subspace $S = \text{span}\{T_1, T_2, T_3\}$ of $T(2, 2)$, where

$$T_1\left(\begin{bmatrix} x_1 \\ x_2 \end{bmatrix}\right) = \begin{bmatrix} x_1 - x_2 \\ 2x_1 \end{bmatrix}, \quad T_2\left(\begin{bmatrix} x_1 \\ x_2 \end{bmatrix}\right) = \begin{bmatrix} x_2 \\ 2x_1 + x_2 \end{bmatrix}, \quad T_3\left(\begin{bmatrix} x_1 \\ x_2 \end{bmatrix}\right) = \begin{bmatrix} 2x_1 - 5x_2 \\ -2x_1 - 3x_2 \end{bmatrix}$$

Solution Let's check if the set is linearly independent. If so, then the dimension is 3 and we are done. If not, then we will identify a linear relation among the vectors that will allow us to remove one.

The zero vector in $T(2, 2)$ is the linear transformation $T_0(\mathbf{x}) = \mathbf{0}$. We need to find the values of c_1, c_2, and c_3 such that

$$c_1 T_1(\mathbf{x}) + c_2 T_2(\mathbf{x}) + c_3 T_3(\mathbf{x}) = T_0(\mathbf{x})$$

is true for all $\mathbf{x}$ in $\mathbf{R}^2$. This is equivalent to the equation

$$c_1 \begin{bmatrix} x_1 - x_2 \\ 2x_1 \end{bmatrix} + c_2 \begin{bmatrix} x_2 \\ 2x_1 + x_2 \end{bmatrix} + c_3 \begin{bmatrix} 2x_1 - 5x_2 \\ -2x_1 - 3x_2 \end{bmatrix} = \begin{bmatrix} 0 \\ 0 \end{bmatrix}$$

which in turn is equivalent to the system

$$c_1(x_1 - x_2) + c_2 x_2 + c_3(2x_1 - 5x_2) = 0$$
$$c_1(2x_1) + c_2(2x_1 + x_2) + c_3(-2x_1 - 3x_2) = 0$$

Reorganizing to separate out x_1 and x_2 gives us

$$x_1(c_1 + 2c_3) + x_2(-c_1 + c_2 - 5c_3) = 0$$
$$x_1(2c_1 + 2c_2 - 2c_3) + x_2(c_2 - 3c_3) = 0$$

Since the system must be satisfied for all values of x_1 and x_2, we require that

$$\begin{aligned} c_1 \quad\quad\quad + 2c_3 &= 0 \\ -c_1 + c_2 - 5c_3 &= 0 \\ 2c_1 + 2c_2 - 2c_3 &= 0 \\ c_2 - 3c_3 &= 0 \end{aligned}$$

Using our standard methods, we can show that this system has infinitely many solutions, among them $c_1 = 2$, $c_2 = -3$, and $c_3 = -1$. Therefore

$$2 T_1(\mathbf{x}) - 3 T_2(\mathbf{x}) - T_3(\mathbf{x}) = T_0(\mathbf{x})$$

or

$$2 T_1(\mathbf{x}) - 3 T_2(\mathbf{x}) = T_3(\mathbf{x})$$

Since T_3 is a linear combination of T_1 and T_2, it follows (see Exercise 52 in Section 7.2) that $S = \text{span}\{T_1, T_2\}$. It is straightforward to verify that T_1 and T_2 are not multiples of each other and so are linearly independent. Hence $\{T_1, T_2\}$ is a basis for S, and therefore $\dim(S) = 2$. ■

In Example 5, we started with a set that spanned a subspace S and removed a vector to form a basis for S. The following theorem formalizes this process, along with the process of expanding a linearly independent set to a basis.

THEOREM 7.14

▶ Theorem 7.14 generalizes Theorem 4.14 in Section 4.2.

Let $\mathcal{V} = \{\mathbf{v}_1, \ldots, \mathbf{v}_m\}$ be a subset of a nontrivial finite dimensional vector space V.

(a) If $\mathcal{V}$ spans V, then either $\mathcal{V}$ is a basis for V or vectors can be removed from $\mathcal{V}$ to form a basis for V.

(b) If $\mathcal{V}$ is linearly independent, then either $\mathcal{V}$ is a basis for V or vectors can be added to $\mathcal{V}$ to form a basis for V.

The proof is left as an exercise. Note that Theorem 7.14 also applies to subspaces of vector spaces.

EXAMPLE 6   Extend the set $V = \{x^2 + 2x + 1,\ x^2 + 6x + 3\}$ to a basis for $\mathbf{P}^2$.

Solution   We know that $\{x^2,\ x,\ 1\}$ is a basis for $\mathbf{P}^2$, so the combined set

$$\{x^2 + 2x + 1,\ x^2 + 6x + 3,\ x^2,\ x,\ 1\} \tag{3}$$

spans $\mathbf{P}^2$. To find a basis that includes V, we need to determine the dependences among the vectors in 3. We start with the equation

$$c_1(x^2 + 2x + 1) + c_2(x^2 + 6x + 3) + c_3 x^2 + c_4 x + c_5 = 0$$

which is equivalent to

$$(c_1 + c_2 + c_3)x^2 + (2c_1 + 6c_2 + c_4)x + (c_1 + 3c_2 + c_5) = 0$$

Setting each coefficient equal to zero yields the linear system

$$
\begin{aligned}
c_1 + c_2 + c_3 \qquad\qquad &= 0 \\
2c_1 + 6c_2 \qquad + c_4 \qquad &= 0 \\
c_1 + 3c_2 \qquad\qquad + c_5 &= 0
\end{aligned}
$$

The augmented matrix and corresponding echelon form are

$$
\begin{bmatrix}
1 & 1 & 1 & 0 & 0 & 0 \\
2 & 6 & 0 & 1 & 0 & 0 \\
1 & 3 & 0 & 0 & 1 & 0
\end{bmatrix}
\sim
\begin{bmatrix}
1 & 1 & 1 & 0 & 0 & 0 \\
0 & 2 & -1 & 0 & 1 & 0 \\
0 & 0 & 0 & 1 & -2 & 0
\end{bmatrix}
$$

Since the leading terms appear in columns 1, 2, and 4 of the echelon matrix, it follows that the vectors associated with c_1, c_2, and c_4 are linearly independent. To see why, note that if the vectors x^2 and 1 were eliminated from 3, then we would have the linear system

$$
\begin{aligned}
c_1 + c_2 \qquad &= 0 \\
2c_1 + 6c_2 + c_4 &= 0 \\
c_1 + 3c_2 \qquad &= 0
\end{aligned}
$$

The augmented matrix and echelon form are as before, only with the third and fifth columns removed. (The row operations are the same as before.)

$$
\begin{bmatrix}
1 & 1 & 0 & 0 \\
2 & 6 & 1 & 0 \\
1 & 3 & 0 & 0
\end{bmatrix}
\sim
\begin{bmatrix}
1 & 1 & 0 & 0 \\
0 & 2 & 0 & 0 \\
0 & 0 & 1 & 0
\end{bmatrix}
$$

We see that the new system has a unique solution, so the set

$$V_1 = \{x^2 + 2x + 1,\ x^2 + 6x + 3,\ x\}$$

is linearly independent. Furthermore, if V_1 is not a basis, then by Theorem 7.14(b) we can add vectors to V_1 to form a basis for $\mathbf{P}^2$. But this would mean that $\dim(\mathbf{P}^2) > 3$, which we know is false. Therefore V_1 is a basis.   ∎

Dimension gives us a way to measure the size of a vector space. Thus if one vector space is contained in another, it seems reasonable that the dimension of the former be smaller than that of latter. This was true in Euclidean space, and also is here.

THEOREM 7.15

▶ Theorem 7.15 is similar to Theorem 4.16 in Section 4.2.

Let V_1 and V_2 be vector spaces with V_1 a subset of V_2, and suppose both have the same definition of addition and scalar multiplication. Then $\dim(V_1) \leq \dim(V_2)$.

Proof If $\dim(V_2) = \infty$, then the theorem is true regardless of the dimension of V_1. (Similar reasoning applies if $\dim(V_1) = 0$.) Now suppose that $\dim(V_2)$ is finite, and let $\mathcal{V}_1$ be a basis for V_1. Since V_1 is a subset of V_2, $\mathcal{V}_1$ is a linearly independent subset of V_2. Thus, by Theorem 7.14(b), $\mathcal{V}_1$ is a basis for V_2 or can be expanded to a basis for V_2. Therefore the number of vectors in $\mathcal{V}_1$ is less than or equal to $\dim(V_2)$. Since $\mathcal{V}_1$ is a basis of V_1, we conclude that $\dim(V_1) \leq \dim(V_2)$. ∎

Note that in Theorem 7.15, V_1 can also be considered a subspace of V_2.

EXAMPLE 7 Show that $\dim(C(\mathbf{R})) = \infty$ without finding a basis for $C(\mathbf{R})$.

Solution We have already shown that $\dim(\mathbf{P}) = \infty$. Since all polynomials are continuous functions, then $\mathbf{P}$ is a subspace of $C(\mathbf{R})$. Hence by Theorem 7.15 we have $\dim(C(\mathbf{R})) = \infty$. ∎

THEOREM 7.16

▶ Theorem 7.16 generalizes Theorem 4.17 in Section 4.2.

Let $\mathcal{V} = \{\mathbf{v}_1, \ldots, \mathbf{v}_m\}$ be a subset of a vector space V with $\dim(V) = n$.

(a) If $m < n$, then $\mathcal{V}$ does not span V.

(b) If $m > n$, then $\mathcal{V}$ is linearly dependent.

The proof is left as an exercise. As with previous theorems in this section, Theorem 7.16 also applies to subspaces.

EXAMPLE 8 Determine by inspection which of the subsets cannot span $\mathbf{R}^{2 \times 2}$ and which must be linearly dependent.

$$\mathcal{V}_1 = \left\{ \begin{bmatrix} 2 & 1 \\ 4 & 3 \end{bmatrix}, \begin{bmatrix} 5 & 0 \\ 1 & 2 \end{bmatrix}, \begin{bmatrix} 8 & 3 \\ 5 & 6 \end{bmatrix} \right\}$$

$$\mathcal{V}_2 = \left\{ \begin{bmatrix} 5 & 9 \\ 0 & 4 \end{bmatrix}, \begin{bmatrix} 7 & 6 \\ 1 & 2 \end{bmatrix}, \begin{bmatrix} 4 & 3 \\ 7 & 8 \end{bmatrix}, \begin{bmatrix} 6 & 2 \\ 7 & 1 \end{bmatrix} \right\}$$

$$\mathcal{V}_3 = \left\{ \begin{bmatrix} 3 & 6 \\ 9 & 0 \end{bmatrix}, \begin{bmatrix} 1 & 4 \\ 3 & 8 \end{bmatrix}, \begin{bmatrix} 3 & 2 \\ 0 & 1 \end{bmatrix}, \begin{bmatrix} 0 & 1 \\ 6 & 6 \end{bmatrix}, \begin{bmatrix} 3 & 2 \\ 5 & 5 \end{bmatrix} \right\}$$

Solution In Example 4, we showed that $\dim(\mathbf{R}^{2 \times 2}) = 4$. Hence by Theorem 7.16(a), $\mathcal{V}_1$ cannot span $\mathbf{R}^{2 \times 2}$ because it has fewer than four vectors. By Theorem 7.16(b), $\mathcal{V}_3$ must be linearly dependent because it has more than four vectors.

Note that Theorem 7.16 cannot tell us if a set spans a vector space or is linearly independent. Hence we cannot conclude that $\mathcal{V}_2$ or $\mathcal{V}_3$ spans $\mathbf{R}^{2 \times 2}$, nor can we conclude that $\mathcal{V}_1$ or $\mathcal{V}_2$ is linearly independent. ∎

THEOREM 7.17

▶ Theorem 7.17 generalizes Theorem 4.15 in Section 4.2.

Let $\mathcal{V} = \{\mathbf{v}_1, \ldots, \mathbf{v}_m\}$ be a subset of a vector space V with $\dim(V) = m < \infty$. If $\mathcal{V}$ is linearly independent *or* spans V, then $\mathcal{V}$ is a basis for V.

Theorem 7.17 also applies to subspaces. This theorem tells us that if a set has the same number of vectors as the dimension of the vector space, then we need only verify either span or linear independence to determine if the set is a basis.

EXAMPLE 9 Show that the set

$$V = \{x^3 - x^2 - 5x + 1,\ 2x^2 + 3x,\ x^3 + 3x^2 - 4x + 2,\ -3x^2 + 5\}$$

is a basis for $\mathbf{P}^3$.

Solution We have $\dim(\mathbf{P}^3) = 4$ and V has four elements, so by Theorem 7.17 we need only show V is linearly independent or spans $\mathbf{P}^3$ to prove that V is a basis for $\mathbf{P}^3$. Let's show that V is linearly independent. Suppose that c_1, c_2, c_3, and c_4 are scalars such that

$$c_1(x^3 - x^2 - 5x + 1) + c_2(2x^2 + 3x) + c_3(x^3 + 3x^2 - 4x + 2) + c_4(-3x^2 + 5) = 0$$

Reorganizing to collect common terms and setting each coefficient equal to zero gives us

$$
\begin{aligned}
c_1 \quad\quad + c_3 \quad\quad &= 0 \\
-c_1 + 2c_2 + 3c_3 - 3c_4 &= 0 \\
-5c_1 + 3c_2 - 4c_3 \quad\quad &= 0 \\
c_1 \quad\quad + 2c_3 + 5c_4 &= 0
\end{aligned}
$$

To determine linear independence, all we need to know is if this system has any nontrivial solutions. Instead of transforming to an augmented matrix and performing row operations, let's define the coefficient matrix

$$
C = \begin{bmatrix}
1 & 0 & 1 & 0 \\
-1 & 2 & 3 & -3 \\
-5 & 3 & -4 & 0 \\
1 & 0 & 2 & 5
\end{bmatrix}
$$

By applying techniques from Chapter 5, we can show that $\det(C) = -59$. Since the determinant of the coefficient matrix C is nonzero, it follows from Theorem 5.7 (the Big Theorem, Version 7) in Section 5.2 that our linear system has only the trivial solution. Therefore V is a linearly independent set and is a basis for $\mathbf{P}^3$. ∎

| EXERCISES |

For Exercises 1–6, determine by inspection if the set V could possibly be a basis for V. Explain your answer.

1. $V = \{x^2 + 7,\ 3x + 5\}$, $V = \mathbf{P}^2$

2. $V = \{x^3 + 3,\ 4x^2 + x - 1,\ 3x^3 + 5,\ x\}$, $V = \mathbf{P}^3$

3. $V = \left\{ \begin{bmatrix} 2 & 1 \\ 0 & 4 \end{bmatrix}, \begin{bmatrix} 0 & 3 \\ 1 & 5 \end{bmatrix}, \begin{bmatrix} 6 & 2 \\ 2 & 1 \end{bmatrix}, \begin{bmatrix} 9 & 0 \\ 0 & 2 \end{bmatrix} \right\}$, $V = \mathbf{R}^{2\times2}$

4. $V = \left\{ \begin{bmatrix} 6 & 0 \\ 0 & 1 \\ 3 & 3 \end{bmatrix}, \begin{bmatrix} 0 & 1 \\ 5 & 5 \\ 4 & 1 \end{bmatrix}, \begin{bmatrix} 1 & 1 \\ 1 & 1 \\ 1 & 1 \end{bmatrix}, \begin{bmatrix} 2 & 2 \\ 1 & 7 \\ 0 & 3 \end{bmatrix} \right\}$, $V = \mathbf{R}^{3\times2}$

5. $V = \{x^4 + 3,\ 4x^3 + x - 1,\ 3x^4 + 5,\ x^2\}$, $V = \mathbf{P}^4$

6. $V = \{x,\ x^3,\ x^5,\ x^7, \ldots\}$, $V = \mathbf{P}$

For Exercises 7–12, determine if the set V is a basis for V.

7. $V = \{2x^2 + x - 3,\ x + 1,\ -5\}$, $V = \mathbf{P}^2$

8. $V = \{x^2 - 5x + 3,\ 3x^2 - 7x + 5,\ x^2 - x + 1\}$, $V = \mathbf{P}^2$

9. $V = \left\{ \begin{bmatrix} 1 & 2 \\ 2 & 1 \end{bmatrix}, \begin{bmatrix} 3 & 1 \\ 0 & 3 \end{bmatrix}, \begin{bmatrix} 2 & 2 \\ 1 & 1 \end{bmatrix}, \begin{bmatrix} 3 & 3 \\ 3 & 4 \end{bmatrix} \right\}$, $V = \mathbf{R}^{2\times2}$

10. $V = \left\{ \begin{bmatrix} 4 & 3 \\ 2 & 1 \end{bmatrix}, \begin{bmatrix} 0 & 1 \\ 2 & 3 \end{bmatrix}, \begin{bmatrix} 0 & 0 \\ 2 & 1 \end{bmatrix}, \begin{bmatrix} 0 & 0 \\ 0 & 1 \end{bmatrix} \right\}$, $V = \mathbf{R}^{2\times2}$

11. $V = \{(1, 0, 0, 0, 0, \ldots),\ (1, -1, 0, 0, 0, \ldots),$ $(1, -1, 1, 0, 0, \ldots),\ (1, -1, 1, -1, 0, \ldots), \ldots\}$, $V = \mathbf{R}^\infty$

12. $V = \{1,\ x + 1,\ x^2 + x + 1,\ x^3 + x^2 + x + 1, \ldots\}$, $V = \mathbf{P}$

For Exercises 13–18, find a basis for the subspace S and determine $\dim(S)$.

13. S is the subspace of $\mathbf{R}^{3\times3}$ consisting of matrices with trace equal to zero. (The *trace* is the sum of the diagonal terms of a matrix.)

14. S is the subspace of $\mathbf{P}^2$ consisting of polynomials with graphs crossing the origin.

15. S is the subspace of $\mathbf{R}^{2\times2}$ consisting of matrices with components that add to zero.

16. S is the subspace of $T(2, 2)$ consisting of linear transformations $T : \mathbf{R}^2 \to \mathbf{R}^2$ such that $T(\mathbf{x}) = a\mathbf{x}$ for some scalar a.

17. S is the subspace of $T(2, 2)$ consisting of linear transformations $T : \mathbf{R}^2 \to \mathbf{R}^2$ such that $T(\mathbf{v}) = \mathbf{0}$ for a specific vector $\mathbf{v}$.

18. S is the subspace of $\mathbf{P}$ consisting of polynomials p such that $p(0) = 0$.

For Exercises 19–20, determine the dimension of the subspace S. Justify your answer.

19. S is the subspace of $C(\mathbf{R})$ consisting of functions f such that $f(k) = 0$ for $k = 0, 1, 2$.

20. S is the subspace of $C(\mathbf{R})$ consisting of functions f such that $f(0) = f(1) = f(2)$.

For Exercises 21–24, extend the linearly independent set $\mathcal{V}$ to a basis for V.

21. $S = \{2x^2 + 1, \ 4x - 3\}, \ V = \mathbf{P}^2$

22. $\mathcal{V} = \{x^3, \ x^2 + x + 1, \ x\}, \ V = \mathbf{P}^3$

23. $\mathcal{V} = \left\{ \begin{bmatrix} 1 & 0 \\ 0 & 1 \end{bmatrix}, \begin{bmatrix} 0 & 1 \\ 1 & 0 \end{bmatrix}, \begin{bmatrix} 1 & 1 \\ 1 & 0 \end{bmatrix} \right\}, \ V = \mathbf{R}^{2 \times 2}$

24. $\mathcal{V} = \left\{ \begin{bmatrix} 1 & 0 \\ 1 & 0 \\ 1 & 0 \end{bmatrix}, \begin{bmatrix} 0 & 1 \\ 0 & 1 \\ 0 & 1 \end{bmatrix}, \begin{bmatrix} 1 & 0 \\ 1 & 0 \\ 0 & 0 \end{bmatrix}, \begin{bmatrix} 0 & 0 \\ 0 & 0 \\ 0 & 1 \end{bmatrix} \right\}, \ V = \mathbf{R}^{3 \times 2}$

For Exercises 25–26, remove vectors from $\mathcal{V}$ to yield a basis for V.

25. $\mathcal{V} = \{x + 1, \ x + 2, \ 2x + 1\}, \ V = \mathbf{P}^1$

26. $\mathcal{V} = \{x^2 + x, \ x + 1, \ x^2 + 1, \ x^2 + x + 1\}, \ V = \mathbf{P}^2$

Ⓒ Exercises 27–30 assume some knowledge of calculus.

27. Let S be the subspace of $C(\mathbf{R})$ consisting of solutions $y(t)$ to the differential equation $y''(t) + y(t) = 0$. (This equation is discussed in Example 10 of Section 7.1.) All solutions to this equation have the form

$$y(t) = c_1 \cos(t) + c_2 \sin(t).$$

Prove that $\dim(S) = 2$.

28. Find a basis for the subspace S of $\mathbf{P}^4$ consisting of polynomials $p(x)$ such that $p'(x) = 0$.

29. Find a basis for the subspace S of $\mathbf{P}^6$ consisting of polynomials $p(x)$ such that $p''(x) = 0$.

30. Determine the dimension of the subspace S of $\mathbf{P}$ consisting of polynomials p such that

$$\int_{-1}^{1} p(x) \, dx = 0.$$

FIND AN EXAMPLE For Exercises 31–36, find an example that meets the given specifications.

31. An infinite dimensional vector space V and a finite dimensional subspace S.

32. A vector space V and a subspace S such that $\dim(S) = 5$.

33. A vector space V and subspace S such that $\dim(V) = 1 + \dim(S)$.

34. A vector space V and subspace S such that $\dim(V) = 2 \cdot \dim(S)$.

35. A vector space V and an infinite subset $\mathcal{V}$ that is linearly independent but does not span V.

36. A vector space V and subspace S such that $\dim(V) = \dim(S)$ but $S \ne V$.

TRUE OR FALSE For Exercises 37–46, determine if the statement is true or false, and justify your answer.

37. The size of a vector space basis varies from one basis to another.

38. There is no linearly independent subset $\mathcal{V}$ of $\mathbf{P}^5$ containing seven elements.

39. No two vector spaces can share the same dimension.

40. If V is a vector space with $\dim(V) = 6$ and S is a subspace of V with $\dim(S) = 6$, then $S = V$.

41. If V is a finite dimensional vector space, then V cannot contain an infinite linearly independent subset $\mathcal{V}$.

42. If V_1 and V_2 are vector spaces and $\dim(V_1) < \dim(V_2)$, then $V_1 \subset V_2$.

43. If $\mathcal{V}$ spans a vector space V, then vectors can be added to $\mathcal{V}$ to produce a basis for V.

44. If V is a finite dimensional vector space, then every subspace of V must also be finite dimensional.

45. If $\{\mathbf{v}_1, \ldots, \mathbf{v}_k\}$ is a basis for a vector space V, then so is $\{c\mathbf{v}_1, \ldots, c\mathbf{v}_k\}$, where c is a scalar.

46. If S_1 is a subspace of a vector space V and $\dim(S_1) = 1$, then the only proper subspace of S_1 is $S_2 = \{\mathbf{0}\}$.

47. Prove that if $\{\mathbf{v}_1, \mathbf{v}_2, \ldots, \mathbf{v}_k\}$ is a basis for a vector space V, then so is $\{\mathbf{v}_1, 2\mathbf{v}_2, \ldots, k\mathbf{v}_k\}$.

48. Prove that $\dim(\mathbf{R}^\infty) = \infty$.

49. Show that

$$\left\{ \begin{bmatrix} 1 & 0 \\ 0 & 0 \end{bmatrix}, \begin{bmatrix} 0 & 1 \\ 0 & 0 \end{bmatrix}, \begin{bmatrix} 0 & 0 \\ 1 & 0 \end{bmatrix}, \begin{bmatrix} 0 & 0 \\ 0 & 1 \end{bmatrix} \right\}$$

is a basis for $\mathbf{R}^{2 \times 2}$.

50. Give a basis for $\mathbf{R}^{n \times m}$, and justify that your set forms a basis. Prove that $\dim(\mathbf{R}^{n \times m}) = mn$.

51. If V is a vector space with $\dim(V) = m$, prove that there exist subspaces $S_0, S_1, \ldots, S_m$ of V such that $\dim(S_k) = k$.

52. Prove that $\dim(T(m, n)) = mn$. (HINT: Recall that if $T : \mathbf{R}^m \rightarrow \mathbf{R}^n$ is a linear transformation, then $T(\mathbf{x}) = A\mathbf{x}$ for some $n \times m$ matrix A.)

53. Prove Theorem 7.11: The set $\{\mathbf{v}_1, \ldots, \mathbf{v}_m\}$ is a basis for a vector space V if and only if the equation

$$c_1 \mathbf{v}_1 + \cdots + c_m \mathbf{v}_m = \mathbf{v}$$

has a unique solution $c_1, \ldots, c_m$ for every $\mathbf{v}$ in V. (HINT: See Theorem 7.9.)

54. Prove Theorem 7.12: Suppose that $\mathcal{V}_1$ and $\mathcal{V}_2$ are both bases of a vector space V. Prove that $\mathcal{V}_1$ and $\mathcal{V}_2$ have the same number of elements. (NOTE: Be sure to address the possibility that $\mathcal{V}_1$ and $\mathcal{V}_2$ both have infinitely many vectors.)

55. Let V_1 and V_2 be vector spaces with V_1 a subset of V_2.

(a) If $\dim(V_2)$ is finite, prove that $\dim(V_1) = \dim(V_2)$ if and only if $V_1 = V_2$.

(b) Give an example showing that (a) need not be true if $\dim(V_2) = \infty$.

56. Prove Theorem 7.14: Let $\mathcal{V} = \{\mathbf{v}_1, \ldots, \mathbf{v}_m\}$ be a subset of a finite dimensional vector space V, and suppose that $\mathcal{V}$ is not a basis of V.

(a) Prove that if $\mathcal{V}$ spans V, then vectors can be removed from $\mathcal{V}$ to form a basis for V.

(b) Prove that if $\mathcal{V}$ is linearly independent, then vectors can be added to $\mathcal{V}$ to form a basis for V.

57. Prove Theorem 7.16: Let $\mathcal{V} = \{\mathbf{v}_1, \ldots, \mathbf{v}_m\}$ be a subset of a vector space V with $\dim(V) = n$.

(a) Prove that if $m < n$, then $\mathcal{V}$ does not span V.

(b) Prove that if $m > n$, then $\mathcal{V}$ is linearly dependent.

58. Prove Theorem 7.17: Let $\mathcal{V} = \{\mathbf{v}_1, \ldots, \mathbf{v}_m\}$ be a subset of a vector space V with $\dim(V) = m$. If $\mathcal{V}$ is linearly independent *or* spans V, then $\mathcal{V}$ is a basis for V.

Orthogonality

The oldest steel bridge in the United States, the Smithfield Street Bridge in Pittsburgh, Pennsylvania is a lenticular truss bridge offering little clearance between its lowest point and the Monongahela River. When transportation officials in Pennsylvania tried to demolish the bridge and replace it with a modern bridge, the Pittsburgh History and Landmarks Foundation acted to preserve the bridge. Instead of demolition, the bridge was renovated to include a widened deck, and a new color and lighting scheme that highlighted its architectural features. Abandoned rail lines were transformed into an extra traffic lane and a light-controlled bus lane was installed for peak traffic hours.

Bridge suggested by Tim Flaherty, Carnegie Mellon University (Johnny Stockshooter/Alamy)

▶ The order of Chapter 7 and Chapter 8 can be reversed if needed.

In this chapter we shift our focus from vector spaces back to Euclidean space $\mathbf{R}^n$. In Section 8.1 we introduce and study the dot product, which provides an algebraic formula for determining if two vectors are perpendicular, or equivalently, *orthogonal*. The familiar notions of perpendicular, angle, and length in $\mathbf{R}^2$ and $\mathbf{R}^3$ still hold here and are extended to higher dimensions by using the dot product. Section 8.2 introduces projections of vectors and the Gram–Schmidt process, which is an algorithm for converting a linearly independent set of vectors into an orthogonal set of vectors. The remaining three sections are applications of orthogonal vectors. Section 8.3 and Section 8.4 focus on matrix factorizations, and Section 8.5 on the problem of fitting functions to data.

8.1 Dot Products and Orthogonal Sets

The *dot product* of two vectors is a form of multiplication of vectors in $\mathbf{R}^n$. Unlike vector addition, which produces a new vector, the dot product of two vectors yields a scalar.

DEFINITION 8.1

Definition Dot Product

Suppose that

$$\mathbf{u} = \begin{bmatrix} u_1 \\ \vdots \\ u_n \end{bmatrix} \quad \text{and} \quad \mathbf{v} = \begin{bmatrix} v_1 \\ \vdots \\ v_n \end{bmatrix}$$

are both in $\mathbf{R}^n$. Then the **dot product** of $\mathbf{u}$ and $\mathbf{v}$ is given by

$$\mathbf{u} \cdot \mathbf{v} = u_1 v_1 + \cdots + u_n v_n$$

▶ An alternative way to define the dot product of $\mathbf{u}$ and $\mathbf{v}$ is with matrix multiplication, by

$$\mathbf{u} \cdot \mathbf{v} = \mathbf{u}^T \mathbf{v}$$

This is discussed in Exercise 71.

EXAMPLE 1 Find $\mathbf{u} \cdot \mathbf{v}$ for $\mathbf{u} = \begin{bmatrix} -1 \\ 3 \\ 2 \end{bmatrix}$ and $\mathbf{v} = \begin{bmatrix} 7 \\ 1 \\ -5 \end{bmatrix}$.

Solution Applying Definition 8.1, we have

$$\mathbf{u} \cdot \mathbf{v} = (-1)(7) + (3)(1) + (2)(-5) = -14 \qquad \blacksquare$$

It is not hard see that $\mathbf{u} \cdot \mathbf{v} = \mathbf{v} \cdot \mathbf{u}$. Theorem 8.2 includes this and other properties of the dot product. Note the similarity to properties of arithmetic of real numbers.

THEOREM 8.2

▶ Note that the properties given in Theorem 8.2 are similar to properties of arithmetic of the real numbers.

Let $\mathbf{u}$, $\mathbf{v}$, and $\mathbf{w}$ be in $\mathbf{R}^n$, and let c be a scalar. Then

(a) $\mathbf{u} \cdot \mathbf{v} = \mathbf{v} \cdot \mathbf{u}$

(b) $(\mathbf{u} + \mathbf{v}) \cdot \mathbf{w} = \mathbf{u} \cdot \mathbf{w} + \mathbf{v} \cdot \mathbf{w}$

(c) $(c\mathbf{u}) \cdot \mathbf{v} = \mathbf{u} \cdot (c\mathbf{v}) = c(\mathbf{u} \cdot \mathbf{v})$

(d) $\mathbf{u} \cdot \mathbf{u} \geq 0$, and $\mathbf{u} \cdot \mathbf{u} = 0$ only when $\mathbf{u} = \mathbf{0}$

The proof of Theorem 8.2 is left as Exercise 62. By combining the properties (b) and (c), it can be shown (see Exercise 63) for $\mathbf{u}_1, \ldots, \mathbf{u}_k$ and $\mathbf{w}$ in $\mathbf{R}^n$ and scalars $c_1, \ldots, c_k$ that

$$(c_1\mathbf{u}_1 + c_2\mathbf{u}_2 + \cdots + c_k\mathbf{u}_k) \cdot \mathbf{w} = c_1(\mathbf{u}_1 \cdot \mathbf{w}) + c_2(\mathbf{u}_2 \cdot \mathbf{w}) + \cdots + c_k(\mathbf{u}_k \cdot \mathbf{w}) \qquad (1)$$

EXAMPLE 2 Suppose that

$$\mathbf{u} = \begin{bmatrix} 2 \\ 1 \\ -3 \\ 2 \end{bmatrix}, \quad \mathbf{v} = \begin{bmatrix} -1 \\ 4 \\ 0 \\ 3 \end{bmatrix}, \quad \mathbf{w} = \begin{bmatrix} 5 \\ 0 \\ 1 \\ 2 \end{bmatrix}$$

and $c = -3$. Show that Theorem 8.2(b) and Theorem 8.2(c) hold for these vectors and this scalar.

Solution Starting with Theorem 8.2(b), we have

$$(\mathbf{u} + \mathbf{v}) \cdot \mathbf{w} = \begin{bmatrix} 1 \\ 5 \\ -3 \\ 5 \end{bmatrix} \cdot \begin{bmatrix} 5 \\ 0 \\ 1 \\ 2 \end{bmatrix} = 5 + 0 - 3 + 10 = 12$$

and

$$\mathbf{u} \cdot \mathbf{w} + \mathbf{v} \cdot \mathbf{w} = (10 + 0 - 3 + 4) + (-5 + 0 + 0 + 6) = 11 + 1 = 12$$

Thus $(\mathbf{u} + \mathbf{v}) \cdot \mathbf{w} = \mathbf{u} \cdot \mathbf{w} + \mathbf{v} \cdot \mathbf{w}$. For Theorem 8.2(c), we compute

$$(-3\mathbf{u}) \cdot \mathbf{v} = \begin{bmatrix} -6 \\ -3 \\ 9 \\ -6 \end{bmatrix} \cdot \begin{bmatrix} -1 \\ 4 \\ 0 \\ 3 \end{bmatrix} = (6 - 12 + 0 - 18) = -24$$

$$\mathbf{u} \cdot (-3\mathbf{v}) = \begin{bmatrix} 2 \\ 1 \\ -3 \\ 2 \end{bmatrix} \cdot \begin{bmatrix} 3 \\ -12 \\ 0 \\ -9 \end{bmatrix} = (6 - 12 + 0 - 18) = -24$$

$$-3(\mathbf{u} \cdot \mathbf{v}) = -3(-2 + 4 + 0 + 6) = -24$$

Therefore $(-3\mathbf{u}) \cdot \mathbf{v} = \mathbf{u} \cdot (-3\mathbf{v}) = -3(\mathbf{u} \cdot \mathbf{v})$. ∎

The dot product can be used to measure distance. Suppose that $\mathbf{x} = \begin{bmatrix} x_1 \\ x_2 \end{bmatrix}$ is a vector in $\mathbf{R}^2$, as shown in Figure 1. If we denote the length of $\mathbf{x}$ by $\|\mathbf{x}\|$, then from the Pythagorean Theorem we know that

$$\|\mathbf{x}\|^2 = x_1^2 + x_2^2$$

By the definition of the dot product, $\mathbf{x} \cdot \mathbf{x} = x_1^2 + x_2^2$, so that we have

$$\|\mathbf{x}\|^2 = \mathbf{x} \cdot \mathbf{x} \quad \Longrightarrow \quad \|\mathbf{x}\| = \sqrt{\mathbf{x} \cdot \mathbf{x}}$$

This suggests a way to extend the definition of "length" to vectors in $\mathbf{R}^n$. In this setting we generally refer to this as the *norm* of a vector.

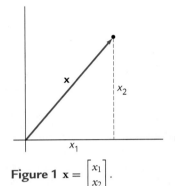

Figure 1 $\mathbf{x} = \begin{bmatrix} x_1 \\ x_2 \end{bmatrix}$.

DEFINITION 8.3

Definition Norm of a Vector

Let $\mathbf{x}$ be a vector in $\mathbf{R}^n$. Then the **norm** (or **length**) of $\mathbf{x}$ is given by

$$\|\mathbf{x}\| = \sqrt{\mathbf{x} \cdot \mathbf{x}}$$

For a scalar c and a vector $\mathbf{x}$, it follows from Theorem 8.2c (see Exercise 64) that

$$\|c\mathbf{x}\| = |c| \|\mathbf{x}\| \tag{2}$$

EXAMPLE 3   Find $\|\mathbf{x}\|$ and $\|-5\mathbf{x}\|$ for $\mathbf{x} = \begin{bmatrix} -3 \\ 1 \\ 4 \end{bmatrix}$.

Solution We have $\mathbf{x} \cdot \mathbf{x} = (-3)^2 + (1)^2 + (4)^2$, so

$$\|\mathbf{x}\| = \sqrt{9 + 1 + 16} = \sqrt{26}$$

By 2,

$$\|-5\mathbf{x}\| = |-5| \|\mathbf{x}\| = 5\sqrt{26}$$ ∎

Among the real numbers, we measure the distance between r and s by computing $|r - s|$. This serves as a model for using norms to define the distance between vectors.

DEFINITION 8.4

Definition Distance Between Vectors

For two vectors $\mathbf{u}$ and $\mathbf{v}$ in $\mathbf{R}^n$, the **distance between u and v** is given by $\|\mathbf{u} - \mathbf{v}\|$.

EXAMPLE 4 Compute the distance between $\mathbf{u} = \begin{bmatrix} -1 \\ 3 \\ 2 \end{bmatrix}$ and $\mathbf{v} = \begin{bmatrix} 7 \\ 1 \\ -5 \end{bmatrix}$.

Solution We have $\mathbf{u} - \mathbf{v} = \begin{bmatrix} -8 \\ 2 \\ 7 \end{bmatrix}$, so that

$$\|\mathbf{u} - \mathbf{v}\| = \sqrt{(-8)^2 + (2)^2 + (7)^2} = \sqrt{117} \qquad \blacksquare$$

Orthogonal Vectors

Suppose that we have two vectors

$$\mathbf{u} = \begin{bmatrix} u_1 \\ u_2 \end{bmatrix} \quad \text{and} \quad \mathbf{v} = \begin{bmatrix} v_1 \\ v_2 \end{bmatrix}$$

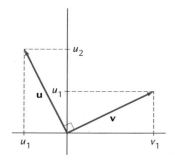

Figure 2 $\mathbf{u}$ and $\mathbf{v}$ are perpendicular vectors.

in $\mathbf{R}^2$ that have equal length and are perpendicular to each other, as shown in Figure 2. Rotating $\mathbf{u}$ by $90°$ gives us $\mathbf{v}$, so that $v_1 = u_2$ and $v_2 = -u_1$. Therefore we have

$$\mathbf{u} \cdot \mathbf{v} = u_1 v_1 + u_2 v_2 = u_1 u_2 - u_2 u_1 = 0$$

Even if the vectors are not the same length, after scaling we can use this argument to show that if two vectors are perpendicular, then $\mathbf{u} \cdot \mathbf{v} = 0$. (See Exercise 72 for another way to show this.) The reverse holds as well: If $\mathbf{u} \cdot \mathbf{v} = 0$, then $\mathbf{u}$ and $\mathbf{v}$ are perpendicular. The same is true in $\mathbf{R}^3$, which suggests using the dot product to extend the notion of perpendicular to higher dimensions. The term *orthogonal* is more commonly used and is equivalent to perpendicular.

DEFINITION 8.5

Definition Orthogonal Vectors

Vectors $\mathbf{u}$ and $\mathbf{v}$ in $\mathbf{R}^n$ are **orthogonal** if $\mathbf{u} \cdot \mathbf{v} = 0$.

EXAMPLE 5 Determine if any pair among $\mathbf{u}$, $\mathbf{v}$, and $\mathbf{w}$ is orthogonal.

$$\mathbf{u} = \begin{bmatrix} 2 \\ -1 \\ 5 \\ -2 \end{bmatrix}, \quad \mathbf{v} = \begin{bmatrix} 3 \\ 2 \\ -4 \\ 0 \end{bmatrix}, \quad \mathbf{w} = \begin{bmatrix} 2 \\ 9 \\ 6 \\ 4 \end{bmatrix}$$

Solution We have

$$\mathbf{u} \cdot \mathbf{v} = (2)(3) + (-1)(2) + (5)(-4) + (-2)(0) = -16 \implies \text{Not Orthogonal}$$

$$\mathbf{u} \cdot \mathbf{w} = (2)(2) + (-1)(9) + (5)(6) + (-2)(4) = 17 \implies \text{Not Orthogonal}$$

$$\mathbf{v} \cdot \mathbf{w} = (3)(2) + (2)(9) + (-4)(6) + (0)(4) = 0 \implies \text{Orthogonal} \qquad \blacksquare$$

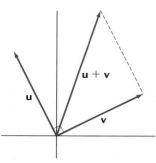

Figure 3 The vectors $\mathbf{u} + \mathbf{v}$ and $\mathbf{v}$ together with the dashed line (which is $\mathbf{u}$ translated) form a right triangle.

Figure 3 shows the orthogonal vectors $\mathbf{u}$ and $\mathbf{v}$ from Figure 2, together with $\mathbf{u} + \mathbf{v}$. Note that the vectors $\mathbf{v}$ and $\mathbf{u} + \mathbf{v}$ together with the dashed line (which is $\mathbf{u}$ translated) form a right triangle. Hence, by the Pythagorean Theorem, we expect

$$\|\mathbf{u} + \mathbf{v}\|^2 = \|\mathbf{u}\|^2 + \|\mathbf{v}\|^2$$

This formulation of the Pythagorean Theorem extends to $\mathbf{R}^n$ and is true exactly when $\mathbf{u}$ and $\mathbf{v}$ are orthogonal—that is, when $\mathbf{u} \cdot \mathbf{v} = 0$.

THEOREM 8.6

(PYTHAGOREAN THEOREM) Suppose that $\mathbf{u}$ and $\mathbf{v}$ are in $\mathbf{R}^n$. Then

$$\|\mathbf{u} + \mathbf{v}\|^2 = \|\mathbf{u}\|^2 + \|\mathbf{v}\|^2 \quad \text{if and only if} \quad \mathbf{u} \cdot \mathbf{v} = 0$$

Proof It is possible to verify the theorem in $\mathbf{R}^2$ using geometric arguments, but that approach is hard to extend to $\mathbf{R}^n$. Instead, we use an algebraic argument that works for any dimension. We have

$$\|\mathbf{u} + \mathbf{v}\|^2 = (\mathbf{u} + \mathbf{v}) \cdot (\mathbf{u} + \mathbf{v})$$
$$= \mathbf{u} \cdot \mathbf{u} + \mathbf{u} \cdot \mathbf{v} + \mathbf{v} \cdot \mathbf{u} + \mathbf{v} \cdot \mathbf{v}$$
$$= \|\mathbf{u}\|^2 + \|\mathbf{v}\|^2 + 2(\mathbf{u} \cdot \mathbf{v})$$

Thus the equality

$$\|\mathbf{u} + \mathbf{v}\|^2 = \|\mathbf{u}\|^2 + \|\mathbf{v}\|^2$$

holds exactly when $\mathbf{u} \cdot \mathbf{v} = 0$. ∎

EXAMPLE 6 Verify the Pythagorean Theorem for the vectors $\mathbf{v}$ and $\mathbf{w}$ in Example 5.

Solution In Example 5 we showed that $\mathbf{v} \cdot \mathbf{w} = 0$, so by the Pythagorean Theorem we expect that $\|\mathbf{v} + \mathbf{w}\|^2 = \|\mathbf{v}\|^2 + \|\mathbf{w}\|^2$. Computing each term, we find

$$\|\mathbf{v} + \mathbf{w}\|^2 = \left\| \begin{bmatrix} 5 \\ 11 \\ 2 \\ 4 \end{bmatrix} \right\|^2 = 5^2 + 11^2 + 2^2 + 4^2 = 166$$

$$\|\mathbf{v}\|^2 = 3^2 + 2^2 + (-4)^2 + 0^2 = 29$$
$$\|\mathbf{w}\|^2 = 2^2 + 9^2 + 6^2 + 4^2 = 137$$

Since $29 + 137 = 166$, we have $\|\mathbf{v} + \mathbf{w}\|^2 = \|\mathbf{v}\|^2 + \|\mathbf{w}\|^2$. ∎

Orthogonal Subspaces

Now that we are acquainted with orthogonal vectors, let's turn to the matter of how sets and vectors can be orthogonal.

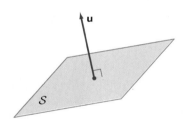

Figure 4 A subspace S and orthogonal vector $\mathbf{u}$.

DEFINITION 8.7

Definition Orthogonal Complement

Let S be a subspace of $\mathbf{R}^n$. A vector $\mathbf{u}$ is **orthogonal** to S if $\mathbf{u} \cdot \mathbf{s} = 0$ for every vector $\mathbf{s}$ in S. The set of all such vectors $\mathbf{u}$ is called the **orthogonal complement** of S and is denoted by $S^{\perp}$.

Figure 4 shows an example in $\mathbf{R}^3$, where we have a subspace S and an orthogonal vector $\mathbf{u}$. Note that the orthogonal complement $S^{\perp}$ consists precisely of the multiples of $\mathbf{u}$, so that $S^{\perp}$ is also a subspace of $\mathbf{R}^3$. This is true for the orthogonal complement of any subspace.

THEOREM 8.8

If S is a subspace of $\mathbf{R}^n$, then so is $S^{\perp}$.

Proof Recall from Definition 4.1, Section 4.1, that a subspace must satisfy three conditions.

(a) Since $\mathbf{0} \cdot \mathbf{s} = 0$ for all $\mathbf{s}$ in S, it follows that $S^{\perp}$ contains $\mathbf{0}$.

(b) Suppose that $\mathbf{u}_1$ and $\mathbf{u}_2$ are in $S^{\perp}$. For any $\mathbf{s}$ in S, we have

▶ $S^{\perp}$ typically is said aloud as "S-perpendicular" or just "S-perp."

$$(\mathbf{u}_1 + \mathbf{u}_2) \cdot \mathbf{s} = \mathbf{u}_1 \cdot \mathbf{s} + \mathbf{u}_2 \cdot \mathbf{s} = 0 + 0 = 0$$

Hence $\mathbf{u}_1 + \mathbf{u}_2$ is in $S^{\perp}$, so that $S^{\perp}$ is closed under addition.

(c) If c is a scalar and $\mathbf{u}$ is in $S^\perp$, then for any $\mathbf{s}$ in S we have

$$(c\mathbf{u}) \cdot \mathbf{s} = c(\mathbf{u} \cdot \mathbf{s}) = c(0) = 0$$

Therefore $c\mathbf{u}$ is also in $S^\perp$ and hence $S^\perp$ is closed under scalar multiplication.

Since parts (a)–(c) of Definition 4.1 hold, $S^\perp$ is a subspace. ∎

How do we find $S^\perp$? Since a nonzero subspace S contains an infinite number of vectors, it appears that determining if $\mathbf{u}$ is in $S^\perp$ will require checking that $\mathbf{u} \cdot \mathbf{s} = 0$ for every $\mathbf{s}$ in S. Fortunately, there is another option. Theorem 8.9 shows that we need only check the basis vectors.

THEOREM 8.9

Let $\mathcal{S} = \{\mathbf{s}_1, \ldots, \mathbf{s}_k\}$ be a basis for a subspace S and $\mathbf{u}$ be a vector. Then

$$\mathbf{u} \cdot \mathbf{s}_1 = 0, \quad \mathbf{u} \cdot \mathbf{s}_2 = 0, \ldots, \quad \mathbf{u} \cdot \mathbf{s}_k = 0$$

if and only if $\mathbf{u}$ is in $S^\perp$.

Proof First, suppose that $\mathbf{u} \cdot \mathbf{s}_1 = 0, \ldots, \mathbf{u} \cdot \mathbf{s}_k = 0$. If $\mathbf{s}$ is in S, then since $\mathcal{S}$ is a basis there exist unique scalars $c_1, \ldots, c_k$ such that

$$\mathbf{s} = c_1\mathbf{s}_1 + \cdots + c_k\mathbf{s}_k$$

Therefore

$$\mathbf{u} \cdot \mathbf{s} = \mathbf{u} \cdot (c_1\mathbf{s}_1 + \cdots + c_k\mathbf{s}_k) = c_1(\mathbf{u} \cdot \mathbf{s}_1) + \cdots + c_k(\mathbf{u} \cdot \mathbf{s}_k) = 0$$

and so $\mathbf{u}$ is in $S^\perp$.

The reverse direction of the proof is easier. If $\mathbf{u}$ is in $S^\perp$, then since each of $\mathbf{s}_1, \ldots, \mathbf{s}_k$ are in S, it follows that $\mathbf{u} \cdot \mathbf{s}_1 = 0, \ldots, \mathbf{u} \cdot \mathbf{s}_k = 0$. ∎

EXAMPLE 7 Let

$$\mathbf{s}_1 = \begin{bmatrix} 2 \\ -1 \\ 3 \\ 0 \end{bmatrix}, \quad \mathbf{s}_2 = \begin{bmatrix} 1 \\ 4 \\ -2 \\ -1 \end{bmatrix}, \quad \text{and} \quad \mathbf{u} = \begin{bmatrix} 3 \\ 6 \\ 0 \\ 5 \end{bmatrix}$$

Suppose that $S = \text{span}\,\{\mathbf{s}_1, \mathbf{s}_2\}$. Determine if $\mathbf{u}$ is in $S^\perp$, and find a basis for $S^\perp$.

Solution To check if $\mathbf{u}$ is in $S^\perp$, by Theorem 8.9 we need only determine if $\mathbf{s}_1 \cdot \mathbf{u} = 0$ and $\mathbf{s}_2 \cdot \mathbf{u} = 0$.

$$\mathbf{s}_1 \cdot \mathbf{u} = 6 - 6 + 0 + 0 = 0$$
$$\mathbf{s}_2 \cdot \mathbf{u} = 3 + 24 + 0 - 5 = 22$$

Since $\mathbf{s}_2 \cdot \mathbf{u} \neq 0$, we know that $\mathbf{u}$ is not in $S^\perp$. To find a basis for $S^\perp$, we start by forming the matrix

$$A = \begin{bmatrix} 2 & -1 & 3 & 0 \\ 1 & 4 & -2 & -1 \end{bmatrix}$$

Note that the *rows* of A are made up of the components of $\mathbf{s}_1$ and $\mathbf{s}_2$, so that if $\mathbf{u}$ is in $\mathbf{R}^4$, then

$$A\mathbf{u} = \begin{bmatrix} \mathbf{s}_1 \cdot \mathbf{u} \\ \mathbf{s}_2 \cdot \mathbf{u} \end{bmatrix}$$

Thus $\mathbf{u}$ is in $S^\perp$ exactly when $A\mathbf{u} = \mathbf{0}$. To solve this equation we use our standard procedure for determining the general solution of a linear system; we find that

$$\mathbf{u} = c_1 \begin{bmatrix} -10 \\ 7 \\ 9 \\ 0 \end{bmatrix} + c_2 \begin{bmatrix} 1 \\ 2 \\ 0 \\ 9 \end{bmatrix}$$

The two vectors in the general solution give a basis for $S^\perp$, so we have

$$S^\perp = \text{span} \left\{ \begin{bmatrix} -10 \\ 7 \\ 9 \\ 0 \end{bmatrix}, \begin{bmatrix} 1 \\ 2 \\ 0 \\ 9 \end{bmatrix} \right\} \quad \blacksquare$$

Orthogonal Sets

We have defined what it means for a vector to be orthogonal to another vector and to a set of vectors. Our next step is to define what it means for a set of vectors to be orthogonal.

DEFINITION 8.10

Definition Orthogonal Set

A set of vectors $\mathcal{V}$ in $\mathbf{R}^n$ form an **orthogonal set** if $\mathbf{v}_i \cdot \mathbf{v}_j = 0$ for all $\mathbf{v}_i$ and $\mathbf{v}_j$ in $\mathcal{V}$ with $i \neq j$.

EXAMPLE 8 Show that $\{\mathbf{v}_1, \mathbf{v}_2, \mathbf{v}_3\}$ is an orthogonal set, where

$$\mathbf{v}_1 = \begin{bmatrix} 1 \\ 4 \\ -1 \end{bmatrix}, \quad \mathbf{v}_2 = \begin{bmatrix} 11 \\ -1 \\ 7 \end{bmatrix}, \quad \mathbf{v}_3 = \begin{bmatrix} 3 \\ -2 \\ -5 \end{bmatrix}$$

Solution We show that $\{\mathbf{v}_1, \mathbf{v}_2, \mathbf{v}_3\}$ is an orthogonal set by showing that each distinct pair of vectors is orthogonal.

$$\mathbf{v}_1 \cdot \mathbf{v}_2 = 11 - 4 - 7 = 0$$
$$\mathbf{v}_1 \cdot \mathbf{v}_3 = 3 - 8 + 5 = 0$$
$$\mathbf{v}_2 \cdot \mathbf{v}_3 = 33 + 2 - 35 = 0$$

Since the dot products are all zero, the set is orthogonal. $\blacksquare$

A useful feature of orthogonal sets of nonzero vectors is that they are linearly independent.

THEOREM 8.11

An orthogonal set of nonzero vectors is linearly independent.

Proof Let $\{\mathbf{v}_1, \ldots, \mathbf{v}_k\}$ be a set of nonzero orthogonal vectors. Suppose that the linear combination

$$c_1\mathbf{v}_1 + \cdots + c_k\mathbf{v}_k = \mathbf{0} \tag{3}$$

To show that $\{\mathbf{v}_1, \ldots, \mathbf{v}_k\}$ is a linearly independent set, we need to show that 3 holds only when $c_1 = \cdots = c_k = 0$. To do so, we start by noting that since $\mathbf{v}_1$ is orthogonal to $\mathbf{v}_2, \ldots, \mathbf{v}_k$,

$$\mathbf{v}_1 \cdot (c_1\mathbf{v}_1 + \cdots + c_k\mathbf{v}_k) = c_1(\mathbf{v}_1 \cdot \mathbf{v}_1) + c_2(\mathbf{v}_1 \cdot \mathbf{v}_2) + \cdots + c_k(\mathbf{v}_1 \cdot \mathbf{v}_k)$$
$$= c_1\|\mathbf{v}_1\|^2 + c_2(0) + \cdots + c_k(0) = c_1\|\mathbf{v}_1\|^2$$

Because $c_1\mathbf{v}_1 + \cdots + c_k\mathbf{v}_k = \mathbf{0}$, it follows that $\mathbf{v}_1 \cdot (c_1\mathbf{v}_1 + \cdots + c_k\mathbf{v}_k) = 0$, and thus

$$c_1\|\mathbf{v}_1\|^2 = 0$$

Since $\mathbf{v}_1 \neq \mathbf{0}$, we know that $\|\mathbf{v}_1\|^2 \neq 0$, and so it must be that $c_1 = 0$. Repeating this argument with each of $\mathbf{v}_2, \ldots, \mathbf{v}_k$ shows that $c_2 = 0, \ldots, c_k = 0$, and hence $\{\mathbf{v}_1, \ldots, \mathbf{v}_k\}$ is a linearly independent set. ∎

Definition Orthogonal Basis

A basis made up of orthogonal vectors is called an **orthogonal basis**. Orthogonal sets are particularly useful for forming a basis, in part because orthogonal vectors are automatically linearly independent, and also because the dot product can be used to form linear combinations.

THEOREM 8.12

Let S be a subspace with an orthogonal basis $\{\mathbf{s}_1, \ldots, \mathbf{s}_k\}$. Then any vector $\mathbf{s}$ in S can be written as

$$\mathbf{s} = c_1\mathbf{s}_1 + \cdots + c_k\mathbf{s}_k$$

where $c_i = \dfrac{\mathbf{s}_i \cdot \mathbf{s}}{\|\mathbf{s}_i\|^2}$ for $i = 1, \ldots, k$.

Proof   The proof is similar to that of Theorem 8.11. We know that there exist unique scalars $c_1, \ldots, c_k$ such that

$$\mathbf{s} = c_1\mathbf{s}_1 + \cdots + c_k\mathbf{s}_k$$

Taking the dot product of $\mathbf{s}_1$ with $\mathbf{s}$, we have

$$\mathbf{s}_1 \cdot \mathbf{s} = \mathbf{s}_1 \cdot (c_1\mathbf{s}_1 + \cdots + c_k\mathbf{s}_k)$$
$$= c_1(\mathbf{s}_1 \cdot \mathbf{s}_1) + c_2(\mathbf{s}_1 \cdot \mathbf{s}_2) + \cdots + c_k(\mathbf{s}_1 \cdot \mathbf{s}_k)$$
$$= c_1(\mathbf{s}_1 \cdot \mathbf{s}_1) + c_2(0) + \cdots + c_k(0)$$
$$= c_1(\mathbf{s}_1 \cdot \mathbf{s}_1) = c_1\|\mathbf{s}_1\|^2$$

Solving for c_1, we find that

$$c_1 = \frac{\mathbf{s}_1 \cdot \mathbf{s}}{\|\mathbf{s}_1\|^2}$$

A similar argument gives the formulas for $c_2, \ldots, c_k$. ∎

EXAMPLE 9   Verify that the set

$$\mathbf{s}_1 = \begin{bmatrix} -2 \\ 1 \\ -1 \end{bmatrix}, \quad \mathbf{s}_2 = \begin{bmatrix} 1 \\ -1 \\ -3 \end{bmatrix}, \quad \mathbf{s}_3 = \begin{bmatrix} 4 \\ 7 \\ -1 \end{bmatrix}$$

forms an orthogonal basis for $\mathbf{R}^3$, and write $\mathbf{s} = \begin{bmatrix} 3 \\ -1 \\ 5 \end{bmatrix}$ as a linear combination of $\mathbf{s}_1$, $\mathbf{s}_2$, and $\mathbf{s}_3$.

Solution We verify orthogonality by computing dot products, which are

$$\mathbf{s}_1 \cdot \mathbf{s}_2 = -2 - 1 + 3 = 0$$
$$\mathbf{s}_1 \cdot \mathbf{s}_3 = -8 + 7 + 1 = 0$$
$$\mathbf{s}_2 \cdot \mathbf{s}_3 = 4 - 7 + 3 = 0$$

By Theorem 8.11 the vectors are linearly independent, and since there are three of them in $\mathbf{R}^3$, the set also spans and hence is a basis for $\mathbf{R}^3$.

To write $\mathbf{s}$ as a linear combination of $\mathbf{s}_1$, $\mathbf{s}_2$, and $\mathbf{s}_3$, we apply Theorem 8.12, starting with the computations

$$c_1 = \frac{\mathbf{s}_1 \cdot \mathbf{s}}{\|\mathbf{s}_1\|^2} = \frac{-6 - 1 - 5}{4 + 1 + 1} = \frac{-12}{6} = -2$$

$$c_2 = \frac{\mathbf{s}_2 \cdot \mathbf{s}}{\|\mathbf{s}_2\|^2} = \frac{3 + 1 - 15}{1 + 1 + 9} = \frac{-11}{11} = -1$$

$$c_3 = \frac{\mathbf{s}_3 \cdot \mathbf{s}}{\|\mathbf{s}_3\|^2} = \frac{12 - 7 - 5}{16 + 49 + 1} = \frac{0}{66} = 0$$

Therefore we have $\mathbf{s} = -2\mathbf{s}_1 - \mathbf{s}_2$. ■

EXERCISES

Exercises 1–8 refer to the vectors $\mathbf{u}_1$ to $\mathbf{u}_8$.

$$\mathbf{u}_1 = \begin{bmatrix} -3 \\ 1 \\ 2 \end{bmatrix}, \ \mathbf{u}_2 = \begin{bmatrix} 1 \\ 1 \\ 1 \end{bmatrix}, \ \mathbf{u}_3 = \begin{bmatrix} 2 \\ 0 \\ -1 \end{bmatrix}, \ \mathbf{u}_4 = \begin{bmatrix} 1 \\ -3 \\ 2 \end{bmatrix},$$

$$\mathbf{u}_5 = \begin{bmatrix} 2 \\ 1 \\ 1 \end{bmatrix}, \ \mathbf{u}_6 = \begin{bmatrix} 0 \\ 3 \\ -1 \end{bmatrix}, \ \mathbf{u}_7 = \begin{bmatrix} 3 \\ -4 \\ -2 \end{bmatrix}, \ \mathbf{u}_8 = \begin{bmatrix} -1 \\ -1 \\ 3 \end{bmatrix}$$

1. Compute the following dot products.

(a) $\mathbf{u}_1 \cdot \mathbf{u}_5$

(b) $\mathbf{u}_3 \cdot (-3\mathbf{u}_2)$

(c) $\mathbf{u}_4 \cdot \mathbf{u}_7$

(d) $2\mathbf{u}_4 \cdot \mathbf{u}_7$

2. Compute the following dot products.

(a) $3\mathbf{u}_7 \cdot \mathbf{u}_3$

(b) $\mathbf{u}_1 \cdot \mathbf{u}_1$

(c) $\mathbf{u}_2 \cdot (-2\mathbf{u}_5)$

(d) $2\mathbf{u}_2 \cdot (-\mathbf{u}_5)$

3. Compute the norms of the given vectors.

(a) $\mathbf{u}_7$

(b) $-\mathbf{u}_7$

(c) $2\mathbf{u}_5$

(d) $-3\mathbf{u}_5$

4. Compute the norms of the given vectors.

(a) $\mathbf{u}_8$

(b) $3\mathbf{u}_8$

(c) $-\mathbf{u}_2$

(d) $-2\mathbf{u}_2$

5. Compute the distance between the given vectors.

(a) $\mathbf{u}_1$ and $\mathbf{u}_2$

(b) $\mathbf{u}_3$ and $\mathbf{u}_8$

(c) $2\mathbf{u}_6$ and $-\mathbf{u}_3$

(d) $-3\mathbf{u}_2$ and $2\mathbf{u}_5$

6. Compute the distance between the given vectors.

(a) $\mathbf{u}_5$ and $\mathbf{u}_1$

(b) $\mathbf{u}_2$ and $\mathbf{u}_8$

(c) $-2\mathbf{u}_3$ and $\mathbf{u}_8$

(d) $4\mathbf{u}_2$ and $-2\mathbf{u}_6$

7. Determine if the given vectors are orthogonal.

(a) $\mathbf{u}_1$ and $\mathbf{u}_3$

(b) $\mathbf{u}_3$ and $\mathbf{u}_4$

(c) $\mathbf{u}_2$ and $\mathbf{u}_5$

(d) $\mathbf{u}_1$ and $\mathbf{u}_8$

8. Determine if the given vectors are orthogonal.

(a) $\mathbf{u}_2$ and $\mathbf{u}_3$

(b) $\mathbf{u}_1$ and $\mathbf{u}_2$

(c) $\mathbf{u}_8$ and $\mathbf{u}_5$

(d) $\mathbf{u}_3$ and $\mathbf{u}_6$

For Exercises 9–12, find all values of a so that $\mathbf{u}$ and $\mathbf{v}$ are orthogonal.

9. $\mathbf{u} = \begin{bmatrix} a \\ 2 \\ -3 \end{bmatrix}, \ \mathbf{v} = \begin{bmatrix} 4 \\ a \\ 3 \end{bmatrix}$

10. $\mathbf{u} = \begin{bmatrix} -1 \\ a \\ 5 \end{bmatrix}$, $\mathbf{v} = \begin{bmatrix} 7 \\ a \\ -2 \end{bmatrix}$

11. $\mathbf{u} = \begin{bmatrix} 2 \\ a \\ -3 \\ -1 \end{bmatrix}$, $\mathbf{v} = \begin{bmatrix} -5 \\ 4 \\ 6 \\ a \end{bmatrix}$

12. $\mathbf{u} = \begin{bmatrix} 1 \\ -5 \\ a \\ 0 \end{bmatrix}$, $\mathbf{v} = \begin{bmatrix} -4 \\ 1 \\ a \\ -2 \end{bmatrix}$

For Exercises 13–16, determine if the given vectors form an orthogonal set.

13. $\mathbf{u}_1 = \begin{bmatrix} 1 \\ -2 \end{bmatrix}$, $\mathbf{u}_2 = \begin{bmatrix} 4 \\ 3 \end{bmatrix}$

14. $\mathbf{u}_1 = \begin{bmatrix} 1 \\ 2 \\ 3 \end{bmatrix}$, $\mathbf{u}_2 = \begin{bmatrix} 5 \\ -4 \\ 1 \end{bmatrix}$, $\mathbf{u}_3 = \begin{bmatrix} 1 \\ 1 \\ -1 \end{bmatrix}$

15. $\mathbf{u}_1 = \begin{bmatrix} 2 \\ 2 \\ -1 \end{bmatrix}$, $\mathbf{u}_2 = \begin{bmatrix} -5 \\ 13 \\ 16 \end{bmatrix}$, $\mathbf{u}_3 = \begin{bmatrix} 5 \\ -4 \\ 2 \end{bmatrix}$

16. $\mathbf{u}_1 = \begin{bmatrix} 1 \\ 2 \\ 0 \\ -1 \end{bmatrix}$, $\mathbf{u}_2 = \begin{bmatrix} 5 \\ 2 \\ 4 \\ 9 \end{bmatrix}$, $\mathbf{u}_3 = \begin{bmatrix} -2 \\ 2 \\ -3 \\ 2 \end{bmatrix}$

In Exercises 17–18, find all values of a (if any) so that the given vectors form an orthogonal set.

17. $\mathbf{u}_1 = \begin{bmatrix} -1 \\ 0 \\ 2 \end{bmatrix}$, $\mathbf{u}_2 = \begin{bmatrix} 4 \\ 3 \\ 2 \end{bmatrix}$, $\mathbf{u}_3 = \begin{bmatrix} 6 \\ a \\ 3 \end{bmatrix}$

18. $\mathbf{u}_1 = \begin{bmatrix} 1 \\ -3 \\ 2 \\ -1 \end{bmatrix}$, $\mathbf{u}_2 = \begin{bmatrix} 4 \\ 2 \\ 1 \\ 0 \end{bmatrix}$, $\mathbf{u}_3 = \begin{bmatrix} -1 \\ 0 \\ a \\ 7 \end{bmatrix}$

In Exercises 19–20, find all values of a and b (if any) so that the given vectors form an orthogonal set.

19. $\mathbf{u}_1 = \begin{bmatrix} 2 \\ 1 \\ -1 \end{bmatrix}$, $\mathbf{u}_2 = \begin{bmatrix} 3 \\ -4 \\ 2 \end{bmatrix}$, $\mathbf{u}_3 = \begin{bmatrix} 2 \\ a \\ b \end{bmatrix}$

20. $\mathbf{u}_1 = \begin{bmatrix} 1 \\ -3 \\ 6 \\ 1 \end{bmatrix}$, $\mathbf{u}_2 = \begin{bmatrix} 2 \\ 1 \\ a \\ -5 \end{bmatrix}$, $\mathbf{u}_3 = \begin{bmatrix} 0 \\ -4 \\ 3 \\ b \end{bmatrix}$

For Exercises 21–24, verify that the Pythagorean Theorem holds for the given orthogonal vectors.

21. $\mathbf{u}_1 = \begin{bmatrix} 3 \\ -1 \end{bmatrix}$, $\mathbf{u}_2 = \begin{bmatrix} 1 \\ 3 \end{bmatrix}$

22. $\mathbf{u}_1 = \begin{bmatrix} 6 \\ 8 \end{bmatrix}$, $\mathbf{u}_2 = \begin{bmatrix} -4 \\ 3 \end{bmatrix}$

23. $\mathbf{u}_1 = \begin{bmatrix} 2 \\ -3 \\ 1 \end{bmatrix}$, $\mathbf{u}_2 = \begin{bmatrix} 4 \\ 3 \\ 1 \end{bmatrix}$

24. $\mathbf{u}_1 = \begin{bmatrix} 5 \\ -2 \\ 4 \\ 2 \end{bmatrix}$, $\mathbf{u}_2 = \begin{bmatrix} 2 \\ 9 \\ 3 \\ -2 \end{bmatrix}$

25. Suppose that $\mathbf{u}_1$ and $\mathbf{u}_2$ are orthogonal vectors, with $\|\mathbf{u}_1\| = 2$ and $\|\mathbf{u}_2\| = 5$. Find $\|3\mathbf{u}_1 + 4\mathbf{u}_2\|$.

26. Suppose that $\mathbf{u}_1$ and $\mathbf{u}_2$ are orthogonal vectors, with $\|\mathbf{u}_1\| = 3$ and $\|\mathbf{u}_2\| = 4$. Find $\|2\mathbf{u}_1 - \mathbf{u}_2\|$.

For Exercises 27–28, determine if $\mathbf{u}$ is orthogonal to the subspace S.

27. $\mathbf{u} = \begin{bmatrix} 2 \\ -3 \\ 1 \end{bmatrix}$, $S = \text{span} \left\{ \begin{bmatrix} 1 \\ 2 \\ -1 \end{bmatrix}, \begin{bmatrix} 3 \\ -1 \\ 1 \end{bmatrix} \right\}$

28. $\mathbf{u} = \begin{bmatrix} 0 \\ 1 \\ 0 \\ 1 \end{bmatrix}$, $S = \text{span} \left\{ \begin{bmatrix} 1 \\ 1 \\ 0 \\ 1 \end{bmatrix}, \begin{bmatrix} 1 \\ 0 \\ 1 \\ 0 \end{bmatrix}, \begin{bmatrix} 1 \\ 1 \\ 1 \\ 1 \end{bmatrix} \right\}$

For Exercises 29–32, find a basis for $S^{\perp}$ for the subspace S.

29. $S = \text{span} \left\{ \begin{bmatrix} 1 \\ -3 \end{bmatrix} \right\}$

30. $S = \text{span} \left\{ \begin{bmatrix} 2 \\ 5 \end{bmatrix} \right\}$

31. $S = \text{span} \left\{ \begin{bmatrix} 1 \\ 1 \\ -2 \end{bmatrix} \right\}$

32. $S = \text{span} \left\{ \begin{bmatrix} -1 \\ 2 \\ 1 \end{bmatrix}, \begin{bmatrix} 2 \\ -3 \\ 2 \end{bmatrix} \right\}$

For Exercises 33–34, show that the given basis for S is orthogonal, and then write $\mathbf{s}$ as a linear combination of the basis vectors.

33. $S = \text{span} \left\{ \begin{bmatrix} 1 \\ 1 \\ 0 \end{bmatrix}, \begin{bmatrix} 1 \\ -1 \\ 4 \end{bmatrix} \right\}$, $\mathbf{s} = \begin{bmatrix} 1 \\ 2 \\ -2 \end{bmatrix}$

34. $S = \text{span} \left\{ \begin{bmatrix} 1 \\ 0 \\ 1 \\ -1 \end{bmatrix}, \begin{bmatrix} 3 \\ 3 \\ 1 \\ 4 \end{bmatrix}, \begin{bmatrix} -2 \\ 3 \\ 1 \\ -1 \end{bmatrix} \right\}$, $\mathbf{s} = \begin{bmatrix} 1 \\ 0 \\ 0 \\ 1 \end{bmatrix}$

FIND AN EXAMPLE For Exercises 35–44, find an example that meets the given specifications.

35. Two vectors $\mathbf{u}$ and $\mathbf{v}$ such that $\mathbf{u} \cdot \mathbf{v} = 12$.

36. A vector that is orthogonal to both $\begin{bmatrix} 1 \\ 0 \end{bmatrix}$ and $\begin{bmatrix} 0 \\ 1 \end{bmatrix}$.

37. A vector $\mathbf{u}$ such that $\|\mathbf{u}\| = 1$ and $\mathbf{u}$ is orthogonal to $\begin{bmatrix} -2 \\ 1 \end{bmatrix}$.

38. An orthogonal basis for $\mathbf{R}^2$ that includes $\begin{bmatrix} 3 \\ 4 \end{bmatrix}$.

39. Two linearly independent vectors that are both orthogonal to $\begin{bmatrix} 2 \\ 0 \\ -1 \end{bmatrix}$.

40. Two vectors in $\mathbf{R}^2$ that are orthogonal but do not span $\mathbf{R}^2$.

41. A subspace S of $\mathbf{R}^3$ that has $\dim(S^\perp) = 2$.

42. Two vectors $\mathbf{u}$ and $\mathbf{v}$ such that $\mathbf{u}$, $\mathbf{v}$, and $\begin{bmatrix} 1 \\ 1 \\ 1 \end{bmatrix}$ form an orthogonal set.

43. Three vectors in $\mathbf{R}^3$ that form an orthogonal set but not an orthogonal basis.

44. A subspace S of $\mathbf{R}^4$ such that $\dim(S) = \dim(S^\perp) = 2$.

TRUE OR FALSE For Exercises 45–56, determine if the statement is true or false, and justify your answer.

45. If $\|\mathbf{u} - \mathbf{v}\| = 3$, then the distance between $2\mathbf{u}$ and $2\mathbf{v}$ is 12.

46. If $\mathbf{u}$ and $\mathbf{v}$ have nonnegative entries, then $\mathbf{u} \cdot \mathbf{v} \geq 0$.

47. $\|\mathbf{u} + \mathbf{v}\| = \|\mathbf{u}\| + \|\mathbf{v}\|$ for all $\mathbf{u}$ and $\mathbf{v}$ in $\mathbf{R}^n$.

48. Suppose that $\{\mathbf{s}_1, \mathbf{s}_2, \mathbf{s}_3\}$ is an orthogonal set and that $c_1, c_2,$ and c_3 are scalars. Then $\{c_1\mathbf{s}_1, c_2\mathbf{s}_2, c_3\mathbf{s}_3\}$ is also an orthogonal set.

49. If S is a one-dimensional subspace of $\mathbf{R}^2$, then so is $S^\perp$.

50. If A is an $n \times n$ matrix and $\mathbf{u}$ is in $\mathbf{R}^n$, then $\|\mathbf{u}\| \leq \|A\mathbf{u}\|$.

51. If $\mathbf{u}_1 \cdot \mathbf{u}_2 = 0$ and $\mathbf{u}_2 \cdot \mathbf{u}_3 = 0$, then $\mathbf{u}_1 \cdot \mathbf{u}_3 = 0$.

52. If A is an $n \times n$ matrix with orthogonal columns, then $A^T A$ is a diagonal matrix.

53. If $\mathbf{u}$ and $\mathbf{v}$ are orthogonal, then the distance between $\mathbf{u}$ and $\mathbf{v}$ is $\sqrt{\|\mathbf{u}\|^2 + \|\mathbf{v}\|^2}$.

54. If $\|\mathbf{u} - \mathbf{v}\| = \|\mathbf{u} + \mathbf{v}\|$, then $\mathbf{u}$ and $\mathbf{v}$ are orthogonal.

55. If $A = \begin{bmatrix} \mathbf{a}_1 & \mathbf{a}_2 \end{bmatrix}$ and $S = \mathrm{span}\{\mathbf{a}_1, \mathbf{a}_2\}$, then $S^\perp = \mathrm{null}(A)$.

56. Even if S is merely a nonempty subset of $\mathbf{R}^n$, the orthogonal complement $S^\perp$ is still a subspace.

57. Prove that Theorem 8.9 is true even if the set $\mathcal{S} = \{\mathbf{s}_1, \ldots, \mathbf{s}_k\}$ only spans the subspace S instead of being a basis for S.

58. Prove that the zero vector $\mathbf{0}$ in $\mathbf{R}^n$ is orthogonal to all vectors in $\mathbf{R}^n$.

59. Prove that the standard basis $\{\mathbf{e}_1, \ldots, \mathbf{e}_n\}$ of $\mathbf{R}^n$ is an orthogonal basis.

60. Prove that if $\mathbf{u}_1$ and $\mathbf{u}_2$ are both orthogonal to $\mathbf{v}$, then so is $\mathbf{u}_1 + \mathbf{u}_2$.

61. Prove that if c_1 and c_2 are scalars and $\mathbf{u}_1$ and $\mathbf{u}_2$ are vectors, then $(c_1\mathbf{u}_1) \cdot (c_2\mathbf{u}_2) = c_1 c_2(\mathbf{u}_1 \cdot \mathbf{u}_2)$.

62. Let $\mathbf{u}$, $\mathbf{v}$, and $\mathbf{w}$ be in $\mathbf{R}^n$, and let c be a scalar. Prove each part of Theorem 8.2.

(a) $\mathbf{u} \cdot \mathbf{v} = \mathbf{v} \cdot \mathbf{u}$

(b) $(\mathbf{u} + \mathbf{v}) \cdot \mathbf{w} = \mathbf{u} \cdot \mathbf{w} + \mathbf{v} \cdot \mathbf{w}$

(c) $(c\mathbf{u}) \cdot \mathbf{v} = \mathbf{u} \cdot (c\mathbf{v}) = c(\mathbf{u} \cdot \mathbf{v})$

(d) $\mathbf{u} \cdot \mathbf{u} \geq 0$, and $\mathbf{u} \cdot \mathbf{u} = 0$ only when $\mathbf{u} = \mathbf{0}$

63. Use the properties of Theorem 8.2 to prove 1, which says that

$$(c_1\mathbf{u}_1 + \cdots + c_k\mathbf{u}_k) \cdot \mathbf{w} = c_1(\mathbf{u}_1 \cdot \mathbf{w}) + \cdots + c_k(\mathbf{u}_k \cdot \mathbf{w})$$

64. Prove 2, that $\|c\mathbf{x}\| = |c|\|\mathbf{x}\|$ for a scalar c and vector $\mathbf{x}$.

65. Prove that if $\mathbf{u} \neq \mathbf{0}$ and $\mathbf{v} = \dfrac{1}{\|\mathbf{u}\|}\mathbf{u}$, then $\|\mathbf{v}\| = 1$.

66. Let $\mathbf{u}$ be a vector in $\mathbf{R}^n$, and then define $T_\mathbf{u} : \mathbf{R}^n \to \mathbf{R}$ by $T_\mathbf{u}(\mathbf{v}) = \mathbf{u} \cdot \mathbf{v}$. Show that $T_\mathbf{u}$ is a linear transformation.

67. Let S be a subspace. Prove that $S \cap S^\perp = \{\mathbf{0}\}$.

68. Prove that

$$\|\mathbf{u} + \mathbf{v}\|^2 + \|\mathbf{u} - \mathbf{v}\|^2 = 2\left(\|\mathbf{u}\|^2 + \|\mathbf{v}\|^2\right).$$

69. For a matrix A, show that $\left(\mathrm{col}(A)\right)^\perp = \mathrm{Null}(A^T)$.

70. Prove that if S is a subspace, then $S = \left(S^\perp\right)^\perp$.

71. Let $\mathbf{u}$ and $\mathbf{v}$ be in $\mathbf{R}^n$ and A be an $n \times n$ matrix.

(a) Explain why $\mathbf{u} \cdot \mathbf{v}$ and $\mathbf{u}^T\mathbf{v}$ are essentially the same.

(b) Show that $(A\mathbf{u}) \cdot \mathbf{v} = \mathbf{u} \cdot \left(A^T\mathbf{v}\right)$.

72. In this problem we show that if $\mathbf{u}$ and $\mathbf{v}$ are perpendicular vectors in $\mathbf{R}^2$, then $\mathbf{u} \cdot \mathbf{v} = 0$. The method of proof is different from the one given at the start of the subsection "Orthogonal Vectors."

(a) Given nonzero vectors $\mathbf{u}$ and $\mathbf{v}$ orthogonal in $\mathbf{R}^2$, show that $\|\mathbf{u}\|, \|\mathbf{v}\|,$ and $\|\mathbf{u} - \mathbf{v}\|$ are the lengths of the sides of a right triangle.

(b) Use the version of the Pythagorean Theorem you learned in high school geometry to show that

$$\|\mathbf{u}\|^2 + \|\mathbf{v}\|^2 = \|\mathbf{u} - \mathbf{v}\|^2$$

(c) Write the equation in (b) in terms of dot products and then simplify to show that $\mathbf{u} \cdot \mathbf{v} = 0$.

Ⓒ For Exercises 73–74, let

$$\mathbf{u}_1 = \begin{bmatrix} 3 \\ -1 \\ 5 \\ 0 \\ 2 \end{bmatrix}, \quad \mathbf{u}_2 = \begin{bmatrix} 7 \\ 4 \\ 0 \\ 2 \\ 8 \end{bmatrix}, \quad \mathbf{u}_3 = \begin{bmatrix} 0 \\ 3 \\ -4 \\ 4 \\ -3 \end{bmatrix}$$

73. Compute each of the following:

(a) $\mathbf{u}_2 \cdot \mathbf{u}_3$

(b) $\|\mathbf{u}_1\|$

(c) $\|2\mathbf{u}_1 + 5\mathbf{u}_3\|$

(d) $\|3\mathbf{u}_1 - 4\mathbf{u}_2 - \mathbf{u}_3\|$

74. Compute each of the following:

(a) $\mathbf{u}_2 \cdot \mathbf{u}_1$

(b) $\|\mathbf{u}_3\|$

(c) $\|3\mathbf{u}_2 + 4\mathbf{u}_3\|$

(d) $\|-2\mathbf{u}_1 + 5\mathbf{u}_2 - 3\mathbf{u}_3\|$

Ⓒ For Exercises 75–76, find a basis for $S^\perp$.

75. $S = \text{span} \left\{ \begin{bmatrix} 2 \\ -1 \\ 3 \\ 5 \end{bmatrix}, \begin{bmatrix} 0 \\ 1 \\ 7 \\ 4 \end{bmatrix} \right\}$

76. $S = \text{span} \left\{ \begin{bmatrix} 6 \\ 0 \\ 2 \\ 5 \\ -1 \end{bmatrix}, \begin{bmatrix} 5 \\ 3 \\ 0 \\ 8 \\ -6 \end{bmatrix} \right\}$

8.2 Projection and the Gram–Schmidt Process

In Section 8.1 we developed the idea of an orthogonal basis. There are numerous applications of orthogonal bases, some explored in the other sections of this chapter. Of course, not every basis is an orthogonal basis. For example, suppose that

$$S = \text{span} \left\{ \begin{bmatrix} 1 \\ 0 \\ 1 \\ 1 \end{bmatrix}, \begin{bmatrix} 0 \\ 2 \\ 0 \\ 3 \end{bmatrix}, \begin{bmatrix} -3 \\ -1 \\ 1 \\ 5 \end{bmatrix} \right\}$$

It can be shown that this basis for S is not orthogonal. However, we can construct an orthogonal basis for S from this basis. The key to understanding how this is done is vector projection, so we start there.

Projection onto Vectors

Suppose that we have two nonzero vectors $\mathbf{u}$ and $\mathbf{v}$ in $\mathbf{R}^2$, as shown in Figure 1a. Draw the line perpendicular to $\mathbf{v}$ that passes through the tip of $\mathbf{u}$ (dashed in Figure 1b). The projection of $\mathbf{u}$ onto $\mathbf{v}$ (denoted $\text{proj}_\mathbf{v}\mathbf{u}$) is the vector parallel to $\mathbf{v}$ with tip at the intersection of $\mathbf{v}$ and the dashed perpendicular line (Figure 1c).

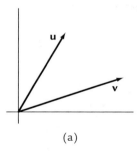
(a)

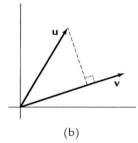
(b)

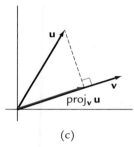
(c)

Figure 1 Constructing $\text{proj}_\mathbf{v}\mathbf{u}$ in $\mathbf{R}^2$.

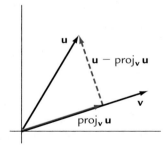

Figure 2 $\mathbf{v}$ and $\mathbf{u} - \text{proj}_\mathbf{v}\mathbf{u}$ are orthogonal.

We can think of $\text{proj}_\mathbf{v}\mathbf{u}$ as the component of $\mathbf{u}$ in the direction of $\mathbf{v}$.

Our next step is to develop a formula for $\text{proj}_\mathbf{v}\mathbf{u}$. Note that the vectors $\mathbf{v}$ and $\mathbf{u} - \text{proj}_\mathbf{v}\mathbf{u}$ are orthogonal to each other (see Figure 2). Therefore

$$\mathbf{v} \cdot (\mathbf{u} - \text{proj}_\mathbf{v}\mathbf{u}) = 0 \tag{1}$$

Since $\text{proj}_\mathbf{v}\mathbf{u}$ is parallel to $\mathbf{v}$, there exists a scalar c such that $\text{proj}_\mathbf{v}\mathbf{u} = c\mathbf{v}$. We can find a formula for c by substituting into (1) and solving for c.

$$\mathbf{v} \cdot (\mathbf{u} - c\mathbf{v}) = 0 \quad \implies \quad \mathbf{v} \cdot \mathbf{u} - c(\mathbf{v} \cdot \mathbf{v}) = 0 \quad \implies \quad c = \frac{\mathbf{v} \cdot \mathbf{u}}{\mathbf{v} \cdot \mathbf{v}} = \frac{\mathbf{v} \cdot \mathbf{u}}{\|\mathbf{v}\|^2}$$

Thus we have

$$\text{proj}_\mathbf{v}\mathbf{u} = c\mathbf{v} = \frac{\mathbf{v} \cdot \mathbf{u}}{\|\mathbf{v}\|^2}\mathbf{v}$$

Although this formula was developed in $\mathbf{R}^2$, it can also be evaluated for vectors in $\mathbf{R}^n$, so we use it to generalize projection to any dimension.

DEFINITION 8.13

Definition Projection Onto a Vector

Let $\mathbf{u}$ and $\mathbf{v}$ be vectors in $\mathbf{R}^n$, with $\mathbf{v}$ nonzero. Then the **projection of u onto v** is given by

$$\operatorname{proj}_{\mathbf{v}}\mathbf{u} = \frac{\mathbf{v} \cdot \mathbf{u}}{\|\mathbf{v}\|^2}\mathbf{v} \tag{2}$$

EXAMPLE 1 Find $\operatorname{proj}_{\mathbf{v}}\mathbf{u}$ for

$$\mathbf{u} = \begin{bmatrix} 7 \\ 14 \\ 4 \end{bmatrix} \quad \text{and} \quad \mathbf{v} = \begin{bmatrix} -2 \\ 5 \\ 1 \end{bmatrix}$$

Solution Applying formula 2, we have

$$\operatorname{proj}_{\mathbf{v}}\mathbf{u} = \frac{\mathbf{v} \cdot \mathbf{u}}{\|\mathbf{v}\|^2}\mathbf{v} = \frac{(-14 + 70 + 4)}{(4 + 25 + 1)}\begin{bmatrix} -2 \\ 5 \\ 1 \end{bmatrix} = \frac{60}{30}\begin{bmatrix} -2 \\ 5 \\ 1 \end{bmatrix} = \begin{bmatrix} -4 \\ 10 \\ 2 \end{bmatrix}$$

Projections onto vectors have several important properties that are summarized in the next theorem. ∎

THEOREM 8.14

Let $\mathbf{u}$ and $\mathbf{v}$ be vectors in $\mathbf{R}^n$ ($\mathbf{v}$ nonzero) and c be a nonzero scalar. Then

(a) $\operatorname{proj}_{\mathbf{v}}\mathbf{u}$ is in span$\{\mathbf{v}\}$.

(b) $\mathbf{u} - \operatorname{proj}_{\mathbf{v}}\mathbf{u}$ is orthogonal to $\mathbf{v}$.

(c) If $\mathbf{u}$ is in span$\{\mathbf{v}\}$, then $\mathbf{u} = \operatorname{proj}_{\mathbf{v}}\mathbf{u}$.

(d) $\operatorname{proj}_{\mathbf{v}}\mathbf{u} = \operatorname{proj}_{c\mathbf{v}}\mathbf{u}$.

Proof We take each part in turn.

(a) Since $\operatorname{proj}_{\mathbf{v}}\mathbf{u} = \dfrac{\mathbf{v} \cdot \mathbf{u}}{\|\mathbf{v}\|^2}\mathbf{v}$ is multiple of $\mathbf{v}$, $\operatorname{proj}_{\mathbf{v}}\mathbf{u}$ must be in span$\{\mathbf{v}\}$.

(b) We verify that $\mathbf{v}$ is orthogonal to $\mathbf{u} - \operatorname{proj}_{\mathbf{v}}\mathbf{u}$ by computing the dot product.

$$\mathbf{v} \cdot (\mathbf{u} - \operatorname{proj}_{\mathbf{v}}\mathbf{u}) = \mathbf{v} \cdot \left(\mathbf{u} - \frac{\mathbf{v} \cdot \mathbf{u}}{\|\mathbf{v}\|^2}\mathbf{v}\right)$$

$$= \mathbf{v} \cdot \mathbf{u} - \frac{\mathbf{v} \cdot \mathbf{u}}{\|\mathbf{v}\|^2}(\mathbf{v} \cdot \mathbf{v})$$

$$= \mathbf{v} \cdot \mathbf{u} - \frac{\mathbf{v} \cdot \mathbf{u}}{\|\mathbf{v}\|^2}\|\mathbf{v}\|^2 = \mathbf{v} \cdot \mathbf{u} - \mathbf{v} \cdot \mathbf{u} = 0$$

(c) If $\mathbf{u}$ is in span$\{\mathbf{v}\}$, then there exists a constant c such that $\mathbf{u} = c\mathbf{v}$. Hence

$$\operatorname{proj}_{\mathbf{v}}\mathbf{u} = \frac{\mathbf{v} \cdot (c\mathbf{v})}{\|\mathbf{v}\|^2}\mathbf{v} = c\frac{\mathbf{v} \cdot \mathbf{v}}{\|\mathbf{v}\|^2}\mathbf{v} = c\mathbf{v} = \mathbf{u}$$

(d) For any nonzero scalar c,

$$\operatorname{proj}_{c\mathbf{v}}\mathbf{u} = \frac{(c\mathbf{v}) \cdot \mathbf{u}}{\|c\mathbf{v}\|^2}(c\mathbf{v}) = \frac{c^2}{|c|^2}\frac{\mathbf{v} \cdot \mathbf{u}}{\|\mathbf{v}\|^2}\mathbf{v} = \frac{\mathbf{v} \cdot \mathbf{u}}{\|\mathbf{v}\|^2}\mathbf{v} = \operatorname{proj}_{\mathbf{v}}\mathbf{u} \quad ∎$$

Projections onto Subspaces

We can extend the idea of projecting onto a vector to projecting onto subspaces. In a sense, we have already taken a step in this direction. Since Theorem 8.14d shows $\text{proj}_\mathbf{v}\mathbf{u} = \text{proj}_{c\mathbf{v}}\mathbf{u}$, we can think of $\text{proj}_\mathbf{v}\mathbf{u}$ as projecting $\mathbf{u}$ onto the subspace $\text{span}\{\mathbf{v}\}$, the line through the origin in the direction of $\mathbf{v}$.

Given a nonzero subspace S and a vector $\mathbf{u}$, we denote the projection of $\mathbf{u}$ onto S by $\text{proj}_S\mathbf{u}$. Although we do not yet have a definition for $\text{proj}_S\mathbf{u}$, its properties should be analogous to those for $\text{proj}_\mathbf{v}\mathbf{u}$ given in Theorem 8.14. In particular, if $\mathbf{u}$ is in S, then we want $\text{proj}_S\mathbf{u} = \mathbf{u}$. At this point, it helps to recall Theorem 8.12 in Section 8.1, which says that if $\{\mathbf{v}_1, \ldots, \mathbf{v}_k\}$ is an orthogonal basis for S, then any $\mathbf{s}$ in S can be expressed

$$\mathbf{s} = \frac{\mathbf{v}_1 \cdot \mathbf{s}}{\|\mathbf{v}_1\|^2}\mathbf{v}_1 + \frac{\mathbf{v}_2 \cdot \mathbf{s}}{\|\mathbf{v}_2\|^2}\mathbf{v}_2 + \cdots + \frac{\mathbf{v}_k \cdot \mathbf{s}}{\|\mathbf{v}_k\|^2}\mathbf{v}_k \tag{3}$$

Given our requirements for $\text{proj}_S\mathbf{u}$, then $\text{proj}_S\mathbf{u}$ must equal the right side of (3) when $\mathbf{u}$ is in S. Moreover, the right side of (3) is also equal to the sum of the projection of $\mathbf{u}$ onto each of $\mathbf{v}_1, \ldots, \mathbf{v}_k$. All of this suggests the following definition.

DEFINITION 8.15

Definition Projection Onto a Subspace

Let S be a nonzero subspace with orthogonal basis $\{\mathbf{v}_1, \ldots, \mathbf{v}_k\}$. Then the **projection of $\mathbf{u}$ onto** S is given by

$$\text{proj}_S\mathbf{u} = \frac{\mathbf{v}_1 \cdot \mathbf{u}}{\|\mathbf{v}_1\|^2}\mathbf{v}_1 + \frac{\mathbf{v}_2 \cdot \mathbf{u}}{\|\mathbf{v}_2\|^2}\mathbf{v}_2 + \cdots + \frac{\mathbf{v}_k \cdot \mathbf{u}}{\|\mathbf{v}_k\|^2}\mathbf{v}_k \tag{4}$$

See Figure 3 for a graphical depiction of projection onto a plane.

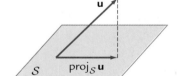

Figure 3 Projection of a vector $\mathbf{u}$ in $\mathbf{R}^3$ onto a two-dimensional subspace S.

EXAMPLE 2 Find $\text{proj}_S\mathbf{u}$ for $S = \text{span}\{\mathbf{v}_1, \mathbf{v}_2\}$, where

$$\mathbf{u} = \begin{bmatrix} 18 \\ -20 \\ 10 \end{bmatrix}, \quad \mathbf{v}_1 = \begin{bmatrix} 4 \\ -1 \\ -5 \end{bmatrix}, \quad \mathbf{v}_2 = \begin{bmatrix} 3 \\ 2 \\ 2 \end{bmatrix}$$

Solution The vectors $\mathbf{v}_1$ and $\mathbf{v}_2$ are orthogonal, so we may apply (4). We have

$$\text{proj}_S\mathbf{u} = \frac{\mathbf{v}_1 \cdot \mathbf{u}}{\|\mathbf{v}_1\|^2}\mathbf{v}_1 + \frac{\mathbf{v}_2 \cdot \mathbf{u}}{\|\mathbf{v}_2\|^2}\mathbf{v}_2 = \frac{42}{42}\begin{bmatrix} 4 \\ -1 \\ -5 \end{bmatrix} + \frac{34}{17}\begin{bmatrix} 3 \\ 2 \\ 2 \end{bmatrix} = \begin{bmatrix} 10 \\ 3 \\ -1 \end{bmatrix} \quad \blacksquare$$

Regarding Definition 8.15:

- If $S = \text{span}\{\mathbf{v}\}$ is a one-dimensional subspace, then (4) reduces to the formula for $\text{proj}_\mathbf{v}\mathbf{u}$.

- We can express $\text{proj}_S\mathbf{u}$ by

$$\text{proj}_S\mathbf{u} = \text{proj}_{\mathbf{v}_1}\mathbf{u} + \text{proj}_{\mathbf{v}_2}\mathbf{u} + \cdots + \text{proj}_{\mathbf{v}_k}\mathbf{u}$$

- The basis $\{\mathbf{v}_1, \ldots, \mathbf{v}_k\}$ for S must be orthogonal in order to apply the formula for $\text{proj}_S\mathbf{u}$.

The next theorem shows that $\text{proj}_S\mathbf{u}$ does not depend on the choice of orthogonal basis for S.

THEOREM 8.16

Let S be a nonzero subspace of $\mathbf{R}^n$ with orthogonal basis $\{\mathbf{v}_1, \ldots, \mathbf{v}_k\}$, and let $\mathbf{u}$ be a vector in $\mathbf{R}^n$. Then

(a) $\text{proj}_S \mathbf{u}$ is in S.

(b) $\mathbf{u} - \text{proj}_S \mathbf{u}$ is orthogonal to S.

(c) if $\mathbf{u}$ is in S, then $\mathbf{u} = \text{proj}_S \mathbf{u}$.

(d) $\text{proj}_S \mathbf{u}$ is independent of the choice of orthogonal basis for S.

▶ This extends Theorem 8.14 to subspaces.

The proof of Theorem 8.16 is given at the end of the section.

EXAMPLE 3 Let

$$\mathbf{u} = \begin{bmatrix} 3 \\ -1 \\ 1 \\ 5 \end{bmatrix}, \quad \mathbf{v}_1 = \begin{bmatrix} 1 \\ 0 \\ -1 \\ 2 \end{bmatrix}, \quad \mathbf{v}_2 = \begin{bmatrix} 2 \\ -1 \\ 2 \\ 0 \end{bmatrix}, \quad \mathbf{w}_1 = \begin{bmatrix} 1 \\ -1 \\ 3 \\ -2 \end{bmatrix}, \quad \mathbf{w}_2 = \begin{bmatrix} 7 \\ -2 \\ 1 \\ 6 \end{bmatrix}$$

It can be shown that $\{\mathbf{v}_1, \mathbf{v}_2\}$ and $\{\mathbf{w}_1, \mathbf{w}_2\}$ both form orthogonal bases for the same subspace S. Show that $\text{proj}_S \mathbf{u}$ is the same for both bases.

Solution Starting with $S = \text{span}\{\mathbf{v}_1, \mathbf{v}_2\}$, we have

$$\text{proj}_S \mathbf{u} = \frac{\mathbf{v}_1 \cdot \mathbf{u}}{\|\mathbf{v}_1\|^2} \mathbf{v}_1 + \frac{\mathbf{v}_2 \cdot \mathbf{u}}{\|\mathbf{v}_2\|^2} \mathbf{v}_2 = \frac{12}{6} \begin{bmatrix} 1 \\ 0 \\ -1 \\ 2 \end{bmatrix} + \frac{9}{9} \begin{bmatrix} 2 \\ -1 \\ 2 \\ 0 \end{bmatrix} = \begin{bmatrix} 4 \\ -1 \\ 0 \\ 4 \end{bmatrix}$$

On the other hand, $S = \text{span}\{\mathbf{w}_1, \mathbf{w}_2\}$ yields

$$\text{proj}_S \mathbf{u} = \frac{\mathbf{w}_1 \cdot \mathbf{u}}{\|\mathbf{w}_1\|^2} \mathbf{w}_1 + \frac{\mathbf{w}_2 \cdot \mathbf{u}}{\|\mathbf{w}_2\|^2} \mathbf{w}_2 = \frac{-3}{15} \begin{bmatrix} 1 \\ -1 \\ 3 \\ -2 \end{bmatrix} + \frac{54}{90} \begin{bmatrix} 7 \\ -2 \\ 1 \\ 6 \end{bmatrix} = \begin{bmatrix} 4 \\ -1 \\ 0 \\ 4 \end{bmatrix}$$

Both bases produce the same projection vector, as promised by Theorem 8.16. ∎

The Gram–Schmidt Process

Now that we know how to find projections of vectors onto subspaces, we are ready to develop a method for finding an orthogonal basis for a subspace. Let's start with a simple case, an arbitrary two-dimensional subspace $S = \text{span}\{\mathbf{s}_1, \mathbf{s}_2\}$ in $\mathbf{R}^n$. Our goal is to find an orthogonal basis for S. Let

$$\mathbf{v}_1 = \mathbf{s}_1$$
$$\mathbf{v}_2 = \mathbf{s}_2 - \text{proj}_{\mathbf{v}_1} \mathbf{s}_2$$

Then $\mathbf{v}_1$ and $\mathbf{v}_2$ are orthogonal by Theorem 8.14b. Moreover, since $\mathbf{v}_1 = \mathbf{s}_1$, it follows that $\text{proj}_{\mathbf{v}_1} \mathbf{s}_2 = \text{proj}_{\mathbf{s}_1} \mathbf{s}_2 = c\mathbf{s}_1$ for some nonzero scalar c. Therefore $\mathbf{v}_2 = \mathbf{s}_2 - c\mathbf{s}_1$, so $\mathbf{v}_1$ and $\mathbf{v}_2$ are both in S. By Theorem 8.11 in Section 8.1, $\mathbf{v}_1$ and $\mathbf{v}_2$ are also linearly independent. Since $\dim(S) = 2$, we may conclude that $\{\mathbf{v}_1, \mathbf{v}_2\}$ is an orthogonal basis for S.

EXAMPLE 4 Let $S = \text{span}\{\mathbf{s}_1, \mathbf{s}_2\}$, where

$$\mathbf{s}_1 = \begin{bmatrix} 1 \\ -2 \\ 3 \end{bmatrix} \quad \text{and} \quad \mathbf{s}_2 = \begin{bmatrix} 3 \\ 5 \\ -7 \end{bmatrix}$$

Find an orthogonal basis for S.

Solution By the above formulas, we define

$$\mathbf{v}_1 = \mathbf{u}_1 = \begin{bmatrix} 1 \\ -2 \\ 3 \end{bmatrix}$$

$$\mathbf{v}_2 = \mathbf{u}_2 - \text{proj}_{\mathbf{v}_1} \mathbf{u}_2 = \begin{bmatrix} 3 \\ 5 \\ -7 \end{bmatrix} - \frac{-28}{14} \begin{bmatrix} 1 \\ -2 \\ 3 \end{bmatrix} = \begin{bmatrix} 5 \\ 1 \\ -1 \end{bmatrix}$$

From the above discussion we know that $\mathbf{v}_1 \cdot \mathbf{v}_2 = 0$ and that $\text{span}\{\mathbf{s}_1, \mathbf{s}_2\} = \text{span}\{\mathbf{v}_1, \mathbf{v}_2\}$. Thus $\{\mathbf{v}_1, \mathbf{v}_2\}$ forms an orthogonal basis for S. ∎

The *Gram–Schmidt process* extends the procedure illustrated in Example 4 and allows us to generate an orthogonal basis for any nonzero subspace.

THEOREM 8.17

▶ Jörgen Gram (1850–1916) was a Danish actuary who worked on the mathematics of accident insurance, and Erhardt Schmidt (1876–1959) was a German mathematician who taught at Berlin University.

(THE GRAM–SCHMIDT PROCESS) Let S be a subspace with basis $\{\mathbf{s}_1, \mathbf{s}_2, \ldots, \mathbf{s}_k\}$. Define $\mathbf{v}_1, \mathbf{v}_2, \ldots, \mathbf{v}_k$, in order, by

$$\mathbf{v}_1 = \mathbf{s}_1$$
$$\mathbf{v}_2 = \mathbf{s}_2 - \text{proj}_{\mathbf{v}_1} \mathbf{s}_2$$
$$\mathbf{v}_3 = \mathbf{s}_3 - \text{proj}_{\mathbf{v}_1} \mathbf{s}_3 - \text{proj}_{\mathbf{v}_2} \mathbf{s}_3$$
$$\mathbf{v}_4 = \mathbf{s}_4 - \text{proj}_{\mathbf{v}_1} \mathbf{s}_4 - \text{proj}_{\mathbf{v}_2} \mathbf{s}_4 - \text{proj}_{\mathbf{v}_3} \mathbf{s}_4$$
$$\vdots \quad \vdots$$
$$\mathbf{v}_k = \mathbf{s}_k - \text{proj}_{\mathbf{v}_1} \mathbf{s}_k - \text{proj}_{\mathbf{v}_2} \mathbf{s}_k - \cdots - \text{proj}_{\mathbf{v}_{k-1}} \mathbf{s}_k$$

Then $\{\mathbf{v}_1, \mathbf{v}_2, \ldots, \mathbf{v}_k\}$ is an orthogonal basis for S.

At each step of the Gram–Schmidt process, the new vector $\mathbf{v}_j$ is orthogonal to the subspace

$$\text{span}\{\mathbf{v}_1, \ldots, \mathbf{v}_{j-1}\} = \text{span}\{\mathbf{s}_1, \ldots, \mathbf{s}_{j-1}\}$$

so we build up our basis for S by adding vectors orthogonal to those already in place, ensuring an orthogonal basis at the end. A proof that the Gram–Schmidt process works can be carried out by induction and is left as an exercise.

Let's return to the problem we encountered at the beginning of the section.

EXAMPLE 5 Find an orthogonal basis for the subspace $S = \text{span}\{\mathbf{s}_1, \mathbf{s}_2, \mathbf{s}_3\}$, where

$$\mathbf{s}_1 = \begin{bmatrix} 1 \\ 0 \\ 1 \\ 1 \end{bmatrix}, \quad \mathbf{s}_2 = \begin{bmatrix} 0 \\ 2 \\ 0 \\ 3 \end{bmatrix}, \quad \mathbf{s}_3 = \begin{bmatrix} -3 \\ -1 \\ 1 \\ 5 \end{bmatrix}$$

Solution The first step of the Gram–Schmidt process is the easiest, setting $\mathbf{v}_1 = \mathbf{s}_1$. Moving to the next step, we let

$$\mathbf{v}_2 = \mathbf{s}_2 - \text{proj}_{\mathbf{v}_1} \mathbf{s}_2 = \mathbf{s}_2 - \frac{\mathbf{v}_1 \cdot \mathbf{s}_2}{\|\mathbf{v}_1\|^2} \mathbf{v}_1 = \begin{bmatrix} 0 \\ 2 \\ 0 \\ 3 \end{bmatrix} - \frac{3}{3} \begin{bmatrix} 1 \\ 0 \\ 1 \\ 1 \end{bmatrix} = \begin{bmatrix} -1 \\ 2 \\ -1 \\ 2 \end{bmatrix}$$

For the last step, we have

$$\mathbf{v}_3 = \mathbf{s}_3 - \text{proj}_{\mathbf{v}_1}\mathbf{s}_3 - \text{proj}_{\mathbf{v}_2}\mathbf{s}_3 = \mathbf{s}_3 - \frac{\mathbf{v}_1 \cdot \mathbf{s}_3}{\|\mathbf{v}_1\|^2}\mathbf{v}_1 - \frac{\mathbf{v}_2 \cdot \mathbf{s}_3}{\|\mathbf{v}_2\|^2}\mathbf{v}_2$$

$$= \begin{bmatrix} -3 \\ -1 \\ 1 \\ 5 \end{bmatrix} - \frac{3}{3}\begin{bmatrix} 1 \\ 0 \\ 1 \\ 1 \end{bmatrix} - \frac{10}{10}\begin{bmatrix} -1 \\ 2 \\ -1 \\ 2 \end{bmatrix} = \begin{bmatrix} -3 \\ -3 \\ 1 \\ 2 \end{bmatrix}$$

This gives us the orthogonal basis

$$\left\{ \begin{bmatrix} 1 \\ 0 \\ 1 \\ 1 \end{bmatrix}, \begin{bmatrix} -1 \\ 2 \\ -1 \\ 2 \end{bmatrix}, \begin{bmatrix} -3 \\ -3 \\ 1 \\ 2 \end{bmatrix} \right\}$$
∎

Orthonormal Bases

In Theorem 8.12 we showed that if S is a subspace with orthogonal basis $\{\mathbf{v}_1, \ldots, \mathbf{v}_k\}$, then any vector $\mathbf{s}$ in S can be expressed

$$\mathbf{s} = \frac{\mathbf{v}_1 \cdot \mathbf{s}}{\|\mathbf{v}_1\|^2}\mathbf{v}_1 + \frac{\mathbf{v}_2 \cdot \mathbf{s}}{\|\mathbf{v}_2\|^2}\mathbf{v}_2 + \cdots + \frac{\mathbf{v}_k \cdot \mathbf{s}}{\|\mathbf{v}_k\|^2}\mathbf{v}_k$$

If each of the vectors $\mathbf{v}_i$ also satisfies $\|\mathbf{v}_i\| = 1$, then this formula simplifies to

$$\mathbf{s} = (\mathbf{v}_1 \cdot \mathbf{s})\mathbf{v}_1 + (\mathbf{v}_2 \cdot \mathbf{s})\mathbf{v}_2 + \cdots + (\mathbf{v}_k \cdot \mathbf{s})\mathbf{v}_k \qquad (5)$$

DEFINITION 8.18

Definition Orthonormal Set

A set of vectors $\{\mathbf{w}_1, \ldots, \mathbf{w}_k\}$ is **orthonormal** if the set is orthogonal and $\|\mathbf{w}_j\| = 1$ for each of $j = 1, 2, \ldots k$.

To obtain an orthonormal basis for a subspace $S = \text{span}\{\mathbf{s}_1, \ldots, \mathbf{s}_k\}$ of dimension k, we first use Gram–Schmidt to find an orthogonal basis $\{\mathbf{v}_1, \ldots, \mathbf{v}_k\}$ for S and then let

$$\mathbf{w}_j = \frac{1}{\|\mathbf{v}_j\|}\mathbf{v}_j \qquad \text{for } j = 1, 2, \ldots k$$

Definition Normalizing

This step is called **normalizing** the vectors. Since each $\mathbf{w}_j$ is a multiple of $\mathbf{v}_j$, the set $\{\mathbf{w}_1, \ldots, \mathbf{w}_k\}$ is orthogonal and $\text{span}\{\mathbf{v}_1, \ldots, \mathbf{v}_k\} = \text{span}\{\mathbf{w}_1, \ldots, \mathbf{w}_k\}$. Furthermore, as

$$\|\mathbf{w}_j\| = \left\| \frac{1}{\|\mathbf{v}_j\|}\mathbf{v}_j \right\| = \frac{1}{\|\mathbf{v}_j\|}\|\mathbf{v}_j\| = 1$$

the set $\{\mathbf{w}_1, \ldots, \mathbf{w}_k\}$ is an orthonormal basis for S.

EXAMPLE 6 Find an orthonormal basis for the subspace S given in Example 5.

Solution We already have the orthogonal basis

$$\mathbf{v}_1 = \begin{bmatrix} 1 \\ 0 \\ 1 \\ 1 \end{bmatrix}, \quad \mathbf{v}_2 = \begin{bmatrix} -1 \\ 2 \\ -1 \\ 2 \end{bmatrix}, \quad \mathbf{v}_3 = \begin{bmatrix} -3 \\ -3 \\ 1 \\ 2 \end{bmatrix}$$

All that remains is to normalize each of $\mathbf{v}_1$, $\mathbf{v}_2$, and $\mathbf{v}_3$ by dividing by their respective lengths. Since $\|\mathbf{v}_1\| = \sqrt{3}$, $\|\mathbf{v}_2\| = \sqrt{10}$, and $\|\mathbf{v}_3\| = \sqrt{23}$, the orthonormal basis is

$$\left\{ \frac{1}{\sqrt{3}} \begin{bmatrix} 1 \\ 0 \\ 1 \\ 1 \end{bmatrix}, \frac{1}{\sqrt{10}} \begin{bmatrix} -1 \\ 2 \\ -1 \\ 2 \end{bmatrix}, \frac{1}{\sqrt{23}} \begin{bmatrix} -3 \\ -3 \\ 1 \\ 2 \end{bmatrix} \right\} \qquad \blacksquare$$

EXAMPLE 7 Let $S = \mathrm{span}\{\mathbf{s}_1, \mathbf{s}_2, \mathbf{s}_3\}$, where

$$\mathbf{s}_1 = \begin{bmatrix} 1 \\ 2 \\ -2 \end{bmatrix}, \quad \mathbf{s}_2 = \begin{bmatrix} 1 \\ 0 \\ -4 \end{bmatrix}, \quad \mathbf{s}_3 = \begin{bmatrix} 5 \\ 2 \\ 0 \end{bmatrix}, \quad \text{and} \quad \mathbf{s} = \begin{bmatrix} 1 \\ 1 \\ -1 \end{bmatrix}$$

Use the Gram–Schmidt process to find an orthonormal basis for S, and then write $\mathbf{s}$ as a linear combination of the orthonormal basis vectors.

Solution We start by finding an orthogonal basis. After setting $\mathbf{v}_1 = \mathbf{s}_1$, we have

$$\mathbf{v}_2 = \mathbf{s}_2 - \mathrm{proj}_{\mathbf{v}_1} \mathbf{s}_2 = \mathbf{s}_2 - \frac{\mathbf{s}_2 \cdot \mathbf{v}_1}{\|\mathbf{v}_1\|^2} \mathbf{v}_1 = \begin{bmatrix} 1 \\ 0 \\ -4 \end{bmatrix} - \frac{9}{9} \begin{bmatrix} 1 \\ 2 \\ -2 \end{bmatrix} = \begin{bmatrix} 0 \\ -2 \\ -2 \end{bmatrix}$$

and

$$\mathbf{v}_3 = \mathbf{s}_3 - \mathrm{proj}_{\mathbf{v}_1} \mathbf{s}_3 - \mathrm{proj}_{\mathbf{v}_2} \mathbf{s}_3 = \mathbf{s}_3 - \frac{\mathbf{s}_3 \cdot \mathbf{v}_1}{\|\mathbf{v}_1\|^2} \mathbf{v}_1 - \frac{\mathbf{s}_3 \cdot \mathbf{v}_2}{\|\mathbf{v}_2\|^2} \mathbf{v}_2$$

$$= \begin{bmatrix} 5 \\ 2 \\ 0 \end{bmatrix} - \frac{9}{9} \begin{bmatrix} 1 \\ 2 \\ -2 \end{bmatrix} - \frac{(-4)}{8} \begin{bmatrix} 0 \\ -2 \\ -2 \end{bmatrix} = \begin{bmatrix} 4 \\ -1 \\ 1 \end{bmatrix}$$

Now that we have an orthogonal basis, we obtain an orthonormal basis by normalizing each of $\mathbf{v}_1$, $\mathbf{v}_2$, and $\mathbf{v}_3$:

$$\mathbf{w}_1 = \frac{1}{\|\mathbf{v}_1\|} \mathbf{v}_1 = \frac{1}{3} \begin{bmatrix} 1 \\ 2 \\ -2 \end{bmatrix}$$

$$\mathbf{w}_2 = \frac{1}{\|\mathbf{v}_2\|} \mathbf{v}_2 = \frac{1}{\sqrt{8}} \begin{bmatrix} 0 \\ -2 \\ -2 \end{bmatrix} = \frac{1}{\sqrt{2}} \begin{bmatrix} 0 \\ -1 \\ -1 \end{bmatrix}$$

$$\mathbf{w}_3 = \frac{1}{\|\mathbf{v}_3\|} \mathbf{v}_3 = \frac{1}{\sqrt{18}} \begin{bmatrix} 4 \\ -1 \\ 1 \end{bmatrix} = \frac{1}{3\sqrt{2}} \begin{bmatrix} 4 \\ -1 \\ 1 \end{bmatrix}$$

To write $\mathbf{s}$ as a linear combination of $\mathbf{w}_1$, $\mathbf{w}_2$, and $\mathbf{w}_3$, we apply the formula in (5). This produces

$$\mathbf{s} = (\mathbf{w}_1 \cdot \mathbf{s})\mathbf{w}_1 + (\mathbf{w}_2 \cdot \mathbf{s})\mathbf{w}_2 + (\mathbf{w}_3 \cdot \mathbf{s})\mathbf{w}_3$$

$$= \left(\frac{5}{3} \right) \mathbf{w}_1 + (0)\, \mathbf{w}_2 + \left(\frac{\sqrt{2}}{3} \right) \mathbf{w}_3 = \frac{5}{3}\mathbf{w}_1 + \frac{\sqrt{2}}{3}\mathbf{w}_3$$

We can check that this is correct by computing

$$\left(\frac{5}{3} \right) \mathbf{w}_1 + \left(\frac{\sqrt{2}}{3} \right) \mathbf{w}_3 = \left(\frac{5}{3} \right) \cdot \frac{1}{3} \begin{bmatrix} 1 \\ 2 \\ -2 \end{bmatrix} + \left(\frac{\sqrt{2}}{3} \right) \cdot \frac{1}{3\sqrt{2}} \begin{bmatrix} 4 \\ -1 \\ 1 \end{bmatrix} = \begin{bmatrix} 1 \\ 1 \\ -1 \end{bmatrix} = \mathbf{s} \qquad \blacksquare$$

Computational Comments

When implemented on a computer, the Gram–Schmidt process can suffer from significant round-off error. As the orthogonal vectors are computed, some dot products $\mathbf{v}_i \cdot \mathbf{v}_j$ might not be close to zero when $|i - j|$ is large. There is a modified version of the Gram–Schmidt process that requires more operations, but it is also more numerically stable and hence is not as prone to loss of orthogonality due to round-off error.

Proof of Theorem 8.16

Proof of Theorem 8.16   Part (a) follows from the definition of $\text{proj}_S\mathbf{u}$, and part (c) follows from Theorem 8.12. For part (b), suppose that $\{\mathbf{v}_1, \ldots, \mathbf{v}_k\}$ is an orthogonal basis for S. Then

$$\mathbf{v}_1 \cdot (\mathbf{u} - \text{proj}_S\mathbf{u}) = \mathbf{v}_1 \cdot \mathbf{u} - \mathbf{v}_1 \cdot \left(\frac{\mathbf{v}_1 \cdot \mathbf{u}}{\|\mathbf{v}_1\|^2}\mathbf{v}_1 + \frac{\mathbf{v}_2 \cdot \mathbf{u}}{\|\mathbf{v}_2\|^2}\mathbf{v}_2 + \cdots + \frac{\mathbf{v}_k \cdot \mathbf{u}}{\|\mathbf{v}_k\|^2}\mathbf{v}_k \right)$$

$$= \mathbf{v}_1 \cdot \mathbf{u} - \left(\frac{\mathbf{v}_1 \cdot \mathbf{u}}{\|\mathbf{v}_1\|^2}(\mathbf{v}_1 \cdot \mathbf{v}_1) + \frac{\mathbf{v}_2 \cdot \mathbf{u}}{\|\mathbf{v}_2\|^2}(\mathbf{v}_1 \cdot \mathbf{v}_2) + \cdots + \frac{\mathbf{v}_k \cdot \mathbf{u}}{\|\mathbf{v}_k\|^2}(\mathbf{v}_1 \cdot \mathbf{v}_k) \right)$$

$$= \mathbf{v}_1 \cdot \mathbf{u} - \left(\frac{\mathbf{v}_1 \cdot \mathbf{u}}{\|\mathbf{v}_1\|^2}\|\mathbf{v}_1\|^2 + \frac{\mathbf{v}_2 \cdot \mathbf{u}}{\|\mathbf{v}_2\|^2}(0) + \cdots + \frac{\mathbf{v}_k \cdot \mathbf{u}}{\|\mathbf{v}_k\|^2}(0) \right)$$

$$= \mathbf{v}_1 \cdot \mathbf{u} - \mathbf{v}_1 \cdot \mathbf{u} = 0$$

The same argument can be used to show that each of

$$\mathbf{v}_2 \cdot (\mathbf{u} - \text{proj}_S\mathbf{u}) = 0, \ldots, \mathbf{v}_k \cdot (\mathbf{u} - \text{proj}_S\mathbf{u}) = 0$$

Thus, by Theorem 8.9, $\mathbf{u} - \text{proj}_S\mathbf{u}$ is orthogonal to S.

To verify part (d), suppose that $\{\tilde{\mathbf{v}}_1, \ldots, \tilde{\mathbf{v}}_k\}$ is another orthogonal basis for S, and let $\text{proj}_{S_v}\mathbf{u}$ and $\text{proj}_{S_{\tilde{v}}}\mathbf{u}$ denote the projections for bases $\{\mathbf{v}_1, \ldots, \mathbf{v}_k\}$ and $\{\tilde{\mathbf{v}}_1, \ldots, \tilde{\mathbf{v}}_k\}$, respectively. By part (b), both $\mathbf{u} - \text{proj}_{S_v}\mathbf{u}$ and $\mathbf{u} - \text{proj}_{S_{\tilde{v}}}\mathbf{u}$ are in $S^{\perp}$, and since $S^{\perp}$ is a subspace, the difference

$$(\mathbf{u} - \text{proj}_{S_v}\mathbf{u}) - (\mathbf{u} - \text{proj}_{S_{\tilde{v}}}\mathbf{u}) = \text{proj}_{S_{\tilde{v}}}\mathbf{u} - \text{proj}_{S_v}\mathbf{u}$$

is in $S^{\perp}$. But by part (a), both $\text{proj}_{S_{\tilde{v}}}\mathbf{u}$ and $\text{proj}_{S_v}\mathbf{u}$ are also in the subspace S, so that $\text{proj}_{S_{\tilde{v}}}\mathbf{u} - \text{proj}_{S_v}\mathbf{u}$ is as well. However, $S \cap S^{\perp} = \{\mathbf{0}\}$ (see Exercise 67 of Section 8.1), which implies $\text{proj}_{S_{\tilde{v}}}\mathbf{u} - \text{proj}_{S_v}\mathbf{u} = \mathbf{0}$. Hence $\text{proj}_{S_{\tilde{v}}}\mathbf{u} = \text{proj}_{S_v}\mathbf{u}$, and therefore $\text{proj}_S\mathbf{u}$ is independent of choice of basis for S. ∎

| EXERCISES |

Exercises 1–6 refer to the vectors given below.

$$\mathbf{u}_1 = \begin{bmatrix} -3 \\ 1 \\ 2 \end{bmatrix}, \quad \mathbf{u}_2 = \begin{bmatrix} 1 \\ 1 \\ 1 \end{bmatrix}, \quad \mathbf{u}_3 = \begin{bmatrix} 2 \\ 0 \\ -1 \end{bmatrix}, \quad \mathbf{u}_4 = \begin{bmatrix} 1 \\ -3 \\ 2 \end{bmatrix}$$

$$\mathbf{u}_5 = \begin{bmatrix} 2 \\ 1 \\ 1 \end{bmatrix}, \quad \mathbf{u}_6 = \begin{bmatrix} 0 \\ 3 \\ -1 \end{bmatrix}, \quad \mathbf{u}_7 = \begin{bmatrix} 3 \\ -4 \\ -2 \end{bmatrix}, \quad \mathbf{u}_8 = \begin{bmatrix} -1 \\ -1 \\ 3 \end{bmatrix}$$

1. Compute the following projections.

(a) $\text{proj}_{\mathbf{u}_3}\mathbf{u}_2$

(b) $\text{proj}_{\mathbf{u}_1}\mathbf{u}_2$

2. Compute the following projections.

(a) $\text{proj}_{\mathbf{u}_5}\mathbf{u}_1$

(b) $\text{proj}_{\mathbf{u}_5}\mathbf{u}_8$

3. Compute $\text{proj}_S\mathbf{u}_2$, where $S = \text{span}\{\mathbf{u}_3, \mathbf{u}_4\}$.

4. Compute $\text{proj}_S\mathbf{u}_8$, where $S = \text{span}\{\mathbf{u}_5, \mathbf{u}_7\}$.

5. Normalize the given vectors.

(a) $\mathbf{u}_1$

(b) $\mathbf{u}_4$

6. Normalize the given vectors.

(a) $\mathbf{u}_3$

(b) $\mathbf{u}_6$

For Exercises 7–14, apply the Gram–Schmidt process to find an orthogonal basis for the given subspace.

7. $S = \text{span}\left\{ \begin{bmatrix} 1 \\ 3 \end{bmatrix}, \begin{bmatrix} 4 \\ 2 \end{bmatrix} \right\}$

8. $S = \text{span}\left\{ \begin{bmatrix} 2 \\ -1 \end{bmatrix}, \begin{bmatrix} 4 \\ 3 \end{bmatrix} \right\}$

9. $S = \text{span}\left\{ \begin{bmatrix} -2 \\ 2 \\ 1 \end{bmatrix}, \begin{bmatrix} 3 \\ 4 \\ -2 \end{bmatrix} \right\}$

10. $S = \text{span}\left\{ \begin{bmatrix} 1 \\ 0 \\ -2 \end{bmatrix}, \begin{bmatrix} 1 \\ 3 \\ 3 \end{bmatrix} \right\}$

11. $S = \text{span}\left\{ \begin{bmatrix} 1 \\ -1 \\ 0 \\ 1 \end{bmatrix}, \begin{bmatrix} 4 \\ 1 \\ 2 \\ 0 \end{bmatrix} \right\}$

12. $S = \text{span}\left\{ \begin{bmatrix} -1 \\ 0 \\ 1 \\ 2 \end{bmatrix}, \begin{bmatrix} 4 \\ 2 \\ 0 \\ -1 \end{bmatrix} \right\}$

13. $S = \text{span}\left\{ \begin{bmatrix} -1 \\ 0 \\ 1 \end{bmatrix}, \begin{bmatrix} 3 \\ 4 \\ 1 \end{bmatrix}, \begin{bmatrix} 4 \\ 1 \\ 6 \end{bmatrix} \right\}$

14. $S = \text{span}\left\{ \begin{bmatrix} 1 \\ 1 \\ 0 \\ -1 \end{bmatrix}, \begin{bmatrix} 1 \\ 3 \\ 0 \\ 1 \end{bmatrix}, \begin{bmatrix} 4 \\ 2 \\ 2 \\ 0 \end{bmatrix} \right\}$

For Exercises 15–22, find $\text{proj}_S \mathbf{u}$.

15. $S = $ subspace in Exercise 7; $\mathbf{u} = \begin{bmatrix} 1 \\ 1 \end{bmatrix}$.

16. $S = $ subspace in Exercise 8; $\mathbf{u} = \begin{bmatrix} -1 \\ 1 \end{bmatrix}$.

17. $S = $ subspace in Exercise 9; $\mathbf{u} = \begin{bmatrix} 1 \\ 0 \\ 2 \end{bmatrix}$.

18. $S = $ subspace in Exercise 10; $\mathbf{u} = \begin{bmatrix} 1 \\ 1 \\ 1 \end{bmatrix}$.

19. $S = $ subspace in Exercise 11; $\mathbf{u} = \begin{bmatrix} 1 \\ -1 \\ 0 \\ 1 \end{bmatrix}$.

20. $S = $ subspace in Exercise 12; $\mathbf{u} = \begin{bmatrix} 0 \\ 1 \\ 1 \\ 0 \end{bmatrix}$.

21. $S = $ subspace in Exercise 13; $\mathbf{u} = \begin{bmatrix} 1 \\ 0 \\ 2 \end{bmatrix}$.

22. $S = $ subspace in Exercise 14; $\mathbf{u} = \begin{bmatrix} 1 \\ 0 \\ 1 \\ 0 \end{bmatrix}$.

For Exercises 23–30, find an orthonormal basis for the given subspace.

23. $S = $ subspace in Exercise 7.

24. $S = $ subspace in Exercise 8.

25. $S = $ subspace in Exercise 9.

26. $S = $ subspace in Exercise 10.

27. $S = $ subspace in Exercise 11.

28. $S = $ subspace in Exercise 12.

29. $S = $ subspace in Exercise 13.

30. $S = $ subspace in Exercise 14.

FIND AN EXAMPLE For Exercises 31–36, find an example that meets the given specifications.

31. Two vectors $\mathbf{u}$ and $\mathbf{v}$ in $\mathbf{R}^2$ with $\text{proj}_\mathbf{v} \mathbf{u} = \mathbf{u}$.

32. Two vectors $\mathbf{u}$ and $\mathbf{v}$ in $\mathbf{R}^3$ with $\text{proj}_\mathbf{v} \mathbf{u} = \mathbf{v}$.

33. Two vectors $\mathbf{u}$ and $\mathbf{v}$ in $\mathbf{R}^2$ with $\text{proj}_\mathbf{v} \mathbf{u} = \mathbf{0}$.

34. A two-dimensional subspace S in $\mathbf{R}^3$ and a vector $\mathbf{u}$ such that $\text{proj}_S \mathbf{u} = \mathbf{u}$.

35. Two nonparallel vectors $\mathbf{u}$ and $\mathbf{v}$ with $\text{proj}_\mathbf{v} \mathbf{u} = \begin{bmatrix} 1 \\ 2 \end{bmatrix}$.

36. A two-dimensional subspace S in $\mathbf{R}^3$ and a vector $\mathbf{u}$ not in S such that $\text{proj}_S \mathbf{u} = \begin{bmatrix} 3 \\ 0 \\ 1 \end{bmatrix}$.

TRUE OR FALSE For Exercises 37–46, determine if the statement is true or false, and justify your answer. Assume S is nontrivial and $\mathbf{u}$ and $\mathbf{v}$ are both nonzero.

37. Every subspace S of $\mathbf{R}^n$ has an orthonormal basis.

38. If $\mathbf{u}$ is in $\mathbf{R}^5$ and S is a three-dimensional subspace of $\mathbf{R}^5$, then $\text{proj}_S \mathbf{u}$ is in $\mathbf{R}^3$.

39. If S is a subspace, then $\text{proj}_S \mathbf{u}$ is in S.

40. If $\mathbf{u}$ and $\mathbf{v}$ are vectors, then $\text{proj}_\mathbf{v} \mathbf{u}$ is a multiple of $\mathbf{u}$.

41. If $\mathbf{u}$ and $\mathbf{v}$ are orthogonal, then $\text{proj}_\mathbf{v} \mathbf{u} = \mathbf{0}$.

42. If $\text{proj}_S \mathbf{u} = \mathbf{u}$, then $\mathbf{u}$ is in S.

43. If $\mathbf{u}$ is in S, then $\text{proj}_{S^\perp} \mathbf{u} = \mathbf{0}$.

44. For a vector $\mathbf{u}$ and a subspace S,

$$\text{proj}_S \left(\text{proj}_S \mathbf{u} \right) = \text{proj}_S \mathbf{u}$$

45. For vectors $\mathbf{u}$ and $\mathbf{v}$,

$$\text{proj}_\mathbf{u} \left(\text{proj}_\mathbf{v} \mathbf{u} \right) = \mathbf{u}$$

46. For every subspace S there exists a nonzero vector $\mathbf{u}$ such that $\text{proj}_S \mathbf{u} = 2\mathbf{u}$.

47. Let $\{\mathbf{v}_1, \ldots, \mathbf{v}_k\}$ be the orthogonal set generated in the course of applying the Gram–Schmidt process to a basis, and define

$$S_j = \text{span}\{\mathbf{v}_1, \ldots, \mathbf{v}_j\} \qquad \text{for } j = 1, \ldots, k$$

(a) Prove that if $i < j$, then S_i is a subspace of S_j.

(b) Prove that if $i < j$, then $S_j^{\perp}$ is a subspace of $S_i^{\perp}$.

48. Suppose that $\{\mathbf{u}_1, \mathbf{u}_2\}$ are linearly independent and that $\mathbf{u}_3$ is in span$\{\mathbf{u}_1, \mathbf{u}_2\}$. Suppose further that the Gram–Schmidt process is applied to $\{\mathbf{u}_1, \mathbf{u}_2, \mathbf{u}_3\}$ to generate a new set $\{\mathbf{v}_1, \mathbf{v}_2, \mathbf{v}_3\}$. What is $\mathbf{v}_3$? Explain your answer.

49. Suppose that $\mathbf{u}$ and $\mathbf{v}$ are nonzero vectors and that S is a subspace. Prove that if $\mathbf{u}$ is in S and $\mathbf{v}$ is in $S^{\perp}$, then $\mathbf{u} + \mathbf{v}$ is not in S or $S^{\perp}$.

50. Suppose that $\{\mathbf{w}_1, \ldots, \mathbf{w}_n\}$ is an orthonormal set and that $\mathbf{x} = c_1\mathbf{w}_1 + \cdots + c_n\mathbf{w}_n$. Prove that

$$\|\mathbf{x}\|^2 = c_1^2 + \cdots + c_n^2$$

51. Let $\mathbf{v} \neq \mathbf{0}$ be a fixed vector in $\mathbf{R}^n$. Prove that $T : \mathbf{R}^n \to \mathbf{R}^n$ given by $T_{\mathbf{v}}(\mathbf{u}) = \text{proj}_{\mathbf{v}}\mathbf{u}$ is a linear transformation.

52. Let S be a nonzero subspace. Prove that $T : \mathbf{R}^n \to \mathbf{R}^n$ given by $T_S(\mathbf{u}) = \text{proj}_S\mathbf{u}$ is a linear transformation.

53. Here we prove that the Gram–Schmidt process works. Suppose that $\{\mathbf{u}_1, \ldots, \mathbf{u}_k\}$ are linearly independent vectors, and that $\{\mathbf{v}_1, \ldots, \mathbf{v}_k\}$ are the vectors generated by the Gram–Schmidt process.

(a) Use induction to show $\{\mathbf{v}_1, \ldots, \mathbf{v}_j\}$ is an orthogonal set for $j = 1, \ldots, k$.

(b) Use induction to show span$\{\mathbf{u}_1, \ldots, \mathbf{u}_j\} = \text{span}\{\mathbf{v}_1, \ldots, \mathbf{v}_j\}$ for $j = 1, \ldots, k$.

(c) Explain why (a) and (b) imply that the Gram–Schmidt process yields an orthogonal basis.

54. Prove that for any nonzero vectors $\mathbf{u}$ and $\mathbf{v}$,

$$\|\mathbf{u}\|^2 = \|\text{proj}_{\mathbf{v}}\mathbf{u}\|^2 + \|\mathbf{u} - \text{proj}_{\mathbf{v}}\mathbf{u}\|^2 \tag{6}$$

(HINT: Apply Theorem 8.14 and Theorem 8.6.)

55. Prove that for any vector $\mathbf{u}$ and nonzero subspace S,

$$\|\mathbf{u}\|^2 = \|\text{proj}_S\mathbf{u}\|^2 + \|\mathbf{u} - \text{proj}_S\mathbf{u}\|^2$$

(HINT: Apply Theorem 8.16 and Theorem 8.6.)

56. If $\mathbf{u}$ and $\mathbf{v}$ are nonzero vectors in $\mathbf{R}^n$, then the angle θ between $\mathbf{u}$ and $\mathbf{v}$ is defined in terms of the formula

$$\cos(\theta) = \frac{\mathbf{u} \cdot \mathbf{v}}{\|\mathbf{u}\| \|\mathbf{v}\|} \tag{7}$$

In this exercise, we use trigonometry to prove that this is true in $\mathbf{R}^2$. The equation in (7) is an extension from $\mathbf{R}^2$ to $\mathbf{R}^n$.

(a) Refer to Figure 4 and use it to explain why

$$\cos(\theta) = \frac{\|\text{proj}_{\mathbf{v}}\mathbf{u}\|}{\|\mathbf{u}\|} = \frac{|\mathbf{u} \cdot \mathbf{v}|}{\|\mathbf{u}\| \|\mathbf{v}\|}$$

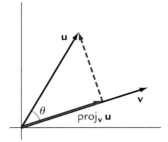

Figure 4 The angle θ between $\mathbf{u}$ and $\mathbf{v}$.

(b) Explain why $|\mathbf{u} \cdot \mathbf{v}| = \mathbf{u} \cdot \mathbf{v}$ in Figure 4. (HINT: $\text{proj}_{\mathbf{v}}\mathbf{u} = c\mathbf{v}$. Is c positive or negative?) Conclude that (7) holds for $\theta < 90°$.

(c) Draw a new diagram with $\mathbf{u}$ and $\mathbf{v}$ arranged so that $\theta > 90°$. Explain why in this case $|\mathbf{u} \cdot \mathbf{v}| = -\mathbf{u} \cdot \mathbf{v}$, and conclude that (7) also holds in this case.

(d) Complete the proof by showing that (7) holds when $\theta = 90°$.

57. In this exercise we prove the *Cauchy–Schwarz inequality*, which states that

$$|\mathbf{u} \cdot \mathbf{v}| \leq \|\mathbf{u}\| \|\mathbf{v}\| \tag{8}$$

for vectors $\mathbf{u}$ and $\mathbf{v}$ in $\mathbf{R}^n$.

(a) Prove that $\|\text{proj}_{\mathbf{v}}\mathbf{u}\| \leq \|\mathbf{u}\|$. (HINT: See (6).)

(b) Use (a) and the definition of projection to show that (8) holds.

(c) Show that $|\mathbf{u} \cdot \mathbf{v}| = \|\mathbf{u}\| \|\mathbf{v}\|$ if and only if $\mathbf{u} = c\mathbf{v}$. Hence there is equality in the Cauchy–Schwarz inequality exactly when $\mathbf{u}$ is a scalar multiple of $\mathbf{v}$.

58. Here we prove that

$$\mathbf{u} = \text{proj}_S\mathbf{u} + \text{proj}_{S^{\perp}}\mathbf{u} \tag{9}$$

for a vector $\mathbf{u}$ and a nontrivial subspace S.

(a) Explain why $\mathbf{u} - \text{proj}_S\mathbf{u}$ and $\text{proj}_{S^{\perp}}\mathbf{u}$ are both in $S^{\perp}$, and use this to prove that

$$\mathbf{u} - \text{proj}_S\mathbf{u} - \text{proj}_{S^{\perp}}\mathbf{u} \tag{10}$$

is in $S^{\perp}$.

(b) Explain why $\mathbf{u} - \text{proj}_{S^{\perp}}\mathbf{u}$ and $\text{proj}_S\mathbf{u}$ are both in S, and use this to prove that the expression in (10) is also in S.

(c) Combine (a) and (b) to show that

$$\mathbf{u} - \text{proj}_S\mathbf{u} - \text{proj}_{S^{\perp}}\mathbf{u} = \mathbf{0}$$

and from this conclude that (9) is true.

Ⓒ In Exercises 59–60, find an orthonormal basis for S.

59. $S = \text{span}\left\{ \begin{bmatrix} 1 \\ 2 \\ -4 \\ -1 \end{bmatrix}, \begin{bmatrix} -3 \\ 0 \\ 5 \\ -2 \end{bmatrix}, \begin{bmatrix} 0 \\ 7 \\ 2 \\ -6 \end{bmatrix} \right\}$

60. $S = \text{span} \left\{ \begin{bmatrix} 2 \\ -1 \\ 0 \\ -2 \\ 3 \end{bmatrix}, \begin{bmatrix} 4 \\ -2 \\ 1 \\ -4 \\ -2 \end{bmatrix}, \begin{bmatrix} 5 \\ -1 \\ 2 \\ 0 \\ -4 \end{bmatrix}, \begin{bmatrix} 4 \\ 0 \\ 2 \\ -3 \\ -3 \end{bmatrix} \right\}$

62. $S = \text{span} \left\{ \begin{bmatrix} 2 \\ 1 \\ 7 \\ 1 \\ 0 \end{bmatrix}, \begin{bmatrix} 3 \\ -2 \\ -5 \\ 0 \\ 4 \end{bmatrix}, \begin{bmatrix} 5 \\ 2 \\ 1 \\ 1 \\ 4 \end{bmatrix} \right\}, \mathbf{u} = \begin{bmatrix} 3 \\ -1 \\ -1 \\ 4 \\ 0 \end{bmatrix}$

Ⓒ In Exercises 61–62, compute $\text{proj}_S \mathbf{u}$.

61. $S = \text{span} \left\{ \begin{bmatrix} 3 \\ -1 \\ -2 \\ 7 \end{bmatrix}, \begin{bmatrix} 2 \\ 1 \\ 6 \\ 3 \end{bmatrix} \right\}, \mathbf{u} = \begin{bmatrix} 3 \\ -1 \\ -1 \\ 4 \end{bmatrix}$

8.3 Diagonalizing Symmetric Matrices and QR Factorization

We start this section by revisiting the problem of diagonalizing matrices. As we discovered in Section 6.4, not all square matrices can be diagonalized, and in general it is not easy to tell if a given matrix can be diagonalized. However, the situation is different if the matrix is symmetric. Let's consider an example.

▶ Recall that a square matrix A is symmetric if $A^T = A$.

EXAMPLE 1 If possible, diagonalize the matrix

$$A = \begin{bmatrix} 1 & -1 & 4 \\ -1 & 4 & -1 \\ 4 & -1 & 1 \end{bmatrix}$$

Solution The characteristic polynomial for A is

$$\det(A - \lambda I) = -\lambda^3 - 6\lambda^2 - 9\lambda - 54 = -(\lambda - 6)(\lambda - 3)(\lambda + 3)$$

so the eigenvalues are $\lambda = 6, 3, -3$. Since the eigenvalues each have multiplicity 1, we know that A is diagonalizable. Following our usual procedure, we find that an eigenvector associated with each eigenvalue is

$$\lambda = 6 \Rightarrow \mathbf{u}_1 = \begin{bmatrix} 1 \\ -1 \\ 1 \end{bmatrix}, \quad \lambda = 3 \Rightarrow \mathbf{u}_2 = \begin{bmatrix} 1 \\ 2 \\ 1 \end{bmatrix}, \quad \lambda = -3 \Rightarrow \mathbf{u}_3 = \begin{bmatrix} -1 \\ 0 \\ 1 \end{bmatrix}$$

Forming the diagonal matrix D and the matrix of eigenvectors P, we have

$$D = \begin{bmatrix} 6 & 0 & 0 \\ 0 & 3 & 0 \\ 0 & 0 & -3 \end{bmatrix} \quad \text{and} \quad P = \begin{bmatrix} 1 & 1 & -1 \\ -1 & 2 & 0 \\ 1 & 1 & 1 \end{bmatrix} \quad ∎$$

We know from previous work that eigenvectors associated with distinct eigenvalues are linearly independent. In Example 1, even more is true: The eigenvectors are orthogonal. This is not a coincidence—it always happens when A is a symmetric matrix.

THEOREM 8.19

If A is a symmetric matrix, then eigenvectors associated with distinct eigenvalues are orthogonal.

▶ Note that $\mathbf{x}_1^T \mathbf{x}_2 = \mathbf{x}_1 \cdot \mathbf{x}_2$. (See Exercise 71 of Section 8.1.)

Proof   Let A be a symmetric matrix, and suppose that $\lambda_1 \neq \lambda_2$ are distinct eigenvalues of A with associated eigenvectors $\mathbf{u}_1$ and $\mathbf{u}_2$, respectively. We now compute $(A\mathbf{u}_1)^T \mathbf{u}_2$ in two different ways. First, we have

$$(A\mathbf{u}_1)^T \mathbf{u}_2 = (\lambda_1 \mathbf{u}_1)^T \mathbf{u}_2 = \lambda_1 \mathbf{u}_1^T \mathbf{u}_2 = \lambda_1 (\mathbf{u}_1 \cdot \mathbf{u}_2)$$

Second, A symmetric means $A = A^T$. Thus

$$(A\mathbf{u}_1)^T \mathbf{u}_2 = (\mathbf{u}_1^T A^T)\mathbf{u}_2 = (\mathbf{u}_1^T A)\mathbf{u}_2 = \mathbf{u}_1^T (A\mathbf{u}_2)$$
$$= \mathbf{u}_1^T (\lambda_2 \mathbf{u}_2) = \lambda_2 (\mathbf{u}_1^T \mathbf{u}_2) = \lambda_2 (\mathbf{u}_1 \cdot \mathbf{u}_2)$$

Hence $\lambda_1 (\mathbf{u}_1 \cdot \mathbf{u}_2) = \lambda_2 (\mathbf{u}_1 \cdot \mathbf{u}_2)$, or equivalently,

$$(\lambda_1 - \lambda_2)(\mathbf{u}_1 \cdot \mathbf{u}_2) = 0$$

Since $\lambda_1 \neq \lambda_2$, then $\mathbf{u}_1 \cdot \mathbf{u}_2 = 0$. Therefore $\mathbf{u}_1$ and $\mathbf{u}_2$ are orthogonal. ■

Returning to Example 1, if we compute $P^T P$ we find that

$$P^T P = \begin{bmatrix} 1 & -1 & 1 \\ 1 & 2 & 1 \\ -1 & 0 & 1 \end{bmatrix} \begin{bmatrix} 1 & 1 & -1 \\ -1 & 2 & 0 \\ 1 & 1 & 1 \end{bmatrix} = \begin{bmatrix} 3 & 0 & 0 \\ 0 & 6 & 0 \\ 0 & 0 & 2 \end{bmatrix}$$

Thus $P^T P$ is a diagonal matrix. To further simplify $P^T P$, we redefine P to have normalized columns (i.e., columns that have length 1),

$$P = \begin{bmatrix} \frac{1}{\sqrt{3}} & \frac{1}{\sqrt{6}} & -\frac{1}{\sqrt{2}} \\ -\frac{1}{\sqrt{3}} & \frac{2}{\sqrt{6}} & 0 \\ \frac{1}{\sqrt{3}} & \frac{1}{\sqrt{6}} & \frac{1}{\sqrt{2}} \end{bmatrix}$$

Since each column is a constant multiple of its predecessor, it is still an eigenvector associated with the same eigenvalue. Thus we could use this definition of P to diagonalize A. Although this choice of P is not as tidy as the previous one, it does have the nice property that

$$P^T P = \begin{bmatrix} \frac{1}{\sqrt{3}} & -\frac{1}{\sqrt{3}} & \frac{1}{\sqrt{3}} \\ \frac{1}{\sqrt{6}} & \frac{2}{\sqrt{6}} & \frac{1}{\sqrt{6}} \\ -\frac{1}{\sqrt{2}} & 0 & \frac{1}{\sqrt{2}} \end{bmatrix} \begin{bmatrix} \frac{1}{\sqrt{3}} & \frac{1}{\sqrt{6}} & -\frac{1}{\sqrt{2}} \\ -\frac{1}{\sqrt{3}} & \frac{2}{\sqrt{6}} & 0 \\ \frac{1}{\sqrt{3}} & \frac{1}{\sqrt{6}} & \frac{1}{\sqrt{2}} \end{bmatrix} = \begin{bmatrix} 1 & 0 & 0 \\ 0 & 1 & 0 \\ 0 & 0 & 1 \end{bmatrix} = I_3$$

Definition Orthogonal Matrix

That is, $P^T P = I_3$, so that $P^T = P^{-1}$. A square matrix with orthonormal columns is called an **orthogonal matrix**, and will always have this property.

THEOREM 8.20

▶ It might be better if an "orthogonal matrix" was called an "orthonormal matrix," but "orthogonal matrix" is standard in linear algebra. There is no special name for a matrix with nonnormal orthogonal columns.

If P is an $n \times n$ orthogonal matrix, then $P^{-1} = P^T$.

Proof   When computing the matrix product $P^T P$, we are just computing the dot products of the columns of P. The diagonal terms of $P^T P$ come from the dot product of a column with itself, with each equal to 1 because of the normality. The nondiagonal terms come from the dot products of distinct columns, and so are zero because the columns are orthogonal. Thus $P^T P = I_n$, and hence $P^{-1} = P^T$. ■

For example, since the following matrix P is orthogonal, we have $P^{-1} = P^T$.

$$P = \begin{bmatrix} 0 & 0 & 0 & 1 \\ 1 & 0 & 0 & 0 \\ 0 & 1 & 0 & 0 \\ 0 & 0 & 1 & 0 \end{bmatrix} \implies P^{-1} = P^T = \begin{bmatrix} 0 & 1 & 0 & 0 \\ 0 & 0 & 1 & 0 \\ 0 & 0 & 0 & 1 \\ 1 & 0 & 0 & 0 \end{bmatrix}$$

Orthogonally Diagonalizable Matrices

Next, we show how to find diagonalizing matrices D and P in the special case where P is an orthogonal matrix.

DEFINITION 8.21

Definition Orthogonally Diagonalizable

A square matrix A is **orthogonally diagonalizable** if there exists an orthogonal matrix P and a diagonal matrix D such that $A = PDP^{-1}$.

Since we can write the symmetric matrix A in Example 1 as $A = PDP^{-1}$ for

$$D = \begin{bmatrix} 6 & 0 & 0 \\ 0 & 3 & 0 \\ 0 & 0 & -3 \end{bmatrix} \quad \text{and} \quad P = \begin{bmatrix} \frac{1}{\sqrt{3}} & \frac{1}{\sqrt{6}} & -\frac{1}{\sqrt{2}} \\ -\frac{1}{\sqrt{3}} & \frac{2}{\sqrt{6}} & 0 \\ \frac{1}{\sqrt{3}} & \frac{1}{\sqrt{6}} & \frac{1}{\sqrt{2}} \end{bmatrix}$$

it follows that A is orthogonally diagonalizable. It is not hard to show that *any* orthogonally diagonalizable matrix must be symmetric.

THEOREM 8.22

Let A be an orthogonally diagonalizable matrix. Then A is symmetric.

Proof If A is orthogonally diagonalizable, then there exists an orthogonal matrix P and diagonal matrix D such that $A = PDP^{-1}$. Using the fact that $P^{-1} = P^T$ (by Theorem 8.20) and $D^T = D$ (because D is diagonal), we have

$$A^T = (PDP^{-1})^T = (PDP^T)^T = (P^T)^T D^T P^T = PDP^T = PDP^{-1} = A$$

Since $A^T = A$, it follows that A is symmetric. ∎

Remarkably, the converse of Theorem 8.22 is also true: If A is a symmetric matrix, then A is orthogonally diagonalizable.

THEOREM 8.23

(SPECTRAL THEOREM) A matrix A is orthogonally diagonalizable if and only if A is symmetric.

A complete proof of the Spectral Theorem is difficult and is not included here. Two consequences of the Spectral Theorem:

- All eigenvalues of a symmetric matrix A are real.
- Each eigenspace of a symmetric matrix A has dimension equal to the multiplicity of the associated eigenvalue.

EXAMPLE 2 Orthogonally diagonalize the symmetric matrix

$$A = \begin{bmatrix} 1 & 3 & -3 & -3 \\ 3 & -3 & 3 & -1 \\ -3 & 3 & 1 & -3 \\ -3 & -1 & -3 & -3 \end{bmatrix}$$

Solution Finding the characteristic polynomial of A (which has degree 4) and then factoring by hand is difficult, but using computer software we find that

$$\det(A - \lambda I) = \lambda^4 + 4\lambda^3 - 48\lambda^2 - 64\lambda + 512 = (\lambda + 8)(\lambda + 4)(\lambda - 4)^2$$

Thus we have eigenvalues $\lambda = -8$, $\lambda = -4$ (both multiplicity 1), and $\lambda = 4$ (multiplicity 2). Our usual methods produce bases for each eigenspace,

$$\lambda = -8 \Rightarrow \left\{ \begin{bmatrix} 1 \\ -1 \\ 1 \\ 1 \end{bmatrix} \right\} ; \quad \lambda = -4 \Rightarrow \left\{ \begin{bmatrix} 0 \\ 1 \\ 0 \\ 1 \end{bmatrix} \right\} ; \quad \lambda = 4 \Rightarrow \left\{ \begin{bmatrix} -1 \\ 0 \\ 1 \\ 0 \end{bmatrix}, \begin{bmatrix} -2 \\ 1 \\ 0 \\ 1 \end{bmatrix} \right\}$$

Since our goal is to orthogonally diagonalize A, we first need an orthogonal basis for each eigenspace. The bases for $\lambda = -8$ and $\lambda = -4$ are fine as is, but we need to apply Gram–Schmidt to the basis for $\lambda = 4$. Designate

$$\mathbf{u}_1 = \begin{bmatrix} -1 \\ 0 \\ 1 \\ 0 \end{bmatrix}, \quad \mathbf{u}_2 = \begin{bmatrix} -2 \\ 1 \\ 0 \\ 1 \end{bmatrix}$$

Setting $\mathbf{v}_1 = \mathbf{u}_1$, the second vector $\mathbf{v}_2$ is given by

$$\mathbf{v}_2 = \mathbf{u}_2 - \frac{\mathbf{v}_1 \cdot \mathbf{u}_2}{\|\mathbf{v}_1\|^2} \mathbf{v}_1 = \begin{bmatrix} -2 \\ 1 \\ 0 \\ 1 \end{bmatrix} - \frac{2}{2} \begin{bmatrix} -1 \\ 0 \\ 1 \\ 0 \end{bmatrix} = \begin{bmatrix} -1 \\ -1 \\ -1 \\ 1 \end{bmatrix}$$

This gives us four orthogonal eigenvectors

$$\begin{bmatrix} 1 \\ -1 \\ 1 \\ 1 \end{bmatrix}, \begin{bmatrix} 0 \\ 1 \\ 0 \\ 1 \end{bmatrix}, \begin{bmatrix} -1 \\ 0 \\ 1 \\ 0 \end{bmatrix}, \begin{bmatrix} -1 \\ -1 \\ -1 \\ 1 \end{bmatrix}$$

Next, we normalize each vector,

$$\begin{bmatrix} \frac{1}{2} \\ -\frac{1}{2} \\ \frac{1}{2} \\ \frac{1}{2} \end{bmatrix}, \begin{bmatrix} 0 \\ \frac{1}{\sqrt{2}} \\ 0 \\ \frac{1}{\sqrt{2}} \end{bmatrix}, \begin{bmatrix} -\frac{1}{\sqrt{2}} \\ 0 \\ \frac{1}{\sqrt{2}} \\ 0 \end{bmatrix}, \begin{bmatrix} -\frac{1}{2} \\ -\frac{1}{2} \\ -\frac{1}{2} \\ \frac{1}{2} \end{bmatrix}$$

Finally, we form D and P,

$$D = \begin{bmatrix} -8 & 0 & 0 & 0 \\ 0 & -4 & 0 & 0 \\ 0 & 0 & 4 & 0 \\ 0 & 0 & 0 & 4 \end{bmatrix} \quad \text{and} \quad P = \begin{bmatrix} \frac{1}{2} & 0 & -\frac{1}{\sqrt{2}} & -\frac{1}{2} \\ -\frac{1}{2} & \frac{1}{\sqrt{2}} & 0 & -\frac{1}{2} \\ \frac{1}{2} & 0 & \frac{1}{\sqrt{2}} & -\frac{1}{2} \\ \frac{1}{2} & \frac{1}{\sqrt{2}} & 0 & \frac{1}{2} \end{bmatrix} \qquad \blacksquare$$

▶ We know the vectors are orthogonal by Theorem 8.19, which says that for a symmetric matrix, eigenvectors associated with distinct eigenvalues are orthogonal.

EXAMPLE 3   Orthogonally diagonalize the matrix $A^T A$ for

$$A = \begin{bmatrix} 1 & 2 \\ 2 & 0 \\ 0 & 2 \end{bmatrix}$$

Solution   We have

$$A^T A = \begin{bmatrix} 1 & 2 & 0 \\ 2 & 0 & 2 \end{bmatrix} \begin{bmatrix} 1 & 2 \\ 2 & 0 \\ 0 & 2 \end{bmatrix} = \begin{bmatrix} 5 & 2 \\ 2 & 8 \end{bmatrix}$$

▶ All matrices of the form $A^T A$ are symmetric. (See Exercise 55 of Section 3.2.)

$A^T A$ is symmetric and hence by the Spectral Theorem orthogonally diagonalizable. The characteristic polynomial is $\lambda^2 - 13\lambda + 36 = (\lambda - 4)(\lambda - 9)$, yielding eigenvalues $\lambda_1 = 9$ and $\lambda_2 = 4$. The corresponding normalized eigenvectors are

$$\lambda = 9 \;\Rightarrow\; \left\{ \begin{bmatrix} \frac{1}{\sqrt{5}} \\ \frac{2}{\sqrt{5}} \end{bmatrix} \right\}; \qquad \lambda = 4 \;\Rightarrow\; \left\{ \begin{bmatrix} \frac{2}{\sqrt{5}} \\ -\frac{1}{\sqrt{5}} \end{bmatrix} \right\}$$

Thus, if we define

$$D = \begin{bmatrix} 9 & 0 \\ 0 & 4 \end{bmatrix} \quad \text{and} \quad P = \begin{bmatrix} \frac{1}{\sqrt{5}} & \frac{2}{\sqrt{5}} \\ \frac{2}{\sqrt{5}} & -\frac{1}{\sqrt{5}} \end{bmatrix}$$

then P is orthogonal and $A^T A = P D P^T$. ∎

Since $A^T A$ is a symmetric matrix, we know from the Spectral Theorem that the eigenvalues are real. In fact, for symmetric matrices of this form, it turns out that the eigenvalues will be nonnegative. Since this result is handy to know in the next section, we state and prove it here.

THEOREM 8.24   If A is a real matrix, then $A^T A$ has nonnegative eigenvalues.

Proof   Suppose that λ is an eigenvalue of $A^T A$ with associated eigenvector $\mathbf{u}$. Then

$$\| A\mathbf{u} \|^2 = (A\mathbf{u}) \cdot (A\mathbf{u}) = (A\mathbf{u})^T (A\mathbf{u}) = (\mathbf{u}^T A^T)(A\mathbf{u})$$

$$= \mathbf{u}^T (A^T A \mathbf{u}) = \mathbf{u}^T (\lambda \mathbf{u}) = \lambda (\mathbf{u}^T \mathbf{u}) = \lambda \| \mathbf{u} \|^2$$

In summary, $\| A\mathbf{u} \|^2 = \lambda \| \mathbf{u} \|^2$. Since both $\| A\mathbf{u} \|^2$ and $\| \mathbf{u} \|^2$ are nonnegative, it must be that λ is nonnegative as well. ∎

QR Factorization

Diagonalizing a matrix is one type of matrix factorization. Diagonalizing is always possible when a matrix is symmetric, but it may or may not be otherwise. Here we consider another type of factorization, which applies to any matrix that has linearly independent columns.

THEOREM 8.25   **(QR FACTORIZATION)**   Let $A = \begin{bmatrix} \mathbf{a}_1 & \cdots & \mathbf{a}_m \end{bmatrix}$ be an $n \times m$ matrix with linearly independent columns. Then A can be factorized as $A = QR$, where $Q = \begin{bmatrix} \mathbf{q}_1 & \cdots & \mathbf{q}_m \end{bmatrix}$ is an $n \times m$ matrix with orthonormal columns and R is an $m \times m$ upper triangular matrix with positive diagonal entries.

Proof   Suppose that $\{\mathbf{q}_1, \ldots, \mathbf{q}_m\}$ is the orthonormal set of vectors we get by applying the Gram–Schmidt process to the set of columns $\{\mathbf{a}_1, \ldots, \mathbf{a}_m\}$. Now define $Q = \begin{bmatrix} \mathbf{q}_1 & \cdots & \mathbf{q}_m \end{bmatrix}$. From the Gram–Schmidt construction, for each $1 \le k \le m$ the vector $\mathbf{a}_k$ is in the span of the orthonormal set $\{\mathbf{q}_1, \ldots, \mathbf{q}_k\}$. Hence by equation (5) in Section 8.2 we have

$$\mathbf{a}_k = (\mathbf{q}_1 \cdot \mathbf{a}_k)\mathbf{q}_1 + (\mathbf{q}_2 \cdot \mathbf{a}_k)\mathbf{q}_2 + \cdots + (\mathbf{q}_k \cdot \mathbf{a}_k)\mathbf{q}_k \tag{1}$$

Now define $r_{ik} = \mathbf{q}_i \cdot \mathbf{a}_k$ for $1 \le k \le m$ and $1 \le i \le k$, and let

$$R = \begin{bmatrix} r_{11} & r_{12} & r_{13} & \cdots & r_{1n} \\ 0 & r_{22} & r_{23} & \cdots & r_{2n} \\ 0 & 0 & r_{33} & \cdots & r_{3n} \\ \vdots & \vdots & \vdots & \ddots & \vdots \\ 0 & 0 & 0 & \cdots & r_{mm} \end{bmatrix}$$

Since $Q = \begin{bmatrix} \mathbf{q}_1 & \cdots & \mathbf{q}_m \end{bmatrix}$, the kth column of the product QR is equal to

$$r_{1k}\mathbf{q}_1 + r_{2k}\mathbf{q}_2 + \cdots + r_{kk}\mathbf{q}_k = \mathbf{a}_k$$

by (1) and the definition of r_{ik}. Since $\mathbf{a}_k$ is the kth column of A, it follows that $A = QR$.

Finally, since $\mathbf{a}_k$ is *not* in span$\{\mathbf{q}_1, \ldots, \mathbf{q}_{k-1}\}$ (why?), it must be that $r_{kk} \ne 0$. If $r_{kk} < 0$, then we replace $\mathbf{q}_k$ with $-\mathbf{q}_k$, which will make $r_{kk} > 0$ while keeping the columns of Q orthonormal and the column space col(Q) unchanged. Hence the diagonal entries of R are positive.   ∎

Before considering an example, we note that once Q has been found we can use matrix multiplication to compute R. Since $A = QR$ we have

$$Q^T A = Q^T Q R = R$$

because $Q^T Q = I_m$. It also can be shown directly that the entries of $Q^T A$ are equal to the entries of R (see Exercise 61).

EXAMPLE 4   Find the QR factorization for

$$A = \begin{bmatrix} 1 & 0 & -3 \\ 0 & 2 & -1 \\ 1 & 0 & 1 \\ 1 & 3 & 5 \end{bmatrix}$$

Solution   The columns of A are the vectors from Examples 5–6 in Section 8.2, where applying the Gram–Schmidt process we found the corresponding orthonormal set

$$\left\{ \frac{1}{\sqrt{3}} \begin{bmatrix} 1 \\ 0 \\ 1 \\ 1 \end{bmatrix}, \frac{1}{\sqrt{10}} \begin{bmatrix} -1 \\ 2 \\ -1 \\ 2 \end{bmatrix}, \frac{1}{\sqrt{23}} \begin{bmatrix} -3 \\ -3 \\ 1 \\ 2 \end{bmatrix} \right\}$$

Therefore we define

$$Q = \begin{bmatrix} \frac{1}{\sqrt{3}} & -\frac{1}{\sqrt{10}} & -\frac{3}{\sqrt{23}} \\ 0 & \frac{2}{\sqrt{10}} & -\frac{3}{\sqrt{23}} \\ \frac{1}{\sqrt{3}} & -\frac{1}{\sqrt{10}} & \frac{1}{\sqrt{23}} \\ \frac{1}{\sqrt{3}} & \frac{2}{\sqrt{10}} & \frac{2}{\sqrt{23}} \end{bmatrix}$$

We find R by computing

$$R = Q^T A = \begin{bmatrix} \frac{1}{\sqrt{3}} & 0 & \frac{1}{\sqrt{3}} & \frac{1}{\sqrt{3}} \\ -\frac{1}{\sqrt{10}} & \frac{2}{\sqrt{10}} & -\frac{1}{\sqrt{10}} & \frac{2}{\sqrt{10}} \\ -\frac{3}{\sqrt{23}} & -\frac{3}{\sqrt{23}} & \frac{1}{\sqrt{23}} & \frac{2}{\sqrt{23}} \end{bmatrix} \begin{bmatrix} 1 & 0 & -3 \\ 0 & 2 & -1 \\ 1 & 0 & 1 \\ 1 & 3 & 5 \end{bmatrix} = \begin{bmatrix} \sqrt{3} & \sqrt{3} & \sqrt{3} \\ 0 & \sqrt{10} & \sqrt{10} \\ 0 & 0 & \sqrt{23} \end{bmatrix}$$

Thus we have the factorization

$$QR = \begin{bmatrix} \frac{1}{\sqrt{3}} & -\frac{1}{\sqrt{10}} & -\frac{3}{\sqrt{23}} \\ 0 & \frac{2}{\sqrt{10}} & -\frac{3}{\sqrt{23}} \\ \frac{1}{\sqrt{3}} & -\frac{1}{\sqrt{10}} & \frac{1}{\sqrt{23}} \\ \frac{1}{\sqrt{3}} & \frac{2}{\sqrt{10}} & \frac{2}{\sqrt{23}} \end{bmatrix} \begin{bmatrix} \sqrt{3} & \sqrt{3} & \sqrt{3} \\ 0 & \sqrt{10} & \sqrt{10} \\ 0 & 0 & \sqrt{23} \end{bmatrix} = \begin{bmatrix} 1 & 0 & -3 \\ 0 & 2 & -1 \\ 1 & 0 & 1 \\ 1 & 3 & 5 \end{bmatrix} = A \quad \blacksquare$$

EXERCISES

In Exercises 1–8, determine if the given matrix is symmetric.

1. $\begin{bmatrix} 1 & -2 \\ 2 & 1 \end{bmatrix}$

2. $\begin{bmatrix} 4 & 3 \\ 3 & 5 \end{bmatrix}$

3. $\begin{bmatrix} 3 & 2 & 1 \\ 2 & 1 & 3 \\ 1 & 3 & 2 \end{bmatrix}$

4. $\begin{bmatrix} 2 & 0 & 1 \\ 0 & 2 & 0 \\ 0 & 0 & 2 \end{bmatrix}$

5. $\begin{bmatrix} 3 & -1 & 4 \\ -1 & 4 & 3 \end{bmatrix}$

6. $\begin{bmatrix} -5 & 2 & -1 \\ 2 & 1 & 0 \\ -1 & 0 & -6 \end{bmatrix}$

7. $\begin{bmatrix} 1 & 7 & -3 \\ 7 & 2 & 4 \\ -3 & 0 & -6 \\ 4 & -6 & -1 \end{bmatrix}$

8. $\begin{bmatrix} 4 & 2 & 0 & -2 \\ 2 & 3 & 4 & 5 \\ 0 & 4 & 2 & 0 \\ -2 & 5 & 0 & 1 \end{bmatrix}$

In Exercises 9–14, determine if the given matrix is orthogonal.

9. $\begin{bmatrix} 1 & -2 \\ 2 & 1 \end{bmatrix}$

10. $\begin{bmatrix} \frac{1}{\sqrt{10}} & \frac{3}{\sqrt{10}} \\ \frac{3}{\sqrt{10}} & -\frac{1}{\sqrt{10}} \end{bmatrix}$

11. $\begin{bmatrix} -\frac{5}{13} & \frac{12}{13} \\ \frac{12}{13} & \frac{5}{13} \end{bmatrix}$

12. $\begin{bmatrix} 1 & 0 & 1 \\ 0 & 1 & 0 \\ -1 & 0 & 1 \end{bmatrix}$

13. $\begin{bmatrix} \frac{1}{2} & \frac{1}{3} & \frac{1}{4} \\ -\frac{1}{2} & \frac{1}{3} & \frac{1}{4} \\ 0 & \frac{1}{3} & -\frac{1}{2} \end{bmatrix}$

14. $\begin{bmatrix} \frac{2}{\sqrt{14}} & \frac{1}{\sqrt{3}} & \frac{4}{\sqrt{42}} \\ \frac{1}{\sqrt{14}} & \frac{1}{\sqrt{3}} & -\frac{5}{\sqrt{42}} \\ -\frac{3}{\sqrt{14}} & \frac{1}{\sqrt{3}} & \frac{1}{\sqrt{42}} \end{bmatrix}$

In Exercises 15–18, the eigenvalues and corresponding eigenvectors for a symmetric matrix A are given. Find matrices D and P of an orthogonal diagonalization of A.

15. $\lambda_1 = 2, \mathbf{u}_1 = \begin{bmatrix} 1 \\ 2 \end{bmatrix}$; $\lambda_2 = -3, \mathbf{u}_2 = \begin{bmatrix} -2 \\ 1 \end{bmatrix}$

16. $\lambda_1 = -1, \mathbf{u}_1 = \begin{bmatrix} 3 \\ 4 \end{bmatrix}$; $\lambda_2 = 1, \mathbf{u}_2 = \begin{bmatrix} -4 \\ 3 \end{bmatrix}$

17. $\lambda_1 = 0, \mathbf{u}_1 = \begin{bmatrix} 1 \\ 1 \\ 1 \end{bmatrix}$; $\lambda_2 = 2, \mathbf{u}_2 = \begin{bmatrix} 1 \\ -1 \\ 0 \end{bmatrix}$;

$\lambda_3 = -1, \mathbf{u}_3 = \begin{bmatrix} -1 \\ -1 \\ 2 \end{bmatrix}$

18. $\lambda_1 = -1, \mathbf{u}_1 = \begin{bmatrix} 2 \\ 1 \\ 0 \end{bmatrix}$; $\lambda_2 = 0, \mathbf{u}_2 = \begin{bmatrix} 1 \\ -2 \\ 1 \end{bmatrix}$;

$\lambda_3 = 3, \mathbf{u}_3 = \begin{bmatrix} -1 \\ 2 \\ 5 \end{bmatrix}$

In Exercises 19–24, the eigenvalues for the symmetric matrix A are given. Find the matrices D and P of an orthogonal diagonalization of A.

19. $A = \begin{bmatrix} 4 & 2 \\ 2 & 1 \end{bmatrix}$, $\lambda = 0, 5$

20. $A = \begin{bmatrix} 3 & 4 \\ 4 & 3 \end{bmatrix}$, $\lambda = -1, 7$

21. $A = \begin{bmatrix} 0 & 1 & 2 \\ 1 & 1 & 1 \\ 2 & 1 & 0 \end{bmatrix}$, $\lambda = -2, 0, 3$

22. $A = \begin{bmatrix} 1 & 2 & 3 \\ 2 & 1 & 3 \\ 3 & 3 & 0 \end{bmatrix}$, $\lambda = -3, -1, 6$

23. $A = \begin{bmatrix} 0 & 0 & 1 \\ 0 & 1 & 0 \\ 1 & 0 & 0 \end{bmatrix}$, $\lambda = -1, 1$

24. $A = \begin{bmatrix} 1 & 1 & 1 \\ 1 & 1 & 1 \\ 1 & 1 & 1 \end{bmatrix}$, $\lambda = 0, 3$

In Exercises 25–28, verify that the eigenvalues of $A^T A$ are non-negative.

25. $A = \begin{bmatrix} 1 & 1 \\ 2 & 1 \\ 1 & 2 \end{bmatrix}$

26. $A = \begin{bmatrix} 0 & 2 \\ 1 & 0 \\ 2 & 1 \end{bmatrix}$

27. $A = \begin{bmatrix} 0 & 2 & 1 \\ 1 & 0 & 0 \end{bmatrix}$

28. $A = \begin{bmatrix} 2 & 2 & 3 \\ 1 & 1 & 0 \end{bmatrix}$

In Exercises 29–32, the given matrix Q has orthogonal columns. Find Q^{-1} without using row operations. (HINT: Exercise 60 could be helpful. If you use it, explain why it works.)

29. $Q = \begin{bmatrix} 1 & 2 \\ -2 & 1 \end{bmatrix}$

30. $Q = \begin{bmatrix} 4 & 5 \\ 5 & -4 \end{bmatrix}$

31. $Q = \begin{bmatrix} 0 & 2 & 0 \\ 1 & 0 & 1 \\ -1 & 0 & 1 \end{bmatrix}$

32. $Q = \begin{bmatrix} 1 & 1 & 5 \\ 2 & 1 & -4 \\ 3 & -1 & 1 \end{bmatrix}$

In Exercises 33–40, find the QR factorization for the matrix A.

33. $A = \begin{bmatrix} 3 & -2 \\ 2 & 3 \end{bmatrix}$

34. $A = \begin{bmatrix} -4 & 3 \\ 2 & 6 \end{bmatrix}$

35. $A = \begin{bmatrix} 1 & 4 \\ 3 & 2 \end{bmatrix}$

36. $A = \begin{bmatrix} 2 & 4 \\ -1 & 3 \end{bmatrix}$

37. $A = \begin{bmatrix} 1 & 2 \\ 1 & -3 \\ 5 & 1 \end{bmatrix}$

38. $A = \begin{bmatrix} 0 & -3 \\ 6 & -3 \\ 9 & 2 \end{bmatrix}$

39. $A = \begin{bmatrix} -2 & 3 \\ 2 & 4 \\ 1 & -2 \end{bmatrix}$

40. $A = \begin{bmatrix} 1 & 1 \\ 0 & 3 \\ -2 & 3 \end{bmatrix}$

FIND AN EXAMPLE For Exercises 41–48, find an example that meets the given specifications.

41. A 2×2 matrix A that has eigenvalues $\lambda_1 = 1$ and $\lambda_2 = 2$ and is orthogonally diagonalizable.

42. A 3×3 matrix A that has eigenvalues $\lambda_1 = 2$, $\lambda_2 = 3$, and $\lambda_3 = 5$ and is orthogonally diagonalizable.

43. A 2×2 matrix A that is orthogonally diagonalizable, has eigenvalues $\lambda_1 = -1$ and $\lambda_2 = 2$, and corresponding eigenvectors

$$\mathbf{u}_1 = \begin{bmatrix} 1 \\ 2 \end{bmatrix}, \quad \mathbf{u}_2 = \begin{bmatrix} -2 \\ 1 \end{bmatrix}$$

44. A 3×3 matrix A that is orthogonally diagonalizable, has eigenvalues $\lambda_1 = -3$, $\lambda_2 = 0$, and $\lambda_3 = 4$, and has corresponding eigenvectors

$$\mathbf{u}_1 = \begin{bmatrix} 1 \\ 0 \\ 2 \end{bmatrix}, \quad \mathbf{u}_2 = \begin{bmatrix} 4 \\ -1 \\ -2 \end{bmatrix}, \quad \mathbf{u}_3 = \begin{bmatrix} 2 \\ 10 \\ -1 \end{bmatrix}$$

45. A 2×2 matrix A that does not have a QR factorization.

46. Two 3×3 matrices A and B that both have a QR factorization, but $A + B$ does not.

47. A 2×2 matrix A that is diagonalizable but not orthogonally diagonalizable.

48. A 2×2 matrix A that is orthogonally diagonalizable but not invertible.

TRUE OR FALSE For Exercises 49–58, determine if the statement is true or false, and justify your answer.

49. If A is a symmetric matrix, then A is diagonalizable.

50. If A is a square matrix, then A is diagonalizable.

51. If A is orthogonally diagonalizable, then $A = A^T$

52. The eigenvalues of a square matrix A are real.

53. $A^T A$ is symmetric for any matrix A.

54. All matrices have a QR factorization.

55. In the QR factorization of a matrix A, the matrix R has columns that span the column space of A.

56. If A and B are orthogonal $n \times n$ matrices, then so is $A + B$.

57. If $A = QR$ is a QR factorization for a matrix A, then R is invertible.

58. If $A^T A = AA^T$ for a square matrix A, then A is an orthogonal matrix.

59. Prove that if A is orthogonal, then $\det(A) = \pm 1$.

60. Suppose $Q = \begin{bmatrix} \mathbf{q}_1 & \cdots & \mathbf{q}_n \end{bmatrix}$ has orthogonal columns. Show that

$$Q^{-1} = \begin{bmatrix} \frac{1}{\|\mathbf{q}_1\|^2} \mathbf{q}_1^T \\ \cdots \\ \frac{1}{\|\mathbf{q}_n\|^2} \mathbf{q}_n^T \end{bmatrix}$$

61. Let A and Q be the matrices in Theorem 8.25. Prove that the entry in position (i, k) of $Q^T A$ is equal to $\mathbf{q}_i \cdot \mathbf{a}_k$.

62. Prove that if A is an orthogonal matrix, then so is A^T.

63. Prove that if A is orthogonally diagonalizable, then so is A^T.

64. Prove that if A and B are orthogonally diagonalizable matrices, then so is $A + B$.

65. Prove that if A is orthogonally diagonalizable, then so is A^2.

66. Suppose that P is an orthogonal matrix. Show that for any vector $\mathbf{x}$, $\| P\mathbf{x} \|^2 = \|\mathbf{x}\|^2$ and therefore any eigenvector of P satisfies $|\lambda| = 1$.

Ⓒ In Exercises 67–70, find an orthogonal diagonalization for the matrix A.

67. $A = \begin{bmatrix} 2 & 1 & 3 \\ 1 & 0 & -4 \\ 3 & -4 & 5 \end{bmatrix}$

68. $A = \begin{bmatrix} -1 & -3 & 0 \\ -3 & 2 & 7 \\ 0 & 7 & 4 \end{bmatrix}$

69. $A = \begin{bmatrix} 2 & 1 & 4 & 0 \\ 1 & 3 & 2 & -5 \\ 4 & 2 & -1 & 3 \\ 0 & -5 & 3 & 2 \end{bmatrix}$

70. $A = \begin{bmatrix} 0 & -8 & 3 & 2 \\ -8 & -1 & 2 & 7 \\ 3 & 2 & 0 & -1 \\ 2 & 7 & -1 & 4 \end{bmatrix}$

Ⓒ In Exercises 71–74, find a QR factorization for the matrix A.

71. $A = \begin{bmatrix} -1 & 3 & 3 \\ 0 & 2 & 4 \\ 1 & 1 & 5 \end{bmatrix}$

72. $A = \begin{bmatrix} 4 & 2 & 0 \\ 1 & 5 & -2 \\ 3 & -3 & 1 \end{bmatrix}$

73. $A = \begin{bmatrix} 1 & 1 & 4 \\ 1 & 3 & 2 \\ 1 & 0 & 2 \\ -1 & 1 & 4 \end{bmatrix}$

74. $A = \begin{bmatrix} 2 & -1 & 1 \\ 4 & 2 & 3 \\ 0 & 3 & -2 \\ 5 & 1 & 0 \end{bmatrix}$

8.4 The Singular Value Decomposition

▶ This section is optional and can be omitted without loss of continuity.

In this section we develop another type of matrix factorization that is a generalization of diagonalization. This new type of matrix factorization is called the *singular value decomposition* (SVD), and it can be applied to any type of matrix, even those that are not square. We start by developing the factorization method and then describe applications to image processing and estimating the rank of a matrix.

Suppose that we have an $n \times m$ matrix A. If $n \geq m$, then the **singular value decomposition** is the factorization of A as the product $A = U \Sigma V^T$, where

Definition Singular Value Decomposition

▶ $0_{(n-m)m}$ is the $(n - m) \times m$ matrix with all entries equal to zero.

- U is an $n \times n$ orthogonal matrix.

- Σ is an $n \times m$ matrix of the form $\Sigma = \begin{bmatrix} D \\ 0_{(n-m)m} \end{bmatrix}$, where D is a diagonal matrix with

$$D = \begin{bmatrix} \sigma_1 & 0 & \cdots & 0 \\ 0 & \sigma_2 & \cdots & 0 \\ \vdots & \vdots & \ddots & \vdots \\ 0 & 0 & \cdots & \sigma_m \end{bmatrix}$$

and $\sigma_1 \geq \sigma_2 \geq \cdots \geq \sigma_m \geq 0$ are the **singular values** of A. The singular values are given by $\sigma_i = \sqrt{\lambda_i}$, where λ_i is an eigenvalue of $A^T A$.

- V is an $m \times m$ orthogonal matrix.

If $n < m$, then $\Sigma = \begin{bmatrix} D_{nn} & 0_{n(m-n)} \end{bmatrix}$ with everything else the same.

THEOREM 8.26 | Every $n \times m$ matrix A has a singular value decomposition.

The proof of this theorem is given at the end of the section. For now, let's look at an example that illustrates how we find the SVD.

EXAMPLE 1 Find the SVD for the 3×2 matrix

$$A = \begin{bmatrix} 1 & 2 \\ 2 & 0 \\ 0 & 2 \end{bmatrix}$$

Solution We find the SVD $A = U\Sigma V^T$ by applying the following sequence of steps.

1. **Orthogonally Diagonalize $A^T A$ to find V.** Since $A^T A$ is symmetric, the Spectral Theorem guarantees that it is orthogonally diagonalizable. Our matrix A appears in Example 3 of Section 8.3, where we showed that the eigenvalues of $A^T A$ are $\lambda_1 = 9$ and $\lambda_2 = 4$ and that the orthogonal diagonalizing matrix is

$$V = \begin{bmatrix} \frac{1}{\sqrt{5}} & \frac{2}{\sqrt{5}} \\ \frac{2}{\sqrt{5}} & -\frac{1}{\sqrt{5}} \end{bmatrix}$$

This is the matrix V in the SVD of A.

▶ By Theorem 8.24, the eigenvalues of $A^T A$ are always nonnegative.

2. **Find Σ.** The singular values for a matrix A are given by $\sigma_i = \sqrt{\lambda_i}$, the square roots of the eigenvalues of $A^T A$. Here we have $\sigma_1 = \sqrt{9} = 3$ and $\sigma_2 = \sqrt{4} = 2$, so that

$$\Sigma = \begin{bmatrix} 3 & 0 \\ 0 & 2 \\ 0 & 0 \end{bmatrix}$$

▶ Recall that col(A) is the column space of A.

3. **Find U.** We determine the columns of U in two steps, one for the columns corresponding to positive singular values, and the other for the columns that form an orthonormal basis for $\left(\text{col}(A)\right)^{\perp}$. (We will see why later.)

▶ $V^T = V^{-1}$ because V is an orthogonal matrix.

3a. **Positive Singular Values.** Our ultimate goal is to find U so that $A = U\Sigma V^T$, or equivalently, $AV = U\Sigma$. Note that the ith column of AV is $A\mathbf{v}_i$, while the ith column of $U\Sigma$ is $\sigma_i \mathbf{u}_i$. Thus, for $AV = U\Sigma$, we must have $A\mathbf{v}_i = \sigma_i \mathbf{u}_i$. When $\sigma_i > 0$, we arrange for this by defining

$$\mathbf{u}_i = \frac{1}{\sigma_i} A\mathbf{v}_i$$

In this example, we have

$$\mathbf{u}_1 = \frac{1}{\sigma_1} A\mathbf{v}_1 = \frac{1}{3} \begin{bmatrix} 1 & 2 \\ 2 & 0 \\ 0 & 2 \end{bmatrix} \begin{bmatrix} \frac{1}{\sqrt{5}} \\ \frac{2}{\sqrt{5}} \end{bmatrix} = \frac{1}{3\sqrt{5}} \begin{bmatrix} 5 \\ 2 \\ 4 \end{bmatrix}$$

$$\mathbf{u}_2 = \frac{1}{\sigma_2} A\mathbf{v}_2 = \frac{1}{2} \begin{bmatrix} 1 & 2 \\ 2 & 0 \\ 0 & 2 \end{bmatrix} \begin{bmatrix} \frac{2}{\sqrt{5}} \\ -\frac{1}{\sqrt{5}} \end{bmatrix} = \frac{1}{\sqrt{5}} \begin{bmatrix} 0 \\ 2 \\ -1 \end{bmatrix}$$

Two important observations:

(a) $\mathbf{u}_1$ and $\mathbf{u}_2$ are orthonormal.

(b) span$\{\mathbf{u}_1, \mathbf{u}_2\}$ = col(A), the column space of A.

This is not a coincidence, but rather a consequence of our method for finding U. We will show why in our proof of Theorem 8.26.

3b. Filling Out U. We now have the first two columns of U. Since the third row of Σ consists of zeros, the product $U\Sigma$ will be the same regardless of our choice of $\mathbf{u}_3$. We want U to be an orthogonal matrix, which we can accomplish by extending span$\{\mathbf{u}_1, \mathbf{u}_2\}$ to an orthonormal basis for $\mathbf{R}^3$. Since span$\{\mathbf{u}_1, \mathbf{u}_2\} = \mathrm{col}(A)$, we proceed by finding an orthonormal basis for $\left(\mathrm{col}(A)\right)^{\perp}$, the orthogonal complement of the column space of A. We do this by noting that $\left(\mathrm{col}(A)\right)^{\perp} = \mathrm{null}(A^T)$ (see Exercise 69 of Section 8.1). The null space of A^T is equal to the set of solutions to $A^T\mathbf{x} = \mathbf{0}$. The augmented matrix and echelon form are

$$\begin{bmatrix} 1 & 2 & 0 & 0 \\ 2 & 0 & 2 & 0 \end{bmatrix} \sim \begin{bmatrix} 1 & 2 & 0 & 0 \\ 0 & -2 & 1 & 0 \end{bmatrix}$$

Back substitution and normalizing the solution yield

$$\mathbf{u}_3 = \begin{bmatrix} -\frac{2}{3} \\ \frac{1}{3} \\ \frac{2}{3} \end{bmatrix}$$

The vector $\mathbf{u}_3$ gives the final column of U. We have

$$U = \begin{bmatrix} \mathbf{u}_1 & \mathbf{u}_2 & \mathbf{u}_3 \end{bmatrix} = \begin{bmatrix} \frac{5}{3\sqrt{5}} & 0 & -\frac{2}{3} \\ \frac{2}{3\sqrt{5}} & \frac{2}{\sqrt{5}} & \frac{1}{3} \\ \frac{4}{3\sqrt{5}} & -\frac{1}{\sqrt{5}} & \frac{2}{3} \end{bmatrix}$$

We can check our work by computing

$$U\Sigma V^T = \begin{bmatrix} \frac{5}{3\sqrt{5}} & 0 & -\frac{2}{3} \\ \frac{2}{3\sqrt{5}} & \frac{2}{\sqrt{5}} & \frac{1}{3} \\ \frac{4}{3\sqrt{5}} & -\frac{1}{\sqrt{5}} & \frac{2}{3} \end{bmatrix} \begin{bmatrix} 3 & 0 \\ 0 & 2 \\ 0 & 0 \end{bmatrix} \begin{bmatrix} \frac{1}{\sqrt{5}} & \frac{2}{\sqrt{5}} \\ \frac{2}{\sqrt{5}} & -\frac{1}{\sqrt{5}} \end{bmatrix} = \begin{bmatrix} 1 & 2 \\ 2 & 0 \\ 0 & 2 \end{bmatrix} = A \quad \blacksquare$$

This procedure will lead to the SVD for any matrix A. We find Σ and V by orthogonally diagonalizing $A^T A$, so we can see that this is always possible. The development of U is not as transparent. The key subtle fact that makes our procedure work is that if $\sigma_1, \ldots, \sigma_k$ are the *positive* singular values of an $n \times m$ matrix A with associated orthonormal eigenvectors $\mathbf{v}_1, \ldots, \mathbf{v}_k$ of $A^T A$, then the set

$$\mathbf{u}_i = \frac{1}{\sigma_i} A\mathbf{v}_i, \quad 1 \leq i \leq k$$

forms an orthonormal basis for $\mathrm{col}(A)$. Finding an orthonormal basis for $\left(\mathrm{col}(A)\right)^{\perp} = \mathrm{null}(A^T)$ allows us to extend the set $\{\mathbf{u}_1, \ldots, \mathbf{u}_k\}$ to an orthonormal basis for $\mathbf{R}^n$ and gives the remaining columns $\mathbf{u}_{k+1}, \ldots, \mathbf{u}_n$ of U. Since rows $k+1, \ldots, n$ of Σ are made up of zeros, the product $U\Sigma$ is independent of $\mathbf{u}_{k+1}, \ldots, \mathbf{u}_n$, and the definition of $\mathbf{u}_1, \ldots, \mathbf{u}_k$ ensures that $AV = U\Sigma$.

▶ Since U and V are orthogonal, so are U^T and V^T (see Exercise 62 of Section 8.3).

Our example covered the case of an $n \times m$ matrix A where $n > m$. The same procedure can be used if $n = m$, but suppose that $n < m$? In this case, we can take transposes. Let $B = A^T$, and suppose $B = U\Sigma V^T$ is the SVD. Since $A = B^T$, we have $A = \left(U\Sigma V^T\right)^T = V\Sigma^T U^T$, the form required by the SVD.

EXAMPLE 2   Find the SVD for the matrix

$$A = \begin{bmatrix} 1 & 2 & 1 & 0 \\ 2 & 0 & 1 & 1 \end{bmatrix}$$

Solution   We start by setting

$$B = A^T = \begin{bmatrix} 1 & 2 \\ 2 & 0 \\ 1 & 1 \\ 0 & 1 \end{bmatrix}$$

Now we apply our algorithm to find the SVD of B.

1. **Orthogonally Diagonalize $B^T B$ to find V.**   We have

$$B^T B = \begin{bmatrix} 6 & 3 \\ 3 & 6 \end{bmatrix}$$

which has eigenvalues and eigenvectors

$$\lambda_1 = 9 \;\Rightarrow\; \mathbf{v}_1 = \begin{bmatrix} \frac{1}{\sqrt{2}} \\ \frac{1}{\sqrt{2}} \end{bmatrix}; \qquad \lambda_2 = 3 \;\Rightarrow\; \mathbf{v}_2 = \begin{bmatrix} -\frac{1}{\sqrt{2}} \\ \frac{1}{\sqrt{2}} \end{bmatrix}$$

Thus the orthogonal diagonalizing matrix is

$$V = \begin{bmatrix} \frac{1}{\sqrt{2}} & -\frac{1}{\sqrt{2}} \\ \frac{1}{\sqrt{2}} & \frac{1}{\sqrt{2}} \end{bmatrix}$$

2. **Find Σ.**   The singular values for a matrix B are $\sigma_1 = 3$ and $\sigma_2 = \sqrt{3}$. Hence

$$\Sigma = \begin{bmatrix} 3 & 0 \\ 0 & \sqrt{3} \\ 0 & 0 \\ 0 & 0 \end{bmatrix}$$

3. **Find U.**   As before, determining the columns of U is performed in two steps.

   3a. Positive Singular Values.   For these we have $\mathbf{u}_i = \frac{1}{\sigma_i} B \mathbf{v}_i$, so that

$$\mathbf{u}_1 = \frac{1}{\sigma_1} B \mathbf{v}_1 = \frac{1}{3} \begin{bmatrix} 1 & 2 \\ 2 & 0 \\ 1 & 1 \\ 0 & 1 \end{bmatrix} \begin{bmatrix} \frac{1}{\sqrt{2}} \\ \frac{1}{\sqrt{2}} \end{bmatrix} = \frac{1}{3\sqrt{2}} \begin{bmatrix} 3 \\ 2 \\ 2 \\ 1 \end{bmatrix}$$

$$\mathbf{u}_2 = \frac{1}{\sigma_2} B \mathbf{v}_2 = \frac{1}{\sqrt{3}} \begin{bmatrix} 1 & 2 \\ 2 & 0 \\ 1 & 1 \\ 0 & 1 \end{bmatrix} \begin{bmatrix} -\frac{1}{\sqrt{2}} \\ \frac{1}{\sqrt{2}} \end{bmatrix} = \frac{1}{\sqrt{6}} \begin{bmatrix} 1 \\ -2 \\ 0 \\ 1 \end{bmatrix}$$

   3b. Filling Out U.   We need two columns to complete U, and we get them from an orthonormal basis for $(\text{col}(B))^{\perp} = \text{null}(B^T)$. Solving $B^T \mathbf{x} = \mathbf{0}$ using our standard procedure and then applying the Gram–Schmidt process gives us

$$\mathbf{u}_3 = \frac{1}{3} \begin{bmatrix} 0 \\ 1 \\ -2 \\ 2 \end{bmatrix}, \qquad \mathbf{u}_4 = \frac{1}{\sqrt{3}} \begin{bmatrix} -1 \\ 0 \\ 1 \\ 1 \end{bmatrix}$$

Combining the four vectors into U yields

$$U = \begin{bmatrix} \mathbf{u}_1 & \mathbf{u}_2 & \mathbf{u}_3 & \mathbf{u}_4 \end{bmatrix} = \begin{bmatrix} \frac{1}{\sqrt{2}} & \frac{1}{\sqrt{6}} & 0 & -\frac{1}{\sqrt{3}} \\ \frac{2}{3\sqrt{2}} & -\frac{2}{\sqrt{6}} & \frac{1}{3} & 0 \\ \frac{2}{3\sqrt{2}} & 0 & -\frac{2}{3} & \frac{1}{\sqrt{3}} \\ \frac{1}{3\sqrt{2}} & \frac{1}{\sqrt{6}} & \frac{2}{3} & \frac{1}{\sqrt{3}} \end{bmatrix}$$

Since $A = B^T$, we have $A = \left(U \Sigma V^T \right)^T = V \Sigma^T U^T$. Checking the calculations, we have

$$V \Sigma^T U^T = \begin{bmatrix} \frac{1}{\sqrt{2}} & -\frac{1}{\sqrt{2}} \\ \frac{1}{\sqrt{2}} & \frac{1}{\sqrt{2}} \end{bmatrix} \begin{bmatrix} 3 & 0 & 0 & 0 \\ 0 & \sqrt{3} & 0 & 0 \end{bmatrix} \begin{bmatrix} \frac{1}{\sqrt{2}} & \frac{2}{3\sqrt{2}} & \frac{2}{3\sqrt{2}} & \frac{1}{3\sqrt{2}} \\ \frac{1}{\sqrt{6}} & -\frac{2}{\sqrt{6}} & 0 & \frac{1}{\sqrt{6}} \\ 0 & \frac{1}{3} & -\frac{2}{3} & \frac{2}{3} \\ -\frac{1}{\sqrt{3}} & 0 & \frac{1}{\sqrt{3}} & \frac{1}{\sqrt{3}} \end{bmatrix}$$

$$= \begin{bmatrix} 1 & 2 & 1 & 0 \\ 2 & 0 & 1 & 1 \end{bmatrix} = A \qquad \blacksquare$$

Image Compression

SVDs can be used to store and transfer digital images efficiently. A digital black-and-white photo can be stored in matrix form, with each entry representing the gray level (the proportion of black to white) for a particular pixel. To simplify the discussion, let's assume that we have a square $n \times n$ matrix A made up of nonnegative entries that are the gray levels for a photo. Such a matrix has n^2 entries, which grows quickly with n and can have significant implications for storage and transmission of digital images.

To work more efficiently, we can take advantage of the fact that pixels near one another in a digital photo frequently have similar gray levels. Hence there can be a lot of redundant information in the image matrix, so that it may be possible to represent the image using much less storage space while still retaining the essential elements. One way to do this is to use the singular value decomposition of the image matrix $A = U \Sigma V^T$. We can use the *outer product expansion* (see Exercise 36) to express

$$A = \sigma_1 \mathbf{u}_1 \mathbf{v}_1^T + \sigma_2 \mathbf{u}_2 \mathbf{v}_2^T + \cdots + \sigma_n \mathbf{u}_n \mathbf{v}_n^T$$

where $\sigma_1 \geq \sigma_2 \geq \cdots \geq \sigma_n$ are the singular values. The terms with the largest singular values often contain most of the "information" in an image, while those associated with the smallest singular values frequently contribute relatively little. We can sometimes discard many—or even most—of the terms and still have a good approximation of the original image, by just taking the first k terms,

$$A_k = \sigma_1 \mathbf{u}_1 \mathbf{v}_1^T + \sigma_2 \mathbf{u}_2 \mathbf{v}_2^T + \cdots + \sigma_k \mathbf{u}_k \mathbf{v}_k^T$$

Figure 1 shows the results of using 2, 7, 14, and 28 of the original 273 singular values from a famous photo of a famous American (See Figure 2 for the original.) Although the image using 28 singular values requires only about 20% of the storage capacity of the original image, it is still fairly good.

Estimating the Rank of a Matrix

In most applications, matrix calculations are carried out on a computer. Unfortunately, the finite precision arithmetic employed by computers can sometimes lead to subtle but critical round-off of matrix entries, which can make it very difficult to determine that

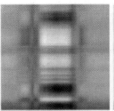

Figure 1 A photo of Abraham Lincoln processed with (from left to right) 2, 7, 14, and 28 of the 273 singular values. (Source: Library of Congress)

Figure 2 Abraham Lincoln image. (Source: Library of Congress)

Definition Machine ϵ
▶ For simplicity in notation we consider only square matrices, but the discussion can be extended to nonsquare matrices.

rank of the matrix. For example, an $n \times n$ matrix A might have true rank n but appear to the computer to have a lower rank, which we refer to as the *numerical rank* of the matrix. Here we briefly describe how to use singular values to find the numerical rank of a matrix.

A computer's sensitivity to round-off depends on the degree of precision used in storing numerical values. The *machine ϵ* provides a measure of this sensitivity. Roughly speaking, the machine ϵ gives an upper bound on the relative error that can occur when representing a number in the computer's floating point memory.

There are different ways to define the numerical rank of an $n \times n$ matrix A. One method employs the singular values $\sigma_1 \geq \sigma_2 \geq \cdots \geq \sigma_n$. Using a machine ϵ as an upper bound of the relative error, we let

$$b = \sigma_1 \cdot \epsilon \cdot n$$

Now define k to be the largest integer such that $\sigma_k \geq b$. Then k is the numerical rank of A.

EXAMPLE 3 Suppose that A is a 4×4 matrix with singular values $\sigma_1 = 3, \sigma_2 = 1$, $\sigma_3 = 10^{-8}$, and $\sigma_4 = 10^{-9}$. If we have $\epsilon = 10^{-10}$, what is the numerical rank of A?

Solution We have the bound
$$b = 3 \cdot 10^{-10} \cdot 4 = 1.2 \times 10^{-9}$$
Since $\sigma_3 \geq b$ but $\sigma_4 < b$, it follows that the numerical rank of A is 3. ∎

Proof of Theorem 8.26

We have verified most of the elements required to show that an SVD always exists. All that remains is to show that the key fact mentioned earlier is true: If $\sigma_1, \ldots, \sigma_k$ are the positive singular values of an $n \times m$ matrix A with associated orthonormal eigenvectors (of $A^T A$) $\mathbf{v}_1, \ldots, \mathbf{v}_k$, then the set

$$\mathbf{u}_i = \frac{1}{\sigma_i} A\mathbf{v}_i, \quad 1 \leq i \leq k$$

forms an orthonormal basis for $\text{col}(A)$. Once this is established, the Rank-Nullity Theorem and Gram–Schmidt process ensure the existence of an orthonormal basis $\{\mathbf{u}_{k+1}, \ldots, \mathbf{u}_n\}$ for $\left(\text{col}(A)\right)^{\perp} = \text{null}(A^T)$, so we are assured that the required orthogonal matrix U can be formed.

First, note that for $1 \leq i, j \leq k$ we have

$$\mathbf{u}_i \cdot \mathbf{u}_j = \left(\frac{1}{\sigma_i} A\mathbf{v}_i\right) \cdot \left(\frac{1}{\sigma_j} A\mathbf{v}_j\right) = \frac{1}{\sigma_i \sigma_j}(A\mathbf{v}_i)^T(A\mathbf{v}_j)$$

$$= \frac{1}{\sigma_i \sigma_j}\mathbf{v}_i^T(A^T A\mathbf{v}_j) = \frac{1}{\sigma_i \sigma_j}\mathbf{v}_i^T(\lambda_j\mathbf{v}_j) = \frac{\lambda_j}{\sigma_i \sigma_j}(\mathbf{v}_i \cdot \mathbf{v}_j)$$

If $i \neq j$, then $\mathbf{v}_i \cdot \mathbf{v}_j = 0$ and so $\mathbf{u}_i \cdot \mathbf{u}_j = 0$. On the other hand, if $i = j$, then $\mathbf{v}_i \cdot \mathbf{v}_i = \frac{\lambda_i}{\sigma_i^2} = 1$ and hence $\mathbf{u}_i \cdot \mathbf{u}_i = 1$. Thus $\{\mathbf{u}_1, \cdots, \mathbf{u}_k\}$ are orthonormal and therefore also linearly independent.

Next, since $\mathbf{u}_i = \frac{1}{\sigma_i} A\mathbf{v}_i$, each $\mathbf{u}_i$ $(1 \leq i \leq k)$ is in col(A). So if we can show that $\dim\big(\mathrm{col}(A)\big) = k$, we are done because $\{\mathbf{u}_1, \cdots, \mathbf{u}_k\}$ is an orthonormal basis for col(A). To see why this is true, we note the following:

(a) rank($A^T A$) = rank(A). (See Exercise 37.)

(b) The eigenvectors $\{\mathbf{v}_1, \ldots, \mathbf{v}_k\}$ associated with the nonzero eigenvalues of $A^T A$ form a basis for col(A). (See Exercise 38.)

Thus, if we have positive singular values $\sigma_1, \ldots, \sigma_k$, then rank($A^T A$) = k, which implies that rank(A) = k. Hence $\dim\big(\mathrm{col}(A)\big) = k$, completing the proof. ∎

| EXERCISES |

Find the singular values for the matrices given in Exercises 1–8.

1. $\begin{bmatrix} 1 & 2 \\ -1 & 2 \end{bmatrix}$

2. $\begin{bmatrix} 1 & 2 \\ -2 & 2 \end{bmatrix}$

3. $\begin{bmatrix} 3 & -1 \\ -1 & 3 \end{bmatrix}$

4. $\begin{bmatrix} 2 & -2 \\ 4 & 1 \end{bmatrix}$

5. $\begin{bmatrix} 1 & 2 \\ 0 & 2 \\ 2 & -1 \end{bmatrix}$

6. $\begin{bmatrix} 3 & 1 \\ -1 & 0 \\ 1 & 2 \end{bmatrix}$

7. $\begin{bmatrix} 1 & 2 & 1 \\ 0 & 1 & -1 \end{bmatrix}$

8. $\begin{bmatrix} -1 & 3 & 2 \\ 1 & 1 & -1 \end{bmatrix}$

Find a singular value decomposition for the matrices given in Exercises 9–16.

9. $\begin{bmatrix} 1 & 2 \\ 2 & 1 \end{bmatrix}$

10. $\begin{bmatrix} 2 & 2 \\ -2 & 1 \end{bmatrix}$

11. $\begin{bmatrix} 2 & 1 \\ -1 & 3 \\ 1 & 0 \end{bmatrix}$

12. $\begin{bmatrix} 1 & -2 \\ 2 & 1 \\ 0 & 2 \end{bmatrix}$

13. $\begin{bmatrix} -1 & 1 & 0 \\ 2 & 2 & 1 \end{bmatrix}$

14. $\begin{bmatrix} 1 & 2 & 0 \\ 2 & -1 & -2 \end{bmatrix}$

15. $\begin{bmatrix} 2 & 2 & 1 & 0 \\ 1 & -1 & 0 & 1 \end{bmatrix}$

16. $\begin{bmatrix} 1 & 3 & 3 & 2 \\ 1 & 2 & -1 & 1 \end{bmatrix}$

For Exercises 17–20, determine the numerical rank of the matrix A.

17. A is a 3×3 matrix with singular values $\sigma_1 = 10$, $\sigma_2 = 6$, $\sigma_3 = 10^{-8}$; $\epsilon = 10^{-9}$.

18. A is a 4×4 matrix with singular values $\sigma_1 = 5$, $\sigma_2 = 3$, $\sigma_3 = 10^{-7}$, $\sigma_4 = 10^{-8}$; $\epsilon = 10^{-8}$.

19. A is a 4×4 matrix with singular values $\sigma_1 = 12$, $\sigma_2 = 4$, $\sigma_3 = 10^{-6}$, $\sigma_4 = 10^{-9}$; $\epsilon = 10^{-7}$.

20. A is a 5×5 matrix with singular values $\sigma_1 = 15$, $\sigma_2 = 8$, $\sigma_3 = 10^{-6}$, $\sigma_4 = 10^{-8}$, $\sigma_5 = 10^{-9}$; $\epsilon = 10^{-7}$.

TRUE OR FALSE For Exercises 21–26, determine if the statement is true or false, and justify your answer.

21. If A is an $n \times m$ matrix, then A has a singular value decomposition only if $n > m$.

22. The singular values of a matrix A are all positive.

23. If A is an invertible matrix with singular value σ, then A^{-1} has singular value σ^{-1}.

24. If A is a square matrix, then the singular value decomposition of A is the same as the diagonalization of A.

25. If A is a square matrix, then $|\det(A)|$ is equal to the product of the singular values of A.

26. The largest singular value of an orthogonal matrix is 1.

27. For the matrix in Exercise 11, compute $\sigma_1 \mathbf{u}_1 \mathbf{v}_1^T$ and $\sigma_1 \mathbf{u}_1 \mathbf{v}_1^T + \sigma_2 \mathbf{u}_2 \mathbf{v}_2^T$, and compare your results to the original matrix.

28. For the matrix in Exercise 12, compute $\sigma_1 \mathbf{u}_1 \mathbf{v}_1^T$ and $\sigma_1 \mathbf{u}_1 \mathbf{v}_1^T + \sigma_2 \mathbf{u}_2 \mathbf{v}_2^T$, and compare your results to the original matrix.

29. For the matrix in Exercise 15, compute $\sigma_1 \mathbf{u}_1 \mathbf{v}_1^T$ and $\sigma_1 \mathbf{u}_1 \mathbf{v}_1^T + \sigma_2 \mathbf{u}_2 \mathbf{v}_2^T$, and compare your results to the original matrix.

30. Prove that if A is a symmetric matrix with eigenvalue λ, then A has singular value $|\lambda|$.

31. Prove that the positive singular values of A and A^T are the same.

32. Prove that if σ is a singular value of A, then there exists a nonzero vector $\mathbf{x}$ such that

$$\sigma = \frac{\|A\mathbf{x}\|}{\|\mathbf{x}\|}$$

For Exercises 33–36, assume that A is an $n \times n$ matrix with SVD $A = U\Sigma V^T$.

33. If A is invertible, find a SVD of A^{-1}.

34. Prove that the columns of U are eigenvectors of AA^T.

35. If P is an orthogonal $n \times n$ matrix, prove that PA has the same singular values as A.

36. Let $U = \begin{bmatrix} \mathbf{u}_1 & \cdots & \mathbf{u}_n \end{bmatrix}$ and $V = \begin{bmatrix} \mathbf{v}_1 & \cdots & \mathbf{v}_n \end{bmatrix}$.

(a) If $n = 2$, show that $A = \sigma_1 \mathbf{u}_1 \mathbf{v}_1^T + \sigma_2 \mathbf{u}_2 \mathbf{v}_2^T$.

(b) For $n \geq 1$, show that $A = \sigma_1 \mathbf{u}_1 \mathbf{v}_1^T + \cdots + \sigma_n \mathbf{u}_n \mathbf{v}_n^T$.

37. Suppose that A is an $m \times n$ matrix. Prove that $\text{rank}(A^T A) = \text{rank}(A)$ by verifying each of the following:

(a) Show that if $\mathbf{x}$ is a solution to $A\mathbf{x} = \mathbf{0}$, then $\mathbf{x}$ is a solution to $A^T A\mathbf{x} = \mathbf{0}$.

(b) Suppose that $\mathbf{x}$ satisfies $A^T A\mathbf{x} = \mathbf{0}$. Show that $A\mathbf{x}$ is in $\left(\text{col}(A)\right)^{\perp}$. Since $A\mathbf{x}$ is also in $\text{col}(A)$ (justify!), show that this means that $A\mathbf{x} = \mathbf{0}$.

(c) Combine (a) and (b) to show that $\text{nullity}(A^T A) = \text{nullity}(A)$, and then apply the Rank–Nullity theorem to conclude that $\text{rank}(A^T A) = \text{rank}(A)$.

38. Prove that the orthogonal eigenvectors $\{\mathbf{v}_1, \ldots, \mathbf{v}_k\}$ associated with the nonzero eigenvalues of $A^T A$ form a basis for $\text{col}(A)$ by verifying each of the following:

(a) Apply Exercise 37 to show that $\dim(\text{col}(A^T A)) = \dim(\text{col}(A))$.

(b) Apply the Spectral Theorem to explain why the orthogonal eigenvectors $\{\mathbf{v}_1, \ldots, \mathbf{v}_k\}$ associated with the nonzero eigenvalues of $A^T A$ form a basis for $\text{col}(A^T A)$.

(c) Combine (a) and (b) to reach the desired conclusion.

Ⓒ Find a singular value decomposition for the matrices given in Exercises 39–42, following the steps illustrated in this section but using computer software to assist with finding the required eigenvalues, eigenvectors, and orthogonal bases.

39. $\begin{bmatrix} 3 & 5 \\ -1 & 2 \end{bmatrix}$

40. $\begin{bmatrix} 2 & -1 & 6 \\ -3 & 0 & 2 \end{bmatrix}$

41. $\begin{bmatrix} -5 & 0 & 2 \\ 1 & -1 & 3 \\ 0 & 4 & 2 \end{bmatrix}$

42. $\begin{bmatrix} 2 & 3 & -1 & 0 \\ 1 & 2 & 1 & 3 \\ -2 & -1 & 1 & 3 \end{bmatrix}$

8.5 Least Squares Regression

▶ This section is optional. However, least squares regression is revisited in optional Section 10.3.

A problem that arises in a wide variety of disciplines is that of finding algebraic formulas to describe data. A simple example involving the relationship between barometric pressure and the boiling point of water is described below.

EXAMPLE 1   The boiling point of water is known to vary depending on the barometric pressure. To determine the relationship between boiling point and pressure, boiling points were found experimentally at several different barometric pressures. The results are summarized in Table 1 and the data plotted in Figure 1. Find a linear equation of the form $T = c_0 + c_1 P$ that will allow us to make accurate predictions of boiling point T for a given barometric pressure P.

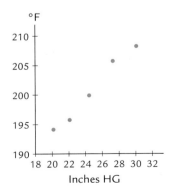

Figure 1 Scatter plot of pressure against boiling point.

Barometric Pressure (inches HG)	20.2	22.1	24.5	27.3	30.1
Boiling Point (°F)	195	197	202	209	212

Table 1   Boiling Point of Water at Different Barometric Pressures

Solution To find the coefficients c_0 and c_1, we could try plugging T and P for each data point into $T = c_0 + c_1 P$, yielding the linear system

$$
\begin{aligned}
c_0 + 20.2c_1 &= 195 \\
c_0 + 22.1c_1 &= 197 \\
c_0 + 24.5c_1 &= 202 \\
c_0 + 27.3c_1 &= 209 \\
c_0 + 30.1c_1 &= 212
\end{aligned}
\tag{1}
$$

Unfortunately, this system cannot have any solutions. If it did, then our points would lie exactly on a line, but we see in Figure 1 that they do not. However, since the points all lie close to a line, the system (1) "almost" has a solution.

In this section we develop an approximation method that gives us a way to change a linear system that has no solutions into a new system that has a solution. Our method is such that we change the system as little as possible, so that the solution to the new system can serve as an approximate solution for the original system. We will return to this example when we have the tools we need to find an answer. ■

Example 1 suggests a more general problem—namely, that of finding an "approximate" solution to a linear system

$$
A\mathbf{x} = \mathbf{y}
\tag{2}
$$

that has no solutions. We solve this problem by changing the vector $\mathbf{y}$ in (2) into a new vector $\hat{\mathbf{y}}$ such that

$$
A\mathbf{x} = \hat{\mathbf{y}}
\tag{3}
$$

has a solution. In order for (3) to have a solution, we must select $\hat{\mathbf{y}}$ from among the vectors in $\text{col}(A)$, the column space of A. We want the systems (2) and (3) to be as similar as possible, so we choose $\hat{\mathbf{y}}$ in order to minimize

▶ The vector $\hat{\mathbf{y}}$ that minimizes (4) also minimizes $\|\mathbf{y} - \hat{\mathbf{y}}\|$. Squaring the distance simplifies calculations by eliminating the square root in the norm formula.

$$
\|\mathbf{y} - \hat{\mathbf{y}}\|^2 = (y_1 - \hat{y}_1)^2 + (y_2 - \hat{y}_2)^2 + \cdots + (y_n - \hat{y}_n)^2
\tag{4}
$$

Thus we want to find $\hat{\mathbf{y}}$ as close as possible to $\mathbf{y}$, subject to the constraint that $\hat{\mathbf{y}}$ is in $\text{col}(A)$. This is depicted in Figure 2.

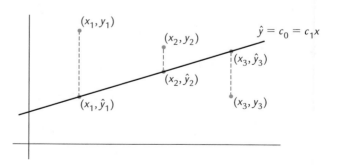

Figure 2 The data points are (x_i, y_i), and $(x_i, \hat{y}_i)$ are the corresponding points on the line $\hat{y} = c_0 + c_1 x$. The sum of the squares of the lengths of the dotted lines equals the expression (4). The coefficients c_0 and c_1 are chosen to make (4) as small as possible.

To find $\hat{\mathbf{y}}$, we use projections developed in Section 8.2. A key property of projections is contained in the next theorem. A proof is given at the end of the section.

THEOREM 8.27

Let $\mathbf{y}$ be a vector and S a subspace. Then the vector closest to $\mathbf{y}$ in S is given by $\hat{\mathbf{y}} = \text{proj}_S\mathbf{y}$.

Definition Least Squares Solution

To sum up, given a linear system $A\mathbf{x} = \mathbf{y}$ that has no solutions, we find an approximate solution by solving $A\mathbf{x} = \hat{\mathbf{y}}$, where $\hat{\mathbf{y}} = \text{proj}_S\mathbf{y}$ and $S = \text{col}(A)$. This approach is called *Least Squares Regression* (or *Linear Regression*), and a solution $\hat{\mathbf{x}}$ to $A\mathbf{x} = \hat{\mathbf{y}}$ is called a **least squares solution**.

EXAMPLE 2 Find the least squares solution $\hat{\mathbf{x}}$ to $A\mathbf{x} = \mathbf{y}$, where

$$A = \begin{bmatrix} \mathbf{a}_1 & \mathbf{a}_2 \end{bmatrix} = \begin{bmatrix} 1 & 4 \\ -3 & 3 \\ 5 & 1 \end{bmatrix} \quad \text{and} \quad \mathbf{y} = \begin{bmatrix} -16 \\ 28 \\ 6 \end{bmatrix}$$

Solution The first step is to find $\hat{\mathbf{y}}$. If $S = \text{col}(A)$, then by Theorem 8.27 the vector $\hat{\mathbf{y}}$ in S that is closest to $\mathbf{y}$ is $\hat{\mathbf{y}} = \text{proj}_S\mathbf{y}$. Since the columns of A are orthogonal, we can apply the projection formula in Definition 8.15 in Section 8.2 to compute

$$\hat{\mathbf{y}} = \text{proj}_S\mathbf{y} = \frac{\mathbf{a}_1 \cdot \mathbf{y}}{\|\mathbf{a}_1\|^2}\mathbf{a}_1 + \frac{\mathbf{a}_2 \cdot \mathbf{y}}{\|\mathbf{a}_2\|^2}\mathbf{a}_2 = \frac{-70}{35}\begin{bmatrix} 1 \\ -3 \\ 5 \end{bmatrix} + \frac{26}{26}\begin{bmatrix} 4 \\ 3 \\ 1 \end{bmatrix} = \begin{bmatrix} 2 \\ 9 \\ -9 \end{bmatrix}$$

We find the least squares solution $\hat{\mathbf{x}}$ by solving the system $A\mathbf{x} = \hat{\mathbf{y}}$, which is

$$\begin{aligned} x_1 + 4x_2 &= 2 \\ -3x_1 + 3x_2 &= 9 \\ 5x_1 + x_2 &= -9 \end{aligned}$$

Using our usual solution methods, we find that $\hat{\mathbf{x}} = \begin{bmatrix} -2 \\ 1 \end{bmatrix}$. ∎

An alternative definition of least squares solution is given below.

DEFINITION 8.28

Definition Least Squares Solution

If A is an $n \times m$ matrix and $\mathbf{y}$ is in $\mathbf{R}^n$, then a **least squares solution** to $A\mathbf{x} = \mathbf{y}$ is a vector $\hat{\mathbf{x}}$ in $\mathbf{R}^m$ such that

$$\|A\hat{\mathbf{x}} - \mathbf{y}\| \leq \|A\mathbf{x} - \mathbf{y}\|$$

for all $\mathbf{x}$ in $\mathbf{R}^m$.

If $A\mathbf{x} = \mathbf{y}$ has a solution $\mathbf{x}_0$, then $\hat{\mathbf{x}} = \mathbf{x}_0$. If A has linearly independent columns, then $\hat{\mathbf{x}}$ will be unique. If not, then there are infinitely many least squares solutions $\hat{\mathbf{x}}$.

The solution method using projection demonstrated in Example 2 requires an orthogonal basis for $S = \text{col}(A)$. The following theorem is more convenient to use because it does not have this requirement.

THEOREM 8.29

▶ The equations in (5) are called the **normal equations** for $A\mathbf{x} = \mathbf{y}$.

▶ When A has linearly independent columns, the formula for $\hat{\mathbf{x}}$ in (6) can be applied. However, for large data sets numerical issues can arise in calculating $(A^T A)^{-1}$ that may make using (5) attractive.

The set of least squares solutions to $A\mathbf{x} = \mathbf{y}$ is equal to the set of solutions to the system

$$A^T A\mathbf{x} = A^T\mathbf{y} \tag{5}$$

If A has linearly independent columns, then there is a unique least squares solution given by

$$\hat{\mathbf{x}} = (A^T A)^{-1} A^T\mathbf{y} \tag{6}$$

Otherwise, there are infinitely many least squares solutions.

Proof Starting with (5), suppose that $\hat{\mathbf{x}}$ is a solution to $A\mathbf{x} = \hat{\mathbf{y}}$, where $\hat{\mathbf{y}} = \text{proj}_S\mathbf{y}$ and $S = \text{col}(A)$. By Theorem 8.16 in Section 8.2, $\mathbf{y} - \hat{\mathbf{y}} = \mathbf{y} - \text{proj}_S\mathbf{y}$ is in $S^\perp$. Since $S^\perp = \left(\text{col}(A)\right)^\perp = \text{null}(A^T)$ (see Exercise 69 of Section 8.1), it follows that $A^T(\mathbf{y} - \hat{\mathbf{y}}) = \mathbf{0}$. As $A\hat{\mathbf{x}} = \hat{\mathbf{y}}$, we have

$$A^T(\mathbf{y} - A\hat{\mathbf{x}}) = \mathbf{0} \implies A^T A\hat{\mathbf{x}} = A^T\mathbf{y}$$

The reasoning also works in the reverse direction, completing the proof of (5).

To prove (6), note that $A^T A$ is invertible if and only if A has linearly independent columns (see Exercise 33). Hence (5) has a unique solution if A has linearly independent columns, with the solution given by (6). Otherwise, $A^T A$ is not invertible, and (5) has infinitely many solutions. ■

EXAMPLE 3 Complete Example 1 by finding the coefficients c_0 and c_1 for the line $T = c_0 + c_1 P$ that best fits the data in Table 1.

Solution We need to find the least squares solution $\hat{\mathbf{c}}$ for the linear system (1), which is equivalent to $A\mathbf{c} = \mathbf{t}$, where

$$A = \begin{bmatrix} 1 & 20.2 \\ 1 & 22.1 \\ 1 & 24.5 \\ 1 & 27.3 \\ 1 & 30.1 \end{bmatrix}, \quad \mathbf{c} = \begin{bmatrix} c_0 \\ c_1 \end{bmatrix}, \quad \text{and} \quad \mathbf{t} = \begin{bmatrix} 195 \\ 197 \\ 202 \\ 209 \\ 212 \end{bmatrix}$$

Although the notation is different than in our general development of least squares solutions, the method of solution is the same. Since the columns of A can be seen to be linearly independent, we have

$$\hat{\mathbf{c}} = \left(A^T A\right)^{-1} A^T\mathbf{t} = \begin{bmatrix} 157.17 \\ 1.845 \end{bmatrix}$$

Therefore the equation that best fits the data is $T = 157.17 + 1.845P$. A graph of the data and line is shown in Figure 3. ■

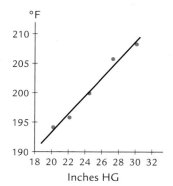

Figure 3 $T = 157.17 + 1.845P$ and five data points.

Fitting Functions to Data

EXAMPLE 4 An economist conducts quarterly surveys to measure consumer confidence. The confidence indices for six consecutive quarters are given in Table 2. Find a cubic polynomial that approximates the data.

Quarter	1	2	3	4	5	6
Confidence Index	5	9	8	4	6	8

Table 2 Quarterly Consumer Confidence Index Data

Solution The data set is displayed in Figure 4. We want to fit to it a cubic polynomial of the form

$$g(t) = c_0 + c_1 t + c_2 t^2 + c_3 t^3$$

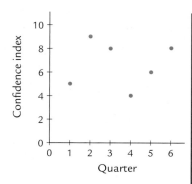

Figure 4 Consumer confidence index data.

For each data point, we plug the quarter number into g and set the result equal to the corresponding index. We get the system of equations

$$
\begin{aligned}
(1, 5) &\implies c_0 + c_1 + c_2 + c_3 = 5 \\
(2, 9) &\implies c_0 + 2c_1 + 4c_2 + 8c_3 = 9 \\
(3, 8) &\implies c_0 + 3c_1 + 9c_2 + 27c_3 = 8 \\
(4, 4) &\implies c_0 + 4c_1 + 16c_2 + 64c_3 = 4 \\
(5, 6) &\implies c_0 + 5c_1 + 25c_2 + 125c_3 = 6 \\
(6, 8) &\implies c_0 + 6c_1 + 36c_2 + 216c_3 = 8
\end{aligned}
$$

The system is equivalent to $A\mathbf{c} = \mathbf{y}$, where

$$
A = \begin{bmatrix} 1 & 1 & 1 & 1 \\ 1 & 2 & 4 & 8 \\ 1 & 3 & 9 & 27 \\ 1 & 4 & 16 & 64 \\ 1 & 5 & 25 & 125 \\ 1 & 6 & 36 & 216 \end{bmatrix}, \quad \mathbf{c} = \begin{bmatrix} c_0 \\ c_1 \\ c_2 \\ c_3 \end{bmatrix}, \quad \mathbf{y} = \begin{bmatrix} 5 \\ 9 \\ 8 \\ 4 \\ 6 \\ 8 \end{bmatrix}
$$

The columns of A can be verified to be linearly independent, so that by Theorem 8.29 we have (some rounding is included)

$$
\hat{\mathbf{c}} = \left(A^T A\right)^{-1} A^T \mathbf{y} = \begin{bmatrix} -5.33 \\ 15.07 \\ -5.02 \\ 0.48 \end{bmatrix}
$$

Hence the best-fitting cubic polynomial is

$$
g(t) = -5.33 + 15.07t - 5.02t^2 + 0.48t^3
$$

A plot of $g(t)$ together with the data is shown in Figure 5. ∎

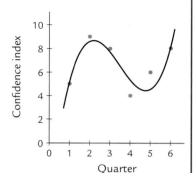

Figure 5 $g(t)$ and the consumer confidence index data.

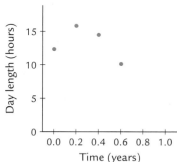

Figure 6 Length of day data. The time $t = 0$ corresponds to March 20, the vernal equinox.

EXAMPLE 5 The times from sunrise to sunset in Vancouver BC on selected days in 2010 are given in Table 3 and plotted in Figure 6. The length of day can be modeled by a function of the form $L(t) = c_1 + c_2 \sin(2\pi t)$, where t is time in years. Find the coefficients c_1 and c_2 that will give the best fit to the data.

Date	Mar 20	June 1	Aug 12	Oct 25
Day Length (hours)	12.17	16.23	12.12	8.18

Table 3 The Time from Sunrise to Sunset on Selected Days in Vancouver, BC

Solution If we let $t = 0$ correspond to March 20, then the remaining dates occur at $t = 0.2$, $t = 0.4$, and $t = 0.6$, respectively. Evaluating $L(t)$ at each of the four times, we obtain the system

$$
\begin{aligned}
(0, 12.17) &\implies c_1 + c_2 \sin(0) = c_1 = 12.17 \\
(0.2, 15.95) &\implies c_1 + c_2 \sin(0.4\pi) = c_1 + 0.951c_2 = 15.95 \\
(0.4, 14.55) &\implies c_1 + c_2 \sin(0.8\pi) = c_1 + 0.588c_2 = 14.55 \\
(0.6, 10.22) &\implies c_1 + c_2 \sin(1.2\pi) = c_1 - 0.588c_2 = 10.22
\end{aligned}
$$

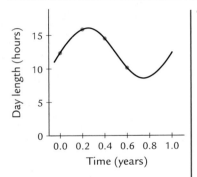

Figure 7 Length of day data and the graph of $L(t)$.

The equivalent system is $A\mathbf{c} = \mathbf{y}$, where

$$A = \begin{bmatrix} 1 & 0 \\ 1 & 0.951 \\ 1 & 0.588 \\ 1 & -0.588 \end{bmatrix}, \quad \mathbf{c} = \begin{bmatrix} c_1 \\ c_2 \end{bmatrix}, \quad \mathbf{y} = \begin{bmatrix} 12.17 \\ 15.95 \\ 14.55 \\ 10.22 \end{bmatrix}$$

The columns of A are linearly independent, so that by Theorem 8.29 we have (some rounding is included)

$$\hat{\mathbf{c}} = \left(A^T A\right)^{-1} A^T \mathbf{y} = \begin{bmatrix} 12.33 \\ 3.75 \end{bmatrix}$$

Therefore the best fit is given by $L(t) = 12.33 + 3.75 \sin(2\pi t)$. A graph of the data and $L(t)$ are shown in Figure 7. ∎

Planetary Orbits Revisited

In Section 1.4 we considered the problem of finding a model that predicts the orbital period (time required to circle the sun) for a planet based on the planet's distance from the sun. We start with the model

$$p = ad^b$$

where p is the orbital period, d is the distance from the sun, and a and b are constants we estimate from the data. To make the model linear, we take the logarithm on both sides of the equation and set $a_1 = \ln(a)$, giving us

$$\ln(p) = a_1 + b\ln(d)$$

Our goal here is to find values for a_1 and b. Previously, we did not have the tools to simultaneously incorporate all of our data into the model, because doing so would have resulted in a system with no solutions. However, now we can find a least squares solution to the system that does incorporate all of the data.

Since the model involves $\ln(p)$ and $\ln(d)$, as a first step we need the logarithms of the periods and distances. This is given in Table 4.

The graph in Figure 8 shows a plot of the points $(\ln(d), \ln(p))$. The points lie very close to a line, suggesting that we are on the right track. Substituting each of the points

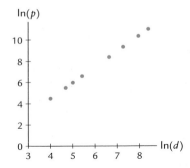

Figure 8 Scatter plot of data $(\ln(d), \ln(p))$.

Planet	Distance ($\times 10^6$ km)	Orbital Period (Days)	$\ln(d)$	$\ln(p)$
Mercury	57.9	88	4.059	4.477
Venus	108.2	224.7	4.684	5.415
Earth	149.6	365.2	5.008	5.900
Mars	227.9	687	5.429	6.532
Jupiter	778.6	4331	6.657	8.374
Saturn	1433.5	10,747	7.268	9.282
Uranus	2871.5	30,589	7.963	10.328
Neptune	4495.1	59,800	8.411	10.999

Table 4 Planetary Orbital Distances and Periods

into the equation $\ln(p) = a_1 + b\ln(d)$ yields the system of equations $A\mathbf{x} = \mathbf{y}$, where

$$
A = \begin{bmatrix} 1 & 4.059 \\ 1 & 4.684 \\ 1 & 5.008 \\ 1 & 5.429 \\ 1 & 6.657 \\ 1 & 7.268 \\ 1 & 7.963 \\ 1 & 8.411 \end{bmatrix}, \quad \mathbf{x} = \begin{bmatrix} a_1 \\ b \end{bmatrix}, \quad \mathbf{y} = \begin{bmatrix} 4.477 \\ 5.415 \\ 5.900 \\ 6.532 \\ 8.374 \\ 9.282 \\ 10.328 \\ 10.999 \end{bmatrix}
$$

Since A has linearly independent columns, we can use (6) to compute

$$
\hat{\mathbf{x}} = \left(A^T A\right)^{-1} A^T \mathbf{y} = \begin{bmatrix} -1.60322 \\ 1.49827 \end{bmatrix}
$$

Therefore $a_1 = -1.60322$ and $b = 1.49827$. Since $a_1 = \ln(a)$, it follows that $a = e^{-1.60322} = 0.201247$. Thus our model is

$$
p = 0.201247 d^{1.49827}
$$

This is consistent with Kepler's third law of motion, which predicts that the exponent should be $3/2$. Table 5 gives the predicted values for the orbital periods. Note that the model provides generally better predictions than those we came up with in Section 1.4.

Proof of Theorem 8.27

Proof of Theorem 8.27 Let $\mathbf{s}$ be a vector in S, and let $\hat{\mathbf{y}} = \text{proj}_S\mathbf{y}$. Then by Theorem 8.16 in Section 8.2, $\mathbf{y} - \hat{\mathbf{y}}$ is in $S^\perp$. Also, both $\hat{\mathbf{y}}$ and $\mathbf{s}$ are in S, so $\hat{\mathbf{y}} - \mathbf{s}$ is in S because S is a subspace. Therefore $\mathbf{y} - \hat{\mathbf{y}}$ and $\hat{\mathbf{y}} - \mathbf{s}$ are orthogonal, so by Theorem 8.6 in Section 8.1 (the Pythagorean Theorem) we have

$$
\|\mathbf{y} - \mathbf{s}\|^2 = \|(\mathbf{y} - \hat{\mathbf{y}}) - (\hat{\mathbf{y}} - \mathbf{s})\|^2 = \|\mathbf{y} - \hat{\mathbf{y}}\|^2 + \|\hat{\mathbf{y}} - \mathbf{s}\|^2
$$

Since $\|\hat{\mathbf{y}} - \mathbf{s}\|^2 \geq 0$, we may conclude that $\|\mathbf{y} - \mathbf{s}\|^2 \geq \|\mathbf{y} - \hat{\mathbf{y}}\|^2$ for all $\mathbf{s}$ in S. Therefore no vector in S is closer to $\mathbf{y}$ than $\hat{\mathbf{y}}$, so $\hat{\mathbf{y}} = \text{proj}_S\mathbf{y}$ is the vector in S that is closest to $\mathbf{y}$. Furthermore, there is equality only when $\hat{\mathbf{y}} = \mathbf{s}$, so $\hat{\mathbf{y}}$ is the unique closest point. ∎

Planet	Distance ($\times 10^6$ km)	Orbital Period (Days)	Predicted Period
Mercury	57.9	88	88.0
Venus	108.2	224.7	224.7
Earth	149.6	365.2	365.1
Mars	227.9	687	685.9
Jupiter	778.6	4331	4322.2
Saturn	1433.5	10,747	10,786.1
Uranus	2871.5	30,589	30,543.0
Neptune	4495.1	59,800	59,775.1

Table 5 Orbital Distances, Periods, and Predicted Periods

EXERCISES

In Exercises 1–4, find the vector in the subspace S closest to $\mathbf{y}$.

1. $\mathbf{y} = \begin{bmatrix} 1 \\ 2 \end{bmatrix}$, $S = \text{span}\left\{ \begin{bmatrix} 1 \\ -1 \end{bmatrix} \right\}$

2. $\mathbf{y} = \begin{bmatrix} 2 \\ -3 \end{bmatrix}$, $S = \text{span}\left\{ \begin{bmatrix} 3 \\ 1 \end{bmatrix} \right\}$

3. $\mathbf{y} = \begin{bmatrix} 1 \\ -1 \\ 2 \end{bmatrix}$, $S = \text{span}\left\{ \begin{bmatrix} 1 \\ 3 \\ -2 \end{bmatrix}, \begin{bmatrix} 5 \\ -1 \\ 1 \end{bmatrix} \right\}$

4. $\mathbf{y} = \begin{bmatrix} 3 \\ 0 \\ 1 \end{bmatrix}$, $S = \text{span}\left\{ \begin{bmatrix} 0 \\ 2 \\ -1 \end{bmatrix}, \begin{bmatrix} 4 \\ -1 \\ -2 \end{bmatrix} \right\}$

In Exercises 5–8, find the normal equations for the given system.

5.
$$2x_1 - x_2 = 4$$
$$x_1 + 2x_2 = 3$$
$$3x_1 - x_2 = 4$$

6.
$$4x_1 + 2x_2 = 3$$
$$-2x_1 - x_2 = 1$$
$$3x_1 + 2x_2 = 0$$

7.
$$x_1 + x_2 - x_3 = 2$$
$$2x_1 - x_2 + 2x_3 = -1$$
$$x_1 + 4x_2 - 5x_3 = 6$$

8.
$$-x_1 - 2x_2 + 2x_3 = -3$$
$$2x_1 + x_2 - 3x_3 = 8$$
$$-x_1 - 5x_2 + 3x_3 = 0$$

In Exercises 9–12, find the least squares solution(s) for the given system.

9.
$$-x_1 - x_2 = 3$$
$$2x_1 + 3x_2 = -1$$
$$-3x_1 + 2x_2 = 2$$

10.
$$x_1 + 2x_2 = -1$$
$$3x_1 - 2x_2 = 1$$
$$-x_1 - 3x_2 = -2$$

11.
$$2x_1 - x_2 - x_3 = 1$$
$$-x_1 + x_2 + 3x_3 = -1$$
$$3x_1 - 2x_2 - 4x_3 = 3$$

12.
$$3x_1 + 2x_2 - x_3 = -2$$
$$2x_1 + 3x_2 - x_3 = 1$$
$$-x_1 - 4x_2 + x_3 = -3$$

13. Find the normal equations for the parabolas that best fit the points $(0, 1)$ and $(2, 5)$, and explain why the system should have infinitely many solutions.

14. Find the normal equations for the cubic polynomials that best fit the points $(0, 1)$, $(1, 4)$, and $(3, -1)$, and explain why the system should have infinitely many solutions.

FIND AN EXAMPLE For Exercises 15–20, find an example that meets the given specifications.

15. A linear system with three equations and two variables that has no solutions and a unique least squares solution.

16. A linear system with four equations and three variables that has no solutions and a unique least squares solution.

17. A linear system with four equations and two variables that has no solutions and infinitely many least squares solutions.

18. A linear system with four equations and three variables that has no solutions and infinitely many least squares solutions.

19. A linear system with three equations and three variables that has a unique solution and a unique least squares solution.

20. A linear system with four equations and three variables that has a unique solution and a unique least squares solution.

TRUE OR FALSE For Exercises 21–28, determine if the statement is true or false, and justify your answer.

21. A least squares solution can be found only for a linear system that has more equations than variables.

22. If a linear system has infinitely many solutions, then it also has infinitely many least squares solutions.

23. If A is an $n \times m$ matrix, then any least squares solution of $A\mathbf{x} = \mathbf{y}$ must be in $\mathbf{R}^m$.

24. The linear system $A\mathbf{x} = \mathbf{y}$ has a unique least squares solution if the columns of A are linearly independent.

25. A least squares solution to $A\mathbf{x} = \mathbf{y}$ is a vector $\hat{\mathbf{x}}$ such that $A\hat{\mathbf{x}}$ is as close as possible to $\mathbf{y}$.

26. The system of normal equations for the linear system $A\mathbf{x} = \mathbf{y}$ has solutions if and only if $A\mathbf{x} = \mathbf{y}$ has solutions.

27. If $\hat{\mathbf{x}}_1$ and $\hat{\mathbf{x}}_2$ are least squares solutions of $A\mathbf{x} = \mathbf{y}$, then so is $\hat{\mathbf{x}}_1 + \hat{\mathbf{x}}_2$.

28. A linear system must be inconsistent in order for there to be infinitely many least squares solutions.

29. Prove that if the matrix A has orthogonal columns, then $A\mathbf{x} = \mathbf{y}$ has a unique least squares solution.

30. Suppose that A is a nonzero matrix and $S = \text{col}(A)$. Prove that if $A\mathbf{x} = \mathbf{y}$ has a solution, then $\mathbf{y} = \text{proj}_S \mathbf{y}$.

31. Prove that if A is an orthogonal matrix, then any least squares solution of $A\mathbf{x} = \mathbf{y}$ is a linear combination of the rows of A.

32. Prove that if $\hat{\mathbf{x}}$ is a least squares solution of $A\mathbf{x} = \mathbf{y}$ and $\mathbf{x}_0$ is in null($A^T A$), then $\hat{\mathbf{x}} + \mathbf{x}_0$ is also a least squares solution of $A\mathbf{x} = \mathbf{y}$.

33. For a matrix A, prove that $A^T A$ is invertible if and only if A has linearly independent columns. (HINT: See Exercise 37 of Section 8.4.)

34. Prove that if A has orthonormal columns, then $\hat{\mathbf{x}} = A^T\mathbf{y}$ is the unique least squares solution to $A\mathbf{x} = \mathbf{y}$.

Ⓒ In Exercises 35–38, find the equation for the line that best fits the given data.

35. $(-2, 1.3)$, $(0, 1.8)$, $(1, 3)$

36. $(-3, -1.6)$, $(-1, -1.9)$, $(1, -2.5)$

37. $(-2, 2.0)$, $(-1, 1.7)$, $(1, 2.6)$, $(3, 2.1)$

38. (1, 3.1), (2, 2.6), (4, 1.9), (5, 2.1)

Ⓒ In Exercises 39–42, find the equation for the parabola that best fits the given data.

39. (−2, 3), (−1, 2), (1, 2.1), (2, 3.4)

40. (−3, −1), (−1, 2), (1, 4), (2, 1)

41. (−2, 0), (−1, 1.5), (0, 2.5), (1, 1.3), (2, −0.2)

42. (0, 4), (1, 3), (2, 1), (3, 2), (4, 5)

Ⓒ In Exercises 43–46, find constants a and b so that the model $y = ae^{bx}$ best fits the given data.

43. (−1, 0.3), (0, 1.3), (1, 3.1), (2, 5.7)

44. (1, 2.1), (3, 3.2), (4, 3.9), (6, 5.8), (9, 10.8)

45. (2, 11.3), (4, 8.2), (5, 7.1), (7, 5.3), (10, 3.2)

46. (−1, 5.1), (0, 1.9), (1, 0.8), (2, 0.3)

Ⓒ In Exercises 47–50, find constants a and b so that the model $y = ax^b$ best fits the given data.

47. (2, 5.4), (4, 13.5), (5, 17.6), (7, 26.0), (9, 40.2)

48. (1, 4.0), (3, 25.2), (4, 42.1), (6, 78.3), (8, 130.4)

49. (2, 26.1), (3, 21.7), (5, 15.8), (7, 12.7), (10, 11.2)

50. (3, 24.1), (4, 18.2), (6, 15.1), (8, 11.9), (9, 10.9)

51. Ⓒ Apply least squares regression to the data for the planets Venus, Earth, and Mars to develop a model to predict orbital period from distance.

52. Ⓒ Apply least squares regression to the data for the planets Mercury, Earth, Jupiter, and Uranus to develop a model to predict orbital period from distance.

53. Ⓒ On January 10, 2010, Nasr Al Niyadi and Omar Al Hegelan parachuted off a platform suspended from a crane attached to the 160th floor of the Burj Khalifa, the tallest building in the world. Suppose that an anvil was dropped from the same platform, located 2205 feet above the ground. Measurements of the elevation of the anvil t seconds after release are given in the table below.

t	Elevation
1	2185 ft
2	2140 ft
3	2055 ft
4	1943 ft

Find the quadratic polynomial that best fits the data, and use it to predict how long it will take for the anvil to hit the ground.

54. Ⓒ Warren invests $100,000 into a fund that combines stocks and bonds. The return varies from year to year. The balance at 5-year intervals is given in the table below.

t	Balance (× 1000)
5	142
10	230
15	314
20	483

Find constants a and b such that the model $y = ae^{bt}$ best fits the data. Use your model to predict when the investment fund balance will reach 1 million dollars.

55. Ⓒ The isotope Polonium-218 is unstable and subject to rapid radioactive decay. The quantity of a sample is measured at various times, with the results in the table below.

t (min)	Mass (g)
2	1.50
4	0.97
6	0.57
8	0.41

Find constants a and b such that the model $y = ae^{bt}$ best fits the data. Use your model to predict the initial size of the sample and the amount that will be present at $t = 15$.

56. Ⓒ Measurements of CO_2 in the atmosphere have been taken regularly over the last 50 years at the Mauna Loa Observatory in Hawaii. In addition to a general upward trend, the CO_2 data also has an annual cyclic behavior. The table below has the monthly measurements (in parts per million) for 2009.

Month	CO_2	Month	CO_2
1	386.92	7	387.74
2	387.41	8	385.91
3	388.77	9	384.77
4	389.46	10	384.38
5	390.18	11	385.99
6	389.43	12	387.27

Find constants a, b, and c such that the model $y = a + bt + c \sin(t\pi/6)$ best fits the data, where t is time in months. Use your model to predict the CO_2 level in January 2020.

Linear Transformations

Opened in 1930, the Old Vicksburg Bridge crosses the mighty Mississippi. Though the bridge is now closed to vehicular traffic, since replaced by a new bridge, it is still used by the railways. The Old Vicksburg Bridge employs a cantilever truss design. Trusses usually apply lower cord tension and upper cord compression to maintain the structural integrity of the bridge, but a cantilever truss is reversed over a portion of the span. Most cantilever truss bridges utilize a *balanced cantilever*, allowing construction to move from a central vertical spar in each direction toward the piers.

Bridge suggested by Seth Oppenheimer, Mississippi State University (Jeff Greenberg/age fotostock)

In Chapter 3 we defined and studied the properties of linear transformations in Euclidean space $\mathbf{R}^n$. In Chapter 7 we developed the concept of a vector space. In this chapter we combine these by extending the definition of linear transformations to vector spaces.

Section 9.1 introduces the definition and basic properties of a linear transformation in the context of a vector space. Most of the definitions, such as one-to-one, onto, kernel, and range, carry over almost word-for-word from $\mathbf{R}^n$. Section 9.2 focuses on a special type of linear transformation called an isomorphism and develops methods for determining when two different vector spaces have the same essential structure. In Section 9.3 we establish matrix representations for linear transformations that are similar to those for linear transformations in Euclidean space. Section 9.4 considers similar matrices, which is a way to group matrices based on their relationship to a linear transformation.

9.1 Definition and Properties

In Section 3.1 we gave the definition for a linear transformation $T : \mathbf{R}^m \to \mathbf{R}^n$. Recall that such a function preserves linear combinations by satisfying

$$T(c_1\mathbf{v}_1 + c_2\mathbf{v}_2) = c_1 T(\mathbf{v}_1) + c_2 T(\mathbf{v}_2)$$

In Chapter 7 we defined the vector space, an extension of Euclidean space that allows vectors to be polynomials, matrices, continuous functions, and other types of mathematical objects.

We start this section by extending the definition of linear transformation to allow domains and codomains that are vector spaces.

DEFINITION 9.1

Definition Linear Transformation

> Let V and W be vector spaces. Then $T : V \to W$ is a **linear transformation** if for all $\mathbf{v}_1$ and $\mathbf{v}_2$ in V and all real scalars c, the function T satisfies
>
> (a) $T(\mathbf{v}_1 + \mathbf{v}_2) = T(\mathbf{v}_1) + T(\mathbf{v}_2)$
>
> (b) $T(c\mathbf{v}_1) = c\,T(\mathbf{v}_1)$

▶ Definition 9.1 generalizes Definition 3.1 in Section 3.1.

Definition Domain, Codomain

For a linear transformation $T : V \to W$, the vector space V is the **domain** and the vector space W is the **codomain**.

EXAMPLE 1 Let $T : \mathbf{R}^2 \to \mathbf{P}^2$ be given by

$$T\left(\begin{bmatrix} a_1 \\ a_2 \end{bmatrix}\right) = a_1 x^2 + (a_1 - a_2)x - a_2.$$

▶ Recall that $\mathbf{P}^n$ denotes the vector space of polynomials with real coefficients that have degree n or less.

Show that T is a linear transformation.

Solution We need to verify that the two conditions given in Definition 9.1 are satisfied. Suppose that $\mathbf{a} = \begin{bmatrix} a_1 \\ a_2 \end{bmatrix}$ and $\mathbf{b} = \begin{bmatrix} b_1 \\ b_2 \end{bmatrix}$. Starting with condition (a), we have

$$\begin{aligned}
T(\mathbf{a} + \mathbf{b}) &= T\left(\begin{bmatrix} a_1 \\ a_2 \end{bmatrix} + \begin{bmatrix} b_1 \\ b_2 \end{bmatrix}\right) \\
&= T\left(\begin{bmatrix} a_1 + b_1 \\ a_2 + b_2 \end{bmatrix}\right) \\
&= (a_1 + b_1)x^2 + \big((a_1 + b_1) - (a_2 + b_2)\big)x - (a_2 + b_2) \\
&= \big(a_1 x^2 + (a_1 - a_2)x - a_2\big) + \big(b_1 x^2 + (b_1 - b_2)x - b_2\big) \\
&= T\left(\begin{bmatrix} a_1 \\ a_2 \end{bmatrix}\right) + T\left(\begin{bmatrix} b_1 \\ b_2 \end{bmatrix}\right) = T(\mathbf{a}) + T(\mathbf{b})
\end{aligned}$$

Thus condition (a) of Definition 9.1 is satisfied. For condition (b), let c be a real scalar. Then we have

$$\begin{aligned}
T(c\mathbf{a}) &= T\left(c\begin{bmatrix} a_1 \\ a_2 \end{bmatrix}\right) = T\left(\begin{bmatrix} ca_1 \\ ca_2 \end{bmatrix}\right) \\
&= ca_1 x^2 + (ca_1 - ca_2)x - ca_2 \\
&= c\big(a_1 x^2 + (a_1 - a_2)x - a_2\big) = cT\left(\begin{bmatrix} a_1 \\ a_2 \end{bmatrix}\right) = cT(\mathbf{a})
\end{aligned}$$

▶ On the left of (1) we have the transformation of a linear combination of two vectors, and on the right we have the same linear combination of the images of the vectors. Hence linear transformations preserve linear combinations.

Hence condition (b) of Definition 9.1 is also satisfied, so T is a linear transformation. ■

Before moving on to the next example, we report the following useful theorem that wraps the two conditions of Definition 9.1 into a single package.

THEOREM 9.2

> $T : V \to W$ is a linear transformation if and only if
>
> $$T(c_1\mathbf{v}_1 + c_2\mathbf{v}_2) = c_1\,T(\mathbf{v}_1) + c_2\,T(\mathbf{v}_2) \tag{1}$$
>
> for all vectors $\mathbf{v}_1$ and $\mathbf{v}_2$ in V and real scalars c_1 and c_2.

The proof follows from applications of Definition 9.1 and is left to Exercise 58. Some consequences of Theorem 9.2:

- The expression (1) can be extended to

$$T(c_1 \mathbf{v}_1 + c_2 \mathbf{v}_2 + \cdots + c_m \mathbf{v}_m) = c_1 T(\mathbf{v}_1) + c_2 T(\mathbf{v}_2) + \cdots + c_m T(\mathbf{v}_m)$$

where $\mathbf{v}_1, \ldots, \mathbf{v}_m$ are in V and $c_1, \ldots, c_m$ are real scalars (see Exercise 54).

- If $\mathbf{0}_V$ denotes the zero vector in V and $\mathbf{0}_W$ the zero vector in W, then $T(\mathbf{0}_V) = \mathbf{0}_W$ (see Exercise 55).

- For any $\mathbf{v}$ in V, we have $T(-\mathbf{v}) = -T(\mathbf{v})$ (see Exercise 56).

EXAMPLE 2   Let $T : \mathbf{P}^2 \to \mathbf{P}^4$ be given by

$$T\big(p(x)\big) = x^2 \, p(x)$$

Show that T is a linear transformation.

Solution   Here we apply Theorem 9.2, so that only one condition needs to be verified.

$$
\begin{aligned}
T\big(c_1 \, p_1(x) + c_2 \, p_2(x)\big) &= x^2 \big(c_1 \, p_1(x) + c_2 \, p_2(x)\big) \\
&= c_1 x^2 \, p_1(x) + c_2 x^2 \, p_2(x) \\
&= c_1 T\big(p_1(x)\big) + c_2 T\big(p_2(x)\big)
\end{aligned}
$$

Therefore Theorem 9.2 is satisfied, so T is a linear transformation.   ■

EXAMPLE 3   Suppose that $T : C[0, 1] \to C[0, 1]$ is defined by

$$T\big(f(x)\big) = \big(f(x)\big)^2$$

Prove that T is *not* a linear transformation.

Solution   All we need to do is show that Definition 9.1 fails to hold for at least one scalar or vector. Given a scalar c and continuous function f, we have

$$T\big(cf(x)\big) = \big(cf(x)\big)^2 = c^2 \big(f(x)\big)^2 = c^2 T\big(f(x)\big)$$

Thus if $c = 2$, then

$$T(2 f(x)) = 4 T(f(x)) \neq 2 T(f(x))$$

Condition (b) of Definition 9.1 is violated, so T is not a linear transformation.   ■

Image, Range, and Kernel

The definitions and notation for image, range, and kernel carry over essentially unchanged from Euclidean space.

Definition Image

- If $\mathbf{v}$ is a vector in V, then $T(\mathbf{v})$ is the **image of v under** T.

- If S is a subspace of V, then $T(S)$ denotes the subset of W consisting of all images of elements of S.

Definition Range

- The **range** of T is denoted range(T) and is the subset of W consisting of all images of elements of V (also can be written $T(V)$).

Definition Kernel

- The **kernel** of T is denoted ker(T) and is the set of all elements $\mathbf{v}$ in V such that $T(\mathbf{v}) = \mathbf{0}_W$.

Note that ker(T) is a subset of V, while range(T) is a subset of W.

Recall that a subspace S is a subset closed under linear combinations. Since linear transformations preserve linear combinations, it makes sense that the image $T(S)$ is also a subspace, as shown in the next theorem.

THEOREM 9.3

▶ Theorem 9.3 generalizes Theorem 4.5 in Section 4.1.

Let $T : V \to W$ be a linear transformation. Then $\ker(T)$ is a subspace of V. If S is a subspace of V, then $T(S)$ is a subspace of W.

Proof We leave the proof that $\ker(T)$ is a subspace of V as Exercise 57 and prove that $T(S)$ is a subspace of W here. Recall that we must verify the three conditions required of a subspace.

(a) As $T(\mathbf{0}_V) = \mathbf{0}_W$ and $\mathbf{0}_V$ must be in S, it follows that $\mathbf{0}_W$ is in $T(S)$.

(b) Suppose that $\mathbf{w}_1$ and $\mathbf{w}_2$ are both in $T(S)$. Then there exist vectors $\mathbf{v}_1$ and $\mathbf{v}_2$ in S such that $T(\mathbf{v}_1) = \mathbf{w}_1$ and $T(\mathbf{v}_2) = \mathbf{w}_2$. Since $\mathbf{v}_1 + \mathbf{v}_2$ must be in S and

$$T(\mathbf{v}_1 + \mathbf{v}_2) = T(\mathbf{v}_1) + T(\mathbf{v}_2) = \mathbf{w}_1 + \mathbf{w}_2$$

it follows that $\mathbf{w}_1 + \mathbf{w}_2$ is also in $T(S)$. Thus S is closed under addition.

(c) Suppose that $\mathbf{w}$ is in $T(S)$ and that c is a scalar. Then there exists a vector $\mathbf{v}$ in S such that $T(\mathbf{v}) = \mathbf{w}$. As $c\mathbf{v}$ is also in S and

$$T(c\mathbf{v}) = c\,T(\mathbf{v}) = c\mathbf{w}$$

we have $c\mathbf{w}$ in $T(S)$. Hence $T(S)$ is also closed under scalar multiplication.

Since (a), (b), and (c) are all satisfied, $T(S)$ is a subspace of W. ∎

EXAMPLE 4 Let $T : \mathbf{R}^{2 \times 2} \to \mathbf{R}$ be given by

$$T(A) = \text{tr}(A)$$

where $\text{tr}(A)$ denotes the trace of A. Determine $\text{range}(T)$ and $\ker(T)$.

Solution Recall that $\text{tr}(A)$ is the sum of the diagonal elements of a square matrix A, so that

$$\text{tr}\left(\begin{bmatrix} a_{11} & a_{12} \\ a_{21} & a_{22} \end{bmatrix} \right) = a_{11} + a_{22}$$

Then T is a linear transformation. (Verifying this is left as Exercise 9.) Note that for any real number r, we have

$$\text{tr}\left(\begin{bmatrix} r & 0 \\ 0 & 0 \end{bmatrix} \right) = r$$

Therefore every real number is the image of an element of $\mathbf{R}^{2 \times 2}$, so we may conclude that $\text{range}(T) = \mathbf{R}$.

Next, $A = \begin{bmatrix} a_{11} & a_{12} \\ a_{21} & a_{22} \end{bmatrix}$ is in $\ker(T)$ if

$$\text{tr}(A) = a_{11} + a_{22} = 0$$

or $a_{22} = -a_{11}$. Therefore $\ker(T)$ is the set of all 2×2 real matrices of the form

$$\begin{bmatrix} a_{11} & a_{12} \\ a_{21} & -a_{11} \end{bmatrix}$$

Since this matrix can be written

$$\begin{bmatrix} a_{11} & a_{12} \\ a_{21} & -a_{11} \end{bmatrix} = \begin{bmatrix} a_{11} & 0 \\ 0 & -a_{11} \end{bmatrix} + \begin{bmatrix} 0 & a_{12} \\ 0 & 0 \end{bmatrix} + \begin{bmatrix} 0 & 0 \\ a_{21} & 0 \end{bmatrix}$$

a basis for ker(T) (which we already know is a subspace) is given by

$$\left\{ \begin{bmatrix} 1 & 0 \\ 0 & -1 \end{bmatrix}, \begin{bmatrix} 0 & 1 \\ 0 & 0 \end{bmatrix}, \begin{bmatrix} 0 & 0 \\ 1 & 0 \end{bmatrix} \right\} \quad \blacksquare$$

One-to-One and Onto Linear Transformations

The definitions of one-to-one and onto for linear transformations carry over almost word-for-word from Chapter 3.

DEFINITION 9.4

Definition One-to-One, Onto

▶ Definition 9.4 generalizes Definition 3.4 in Section 3.1.

▶ Theorem 9.5 generalizes Theorem 3.5 in Section 3.1.

Let $T : V \to W$ be a linear transformation. Then

(a) T is **one-to-one** if for each $\mathbf{w}$ in W there is *at most* one $\mathbf{v}$ in V such that $T(\mathbf{v}) = \mathbf{w}$.

(b) T is **onto** if for each $\mathbf{w}$ in W there is *at least* one $\mathbf{v}$ in V such that $T(\mathbf{v}) = \mathbf{w}$.

One way to determine if a linear transformation is one-to-one is to find ker(T) and then apply the next theorem.

THEOREM 9.5

Let $T : V \to W$ be a linear transformation. Then T is one-to-one if and only if ker(T) = $\{\mathbf{0}_V\}$.

The proof is similar to that of Theorem 3.5 in Section 3.1 and is left as Exercise 59.

EXAMPLE 5   Let $T : \mathbf{R}^{2 \times 2} \to \mathbf{P}^2$ be given by

$$T\left(\begin{bmatrix} a & b \\ c & d \end{bmatrix} \right) = (a - d)x^2 - bx + c$$

Then T is a linear transformation (see Exercise 8). Determine if T is onto or one-to-one.

Solution A typical element of $\mathbf{P}^2$ has the form $h(x) = ex^2 + fx + g$. Thus, for T to be onto, we need to be able to find a solution to

$$T\left(\begin{bmatrix} a & b \\ c & d \end{bmatrix} \right) = (a - d)x^2 - bx + c = ex^2 + fx + g$$

Comparing coefficients yields the linear system

$$a - d = e$$
$$-b = f$$
$$c = g$$

This system has infinitely many solutions, among them

$$a = e, \quad b = -f, \quad c = g, \quad d = 0$$

Hence T is onto. Moreover, for the special case where $e = f = g = 0$, a solution to the system is $a = d = 1$, $b = c = 0$. Therefore

$$T\left(\begin{bmatrix} 1 & 0 \\ 0 & 1 \end{bmatrix} \right) = 0$$

so that ker(T) is nontrivial. Hence T is not one-to-one by Theorem 9.5. $\blacksquare$

The next theorem shows how a linear transformation $T : V \to W$ can relate linearly independent sets in V and W.

THEOREM 9.6

Let $T : V \to W$ be a linear transformation. Suppose that $\mathcal{V} = \{\mathbf{v}_1, \ldots, \mathbf{v}_m\}$ is a subset of V, $\mathcal{W} = \{\mathbf{w}_1, \ldots, \mathbf{w}_m\}$ is a subset of W, and $T(\mathbf{v}_i) = \mathbf{w}_i$ for $i = 1, \ldots, m$. If $\mathcal{W}$ is linearly independent, then so is $\mathcal{V}$.

Proof Suppose that

$$c_1 \mathbf{v}_1 + \cdots + c_m \mathbf{v}_m = \mathbf{0}_V \tag{2}$$

Since

$$T(c_1 \mathbf{v}_1 + \cdots + c_m \mathbf{v}_m) = c_1 T(\mathbf{v}_1) + \cdots + c_m T(\mathbf{v}_m) = c_1 \mathbf{w}_1 + \cdots + c_m \mathbf{w}_m$$

and $T(\mathbf{0}_V) = \mathbf{0}_W$, we have

$$c_1 \mathbf{w}_1 + \cdots + c_m \mathbf{w}_m = \mathbf{0}_W$$

As $\mathcal{W}$ is a linearly independent set, it must be that $c_1 = \cdots = c_m = 0$, so that (2) has only the trivial solution. Hence $\mathcal{V}$ is also linearly independent. ∎

Note that the reverse is not always true. Just because $\mathcal{V}$ is linearly independent does *not* guarantee that $\mathcal{W}$ is linearly independent (see Exercise 37). However, if T is one-to-one and $\mathcal{V}$ is a linearly independent set, then that is enough to ensure that $\mathcal{W}$ is also linearly independent (see Exercise 61).

EXAMPLE 6 Let $T : C[0, 1] \to \mathbf{R}^2$ be defined by

$$T(f) = \begin{bmatrix} f(0) \\ f(1) \end{bmatrix}$$

Use T to prove that the set $\{\cos(x\pi/2), \sin(x\pi/2)\}$ is linearly independent.

Solution It is shown in Exercise 6 that T is a linear transformation. Next, note that

$$T\big(\cos(x\pi/2)\big) = \begin{bmatrix} \cos(0) \\ \cos(\pi/2) \end{bmatrix} = \begin{bmatrix} 1 \\ 0 \end{bmatrix}$$

$$T\big(\sin(x\pi/2)\big) = \begin{bmatrix} \sin(0) \\ \sin(\pi/2) \end{bmatrix} = \begin{bmatrix} 0 \\ 1 \end{bmatrix}$$

Since the set $\left\{ \begin{bmatrix} 1 \\ 0 \end{bmatrix}, \begin{bmatrix} 0 \\ 1 \end{bmatrix} \right\}$ is linearly independent, then by Theorem 9.6 the set $\{\cos(x\pi/2), \sin(x\pi/2)\}$ is also linearly independent. ∎

Our next theorem relates the dimensions of V, $\ker(T)$, and $\mathrm{range}(T)$ for a linear transformation $T : V \to W$.

THEOREM 9.7

▶ Theorem 9.7 generalizes Theorem 4.23 in Section 4.3.

Let $T : V \to W$ be a linear transformation, with V and W finite dimensional. Then

$$\dim(V) = \dim\big(\ker(T)\big) + \dim\big(\mathrm{range}(T)\big) \tag{3}$$

Proof We start by letting $\{\mathbf{v}_1, \ldots, \mathbf{v}_k\}$ be a basis for $\ker(T)$, so that $\dim\big(\ker(T)\big) = k$. Now extend this set to a basis $\{\mathbf{v}_1, \ldots, \mathbf{v}_k, \mathbf{v}_{k+1}, \ldots, \mathbf{v}_m\}$ for V. Hence $\dim(V) = m$. For $i = k + 1, \ldots, m$, let $\mathbf{w}_i = T(\mathbf{v}_i)$. Our goal is to show that $\{\mathbf{w}_{k+1}, \ldots, \mathbf{w}_m\}$ is a basis for $\mathrm{range}(T)$.

Given real scalars $c_1, \ldots, c_m$, we have

$$T(c_1\mathbf{v}_1 + \cdots + c_k\mathbf{v}_k + c_{k+1}\mathbf{v}_{k+1} + \cdots + c_m\mathbf{v}_m)$$
$$= c_1 T(\mathbf{v}_1) + \cdots + c_k T(\mathbf{v}_k) + c_{k+1} T(\mathbf{v}_{k+1}) + \cdots + c_m T(\mathbf{v}_m)$$
$$= c_1\mathbf{0}_W + \cdots + c_k\mathbf{0}_W + c_{k+1}\mathbf{w}_{k+1} + \cdots + c_m\mathbf{w}_m$$
$$= c_{k+1}\mathbf{w}_{k+1} + \cdots + c_m\mathbf{w}_m$$

▶ The cases $\dim(V) = 0$ and $\dim(\ker(T)) = 0$ are left to the reader.

As range(T) is the set of all such elements, it follows that $\{\mathbf{w}_{k+1}, \ldots, \mathbf{w}_m\}$ spans range(T). Moreover, if

$$c_{k+1}\mathbf{w}_{k+1} + \cdots + c_m\mathbf{w}_m = \mathbf{0}_W$$

then $c_{k+1}\mathbf{v}_{k+1} + \cdots + c_m\mathbf{v}_m$ is in ker(T). Since $\mathbf{v}_{k+1}, \ldots, \mathbf{v}_m$ are linearly independent and none are in ker(T), this implies $c_{k+1} = \cdots = c_m = 0$. Hence $\{\mathbf{w}_{k+1}, \ldots, \mathbf{w}_m\}$ is also linearly independent and thus is a basis for range(T). Therefore $\dim(\text{range}(T)) = m - k$, and so (3) is true. ■

For example, recall $T : \mathbf{R}^{2\times2} \to \mathbf{R}$ given by $T(A) = \text{tr}(A)$ in Example 4. There, it is shown that a basis for ker(T) is

$$\left\{ \begin{bmatrix} 1 & 0 \\ 0 & -1 \end{bmatrix}, \begin{bmatrix} 0 & 1 \\ 0 & 0 \end{bmatrix}, \begin{bmatrix} 0 & 0 \\ 1 & 0 \end{bmatrix} \right\}$$

so that $\dim(\ker(T)) = 3$. Since range(T) = $\mathbf{R}$, we have $\dim(\text{range}(T)) = 1$. We also have $\dim(\mathbf{R}^{2\times2}) = 4$, which is exactly as predicted by Theorem 9.7.

Another application of Theorem 9.7 is given in the next theorem.

THEOREM 9.8

Let $T : V \to W$ be a linear transformation, with V and W finite dimensional.

(a) If T is onto, then $\dim(V) \geq \dim(W)$.

(b) If T is one-to-one, then $\dim(V) \leq \dim(W)$.

Proof If T is onto, then range(T) = W. Therefore $\dim(\text{range}(T)) = \dim(W)$, so that by Theorem 9.7,

$$\dim(V) = \dim(\ker(T)) + \dim(W)$$

Since $\dim(\ker(T)) \geq 0$, it follows that $\dim(V) \geq \dim(W)$, so (a) is true.

For part (b), if T is one-to-one, then by Theorem 9.5 we have ker(T) = $\{\mathbf{0}_V\}$. Hence $\dim(\ker(T)) = 0$, and so by Theorem 9.7,

$$\dim(V) = \dim(\text{range}(T)) \leq \dim(W)$$

because range(T) is a subset of W. Hence (b) is also true. ■

| EXERCISES |

1. Let $T : V \to \mathbf{R}^2$ be a linear transformation satisfying

$$T(\mathbf{v}_1) = \begin{bmatrix} 1 \\ 2 \end{bmatrix}, \quad T(\mathbf{v}_2) = \begin{bmatrix} -3 \\ 1 \end{bmatrix}$$

Find $T(\mathbf{v}_2 - 2\mathbf{v}_1)$.

2. Let $T : V \to \mathbf{P}^2$ be a linear transformation satisfying

$$T(\mathbf{v}_1) = x^2 + 1$$
$$T(\mathbf{v}_2) = -x^2 + 3x$$
$$T(\mathbf{v}_3) = 4x - 2$$

Find $T(2\mathbf{v}_1 + \mathbf{v}_2 - 3\mathbf{v}_3)$.

3. Let $T : \mathbf{P}^2 \to \mathbf{R}^2$ be a linear transformation satisfying

$$T(x^2 + 1) = \begin{bmatrix} -1 \\ 3 \end{bmatrix}, \quad T(4x + 3) = \begin{bmatrix} 2 \\ 1 \end{bmatrix}$$

Find $T(2x^2 - 4x - 1)$. (HINT: Write $2x^2 - 4x - 1$ as a linear combination of $x^2 + 1$ and $4x + 3$.)

4. Let $T : \mathbf{R}^{2\times 2} \to \mathbf{P}^2$ be a linear transformation satisfying

$$T\left(\begin{bmatrix} 1 & -2 \\ -1 & 3 \end{bmatrix}\right) = x^2 - x + 3$$

$$T\left(\begin{bmatrix} 2 & 4 \\ 1 & -1 \end{bmatrix}\right) = 3x^2 + 4x - 1$$

Find $T\left(\begin{bmatrix} 7 & 2 \\ -1 & 7 \end{bmatrix}\right)$. (HINT: Write $\begin{bmatrix} 7 & 2 \\ -1 & 7 \end{bmatrix}$ as a linear combination of $\begin{bmatrix} 1 & -2 \\ -1 & 3 \end{bmatrix}$ and $\begin{bmatrix} 2 & 4 \\ 1 & -1 \end{bmatrix}$.)

For Exercises 5–10, prove that the given function is a linear transformation.

5. $T : \mathbf{P}^2 \to \mathbf{P}^2$ with

$$T(ax^2 + bx + c) = cx^2 + bx + a$$

6. $T : C[0, 1] \to \mathbf{R}^2$ with $T(f) = \begin{bmatrix} f(0) \\ f(1) \end{bmatrix}$

7. $T : \mathbf{P}^n \to C[0, 1]$ with $T(p(x)) = e^x p(x)$

8. $T : \mathbf{R}^{2\times 2} \to \mathbf{P}^2$ with

$$T\left(\begin{bmatrix} a & b \\ c & d \end{bmatrix}\right) = (a - d)x^2 - bx + c$$

9. $T : \mathbf{R}^{2\times 2} \to \mathbf{R}$ with $T(A) = \operatorname{tr}(A)$ (the trace of A).

10. $T : V \to W$ with $T(\mathbf{v}) = \mathbf{0}_W$ for all $\mathbf{v}$.

For Exercises 11–22, determine if the given function is a linear transformation. Be sure to completely justify your answer.

11. $T : \mathbf{R}^{n'} \to \mathbf{R}^n$ with $T(\mathbf{v}) = -4\mathbf{v}$

12. $T : \mathbf{R}^2 \to \mathbf{R}^2$ with

$$T\left(\begin{bmatrix} a \\ b \end{bmatrix}\right) = -\begin{bmatrix} a \\ b \end{bmatrix} + \begin{bmatrix} 3 \\ 2 \end{bmatrix}$$

13. $T : \mathbf{R}^2 \to \mathbf{R}^2$ with $T\left(\begin{bmatrix} a \\ b \end{bmatrix}\right) = \begin{bmatrix} b \\ a \end{bmatrix}$

14. $T : \mathbf{R}^n \to \mathbf{R}^n$ with $T(\mathbf{v}) = \begin{bmatrix} 0 \\ \vdots \\ 0 \end{bmatrix}$

15. $T : \mathbf{P}^2 \to \mathbf{R}^2$ with $T(ax^2 + bx + c) = \begin{bmatrix} a - b \\ b + c \end{bmatrix}$

16. $T : \mathbf{R}^{2\times 2} \to \mathbf{R}$ with $T(A) = \det(A)$

17. $T : \mathbf{R}^{n\times n} \to \mathbf{R}^{n\times n}$ with $T(A) = A^T$

18. $T : C[0, 1] \to \mathbf{R}$ with $T(f) = f(2)$

19. $T : \mathbf{R}^{n\times n} \to \mathbf{R}$ with $T(A) = \operatorname{tr}(A)$

20. $T : C[0, 1] \to C[0, 1]$ with $T(f) = x + f(x)$

21. $T : \mathbf{R}^{3\times 2} \to \mathbf{R}^{2\times 2}$ with $T(A) = A^T A$

22. $T : \mathbf{R}^2 \to C[0, 1]$ with $T\left(\begin{bmatrix} a \\ b \end{bmatrix}\right) = ae^{bx}$

In Exercises 23–26, describe the kernel and range of the given linear transformation.

23. $T : \mathbf{P}^1 \to \mathbf{R}$ with $T(ax + b) = a - b$

24. $T : \mathbf{P}^1 \to \mathbf{P}^2$ with $T(f) = xf(x)$

25. $T : \mathbf{P}^2 \to \mathbf{R}^{2\times 2}$ with

$$T(ax^2 + bx + c) = \begin{bmatrix} a & b \\ b & c \end{bmatrix}$$

26. $T : C[0, 1] \to \mathbf{P}^2$ with

$$T(f) = f(0)x^2 + f(1)$$

In Exercises 27–30, determine if the given linear transformation is onto and/or one-to-one.

27. $T : \mathbf{P}^2 \to \mathbf{R}^2$ with $T(f) = \begin{bmatrix} f(1) \\ f(2) \end{bmatrix}$

28. $T : \mathbf{P}^2 \to \mathbf{P}^3$ with $T(f) = xf(x)$

29. $T : C[0, 1] \to \mathbf{R}$ with $T(f) = f(1)$

30. $T : \mathbf{R}^2 \to \mathbf{P}^2$ with

$$T\left(\begin{bmatrix} a \\ b \end{bmatrix}\right) = ax^2 + (b - a)x + (a - b)$$

FIND AN EXAMPLE For Exercises 31–38, find an example of vector spaces V and W and a function $T : V \to W$ that meets the given specifications.

31. T is a linear transformation that is one-to-one but not onto.

32. T is a linear transformation that is onto but not one-to-one.

33. T is a linear transformation that is neither onto nor one-to-one.

34. T is a linear transformation that is both onto and one-to-one.

35. T is a linear transformation such that $\dim\big(\ker(T)\big) = 1$ and $\dim\big(\operatorname{range}(T)\big) = 3$.

36. T is a linear transformation such that $\dim\big(\ker(T)\big) = 4$ and $\dim\big(\operatorname{range}(T)\big) = 2$.

37. T is a linear transformation such that, for any set of linearly independent vectors $\{\mathbf{v}_1, \ldots, \mathbf{v}_k\}$ the set $\{T(\mathbf{v}_1), \ldots, T(\mathbf{v}_k)\}$ is linearly dependent.

38. T satisfies condition (b) but not condition (a) of Definition 9.1.

TRUE OR FALSE For Exercises 39–48, determine if the statement is true or false, and justify your answer.

39. If $T : V \to W$ is a linear transformation, then $T(\mathbf{v}_1 - \mathbf{v}_2) = T(\mathbf{v}_1) - T(\mathbf{v}_2)$.

40. If $T : V \to W$ is a linear transformation, then $T(\mathbf{v}) = \mathbf{0}_W$ implies that $\mathbf{v} = \mathbf{0}_V$.

41. If $T : V \to W$ is a linear transformation and S is a subspace of V, then $T(S)$ is a subspace of W.

42. If $T : V \to W$ is a linear transformation, then $\dim\big(\ker(T)\big) \leq \dim\big(\operatorname{range}(T)\big)$.

43. If $T : V \to W$ is a linear transformation and $\{\mathbf{v}_1, \dots, \mathbf{v}_k\}$ is a linearly independent set, then so is $\{T(\mathbf{v}_1), \dots, T(\mathbf{v}_k)\}$.

44. If $T : V \to W$ is a linear transformation and $\{\mathbf{v}_1, \dots, \mathbf{v}_k\}$ is a linearly dependent set, then so is $\{T(\mathbf{v}_1), \dots, T(\mathbf{v}_k)\}$.

45. If $T : \mathbf{R}^{2 \times 2} \to \mathbf{P}^6$, then is it impossible for T to be onto.

46. If $T : \mathbf{P}^4 \to \mathbf{R}^6$, then it is impossible for T to be one-to-one.

47. Let $T : V \to W$ be a linear transformation and $\mathbf{w}$ a nonzero vector in W. Then the set of all $\mathbf{v}$ in V such that $T(\mathbf{v}) = \mathbf{w}$ forms a subspace.

48. For every pair of vector spaces V and W, it is always possible to define a linear transformation $T : V \to W$.

49. Let $T : \mathbf{R}^5 \to C[0, 1]$ be a linear transformation, and suppose that $\dim(\ker(T)) = 2$. What is $\dim(\text{range}(T))$?

50. Let $T : \mathbf{R}^{4 \times 3} \to \mathbf{P}$ be a one-to-one linear transformation. What is $\dim(\text{range}(T))$?

51. Prove that if $T : V \to W$ is a linear transformation with $T(\mathbf{v}_1) = T(\mathbf{v}_2)$, then $\mathbf{v}_1 - \mathbf{v}_2$ is in $\ker(T)$.

52. Prove that if $T : V \to W$ is an onto and one-to-one linear transformation, and both V and W are of finite dimension, then $\dim(V) = \dim(W)$.

53. Suppose that $T_1 : V \to W$ and $T_2 : V \to W$ are both linear transformations. Prove that $T_1 + T_2$ is also a linear transformation from V to W.

54. Prove the extended version of Theorem 9.2: If $T : V \to W$ is a linear transformation, then

$$T(c_1\mathbf{v}_1 + c_2\mathbf{v}_2 + \cdots + c_m\mathbf{v}_m)$$
$$= c_1 T(\mathbf{v}_1) + c_2 T(\mathbf{v}_2) + \cdots + c_m T(\mathbf{v}_m),$$

where $\mathbf{v}_1, \dots, \mathbf{v}_m$ are in V and $c_1, \dots, c_m$ are real scalars.

55. Prove that if $T : V \to W$ is a linear transformation, then $T(\mathbf{0}_V) = \mathbf{0}_W$.

56. Prove that if $T : V \to W$ is a linear transformation, then for any $\mathbf{v}$ in V we have $T(-\mathbf{v}) = -T(\mathbf{v})$.

57. Prove part of Theorem 9.3: Let $T : V \to W$. Then $\ker(T)$ is a subspace of V.

58. Prove Theorem 9.2: $T : V \to W$ is a linear transformation if and only if

$$T(c_1\mathbf{v}_1 + c_2\mathbf{v}_2) = c_1 T(\mathbf{v}_1) + c_2 T(\mathbf{v}_2)$$

for all vectors $\mathbf{v}_1$ and $\mathbf{v}_2$ in V and real scalars c_1 and c_2.

59. Prove Theorem 9.5: Let $T : V \to W$ be a linear transformation. Then T is one-to-one if and only if $\ker(T) = \{\mathbf{0}_V\}$. (HINT: Theorem 3.5 is similar.)

60. Suppose that $T_1 : V \to W$ and $T_2 : W \to Y$ are both linear transformations. Prove that $T_2\big(T_1(\mathbf{v})\big)$ is also a linear transformation from V to Y.

61. Prove a partial converse of Theorem 9.6: Let $T : V \to W$ be a one-to-one linear transformation, with $\mathcal{V} = \{\mathbf{v}_1, \dots, \mathbf{v}_m\}$ a subset of V, and $\mathcal{W} = \{\mathbf{w}_1, \dots, \mathbf{w}_m\}$ a subset of W. Suppose that $T(\mathbf{v}_i) = \mathbf{w}_i$ for $i = 1, \dots, m$. If $\mathcal{V}$ is linearly independent, then so is $\mathcal{W}$.

62. Prove Theorem 9.8, but with the condition that V and W are finite dimensional removed.

(Calculus required) For Exercises 63–68, some basic knowledge of calculus is required.

63. Let $C^1(a, b)$ denote the set of functions that are continuously differentiable on the interval (a, b). Prove that $T : C^1(a, b) \to C(a, b)$ given by

$$T(f) = f'(x)$$

is a linear transformation.

64. Let $T : C[a, b] \to \mathbf{R}$ be given by

$$T(f) = \int_a^b f(x)\, dx.$$

Prove that T is a linear transformation.

65. Determine if $T : \mathbf{P}^4 \to \mathbf{P}^2$ with $T(p) = p''(x)$ is a linear transformation.

66. Determine if $T : \mathbf{P}^3 \to \mathbf{R}$ with

$$T(p) = \int_a^b x p(x)\, dx$$

is a linear transformation.

67. Determine if $T : \mathbf{P}^4 \to \mathbf{P}^5$ with $T(p) = \big(x^2 p(x)\big)'$ is a linear transformation.

68. Determine if $T : \mathbf{P}^3 \to \mathbf{R}$ with

$$T(p) = \int_a^b e^x - x^2 p(x)\, dx$$

is a linear transformation.

9.2 Isomorphisms

At the beginning of Section 7.1, we compared the features of $\mathbf{R}^3$ and $\mathbf{P}^2$ and concluded that these two vector spaces have a lot in common. In fact, many superficially different vector spaces are essentially the same in important ways. In this section we define precisely what it means for two vector spaces to be essentially the same, and determine which vector spaces are essentially the same.

Definition of Isomorphism

Our mechanism for establishing when two vector spaces are essentially the same is through a special type of linear transformation called an *isomorphism*.

DEFINITION 9.9

Definition Isomorphism, Isomorphic

A linear transformation $T : V \rightarrow W$ is an **isomorphism** if T is both one-to-one and onto. If such an isomorphism exists, then we say that V and W are **isomorphic** vector spaces.

▶ The word *isomorphism* has Greek origins and means "same structure."

Regarding isomorphisms and isomorphic vector spaces:

- Some pairs of vector spaces are isomorphic, while others are not. For instance, $\mathbf{R}^3$ and $\mathbf{P}^2$ are isomorphic (see Example 1 below), while $\mathbf{R}^{2 \times 3}$ and $C[0, 1]$ are not. (We will explain why later.)

- If $T : V \rightarrow W$ is an isomorphism, then there will also exist an isomorphism $S : W \rightarrow V$. (This is developed in more detail later in this section.) Thus the notion of isomorphic is symmetric. If V and W are isomorphic, then W and V are also isomorphic.

- The requirement that $T : V \rightarrow W$ be both onto and one-to-one ensures that there is an exact correspondence between the elements of V and W, called a **one-to-one correspondence**: Every $\mathbf{v}$ in V is paired up with a specific $\mathbf{w}$ in W by $T(\mathbf{v}) = \mathbf{w}$. Matching up elements of V and W in this manner is one part of establishing that V and W are essentially the same.

- Since T is a linear transformation, addition and scalar multiplication work the same between corresponding elements of V and W. For instance, if $T(\mathbf{v}_i) = \mathbf{w}_i$ for $i = 1, 2, 3$ and $\mathbf{v}_1 + \mathbf{v}_2 = \mathbf{v}_3$, then

$$\mathbf{w}_1 + \mathbf{w}_2 = T(\mathbf{v}_1) + T(\mathbf{v}_2) = T(\mathbf{v}_1 + \mathbf{v}_2) = T(\mathbf{v}_3) = \mathbf{w}_3$$

That is, $\mathbf{v}_1 + \mathbf{v}_2 = \mathbf{v}_3$ implies that $\mathbf{w}_1 + \mathbf{w}_2 = \mathbf{w}_3$, so that addition works the same in V and W. This principle also holds for scalar multiplication.

Since an isomorphism between two vector spaces V and W matches up vectors and preserves the respective arithmetic operations, from the standpoint of vector spaces V and W are essentially the same.

Let's consider some examples.

EXAMPLE 1 Show that $\mathbf{R}^3$ and $\mathbf{P}^2$ are isomorphic.

▶ To establish that V and W are isomorphic, we need only find one isomorphism $T : V \rightarrow W$.

Solution To show that two vector spaces V and W are isomorphic, we need to find an isomorphism $T : V \rightarrow W$. That is, T must be a linear transformation that is onto and one-to-one.

Since our goal is to show that $\mathbf{R}^3$ and $\mathbf{P}^2$ are isomorphic, we choose a linear transformation $T : \mathbf{R}^3 \rightarrow \mathbf{P}^2$ that is as simple as possible while still meeting the requirements of an isomorphism. Often something obvious makes a good choice. Here we try

$$T \left(\begin{bmatrix} a_0 \\ a_1 \\ a_2 \end{bmatrix} \right) = a_2 x^2 + a_1 x + a_0$$

We have

$$T\left(\begin{bmatrix} a_0 \\ a_1 \\ a_2 \end{bmatrix} + \begin{bmatrix} b_0 \\ b_1 \\ b_2 \end{bmatrix}\right) = T\left(\begin{bmatrix} a_0 + b_0 \\ a_1 + b_1 \\ a_2 + b_2 \end{bmatrix}\right) = (a_2 + b_2)x^2 + (a_1 + b_1)x + (a_0 + b_0)$$

$$= (a_2 x^2 + a_1 x + a_0) + (b_2 x^2 + b_1 x + b_0) = T\left(\begin{bmatrix} a_0 \\ a_1 \\ a_2 \end{bmatrix}\right) + T\left(\begin{bmatrix} b_0 \\ b_1 \\ b_2 \end{bmatrix}\right)$$

and

$$T\left(c\begin{bmatrix} a_0 \\ a_1 \\ a_2 \end{bmatrix}\right) = T\left(\begin{bmatrix} ca_0 \\ ca_1 \\ ca_2 \end{bmatrix}\right) = ca_2 x^2 + ca_1 x + ca_0 = c(a_2 x^2 + a_1 x + a_0) = cT\left(\begin{bmatrix} a_0 \\ a_1 \\ a_2 \end{bmatrix}\right)$$

so that T is a linear transformation. We can see from the definition of T that every vector $a_2 x^2 + a_1 x + a_0$ in $\mathbf{P}^2$ is matched up by T with exactly one vector $\begin{bmatrix} a_0 \\ a_1 \\ a_2 \end{bmatrix}$ in $\mathbf{R}^3$, and that every vector in $\mathbf{R}^3$ is matched up with a vector in $\mathbf{P}^2$. Therefore T is one-to-one and onto, and thus is an isomorphism. This proves that $\mathbf{R}^3$ and $\mathbf{P}^2$ are isomorphic. ∎

It is possible for a subspace of one vector space to be isomorphic to a vector space or even another subspace.

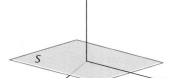

Figure 1 The subspace S in $\mathbf{R}^3$.

EXAMPLE 2 Show that $\mathbf{R}^2$ and the subspace

$$S = \text{span}\left\{\begin{bmatrix} 1 \\ 2 \\ 0 \end{bmatrix}, \begin{bmatrix} 3 \\ 1 \\ 1 \end{bmatrix}\right\} \tag{1}$$

of $\mathbf{R}^3$ are isomorphic.

Solution The subspace S is a plane in $\mathbf{R}^3$ (Figure 1), so it resembles the coordinate plane $\mathbf{R}^2$. Hence it seems plausible that the two would be isomorphic. But this is not a proof.

To prove that S and $\mathbf{R}^2$ are isomorphic, we need to find an isomorphism. Let $T : \mathbf{R}^2 \to S$ be given by $T(\mathbf{x}) = A\mathbf{x}$, where

$$A = \begin{bmatrix} 1 & 3 \\ 2 & 1 \\ 0 & 1 \end{bmatrix}$$

Then T is a linear transformation. Since $S = \text{col}(A)$, the column space of A, it follows that T is onto. Furthermore, since the columns of A are linearly independent, T is one-to-one by Theorem 3.6 in Section 3.1. Therefore T is an isomorphism, and S and $\mathbf{R}^2$ are isomorphic. ∎

In both of the examples we have considered, the isomorphic vector spaces had the same dimension. This is not a coincidence.

THEOREM 9.10

Suppose that finite dimensional vector spaces V and W are isomorphic. Then $\dim(V) = \dim(W)$.

Proof Since V and W are isomorphic, there exists an isomorphism $T : V \to W$. Now recall from Theorem 9.8 that

- If T is onto, then $\dim(V) \geq \dim(W)$.

- If T is one-to-one, then $\dim(V) \leq \dim(W)$.

Since T is onto and one-to-one, both inequalities must hold. The only way this can happen is if $\dim(V) = \dim(W)$. ∎

Theorem 9.10 provides a quick and easy way to show that two vector spaces are *not* isomorphic, because

If $\dim(V) \neq \dim(W)$, then V and W are not isomorphic.

EXAMPLE 3 Show that $\mathbf{R}^{3\times 2}$ and $\mathbf{P}^4$ are not isomorphic.

Solution Since $\dim(\mathbf{R}^{3\times 2}) = 6$ and $\dim(\mathbf{P}^4) = 5$, these two vector spaces cannot be isomorphic. ∎

Theorem 9.10 also holds for infinite-dimensional vector spaces. If V and W are isomorphic and one is infinite dimensional, then so is the other. Such vector spaces can lead to interesting and counter-intuitive results.

EXAMPLE 4 Let $\mathbf{P}_e$ denote the set of polynomials with real coefficients and only even-powered terms, and let $\mathbf{P}$ be the set of all polynomials with real coefficients. Then $\mathbf{P}_e$ is a subspace of $\mathbf{P}$. Is $\mathbf{P}_e$ isomorphic to $\mathbf{P}$?

Solution Viewed one way, it seems unlikely that $\mathbf{P}_e$ is isomorphic to $\mathbf{P}$. After all, $\mathbf{P}_e$ is a proper subspace of $\mathbf{P}$, so how could a one-to-one correspondence—required of an isomorphism—possibly exist between these two sets? But intuition can be misleading when it comes to infinite-dimensional vector spaces.

Suppose that we define $T : \mathbf{P}_e \to \mathbf{P}$ by

$$T(a_n x^{2n} + a_{n-1} x^{2(n-1)} + \cdots + a_1 x^2 + a_0) = a_n x^n + a_{n-1} x^{n-1} + \cdots + a_1 x + a_0$$

It is not hard to verify that T is a linear transformation, and that T is also onto and one-to-one. Therefore, perhaps surprisingly, $\mathbf{P}_e$ and $\mathbf{P}$ are isomorphic. ∎

Thus far we know that if V and W are isomorphic, then $\dim(V) = \dim(W)$. Now we consider the reverse direction. Do the dimensions of V and W tell us anything about whether or not V and W are isomorphic?

THEOREM 9.11

Let V and W be finite-dimensional vector spaces with $\dim(V) = \dim(W) = m$, where $m > 0$. Suppose that

$$\mathcal{V} = \{\mathbf{v}_1, \ldots, \mathbf{v}_m\} \quad \text{and} \quad \mathcal{W} = \{\mathbf{w}_1, \ldots, \mathbf{w}_m\}$$

are bases for V and W, respectively. Now define $T : V \to W$ as follows: For $\mathbf{v}$ in V, let $c_1, \ldots, c_m$ be such that $\mathbf{v} = c_1 \mathbf{v}_1 + \cdots + c_m \mathbf{v}_m$. Then set

$$T(\mathbf{v}) = T(c_1 \mathbf{v}_1 + \cdots + c_m \mathbf{v}_m) = c_1 \mathbf{w}_1 + \cdots + c_m \mathbf{w}_m$$

Then T is an isomorphism, and V and W are isomorphic vector spaces.

The proof of Theorem 9.11 is not hard, but it is a bit long and so is given at the end of the section. Here we report the significant implication of this theorem.

THEOREM 9.12

Finite dimensional vector spaces V and W are isomorphic if and only if $\dim(V) = \dim(W)$.

Proof By Theorem 9.11, if two finite-dimensional vector spaces V and W have the same dimension, then they are isomorphic. The converse follows from Theorem 9.10, which tells us that two finite-dimensional isomorphic vector spaces must have the same dimension. ■

If we think of dimension as giving a measure of the size of a vector space, then Theorem 9.12 tells us that size is *all* that matters when determining if two vector spaces are isomorphic. Now we can easily answer questions such as:

- Are $\mathbf{P}^7$ and $\mathbf{R}^{4\times 2}$ isomorphic? (Yes. Both have dimension 8.)
- Are $\mathbf{R}^{2\times 3}$ and $C[0, 1]$ isomorphic? (No. $\dim(\mathbf{R}^{2\times 3}) = 6$ and $\dim(C[0, 1]) = \infty$.)
- Are $\mathbf{R}^5$ and $S = \text{span}\{\cos(x), \sin(x), \cos(2x), \sin(2x)\}$ in $C[0, 1]$ isomorphic? (No. $\dim(\mathbf{R}^5) = 5$ and $\dim(S) \le 4$.)
- Are $\mathbf{R}^{7\times 5}$ and $\mathbf{R}^{35}$ isomorphic? (Yes. Both have dimension 35.)
- Are $\mathbf{R}^\infty$ and $\mathbf{P}$ isomorphic? (Maybe. We cannot tell because Theorem 9.12 does not apply to a pair of vector spaces having infinite dimension. Not every question is easily answered.)

Another consequence of Theorem 9.12 is contained in the next theorem.

THEOREM 9.13

If V is a vector space and $\dim(V) = n$, then V is isomorphic to $\mathbf{R}^n$.

In a way, this brings our development of vector spaces full circle. Since all n-dimensional vector spaces are isomorphic to n-dimensional Euclidean space, it is not so surprising that our Euclidean space results carried over so readily to vector spaces.

Inverses

We now revisit the notion of inverse functions, first treated in Section 3.3. Here is a definition, updated from earlier to our current more general setting.

DEFINITION 9.14

Definition Inverse, Invertible

A linear transformation $T : V \to W$ is **invertible** if T is one-to-one and onto. When T is invertible, the **inverse** function $T^{-1} : W \to V$ is defined by

$$T^{-1}(\mathbf{w}) = \mathbf{v} \quad \text{if and only if} \quad T(\mathbf{v}) = \mathbf{w}$$

Much of the development of Section 3.3 carries over directly to vector spaces. The main points are:

- A linear transformation T is invertible exactly when T is an isomorphism.
- If a linear transformation T is invertible, then the inverse T^{-1} is unique.
- If $T : V \to W$ is an isomorphism, then $T^{-1} : W \to V$ is also an isomorphism.

The proofs of these properties are left as exercises. Let's consider an example.

EXAMPLE 5 Let $T : \mathbf{P}^1 \to \mathbf{R}^2$ be the linear transformation given by

$$T(p(x)) = \begin{bmatrix} p(0) \\ p(1) \end{bmatrix}$$

Show that T is an isomorphism by showing that T is one-to-one and onto, and find T^{-1}.

Solution Elements of $\mathbf{P}^1$ are all polynomials of the form $p(x) = ax + b$, so that $p(0) = b$ and $p(1) = a + b$. Thus

$$T(ax + b) = \begin{bmatrix} b \\ a + b \end{bmatrix}$$

For any real numbers c and d, the vector equation

$$\begin{bmatrix} b \\ a + b \end{bmatrix} = \begin{bmatrix} c \\ d \end{bmatrix}$$

has unique solution $a = d - c$ and $b = c$. Therefore T is one-to-one and onto. Moreover, the unique solution also shows us how to define $T^{-1} : \mathbf{R}^2 \to \mathbf{P}^1$. Since $a = d - c$ and $b = c$, it follows that

$$T\big((d - c)x + c\big) = \begin{bmatrix} c \\ d \end{bmatrix} \quad \Longrightarrow \quad T^{-1}\left(\begin{bmatrix} c \\ d \end{bmatrix}\right) = (d - c)x + c$$

We can check that this is correct by computing

$$T^{-1}\big(T(ax + b)\big) = T^{-1}\left(\begin{bmatrix} b \\ a + b \end{bmatrix}\right) = ((a + b) - b)x + b = ax + b$$

and

$$T\left(T^{-1}\left(\begin{bmatrix} c \\ d \end{bmatrix}\right)\right) = T\big((d - c)x + c\big) = \begin{bmatrix} c \\ (d - c) + c \end{bmatrix} = \begin{bmatrix} c \\ d \end{bmatrix} \quad \blacksquare$$

Proof of Theorem 9.11

Proof of Theorem 9.11 First, since $\mathcal{V} = \{\mathbf{v}_1, \ldots, \mathbf{v}_m\}$ is a basis for V, for any $\mathbf{v}$ there is exactly one set of scalars $c_1, \ldots, c_m$ such that $\mathbf{v} = c_1\mathbf{v}_1 + \cdots + c_m\mathbf{v}_m$. Therefore T is actually a well-defined function.

Second, suppose that $\mathbf{u}$ is also in V, with $\mathbf{u} = d_1\mathbf{v}_1 + \cdots + d_m\mathbf{v}_m$. For $\mathbf{v} = c_1\mathbf{v}_1 + \cdots + c_m\mathbf{v}_m$ as above and scalars a and b, we have

$$\begin{aligned} T(a\mathbf{v} + b\mathbf{u}) &= T\big((ac_1 + bd_1)\mathbf{v}_1 + \cdots + (ac_m + bd_m)\mathbf{v}_m\big) \\ &= (ac_1 + bd_1)\mathbf{w}_1 + \cdots + (ac_m + bd_m)\mathbf{w}_m \\ &= a(c_1\mathbf{w}_1 + \cdots + c_m\mathbf{w}_m) + b(d_1\mathbf{w}_1 + \cdots + d_m\mathbf{w}_m) = aT(\mathbf{v}) + bT(\mathbf{u}). \end{aligned}$$

Hence T is a linear transformation.

Third, since $\mathcal{W} = \{\mathbf{w}_1, \ldots, \mathbf{w}_m\}$ is a basis, every element $\mathbf{w}$ in W can be expressed in the form $\mathbf{w} = c_1\mathbf{w}_1 + \cdots + c_m\mathbf{w}_m$ for unique $c_1, \ldots, c_m$. For such a $\mathbf{w}$, we see that $\mathbf{v} = c_1\mathbf{v}_1 + \cdots + c_m\mathbf{v}_m$ satisfies $T(\mathbf{v}) = \mathbf{w}$, and therefore T is onto.

Finally, as $\mathcal{W}$ is a basis, the only linear combination $c_1\mathbf{w}_1 + \cdots + c_m\mathbf{w}_m = \mathbf{0}_W$ is when $c_1 = \cdots = c_m = 0$. Therefore the only $\mathbf{v}$ in V such that $T(\mathbf{v}) = \mathbf{0}_W$ is $\mathbf{v} = \mathbf{0}_V$, which implies that T is also one-to-one. Thus T is an isomorphism, and therefore V and W are isomorphic vector spaces. $\blacksquare$

| EXERCISES |

For Exercises 1–6, use dimensions when possible to determine if the given vector spaces are isomorphic. If not possible, explain why.

1. $V = \mathbf{R}^8$ and $W = \mathbf{P}^9$

2. $V = \mathbf{R}^{5 \times 3}$ and $W = \mathbf{R}^{15}$

3. $V = \mathbf{R}^{3 \times 6}$ and $W = \mathbf{P}^{17}$

4. $V = \mathbf{R}^\infty$ and $\mathbf{P}^{20}$

5. $V = \mathbf{R}^{13}$ and $C[0, 1]$

6. $V = \mathbf{R}^\infty$ and $C[0, 1]$

For Exercises 7–10, prove that the given function is an isomorphism.

7. $T : \mathbf{R}^3 \to \mathbf{P}^2$ with $T \left(\begin{bmatrix} a \\ b \\ c \end{bmatrix} \right) = cx^2 + bx + a$

8. $T : \mathbf{P}^1 \to \mathbf{R}^2$ with $T(p) = \begin{bmatrix} p(-2) \\ p(1) \end{bmatrix}$

9. $T : \mathbf{P}^3 \to \mathbf{R}^{2\times 2}$ with $T(ax^3 + bx^2 + cx + d) =$
$\begin{bmatrix} (a+b+c+d) & (a+b+c) \\ (a+b) & a \end{bmatrix}$

10. $T : \mathbf{R}^{2\times 2} \to \mathbf{R}^{2\times 2}$ with $T(A) = A^T$

For Exercises 11–14, determine if the given linear transformation is an isomorphism.

11. $T : \mathbf{P}^1 \to \mathbf{R}^2$ with $T(ax + b) = \begin{bmatrix} a + b \\ b - a \end{bmatrix}$

12. $T : \mathbf{P} \to \mathbf{P}$ with $T(f) = xp(x)$

13. $T : C(\mathbf{R}) \to \mathbf{R}^\infty$ with $T(f) = (f(1), f(2), f(3), \ldots)$

14. $T : \mathbf{R}^{2\times 2} \to \mathbf{P}^3$ with
$$T(A) = \operatorname{tr}(A)x^3 + a_{11}x^2 + a_{21}x - a_{12}$$

For Exercises 15–18, find T^{-1} for the given isomorphism T.

15. $T : \mathbf{P}^1 \to \mathbf{R}^2$ with $T(ax + b) = \begin{bmatrix} 2b \\ a - b \end{bmatrix}$

16. $T : \mathbf{R}^{2\times 2} \to \mathbf{R}^{2\times 2}$ with $T(A) = A^T$

17. $T : \mathbf{P}^2 \to \mathbf{P}^2$ with $T(ax^2 + bx + c) = cx^2 - bx + a$

18. $T : \mathbf{P}^3 \to \mathbf{R}^{2\times 2}$ with
$$T(ax^3 + bx^2 + cx + d) = \begin{bmatrix} -a & c \\ -d & b \end{bmatrix}$$

19. Let S be the subspace of $\mathbf{R}^3$ given by
$$S = \operatorname{span} \left\{ \begin{bmatrix} 1 \\ 0 \\ 0 \end{bmatrix}, \begin{bmatrix} 0 \\ 1 \\ 0 \end{bmatrix} \right\}$$

Show that $T : \mathbf{R}^2 \to S$ given by
$$T \left(\begin{bmatrix} a_1 \\ a_2 \end{bmatrix} \right) = \begin{bmatrix} a_1 \\ a_2 \\ 0 \end{bmatrix}$$

is an isomorphism.

20. Let S be the subspace of $\mathbf{R}^4$ given by
$$S = \operatorname{span} \left\{ \begin{bmatrix} 1 \\ 0 \\ 0 \\ 0 \end{bmatrix}, \begin{bmatrix} 0 \\ 0 \\ 0 \\ 2 \end{bmatrix} \right\}$$

Show that $T : \mathbf{R}^2 \to S$ given by
$$T \left(\begin{bmatrix} a_1 \\ a_2 \end{bmatrix} \right) = \begin{bmatrix} a_1 \\ 0 \\ 0 \\ a_2 \end{bmatrix}$$

is an isomorphism.

21. Let $\mathbf{P}_e$ be the subspace of $\mathbf{P}$ defined in Example 4. Show that $T : \mathbf{P}_e \to \mathbf{P}$ given by
$$T(a_n x^{2n} + a_{n-1} x^{2(n-1)} + \cdots + a_1 x^2 + a_0)$$
$$= a_n x^n + a_{n-1} x^{n-1} + \cdots + a_1 x + a_0$$

is an isomorphism.

22. Let $\mathbf{P}_o$ be the subspace of $\mathbf{P}$ consisting of polynomials with only odd-powered terms, and the zero polynomial. Define $T : \mathbf{P}_o \to \mathbf{P}$ by
$$T(a_n x^{2n+1} + a_{n-1} x^{2n-1} + \cdots + a_1 x^3 + a_0 x)$$
$$= a_n x^n + a_{n-1} x^{n-1} + \cdots + a_1 x + a_0$$

and $T(0) = 0$ for the zero polynomial. Is T a linear transformation?

FIND AN EXAMPLE For Exercises 23–30, find an example that meets the given specifications. Prove your claim.

23. An isomorphism $T : \mathbf{R}^5 \to \mathbf{P}^4$.

24. An isomorphism $T : \mathbf{R}^{2\times 3} \to \mathbf{R}^6$.

25. An isomorphism $T : \mathbf{R}^{2\times 2} \to \mathbf{P}^3$.

26. An isomorphism $T : \mathbf{R}^4 \to S$, where S is the subspace of $\mathbf{P}^6$ of polynomials that have only terms with even-powered exponents.

27. A subspace of $\mathbf{R}^{2\times 3}$ that is isomorphic to $\mathbf{P}^3$.

28. A subspace of $\mathbf{P}$ that is isomorphic to $\mathbf{R}^4$.

29. A subspace of $\mathbf{R}^\infty$ that is isomorphic to $\mathbf{P}$.

30. A proper subspace of $\mathbf{R}^\infty$ that is isomorphic to $\mathbf{R}^\infty$.

TRUE OR FALSE For Exercises 31–44, determine if the statement is true or false, and justify your answer.

31. Every linear transformation is also an isomorphism.

32. Every finite dimensional vector space V is isomorphic to $\mathbf{R}^n$ for some n.

33. If $T : \mathbf{R}^n \to \mathbf{R}^n$ is a one-to-one linear transformation, then T is an isomorphism.

34. There exists a subspace of $\mathbf{P}^{10}$ that is isomorphic to $\mathbf{R}^{3\times 3}$.

35. If V and W are isomorphic, then there is a unique linear transformation $T : V \to W$ that is an isomorphism.

36. Every three-dimensional subspace of a vector space V is isomorphic to $\mathbf{P}^2$.

37. If $T_1 : V \to W$ and $T_2 : V \to W$ are isomorphisms, then so is $T_1 + T_2$.

38. If $T : V \to W$ is an isomorphism and $\{\mathbf{v}_1, \mathbf{v}_2, \mathbf{v}_3\}$ is a basis for V, then $\{T(\mathbf{v}_1), T(\mathbf{v}_2), T(\mathbf{v}_3)\}$ is a basis for W.

39. If $T : \mathbf{P} \to \mathbf{P}$ is a linear transformation and range$(T) = \mathbf{P}$, then T is an isomorphism.

40. If $T : \mathbf{P} \to \mathbf{P}$ is defined by $T(p(x)) = p(2x + 1)$, then T is an isomorphism.

41. (Calculus required) Let $C^\infty(a, b)$ denote the set of functions that have an infinite number of continuous derivatives on the interval (a, b). Then $T : C^\infty(a, b) \to C^\infty(a, b)$ given by $T(f) = f'(x)$ is an isomorphism.

42. (Calculus required) Let $T : C[a, b] \to \mathbf{R}$ be given by

$$T(f) = \int_a^b f(x)\, dx$$

Then T is an isomorphism.

43. (Calculus required) The linear transformation $T : \mathbf{P}^4 \to \mathbf{P}^4$ with $T(p) = xp'(x)$ is an isomorphism.

44. (Calculus required) If $T : \mathbf{P}^1 \to \mathbf{P}^1$ with

$$T(p) = x - \int_a^b p(x)\, dx$$

then T is an isomorphism.

45. Prove that a linear tranformation T is an isomorphism if and only if T has an inverse.

46. Prove that if the linear transformation T has an inverse, then it is unique.

47. Prove that if $T : V \to W$ is an isomorphism, then $T^{-1} : W \to V$ is also an isomorphism.

48. Prove that if a linear transformation $T : V \to W$ is either onto *or* one-to-one, and $\dim(V) = \dim(W)$ are both finite, then T is an isomorphism.

49. Suppose that $T : V \to W$ is a one-to-one linear transformation. Prove that V and range(T) are isomorphic.

50. Suppose that $T_1 : V \to W$ and $T_2 : W \to Y$ are isomorphisms. Prove that V and Y are isomorphic vector spaces.

9.3 The Matrix of a Linear Transformation

In Section 3.1 we defined the linear transformation $T : \mathbf{R}^m \to \mathbf{R}^n$ and showed that it has the form $T(\mathbf{x}) = A\mathbf{x}$ for an $n \times m$ matrix A. If V and W are finite-dimensional vector spaces, then we can establish a similar connection between linear transformations $T : V \to W$ and matrices. Before getting to that, we first revisit coordinate vectors to see how they are applied to vector spaces.

Coordinate Vectors

To establish a connection between linear transformations and vector spaces, we need a way to express vectors in the form of elements of $\mathbf{R}^m$. This can be done using coordinate vectors as defined below.

DEFINITION 9.15

Definition Coordinate Vector

Let V be a vector space with basis $\mathcal{G} = \{\mathbf{g}_1, \ldots, \mathbf{g}_m\}$. For each $\mathbf{v} = c_1\mathbf{g}_1 + \cdots + c_m\mathbf{g}_m$ in V, we define the **coordinate vector of v with respect to** $\mathcal{G}$ by

$$\mathbf{v}_\mathcal{G} = \begin{bmatrix} c_1 \\ \vdots \\ c_m \end{bmatrix}_\mathcal{G}$$

▶ In this section and the next, we will sometimes need to refer to more than one basis for a vector space. To avoid the appearance of favoring one basis over another (and additional cluttering subscripts), we depart from our customary use of $\mathcal{V}$ and $\mathcal{W}$ to represent bases for V and W, respectively.

Regarding Definition 9.15:

- Although $\mathbf{v}_\mathcal{G}$ is expressed in terms of a vector in Euclidean space, it represents a vector in V.

- The choice of basis matters. Different bases will yield different coordinate vectors for the same vector $\mathbf{v}$.

- The order of the basis vectors matters. Our convention is to read a list of basis vectors from left to right, taking them in that order.

- Since $\mathcal{G}$ is a basis for V, the scalars $c_1, \ldots, c_m$ are unique, so that for each $\mathbf{v}$ there is exactly one coordinate vector with respect to a given basis $\mathcal{G}$.

EXAMPLE 1 Two bases for $\mathbf{P}^2$ are

$$\mathcal{G} = \{x^2, x, 1\} \quad \text{and} \quad \mathcal{H} = \{x^2 + 2x - 4, x + 1, x^2 - x\}$$

Find the coordinate vector of $\mathbf{v} = x^2 - 6x + 2$ with respect to $\mathcal{G}$ and with respect to $\mathcal{H}$.

Solution Starting with $\mathcal{G}$, it is not difficult to see that

$$\mathbf{v} = (1)x^2 + (-6)x + (2) \quad \Longrightarrow \quad \mathbf{v}_{\mathcal{G}} = \begin{bmatrix} 1 \\ -6 \\ 2 \end{bmatrix}_{\mathcal{G}}$$

For the basis $\mathcal{H}$, we need to find scalars a, b, and c such that

$$\mathbf{v} = x^2 - 6x + 2 = a(x^2 + 2x - 4) + b(x + 1) + c(x^2 - x)$$

Using our standard methods, it can be shown that $a = -1$, $b = -2$, and $c = 2$. Hence

$$\mathbf{v} = (-1)(x^2 + 2x - 4) + (-2)(x + 1) + (2)(x^2 - x) \quad \Longrightarrow \quad \mathbf{v}_{\mathcal{H}} = \begin{bmatrix} -1 \\ -2 \\ 2 \end{bmatrix}_{\mathcal{H}} \quad \blacksquare$$

EXAMPLE 2 Find the coordinate vector of $\mathbf{v} = \begin{bmatrix} 3 & 2 \\ 5 & 7 \end{bmatrix}$ with respect to the basis for $\mathbf{R}^{2\times2}$ given by

$$\mathcal{G} = \left\{ \begin{bmatrix} 1 & 0 \\ 0 & 0 \end{bmatrix}, \begin{bmatrix} 0 & 1 \\ 0 & 0 \end{bmatrix}, \begin{bmatrix} 0 & 0 \\ 1 & 0 \end{bmatrix}, \begin{bmatrix} 0 & 0 \\ 0 & 1 \end{bmatrix} \right\}$$

Solution The set $\mathcal{G}$ is the standard basis for $\mathbf{R}^{2\times2}$ and can be especially easy to use. We see that

$$\mathbf{v} = \begin{bmatrix} 3 & 2 \\ 5 & 7 \end{bmatrix} = 3\begin{bmatrix} 1 & 0 \\ 0 & 0 \end{bmatrix} + 2\begin{bmatrix} 0 & 1 \\ 0 & 0 \end{bmatrix} + 5\begin{bmatrix} 0 & 0 \\ 1 & 0 \end{bmatrix} + 7\begin{bmatrix} 0 & 0 \\ 0 & 1 \end{bmatrix} \quad \Longrightarrow \quad \mathbf{v}_{\mathcal{G}} = \begin{bmatrix} 3 \\ 2 \\ 5 \\ 7 \end{bmatrix}_{\mathcal{G}} \quad \blacksquare$$

EXAMPLE 3 Suppose that $\mathcal{G} = \{\sin(x), \cos(x), e^{-x}\}$ is a basis for a subspace of $C[0, 1]$. Find $\mathbf{v}$ if

$$\mathbf{v}_{\mathcal{G}} = \begin{bmatrix} 5 \\ -3 \\ 2 \end{bmatrix}_{\mathcal{G}}$$

Solution Here we are given the coordinate vector $\mathbf{v}_{\mathcal{G}}$ and need to extract $\mathbf{v}$. All we have to do is multiply the basis vectors by the scalars given in $\mathbf{v}_{\mathcal{G}}$. We have

$$\mathbf{v} = 5\sin(x) - 3\cos(x) + 2e^{-x} \quad \blacksquare$$

Transformation Matrices

Now that we have an understanding of coordinate vectors, let's consider how matrices can be used to represent linear transformations. Suppose that $T : V \to W$ is a linear transformation, and that $\mathcal{G} = \{\mathbf{g}_1, \ldots, \mathbf{g}_m\}$ and $\mathcal{Q} = \{\mathbf{q}_1, \ldots, \mathbf{q}_n\}$ are bases of V and W, respectively. Our goal is to find a matrix A such that

$$\left[T(\mathbf{v})\right]_{\mathcal{Q}} = A\mathbf{v}_{\mathcal{G}}$$

where

$$\mathbf{v}_{\mathcal{G}} = \text{coordinate vector of } \mathbf{v} \text{ with respect to } \mathcal{G}$$
$$A\mathbf{v}_{\mathcal{G}} = \text{coordinate vector of } T(\mathbf{v}) \text{ with respect to } \mathcal{Q}$$

If $A = \begin{bmatrix} \mathbf{a}_1 & \cdots & \mathbf{a}_m \end{bmatrix}$ and $\mathbf{v} = c_1\mathbf{g}_1 + \cdots + c_m\mathbf{g}_m = \begin{bmatrix} c_1 \\ \vdots \\ c_m \end{bmatrix}_\mathcal{G}$, then

$$A\mathbf{v}_\mathcal{G} = A\begin{bmatrix} c_1 \\ \vdots \\ c_m \end{bmatrix}_\mathcal{G} = c_1\mathbf{a}_1 + \cdots + c_m\mathbf{a}_m$$

On the other hand, we also have

$$T(\mathbf{v}) = c_1 T(\mathbf{g}_1) + \cdots + c_m T(\mathbf{g}_m)$$

Thus, in order for $A\mathbf{v}_\mathcal{G} = [T(\mathbf{v})]_\mathcal{Q}$, we should set $\mathbf{a}_i = [T(\mathbf{g}_i)]_\mathcal{Q}$ for each $i = 1, \ldots, m$. This brings us to our next definition.

DEFINITION 9.16

Definition Matrix of a Linear Transformation

Let $T : V \to W$ be a linear transformation, $\mathcal{G} = \{\mathbf{g}_1, \ldots, \mathbf{g}_m\}$ a basis of V, and let $\mathcal{Q} = \{\mathbf{q}_1, \ldots, \mathbf{q}_n\}$ a basis of W. If $A = \begin{bmatrix} \mathbf{a}_1 & \cdots & \mathbf{a}_m \end{bmatrix}$ with

$$\mathbf{a}_i = [T(\mathbf{g}_i)]_\mathcal{Q}$$

for each $i = 1, \ldots, m$, then A is the **matrix of T with respect to $\mathcal{G}$ and $\mathcal{Q}$**.

Definition Transformation Matrix

The matrix A in Definition 9.16 is also called a **transformation matrix**.

EXAMPLE 4 Let $T : \mathbf{P}^2 \to \mathbf{P}^1$ be given by

$$T(ax^2 + bx + c) = (2a + c - 3b)x + (c + 4b + 3a)$$

and let $\mathcal{G} = \{x^2, x, 1\}$ be a basis for $\mathbf{P}^2$ and $\mathcal{Q} = \{x, 1\}$ a basis for $\mathbf{P}^1$. Find the matrix of T with respect to $\mathcal{G}$ and $\mathcal{Q}$, and then use it to compute $T(2x^2 - 4x + 1)$.

Solution Finding A requires us to compute $[T(\mathbf{g}_i)]_\mathcal{Q}$ for each basis vector of $\mathcal{G}$. We have

$$T(x^2) = 2x + 3 \implies [T(x^2)]_\mathcal{Q} = \begin{bmatrix} 2 \\ 3 \end{bmatrix}_\mathcal{Q}$$

$$T(x) = -3x + 4 \implies [T(x)]_\mathcal{Q} = \begin{bmatrix} -3 \\ 4 \end{bmatrix}_\mathcal{Q}$$

$$T(1) = x + 1 \implies [T(1)]_\mathcal{Q} = \begin{bmatrix} 1 \\ 1 \end{bmatrix}_\mathcal{Q}$$

Therefore, by Definition 9.16,

$$A = \begin{bmatrix} [T(x^2)]_\mathcal{Q} & [T(x)]_\mathcal{Q} & [T(1)]_\mathcal{Q} \end{bmatrix} = \begin{bmatrix} 2 & -3 & 1 \\ 3 & 4 & 1 \end{bmatrix}$$

Now let's use A to compute $T(2x^2 - 4x + 1)$. We have $2x^2 - 4x + 1 = \begin{bmatrix} 2 \\ -4 \\ 1 \end{bmatrix}_\mathcal{G}$, so that

$$T(2x^2 - 4x + 1) = A\begin{bmatrix} 2 \\ -4 \\ 1 \end{bmatrix}_\mathcal{G} = \begin{bmatrix} 17 \\ -9 \end{bmatrix}_\mathcal{Q} = 17x - 9$$

To check our work, let's compute directly,

$$T(2x^2 - 4x + 1) = \big(2(2) + (1) - 3(-4)\big)x + \big((1) + 4(-4) + 3(2)\big) = 17x - 9 \quad \blacksquare$$

In Example 4 we used the standard bases for $\mathbf{P}^2$ and $\mathbf{P}^1$. While this simplified computations, there is no reason why other bases cannot be used. Let's try Example 4 again, this time with different choices for $\mathcal{G}$ and $\mathcal{Q}$.

EXAMPLE 5   Repeat Example 4, this time with bases

$$\mathcal{G} = \{x^2 - 2,\ 2x - 1,\ x^2 + 4x - 1\} \quad \text{and} \quad \mathcal{Q} = \{2x + 1,\ 5x + 3\}$$

Solution As before, the columns of our transformation matrix A are given by $[T(\mathbf{g}_i)]_{\mathcal{Q}}$. Starting with $\mathbf{g}_1$, we have $T(x^2 - 2) = 1$. Finding the coordinate vector with respect to $\mathcal{Q}$ requires more work than before. Here we need to find scalars a and b such that

$$1 = a(2x + 1) + b(5x + 3)$$

Applying our usual methods yields the solution $a = -5$ and $b = 2$. Hence

$$[T(x^2 - 2)]_{\mathcal{Q}} = \begin{bmatrix} -5 \\ 2 \end{bmatrix}_{\mathcal{Q}}$$

Applying the same procedure to the other basis vectors yields

$$T(2x - 1) = -7x + 7 = -56(2x + 1) + 21(5x + 3) \implies [T(2x - 1)]_{\mathcal{Q}} = \begin{bmatrix} -56 \\ 21 \end{bmatrix}_{\mathcal{Q}}$$

and

$$T(x^2 + 4x - 1) = -11x + 18 = -123(2x + 1) + 47(5x + 3)$$
$$\implies [T(x^2 + 4x - 1)]_{\mathcal{Q}} = \begin{bmatrix} -123 \\ 47 \end{bmatrix}_{\mathcal{Q}}$$

Therefore this time the transformation matrix is

$$B = \begin{bmatrix} -5 & -56 & -123 \\ 2 & 21 & 47 \end{bmatrix}$$

To test this out by computing $T(2x^2 - 4x + 1)$, we need to determine the coordinate vector of $2x^2 - 4x + 1$ with respect to $\mathcal{G}$. To do so, we need to find scalars a, b, and c such that

$$2x^2 - 4x + 1 = a(x^2 - 2) + b(2x - 1) + c(x^2 + 4x - 1)$$

▶ Here we use B to denote the transformation matrix to avoid confusing this matrix with the matrix A found in Example 4. In Section 9.4 we will see that there is a relationship between two transformation matrices A and B representing the same linear transformation with respect to different bases.

Our usual methods produce $a = 1$, $b = -4$, and $c = 1$, so that $2x^2 - 4x + 1 = \begin{bmatrix} 1 \\ -4 \\ 1 \end{bmatrix}_{\mathcal{G}}$. Hence

$$T(2x^2 - 4x + 1) = B \begin{bmatrix} 1 \\ -4 \\ 1 \end{bmatrix}_{\mathcal{G}} = \begin{bmatrix} 96 \\ -35 \end{bmatrix}_{\mathcal{Q}} = 96(2x + 1) - 35(5x + 3) = 17x - 9$$

which agrees with our earlier work. ∎

Inverses

Let $T : V \to W$ be an isomorphism of finite-dimensional vector spaces. By Theorem 9.10, we know that $\dim(V) = \dim(W) = m$ for some m. Now suppose that A is the transformation matrix of T with respect to bases $\mathcal{G}$ and $\mathcal{Q}$, and that $T(\mathbf{v}) = \mathbf{w}$. Then A is an $m \times m$ matrix with

$$A\mathbf{v}_{\mathcal{G}} = \mathbf{w}_{\mathcal{Q}} \tag{1}$$

As T is an isomorphism, T must be onto and one-to-one, so that by the Big Theorem (Version 3 or later), A is an invertible matrix. Multiplying by A^{-1} on both sides of 1 yields

$$A^{-1}\mathbf{w}_Q = \mathbf{v}_G$$

Thus A^{-1} reverses T, so we can conclude that A^{-1} is the matrix of T^{-1} with respect to Q and G. (Note that the domain and codomain reverse when switching from T to T^{-1}.)

EXAMPLE 6 In Example 5 of Section 9.2, it is shown that the linear transformation $T : \mathbf{P}^1 \rightarrow \mathbf{R}^2$ given by

$$T(p) = \begin{bmatrix} p(0) \\ p(1) \end{bmatrix}$$

is an isomorphism. Let $G = \{x, 1\}$ and $Q = \{\mathbf{e}_1, \mathbf{e}_2\}$ be bases for $\mathbf{P}^1$ and $\mathbf{R}^2$, respectively. Find the matrix of T^{-1} with respect to G and Q.

Solution We have

$$T(x) = \begin{bmatrix} 0 \\ 1 \end{bmatrix} \implies [T(x)]_Q = \begin{bmatrix} 0 \\ 1 \end{bmatrix}_Q$$

$$T(1) = \begin{bmatrix} 1 \\ 1 \end{bmatrix} \implies [T(1)]_Q = \begin{bmatrix} 1 \\ 1 \end{bmatrix}_Q$$

Therefore

$$A = \begin{bmatrix} 0 & 1 \\ 1 & 1 \end{bmatrix}$$

is the matrix of T with respect to G and Q. By the preceding comments, the matrix of T^{-1} with respect to Q and G is given by

$$A^{-1} = \begin{bmatrix} -1 & 1 \\ 1 & 0 \end{bmatrix}$$

For a typical vector $\mathbf{w} = \begin{bmatrix} c \\ d \end{bmatrix}$ in $\mathbf{R}^2$, we have $\mathbf{w} = c\mathbf{e}_1 + d\mathbf{e}_2$, so that

$$T^{-1}(\mathbf{w}) = A^{-1} \begin{bmatrix} c \\ d \end{bmatrix}_Q = \begin{bmatrix} -1 & 1 \\ 1 & 0 \end{bmatrix} \begin{bmatrix} c \\ d \end{bmatrix}_Q = \begin{bmatrix} -c + d \\ c \end{bmatrix}_G = (d - c)x + c$$

which matches T^{-1} found in Example 5 of Section 9.2. ∎

EXERCISES

For Exercises 1–4, find $\mathbf{v}$ given the coordinate vector $\mathbf{v}_G$ with respect to the basis G.

1. $\mathbf{v}_G = \begin{bmatrix} -4 \\ 1 \end{bmatrix}_G$; $G = \left\{ \begin{bmatrix} 3 \\ 2 \end{bmatrix}, \begin{bmatrix} 1 \\ 4 \end{bmatrix} \right\}$

2. $\mathbf{v}_G = \begin{bmatrix} 2 \\ 5 \end{bmatrix}_G$; $G = \{-3x + 1, 2x - 4\}$

3. $\mathbf{v}_G = \begin{bmatrix} -1 \\ 0 \\ 3 \end{bmatrix}_G$; $G = \left\{ x^2 - x + 3, 3x^2 + 4, -5x - 2 \right\}$

4. $\mathbf{v}_G = \begin{bmatrix} 1 \\ 2 \\ 3 \\ 1 \end{bmatrix}_G$; $G = \left\{ \begin{bmatrix} 1 & 1 \\ 2 & 1 \end{bmatrix}, \begin{bmatrix} 1 & 2 \\ 2 & 2 \end{bmatrix}, \begin{bmatrix} 0 & 2 \\ 1 & 0 \end{bmatrix}, \begin{bmatrix} 2 & 1 \\ 0 & 3 \end{bmatrix} \right\}$

For Exercises 5–12, find the coordinate vector of $\mathbf{v}$ with respect to the given basis G.

5. $\mathbf{v} = \begin{bmatrix} 8 \\ 9 \end{bmatrix}$; $G = \left\{ \begin{bmatrix} 2 \\ 0 \end{bmatrix}, \begin{bmatrix} 0 \\ 3 \end{bmatrix} \right\}$

6. $\mathbf{v} = -14x + 15$; $G = \{2x, 5\}$

7. $\mathbf{v} = 12x^2 - 15x + 30$; $G = \left\{ 4x^2, 3x, 5 \right\}$

8. $\mathbf{v} = \begin{bmatrix} -6 & 6 \\ 20 & -7 \end{bmatrix}$;

$$\mathcal{G} = \left\{ \begin{bmatrix} 2 & 0 \\ 0 & 0 \end{bmatrix}, \begin{bmatrix} 0 & 3 \\ 0 & 0 \end{bmatrix}, \begin{bmatrix} 0 & 0 \\ 5 & 0 \end{bmatrix}, \begin{bmatrix} 0 & 0 \\ 0 & 1 \end{bmatrix} \right\}$$

9. $\mathbf{v} = \begin{bmatrix} 5 \\ -5 \end{bmatrix}$; $\mathcal{G} = \left\{ \begin{bmatrix} 1 \\ 2 \end{bmatrix}, \begin{bmatrix} 3 \\ 1 \end{bmatrix} \right\}$

10. $\mathbf{v} = -11x$; $\mathcal{G} = \{-3x + 2, 2x - 5\}$

11. $\mathbf{v} = 9x + 1$; $\mathcal{G} = \{x^2 - 1, 2x + 1, 3x^2 - 1\}$

12. $\mathbf{v} = \begin{bmatrix} 0 & 3 \\ 1 & -1 \end{bmatrix}$;

$$\mathcal{G} = \left\{ \begin{bmatrix} 1 & 0 \\ 1 & 0 \end{bmatrix}, \begin{bmatrix} 0 & 2 \\ 1 & 1 \end{bmatrix}, \begin{bmatrix} 1 & -1 \\ 0 & 1 \end{bmatrix}, \begin{bmatrix} 0 & 0 \\ 1 & 1 \end{bmatrix} \right\}$$

For Exercises 13–18, A is the matrix of linear transformation $T : V \to W$ with respect to bases $\mathcal{G}$ and $\mathcal{Q}$, respectively. Find $T(\mathbf{v})$ for the given $\mathbf{v}_{\mathcal{G}}$.

13. $A = \begin{bmatrix} 1 & 3 \\ 2 & -1 \end{bmatrix}$; $\mathbf{v}_{\mathcal{G}} = \begin{bmatrix} 4 \\ -3 \end{bmatrix}_{\mathcal{G}}$; $\mathcal{Q} = \left\{ \begin{bmatrix} 1 \\ 1 \end{bmatrix}, \begin{bmatrix} 2 \\ 3 \end{bmatrix} \right\}$

14. $A = \begin{bmatrix} 4 & 0 \\ 3 & 1 \end{bmatrix}$; $\mathbf{v}_{\mathcal{G}} = \begin{bmatrix} 2 \\ 2 \end{bmatrix}_{\mathcal{G}}$; $\mathcal{Q} = \{3x - 2, x + 5\}$

15. $A = \begin{bmatrix} 1 & 1 & 2 \\ 0 & 1 & 3 \\ 0 & 1 & 1 \end{bmatrix}$; $\mathbf{v}_{\mathcal{G}} = \begin{bmatrix} 1 \\ 3 \\ 4 \end{bmatrix}_{\mathcal{G}}$;

$\mathcal{Q} = \{x^2 - 2x, x^2 + x + 4, 3x + 1\}$

16. $A = \begin{bmatrix} -1 & 3 & 1 \\ 4 & 0 & 1 \\ 1 & -1 & 0 \end{bmatrix}$; $\mathbf{v}_{\mathcal{G}} = \begin{bmatrix} 2 \\ 0 \\ -1 \end{bmatrix}_{\mathcal{G}}$;

$\mathcal{Q} = \left\{ \begin{bmatrix} 3 \\ 1 \\ -2 \end{bmatrix}, \begin{bmatrix} 2 \\ 4 \\ 0 \end{bmatrix}, \begin{bmatrix} 1 \\ 2 \\ 3 \end{bmatrix} \right\}$

17. $A = \begin{bmatrix} 0 & 4 & 3 \\ 2 & -1 & -3 \\ 2 & -2 & -1 \end{bmatrix}$; $\mathbf{v}_{\mathcal{G}} = \begin{bmatrix} 2 \\ -3 \\ 1 \end{bmatrix}_{\mathcal{G}}$;

$\mathcal{Q} = \{\sin(x), \cos(x), e^{-x}\}$

18. $A = \begin{bmatrix} 0 & 0 & 1 & 1 \\ 1 & 0 & 1 & 0 \\ 2 & 0 & 0 & 1 \\ 1 & 1 & 0 & 0 \end{bmatrix}$; $\mathbf{v}_{\mathcal{G}} = \begin{bmatrix} 3 \\ 2 \\ 1 \\ -1 \end{bmatrix}_{\mathcal{G}}$;

$\mathcal{Q} = \left\{ \begin{bmatrix} 1 & 0 \\ 1 & 2 \end{bmatrix}, \begin{bmatrix} 2 & -1 \\ 0 & 2 \end{bmatrix}, \begin{bmatrix} 1 & 4 \\ 2 & 1 \end{bmatrix}, \begin{bmatrix} 3 & 1 \\ 2 & 4 \end{bmatrix} \right\}$

For Exercises 19–26, find the matrix A of the linear transformation $T : V \to W$ with respect to bases $\mathcal{G}$ and $\mathcal{Q}$, respectively.

19. $T\left(\begin{bmatrix} a \\ b \end{bmatrix} \right) = bx - a$; $\mathcal{G} = \{\mathbf{e}_1, \mathbf{e}_2\}$; $\mathcal{Q} = \{x, 1\}$

20. $T(ax + b) = ax^2 + (a + b)x - b$; $\mathcal{G} = \{x, 1\}$; $\mathcal{Q} = \{x^2, x, 1\}$

21. $T(f(x)) = f'(x)$; $\mathcal{G} = \{x^2, x, 1\}$; $\mathcal{Q} = \{x, 1\}$

22. $T(f(x)) = xf'(x) + f(0)$; $\mathcal{G} = \{x^2 + 3, x - 2, x^2 + x\}$; $\mathcal{Q} = \{x^2, x, 1\}$

23. $T\left(\begin{bmatrix} a \\ b \end{bmatrix} \right) = \begin{bmatrix} b - a \\ a + 2b \end{bmatrix}$; $\mathcal{G} = \{\mathbf{e}_1, \mathbf{e}_2\}$; $\mathcal{Q} = \left\{ \begin{bmatrix} 1 \\ 1 \end{bmatrix}, \begin{bmatrix} 1 \\ 2 \end{bmatrix} \right\}$

24. $T(ax + b) = (b - a)x - (a + b)$; $\mathcal{G} = \{x, 1\}$; $\mathcal{Q} = \{x + 1, x + 2\}$

25. $T\left(\begin{bmatrix} a \\ b \end{bmatrix} \right) = -bx + a + b$; $\mathcal{G} = \left\{ \begin{bmatrix} 2 \\ 1 \end{bmatrix}, \begin{bmatrix} 3 \\ 2 \end{bmatrix} \right\}$;

$\mathcal{Q} = \{5x + 3, 2x + 1\}$

26. $T\left(\begin{bmatrix} a & b \\ c & d \end{bmatrix} \right) = \begin{bmatrix} a + b \\ b + c \\ c + d \end{bmatrix}$;

$\mathcal{G} = \left\{ \begin{bmatrix} 1 & 0 \\ 0 & 0 \end{bmatrix}, \begin{bmatrix} 1 & 1 \\ 0 & 0 \end{bmatrix}, \begin{bmatrix} 1 & 1 \\ 1 & 0 \end{bmatrix}, \begin{bmatrix} 1 & 1 \\ 1 & 1 \end{bmatrix} \right\}$;

$\mathcal{Q} = \left\{ \begin{bmatrix} 1 \\ 0 \\ 0 \end{bmatrix}, \begin{bmatrix} 1 \\ 1 \\ 0 \end{bmatrix}, \begin{bmatrix} 1 \\ 1 \\ 1 \end{bmatrix} \right\}$

For Exercises 27–30, suppose that

$$A = \begin{bmatrix} a & b & c \\ d & e & f \end{bmatrix}$$

is the matrix of $T : V \to W$ with respect to bases $\mathcal{G} = \{\mathbf{g}_1, \mathbf{g}_2, \mathbf{g}_3\}$ and $\mathcal{Q} = \{\mathbf{q}_1, \mathbf{q}_2\}$, respectively. Find the matrix of T with respect to the given bases $\mathcal{H}$ and $\mathcal{R}$.

27. (a) $\mathcal{H} = \{\mathbf{g}_1, \mathbf{g}_2, \mathbf{g}_3\}$, $\mathcal{R} = \{2\mathbf{q}_1, \mathbf{q}_2\}$

(b) $\mathcal{H} = \{3\mathbf{g}_1, \mathbf{g}_2, \mathbf{g}_3\}$, $\mathcal{R} = \{\mathbf{q}_1, \mathbf{q}_2\}$

28. (a) $\mathcal{H} = \{\mathbf{g}_3, \mathbf{g}_2, \mathbf{g}_1\}$, $\mathcal{R} = \{\mathbf{q}_1, \mathbf{q}_2\}$

(b) $\mathcal{H} = \{\mathbf{g}_1, \mathbf{g}_2, \mathbf{g}_3\}$, $\mathcal{R} = \{\mathbf{q}_2, \mathbf{q}_1\}$

29. (a) $\mathcal{H} = \{\mathbf{g}_3, \mathbf{g}_1, \mathbf{g}_2\}$, $\mathcal{R} = \{\mathbf{q}_1, \mathbf{q}_2\}$

(b) $\mathcal{H} = \{\mathbf{g}_1, \mathbf{g}_3, \mathbf{g}_2\}$, $\mathcal{R} = \{\mathbf{q}_2, \mathbf{q}_1\}$

30. (a) $\mathcal{H} = \{\mathbf{g}_1, 5\mathbf{g}_2, \mathbf{g}_3\}$, $\mathcal{R} = \{\mathbf{q}_1, 2\mathbf{q}_2\}$

(b) $\mathcal{H} = \{\mathbf{g}_2, \mathbf{g}_1, 3\mathbf{g}_3\}$, $\mathcal{R} = \{4\mathbf{q}_2, \mathbf{q}_1\}$

31. Suppose that $T : \mathbf{P}^1 \to \mathbf{P}^1$ has matrix $A = \begin{bmatrix} 1 & 3 \\ 0 & 1 \end{bmatrix}$ with respect to the basis $\mathcal{G} = \{x + 1, x - 1\}$ for the domain and $\mathcal{Q} = \{1, x\}$ for the codomain. Use the inverse of A to find $T^{-1}(x)$.

32. Suppose that $T : \mathbf{P}^1 \to \mathbf{P}^1$ has matrix $A = \begin{bmatrix} 2 & 7 \\ 1 & 3 \end{bmatrix}$ with respect to the basis $\mathcal{G} = \{2x + 1, 3x - 1\}$ for the domain and $\mathcal{Q} = \{1, x + 3\}$ for the codomain. Use the inverse of A to find $T^{-1}(x + 2)$.

33. Suppose that $T : \mathbf{R}^2 \to \mathbf{P}^1$ has matrix $A = \begin{bmatrix} 2 & 1 \\ 3 & 2 \end{bmatrix}$ with respect to the basis $\mathcal{G} = \left\{ \begin{bmatrix} 1 \\ 3 \end{bmatrix}, \begin{bmatrix} 2 \\ 1 \end{bmatrix} \right\}$ for the domain and $\mathcal{Q} = \{x, 2x + 1\}$ for the codomain. Use the inverse of A to find $T^{-1}(x + 1)$.

34. Suppose that $T : \mathbf{R}^2 \to \mathbf{P}^1$ has matrix $A = \begin{bmatrix} 1 & 3 \\ 1 & 4 \end{bmatrix}$ with respect to the basis $\mathcal{G} = \left\{ \begin{bmatrix} 2 \\ 5 \end{bmatrix}, \begin{bmatrix} 1 \\ 3 \end{bmatrix} \right\}$ for the domain and $\mathcal{Q} = \{3x, 2x - 1\}$ for the codomain. Use the inverse of A to find $T^{-1}(x - 2)$.

FIND AN EXAMPLE For Exercises 35–40, find an example that meets the given specifications. Prove your claim.

35. An element $\mathbf{v}$ and basis $\mathcal{G}$ of $\mathbf{P}^2$ such that

$$\mathbf{v}_\mathcal{G} = \begin{bmatrix} 2 \\ 0 \\ -3 \end{bmatrix}_\mathcal{G}$$

36. An element $\mathbf{v}$ and basis $\mathcal{G}$ of $\mathbf{R}^{2\times 2}$ such that

$$\mathbf{v}_\mathcal{G} = \begin{bmatrix} 1 \\ -7 \\ 4 \\ -4 \end{bmatrix}_\mathcal{G}$$

37. A basis $\mathcal{G}$ of $\mathbf{R}^2$ such that

$$\begin{bmatrix} -3 \\ 4 \end{bmatrix}_\mathcal{G} = \begin{bmatrix} 7 \\ 5 \end{bmatrix}$$

38. A basis $\mathcal{G}$ of $\mathbf{P}^1$ such that

$$\begin{bmatrix} 3 \\ -5 \end{bmatrix}_\mathcal{G} = 4x - 11$$

39. Vector spaces V and W, bases for each, and a linear transformation $T : V \to W$ that has matrix

$$A = \begin{bmatrix} 2 & 1 & 2 \\ 0 & 3 & 1 \end{bmatrix}$$

with respect to the bases.

40. Vector spaces V and W, bases for each, and a linear transformation $T : V \to W$ that has matrix

$$A = \begin{bmatrix} 6 & 1 \\ 3 & 3 \\ -2 & 5 \end{bmatrix}$$

with respect to the bases.

TRUE OR FALSE For Exercises 41–46, determine if the statement is true or false, and justify your answer.

41. The matrix of any linear transformation between finite-dimensional vector spaces must be square.

42. The matrix of a linear transformation $T : V \to W$ is unique.

43. If $\begin{bmatrix} a \\ b \end{bmatrix}_\mathcal{G}$ is the coordinate vector of a vector with respect to a basis $\mathcal{G}$, then $\begin{bmatrix} 2a \\ -3b \end{bmatrix}_\mathcal{G}$ is the coordinate vector with respect to $\mathcal{G}$ of some vector.

44. If $\mathbf{v} = \begin{bmatrix} a \\ b \end{bmatrix}_\mathcal{G}$ for $\mathcal{G} = \{\mathbf{g}_1, \mathbf{g}_2\}$, then $\mathbf{v} = \begin{bmatrix} b \\ a \end{bmatrix}_{\tilde{\mathcal{G}}}$ for $\tilde{\mathcal{G}} = \{\mathbf{g}_2, \mathbf{g}_1\}$.

45. If $\mathbf{0}$ is the zero vector of a finite-dimensional vector space V, then

$$\mathbf{0}_V = \begin{bmatrix} 0 \\ \vdots \\ 0 \end{bmatrix}_\mathcal{G}$$

for every basis $\mathcal{G}$ of V.

46. If $\mathcal{G}$ and $\mathcal{H}$ are two distinct bases for a finite-dimensional vector space V, then $\mathbf{v}_\mathcal{G}$ and $\mathbf{v}_\mathcal{H}$ cannot have the same entries for any element $\mathbf{v}$ of V.

47. Let $\mathcal{G}$ be a basis for a vector space V of dimension m. Show that a set of vectors $\{\mathbf{v}_1, \ldots, \mathbf{v}_k\}$ is linearly independent in V if and only if the coordinate vectors $\{[\mathbf{v}_1]_\mathcal{G}, \ldots, [\mathbf{v}_k]_\mathcal{G}\}$ are linearly independent in $\mathbf{R}^m$.

48. Let $\mathcal{G}$ be a basis for a vector space V, and suppose that $\mathbf{v}_\mathcal{G} = \mathbf{w}_\mathcal{G}$. Prove that $\mathbf{v} = \mathbf{w}$.

49. Let $\mathcal{G}$ be a basis for a vector space V of dimension m. Show that a linear combination of vectors $\mathbf{v}_1, \ldots, \mathbf{v}_k$ is equal to $\mathbf{v}$ in V if and only if there is a linear combination of the coordinate vectors $[\mathbf{v}_1]_\mathcal{G}, \ldots, [\mathbf{v}_k]_\mathcal{G}$ that is equal to the coordinate vector $\mathbf{v}_\mathcal{G}$ in $\mathbf{R}^m$.

50. Suppose that A is the matrix of linear transformation $T : V \to W$ with respect to bases $\mathcal{G}$ and $\mathcal{Q}$, respectively.

(a) Show that $\mathbf{v}$ is in the kernel of T if and only if $\mathbf{v}_\mathcal{G}$ is in the null space of A.

(b) Show that $\mathbf{w}$ is in the range of T if and only if $\mathbf{w}_\mathcal{Q}$ is in the column space of A.

51. Suppose that A is the matrix of linear transformation $T : V \to V$ with respect to basis $\mathcal{G}$ for both the domain and codomain. Let $T^2(\mathbf{v}) = T \circ T(\mathbf{v}) = T(T(\mathbf{v}))$ denote the composition of T with itself.

(a) Show that A^2 is the matrix of the linear transformation $T^2 : V \to V$ with respect to basis $\mathcal{G}$ for both the domain and codomain.

(b) If T^n denotes the n-fold composition of T with itself, then show that A^n is the matrix of the linear transformation $T^n : V \to V$ with respect to basis $\mathcal{G}$ for both the domain and codomain.

9.4 Similarity

In this section we continue our exploration of matrix representatives of linear transformations, now focusing on the special case $T : V \to V$, where the same basis $\mathcal{G}$ is used for both the domain and codomain. Let's start with an example.

EXAMPLE 1 Let $T : \mathbf{R}^3 \to \mathbf{R}^3$ be given by $T(\mathbf{x}) = A\mathbf{x}$, where

$$A = \begin{bmatrix} 1 & 0 & 1 \\ 0 & 1 & -4 \\ 2 & 1 & -1 \end{bmatrix}$$

and $\mathbf{x}$ is with respect to the standard basis. Find the matrix B of T with respect to the basis

$$\mathcal{G} = \left\{ \begin{bmatrix} 2 \\ -4 \\ 1 \end{bmatrix}, \begin{bmatrix} -1 \\ 3 \\ 0 \end{bmatrix}, \begin{bmatrix} 0 \\ 1 \\ 1 \end{bmatrix} \right\} = \{\mathbf{g}_1, \mathbf{g}_2, \mathbf{g}_3\}$$

Solution Even though we are working with a linear transformation from a vector space to itself, and the same basis is being used for both the domain and codomain, we still follow the same procedure as we used in Section 9.3. For each $\mathbf{g}_i$ in $\mathcal{G}$, we need to find the coordinate vector of $T(\mathbf{g}_i)$ with respect to $\mathcal{G}$. Starting with $\mathbf{g}_1$, we have

$$T(\mathbf{g}_1) = T\left(\begin{bmatrix} 2 \\ -4 \\ 1 \end{bmatrix} \right) = A \begin{bmatrix} 2 \\ -4 \\ 1 \end{bmatrix} = \begin{bmatrix} 3 \\ -8 \\ -1 \end{bmatrix}$$

To determine $[T(\mathbf{g}_1)]_{\mathcal{G}}$, we need to find c_1, c_2, and c_3 such that

$$\begin{bmatrix} 3 \\ -8 \\ -1 \end{bmatrix} = c_1 \begin{bmatrix} 2 \\ -4 \\ 1 \end{bmatrix} + c_2 \begin{bmatrix} -1 \\ 3 \\ 0 \end{bmatrix} + c_3 \begin{bmatrix} 0 \\ 1 \\ 1 \end{bmatrix} = \begin{bmatrix} 2 & -1 & 0 \\ -4 & 3 & 1 \\ 1 & 0 & 1 \end{bmatrix} \begin{bmatrix} c_1 \\ c_2 \\ c_3 \end{bmatrix}$$

Setting $S = \begin{bmatrix} \mathbf{g}_1 & \mathbf{g}_2 & \mathbf{g}_3 \end{bmatrix} = \begin{bmatrix} 2 & -1 & 0 \\ -4 & 3 & 1 \\ 1 & 0 & 1 \end{bmatrix}$, we have $S^{-1} = \begin{bmatrix} 3 & 1 & -1 \\ 5 & 2 & -2 \\ -3 & -1 & 2 \end{bmatrix}$, so that

$$\begin{bmatrix} c_1 \\ c_2 \\ c_3 \end{bmatrix} = S^{-1} \begin{bmatrix} 3 \\ -8 \\ -1 \end{bmatrix} = \begin{bmatrix} 3 & 1 & -1 \\ 5 & 2 & -2 \\ -3 & -1 & 2 \end{bmatrix} \begin{bmatrix} 3 \\ -8 \\ -1 \end{bmatrix} = \begin{bmatrix} 2 \\ 1 \\ -3 \end{bmatrix}$$

Therefore $[T(\mathbf{g}_1)]_{\mathcal{G}} = \begin{bmatrix} 2 \\ 1 \\ -3 \end{bmatrix}_{\mathcal{G}}$. Since we have S^{-1}, computing $[T(\mathbf{g}_2)]_{\mathcal{G}}$ and $[T(\mathbf{g}_3)]_{\mathcal{G}}$ is easier. They are

$$T(\mathbf{g}_2) = \begin{bmatrix} -1 \\ 3 \\ 1 \end{bmatrix} \implies [T(\mathbf{g}_2)]_{\mathcal{G}} = S^{-1} \begin{bmatrix} -1 \\ 3 \\ 1 \end{bmatrix} = \begin{bmatrix} -1 \\ -1 \\ 2 \end{bmatrix}_{\mathcal{G}}$$

$$T(\mathbf{g}_3) = \begin{bmatrix} 1 \\ -3 \\ 0 \end{bmatrix} \implies [T(\mathbf{g}_3)]_{\mathcal{G}} = S^{-1} \begin{bmatrix} 1 \\ -3 \\ 0 \end{bmatrix} = \begin{bmatrix} 0 \\ -1 \\ 0 \end{bmatrix}_{\mathcal{G}}$$

Thus the matrix B of T with respect to the basis $\mathcal{G}$ is

$$B = \begin{bmatrix} 2 & -1 & 0 \\ 1 & -1 & -1 \\ -3 & 2 & 0 \end{bmatrix} \qquad \blacksquare$$

Looking back at our work, we see that since $T(\mathbf{x}) = A\mathbf{x}$, computing each column $\mathbf{b}_i$ of B amounted to first computing $T(\mathbf{g}_i) = A\mathbf{g}_i$, and from this computing

$S^{-1}\big(T(\mathbf{g}_i)\big) = S^{-1}A\mathbf{g}_i$. Thus we have

$$\mathbf{b}_i = S^{-1}A\mathbf{g}_i \quad \text{for } i = 1, 2, 3$$

But since $B = \begin{bmatrix} \mathbf{b}_1 & \mathbf{b}_2 & \mathbf{b}_3 \end{bmatrix}$ and $S = \begin{bmatrix} \mathbf{g}_1 & \mathbf{g}_2 & \mathbf{g}_3 \end{bmatrix}$, it follows that

$$B = S^{-1}AS$$

This form is reminiscent of the diagonalization of matrices, and it results from an underlying change of basis that is taking place. Specifically, recall from Section 6.3 that S is the change of basis matrix from $\mathcal{G}$ to the standard basis for $\mathbf{R}^3$. Thus we can think of the product $S^{-1}AS\mathbf{v}_\mathcal{G}$ as doing the following:

(a) Multiplying S by $\mathbf{v}_\mathcal{G}$ converts $\mathbf{v}_\mathcal{G}$ from the coordinate vector with respect to $\mathcal{G}$ to the standard basis.

(b) Multiplying A by $S\mathbf{v}_\mathcal{G}$ performs the linear transformation T.

(c) Multiplying S^{-1} by $AS\mathbf{v}_\mathcal{G}$ converts from the standard basis back to the coordinate vector with respect to $\mathcal{G}$.

The same change of basis can be performed between two bases of any finite-dimensional vector space.

Change of Basis

Suppose that $\mathcal{G} = \{\mathbf{g}_1, \ldots, \mathbf{g}_m\}$ and $\mathcal{H} = \{\mathbf{h}_1, \ldots, \mathbf{h}_m\}$ are bases for a vector space V. Then for each $\mathbf{g}_i$, there exists a unique set of scalars $s_{1i}, \ldots, s_{mi}$ such that

$$\mathbf{g}_i = s_{1i}\mathbf{h}_1 + \cdots + s_{mi}\mathbf{h}_m$$

Now set

$$S = \begin{bmatrix} s_{11} & s_{12} & \cdots & s_{1m} \\ s_{21} & s_{22} & \cdots & s_{2m} \\ \vdots & \vdots & \ddots & \vdots \\ s_{m1} & s_{m2} & \cdots & s_{mm} \end{bmatrix} \tag{1}$$

Then for $\mathbf{v} = a_1\mathbf{g}_1 + \cdots + a_m\mathbf{g}_m$, we have

$$\begin{aligned}
\begin{bmatrix} a_1 \\ \vdots \\ a_m \end{bmatrix}_\mathcal{G} &= a_1\mathbf{g}_1 + \cdots + a_m\mathbf{g}_m \\
&= a_1(s_{11}\mathbf{h}_1 + \cdots + s_{m1}\mathbf{h}_m) + \cdots + a_m(s_{1m}\mathbf{h}_1 + \cdots + s_{mm}\mathbf{h}_m) \\
&= (a_1 s_{11} + \cdots + a_m s_{1m})\mathbf{h}_1 + \cdots + (a_1 s_{m1} + \cdots + a_m s_{mm})\mathbf{h}_m \\
&= \left[S \begin{bmatrix} a_1 \\ \vdots \\ a_m \end{bmatrix} \right]_\mathcal{H}
\end{aligned}$$

Therefore

$$\mathbf{v} = \begin{bmatrix} a_1 \\ \vdots \\ a_m \end{bmatrix}_\mathcal{G} = \left[S \begin{bmatrix} a_1 \\ \vdots \\ a_m \end{bmatrix} \right]_\mathcal{H}$$

Thus multiplication by the matrix S changes the coordinate vector with respect to $\mathcal{G}$ into the coordinate vector with respect to $\mathcal{H}$.

DEFINITION 9.17

Definition Change of Basis Matrix

The matrix S in 1 is called the **change of basis matrix** from $\mathcal{G}$ to $\mathcal{H}$.

Note that the change can be reversed with the inverse matrix: S^{-1} converts a coordinate vector with respect to $\mathcal{H}$ into the equivalent coordinate vector with respect to $\mathcal{G}$.

EXAMPLE 2 Find the change of basis matrix S from basis $\mathcal{G}$ to basis $\mathcal{H}$ of $\mathbf{P}^1$, where

$$\mathcal{G} = \{2x + 5, \, x + 3\}, \quad \mathcal{H} = \{2x - 1, \, x - 1\}$$

Then use S to find $\mathbf{v}_{\mathcal{H}}$ for $\mathbf{v} = \begin{bmatrix} -1 \\ 3 \end{bmatrix}_{\mathcal{G}}$, and find the change of basis matrix from $\mathcal{H}$ to $\mathcal{G}$.

Solution Starting with the first vector in $\mathcal{G}$, we have

$$2x + 5 = s_{11}(2x - 1) + s_{21}(x - 1)$$

Regrouping and comparing coefficients yields the linear system

$$2s_{11} + s_{21} = 2$$
$$-s_{11} - s_{21} = 5$$

which has the unique solution $s_{11} = 7$, $s_{21} = -12$. The second vector $x + 3$ gives rise to the equation

$$x + 3 = s_{12}(2x - 1) + s_{22}(x - 1)$$

Solving the equivalent linear system yields the unique solution $s_{12} = 4$, $s_{22} = -7$. Therefore

$$S = \begin{bmatrix} 7 & 4 \\ -12 & -7 \end{bmatrix}$$

and so

$$\mathbf{v}_{\mathcal{H}} = S\mathbf{v}_{\mathcal{G}} = \begin{bmatrix} 7 & 4 \\ -12 & -7 \end{bmatrix} \begin{bmatrix} -1 \\ 3 \end{bmatrix}_{\mathcal{G}} = \begin{bmatrix} 5 \\ -9 \end{bmatrix}_{\mathcal{H}}$$

Thus we can conclude that

$$-(2x + 5) + 3(x + 3) = 5(2x - 1) - 9(x - 1)$$

(Both are equal to $x + 4$.) The change of basis matrix from $\mathcal{H}$ to $\mathcal{G}$ is given by

$$S^{-1} = \begin{bmatrix} 7 & 4 \\ -12 & -7 \end{bmatrix}$$

Note that $S^{-1} = S$, so the change of basis matrix is the same from $\mathcal{H}$ to $\mathcal{G}$ as it is from $\mathcal{G}$ to $\mathcal{H}$. This is not typical, but it is possible. ∎

Transformation Matrices Revisited

Now that we know how change of basis matrices are defined for a general vector space V, we can combine this with what we learned in Section 9.3 about transformation matrices.

THEOREM 9.18

Let $T : V \to V$ be a linear transformation. Suppose that A and B are the matrices of T with respect to the bases $\mathcal{G}$ and $\mathcal{H}$, respectively, and let S be the change of basis matrix from $\mathcal{G}$ to $\mathcal{H}$. Then $A = S^{-1}BS$.

Proof For a typical vector $\mathbf{v}$ in V, the product $A\mathbf{v}_\mathcal{G}$ produces $[T(\mathbf{v})]_\mathcal{G}$, the coordinate vector of $T(\mathbf{v})$ with respect to the basis $\mathcal{G}$. On the other hand,

(a) $S\mathbf{v}_\mathcal{G}$ converts $\mathbf{v}_\mathcal{G}$ to $\mathbf{v}_\mathcal{H}$, the coordinate vector with respect to $\mathcal{H}$.

(b) $B(S\mathbf{v}_\mathcal{G})$ yields $[T(\mathbf{v})]_\mathcal{H}$, the coordinate vector of $T(\mathbf{v})$ with respect to the basis $\mathcal{H}$.

(c) $S^{-1}(BS\mathbf{v}_\mathcal{G})$ converts $[T(\mathbf{v})]_\mathcal{H}$ back to $[T(\mathbf{v})]_\mathcal{G}$.

Therefore $A\mathbf{v}_\mathcal{G}$ and $S^{-1}BS\mathbf{v}_\mathcal{G}$ are the same for all $\mathbf{v}$, and hence $A = S^{-1}BS$. ∎

Two matrices A and B related as in Theorem 9.18 go by a special name.

DEFINITION 9.19
Definition Similar Matrices, Similarity Transformation

A square matrix A is **similar** to matrix B if there exists an invertible matrix S such that $A = S^{-1}BS$. The change from B to A is called a **similarity transformation**.

Note that if $A = S^{-1}BS$, then $B = SAS^{-1}$. Hence setting $R = S^{-1}$ gives

$$R^{-1}AR = (S^{-1})^{-1}AS^{-1} = SAS^{-1} = B$$

so that B is also similar to A. Thus it makes sense to simply say that A and B are similar matrices.

EXAMPLE 3 Let matrices

$$A = \begin{bmatrix} 4 & 1 & -3 \\ 0 & 2 & 5 \\ 1 & -1 & 3 \end{bmatrix}, \quad B = \begin{bmatrix} 4 & 3 & -8 \\ -109 & -16 & 151 \\ -16 & -1 & 21 \end{bmatrix}, \quad S = \begin{bmatrix} 1 & 3 & -2 \\ 2 & 7 & 2 \\ 1 & 3 & -1 \end{bmatrix}$$

Show that A and B are similar matrices with similarity transformation matrix S.

Solution First, we note that $\det(S) = 1$, so that S is an invertible matrix. To show that A and B are similar, we shall show that $SA = BS$, which saves us the trouble of computing S^{-1}. We have

$$SA = \begin{bmatrix} 1 & 3 & -2 \\ 2 & 7 & 2 \\ 1 & 3 & -1 \end{bmatrix} \begin{bmatrix} 4 & 1 & -3 \\ 0 & 2 & 5 \\ 1 & -1 & 3 \end{bmatrix} = \begin{bmatrix} 2 & 9 & 6 \\ 10 & 14 & 35 \\ 3 & 8 & 9 \end{bmatrix}$$

and

$$BS = \begin{bmatrix} 4 & 3 & -8 \\ -109 & -16 & 151 \\ -16 & -1 & 21 \end{bmatrix} \begin{bmatrix} 1 & 3 & -2 \\ 2 & 7 & 2 \\ 1 & 3 & -1 \end{bmatrix} = \begin{bmatrix} 2 & 9 & 6 \\ 10 & 14 & 35 \\ 3 & 8 & 9 \end{bmatrix}$$

Since $SA = BS$ and S is invertible, then $A = S^{-1}BS$ and hence A and B are similar matrices. ∎

THEOREM 9.20

Two matrices A and B are similar if and only if A and B are the transformation matrices of linear transformation $T : V \to V$ with respect to different bases of V.

Proof Combining Theorem 9.18 and Definition 9.19, we have that if A and B are transformation matrices for the same linear transformation T but with respect to different bases, then A and B are similar.

Now suppose that A and B are similar matrices with $A = S^{-1}BS$ and $S = \begin{bmatrix} \mathbf{s}_1 & \cdots & \mathbf{s}_m \end{bmatrix}$. Then $\mathcal{G} = \{\mathbf{s}_1, \ldots, \mathbf{s}_m\}$ is a basis for $\mathbf{R}^m$ because S is invertible.

Let $T : \mathbf{R}^m \to \mathbf{R}^m$ be given by $T(\mathbf{v}) = B\mathbf{v}$. Then B is the transformation matrix with respect to the standard basis, and as S is the change of basis matrix from $\mathcal{G}$ to the standard basis, it follows that A is the transformation matrix of T with respect to $\mathcal{G}$. Hence A and B are both transformation matrices for T. ■

Note that given two matrices A and B, generally there is no simple way to determine if they are similar. One approach that is sure to work is illustrated in the next example.

EXAMPLE 4 Determine if the given matrices A and B are similar.

$$A = \begin{bmatrix} -2 & -9 \\ 2 & 7 \end{bmatrix}, \quad B = \begin{bmatrix} 3 & -1 \\ -2 & 2 \end{bmatrix}$$

Solution A and B are similar if there exists an invertible matrix S such that $A = S^{-1}BS$, or equivalently, $SA = BS$. Letting $S = \begin{bmatrix} s_{11} & s_{12} \\ s_{21} & s_{22} \end{bmatrix}$ and multiplying out SA and BS, we have

$$\begin{bmatrix} -2s_{11} + 2s_{12} & -9s_{11} + 7s_{12} \\ -2s_{21} + 2s_{22} & -9s_{21} + 7s_{22} \end{bmatrix} = \begin{bmatrix} 3s_{11} - s_{21} & 3s_{12} - s_{22} \\ -2s_{11} + 2s_{21} & -2s_{12} + 2s_{22} \end{bmatrix}$$

Setting the components equal to one another yields the homogeneous linear system

$$\begin{array}{rcl} 5s_{11} - 2s_{12} - s_{21} & = & 0 \\ 9s_{11} - 4s_{12} \quad\quad - s_{22} & = & 0 \\ 2s_{11} \quad\quad - 4s_{21} + 2s_{22} & = & 0 \\ 2s_{12} - 9s_{21} + 5s_{22} & = & 0 \end{array}$$

The system has the trivial solution, but since we require S to be invertible, we are not interested in that solution. However, there are also nontrivial solutions—for instance, $s_{11} = 1$, $s_{12} = 2$, $s_{21} = 1$, and $s_{22} = 1$. These give us

$$S = \begin{bmatrix} 1 & 2 \\ 1 & 1 \end{bmatrix}$$

As $\det(S) = -1 \neq 0$, it follows that S is invertible. We can check that $A = S^{-1}BS$ by computing

$$SA = \begin{bmatrix} 1 & 2 \\ 1 & 1 \end{bmatrix} \begin{bmatrix} -2 & -9 \\ 2 & 7 \end{bmatrix} = \begin{bmatrix} 2 & 5 \\ 0 & -2 \end{bmatrix}$$

and

$$BS = \begin{bmatrix} 3 & -1 \\ -2 & 2 \end{bmatrix} \begin{bmatrix} 1 & 2 \\ 1 & 1 \end{bmatrix} = \begin{bmatrix} 2 & 5 \\ 0 & -2 \end{bmatrix}$$

Hence $SA = BS$, so that A and B are similar. ■

Note that in Example 4, any nonzero scalar multiple of S would also serve as a solution to $SA = BS$, showing us that such matrices S are not unique.

Two matrices that are similar have several interesting properties in common.

THEOREM 9.21

If A and B are similar matrices, then they have the same characteristic polynomial and the same eigenvalues (including multiplicities), and $\det(A) = \det(B)$.

Proof If A and B are similar matrices, then there exists an invertible matrix S such that $A = S^{-1}BS$. Hence

$$
\begin{aligned}
\det(A - \lambda I) &= \det(S^{-1}BS - \lambda I) \\
&= \det(S^{-1}BS - S^{-1}(\lambda I)S) \\
&= \det(S^{-1}(B - \lambda I)S) \\
&= \det(S^{-1})\det(B - \lambda I)\det(S) \\
&= \det(B - \lambda I)
\end{aligned}
$$

▶ Here we use the fact that $\det(CD) = \det(C)\det(D)$ for $n \times n$ matrices C and D, and if C is invertible then $\det(C^{-1}) = \left(\det(C)\right)^{-1}$.

Thus the characteristic polynomials of A and B are the same, and therefore the eigenvalues (including multiplicities) are also the same. Setting $\lambda = 0$ above shows that $\det(A) = \det(B)$, completing the proof. ■

The last part of Theorem 9.21 tells us that if $\det(A) \neq \det(B)$, then A and B are not similar. However, note that if $\det(A) = \det(B)$, then we cannot draw any conclusion.

EXAMPLE 5 Use determinants to try to determine if the pairs of matrices A and B are similar.

(a) $A = \begin{bmatrix} 3 & 2 \\ 4 & 5 \end{bmatrix}, \quad B = \begin{bmatrix} 5 & 1 \\ 2 & 1 \end{bmatrix}$

(b) $A = \begin{bmatrix} 1 & 0 & 3 \\ 2 & 1 & 3 \\ 0 & 2 & 2 \end{bmatrix}, \quad B = \begin{bmatrix} 2 & 3 & 0 \\ 4 & 1 & 2 \\ 3 & 0 & 1 \end{bmatrix}$

Solution For the matrices in (a), we have $\det(A) = 7$ and $\det(B) = 4$. Since the determinants differ, the two matrices cannot be similar.

For (b), both $\det(A) = 8$ and $\det(B) = 8$. Thus determinants tell us nothing about whether or not the two matrices are similar. ■

Although determinants are not helpful for part (b), we can apply another part of Theorem 9.21. If A and B are similar, then they have the same eigenvalues. For the matrices in part (b), it can be shown that they have different eigenvalues and so cannot be similar matrices.

Computational Comments

There are a number of algorithms for estimating eigenvalues that exploit the fact that similar matrices have the same eigenvalues. The popular *QR algorithm* produces a sequence of similar matrices that become successively closer to upper triangular. It starts by producing the QR factorization (see Section 8.4) $A = Q_1 R_1$, where Q is an orthogonal matrix and R is upper triangular. Next, we let $A_1 = R_1 Q_1$, so that

$$A_1 = R_1 Q_1 = Q_1^{-1} Q_1 R_1 Q_1 = Q_1^{-1} A Q_1 \tag{2}$$

Thus A and A_1 are similar matrices and hence by Theorem 9.21 have the same eigenvalues. For $i > 1$, we let $A_i = Q_{i+1} R_{i+1}$ be the QR factorization for A_i and then define $A_{i+1} = R_{i+1} Q_{i+1}$. By the same reasoning as in 2, A_i and A_{i+1} are similar matrices. Therefore the sequence $A, A_1, A_2, \ldots$ of matrices all have the same eigenvalues. Under certain conditions, the sequence of matrices converges to a triangular matrix that has eigenvalues along the diagonal.

Another (older) algorithm called *Jacobi's Method* is applicable to symmetric matrices A. This method resembles matrix diagonalization, starting with $A_1 = A$ and setting

$$A_{i+1} = P_i^{-1} A_i P_i \quad \text{for } i = 1, 2, \ldots$$

Note that P_i is not the same as the orthogonal matrix found when diagonalizing a symmetric matrix, and A_i is not diagonal. (How P_i and A_i are defined is beyond the scope of this discussion.) However, the sequence of matrices $A_1, A_2, \ldots$ are all similar, and they converge to a diagonal matrix with the eigenvalues of A on the diagonal.

EXERCISES

For Exercises 1–8, find the change of basis matrix from $\mathcal{G}$ to $\mathcal{H}$.

1. $\mathcal{G} = \{2x - 1, 5x + 4\}$, $\mathcal{H} = \{x, 1\}$

2. $\mathcal{G} = \{x^2 - 7x + 5, 3x^2 + 1, 7x - 3\}$, $\mathcal{H} = \{x^2, x, 1\}$

3. $\mathcal{G} = \left\{ \begin{bmatrix} 3 & 2 \\ 1 & 0 \end{bmatrix}, \begin{bmatrix} 4 & 0 \\ 0 & 2 \end{bmatrix}, \begin{bmatrix} 1 & 7 \\ 5 & 1 \end{bmatrix}, \begin{bmatrix} 0 & 6 \\ 2 & 3 \end{bmatrix} \right\}$,

$\mathcal{H} = \left\{ \begin{bmatrix} 1 & 0 \\ 0 & 0 \end{bmatrix}, \begin{bmatrix} 0 & 1 \\ 0 & 0 \end{bmatrix}, \begin{bmatrix} 0 & 0 \\ 1 & 0 \end{bmatrix}, \begin{bmatrix} 0 & 0 \\ 0 & 1 \end{bmatrix} \right\}$

4. $\mathcal{G} = \{x + 3, 4x + 1\}$, $\mathcal{H} = \{1, x\}$

5. $\mathcal{G} = \{x - 2, x^2 + 9x, x^2 - x - 1\}$, $\mathcal{H} = \{x, 1, x^2\}$

6. $\mathcal{G} = \left\{ \begin{bmatrix} 6 & 1 \\ 0 & 0 \end{bmatrix}, \begin{bmatrix} 5 & 2 \\ 3 & 8 \end{bmatrix}, \begin{bmatrix} 2 & 5 \\ 9 & 7 \end{bmatrix}, \begin{bmatrix} 1 & 3 \\ 2 & 0 \end{bmatrix} \right\}$

$\mathcal{H} = \left\{ \begin{bmatrix} 0 & 1 \\ 0 & 0 \end{bmatrix}, \begin{bmatrix} 1 & 0 \\ 0 & 0 \end{bmatrix}, \begin{bmatrix} 0 & 0 \\ 0 & 1 \end{bmatrix}, \begin{bmatrix} 0 & 0 \\ 1 & 1 \end{bmatrix} \right\}$

7. $\mathcal{G} = \{7x + 4, 3x + 2\}$, $\mathcal{H} = \{2x + 1, 5x + 3\}$

8. $\mathcal{G} = \{x^2 + 2x + 1, x^2 + 1, x^2 - x + 2\}$
$\mathcal{H} = \{x^2 + x + 1, x + 1, 1\}$

For Exercises 9–12, B is the matrix of $T : V \to V$ with respect to a basis $\mathcal{H}$, and S is the change of basis matrix from a basis $\mathcal{G}$ to $\mathcal{H}$. Find the matrix A of T with respect to $\mathcal{G}$.

9. $B = \begin{bmatrix} 2 & 3 \\ -4 & 1 \end{bmatrix}$, $S = \begin{bmatrix} 1 & 3 \\ 2 & 7 \end{bmatrix}$

10. $B = \begin{bmatrix} 5 & 0 \\ 1 & 7 \end{bmatrix}$, $S = \begin{bmatrix} 4 & 3 \\ 3 & 2 \end{bmatrix}$

11. $B = \begin{bmatrix} 1 & 0 & 2 \\ 0 & 1 & 1 \\ 1 & 2 & -1 \end{bmatrix}$, $S = \begin{bmatrix} 1 & 0 & 0 \\ 2 & 1 & 0 \\ 1 & 3 & 1 \end{bmatrix}$

12. $B = \begin{bmatrix} 3 & 1 & 1 \\ 1 & 4 & 1 \\ 4 & 2 & 3 \end{bmatrix}$, $S = \begin{bmatrix} 0 & 1 & 0 \\ 1 & 0 & 1 \\ 1 & 0 & 0 \end{bmatrix}$

For Exercises 13–16, B is the matrix of $T : V \to V$ with respect to the basis $\mathcal{H}$. Find the matrix A of T with respect to $\mathcal{G}$.

13. $B = \begin{bmatrix} 1 & 2 \\ 1 & 1 \end{bmatrix}$, $\mathcal{H} = \{x, 1\}$, $\mathcal{G} = \{3x + 1, 2x + 1\}$

14. $B = \begin{bmatrix} 1 & 2 & 1 \\ 0 & 1 & 1 \\ 1 & 0 & 2 \end{bmatrix}$, $\mathcal{H} = \{x^2, x, 1\}$,

$\mathcal{G} = \{x^2 + x + 1, x + 1, 1\}$

15. $B = \begin{bmatrix} 2 & 2 \\ 4 & 1 \end{bmatrix}$, $\mathcal{H} = \{x + 3, 2x + 5\}$,

$\mathcal{G} = \{2x - 1, -3x + 2\}$

16. $B = \begin{bmatrix} 3 & -1 & 0 \\ 2 & -1 & 1 \\ 1 & 1 & 4 \end{bmatrix}$, $\mathcal{H} = \{1, x^2 + 1, x - 1\}$,

$\mathcal{G} = \{x - 3, x^2 - 2x + 4, 1\}$

For Exercises 17–20, determine if A and B are similar matrices.

17. $A = \begin{bmatrix} 1 & 3 \\ 2 & 5 \end{bmatrix}$, $B = \begin{bmatrix} 2 & 1 \\ 3 & 2 \end{bmatrix}$

18. $A = \begin{bmatrix} 2 & 1 \\ -1 & 0 \end{bmatrix}$, $B = \begin{bmatrix} 1 & 0 \\ 1 & 1 \end{bmatrix}$

19. $A = \begin{bmatrix} 1 & -1 & 3 \\ 1 & -3 & -3 \\ 0 & 1 & 2 \end{bmatrix}$, $B = \begin{bmatrix} 1 & 1 & 2 \\ 1 & 0 & 1 \\ 0 & 1 & -1 \end{bmatrix}$

20. $A = \begin{bmatrix} 1 & 2 & 1 \\ 3 & 1 & -1 \\ 0 & 1 & -2 \end{bmatrix}$, $B = \begin{bmatrix} 1 & 1 & -1 \\ -2 & 3 & 1 \\ 1 & 0 & -2 \end{bmatrix}$

FIND AN EXAMPLE For Exercises 21–24, find an example that meets the given specifications. Prove your claim.

21. A vector space V with bases $\mathcal{G}$ and $\mathcal{H}$ related by the matrix S in (1) for

$$S = \begin{bmatrix} 3 & 4 \\ 2 & 3 \end{bmatrix}$$

22. A vector space V with bases $\mathcal{G}$ and $\mathcal{H}$ related by the matrix S in (1) for

$$S = \begin{bmatrix} 1 & 3 & -2 \\ 1 & 4 & 4 \\ 2 & 6 & -3 \end{bmatrix}$$

23. Two similar matrices A and B that are related by

$$S = \begin{bmatrix} 5 & 2 \\ 8 & 3 \end{bmatrix}$$

24. Two similar matrices A and B that are related by

$$S = \begin{bmatrix} 3 & 4 & -2 \\ 2 & 3 & -4 \\ 3 & 4 & 6 \end{bmatrix}$$

TRUE OR FALSE For Exercises 25–34, determine if the statement is true or false, and justify your answer.

25. If $A = S^{-1}BS$, then A and B are similar.

26. The matrix of a linear transformation $T : V \to V$ is unique for a fixed basis $\mathcal{G}$ of V.

27. If A and B have the same rank, then they are similar.

28. Two similar matrices have the same eigenvectors.

29. If A and B are not similar and B and C are not similar, then A and C are not similar.

30. If A, B, and C are similar, then AB and BC are similar.

31. If there exists a matrix S such that $SA = BS$, then A and B are similar matrices.

32. For every A there exists a distinct B such that A and B are similar matrices.

33. If A and B are similar matrices, then $\text{null}(A) = \text{null}(B)$.

34. If A and B are similar matrices, then AB and BA are similar matrices.

35. Suppose that A and B are similar matrices, related by $A = S_1^{-1}BS_1$, and that B and C are also similar matrices, related by $B = S_2^{-1}CS_2$. Find the matrix D that relates A and C by $A = D^{-1}CD$.

36. Prove that similarity of matrices is transitive: if A is similar to B and B is similar to C, then A is similar to C.

37. Suppose that A and B are both diagonalizable matrices that have the same eigenvalues, including multiplicities. Prove that A and B are similar matrices.

38. Prove that if A and B are similar matrices, then so are A^k and B^k.

39. Prove that if A and B are similar matrices, then so are A^T and B^T.

40. Suppose that A and B are similar matrices and that A is invertible. Prove that B is also invertible and that A^{-1} and B^{-1} are also similar.

Ⓒ For Exercises 41–44, determine if the given matrices are similar.

41. $A = \begin{bmatrix} 1 & -2 & 4 \\ 5 & 1 & 2 \\ 0 & 1 & -3 \end{bmatrix}$, $B = \begin{bmatrix} 1 & -1 & 1 \\ 5 & 0 & 0 \\ 0 & 1 & 2 \end{bmatrix}$

42. $A = \begin{bmatrix} 3 & 2 & -2 \\ 1 & 4 & 0 \\ -2 & 1 & -1 \end{bmatrix}$, $B = \begin{bmatrix} 1 & 3 & -1 \\ 3 & 3 & 1 \\ -2 & 1 & 2 \end{bmatrix}$

43. $A = \begin{bmatrix} 1 & 0 & 1 & 3 \\ -1 & 2 & 4 & 1 \\ 2 & 3 & -1 & 0 \\ 0 & 2 & -2 & -2 \end{bmatrix}$, $B = \begin{bmatrix} 1 & 2 & 1 & 1 \\ -3 & -3 & 4 & -1 \\ 2 & 5 & 2 & -1 \\ 0 & 2 & 0 & 0 \end{bmatrix}$

44. $A = \begin{bmatrix} 2 & 1 & -1 & 2 \\ 3 & 0 & 1 & 0 \\ -1 & 2 & 4 & 1 \\ 0 & 0 & 3 & -1 \end{bmatrix}$, $B = \begin{bmatrix} 4 & 0 & 1 & -3 \\ 2 & 3 & 4 & -4 \\ 1 & 0 & 0 & 2 \\ -2 & -1 & 0 & 2 \end{bmatrix}$

Inner Product Spaces

In Chapter 8 we introduced dot products, which provided an algebraic way to determine when vectors in Euclidean space $\mathbf{R}^n$ are orthogonal. There we also developed applications of the dot product, including projections of vectors, the Gram–Schmidt process, and orthonormal bases.

In this chapter we introduce inner products, which extend the dot product from Euclidean space to vector spaces. In Section 10.1 we define the inner product, provide a number of examples of inner products, and give some results that are generalizations of those in Chapter 8. In Section 10.2 we develop the Gram–Schmidt process in the context of an inner product space. Section 10.3 is devoted to a few applications of inner products.

In previous chapters, topics involving calculus were separated from other material. Since some of the most important examples of inner products involve calculus, examples from calculus are more fully integrated into this chapter.

10.1 Inner Products

Since the dot product proved so useful in Euclidean space $\mathbf{R}^n$, we would like to extend the dot product to a similar product in other vector spaces. The four properties of dot products given in Theorem 8.2 of Section 8.1 are really what make the dot product useful. For instance, much of the development of the Gram–Schmidt process relies on these properties. Hence it makes sense that a generalized product on a vector space should have these same properties. In fact, it makes so much sense that we adapt the properties in Theorem 8.2 as a definition.

DEFINITION 10.1

Definition Inner Product, Inner Product Space

Let $\mathbf{u}$, $\mathbf{v}$, and $\mathbf{w}$ be elements of a vector space V, and let c be a scalar. An **inner product** on V is a function that takes two vectors in V as input and produces a scalar as output. An inner product function is denoted by $\langle \mathbf{u}, \mathbf{v} \rangle$ and must satisfy the following conditions:

(a) $\langle \mathbf{u}, \mathbf{v} \rangle = \langle \mathbf{v}, \mathbf{u} \rangle$

(b) $\langle \mathbf{u} + \mathbf{v}, \mathbf{w} \rangle = \langle \mathbf{u}, \mathbf{w} \rangle + \langle \mathbf{v}, \mathbf{w} \rangle$

(c) $\langle c\mathbf{u}, \mathbf{v} \rangle = \langle \mathbf{u}, c\mathbf{v} \rangle = c\langle \mathbf{u}, \mathbf{v} \rangle$

(d) $\langle \mathbf{u}, \mathbf{u} \rangle \geq 0$, and $\langle \mathbf{u}, \mathbf{u} \rangle = 0$ only when $\mathbf{u} = \mathbf{0}$

A vector space V with an inner product defined on it is called an **inner product space**.

Since this definition is guided by properties of the dot product, it follows that the dot product is an inner product on $\mathbf{R}^n$. But this is only one of many inner products on vector spaces. In fact, we can modify the usual dot product to produce a "weighted dot product" on $\mathbf{R}^n$.

▶ Here $\mathbf{u} = \begin{bmatrix} u_1 \\ \vdots \\ u_n \end{bmatrix}$, $\mathbf{v} = \begin{bmatrix} v_1 \\ \vdots \\ v_n \end{bmatrix}$

EXAMPLE 1 Let $t_1, \ldots, t_n$ be positive scalars, which are the "weights." Show that

$$\langle \mathbf{u}, \mathbf{v} \rangle = t_1 u_1 v_1 + t_2 u_2 v_2 + \cdots + t_n u_n v_n \tag{1}$$

is an inner product on $\mathbf{R}^n$.

Solution A function taking two vectors as input and producing a scalar as output is an inner product if it satisfies conditions (a)–(d) in Definition 10.1. We verify (a) and (d) here, leaving (b) and (c) as an exercise.

To show that condition (a) is met, note that

$$\begin{aligned} \langle \mathbf{u}, \mathbf{v} \rangle &= t_1 u_1 v_1 + t_2 u_2 v_2 + \cdots + t_n u_n v_n \\ &= t_1 v_1 u_1 + t_2 v_2 u_2 + \cdots + t_n v_n u_n = \langle \mathbf{v}, \mathbf{u} \rangle \end{aligned}$$

To verify (d), we start by computing

$$\langle \mathbf{u}, \mathbf{u} \rangle = t_1 u_1^2 + t_2 u_2^2 + \cdots + t_n u_n^2$$

Since $u_i^2 \geq 0$ for $i = 1, \ldots, n$ and the weights $t_1, \ldots, t_n$ are all positive, we have $\langle \mathbf{u}, \mathbf{u} \rangle \geq 0$. Furthermore, because the weights are positive, the only way that $\langle \mathbf{u}, \mathbf{u} \rangle = 0$ is if $u_1 = \cdots = u_n = 0$, which implies $\mathbf{u} = \mathbf{0}$. Hence condition (d) holds and the weighted dot product is an inner product. ∎

EXAMPLE 2 Let $\mathbf{u} = \begin{bmatrix} 1 \\ 3 \\ -2 \end{bmatrix}$ and $\mathbf{v} = \begin{bmatrix} 4 \\ -1 \\ 1 \end{bmatrix}$ be in $\mathbf{R}^3$. Compute $\langle \mathbf{u}, \mathbf{v} \rangle$ using the weighted dot product defined in Example 1 with weights $t_1 = 2$, $t_2 = 3$, and $t_3 = 1$.

Solution We have

$$\begin{aligned} \langle \mathbf{u}, \mathbf{v} \rangle &= t_1 u_1 v_1 + t_2 u_2 v_2 + t_3 u_3 v_3 \\ &= (2)(1)(4) + (3)(3)(-1) + (1)(-2)(1) = -3 \end{aligned}$$ ∎

EXAMPLE 3 Let $p(x)$ and $q(x)$ be polynomials in $\mathbf{P}^n$, and suppose $x_0, x_1, \ldots, x_n$ are $n + 1$ distinct real numbers. Prove that

$$\langle p, q \rangle = p(x_0)q(x_0) + p(x_1)q(x_1) + \cdots + p(x_n)q(x_n) \tag{2}$$

is an inner product on $\mathbf{P}^n$.

Solution Properties (a)–(c) of Definition 10.1 follow readily from the properties of real numbers and are left as an exercise. For (d) we have

$$\langle p, p \rangle = \left(p(x_0) \right)^2 + \left(p(x_1) \right)^2 + \cdots + \left(p(x_n) \right)^2$$

Since each term on the right must be nonnegative, it follows that $\langle p, p \rangle \geq 0$ for any polynomial p in $\mathbf{P}^n$. Moreover, $\langle p, p \rangle = 0$ exactly when $p(x_0) = p(x_1) = \cdots = p(x_n) = 0$. But the only way a polynomial of degree n or less can have $n + 1$ distinct roots is if it is the zero polynomial. Thus property (d) is also satisfied, so that (2) defines an inner product. ■

EXAMPLE 4 Suppose that $p(x) = x^2 - 3x + 2$ and $q(x) = 2x^2 + 4x - 1$ are in $\mathbf{P}^2$, and that $x_0 = -1$, $x_1 = 1$, and $x_2 = 4$. Compute $\langle p, q \rangle$ using the inner product defined in Example 3.

Solution We have

$$\langle p, q \rangle = p(-1)q(-1) + p(1)q(1) + p(4)q(4) = (6)(-3) + (0)(5) + (6)(47) = 264$$

■

Note that a weighted version of this inner product (2) can also be defined (see Exercise 53).

EXAMPLE 5 Let f and g be two continuous functions in $C[-1, 1]$. Show that

$$\langle f, g \rangle = \int_{-1}^{1} f(x)g(x) \, dx \tag{3}$$

defines an inner product on $C[-1, 1]$.

Solution Properties (a)–(c) of Definition 10.1 follow readily from basic properties of the definite integral. If f is in $C[-1, 1]$, then $\left(f(x) \right)^2 \geq 0$ for all x in $[-1, 1]$, so that

$$\langle f, f \rangle = \int_{-1}^{1} \left(f(x) \right)^2 \, dx \geq 0$$

The second part of property (d) follows from the more subtle but plausible fact from real analysis that for a continuous function $f(x)$,

$$\int_{-1}^{1} \left(f(x) \right)^2 \, dx > 0$$

▶ $z(x) = 0$ for all x in $[-1, 1]$.

except when $f(x) = z(x)$, the identically zero function. Therefore this is the only function for which $\langle f, f \rangle = 0$, and hence (3) gives an inner product. ■

EXAMPLE 6 Let $f(x) = x^2 + 4x$ and $g(x) = 5x^2 - 3$. Evaluate $\langle f, g \rangle$ using the inner product defined in Example 5.

Solution The inner product of $f(x)$ and $g(x)$ is

$$\langle f, g \rangle = \int_{-1}^{1} (x^2 + 4x)(5x^2 - 3)\, dx = \int_{-1}^{1} \left(5x^4 + 20x^3 - 3x^2 - 12x\right) dx = 0 \quad \blacksquare$$

Orthogonality and Norms

In Chapter 8 two vectors $\mathbf{u}$ and $\mathbf{v}$ in $\mathbf{R}^n$ were defined to be orthogonal if $\mathbf{u} \cdot \mathbf{v} = 0$. We extend this to inner products in a natural way.

DEFINITION 10.2

Definition Orthogonal Vectors

Two vectors $\mathbf{u}$ and $\mathbf{v}$ in an inner product space V are **orthogonal** if and only if $\langle \mathbf{u}, \mathbf{v} \rangle = 0$.

For instance, in Example 6 we showed that the vectors $f(x) = x^2 + 4x$ and $g(x) = 5x^2 - 3$ are orthogonal with respect to the inner product given in Example 5.

EXAMPLE 7 Which pairs among $p_1(x) = x^2 - 5x + 4$, $p_2(x) = x^2 - x - 2$, and $p_3(x) = 3x^2 - x - 4$ are orthogonal with respect to the inner product on $\mathbf{P}^2$ in Example 4?

Solution We have

$$\langle p_1, p_2 \rangle = p_1(-1)p_2(-1) + p_1(1)p_2(1) + p_1(4)p_2(4) = 0$$
$$\langle p_1, p_3 \rangle = p_1(-1)p_3(-1) + p_1(1)p_3(1) + p_1(4)p_3(4) = 0$$
$$\langle p_2, p_3 \rangle = p_2(-1)p_3(-1) + p_2(1)p_3(1) + p_2(4)p_3(4) = 404$$

Hence $p_1(x)$ and $p_2(x)$ are orthogonal, $p_1(x)$ and $p_3(x)$ are orthogonal, but $p_2(x)$ and $p_3(x)$ are not. $\quad \blacksquare$

EXAMPLE 8 If $V = C[-1, 1]$, then we can show that

$$\langle f, g \rangle = \int_{-1}^{1} (x^2 + 1)\, f(x)g(x)\, dx$$

is an inner product on V, a weighted version of the inner product in Example 5 (see Exercise 54). Compute $\langle f, g \rangle$ for $f(x)$ and $g(x)$ in Example 6.

Solution Here we have

$$\langle f, g \rangle = \int_{-1}^{1} (x^2 + 1)(x^2 + 4x)(5x^2 - 3)\, dx = \frac{8}{35}$$

so that $f(x)$ and $g(x)$ are not orthogonal with respect to this inner product. $\quad \blacksquare$

In Section 10.1 we defined the norm (length) of a vector in $\mathbf{R}^n$ in terms of the dot product. Here we define the norm in terms of an inner product. Just as in Euclidean space, the norm on a vector space gives us a way to define the length of each vector in the space (the norm of the vector) and to measure the distance between vectors (the norm of the difference).

DEFINITION 10.3

Definition Norm

Let $\mathbf{v}$ be a vector in an inner product space V. Then the **norm** of $\mathbf{v}$ is given by

$$\|\mathbf{v}\| = \sqrt{\langle \mathbf{v}, \mathbf{v} \rangle}$$

EXAMPLE 9 If A and B are matrices in $\mathbf{R}^{3\times3}$, then it can be shown that (see Exercise 57)

$$\langle A, B \rangle = \text{tr}(A^T B)$$

▶ Recall that $\text{tr}(C)$ denotes the *trace* of C, the sum of the diagonal entries of C.

is an inner product. Compute $\|A\|$ for $A = \begin{bmatrix} 1 & -2 & -1 \\ 3 & 2 & 0 \\ -1 & 1 & 1 \end{bmatrix}$.

Solution We have

$$\|A\|^2 = \langle A, A \rangle = \text{tr}(A^T A) = \text{tr}\left(\begin{bmatrix} 11 & 3 & -2 \\ 3 & 9 & 3 \\ -2 & 3 & 2 \end{bmatrix} \right) = 22$$

Hence $\|A\| = \sqrt{22} \approx 4.69$. ■

EXAMPLE 10 Suppose that the vector space $\mathbf{P}^1$ has inner product

$$\langle p, q \rangle = p(0)q(0) + 3p(1)q(1) + 2p(3)q(3)$$

Determine which of $p(x) = 3x - 2$ and $q(x) = -2x + 4$ is longer, and find the distance between the two vectors.

Solution The length of each vector is given by the norm, so that

$$\|p\| = \sqrt{\langle p, p \rangle} = \sqrt{(-2)^2 + 3(1)^2 + 2(7)^2} = \sqrt{105} \approx 10.247$$
$$\|q\| = \sqrt{\langle q, q \rangle} = \sqrt{(4)^2 + 3(2)^2 + 2(-2)^2} = \sqrt{36} = 6$$

Since $\|p\| > \|q\|$, p is longer than q. (Remember that the results might be different with another inner product.) The distance between our vectors is given by $\|p - q\|$. Since $p(x) - q(x) = 5x - 6$, we have

$$\|p - q\| = \sqrt{\langle p - q, p - q \rangle} = \sqrt{(-6)^2 + 3(-1)^2 + 2(9)^2} = \sqrt{201} \approx 14.177 \quad ■$$

In Euclidean space we formulated the Pythagorean Theorem in terms of norms. We can do the same with inner product spaces.

THEOREM 10.4

▶ Theorem 10.4 generalizes Theorem 8.6 in Section 8.1.

(PYTHAGOREAN THEOREM) Let $\mathbf{u}$ and $\mathbf{v}$ be vectors in an inner product space V. Then $\mathbf{u}$ and $\mathbf{v}$ are orthogonal if and only if

$$\|\mathbf{u}\|^2 + \|\mathbf{v}\|^2 = \|\mathbf{u} + \mathbf{v}\|^2 \tag{4}$$

Proof The proof only uses properties from the definition of an inner product, without any reference to a specific inner product. We have

$$\begin{aligned} \|\mathbf{u} + \mathbf{v}\|^2 &= \langle \mathbf{u} + \mathbf{v}, \mathbf{u} + \mathbf{v} \rangle \\ &= \langle \mathbf{u}, \mathbf{u} \rangle + \langle \mathbf{u}, \mathbf{v} \rangle + \langle \mathbf{v}, \mathbf{u} \rangle + \langle \mathbf{v}, \mathbf{v} \rangle \\ &= \|\mathbf{u}\|^2 + \|\mathbf{v}\|^2 + 2\langle \mathbf{u}, \mathbf{v} \rangle \end{aligned}$$

Since $\mathbf{u}$ and $\mathbf{v}$ are orthogonal if and only if $\langle \mathbf{u}, \mathbf{v} \rangle = 0$, then $\mathbf{u}$ and $\mathbf{v}$ are orthogonal if and only if (4) is true. ■

EXAMPLE 11 Verify the Pythagorean Theorem for the vectors $p_1(x) = x^2 - 5x + 4$ and $p_2(x) = x^2 - x - 2$ in the inner product space given in Example 4.

Solution We saw in Example 7 that $\langle p_1, p_2 \rangle = 0$, so by the Pythagorean Theorem we expect that $\| p_1 \|^2 + \| p_2 \|^2 = \| p_1 + p_2 \|^2$. To verify this, we compute

$$\| p_1 \|^2 = \langle p_1, p_1 \rangle = \left(p_1(-1) \right)^2 + \left(p_1(1) \right)^2 + \left(p_1(4) \right)^2 = 100$$
$$\| p_2 \|^2 = \langle p_2, p_2 \rangle = \left(p_2(-1) \right)^2 + \left(p_2(1) \right)^2 + \left(p_2(4) \right)^2 = 104$$
$$\| p_1 + p_2 \|^2 = \langle p_1 + p_2, p_1 + p_2 \rangle$$
$$= \left(p_1(-1) + p_2(-1) \right)^2 + \left(p_1(1) + p_2(1) \right)^2 + \left(p_1(4) + p_2(4) \right)^2 = 204$$

Therefore $\| p_1 \|^2 + \| p_2 \|^2 = \| p_1 + p_2 \|^2$. ∎

Projection and Inequalities

When studying dot products in Euclidean space, we developed the projection of one vector onto another and extended this to the projection of a vector onto a subspace. Here we shall generalize projections onto a vector in a vector space. We treat projections onto subspaces in the next section.

In Section 8.2, the formula for the projection of one vector onto another was defined in terms of dot products. Hence it is reasonable to generalize projections to inner product spaces by changing the dot products into inner products.

DEFINITION 10.5

Definition Projection onto a Vector

Let $\mathbf{u}$ and $\mathbf{v}$ be vectors in an inner product space V, with $\mathbf{v}$ nonzero. Then the **projection of $\mathbf{u}$ onto $\mathbf{v}$** is given by

$$\text{proj}_{\mathbf{v}} \mathbf{u} = \frac{\langle \mathbf{v}, \mathbf{u} \rangle}{\langle \mathbf{v}, \mathbf{v} \rangle} \mathbf{v} = \frac{\langle \mathbf{v}, \mathbf{u} \rangle}{\| \mathbf{v} \|^2} \mathbf{v} \tag{5}$$

EXAMPLE 12 Determine $\text{proj}_{\mathbf{v}} \mathbf{u}$ for the vectors $\mathbf{u}$, $\mathbf{v}$, and the inner product space given in Example 2.

Solution In Example 2 we showed that $\langle \mathbf{v}, \mathbf{u} \rangle = -3$. The inner product of $\mathbf{v}$ with itself is

$$\langle \mathbf{v}, \mathbf{v} \rangle = v_1^2 t_1 + v_2^2 t_2 + v_3^2 t_3 = (4)^2(2) + (-1)^2(3) + (1)^2(1) = 36$$

Therefore

$$\text{proj}_{\mathbf{v}} \mathbf{u} = \frac{\langle \mathbf{v}, \mathbf{u} \rangle}{\langle \mathbf{v}, \mathbf{v} \rangle} \mathbf{v} = \frac{-3}{36} \begin{bmatrix} 4 \\ -1 \\ 1 \end{bmatrix} = -\frac{1}{12} \begin{bmatrix} 4 \\ -1 \\ 1 \end{bmatrix} = \begin{bmatrix} -\frac{1}{3} \\ \frac{1}{12} \\ -\frac{1}{12} \end{bmatrix}$$ ∎

Our new definition for projection will be a disappointment if it does not have properties similar to those of the Euclidean space version of projection. Happily, everything carries over to inner product spaces with no significant changes.

THEOREM 10.6

▶ Theorem 10.6 generalizes Theorem 8.14 in Section 8.2.

Let $\mathbf{u}$ and $\mathbf{v}$ be vectors in an inner product space V, with $\mathbf{v}$ nonzero, and let c be a nonzero scalar. Then

(a) $\text{proj}_{\mathbf{v}} \mathbf{u}$ is in span$\{\mathbf{v}\}$.

(b) $\mathbf{u} - \text{proj}_{\mathbf{v}} \mathbf{u}$ is orthogonal to $\mathbf{v}$.

(c) If $\mathbf{u}$ is in span$\{\mathbf{v}\}$, then $\mathbf{u} = \text{proj}_{\mathbf{v}} \mathbf{u}$.

(d) $\text{proj}_{\mathbf{v}} \mathbf{u} = \text{proj}_{c\mathbf{v}} \mathbf{u}$.

Proof   We give the proofs of parts (a) and (b) here and leave the proofs of parts (c) and (d) as an exercise. To prove (a), we note that since

$$\text{proj}_\mathbf{v}\mathbf{u} = \frac{\langle \mathbf{v}, \mathbf{u} \rangle}{\langle \mathbf{v}, \mathbf{v} \rangle}\mathbf{v}$$

then $\text{proj}_\mathbf{v}\mathbf{u}$ is a scalar multiple of $\mathbf{v}$ and so is in span$\{\mathbf{v}\}$.

For part (b), we have

$$\langle \mathbf{u} - \text{proj}_\mathbf{v}\mathbf{u}, \mathbf{v} \rangle = \langle \mathbf{u}, \mathbf{v} \rangle - \langle \text{proj}_\mathbf{v}\mathbf{u}, \mathbf{v} \rangle$$

$$= \langle \mathbf{u}, \mathbf{v} \rangle - \left\langle \frac{\langle \mathbf{v}, \mathbf{u} \rangle}{\langle \mathbf{v}, \mathbf{v} \rangle}\mathbf{v}, \mathbf{v} \right\rangle$$

$$= \langle \mathbf{u}, \mathbf{v} \rangle - \frac{\langle \mathbf{v}, \mathbf{u} \rangle}{\langle \mathbf{v}, \mathbf{v} \rangle}\langle \mathbf{v}, \mathbf{v} \rangle = \langle \mathbf{u}, \mathbf{v} \rangle - \langle \mathbf{v}, \mathbf{u} \rangle = 0$$

because $\langle \mathbf{u}, \mathbf{v} \rangle = \langle \mathbf{v}, \mathbf{u} \rangle$. Therefore $\mathbf{u} - \text{proj}_\mathbf{v}\mathbf{u}$ is orthogonal to $\mathbf{v}$.   ■

Combining (a) and (b) of Theorem 10.6 tells us that $\text{proj}_\mathbf{v}\mathbf{u}$ and $\mathbf{u} - \text{proj}_\mathbf{v}\mathbf{u}$ are orthogonal (see Exercise 64), so that by the Pythagorean Theorem,

$$\|\mathbf{u}\|^2 = \|\text{proj}_\mathbf{v}\mathbf{u}\|^2 + \|\mathbf{u} - \text{proj}_\mathbf{v}\mathbf{u}\|^2$$

Since $\|\mathbf{u} - \text{proj}_\mathbf{v}\mathbf{u}\|^2 \geq 0$, it follows that

$$\|\text{proj}_\mathbf{v}\mathbf{u}\| \leq \|\mathbf{u}\| \tag{6}$$

This inequality is useful for proving the next theorem.

THEOREM 10.7

▶ Theorem 10.7 generalizes Exercise 57 in Section 8.2.

(THE CAUCHY–SCHWARZ INEQUALITY)   For all $\mathbf{u}$ and $\mathbf{v}$ in an inner product space V,

$$|\langle \mathbf{u}, \mathbf{v} \rangle| \leq \|\mathbf{u}\|\|\mathbf{v}\| \tag{7}$$

Proof   First, if either $\mathbf{u} = \mathbf{0}$ or $\mathbf{v} = \mathbf{0}$, then both sides of (7) are equal to 0 and we are done. So let's assume that both $\mathbf{u}$ and $\mathbf{v}$ are nonzero vectors. Then

$$\|\text{proj}_\mathbf{v}\mathbf{u}\| = \left\| \frac{\langle \mathbf{v}, \mathbf{u} \rangle}{\|\mathbf{v}\|^2}\mathbf{v} \right\| = \frac{|\langle \mathbf{v}, \mathbf{u} \rangle|}{\|\mathbf{v}\|^2}\|\mathbf{v}\| = \frac{|\langle \mathbf{v}, \mathbf{u} \rangle|}{\|\mathbf{v}\|}$$

Combining this with (6) yields the inequality

$$\frac{|\langle \mathbf{v}, \mathbf{u} \rangle|}{\|\mathbf{v}\|} \leq \|\mathbf{u}\|$$

Hence (7) holds, completing the proof.   ■

EXAMPLE 13   Verify that the Cauchy–Schwarz inequality holds for the inner product given in Example 9 when applied to the matrices

$$A = \begin{bmatrix} 2 & 0 & -1 \\ 1 & 1 & 2 \\ -2 & -1 & 0 \end{bmatrix} \quad \text{and} \quad B = \begin{bmatrix} 1 & 0 & 1 \\ -2 & 1 & 2 \\ 0 & 1 & 1 \end{bmatrix}$$

Solution We have

$$\langle A, B \rangle = \text{tr}(A^T B) = \text{tr}\left(\begin{bmatrix} 0 & -1 & 2 \\ -2 & 0 & 1 \\ -5 & 2 & 3 \end{bmatrix}\right) = 3$$

$$\|A\| = \sqrt{\langle A, A \rangle} = \sqrt{\text{tr}(A^T A)} = \sqrt{\text{tr}\left(\begin{bmatrix} 9 & 3 & 0 \\ 3 & 2 & 2 \\ 0 & 2 & 5 \end{bmatrix}\right)} = \sqrt{16}$$

$$\|B\| = \sqrt{\langle B, B \rangle} = \sqrt{\text{tr}(B^T B)} = \sqrt{\text{tr}\left(\begin{bmatrix} 5 & -2 & -3 \\ -2 & 2 & 3 \\ -3 & 3 & 6 \end{bmatrix}\right)} = \sqrt{13}$$

Since $3 < \sqrt{16}\sqrt{13} \approx 14.422$, the Cauchy–Schwarz inequality is verified for this pair of matrices. ■

The Cauchy–Schwarz inequality makes it easy to prove a second important inequality.

THEOREM 10.8

(THE TRIANGLE INEQUALITY) For all **u** and **v** in an inner product space V,

$$\|\mathbf{u} + \mathbf{v}\| \leq \|\mathbf{u}\| + \|\mathbf{v}\| \tag{8}$$

Proof We have

$$\begin{aligned}
\|\mathbf{u} + \mathbf{v}\|^2 &= \langle \mathbf{u} + \mathbf{v}, \mathbf{u} + \mathbf{v} \rangle \\
&= \langle \mathbf{u}, \mathbf{u} \rangle + \langle \mathbf{u}, \mathbf{v} \rangle + \langle \mathbf{v}, \mathbf{u} \rangle + \langle \mathbf{v}, \mathbf{v} \rangle \\
&= \|\mathbf{u}\|^2 + \|\mathbf{v}\|^2 + 2\langle \mathbf{u}, \mathbf{v} \rangle \\
&\leq \|\mathbf{u}\|^2 + \|\mathbf{v}\|^2 + 2\|\mathbf{u}\|\|\mathbf{v}\| \quad \text{(By Cauchy–Schwarz inequality)} \\
&= \left(\|\mathbf{u}\| + \|\mathbf{v}\|\right)^2
\end{aligned}$$

Taking square roots on both sides yields (8) and completes the proof. ■

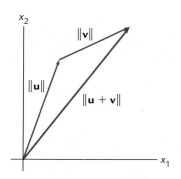

Figure 1 $\|\mathbf{u} + \mathbf{v}\| \leq \|\mathbf{u}\| + \|\mathbf{v}\|$.

If we think of **u** and **v** placed tip-to-tail to form two sides of a triangle (see Figure 1), then $\mathbf{u} + \mathbf{v}$ gives the third side. It is geometrically evident in $\mathbf{R}^2$ that the sum of the lengths of two sides of a triangle must be at least as great as the length of the third side. The triangle inequality tells us that the same is true in any inner product space.

EXERCISES

For Exercises 1–8, compute the indicated inner product.

1. $\langle \mathbf{u}, \mathbf{v} \rangle$ for $\mathbf{u} = \begin{bmatrix} 1 \\ 2 \\ 1 \end{bmatrix}$, $\mathbf{v} = \begin{bmatrix} 3 \\ 4 \\ 2 \end{bmatrix}$, and the inner product given in Example 2.

2. $\langle \mathbf{u}, \mathbf{v} \rangle$ for $\mathbf{u} = \begin{bmatrix} 2 \\ 5 \\ 2 \end{bmatrix}$, $\mathbf{v} = \begin{bmatrix} 1 \\ 3 \\ 0 \end{bmatrix}$, and the inner product given in Example 2 with weights $t_1 = 3$, $t_2 = 1$, and $t_3 = 4$.

3. $\langle p, q \rangle$ for $p(x) = 3x + 2$, $q(x) = -x + 1$, and the inner product given in Example 3 with $x_0 = -1$, $x_1 = 0$, and $x_2 = 2$.

4. $\langle p, q \rangle$ for $p(x) = x^2 + 1$, $q(x) = 2x - 3$, and the inner product given in Example 3 with $x_0 = -1$, $x_1 = 1$, $x_2 = 2$, and $x_3 = 5$.

5. $\langle f, g \rangle$ for $f(x) = x + 3$, $g(x) = x^2$, and the inner product given in Example 5.

6. $\langle f, g \rangle$ for $f(x) = x$, $g(x) = e^x$, and the inner product given in Example 5.

7. $\langle A, B \rangle = \text{tr}(A^T B)$ for $A = \begin{bmatrix} 2 & -1 \\ 3 & 4 \end{bmatrix}$, $B = \begin{bmatrix} 5 & 2 \\ -3 & -2 \end{bmatrix}$.

8. $\langle A, B \rangle = \text{tr}(A^T B)$ for $A = \begin{bmatrix} 7 & -3 \\ 2 & 6 \end{bmatrix}$, $B = \begin{bmatrix} 4 & 2 \\ 0 & 5 \end{bmatrix}$.

9. Suppose that $\mathbf{u} = \begin{bmatrix} 1 \\ 0 \\ -1 \end{bmatrix}$ and $\mathbf{v} = \begin{bmatrix} 2 \\ 1 \\ 2 \end{bmatrix}$ are orthogonal with respect to the inner product given in Example 2 with weights $t_1 = 3$, $t_2 = 1$, and $t_3 = a$. What are the possible value(s) of a?

10. Suppose that $\mathbf{u} = \begin{bmatrix} 3 \\ 2 \\ 2 \end{bmatrix}$ and $\mathbf{v} = \begin{bmatrix} 5 \\ -1 \\ 2 \end{bmatrix}$ are orthogonal with respect to the inner product given in Example 2 with weights $t_1 = 1$, $t_2 = b$, and $t_3 = 2$. What are the possible value(s) of b?

11. Suppose that $p(x) = x + 2$, $q(x) = -3x + 1$ are orthogonal with respect to the inner product given in Example 3 with $x_0 = -1$, $x_1 = a$, and $x_2 = 2$. What are the possible value(s) of a?

12. Suppose that $p(x) = x^2 - 3x - 1$, $q(x) = x + 2$ are orthogonal with respect to the inner product given in Example 3 with $x_0 = -2$, $x_1 = 0$, $x_2 = 1$, and $x_3 = a$. What are the possible value(s) of a?

13. Suppose that $f(x) = 2x$ and $g(x) = x + b$. For what value(s) of b are f and g orthogonal with respect to the inner product in Example 5?

14. Suppose that $f(x) = x^2$ and $g(x) = x + b$. For what value(s) of b are f and g orthogonal with respect to the inner product in Example 5?

For Exercises 15–22, compute the norm with respect to the indicated inner product.

15. $\left\| \begin{bmatrix} 1 \\ -3 \\ 2 \end{bmatrix} \right\|$ for the inner product given in Example 2.

16. $\left\| \begin{bmatrix} 2 \\ 0 \\ -5 \end{bmatrix} \right\|$ for the inner product given in Example 2 with weights $t_1 = 1$, $t_2 = 5$, and $t_3 = 2$.

17. $\|3x - 5\|$ for the inner product given in Example 3 with $x_0 = -2$, $x_1 = 1$, and $x_2 = 4$.

18. $\|-x^2 + x - 4\|$ for the inner product given in Example 3 with $x_0 = 0$, $x_1 = 3$, $x_2 = 2$, and $x_3 = 6$.

19. $\|x^3\|$ for the inner product given in Example 5.

20. $\|xe^{-x}\|$ for the inner product given in Example 5.

21. $\|A\|$ for $A = \begin{bmatrix} 3 & -1 \\ 2 & 0 \end{bmatrix}$ and $\langle A, B \rangle = \text{tr}(A^T B)$.

22. $\|A\|$ for $A = \begin{bmatrix} 2 & 3 & 0 \\ 1 & -3 & -1 \\ 2 & 5 & 2 \end{bmatrix}$ and $\langle A, B \rangle = \text{tr}(A^T B)$.

For Exercises 23–30, compute the indicated projection with respect to the given inner product.

23. $\text{proj}_\mathbf{u}\mathbf{v}$ for $\mathbf{u} = \begin{bmatrix} 1 \\ 2 \\ 1 \end{bmatrix}$, $\mathbf{v} = \begin{bmatrix} 3 \\ 4 \\ 2 \end{bmatrix}$, and the inner product given in Example 2.

24. $\text{proj}_\mathbf{u}\mathbf{v}$ for $\mathbf{u} = \begin{bmatrix} 2 \\ 5 \\ 2 \end{bmatrix}$, $\mathbf{v} = \begin{bmatrix} 1 \\ 3 \\ 0 \end{bmatrix}$, and the inner product given in Example 2 with weights $t_1 = 3$, $t_2 = 1$, and $t_3 = 4$.

25. $\text{proj}_p q$ for $p(x) = 3x + 2$, $q(x) = -x + 1$, and the inner product given in Example 3 with $x_0 = -1$, $x_1 = 0$, and $x_2 = 2$.

26. $\text{proj}_p q$ for $p(x) = x^2 + 1$, $q(x) = 2x - 3$, and the inner product given in Example 3 with $x_0 = -1$, $x_1 = 1$, $x_2 = 2$, and $x_3 = 5$.

27. $\text{proj}_f g$ for $f(x) = x$, $g(x) = x^2$, and the inner product given in Example 5.

28. $\text{proj}_f g$ for $f(x) = \sin(x)$, $g(x) = 1 - x^2$, and the inner product given in Example 5.

29. $\text{proj}_A B$ for $A = \begin{bmatrix} 2 & -1 \\ 1 & 0 \end{bmatrix}$, $B = \begin{bmatrix} 2 & 3 \\ 0 & -2 \end{bmatrix}$, and $\langle A, B \rangle = \text{tr}(A^T B)$.

30. $\text{proj}_A B$ for $A = \begin{bmatrix} 3 & 4 \\ -1 & -3 \end{bmatrix}$, $B = \begin{bmatrix} 1 & 5 \\ -2 & 1 \end{bmatrix}$, and $\langle A, B \rangle = \text{tr}(A^T B)$.

FIND AN EXAMPLE For Exercises 31–40, find an example that meets the given specifications.

31. An orthogonal basis for $\mathbf{R}^2$ with respect to the inner product $\langle \mathbf{u}, \mathbf{v} \rangle = 3u_1 v_1 + 2u_2 v_2$.

32. An orthogonal basis for $\mathbf{P}^1$ with respect to the inner product $\langle p, q \rangle = \int_0^1 p(x)q(x)\,dx$.

33. A vector $\mathbf{u}$ such that $\|\mathbf{u}\| = 2$ and $\mathbf{u}$ is orthogonal to $\begin{bmatrix} 2 \\ 3 \end{bmatrix}$ with respect to a weighted dot product of the form $\langle \mathbf{u}, \mathbf{v} \rangle = t_1 u_1 v_1 + t_2 u_2 v_2$.

34. An inner product on $\mathbf{P}^2$ such that $\|p\| = 3$ for $p(x) = x^2 - 4x + 3$.

35. A nonidentity matrix A such that $\langle \mathbf{u}, \mathbf{v} \rangle = \mathbf{u}^T A\mathbf{v}$ is an inner product on $\mathbf{R}^3$.

36. A nonidentity matrix A such that $\langle \mathbf{u}, \mathbf{v} \rangle = \mathbf{u}^T A\mathbf{v}$ is *not* an inner product on $\mathbf{R}^3$.

37. An inner product of your creation on $\mathbf{P}^2$.

38. An inner product of your own creation on $C[0, 1]$.

39. A function $\langle p, q \rangle$ that is *almost* an inner product on $\mathbf{P}^2$: It satisfies (a)–(c) of Definition 10.1, but not (d).

40. A function $\langle A, B \rangle$ that is a poor attempt at an inner product on $\mathbf{R}^{3\times3}$: It satisfies (a) of Definition 10.1, but not (b)–(d).

TRUE OR FALSE For Exercises 41–50, determine if the statement is true or false, and justify your answer. Here $\mathbf{u}$ and $\mathbf{v}$ are vectors in an inner product space V.

41. If $\langle \mathbf{u}, \mathbf{v} \rangle = 3$, then $\langle 2\mathbf{u}, -4\mathbf{v} \rangle = -24$.

42. $\|\mathbf{u} + \mathbf{v}\|^2 = \|\mathbf{u}\|^2 + \|\mathbf{v}\|^2$ for all $\mathbf{u}$ and $\mathbf{v}$ in V.

43. If $\mathbf{u}$ and $\mathbf{v}$ are orthogonal with $\|\mathbf{u}\| = 3$ and $\|\mathbf{v}\| = 4$, then $\|\mathbf{u} + \mathbf{v}\| = 5$.

44. If $\mathbf{u} = c\mathbf{v} \neq \mathbf{0}$ for a scalar c, then $\mathbf{u} = \text{proj}_\mathbf{u}\mathbf{v}$.

45. If $\{\mathbf{u}, \mathbf{v}\}$ is an orthogonal set and c_1 and c_2 are scalars, then $\{c_1\mathbf{u}, c_2\mathbf{v}\}$ is also an orthogonal set.

46. $-\|\mathbf{u}\|\|\mathbf{v}\| \leq \langle \mathbf{u}, \mathbf{v} \rangle$ for all $\mathbf{u}$ and $\mathbf{v}$ in V.

47. $\|\mathbf{u} - \mathbf{v}\| \leq \|\mathbf{u}\| - \|\mathbf{v}\|$ for all $\mathbf{u}$ and $\mathbf{v}$ in V.

48. $\langle f, g \rangle = \int_{-1}^{1} x f(x) g(x) \, dx$ is an inner product on $C[-1, 1]$.

49. $\langle p, q \rangle = p(x_0)q(x_0) + p(x_1)q(x_1)$ is an inner product on $\mathbf{P}^2$ when $x_0 \neq x_1$.

50. If $T : V \rightarrow \mathbf{R}$ is a linear transformation, then $\langle \mathbf{u}, \mathbf{v} \rangle = T(\mathbf{u}) \cdot T(\mathbf{v})$ is an inner product.

51. Complete Example 1. Prove that the weighted dot product on $\mathbf{R}^n$ given by

$$\langle \mathbf{u}, \mathbf{v} \rangle = t_1 u_1 v_1 + t_2 u_2 v_2 + \cdots + t_n u_n v_n$$

where $t_1, t_2, \ldots, t_n$ are all positive, is an inner product

52. Complete Example 3. Show that properties (a)–(c) of Definition 10.1 are true for the inner product

$$\langle p, q \rangle = p(x_0)q(x_0) + \cdots + p(x_n)q(x_n)$$

53. A weighted version of the inner product given in Example 3 is defined as follows: For $p(x)$ and $q(x)$ in $\mathbf{P}^n$ and distinct real numbers $x_0, x_1, \ldots, x_n$, let

$$\langle p, q \rangle = t(x_0) p(x_0) q(x_0) + \cdots + t(x_n) p(x_n) q(x_n)$$

where $t(x)$ takes positive values on $x_0, \ldots, x_n$. Show that $\langle p, q \rangle$ is an inner product on $\mathbf{P}^n$.

54. Let f and g be continuous functions in $C[-1, 1]$. Show that the weighted version of (3) given by

$$\langle f, g \rangle = \int_{-1}^{1} t(x) f(x) g(x) \, dx$$

where $t(x) > 0$ is continuous for all x in $[-1, 1]$, defines an inner product on $C[-1, 1]$.

55. Let f and g be continuous functions in $C[-\pi, \pi]$. Show that

$$\langle f, g \rangle = \frac{1}{\pi} \int_{-\pi}^{\pi} f(x) g(x) \, dx$$

defines an inner product on $C[-\pi, \pi]$.

56. Complete the proof of Theorem 10.6, by showing that parts (c) and (d) are true.

57. Prove that $\langle A, B \rangle = \text{tr}(A^T B)$ is an inner product on $\mathbf{R}^{3 \times 3}$.

58. For nonzero vectors $\mathbf{u}$ and $\mathbf{v}$, show that there is equality in the Cauchy–Schwarz inequality exactly when $\mathbf{u} = c\mathbf{v}$ for some scalar c. (HINT: The key lies with the inequality (6).)

For Exercises 59–68, $\mathbf{u}$, $\mathbf{v}$, and $\mathbf{w}$ (and their subscripted associates) are vectors in an inner product space V.

59. Prove that $\|c\mathbf{v}\| = |c| \|\mathbf{v}\|$ for every $\mathbf{v}$ and scalar c.

60. Prove that $\|\mathbf{v}\| \geq 0$ for every $\mathbf{v}$ in V, with equality holding only for $\mathbf{v} = \mathbf{0}$.

61. Prove that

$$\langle c_1 \mathbf{u}_1 + \cdots + c_k \mathbf{u}_k, \mathbf{w} \rangle = c_1 \langle \mathbf{u}_1, \mathbf{w} \rangle + \cdots + c_k \langle \mathbf{u}_k, \mathbf{w} \rangle$$

62. Prove that the zero vector $\mathbf{0}$ is orthogonal to all vectors in V.

63. Prove that if $\mathbf{v} \neq \mathbf{0}$, then $\mathbf{w} = \dfrac{1}{\|\mathbf{v}\|} \mathbf{v}$ satisfies $\|\mathbf{w}\| = 1$.

64. Prove that $\text{proj}_\mathbf{v} \mathbf{u}$ and $\mathbf{u} - \text{proj}_\mathbf{v} \mathbf{u}$ are orthogonal.

65. For a fixed $\mathbf{v}$ in V, define $T_\mathbf{v} : V \rightarrow \mathbf{R}$ by $T_\mathbf{v}(\mathbf{u}) = \langle \mathbf{u}, \mathbf{v} \rangle$. Show that $T_\mathbf{v}$ is a linear transformation.

66. For a fixed $\mathbf{v}$ in V, define $T_\mathbf{v} : V \rightarrow V$ by $T_\mathbf{v}(\mathbf{u}) = \text{proj}_\mathbf{v} \mathbf{u}$. Show that $T_\mathbf{v}$ is a linear transformation.

67. Prove that if $\mathbf{u}$ and $\mathbf{v}$ are orthogonal, then the distance between $\mathbf{u}$ and $\mathbf{v}$ is $\sqrt{\|\mathbf{u}\|^2 + \|\mathbf{v}\|^2}$.

68. Prove that

$$\|\mathbf{u} + \mathbf{v}\|^2 + \|\mathbf{u} - \mathbf{v}\|^2 = 2\left(\|\mathbf{u}\|^2 + \|\mathbf{v}\|^2\right)$$

If S is a subspace of a finite-dimensional inner product space V, a vector $\mathbf{v}$ is **orthogonal** to S if $\langle \mathbf{v}, \mathbf{s} \rangle = 0$ for every vector $\mathbf{s}$ in S. The set of all such vectors $\mathbf{v}$ is called the **orthogonal complement** of S and is denoted by $S^\perp$. In Exercises 69–72, prove that the statement involving $S^\perp$ is true.

69. If S is a subspace, then so is $S^\perp$.

70. If S is a subspace, then $\left(S^\perp\right)^\perp = S$.

71. If $\mathbf{s}$ is in S and $\mathbf{s}^\perp$ is in $S^\perp$, then

$$\|\mathbf{s} \pm \mathbf{s}^\perp\|^2 = \|\mathbf{s}\|^2 + \|\mathbf{s}^\perp\|^2.$$

72. If S is a subspace, then $S \cap S^\perp = \{\mathbf{0}\}$.

10.2 The Gram–Schmidt Process Revisited

▶ Recall that the Gram–Schmidt process allows us to transform a basis into an orthogonal basis. See Section 8.2 for details.

Our main goal for this section is to develop a version of the Gram–Schmidt process for an inner product space. Before we can do that, we need to carry over some concepts from Euclidean space, starting with the definition of an orthonormal set.

DEFINITION 10.9
Definition Orthogonal Set

The vectors $\{\mathbf{v}_1, \ldots, \mathbf{v}_k\}$ in an inner product space V form an **orthogonal set** if $\langle \mathbf{v}_i, \mathbf{v}_j \rangle = 0$ for $i \neq j$.

EXAMPLE 1   Let V be the inner product space consisting of vectors in $\mathbf{R}^3$ and the weighted dot product $\langle \mathbf{u}, \mathbf{v} \rangle = 2u_1v_1 + 3u_2v_2 + u_3v_3$. Show that the vectors

$$\mathbf{v}_1 = \begin{bmatrix} 4 \\ 5 \\ 1 \end{bmatrix}, \quad \mathbf{v}_2 = \begin{bmatrix} -3 \\ 2 \\ -6 \end{bmatrix}, \quad \mathbf{v}_3 = \begin{bmatrix} 16 \\ -7 \\ -23 \end{bmatrix}$$

form an orthogonal set.

Solution   The inner products of each pair are

$$\langle \mathbf{v}_1, \mathbf{v}_2 \rangle = (2)(4)(-3) + (3)(5)(2) + (1)(-6) = 0$$
$$\langle \mathbf{v}_1, \mathbf{v}_3 \rangle = (2)(4)(16) + (3)(5)(-7) + (1)(-23) = 0$$
$$\langle \mathbf{v}_2, \mathbf{v}_3 \rangle = (2)(-3)(16) + (3)(2)(-7) + (-6)(-23) = 0$$

and therefore the set $\{\mathbf{v}_1, \mathbf{v}_2, \mathbf{v}_3\}$ is orthogonal.   ∎

EXAMPLE 2   Let $V = C[-\pi, \pi]$ be the inner product space of continuous functions on $[-\pi, \pi]$ with respect to the inner product

$$\langle f, g \rangle = \frac{1}{\pi} \int_{-\pi}^{\pi} f(x)g(x)\, dx$$

Show that the set $\{1, \cos(x), \sin(x)\}$ is orthogonal.

Solution   As in Example 1, we need to compute the inner products of the three possible pairs of vectors:

$$\langle 1, \cos(x) \rangle = \frac{1}{\pi} \int_{-\pi}^{\pi} (1)(\cos(x))\, dx = \frac{1}{\pi}\left(\sin(\pi) - \sin(-\pi)\right) = 0$$

$$\langle 1, \sin(x) \rangle = \frac{1}{\pi} \int_{-\pi}^{\pi} (1)(\sin(x))\, dx = -\frac{1}{\pi}\left(\cos(\pi) - \cos(-\pi)\right) = 0$$

$$\langle \cos(x), \sin(x) \rangle = \frac{1}{\pi} \int_{-\pi}^{\pi} (\cos(x))(\sin(x))\, dx = \frac{1}{\pi} \int_{-\pi}^{\pi} \frac{1}{2}\sin(2x)\, dx$$

$$= -\frac{1}{4\pi}\left(\cos(2\pi) - \cos(-2\pi)\right) = 0$$

▶ Recall the identity
$2\cos(x)\sin(x) = \sin(2x)$.

Thus our set of vectors is orthogonal.   ∎

When studying orthogonality in Euclidean space, we saw that an orthogonal set of nonzero vectors must be linearly independent. The same is true of such a set in an inner product space.

THEOREM 10.10

▶ Theorem 10.10 generalizes
Theorem 8.11 in Section 8.1.

Let $\mathcal{V} = \{\mathbf{v}_1, \ldots, \mathbf{v}_m\}$ be an orthogonal set of nonzero vectors in an inner product space V. Then $\mathcal{V}$ is linearly independent.

The proof is left as an exercise. An interesting consequence of Theorem 10.10 is that if a given set of nonzero vectors is orthogonal with respect to *just one* inner product, then the set must be linearly independent. We also can flip this around: If the set is linearly dependent, then it cannot be orthogonal with respect to *any* inner product, no matter how cleverly selected.

An orthogonal set of vectors that forms a basis for an inner product space is called an **orthogonal basis**.

Definition Orthogonal Basis

EXAMPLE 3 An important class of orthogonal polynomials are the *Legendre polynomials*. There is an infinite sequence of them—the first four are

$$p_0(x) = 1, \quad p_1(x) = x, \quad p_2(x) = \tfrac{1}{2}(3x^2 - 1), \quad p_3(x) = \tfrac{1}{2}(5x^3 - 3x)$$

(See Figure 1.) Show that this set of polynomials is a basis for $\mathbf{P}^3$ by showing they are an orthogonal set with respect to the inner product

$$\langle p, q \rangle = \int_{-1}^{1} p(x)q(x) \, dx$$

Solution The inner products of each pair of polynomials are

$$\langle p_0, p_1 \rangle = \int_{-1}^{1} x \, dx = 0 \qquad \langle p_1, p_2 \rangle = \int_{-1}^{1} \frac{1}{2}(3x^3 - x) \, dx = 0$$

$$\langle p_0, p_2 \rangle = \int_{-1}^{1} \frac{1}{2}(3x^2 - 1) \, dx = 0 \quad \langle p_1, p_3 \rangle = \int_{-1}^{1} \frac{1}{2}(5x^4 - 3x^2) \, dx = 0$$

$$\langle p_0, p_3 \rangle = \int_{-1}^{1} \frac{1}{2}(5x^3 - 3x) \, dx = 0 \quad \langle p_2, p_3 \rangle = \int_{-1}^{1} \frac{1}{4}(15x^5 - 14x^3 + 3x) \, dx = 0$$

Therefore the set $\{p_0(x), p_1(x), p_2(x), p_3(x)\}$ is orthogonal and hence by Theorem 10.10 is linearly independent. Since $\dim(\mathbf{P}^3) = 4$, it follows that this set is a basis for $\mathbf{P}^3$. ■

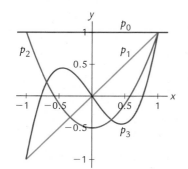

Figure 1 Graphs of the first four Legendre polynomials

An orthogonal basis is handy because the inner product can be used to easily determine how to express vectors as a linear combination of basis vectors. The next theorem shows how this is accomplished.

THEOREM 10.11

Let $\mathcal{V} = \{\mathbf{v}_1, \ldots, \mathbf{v}_k\}$ be an orthogonal basis for an inner product space V. Then any vector $\mathbf{v}$ in V can be written as

$$\mathbf{v} = s_1\mathbf{v}_1 + \cdots + s_k\mathbf{v}_k$$

where $s_i = \dfrac{\langle \mathbf{v}_i, \mathbf{v} \rangle}{\langle \mathbf{v}_i, \mathbf{v}_i \rangle} = \dfrac{\langle \mathbf{v}_i, \mathbf{v} \rangle}{\|\mathbf{v}_i\|^2}$ for $i = 1, \ldots, k$.

▶ Theorem 10.11 generalizes Theorem 8.12 in Section 8.1.

The proof is left as an exercise.

EXAMPLE 4 Use Theorem 10.11 to write $p(x) = 10x^3 + 3x^2 - 11x + 2$ as a linear combination of the Legendre polynomials in Example 3.

Solution We know that $\{p_0(x), p_1(x), p_2(x), p_3(x)\}$ is a basis for $\mathbf{P}^3$, so $p(x)$ must be a linear combination of these vectors. Since the basis is orthogonal, we can apply Theorem 10.11 to find the scalars. The squares of the norms of p_0, p_1, p_2, and p_3 are

$$\langle p_0, p_0 \rangle = \int_{-1}^{1} \left(p_0(x)\right)^2 dx = \int_{-1}^{1} (1)^2 \, dx = 2$$

$$\langle p_1, p_1 \rangle = \int_{-1}^{1} \left(p_1(x)\right)^2 dx = \int_{-1}^{1} (x)^2 \, dx = 2/3$$

$$\langle p_2, p_2 \rangle = \int_{-1}^{1} \left(p_2(x)\right)^2 dx = \int_{-1}^{1} \left(\tfrac{1}{2}(3x^2 - 1)\right)^2 dx = 2/5$$

$$\langle p_3, p_3 \rangle = \int_{-1}^{1} \left(p_3(x)\right)^2 dx = \int_{-1}^{1} \left(\tfrac{1}{2}(5x^3 - 3x)\right)^2 dx = 2/7$$

It is left to the reader to verify the other required inner products, namely,

$$\langle p_0, p \rangle = 6, \quad \langle p_1, p \rangle = -10/3, \quad \langle p_2, p \rangle = 4/5, \quad \langle p_3, p \rangle = 8/7$$

By Theorem 10.11 we have

$$p(x) = \frac{\langle p_0, p \rangle}{\langle p_0, p_0 \rangle} p_0(x) + \frac{\langle p_1, p \rangle}{\langle p_1, p_1 \rangle} p_1(x) + \frac{\langle p_2, p \rangle}{\langle p_2, p_2 \rangle} p_2(x) + \frac{\langle p_3, p \rangle}{\langle p_3, p_3 \rangle} p_3(x)$$

$$= \frac{6}{2} p_0(x) + \frac{-10/3}{2/3} p_1(x) + \frac{4/5}{2/5} p_2(x) + \frac{8/7}{2/7} p_3(x)$$

$$= 3 p_0(x) - 5 p_1(x) + 2 p_2(x) + 4 p_3(x)$$

We can check our calculations by computing

$$3 p_0(x) - 5 p_1(x) + 2 p_2(x) + 4 p_3(x)$$
$$= 3(1) - 5(x) + 2\left(\tfrac{1}{2}(3x^2 - 1)\right) + 4\left(\tfrac{1}{2}(5x^3 - 3x)\right)$$
$$= 10x^3 + 3x^2 - 11x + 2 = p(x) \qquad \blacksquare$$

Orthonormal Sets

The formula for the scalars s_i given in Theorem 10.11 is simplified if $\|v_i\| = 1$. In this case, the vectors are said to be **normal**, and such a set of vectors is called an **orthonormal basis**. When we have an orthonormal basis, Theorem 10.11 can be simplified to the following form.

Definition Normal Vector, Orthonormal Basis

THEOREM 10.12

Let $\mathcal{V} = \{v_1, \ldots, v_k\}$ be an orthonormal basis for an inner product space V. Then any vector v in V can be written as

$$v = \langle v_i, v \rangle v_1 + \cdots + \langle v_k, v \rangle v_k$$

The proof of Theorem 10.12 is left as an exercise.

EXAMPLE 5 Let S be the subspace of $C[-\pi, \pi]$ with basis $\{1, \cos(x), \sin(x)\}$. Convert this to an orthonormal basis with respect to the inner product

$$\langle f, g \rangle = \frac{1}{\pi} \int_{-\pi}^{\pi} f(x)g(x) \, dx$$

Then express $f(x) = \sin^2(x/2)$ (which is in S) as a linear combination of the orthonormal basis functions.

Solution In Example 2 we showed that the basis functions are orthogonal, so all that remains is to scale each so that they have norm 1. For any norm, if $v \neq 0$ and $w = \frac{1}{\|v\|} v$, then $\|w\| = 1$ (see Exercise 63 in Section 10.1). Hence we can convert an orthogonal basis to an orthonormal basis by dividing each vector by its norm. For our basis vectors we have

▶ The details of evaluating the integrals in this example are left to the reader.

$$\|1\|^2 = \frac{1}{\pi} \int_{-\pi}^{\pi} 1^2 \, dx = 2 \qquad \Longrightarrow \qquad g_1(x) = \frac{1}{\sqrt{2}} 1 = \frac{1}{\sqrt{2}}$$

$$\|\cos(x)\|^2 = \frac{1}{\pi} \int_{-\pi}^{\pi} \cos^2(x) \, dx = 1 \qquad \Longrightarrow \qquad g_2(x) = \frac{1}{\sqrt{1}} \cos(x) = \cos(x)$$

$$\|\sin(x)\|^2 = \frac{1}{\pi} \int_{-\pi}^{\pi} \sin^2(x) \, dx = 1 \qquad \Longrightarrow \qquad g_3(x) = \frac{1}{\sqrt{1}} \sin(x) = \sin(x)$$

Then $\{g_1(x), g_2(x), g_3(x)\}$ is an orthonormal basis for S. To express $f(x) = \sin^2(x/2)$ as a linear combination of the basis functions, we compute the inner products

$$\langle g_1, f \rangle = \frac{1}{\pi} \int_{-\pi}^{\pi} \frac{1}{\sqrt{2}} \sin^2(x/2)\, dx = \frac{1}{\sqrt{2}}$$

$$\langle g_2, f \rangle = \frac{1}{\pi} \int_{-\pi}^{\pi} \cos(x) \sin^2(x/2)\, dx = -\frac{1}{2}$$

$$\langle g_3, f \rangle = \frac{1}{\pi} \int_{-\pi}^{\pi} \sin(x) \sin^2(x/2)\, dx = 0$$

Therefore, by Theorem 10.12,

$$\sin^2(x/2) = \frac{1}{\sqrt{2}} g_1(x) - \frac{1}{2} g_2(x) + 0 g_3(x)$$
$$= \frac{1}{\sqrt{2}} \left(\frac{1}{\sqrt{2}} \right) - \frac{1}{2} \cos(x) = \frac{1 - \cos(x)}{2}$$

which agrees with the half-angle formula for sine. ∎

Projections Onto Subspaces

Theorem 10.11 provides a formula for expressing a vector $\mathbf{v}$ in a vector space V as a linear combination of orthogonal basis vectors $\{\mathbf{v}_1, \ldots, \mathbf{v}_m\}$. Similar to our approach in Section 8.2, we use the formula from Theorem 10.11 to serve as a guide for the formula for projecting a vector onto a subspace.

DEFINITION 10.13

Definition Projection onto a Subspace

Let S be a subspace of an inner product space V, and suppose that S has orthogonal basis $\{\mathbf{v}_1, \ldots, \mathbf{v}_k\}$. Then the **projection of $\mathbf{v}$ onto** S is given by

$$\text{proj}_S \mathbf{v} = \frac{\langle \mathbf{v}_1, \mathbf{v} \rangle}{\|\mathbf{v}_1\|^2} \mathbf{v}_1 + \frac{\langle \mathbf{v}_2, \mathbf{v} \rangle}{\|\mathbf{v}_2\|^2} \mathbf{v}_2 + \cdots + \frac{\langle \mathbf{v}_k, \mathbf{v} \rangle}{\|\mathbf{v}_k\|^2} \mathbf{v}_k \qquad (1)$$

As in Euclidean space, the following are true about projections onto inner product subspaces:

- If $S = \text{span}\{\mathbf{v}_1\}$ is a one-dimensional subspace, then (1) reduces to the formula for $\text{proj}_{\mathbf{v}_1} \mathbf{v}$.
- The basis $\{\mathbf{v}_1, \ldots, \mathbf{v}_k\}$ for S must be orthogonal in order to apply the formula for $\text{proj}_S \mathbf{v}$. If the basis is orthonormal, then (1) reduces to

$$\text{proj}_S \mathbf{v} = \langle \mathbf{v}_1, \mathbf{v} \rangle \mathbf{v}_1 + \langle \mathbf{v}_2, \mathbf{v} \rangle \mathbf{v}_2 + \cdots + \langle \mathbf{v}_k, \mathbf{v} \rangle \mathbf{v}_k \qquad (2)$$

- The vector $\text{proj}_S \mathbf{v}$ does not depend on the choice of orthogonal basis for S.

THEOREM 10.14

▶ Theorem 10.14 generalizes Theorem 8.16 in Section 8.2.

Let S be a nonzero finite-dimensional subspace of an inner product space V, and $\mathbf{v}$ a vector in V. Then

(a) $\text{proj}_S \mathbf{v}$ is in S.

(b) $\mathbf{v} - \text{proj}_S \mathbf{v}$ is orthogonal to S.

(c) If $\mathbf{v}$ is in S, then $\mathbf{v} = \text{proj}_S \mathbf{v}$.

(d) $\text{proj}_S \mathbf{v}$ is independent of the choice of orthogonal basis for S.

The proof of Theorem 10.14 is similar to that of Theorem 8.16 in Section 8.2 and is left as an exercise.

EXAMPLE 6 Let $S = \text{span}\{\mathbf{v}_1, \mathbf{v}_2\}$, where

$$\mathbf{v}_1 = \begin{bmatrix} 4 \\ 5 \\ 1 \end{bmatrix}, \quad \mathbf{v}_2 = \begin{bmatrix} -3 \\ 2 \\ -6 \end{bmatrix}, \quad \mathbf{v} = \begin{bmatrix} 3 \\ 25 \\ 33 \end{bmatrix}$$

Find $\text{proj}_S\mathbf{v}$ using the inner product $\langle \mathbf{u}, \mathbf{v} \rangle = 2u_1v_1 + 3u_2v_2 + u_3v_3$ from Example 1.

Solution In Example 1 we showed that $\mathbf{v}_1$ and $\mathbf{v}_2$ are orthogonal with respect to our inner product, so all that remains is to apply the formula (1). To do so, we need

$$\langle \mathbf{v}_1, \mathbf{v} \rangle = (2)(4)(3) + (3)(5)(25) + (1)(1)(33) = 432$$
$$\langle \mathbf{v}_2, \mathbf{v} \rangle = (2)(-3)(3) + (3)(2)(25) + (1)(-6)(33) = -66$$
$$\|\mathbf{v}_1\|^2 = \langle \mathbf{v}_1, \mathbf{v}_1 \rangle = (2)(4)^2 + (3)(5)^2 + (1)(1)^2 = 108$$
$$\|\mathbf{v}_2\|^2 = \langle \mathbf{v}_2, \mathbf{v}_2 \rangle = (2)(-3)^2 + (3)(2)^2 + (1)(-6)^2 = 66$$

Therefore

$$\text{proj}_S\mathbf{v} = \frac{432}{108}\mathbf{v}_1 + \frac{-66}{66}\mathbf{v}_2 = 4\mathbf{v}_1 - \mathbf{v}_2 = 4\begin{bmatrix} 4 \\ 5 \\ 1 \end{bmatrix} - \begin{bmatrix} -3 \\ 2 \\ -6 \end{bmatrix} = \begin{bmatrix} 19 \\ 18 \\ 10 \end{bmatrix} \quad \blacksquare$$

EXAMPLE 7 Let $S = \text{span}\{1/\sqrt{2}, \cos(x), \sin(x)\}$ be a subspace of $C[-\pi, \pi]$. Find the projection of $h(x) = x^2$ onto S with respect to the inner product

$$\langle f, g \rangle = \frac{1}{\pi} \int_{-\pi}^{\pi} f(x)g(x) \, dx$$

Solution In Example 5 we showed the basis of S is orthonormal. Hence we can compute the projection using the simplified formula (2), which only requires the inner products

$$\langle 1/\sqrt{2}, h \rangle = \frac{1}{\pi} \int_{-\pi}^{\pi} \frac{1}{\sqrt{2}} \cdot x^2 \, dx = \frac{\sqrt{2}\pi^2}{3}$$

$$\langle \cos(x), h \rangle = \frac{1}{\pi} \int_{-\pi}^{\pi} \cos(x) \cdot x^2 \, dx = -4$$

$$\langle \sin(x), h \rangle = \frac{1}{\pi} \int_{-\pi}^{\pi} \sin(x) \cdot x^2 \, dx = 0$$

Therefore the projection is given by

$$\text{proj}_S h = \frac{\sqrt{2}\pi^2}{3}\left(\frac{1}{\sqrt{2}}\right) - 4\cos(x) = \frac{\pi^2}{3} - 4\cos(x)$$

A graph of $h(x) = x^2$ together with $\text{proj}_S h$ is given in Figure 2. Note that the two graphs are fairly close together. $\quad \blacksquare$

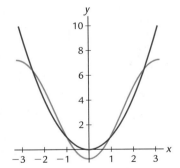

Figure 2 $h(x) = x^2$ (blue) and $\text{proj}_S h$ (red) from Example 7.

The Gram–Schmidt Process

We are now ready to recast the Gram–Schmidt process in the setting of an inner product space. The goal of the Gram–Schmidt process is the same here as in Euclidean space, to convert a basis into an orthogonal basis. (Which can then be made into an orthonormal basis by normalizing each vector.)

The basic procedure for implementing the Gram–Schmidt process in an inner product space is exactly the same as in Euclidean space. The only computational change is that we replace the dot products with inner products when computing projections.

THEOREM 10.15

(THE GRAM–SCHMIDT PROCESS) Let S be a subspace with basis $\{\mathbf{s}_1, \mathbf{s}_2, \ldots, \mathbf{s}_k\}$. Define $\mathbf{v}_1, \mathbf{v}_2, \ldots, \mathbf{v}_k$, in order, by

$$\mathbf{v}_1 = \mathbf{s}_1$$
$$\mathbf{v}_2 = \mathbf{s}_2 - \text{proj}_{\mathbf{v}_1} \mathbf{s}_2$$
$$\mathbf{v}_3 = \mathbf{s}_3 - \text{proj}_{\mathbf{v}_1} \mathbf{s}_3 - \text{proj}_{\mathbf{v}_2} \mathbf{s}_3$$
$$\mathbf{v}_4 = \mathbf{s}_4 - \text{proj}_{\mathbf{v}_1} \mathbf{s}_4 - \text{proj}_{\mathbf{v}_2} \mathbf{s}_4 - \text{proj}_{\mathbf{v}_3} \mathbf{s}_4$$
$$\vdots \qquad \vdots$$
$$\mathbf{v}_k = \mathbf{s}_k - \text{proj}_{\mathbf{v}_1} \mathbf{s}_k - \text{proj}_{\mathbf{v}_2} \mathbf{s}_k - \cdots - \text{proj}_{\mathbf{v}_{k-1}} \mathbf{s}_k$$

Then $\{\mathbf{v}_1, \mathbf{v}_2, \ldots, \mathbf{v}_k\}$ is an orthogonal basis for S.

The proof is left as an exercise. After using Gram–Schmidt to find an orthogonal basis $\{\mathbf{v}_1, \ldots, \mathbf{v}_k\}$, we can find an orthonormal basis $\{\mathbf{w}_1, \ldots, \mathbf{w}_k\}$ by setting

$$\mathbf{w}_i = \frac{1}{\|\mathbf{v}_i\|} \mathbf{v}_i$$

for each $i = 1, \ldots, k$.

EXAMPLE 8   Starting with the vectors

$$\mathbf{s}_1 = \begin{bmatrix} 1 \\ 1 \\ 1 \end{bmatrix}, \quad \mathbf{s}_2 = \begin{bmatrix} 1 \\ 1 \\ 0 \end{bmatrix}, \quad \mathbf{s}_3 = \begin{bmatrix} 1 \\ 0 \\ 0 \end{bmatrix}$$

implement the Gram–Schmidt process to find basis for $\mathbf{R}^3$ that is orthogonal with respect to the inner product $\langle \mathbf{u}, \mathbf{v} \rangle = 2u_1 v_1 + 3u_2 v_2 + u_3 v_3$ from Example 1.

Solution The first step requires no computation: $\mathbf{v}_1 = \mathbf{s}_1$. For the second vector, we have

$$\mathbf{v}_2 = \mathbf{s}_2 - \text{proj}_{\mathbf{v}_1} \mathbf{s}_2 = \mathbf{s}_2 - \frac{\langle \mathbf{v}_1, \mathbf{s}_2 \rangle}{\langle \mathbf{v}_1, \mathbf{v}_1 \rangle} \mathbf{v}_1 = \begin{bmatrix} 1 \\ 1 \\ 0 \end{bmatrix} - \frac{5}{6} \begin{bmatrix} 1 \\ 1 \\ 1 \end{bmatrix} = \begin{bmatrix} 1/6 \\ 1/6 \\ -5/6 \end{bmatrix}$$

Since any multiple of an orthogonal vector is still orthogonal, we multiply v_2 by 6 to clear the fractions and make future computations easier. This gives us

$$\mathbf{v}_2 = \begin{bmatrix} 1 \\ 1 \\ -5 \end{bmatrix}$$

Finally, for $\mathbf{v}_3$ we have

$$\mathbf{v}_3 = \mathbf{s}_3 - \text{proj}_{\mathbf{v}_1} \mathbf{s}_3 - \text{proj}_{\mathbf{v}_2} \mathbf{s}_3$$

$$= \mathbf{s}_3 - \frac{\langle \mathbf{v}_1, \mathbf{s}_3 \rangle}{\langle \mathbf{v}_1, \mathbf{v}_1 \rangle} \mathbf{v}_1 - \frac{\langle \mathbf{v}_2, \mathbf{s}_3 \rangle}{\langle \mathbf{v}_2, \mathbf{v}_2 \rangle} \mathbf{v}_2$$

$$= \begin{bmatrix} 1 \\ 0 \\ 0 \end{bmatrix} - \frac{2}{6} \begin{bmatrix} 1 \\ 1 \\ 1 \end{bmatrix} - \frac{2}{30} \begin{bmatrix} 1 \\ 1 \\ -5 \end{bmatrix} = \begin{bmatrix} 3/5 \\ -2/5 \\ 0 \end{bmatrix}$$

For the sake of consistency, we again clear fractions by multiplying $\mathbf{v}_3$ by 5. This leaves us with the orthogonal basis

$$\left\{ \begin{bmatrix} 1 \\ 1 \\ 1 \end{bmatrix}, \begin{bmatrix} 1 \\ 1 \\ -5 \end{bmatrix}, \begin{bmatrix} 3 \\ -2 \\ 0 \end{bmatrix} \right\} \qquad \blacksquare$$

EXAMPLE 9   The monomials $\{1, x, x^2, x^3\}$ give a basis for $\mathbf{P}^3$. Apply the Gram–Schmidt process to find a basis that is orthonormal with respect to the inner product

$$\langle p, q \rangle = \int_{-1}^{1} p(x)q(x)\, dx$$

Solution   We are asked for an orthonormal basis, but we start by finding an orthogonal basis, and then normalize at the end. Let $p_1(x) = 1$, $p_2(x) = x$, $p_3(x) = x^2$, and $p_4(x) = x^3$. The first basis vector is $q_1(x) = p_1(x) = 1$. For the second basis vector, we have

$$q_2(x) = p_2(x) - \text{proj}_{q_1} p_2 = x - \frac{\langle q_1, p_2 \rangle}{\langle q_1, q_1 \rangle} q_1 = x - \frac{0}{2/3}(1) = x$$

We have $q_2(x) = p_2(x)$ because $p_2(x)$ is orthogonal to $p_1(x)$, which is why $\text{proj}_{q_1} p_2 = 0$. Proceeding to $q_3(x)$,

$$q_3(x) = p_3(x) - \text{proj}_{q_1} p_3 - \text{proj}_{q_2} p_3$$

$$= p_3(x) - \frac{\langle q_1, p_3 \rangle}{\langle q_1, q_1 \rangle} q_1 - \frac{\langle q_2, p_3 \rangle}{\langle q_2, q_2 \rangle} q_2$$

$$= x^2 - \frac{2/3}{2}(1) - \frac{0}{2/3}(x) = x^2 - \tfrac{1}{3} = \tfrac{1}{3}(3x^2 - 1)$$

For the last polynomial, $q_4(x)$, we have

$$q_4(x) = p_4(x) - \text{proj}_{q_1} p_4 - \text{proj}_{q_2} p_4 - \text{proj}_{q_3} p_4$$

$$= p_4(x) - \frac{\langle q_1, p_4 \rangle}{\langle q_1, q_1 \rangle} q_1 - \frac{\langle q_2, p_4 \rangle}{\langle q_2, q_2 \rangle} q_2 - \frac{\langle q_3, p_4 \rangle}{\langle q_3, q_3 \rangle} q_3$$

$$= x^3 - \frac{0}{2}(1) - \frac{2/5}{2/3}(x) - \frac{0}{8/45}\left(\tfrac{1}{3}(3x^2 - 1)\right)$$

$$= x^3 - \tfrac{3}{5}x = \tfrac{1}{5}(5x^3 - 3x)$$

▶ These orthogonal polynomials are familiar—they are all scalar multiples of the Legendre polynomials introduced earlier. One way to define Legendre polynomials is in terms of the result of applying the Gram–Schmidt process to $1, x, x^2, x^3, \ldots$ and scaling.

The last step is to normalize each polynomial to produce an orthonormal basis. While implementing the Gram–Schmidt process, we computed

$$\|q_1\|^2 = \langle q_1, q_1 \rangle = 2, \quad \|q_2\|^2 = \langle q_2, q_2 \rangle = \frac{2}{3}, \quad \|q_3\|^2 = \langle q_3, q_3 \rangle = \frac{8}{45}$$

We also have

$$\|q_4\|^2 = \langle q_4, q_4 \rangle = \int_{-1}^{1} \left(\tfrac{1}{5}(5x^3 - 3x)\right)^2 dx = \frac{8}{175}$$

Therefore the orthonormal polynomials are

$$r_1(x) = \frac{1}{\|q_1\|} q_1(x) = \frac{1}{\sqrt{2}}$$

$$r_2(x) = \frac{1}{\|q_2\|} q_2(x) = \sqrt{\frac{3}{2}}\, x$$

$$r_3(x) = \frac{1}{\|q_3\|} q_3(x) = \sqrt{\frac{45}{8}} \cdot \frac{1}{3}(3x^2 - 1) = \sqrt{\frac{5}{8}}(3x^2 - 1)$$

$$r_4(x) = \frac{1}{\|q_4\|} q_4(x) = \sqrt{\frac{175}{8}} \cdot \frac{1}{5}(5x^3 - 3x) = \sqrt{\frac{7}{8}}(5x^3 - 3x)$$

The orthonormal basis is

$$\left\{ \frac{1}{\sqrt{2}}, \ \sqrt{\frac{3}{2}}x, \ \sqrt{\frac{5}{8}}(3x^2 - 1), \ \sqrt{\frac{7}{8}}(5x^3 - 3x) \right\}$$ ∎

EXAMPLE 10 Use the orthonormal basis found in Example 9 to find the projection of $f(x) = \sin(x) + \cos(x)$ onto $\mathbf{P}^3$.

Solution Since we have an orthonormal basis $\{r_1(x), r_2(x), r_3(x), r_4(x)\}$, we can apply (2) to find $\text{proj}_{\mathbf{P}^3} f$. The required inner products are

▶ The exact values of these integrals are somewhat complicated, so decimal approximations are reported.

$$\langle r_1, f \rangle = \int_{-1}^{1} \frac{1}{\sqrt{2}}(\sin(x) + \cos(x)) \, dx \approx 1.1900$$

$$\langle r_2, f \rangle = \int_{-1}^{1} \sqrt{\frac{3}{2}}x(\sin(x) + \cos(x)) \, dx \approx 0.7377$$

$$\langle r_3, f \rangle = \int_{-1}^{1} \sqrt{\frac{5}{8}}(3x^2 - 1)(\sin(x) + \cos(x)) \, dx \approx -0.1962$$

$$\langle r_4, f \rangle = \int_{-1}^{1} \sqrt{\frac{7}{8}}(5x^3 - 3x)(\sin(x) + \cos(x)) \, dx \approx -0.0337$$

Hence we have

$$\text{proj}_{\mathbf{P}^3} f \approx 1.19\frac{1}{\sqrt{2}} + 0.7377\sqrt{\frac{3}{2}}x - 0.1962\sqrt{\frac{5}{8}}(3x^2 - 1) - 0.0337\sqrt{\frac{7}{8}}(5x^3 - 3x)$$

$$\approx -0.1576x^3 - 0.4653x^2 + 0.9981x + 0.9966$$

The graph of $f(x) = \sin(x) + \cos(x)$ and $\text{proj}_{\mathbf{P}^3} f$ are shown in Figure 3(a). The two graphs are virtually indistinguishable because the difference between $f(x)$ and $\text{proj}_{\mathbf{P}^3} f$ on $[-1, 1]$ is quite small. The graph of the difference $f(x) - \text{proj}_{\mathbf{P}^3} f$ in Figure 3(b) is more revealing. Among the polynomials of degree 3, the projection provides an excellent approximation to $f(x) = \sin(x) + \cos(x)$.

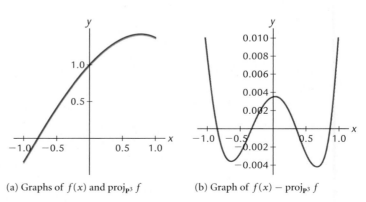

(a) Graphs of $f(x)$ and $\text{proj}_{\mathbf{P}^3} f$ (b) Graph of $f(x) - \text{proj}_{\mathbf{P}^3} f$

Figure 3 $f(x)$ and $\text{proj}_{\mathbf{P}^3} f$. ∎

EXERCISES

1. Convert the set $\{\mathbf{v}_1, \mathbf{v}_2, \mathbf{v}_3\}$ from Example 1 into an orthonormal set with respect to the inner product $\langle \mathbf{u}, \mathbf{v} \rangle = 2u_1v_1 + 3u_2v_2 + u_3v_3$.

2. Verify that the set

$$\{x + 1, \ -9x + 5, \ 6x^2 - 6x + 1\}$$

is orthogonal with respect to the inner product

$$\langle p, q \rangle = \int_0^1 p(x)q(x) \, dx$$

and then make the set orthonormal.

In Exercises 3–4, determine the values of a (if any) that will make the given set of vectors orthogonal in $\mathbf{R}^3$ with respect to the weighted dot product with $t_1 = 2$, $t_2 = 3$, and $t_3 = 1$. If possible, normalize the vectors to make the set orthonormal.

3. $\left\{ \begin{bmatrix} 2 \\ -1 \\ 1 \end{bmatrix}, \begin{bmatrix} 2 \\ 1 \\ a \end{bmatrix}, \begin{bmatrix} 1 \\ 2 \\ 2 \end{bmatrix} \right\}$

4. $\left\{ \begin{bmatrix} 1 \\ 1 \\ a \end{bmatrix}, \begin{bmatrix} -5 \\ 1 \\ 7 \end{bmatrix}, \begin{bmatrix} 3 \\ a \\ 6 \end{bmatrix} \right\}$

In Exercises 5–6, determine the values of a (if any) that will make the given set of vectors orthogonal in $\mathbf{P}^2$ with respect to the inner product

$$\langle p, q \rangle = p(-1)q(-1) + p(0)q(0) + p(2)q(2)$$

If possible, normalize the vectors to make the set orthonormal.

5. $\left\{ x^2 + x, \ x^2 + ax - 2, \ x^2 - 2x \right\}$

6. $\left\{ 3x^2 - 2x - 1, \ ax^2 + x - 1, \ 5x^2 + ax - 9 \right\}$

7. Use Theorem 10.11 to write $\mathbf{v} = \begin{bmatrix} 1 \\ 34 \\ 22 \end{bmatrix}$ as a linear combination of the vectors given in Example 1.

8. Use Theorem 10.11 to write $f(x) = 12x^2 - 6x - 6$ as a linear combination of the vectors given in Exercise 2.

9. Let $\mathbf{v} = \begin{bmatrix} 1 \\ 0 \\ -1 \end{bmatrix}$. Find $\text{proj}_S \mathbf{v}$ for the subspace S spanned by the vectors $\mathbf{v}_1$ and $\mathbf{v}_2$ and the inner product in Example 1.

10. Let $f(x) = x$. Find $\text{proj}_S f$ for the inner product and subspace S spanned by the functions in Exercise 2.

11. Let $f(x) = x$. Find $\text{proj}_S f$ for the inner product and subspace S in Example 7.

12. Find $\text{proj}_S f$ for $f(x) = e^x$, where $S = \text{span}\{1, x\}$ and the inner product is

$$\langle f, g \rangle = \int_{-1}^1 f(x)g(x) \, dx$$

In Exercises 13–18, use the Gram–Schmidt process to convert the given set of vectors to an orthogonal basis with respect to the given inner product.

13. The set $\left\{ \begin{bmatrix} 1 \\ -1 \\ 0 \end{bmatrix}, \begin{bmatrix} 2 \\ 0 \\ 1 \end{bmatrix} \right\}$ with respect to the inner product given in Example 1.

14. The set $\left\{ \begin{bmatrix} 0 \\ 1 \\ 0 \end{bmatrix}, \begin{bmatrix} 2 \\ 1 \\ 1 \end{bmatrix}, \begin{bmatrix} 1 \\ 0 \\ 1 \end{bmatrix} \right\}$ with respect to the inner product given in Example 1.

15. The set $\{1, x^2\}$ with respect to the inner product given in Exercise 2.

16. The set $\{x^2 + 1, 4x, -3\}$ with respect to the inner product given in Exercise 2.

17. The set $\{x, 1\}$ with respect to the inner product given in Exercises 5–6.

18. The set $\{2x + 1, x^2, 3\}$ with respect to the inner product given in Exercises 5–6.

FIND AN EXAMPLE For Exercises 19–24, find an example that meets the given specifications.

19. An orthogonal basis for $\mathbf{R}^2$ with respect to the inner product $\langle \mathbf{u}, \mathbf{v} \rangle = 3u_1v_1 + 2u_2v_2$ that includes $\mathbf{u}_1 = \begin{bmatrix} 1 \\ 2 \end{bmatrix}$.

20. An orthogonal basis for $\mathbf{R}^3$ with respect to the inner product $\langle \mathbf{u}, \mathbf{v} \rangle = u_1v_1 + 3u_2v_2 + 2u_3v_3$ that includes $\mathbf{u}_1 = \begin{bmatrix} -1 \\ 2 \\ 1 \end{bmatrix}$.

21. An orthogonal basis for $\mathbf{P}^1$ with respect to the inner product $\langle p, q \rangle = \int_0^1 p(x)q(x) \, dx$ that contains $p_1(x) = 3x + 1$.

22. An orthogonal basis for $\mathbf{P}^2$ with respect to the inner product $\langle p, q \rangle = \int_{-1}^1 p(x)q(x) \, dx$ that contains $p_1(x) = x^2 + 4x - 1$.

23. Three distinct functions in $C[-\pi, \pi]$ that are orthogonal, but cannot be made orthonormal, with respect to the inner product

$$\langle f, g \rangle = \frac{1}{\pi} \int_{-\pi}^\pi f(x)g(x) \, dx$$

24. A basis for $\mathbf{R}^n$ that is orthogonal with respect to any weighted dot product.

TRUE OR FALSE For Exercises 25–32, determine if the statement is true or false, and justify your answer.

25. If $\{\mathbf{v}_1, \mathbf{v}_2, \mathbf{v}_3\}$ is an orthonormal set in an inner product space V, then so is $\{c_1\mathbf{v}_1, c_2\mathbf{v}_2, c_3\mathbf{v}_3\}$, where c_1, c_2, and c_3 are nonzero scalars.

26. The set $\{1, \cos(2x), \sin(2x)\}$ is orthogonal in $C[-\pi, \pi]$ with respect to the inner product

$$\langle f, g \rangle = \frac{1}{\pi} \int_{-\pi}^{\pi} f(x)g(x) \, dx$$

27. Any finite linearly independent set in an inner product space V can be converted to an orthonormal set by applying the Gram–Schmidt process.

28. Every set of orthogonal vectors is linearly independent.

29. If $\mathbf{v}$ is a nonzero vector in an inner product space V and S is a nonzero finite-dimensional subspace of V, then $\text{proj}_S \mathbf{v} \neq \mathbf{0}$.

30. If a set of vectors in an inner product space V is linearly dependent, then the set cannot be orthogonal.

31. If the Gram–Schmidt process is applied to a linearly dependent set, then one of the vectors produced will be the zero vector $\mathbf{0}$.

32. If $\mathcal{V} = \{\mathbf{v}_1, \mathbf{v}_2, \mathbf{v}_3, \mathbf{v}_4\}$ is a set of nonzero vectors in $\mathbf{R}^3$, then $\mathcal{V}$ is not an orthogonal set with respect to any inner product.

33. Prove Theorem 10.10. (HINT: See the proof of Theorem 8.11.)

34. Prove Theorem 10.11. (HINT: See the proof of Theorem 8.12.)

35. Apply Theorem 10.11 to prove Theorem 10.12.

36. Prove Theorem 10.14. (HINT: See the proof of Theorem 8.16.)

For Exercises 37–42, $\mathbf{u}$ and $\mathbf{v}$ (and their subscripted relatives) are vectors in an inner product space V, and S is a nonzero finite-dimensional subspace of V.

37. Prove that $\text{proj}_S \mathbf{u} = \text{proj}_S \left(\text{proj}_S \mathbf{u} \right)$.

38. Prove that $T : V \rightarrow V$ given by $T(\mathbf{u}) = \text{proj}_S \mathbf{u}$ is a linear transformation.

39. Suppose that $\mathbf{u}$ is in S, and that $S^\perp$ is nonzero.

(a) What is $\text{proj}_S \mathbf{u}$?

(b) What is $\text{proj}_{S^\perp} \mathbf{u}$?

40. Let $\{\mathbf{u}_1, \mathbf{u}_2\}$ be nonzero vectors, and define

$$\mathbf{v}_1 = \mathbf{u}_1, \quad \mathbf{v}_2 = \mathbf{u}_2 - \text{proj}_{\mathbf{v}_1} \mathbf{u}_2$$

Prove that $\text{span}\{\mathbf{u}_1, \mathbf{u}_2\} = \text{span}\{\mathbf{v}_1, \mathbf{v}_2\}$.

41. Let $\{\mathbf{v}_1, \ldots, \mathbf{v}_k\}$ be an orthonormal basis of V. Prove that for any $\mathbf{v}$ in V, we have

$$\|\mathbf{v}\|^2 = \langle \mathbf{v}_1, \mathbf{v} \rangle^2 + \cdots + \langle \mathbf{v}_k, \mathbf{v} \rangle^2$$

42. Here we prove that the Gram–Schmidt process works. Suppose that $\{\mathbf{u}_1, \ldots, \mathbf{u}_k\}$ are linearly independent vectors, and that $\{\mathbf{v}_1, \ldots, \mathbf{v}_k\}$ are as defined in the statement of the Gram–Schmidt process.

(a) Use induction to show $\{\mathbf{v}_1, \ldots, \mathbf{v}_j\}$ is an orthogonal set for $j = 1, \ldots, k$.

(b) Use induction to show $\text{span}\{\mathbf{u}_1, \ldots, \mathbf{u}_j\} = \text{span}\{\mathbf{v}_1, \ldots, \mathbf{v}_j\}$ for $j = 1, \ldots, k$.

(c) Explain why (a) and (b) imply that Gram–Schmidt yields an orthogonal basis.

10.3 Applications of Inner Products

▶ This section is optional and can be omitted without loss of continuity.

In this section we consider a few applications of inner products. These applications exploit our ability to do the following: Given a vector $\mathbf{v}$ in an inner product space V, we can use the projection to find the vector $\mathbf{s}$ in a subspace S of V that is closest to $\mathbf{v}$.

A vector $\mathbf{s}$ is "closest" to $\mathbf{v}$ when the norm of their difference $\|\mathbf{v} - \mathbf{s}\|$ is as small as possible. We encountered this when fitting lines to data in Euclidean space, where we found the required vector by using projections. The same approach works here, by applying this key theorem.

THEOREM 10.16

▶ Theorem 10.16 generalizes Theorem 8.29 in Section 8.5.

Let S be a finite-dimensional subspace of an inner product space V, and suppose that $\mathbf{v}$ is in V. Then the closest vector in S to $\mathbf{v}$ is given by $\text{proj}_S \mathbf{v}$. That is,

$$\|\mathbf{v} - \text{proj}_S \mathbf{v}\| \leq \|\mathbf{v} - \mathbf{s}\|$$

for all $\mathbf{s}$ in S, with equality holding exactly when $\mathbf{s} = \text{proj}_S \mathbf{v}$.

Proof The proof is similar to that of Theorem 8.29. If $\mathbf{s}$ is in S, then since $\text{proj}_S \mathbf{v}$ is also in S, the difference $\text{proj}_S \mathbf{v} - \mathbf{s}$ must be in S. On the other hand, $\mathbf{v} - \text{proj}_S \mathbf{v}$ is in $S^\perp$ (by Theorem 10.14 in Section 10.2). Therefore by the Pythagorean theorem (Theorem 10.4 in Section 10.1), we have

$$\|(\mathbf{v} - \text{proj}_S \mathbf{v}) - (\text{proj}_S \mathbf{v} - \mathbf{s})\|^2 = \|\mathbf{v} - \text{proj}_S \mathbf{v}\|^2 + \|\text{proj}_S \mathbf{v} - \mathbf{s}\|^2$$

As $(\mathbf{v} - \text{proj}_S\mathbf{v}) - (\text{proj}_S\mathbf{v} - \mathbf{s}) = \mathbf{v} - \mathbf{s}$, it follows that

$$\|\mathbf{v} - \mathbf{s}\|^2 = \|\mathbf{v} - \text{proj}_S\mathbf{v}\|^2 + \|\text{proj}_S\mathbf{v} - \mathbf{s}\|^2$$

Since $\|\text{proj}_S\mathbf{v} - \mathbf{s}\|^2 \geq 0$, we have $\|\mathbf{v} - \mathbf{s}\| \geq \|\mathbf{v} - \text{proj}_S\mathbf{v}\|$. Furthermore, there is equality in this inequality exactly when $\|\text{proj}_S\mathbf{v} - \mathbf{s}\| = 0$. That is, when $\mathbf{s} = \text{proj}_S\mathbf{v}$. ■

Weighted Least Squares Regression

In Section 8.5 we used projection onto a subspace to fit a line to a data set of the form $(x_1, y_1), \ldots, (x_n, y_n)$. There we treated each data point as being equally important, but now suppose that we view some points as more important than others. For instance, those that have the most extreme x-coordinates could be viewed as potential outliers. We might want to adjust our inner product so that these points have less influence on the model than those near the "center" of the data set.

We adopt the same notation as in Section 8.5. Given a line $\hat{y} = c_0 + c_1 x$, for each data point (x_i, y_i) we define $\hat{y}_i = c_0 + c_1 x_i$. The goal of ordinary **least squares regression** is to select c_0 and c_1 so that

$$(y_1 - \hat{y}_1)^2 + (y_2 - \hat{y}_2)^2 + \cdots + (y_n - \hat{y}_n)^2$$

is as small as possible. With **weighted least squares regression**, we minimize the expression

$$t_1(y_1 - \hat{y}_1)^2 + t_2(y_2 - \hat{y}_2)^2 + \cdots + t_n(y_n - \hat{y}_n)^2 \tag{1}$$

where $t_1, t_2, \ldots, t_n$ are the positive **weights**. If we set

$$\mathbf{y} = \begin{bmatrix} y_1 \\ y_2 \\ \vdots \\ y_n \end{bmatrix} \quad \text{and} \quad \hat{\mathbf{y}} = \begin{bmatrix} \hat{y}_1 \\ \hat{y}_2 \\ \vdots \\ \hat{y}_n \end{bmatrix}$$

then (1) is equal to $\|\mathbf{y} - \hat{\mathbf{y}}\|^2$, where the norm is with respect to the weighted dot product with weights $t_1, t_2, \ldots, t_n$. Since $\hat{y} = c_0 + c_1 x$, if we define

$$A = \begin{bmatrix} 1 & x_1 \\ 1 & x_2 \\ \vdots & \vdots \\ 1 & x_n \end{bmatrix} \quad \text{and} \quad \mathbf{x} = \begin{bmatrix} c_0 \\ c_1 \end{bmatrix}$$

then we need a solution to $\hat{\mathbf{y}} = A\mathbf{x}$, where $\hat{\mathbf{y}}$ is the vector in $S = \text{col}(A)$ (the column space of A) that is closest to $\mathbf{y}$. Thus, by Theorem 10.16, we should set

$$\hat{\mathbf{y}} = \text{proj}_S\mathbf{y}.$$

Let's consider an example that shows how this process works.

EXAMPLE 1   Use least squares regression to fit a line to the data set (scatter plot shown in Figure 1)

$$(-6, 2.9), \quad (-3, 1.5), \quad (-2, 2), \quad (2, 2.7), \quad (3, 3.3), \quad (6, 1.1)$$

Then use weighted least squares regression with weights designed to emphasize the four points in the middle of the data set.

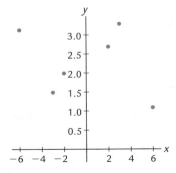

Figure 1 Scatter plot of data for Example 1.

Solution Instead of using the formula for least squares regression developed in Section 8.5, we will use projection onto a subspace to find the equation for the regression line because that method generalizes to weighted least squares regression. Let $A = \begin{bmatrix} \mathbf{a}_1 & \mathbf{a}_2 \end{bmatrix}$, where

$$\mathbf{a}_1 = \begin{bmatrix} 1 \\ 1 \\ 1 \\ 1 \\ 1 \\ 1 \end{bmatrix} \quad \text{and} \quad \mathbf{a}_2 = \begin{bmatrix} -6 \\ -3 \\ -2 \\ 2 \\ 3 \\ 6 \end{bmatrix}$$

▶ The 1's in $\mathbf{a}_1$ are multiplied times the constant term in the regression equation.

We need to find $\hat{\mathbf{y}} = \text{proj}_S \mathbf{y}$, where $S = \text{span}\{\mathbf{a}_1, \mathbf{a}_2\}$. We have

$$\mathbf{a}_1 \cdot \mathbf{a}_2 = (1)(-6) + (1)(-3) + (1)(-2) + (1)(6) + (1)(3) + (1)(2) = 0$$

so that $\{\mathbf{a}_1, \mathbf{a}_2\}$ is an orthogonal basis of S with respect to the dot product. Therefore we can apply the projection formula

$$\hat{\mathbf{y}} = \text{proj}_S \mathbf{y} = \frac{\mathbf{a}_1 \cdot \mathbf{y}}{\mathbf{a}_1 \cdot \mathbf{a}_1} \mathbf{a}_1 + \frac{\mathbf{a}_2 \cdot \mathbf{y}}{\mathbf{a}_2 \cdot \mathbf{a}_2} \mathbf{a}_2$$

$$= \frac{13.5}{6} \mathbf{a}_1 + \frac{-4}{98} \mathbf{a}_2 \approx 2.25 \mathbf{a}_1 - 0.0408 \mathbf{a}_2$$

Since $A\mathbf{x} = c_0 \mathbf{a}_1 + c_1 \mathbf{a}_2$ and $\hat{\mathbf{y}} = 2.25 \mathbf{a}_1 - 0.0408 \mathbf{a}_2$, the solution to $A\mathbf{x} = \hat{\mathbf{y}}$ is $c_0 = 2.25$ and $c_1 = -0.0408$. Hence the least squares regression line is $\hat{y} = 2.25 - 0.0408x$. A graph of the line together with the data is shown in Figure 2. The central four points lie roughly on a line, but the fit is poor due to the influence of the extreme points $(-6, 2.9)$ and $(6, 1.1)$.

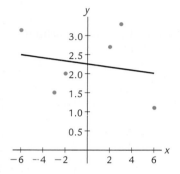

Figure 2 Scatter plot of data and $\hat{y} = 2.25 - 0.0408x$.

We now repeat the analysis, but this time using a weighted dot product. We can diminish the "pull" of the extreme points with the weights

$$\mathbf{t} = (t_1, t_2, t_3, t_4, t_5, t_6) = (1, 5, 5, 5, 5, 1)$$

This gives the four data points closest to the middle 5 times the weight of the outer two points. It is not hard to verify that the column vectors $\mathbf{a}_1$ and $\mathbf{a}_2$ satisfy $\langle \mathbf{a}_1, \mathbf{a}_2 \rangle = 0$ with respect to this weighted dot product. Thus we can use them for the projection function,

$$\hat{\mathbf{y}} = \text{proj}_S \mathbf{y} = \frac{\langle \mathbf{a}_1, \mathbf{y} \rangle}{\langle \mathbf{a}_1, \mathbf{a}_1 \rangle} \mathbf{a}_1 + \frac{\langle \mathbf{a}_2, \mathbf{y} \rangle}{\langle \mathbf{a}_2, \mathbf{a}_2 \rangle} \mathbf{a}_2$$

The required inner products are

$$\langle \mathbf{a}_1, \mathbf{y} \rangle = 51.5, \quad \langle \mathbf{a}_2, \mathbf{y} \rangle = 23.2, \quad \langle \mathbf{a}_1, \mathbf{a}_1 \rangle = 22, \quad \langle \mathbf{a}_2, \mathbf{a}_2 \rangle = 202$$

Hence the projection is

$$\hat{\mathbf{y}} = \text{proj}_S \mathbf{y} = \frac{51.5}{22} \mathbf{a}_1 + \frac{23.2}{202} \mathbf{a}_2 \approx 2.34 \mathbf{a}_1 + 0.115 \mathbf{a}_2$$

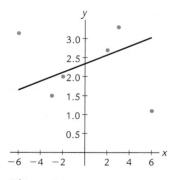

Figure 3 Scatter plot of data and $\hat{y} = 2.34 - 0.115x$.

By the same reasoning as before, the equation of the weighted least squares regression line is $\hat{y} = 2.34 + 0.115x$. The data and graph of this line are shown in Figure 3. Although the line has positive slope and is an improvement over ordinary least squares regression, it still does not fit the central data very well. Two more lines, with weights even more extreme to further diminish the effects of the extreme points, are shown in Figure 4.

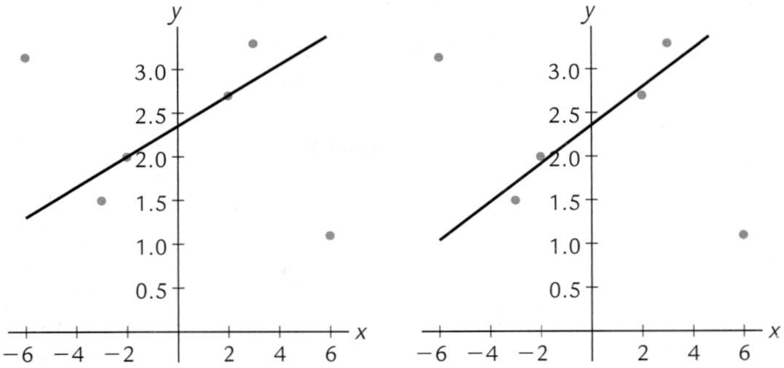

(a) **t** = (1, 10, 10, 10, 10, 1)
$\hat{y} = 2.36 + 0.172x$

(b) **t** = (1, 25, 25, 25, 25, 1)
$\hat{y} = 2.37 + 0.221x$

Figure 4 The graphs of weighted least squares regression lines with weights as shown.

The line in (b) of Figure 4 fits the central data fairly well, but the weights are so extreme that we are close to simply discarding the two outside points. In practice, one would typically make decisions about weights before collecting data, and it might happen that a linear equation is not appropriate for describing the data. ■

Fourier Approximations

We now return to the vector space $V = C[-\pi, \pi]$ together with the inner product

$$\langle f, g \rangle = \frac{1}{\pi} \int_{-\pi}^{\pi} f(x)g(x) \, dx \qquad (2)$$

In Example 2 (Section 10.2), we showed that the set $\{1, \cos(x), \sin(x)\}$ forms an orthogonal set in V. This can be expanded to a larger orthogonal set.

THEOREM 10.17

For each integer $k \geq 1$, the set

$$\{1, \cos(x), \cos(2x), \ldots, \cos(kx), \sin(x), \sin(2x), \ldots, \sin(kx)\} \qquad (3)$$

is orthogonal in $V = C[-\pi, \pi]$ with the inner product (2).

Proof   The proof involves computing a number of definite integrals to verify the orthogonality. Two are given here, and the rest are left as exercises. Starting with 1 and $\cos(2k\pi x)$, we have

$$\langle 1, \cos(kx) \rangle = \frac{1}{\pi} \int_{-\pi}^{\pi} \cos(kx) \, dx = \frac{1}{k\pi} \left[\sin(kx) \Big|_{-\pi}^{\pi} \right] = 0$$

▶ f is an *odd function* if $f(-x) = -f(x)$. If f is odd, then for any b we have

$$\int_{-b}^{b} f(x) \, dx = 0$$

so 1 and $\cos(kx)$ are orthogonal for any $k \geq 1$.

Since the product $\sin(jx)\cos(kx)$ is an odd function for any positive integers j and k, we have

$$\langle \sin(jx), \cos(kx) \rangle = \frac{1}{\pi} \int_{-\pi}^{\pi} \sin(jx) \cos(kx) \, dx = 0$$

As noted above, the remaining integrals are left as exercises. ■

Now let F_n denote the subspace of $V = C[-\pi, \pi]$ spanned by the orthogonal basis given in Theorem 10.17. For any f in V, the best approximation in F_n to f is given by

$$f_n(x) = \text{proj}_{F_n} f = a_0 + a_1 \cos(x) + \cdots + a_n \cos(nx) + b_1 \sin(x) + \cdots + b_n \sin(nx) \qquad (4)$$

Definition *n*th-Order Fourier Approximation

The function $f_n(x)$ is called the **nth-order Fourier approximation** of f. Since the basis functions of F_n are orthogonal, from the projection formula we have

$$a_k = \frac{\langle f, \cos(kx) \rangle}{\langle \cos(kx), \cos(kx) \rangle} \quad (k \geq 1)$$

$$b_k = \frac{\langle f, \sin(kx) \rangle}{\langle \sin(kx), \sin(kx) \rangle} \quad (k \geq 1)$$

For $k \geq 1$ we have $\langle \cos(kx), \cos(kx) \rangle = \langle \sin(kx), \sin(kx) \rangle = 1$ (see Exercises 37–38), so that the formulas for a_k and b_k simplify to

$$a_k = \frac{1}{\pi} \int_{-\pi}^{\pi} f(x) \cos(kx) \, dx \quad \text{and} \quad b_k = \frac{1}{\pi} \int_{-\pi}^{\pi} f(x) \sin(kx) \, dx \qquad (5)$$

Since $\langle 1, 1 \rangle = 2$, the constant term is

$$a_0 = \frac{1}{2\pi} \int_{-\pi}^{\pi} f(x) \, dx$$

Definition Fourier Coefficients

The a_k's and b_k's are called the **Fourier coefficients** of f.

EXAMPLE 2 Find the Fourier coefficients for $f(x) = x$ on $[-\pi, \pi]$.

Solution We start with

$$a_0 = \frac{1}{2\pi} \int_{-\pi}^{\pi} f(x) \, dx = \frac{1}{2\pi} \int_{-\pi}^{\pi} x \, dx = 0$$

because x is an odd function. As $x \cos(kx)$ is also an odd function, then by the same reasoning we have

$$a_k = \frac{1}{\pi} \int_{-\pi}^{\pi} x \cos(kx) \, dx = 0$$

For $k \geq 1$, an application of integration by parts (the details are left as an exercise) gives us

$$b_k = \frac{1}{\pi} \int_{-\pi}^{\pi} x \sin(kx) \, dx = \frac{2}{k}(-1)^{k+1}$$

Therefore our *n*th-order Fourier approximation to $f(x) = x$ is given by

$$f_n(x) = \sum_{k=1}^{n} \frac{2}{k}(-1)^{k+1} \sin(kx)$$

$$= 2\sin(x) - \sin(2x) + \frac{2}{3}\sin(3x) - \cdots + \frac{2}{n}(-1)^{n+1} \sin(nx)$$

The graphs of f_n for $n = 2, 4, 6$, and 8 are given in Figure 5.

If the Fourier coefficients decrease in size sufficiently quickly, then we can extend the *n*th-order Fourier approximation f_n to a **Fourier series**

$$a_0 + \sum_{k=1}^{\infty} \left(a_k \cos(kx) + b_k \sin(kx) \right)$$

Under the right conditions, the infinite series is equal to $f(x)$. The theory of Fourier series is covered in more advanced mathematics courses. ∎

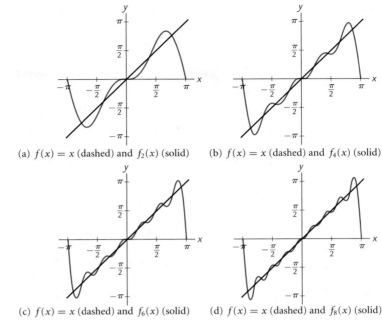

(a) $f(x) = x$ (dashed) and $f_2(x)$ (solid)

(b) $f(x) = x$ (dashed) and $f_4(x)$ (solid)

(c) $f(x) = x$ (dashed) and $f_6(x)$ (solid)

(d) $f(x) = x$ (dashed) and $f_8(x)$ (solid)

Figure 5 The graph of $f(x) = x$ and $f_n(x)$ for $n = 2, 4, 6,$ and 8. Note that the approximation improves with larger n.

Discrete Fourier Transforms

Often in applications we would like to find a Fourier approximation for a function f but do not have a formula for the function. Instead, we might know only values of $f(x)$ at discrete values of x, so we cannot directly apply the formulas given in (5) to find the Fourier coefficients.

▶ Here we assume that f has domain $[-\pi, \pi]$. A change of variables can be used to accommodate other domains.

We use a numerical integration technique to get around this problem. Suppose that for a function g, all we know are function values at n points evenly distributed in $[-\pi, \pi]$,

$$g\left(\frac{2\pi}{n} - \pi\right), \ g\left(\frac{4\pi}{n} - \pi\right), \ g\left(\frac{6\pi}{n} - \pi\right), \ \ldots, g\left(\frac{2n\pi}{n} - \pi\right) = g(\pi)$$

Then we can approximate the definite integral of g with the numerical integration formula

▶ (6) is just one of many numerical integration formulas.

$$\frac{1}{\pi} \int_{-\pi}^{\pi} g(x) \, dx \approx \frac{2}{n} \sum_{j=1}^{n} g\left(\frac{2j\pi}{n} - \pi\right) \tag{6}$$

To find a Fourier approximation for a function f with values known only at the discrete points

$$f\left(\frac{2\pi}{n} - \pi\right), \ f\left(\frac{4\pi}{n} - \pi\right), \ f\left(\frac{6\pi}{n} - \pi\right), \ \ldots, f\left(\frac{2n\pi}{n} - \pi\right) = f(\pi)$$

we use the numerical integration formula (6) to approximate a_k and b_k given by the definite integrals in (5) with the approximations

$$c_0 = \frac{1}{n} \sum_{j=1}^{n} f\left(\frac{2j\pi}{n} - \pi\right)$$

$$c_k = \frac{2}{n} \sum_{j=1}^{n} f\left(\frac{2j\pi}{n} - \pi\right) \cos\left(\frac{2jk\pi}{n} - k\pi\right) \qquad (k \geq 1) \tag{7}$$

$$d_k = \frac{2}{n} \sum_{j=1}^{n} f\left(\frac{2j\pi}{n} - \pi\right) \sin\left(\frac{2jk\pi}{n} - k\pi\right) \qquad (k \geq 1)$$

Definition Discrete Fourier
Coefficients

The c_k's and d_k's are called **discrete Fourier coefficients**. In general, the larger the number of function values that are known (that is, the larger the value of n), the closer c_k is to a_k and d_k is to b_k. We examine this computationally in the next example.

EXAMPLE 3 Let $f(x) = x$. Compare the values of the Fourier coefficients a_k and b_k of f with the discrete Fourier coefficients c_k and d_k.

Solution In Example 5 we determined that $a_k = 0$ for each $k \geq 0$, so we expect that c_k should get smaller as n (the number of discrete function values) gets larger. Table 1 gives the values of c_k for $0 \leq k \leq 10$ and $n = 50$, $n = 100$, $n = 500$, and $n = 1000$. We can see that the values of c_k get smaller, and hence closer to $a_k = 0$, as n gets larger.

▶ Other than sign, the values of c_k are the same for each k. This is not typical and an idiosyncrasy of this particular example.

k	a_k	c_k $(n = 50)$	c_k $(n = 100)$	c_k $(n = 500)$	c_k $(n = 1000)$
0	0	−0.125664	−0.062832	−0.012566	−0.006283
1	0	0.125664	0.062832	0.012566	0.006283
2	0	−0.125664	−0.062832	−0.012566	−0.006283
3	0	0.125664	0.062832	0.012566	0.006283
4	0	−0.125664	−0.062832	−0.012566	−0.006283
5	0	0.125664	0.062832	0.012566	0.006283
6	0	−0.125664	−0.062832	−0.012566	−0.006283
7	0	0.125664	0.062832	0.012566	0.006283
8	0	−0.125664	−0.062832	−0.012566	−0.006283
9	0	0.125664	0.062832	0.012566	0.006283
10	0	−0.125664	−0.062832	−0.012566	−0.006283

Table 1 The Values of a_k and c_k for $f(x) = x$

For $f(x) = x$ we have $b_k = \frac{2}{k}(-1)^{k+1}$ for $k \geq 1$. Table 2 gives the values of b_k for $1 \leq k \leq 10$ and $n = 50$, $n = 100$, $n = 500$, and $n = 1000$. The table values suggest that d_k is getting closer to b_k as n gets larger.

k	b_k	d_k $(n = 50)$	d_k $(n = 100)$	d_k $(n = 500)$	d_k $(n = 1000)$
1	2.00000	1.99737	1.99934	1.99997	1.99999
2	−1.00000	−0.99473	−0.99868	−0.99995	−0.99999
3	0.66667	0.65875	0.66469	0.66659	0.66665
4	−0.50000	−0.48943	−0.49737	−0.49990	−0.49997
5	0.40000	0.38675	0.39671	0.39987	0.39997
6	−0.33333	−0.31739	−0.32938	−0.33318	−0.33329
7	0.28571	0.26705	0.28109	0.28553	0.28567
8	−0.25000	−0.22858	−0.24471	−0.24979	−0.24995
9	0.22222	0.19801	0.21627	0.22199	0.22216
10	−0.20000	−0.17296	−0.19338	−0.19974	−0.19993

Table 2 The Values of b_k and d_k for $f(x) = x$

■

Definition nth-Order Discrete
Fourier Approximation

The **nth-order discrete Fourier approximation** is defined by

$$g_n(x) = c_0 + c_1 \cos(x) + \cdots + c_n \cos(nx) + d_1 \sin(x) + \cdots + d_n \sin(nx).$$

The only difference between $f_n(x)$ and $g_n(x)$ is in how the coefficients are computed.

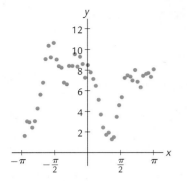

Figure 6 Data set for Example 4.

EXAMPLE 4 Find discrete Fourier approximations for the data set shown in Figure 6.

Solution There are 50 points shown in Figure 6. We have no formula for the function that generated the data, so finding the exact Fourier coefficients is out of the question. But we can use the discrete Fourier approximation and the formulas in (7) to compute approximations. The functions $g_2(x)$, $g_4(x)$, $g_8(x)$, and $g_{12}(x)$ are shown with the data in Figure 7. ■

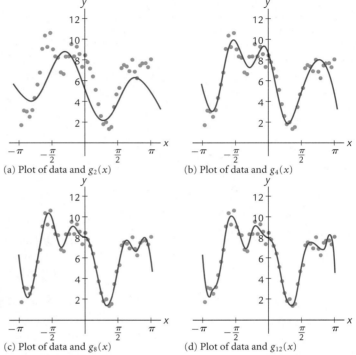

(a) Plot of data and $g_2(x)$ (b) Plot of data and $g_4(x)$

(c) Plot of data and $g_8(x)$ (d) Plot of data and $g_{12}(x)$

Figure 7 Scatter plot of data for Example 4 and plot of discrete Fourier approximations of degree 2, 4, 8, and 12.

Computational Comment

Calculating coefficients for the discrete Fourier approximation can be computationally intensive. To address this, various methods known collectively as *fast Fourier transforms* (FFT) have been developed to improve computational efficiency. A popular FFT method developed by J. W. Cooley and J. W. Tukey works recursively by factoring n, the number of points, as $n = n_1 n_2$ and then focusing on the smaller n_1 and n_2. Interestingly, there is evidence that Gauss knew about this method in the early 1800s. (There seems to be little about linear algebra that was unknown to Gauss.)

| EXERCISES |

1. Find the weighted least-squares line for the data set $\{(-2, 0), (-1, 2), (1, 3), (2, 5)\}$, with the inner two points weighted twice as much as the outer two points.

2. Find the weighted least-squares line for the data set $\{(-2, -2), (-1, 0), (1, 3), (2, 6)\}$, with the inner two points weighted three times as much as the outer two points.

Exercises 3–4 refer to the points shown in Figure 8.

3. Suppose that a line ℓ_1 is fitted to the points shown using ordinary least squares regression, and then a second line ℓ_2 is fitted using weighted least squares regression, with the two extreme points having half the weight of the others. How would you expect the slope of ℓ_1 to compare to that of ℓ_2? Explain your answer.

Figure 8

4. Suppose that a line ℓ_1 is fitted to the points shown using ordinary least squares regression, and then a second line ℓ_2 is fitted using weighted least squares regression, with the two extreme points having triple the weight of the others. How would you expect the slope of ℓ_1 to compare to that of ℓ_2? Explain your answer.

5. Suppose that all of the points in a weighted least-squares approximation have their weights tripled. Will it change the equation of the resulting line? Explain your answer.

6. Suppose that all of the points in a weighted least-squares approximation have their weights increased by a factor of 10. Will it change the equation of the resulting line? Explain your answer.

7. Find the Fourier approximation f_2 for
$$f(x) = \begin{cases} 1 & \text{if } -\frac{\pi}{2} \le x < \frac{\pi}{2} \\ 0 & \text{otherwise} \end{cases}$$

8. Find the Fourier approximation f_2 for
$$f(x) = \begin{cases} 1 & \text{if } -\pi \le x < -\frac{\pi}{2} \text{ and } \frac{\pi}{2} \le x < \pi \\ 0 & \text{otherwise} \end{cases}$$

9. Find the Fourier approximation f_2 for
$$f(x) = \begin{cases} 0 & \text{if } -\pi \le x < 0 \\ 1 & \text{if } 0 \le x < \pi \end{cases}$$

10. Find the Fourier approximation f_2 for
$$f(x) = \begin{cases} 1 & \text{if } -\pi \le x < 0 \\ 0 & \text{if } 0 \le x < \pi \end{cases}$$

11. Find the Fourier approximation f_2 for $f(x) = x + 1$.

12. Find the Fourier approximation f_2 for $f(x) = 3 - 2x$.

13. Find the Fourier approximation f_2 for $f(x) = x^2$.

14. Find the Fourier approximation f_2 for $f(x) = |x|$.

In Exercises 15–18, find the Fourier coefficients for the given function without performing any integrals.

15. $f(x) = \cos(2x) - \sin(3x)$

16. $f(x) = -2 + 3\cos(2x) - 4\sin(4x)$

17. $f(x) = 1 + \sin^2(4x)$

18. $f(x) = 1 - \cos^2(6x)$

19. Find the discrete Fourier approximation $g_1(x)$ for $f(x)$ based on the table information.

x	0	π
$f(x)$	1	2

20. Find the discrete Fourier approximation $g_2(x)$ for $f(x)$ based on the table information.

x	0	π
$f(x)$	-1	3

21. Find the discrete Fourier approximation $g_1(x)$ for $f(x)$ based on the table information.

x	$-\frac{\pi}{2}$	0	$\frac{\pi}{2}$	π
$f(x)$	0	1	3	-2

22. Find the discrete Fourier approximation $g_2(x)$ for $f(x)$ based on the table information.

x	$-\frac{\pi}{2}$	0	$\frac{\pi}{2}$	π
$f(x)$	-2	-1	0	2

FIND AN EXAMPLE For Exercises 23–28, find an example that meets the given specifications.

23. A data set such that the ordinary least squares regression line has slope zero, but the weighted least squares regression line, with triple weight on the right-most data point, has a negative slope.

24. A data set such that the ordinary least squares regression line has slope zero, but the weighted least squares regression line, with triple weight on the right-most and left-most data points, has a negative slope.

25. A function $f(x)$ such that the Fourier coefficients are all zero except for a_0 which is nonzero.

26. A function $f(x)$ such that the Fourier coefficients $b_1 = b_2 = b_3 = \cdots = 0$.

27. A function $f(x)$ such that the Fourier coefficients $a_1 = a_2 = a_3 = \cdots = 0$ and $a_0 = 1$.

28. A function $f(x)$ such that the Fourier coefficients $b_1 = b_2 = b_3 = \cdots = 0$ and $a_0 = -2$.

TRUE OR FALSE For Exercises 29–34, determine if the statement is true or false, and justify your answer.

29. In weighted least squares regression, the weights must all be positive.

30. In weighted least squares regression, the weights must all be greater than or equal to one.

31. Weighted least squares can only be applied to data sets where the corresponding matrix A has orthogonal columns.

32. The Fourier approximation can only be applied to positive functions.

33. If it is possible to compute Fourier coefficients for a function f, then it is also possible to compute discrete Fourier coefficients for f.

34. In general, the higher the number of discrete function values used, the better the approximation given by the discrete Fourier approximation.

In Exercises 35–38, we evaluate the remaining integrals required to show that the set given in Theorem 10.17 is orthogonal on $C[-\pi, \pi]$ with respect to the inner product

$$\langle f, g \rangle = \frac{1}{\pi} \int_{-\pi}^{\pi} f(x)g(x) \, dx$$

We also find the norms of these functions.

35. Show that $\langle 1, \sin(kx) \rangle = 0$ for all integers $k \geq 1$.

36. Show that $\|1\| = \sqrt{2}$.

37. Use the identity $2\sin^2(a) = 1 - \cos(2a)$ to show that $\|\sin(kx)\| = 1$ for all integers $k \geq 1$.

38. Use the identity $2\cos^2(a) = 1 + \cos(2a)$ to show that $\|\cos(kx)\| = 1$ for all integers $k \geq 1$.

39. Use integration by parts to show that

$$\frac{1}{\pi} \int_{-\pi}^{\pi} x \cos(kx) \, dx = 0 \quad (k \geq 1)$$

40. Use integration by parts to show that

$$\frac{1}{\pi} \int_{-\pi}^{\pi} x \sin(kx) \, dx = \frac{2}{k}(-1)^{k+1} \quad (k \geq 1)$$

41. Use the trigonometric identities

$$\cos(a - b) = \cos(a)\cos(b) + \sin(a)\sin(b)$$
$$\sin(a - b) = \sin(a)\cos(b) - \sin(b)\cos(a)$$

to prove that the formulas for c_k and d_k in (7) can be simplified to

$$c_k = \frac{2}{n} \sum_{j=1}^{n} (-1)^k f\left(\frac{2j\pi}{n} - \pi\right) \cos\left(\frac{2jk\pi}{n}\right) \quad (k \geq 1)$$

$$d_k = \frac{2}{n} \sum_{j=1}^{n} (-1)^k f\left(\frac{2j\pi}{n} - \pi\right) \sin\left(\frac{2jk\pi}{n}\right) \quad (k \geq 1)$$

42. $\boxed{C}$ Discrete values of f are given in the table below. Use this information to find the discrete Fourier approximation $g_3(x)$.

x	$-\frac{3\pi}{4}$	$-\frac{\pi}{2}$	$-\frac{\pi}{4}$	0	$\frac{\pi}{4}$	$\frac{\pi}{2}$	$\frac{3\pi}{4}$	π
$f(x)$	3.1	3.5	3.3	3.0	2.7	2.6	2.8	3.0

43. $\boxed{C}$ Discrete values of f are given in the table below. Use this information to find the discrete Fourier approximation $g_5(x)$.

x	$-\frac{3\pi}{4}$	$-\frac{\pi}{2}$	$-\frac{\pi}{4}$	0	$\frac{\pi}{4}$	$\frac{\pi}{2}$	$\frac{3\pi}{4}$	π
$f(x)$	2.4	2.8	3.0	3.5	2.9	2.6	2.4	2.1

44. $\boxed{C}$ Suppose that $f(x) = x^2 + 1$. Find $f_3(x)$. Then generate a list of values of f corresponding to $x = 0.02, 0.04, \ldots, 1.0$, and use these to find $g_3(x)$.

45. $\boxed{C}$ Suppose that $f(x) = e^x$. Find $f_3(x)$. Then generate a list of values of f corresponding to $x = 0.02, 0.04, \ldots, 1.0$, and use these to find $g_3(x)$.

Additional Topics and Applications

The Quebec Bridge, opened in 1919, crosses the St. Lawrence River at Levis and west of Quebec City. At 3,239 feet in length, it maintains its status as the longest cantilever bridge in the world. The bridge plays a notoriously important role in bridge engineering, as it collapsed twice during construction. The collapse of 1907 resulted in the death of 75 workers; the collapse of 1916 caused 13 deaths. The disasters raised questions about the control held by one engineer on such a project. Engineers, worried about potentially strict governmental intervention, organized to create professional engineering societies that served both to educate their members and to administer licensing requirements and exams.

Bridge suggested by Pont de Quebec (Martin St-Amant-Wikipedia-CC-BY-SA-3.0)

▶ The sections in this chapter are optional.

11.1 Quadratic Forms

In Chapter 3 we showed that every linear transformation $T : \mathbf{R}^m \rightarrow \mathbf{R}^n$ has the form $T(\mathbf{x}) = A\mathbf{x}$ for some $n \times m$ matrix A. In this section we study a function $Q : \mathbf{R}^n \rightarrow \mathbf{R}$ called a *quadratic form* that also can be defined in terms of matrix and vector multiplication. Such functions arise naturally in a variety of disciplines, including engineering (control theory), physics, economics, and mathematics.

DEFINITION 11.1

Definition Quadratic Form

Definition Matrix of the Quadratic Form

A **quadratic form** is a function $Q : \mathbf{R}^n \rightarrow \mathbf{R}$ that has the form

$$Q(\mathbf{x}) = \mathbf{x}^T A \mathbf{x} \tag{1}$$

where A is an $n \times n$ symmetric matrix called the **matrix of the quadratic form**.

▶ We interpret a quadratic form $\mathbf{x}^T A\mathbf{x}$ as a scalar instead of as a 1×1 matrix.

EXAMPLE 1 Suppose that $\mathbf{x} = \begin{bmatrix} x_1 \\ x_2 \end{bmatrix}$. Evaluate $Q(\mathbf{x}) = \mathbf{x}^T A\mathbf{x}$ for each of the matrices

$$(a)\ A = \begin{bmatrix} 2 & 0 \\ 0 & 5 \end{bmatrix} \qquad (b)\ A = \begin{bmatrix} 3 & 1 \\ 1 & -2 \end{bmatrix}$$

Solution (a) $\mathbf{x}^T A\mathbf{x} = \begin{bmatrix} x_1 & x_2 \end{bmatrix} \begin{bmatrix} 2 & 0 \\ 0 & 5 \end{bmatrix} \begin{bmatrix} x_1 \\ x_2 \end{bmatrix} = \begin{bmatrix} x_1 & x_2 \end{bmatrix} \begin{bmatrix} 2x_1 \\ 5x_2 \end{bmatrix} = 2x_1^2 + 5x_2^2.$

(b) Since A is not diagonal, this quadratic form is a bit more complicated. We have

$$\mathbf{x}^T A\mathbf{x} = \begin{bmatrix} x_1 & x_2 \end{bmatrix} \begin{bmatrix} 3 & 1 \\ 1 & -2 \end{bmatrix} \begin{bmatrix} x_1 \\ x_2 \end{bmatrix}$$

$$= \begin{bmatrix} x_1 & x_2 \end{bmatrix} \begin{bmatrix} 3x_1 + x_2 \\ x_1 - 2x_2 \end{bmatrix}$$

$$= x_1(3x_1 + x_2) + x_2(x_1 - 2x_2) = 3x_1^2 + 2x_1 x_2 - 2x_2^2 \quad \blacksquare$$

Note that in both parts, the coefficients on x_1^2 and x_2^2 come directly from the diagonal entries of A. In (b), we see that the coefficient on the *cross-product* term $x_1 x_2$ is the sum of the nondiagonal matrix entries. (This is also happening in part (a), but is not visible because the nondiagonal entries are 0.) These observations generalize so that it is not too hard to construct the matrix of a quadratic form from the equation.

EXAMPLE 2 Suppose that Q is the quadratic form

$$Q(\mathbf{x}) = 3x_1^2 - 7x_3^2 - 4x_1 x_2 + 10x_2 x_3$$

Directly compute $Q(\mathbf{x}_0)$ for $\mathbf{x}_0 = \begin{bmatrix} 1 \\ 3 \\ -2 \end{bmatrix}$. Then find the 3×3 matrix A of the quadratic form and use it to recompute $Q(\mathbf{x}_0)$ by applying the formula $Q(\mathbf{x}) = \mathbf{x}^T A\mathbf{x}$.

Solution For our given $\mathbf{x}_0$ we have $x_1 = 1$, $x_2 = 3$, and $x_3 = -2$. Hence

$$Q(\mathbf{x}_0) = 3(1)^2 - 7(-2)^2 - 4(1)(3) + 10(3)(-2) = -97$$

To find the matrix A of this quadratic form, we start by noting that the terms $3x_1^2$ and $-7x_3^2$ indicate that there should be a 3 and -7 in the first and third diagonal entries, respectively. Since there is no x_2^2 term, the second diagonal entry is 0. Thus the diagonal portion of A is

$$A = \begin{bmatrix} 3 & \bullet & \bullet \\ \bullet & 0 & \bullet \\ \bullet & \bullet & -7 \end{bmatrix}$$

The coefficient -4 on the cross-product term $-4x_1 x_2$ should be evenly split across the $(1, 2)$ and $(2, 1)$ entries to ensure that A is symmetric. Similarly, we evenly split the coefficient 10 from $10x_2 x_3$ across the $(2, 3)$ and $(3, 2)$ entries. Since that accounts for all of the terms of Q, any other entries of A should be zero. Therefore we end up with

$$A = \begin{bmatrix} 3 & -2 & 0 \\ -2 & 0 & 5 \\ 0 & 5 & -7 \end{bmatrix}$$

Now we test this by computing $\mathbf{x}_0^T A \mathbf{x}_0$. We have

$$\mathbf{x}_0^T A \mathbf{x}_0 = \begin{bmatrix} 1 & 3 & -2 \end{bmatrix} \begin{bmatrix} 3 & -2 & 0 \\ -2 & 0 & 5 \\ 0 & 5 & -7 \end{bmatrix} \begin{bmatrix} 1 \\ 3 \\ -2 \end{bmatrix} = \begin{bmatrix} 1 & 3 & -2 \end{bmatrix} \begin{bmatrix} -3 \\ -12 \\ 29 \end{bmatrix} = -97 \quad ∎$$

Quadratic forms can be easier to apply when there are no cross-product terms to complicate things. Happily, we can use a change of variables to arrange for this.

<table>
<tr><td>

THEOREM 11.2

▶ The "principal axes" in Theorem 11.2 are the columns of P, which are eigenvectors of A. The name will be explained shortly.

</td><td>

(PRINCIPAL AXES THEOREM)   If A is a symmetric matrix, then there exists an orthogonal matrix P such that the transformation $\mathbf{y} = P^T \mathbf{x}$ changes the quadratic form $\mathbf{x}^T A \mathbf{x}$ into the quadratic form $\mathbf{y}^T D \mathbf{y}$ (where D is diagonal) that has no cross-product terms.

</td></tr>
</table>

Proof   Since A is a symmetric matrix, by the Spectral Theorem (Section 8.3) A can be diagonalized as $A = PDP^{-1}$, where P is an orthogonal matrix with eigenvectors of A for columns and D is a diagonal matrix with eigenvalues of A for diagonal entries. Since P is orthogonal, we have $P^{-1} = P^T$, so that

$$A = PDP^T \quad \Longrightarrow \quad D = P^T A P$$

If we set $\mathbf{y} = P^{-1}\mathbf{x} = P^T\mathbf{x}$, then $\mathbf{x} = P\mathbf{y}$ so that

$$\mathbf{x}^T A \mathbf{x} = (P\mathbf{y})^T A (P\mathbf{y}) = \mathbf{y}^T P^T A P \mathbf{y} = \mathbf{y}^T (P^T A P) \mathbf{y} = \mathbf{y}^T D \mathbf{y}$$

Hence the matrix of the quadratic form is diagonal with respect to the change of variables $\mathbf{y} = P^{-1}\mathbf{x}$. ∎

EXAMPLE 3   Find a change of variables to express the quadratic form with matrix

$$A = \begin{bmatrix} 5 & 2 \\ 2 & 8 \end{bmatrix}$$

as a quadratic form with no cross-product terms.

Solution   The quadratic form with matrix A is

$$Q(\mathbf{x}) = 5x_1^2 + 8x_2^2 + 4x_1 x_2 \quad (2)$$

To eliminate the cross-product term, we need to find a matrix P that orthogonally diagonalizes A. Computing the roots of the characteristic polynomial reveals that the eigenvalues of this matrix are $\lambda_1 = 9$ and $\lambda_2 = 4$. The corresponding normalized eigenvectors are

$$\lambda_1 = 9 \implies \mathbf{p}_1 = \tfrac{1}{\sqrt{5}} \begin{bmatrix} 1 \\ 2 \end{bmatrix} \qquad \lambda_2 = 4 \implies \mathbf{p}_2 = \tfrac{1}{\sqrt{5}} \begin{bmatrix} -2 \\ 1 \end{bmatrix}$$

▶ In this chapter the details of computing eigenvalues and eigenvectors are often left to the reader.

Therefore we have

$$D = \begin{bmatrix} \lambda_1 & 0 \\ 0 & \lambda_2 \end{bmatrix} = \begin{bmatrix} 9 & 0 \\ 0 & 4 \end{bmatrix} \quad \text{and} \quad P = \begin{bmatrix} \mathbf{p}_1 & \mathbf{p}_2 \end{bmatrix} = \begin{bmatrix} \frac{1}{\sqrt{5}} & -\frac{2}{\sqrt{5}} \\ \frac{2}{\sqrt{5}} & \frac{1}{\sqrt{5}} \end{bmatrix}$$

Hence for $\mathbf{y} = \begin{bmatrix} y_1 \\ y_2 \end{bmatrix} = P^T\mathbf{x}$,

$$Q(\mathbf{y}) = 9y_1^2 + 4y_2^2 \quad (3)$$

We can test out the two versions of the quadratic form by starting with a specific vector—say, $\mathbf{x}_0 = \begin{bmatrix} 2 \\ 1 \end{bmatrix}$. Evaluating (2) directly, we find that

$$Q\left(\begin{bmatrix} 2 \\ 1 \end{bmatrix}\right) = 5(2)^2 + 8(1)^2 + 4(2)(1) = 36$$

For this choice of $\mathbf{x}_0$, the corresponding $\mathbf{y}_0$ is

$$\mathbf{y}_0 = P^T\mathbf{x}_0 = \begin{bmatrix} \frac{1}{\sqrt{5}} & \frac{2}{\sqrt{5}} \\ -\frac{2}{\sqrt{5}} & \frac{1}{\sqrt{5}} \end{bmatrix}\begin{bmatrix} 2 \\ 1 \end{bmatrix} = \begin{bmatrix} \frac{4}{\sqrt{5}} \\ -\frac{3}{\sqrt{5}} \end{bmatrix}$$

Therefore from (3) we have

$$9y_1^2 + 4y_2^2 = 9\left(\frac{4}{\sqrt{5}}\right)^2 + 4\left(-\frac{3}{\sqrt{5}}\right)^2 = 9\left(\frac{16}{5}\right) + 4\left(\frac{9}{5}\right) = \frac{180}{5} = 36 \quad \blacksquare$$

Geometry of Quadratic Forms

Definition Principal Axes

In the Principal Axes Theorem, the columns of P—which are eigenvectors of A—are the **principal axes** of the quadratic form $\mathbf{x}^T A\mathbf{x}$. The use of the word "axes" makes more sense when we view quadratic forms geometrically. Consider the set of vectors $\mathbf{x}$ in $\mathbf{R}^2$ that satisfy the equation

$$\mathbf{x}^T A\mathbf{x} = c \tag{4}$$

where c is a fixed constant and A is an invertible 2×2 symmetric matrix. It turns out that graph of the solution set can be one of the following: an ellipse (including circles), a hyperbola, two intersecting lines, a single point, or the empty set. Here we focus on the ellipse and hyperbola.

If $A = \begin{bmatrix} a & 0 \\ 0 & b \end{bmatrix}$ is diagonal, then (4) is equivalent to

$$ax_1^2 + bx_2^2 = c$$

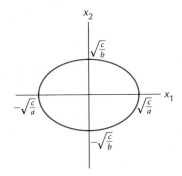

Figure 1 Graph of the solutions to $ax_1^2 + bx_2^2 = c$. The axis intercepts are as shown.

Definition Standard Position

When a, b, and c are all positive, the graph of the solution set is the ellipse in Figure 1. The graph of the solution set of a quadratic form with no cross-product terms is said to be in **standard position**. Since A is diagonal, the eigenvectors of A point in the direction of the coordinate axes, which coincide with the major and minor axes of the ellipse. This carries over to quadratic forms that have cross-product terms.

EXAMPLE 4 Graph the set of solutions to the quadratic form
$$3x_1^2 + 6x_2^2 + 4x_1x_2 = 40$$

Solution The quadratic form $3x_1^2 + 6x_2^2 + 4x_1x_2$ has matrix $A = \begin{bmatrix} 3 & 2 \\ 2 & 6 \end{bmatrix}$. The eigenvalues and normalized eigenvectors of A are

$$\lambda_1 = 2 \implies \mathbf{p}_1 = \frac{1}{\sqrt{5}}\begin{bmatrix} 2 \\ -1 \end{bmatrix}, \qquad \lambda_2 = 7 \implies \mathbf{p}_2 = \frac{1}{\sqrt{5}}\begin{bmatrix} 1 \\ 2 \end{bmatrix}$$

Now let

$$D = \begin{bmatrix} \lambda_1 & 0 \\ 0 & \lambda_2 \end{bmatrix} = \begin{bmatrix} 2 & 0 \\ 0 & 7 \end{bmatrix} \quad \text{and} \quad P = \begin{bmatrix} \mathbf{p}_1 & \mathbf{p}_2 \end{bmatrix} = \begin{bmatrix} \frac{2}{\sqrt{5}} & \frac{1}{\sqrt{5}} \\ -\frac{1}{\sqrt{5}} & \frac{2}{\sqrt{5}} \end{bmatrix}$$

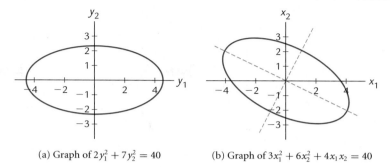

(a) Graph of $2y_1^2 + 7y_2^2 = 40$ (b) Graph of $3x_1^2 + 6x_2^2 + 4x_1x_2 = 40$

Figure 2 In (a), the graph of $\mathbf{y}^T D\mathbf{y} = 40$, which is in standard position. Rotating this graph to align the ellipse axes (dashed) with the eigenvectors $\mathbf{p}_1$ and $\mathbf{p}_2$ of A gives the graph of $\mathbf{x}^T A\mathbf{x} = 40$ in (b).

Then $\mathbf{y}^T D\mathbf{y} = 40$ is equivalent to $2y_1^2 + 7y_2^2 = 40$. Since there are no cross-product terms, we can use Figure 1 as a model for the graph of $\mathbf{y}^T D\mathbf{y} = 40$, which is given in Figure 2(a).

By the Principal Axes Theorem, $\mathbf{x}^T A\mathbf{x} = c$ and $\mathbf{y}^T D\mathbf{y} = c$ are the same equation when $\mathbf{y} = P^T\mathbf{x}$, or equivalently, $\mathbf{x} = P\mathbf{y}$. Therefore the graph of all $\mathbf{x}$ that satisfy $\mathbf{x}^T A\mathbf{x} = c$ is the same as the graph of all $\mathbf{y}$ that satisfy $\mathbf{y}^T D\mathbf{y} = c$ after applying the transformation $\mathbf{x} = P\mathbf{y}$. Since P is an orthogonal matrix, this transformation is a rotation or reflection (see Exercise 46), so the graph of $\mathbf{x}^T A\mathbf{x} = c$ is a rotation or reflection of the graph of $\mathbf{y}^T D\mathbf{y} = c$. Furthermore, since $\mathbf{e}_1$ and $\mathbf{e}_2$ are parallel to the major and minor axes of the ellipse $\mathbf{y}^T D\mathbf{y} = c$, then $P(\mathbf{e}_1) = \mathbf{p}_1$ and $P(\mathbf{e}_2) = \mathbf{p}_2$ are parallel to the axes of $\mathbf{x}^T A\mathbf{x} = c$. The graph is given in Figure 2(b). ■

▶ Recall that $\{\mathbf{e}_1, \mathbf{e}_2\}$ are the standard basis for $\mathbf{R}^2$,

$$\mathbf{e}_1 = \begin{bmatrix} 1 \\ 0 \end{bmatrix}, \quad \mathbf{e}_2 = \begin{bmatrix} 0 \\ 1 \end{bmatrix}$$

Summing up the solution to Example 4, we did the following:

- Found D and P for A.

- Graphed $\mathbf{y}^T D\mathbf{y} = c$, which is not difficult, because it is in standard form.

- Rotated the graph so that the axes of symmetry align with the principal axes of A to graph $\mathbf{x}^T A\mathbf{x} = c$.

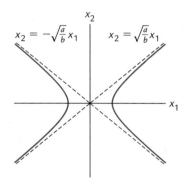

$$x_2 = -\sqrt{\tfrac{a}{b}}\,x_1 \qquad x_2 = \sqrt{\tfrac{a}{b}}\,x_1$$

Figure 3 Graph of the solutions to $ax_1^2 - bx_2^2 = c$. The asymptotes (dashed) have equations shown.

Turning to hyperbolas, if a, b, and c are still positive, then the graph of $ax_1^2 - bx_2^2 = c$ is a hyperbola in standard position with asymptotes $x_2 = \pm\sqrt{\tfrac{a}{b}}\,x_1$ (Figure 3). We can use the same approach as in Example 4 to graph hyperbolas that are not in standard position.

EXAMPLE 5 Graph the set of solutions to the quadratic form

$$4x_1^2 - x_2^2 + 12x_1x_2 = 10$$

Solution Here the matrix of the quadratic form is $A = \begin{bmatrix} 4 & 6 \\ 6 & -1 \end{bmatrix}$, which has eigenvalues and normalized eigenvectors

$$\lambda_1 = 8 \;\Rightarrow\; \mathbf{p}_1 = \tfrac{1}{\sqrt{13}}\begin{bmatrix} 3 \\ 2 \end{bmatrix}, \qquad \lambda_2 = -5 \;\Rightarrow\; \mathbf{p}_2 = \tfrac{1}{\sqrt{13}}\begin{bmatrix} -2 \\ 3 \end{bmatrix}$$

Hence the matrices D and P from the Principal Axes Theorem are

$$D = \begin{bmatrix} 8 & 0 \\ 0 & -5 \end{bmatrix} \quad \text{and} \quad P = \begin{bmatrix} \frac{3}{\sqrt{13}} & -\frac{2}{\sqrt{13}} \\ \frac{2}{\sqrt{13}} & \frac{3}{\sqrt{13}} \end{bmatrix}$$

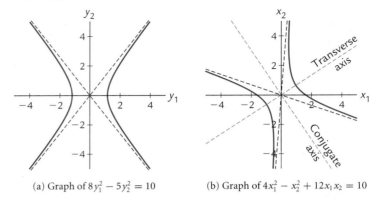

(a) Graph of $8y_1^2 - 5y_2^2 = 10$ (b) Graph of $4x_1^2 - x_2^2 + 12x_1 x_2 = 10$

Figure 4 In (a), the graph of $8y_1^2 - 5y_2^2 = 10$, which is in standard position. Rotating this graph to align the axes of symmetry with the eigenvectors of A gives the graph of $4x_1^2 - x_2^2 + 12x_1 x_2 = 10$ in (b).

The equation $8y_1^2 - 5y_2^2 = 10$ is in standard form, and the graph is a hyperbola with asymptotes $y_2 = \pm\sqrt{\frac{8}{5}}\, y_1$ (Figure 4(a)). Hyperbolas also have two axes of symmetry, the transverse axis (the x-axis in Figure 4(a)) and the conjugate axis (the y-axis in Figure 4(a)).

As with ellipses, we get the graph of $\mathbf{x}^T A\mathbf{x} = 10$ by applying the transformation $\mathbf{x} = P\mathbf{y}$ to the graph of $\mathbf{y}^T D\mathbf{y} = 10$. The axes of symmetry of the graph of $\mathbf{y}^T D\mathbf{y} = 10$ are rotated to align with the eigenvectors of A, yielding the graph in Figure 4(b). ■

Types of Quadratic Forms

We can classify a quadratic form $Q(\mathbf{x}) = \mathbf{x}^T A\mathbf{x}$ based on the values of $Q(\mathbf{x})$ as $\mathbf{x}$ ranges over different possibilities in $\mathbf{R}^n$.

DEFINITION 11.3

Definition Positive Definite, Negative Definite, Indefinite, Positive Semidefinite, Negative Semidefinite

Let $Q(\mathbf{x}) = \mathbf{x}^T A\mathbf{x}$ be a quadratic form.

(a) Q is **positive definite** if $Q(\mathbf{x}) > 0$ for all nonzero vectors $\mathbf{x}$ in $\mathbf{R}^n$, and Q is **positive semidefinite** if $Q(\mathbf{x}) \geq 0$ for all $\mathbf{x}$ in $\mathbf{R}^n$.

(b) Q is **negative definite** if $Q(\mathbf{x}) < 0$ for all nonzero vectors $\mathbf{x}$ in $\mathbf{R}^n$, and Q is **negative semidefinite** if $Q(\mathbf{x}) \leq 0$ for all $\mathbf{x}$ in $\mathbf{R}^n$.

(c) Q is **indefinite** if $Q(\mathbf{x})$ is positive for some $\mathbf{x}$'s in $\mathbf{R}^n$ and negative for others.

The graphs in Figure 5 show quadratic forms that are positive definite, negative definite, and indefinite.

It might seem difficult to classify a quadratic form, but it turns out that the eigenvalues of the matrix of a quadratic form tell the story.

THEOREM 11.4

Let A be an $n \times n$ symmetric matrix, and suppose that $Q(\mathbf{x}) = \mathbf{x}^T A\mathbf{x}$. Then

(a) Q is positive definite exactly when A has only positive eigenvalues.

(b) Q is negative definite exactly when A has only negative eigenvalues.

(c) Q is indefinite exactly when A has positive and negative eigenvalues.

Proof Since A is a symmetric matrix, there exist matrices P and D such that $P^T A P = D$, where the columns of P are orthonormal eigenvectors of A and the diagonal entries

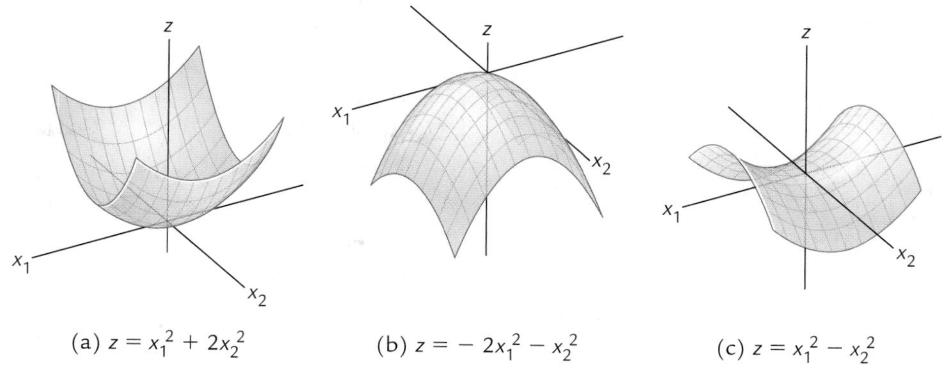

(a) $z = x_1^2 + 2x_2^2$ (b) $z = -2x_1^2 - x_2^2$ (c) $z = x_1^2 - x_2^2$

Figure 5 Plots of (a) positive definite, (b) negative definite, and (c) indefinite quadratic forms.

of D are the eigenvalues $\lambda_1, \ldots, \lambda_n$ of A. Since P is invertible, for a given $\mathbf{x}$ we can define $\mathbf{y} = P^{-1}\mathbf{x}$, so that $\mathbf{x} = P\mathbf{y}$. Then

$$Q(\mathbf{x}) = \mathbf{x}^T A\mathbf{x} = (P\mathbf{y})^T A(P\mathbf{y})$$
$$= \mathbf{y}^T (P^T AP)\mathbf{y} = \mathbf{y}^T D\mathbf{y} = \lambda_1 y_1^2 + \cdots + \lambda_n y_n^2$$

If the eigenvalues are all positive, then $Q(\mathbf{x}) > 0$ except when $\mathbf{y} = \mathbf{0}$, which implies $\mathbf{x} = \mathbf{0}$. Hence Q is positive definite. On the other hand, suppose that A has a nonpositive eigenvalue—say, $\lambda_1 \leq 0$. If $\mathbf{y}$ has $y_1 = 1$ and the other components are 0, then for the corresponding $\mathbf{x} \neq \mathbf{0}$ we have

$$Q(\mathbf{x}) = \lambda_1 \leq 0$$

so that Q is not positive definite. This proves part (a). The other parts are similar and left as an exercise. ■

EXAMPLE 6   Determine if $Q(\mathbf{x}) = x_2^2 + 2x_1 x_2 + 4x_1 x_3 + 2x_2 x_3$ is positive definite, negative definite, or indefinite.

Solution The matrix of this quadratic form is

$$A = \begin{bmatrix} 0 & 1 & 2 \\ 1 & 1 & 1 \\ 2 & 1 & 0 \end{bmatrix}$$

The eigenvalues of A are $\lambda_1 = -2$, $\lambda_2 = 0$, and $\lambda_3 = 3$. Thus, by Theorem 11.4, Q is indefinite. ■

Applying Theorem 11.4 requires knowing the eigenvalues of A, which can sometimes be hard to find. In the next section we will see how to use determinants to accomplish the same thing.

| EXERCISES |

For Exercises 1–4, evaluate $Q(\mathbf{x})$ for the given $\mathbf{x}_0$.

1. $Q(\mathbf{x}) = x_1^2 - 5x_2^2 + 6x_1 x_2$; $\mathbf{x}_0 = \begin{bmatrix} 3 \\ 1 \end{bmatrix}$

2. $Q(\mathbf{x}) = x_1^2 + 4x_2^2 - 3x_3^2 + 6x_1 x_2$; $\mathbf{x}_0 = \begin{bmatrix} 1 \\ 0 \\ 2 \end{bmatrix}$

3. $Q(\mathbf{x}) = 3x_1^2 + x_2^2 - x_3^2 + 6x_1 x_3$; $\mathbf{x}_0 = \begin{bmatrix} 0 \\ 3 \\ 1 \end{bmatrix}$

4. $Q(\mathbf{x}) = -2x_1^2 + 7x_2^2 - 8x_1 x_2 - 4x_1 x_3 + 2x_2 x_3$; $\mathbf{x}_0 = \begin{bmatrix} 0 \\ 3 \\ 1 \end{bmatrix}$

For Exercises 5–12, find a formula for the quadratic form with the given matrix A.

5. $A = \begin{bmatrix} 4 & 0 \\ 0 & 1 \end{bmatrix}$

6. $A = \begin{bmatrix} -3 & 0 \\ 0 & 7 \end{bmatrix}$

7. $A = \begin{bmatrix} 1 & 3 \\ 3 & 2 \end{bmatrix}$

8. $A = \begin{bmatrix} 5 & -2 \\ -2 & 0 \end{bmatrix}$

9. $A = \begin{bmatrix} 1 & 0 & 0 \\ 0 & 3 & 0 \\ 0 & 0 & -2 \end{bmatrix}$

10. $A = \begin{bmatrix} 2 & 1 & -3 \\ 1 & 0 & 0 \\ -3 & 0 & 5 \end{bmatrix}$

11. $A = \begin{bmatrix} 2 & 0 & 0 & 0 \\ 0 & 1 & 0 & 0 \\ 0 & 0 & 2 & 0 \\ 0 & 0 & 0 & 3 \end{bmatrix}$

12. $A = \begin{bmatrix} 5 & 0 & 1 & 0 \\ 0 & 2 & 0 & 0 \\ 1 & 0 & 0 & 3 \\ 0 & 0 & 3 & 4 \end{bmatrix}$

For Exercises 13–18, find a matrix A such that $Q(\mathbf{x}) = \mathbf{x}^T A\mathbf{x}$.

13. $Q(\mathbf{x}) = x_1^2 - 5x_2^2 + 6x_1 x_2$

14. $Q(\mathbf{x}) = x_1^2 + 4x_2^2 - 3x_3^2 + 6x_1 x_2$

15. $Q(\mathbf{x}) = 3x_1^2 + x_2^2 - x_3^2 + 6x_1 x_3$

16. $Q(\mathbf{x}) = -2x_1^2 + 7x_2^2 - 8x_1 x_2 - 4x_1 x_3 + 2x_2 x_3$

17. $Q(\mathbf{x}) = 5x_1^2 - x_2^2 + 3x_3^2 + 6x_1 x_3 - 12x_2 x_3$

18. $Q(\mathbf{x}) = x_2^2 + x_3^2 + 2x_1 x_2 - 4x_1 x_3 - 8x_2 x_3$

In Exercises 19–26, determine if the quadratic form $Q(\mathbf{x}) = \mathbf{x}^T A\mathbf{x}$ is positive definite, negative definite, indefinite, or none of these.

19. $A = \begin{bmatrix} 1 & 2 \\ 2 & 1 \end{bmatrix}$

20. $A = \begin{bmatrix} 5 & 1 \\ 1 & 1 \end{bmatrix}$

21. $A = \begin{bmatrix} 2 & 2 \\ 2 & -1 \end{bmatrix}$

22. $A = \begin{bmatrix} 5 & 8 \\ 8 & -1 \end{bmatrix}$

23. $A = \begin{bmatrix} 0 & 1 & 0 \\ 1 & 0 & 0 \\ 0 & 0 & 1 \end{bmatrix}$

24. $A = \begin{bmatrix} 0 & 0 & 1 \\ 0 & 1 & 0 \\ 1 & 0 & 1 \end{bmatrix}$

25. $A = \begin{bmatrix} 0 & 0 & 1 & 0 \\ 0 & 1 & 0 & 0 \\ 1 & 0 & 0 & 0 \\ 0 & 0 & 0 & 1 \end{bmatrix}$

26. $A = \begin{bmatrix} 1 & 1 & 0 & 0 \\ 1 & 1 & 0 & 0 \\ 0 & 0 & 1 & 1 \\ 0 & 0 & 1 & 1 \end{bmatrix}$

FIND AN EXAMPLE For Exercises 27–34, find an example that meets the given specifications.

27. A quadratic form $Q(\mathbf{x})$ and a constant c such that $Q(\mathbf{x}) = c$ has no solutions.

28. A quadratic form $Q(\mathbf{x})$ and a constant c such that $Q(\mathbf{x}) = c$ has exactly one solution.

29. A quadratic form $Q(\mathbf{x})$ and a constant c such that the graph of $Q(\mathbf{x}) = c$ is two intersecting lines.

30. A quadratic form $Q : \mathbf{R} \to \mathbf{R}$.

31. A quadratic form $Q(\mathbf{x})$ that is also a linear transformation.

32. A quadratic form $Q : \mathbf{R}^4 \to \mathbf{R}$ that is indefinite.

33. A quadratic form $Q : \mathbf{R}^3 \to \mathbf{R}$ that is positive semidefinite but not positive definite.

34. A quadratic form $Q : \mathbf{R}^2 \to \mathbf{R}$ that is negative semidefinite but not negative definite.

TRUE OR FALSE For Exercises 35–40, determine if the statement is true or false, and justify your answer.

35. The matrix A of a quadratic form Q must be symmetric.

36. If Q is a quadratic form, then $Q(\mathbf{x}) = Q(-\mathbf{x})$.

37. If Q is a quadratic form, then

$$Q(\mathbf{x}_1 + \mathbf{x}_2) = Q(\mathbf{x}_1) + Q(\mathbf{x}_2)$$

38. If A is a diagonal matrix and the matrix of a quadratic form Q, then Q is positive definite.

39. If $Q_1(\mathbf{x})$ and $Q_2(\mathbf{x})$ are quadratic forms, then so is $Q_1(\mathbf{x}) + Q_2(\mathbf{x})$.

40. If A is the matrix of the quadratic form $Q(\mathbf{x})$, then cA is the matrix of the quadratic form $c\,Q(\mathbf{x})$.

41. Prove that a quadratic form $Q : \mathbf{R} \to \mathbf{R}$ cannot be indefinite.

42. Prove that if $Q(\mathbf{x}) = \mathbf{x}^T A\mathbf{x}$ and $Q_T(\mathbf{x}) = \mathbf{x}^T A^T\mathbf{x}$, then $Q(\mathbf{x}) = Q_T(\mathbf{x})$.

43. Show that $Q(\mathbf{x}) = \|\mathbf{x}\|^2$ is a quadratic form and give the matrix of the quadratic form. Then show that Q is positive definite.

44. Show that $Q(\mathbf{x}) = \|A\mathbf{x}\|^2$ is a quadratic form and give the matrix of the quadratic form. Then show that Q is positive definite if and only if $\text{null}(A) = \{\mathbf{0}\}$.

45. Prove that $Q(\mathbf{0}) = \mathbf{0}$ for every quadratic form Q.

46. In this exercise we show that if P is an orthogonal 2×2 matrix, then the transformation $P\mathbf{y} = \mathbf{x}$ is a rotation or reflection.

(a) Prove that $\|P\mathbf{y}\| = \|\mathbf{y}\|$ for $\mathbf{y}$ in $\mathbf{R}^2$.

(b) Prove that if $\mathbf{x}$ and $\mathbf{y}$ are in $\mathbf{R}^2$, then the angle between $\mathbf{x}$ and $\mathbf{y}$ is the same as the angle between $P\mathbf{x}$ and $P\mathbf{y}$. HINT: Recall the formula

$$\cos(\theta) = \frac{\mathbf{x} \cdot \mathbf{y}}{\|\mathbf{x}\| \|\mathbf{y}\|}$$

(c) Combine (a) and (b) to explain why $P\mathbf{y} = \mathbf{x}$ is a rotation or reflection.

11.2 Positive Definite Matrices

Definition 11.3 in Section 11.1 gives, for a quadratic form, the meaning of positive definite, positive semidefinite, and so on. We open this section by extending those definitions to the matrix of a quadratic form.

DEFINITION 11.5

Definition **Positive Definite**

A symmetric $n \times n$ matrix A is **positive definite** if the corresponding quadratic form $Q(\mathbf{x}) = \mathbf{x}^T A \mathbf{x}$ is positive definite. Analogous definitions apply for **negative definite** and **indefinite**.

By Theorem 11.4 in Section 11.1, we know that one way to determine if a symmetric matrix A is positive definite is to examine the eigenvalues of A. If all the eigenvalues are positive, then A is positive definite. The shortcoming of this approach is that it can be difficult to find the eigenvalues, so it would be useful to have another method available. The next theorem gives us a start in this direction.

THEOREM 11.6

If A is a symmetric positive definite matrix, then A is nonsingular and $\det(A) > 0$.

Proof   Since A is positive definite, by Theorem 11.4 in Section 11.1, all of the eigenvalues of A are positive and hence nonzero. Thus, by The Big Theorem, Version 8 (Section 6.1), A is nonsingular. Next, recall that

$$\det(A) = \lambda_1 \lambda_2 \cdots \lambda_n$$

where $\lambda_1, \lambda_2, \ldots, \lambda_n$ are the eigenvalues of A (see Exercise 66 in Section 6.1). Since the eigenvalues are positive, so is their product, and therefore $\det(A) > 0$. ∎

EXAMPLE 1   Show that Theorem 11.6 holds for the matrix

$$A = \begin{bmatrix} 3 & -1 & 0 \\ -1 & 3 & -1 \\ 0 & -1 & 3 \end{bmatrix}$$

Solution   The eigenvalues of A are

$$\lambda_1 = 3 - \sqrt{2} \approx 1.586, \qquad \lambda_2 = 3, \qquad \lambda_3 = 3 + \sqrt{2} \approx 4.414$$

Since the eigenvalues are all positive, by Theorem 11.4 in Section 11.1 the associated quadratic form, and hence the matrix A, is positive definite. We also have $\det(A) = 21 > 0$. This implies A is nonsingular and shows that Theorem 11.6 is true for this matrix. ∎

▶ Suppose that

$$A = \begin{bmatrix} -2 & 0 \\ 0 & -1 \end{bmatrix}$$

Then $\det(A) = 2$, but A has negative eigenvalues -2 and -1 and so is not positive definite.

It would be nice if the converse of Theorem 11.6 was true, so that $\det(A) > 0$ would be enough to guarantee that A is positive definite, but that is not true. (See the example in the margin.) However, let's not abandon determinants yet. If

$$A = \begin{bmatrix} a_{11} & a_{12} & \cdots & a_{1n} \\ a_{21} & a_{22} & \cdots & a_{2n} \\ \vdots & \vdots & \ddots & \vdots \\ a_{n1} & a_{n2} & \cdots & a_{nn} \end{bmatrix}$$

Definition Leading Principal Submatrix

then the **leading principal submatrices** of A are given by

$$A_1 = \begin{bmatrix} a_{11} \end{bmatrix}, \quad A_2 = \begin{bmatrix} a_{11} & a_{12} \\ a_{21} & a_{22} \end{bmatrix}, \quad A_3 = \begin{bmatrix} a_{11} & a_{12} & a_{13} \\ a_{21} & a_{22} & a_{23} \\ a_{31} & a_{32} & a_{33} \end{bmatrix}$$

and so on through $A_n = A$.

THEOREM 11.7

A symmetric positive definite matrix A has leading principal submatrices A_1, A_2, ..., A_n that are also positive definite.

Proof Let $1 \leq m \leq n$, and suppose that $\mathbf{x}_m \neq \mathbf{0}$ is in $\mathbf{R}^m$. If we set

$$\mathbf{x} = \begin{bmatrix} \mathbf{x}_m \\ 0 \\ \vdots \\ 0 \end{bmatrix} \text{ in } \mathbf{R}^n$$

then we have

$$\mathbf{x}_m^T A_m \mathbf{x}_m = \mathbf{x}^T A \mathbf{x} > 0$$

because A is positive definite. Since this works for any nonzero $\mathbf{x}_m$, it follows that A_m is positive definite. ∎

EXAMPLE 2 Show that the leading principal submatrices of the positive definite matrix A in Example 1 are also positive definite.

Solution Since $A_1 = \begin{bmatrix} 3 \end{bmatrix}$ has associated quadratic form $Q(x) = 3x^2$ that is positive definite, then A_1 is positive definite. The matrix

$$A_2 = \begin{bmatrix} 3 & -1 \\ -1 & 3 \end{bmatrix}$$

has positive eigenvalues $\lambda_1 = 2$ and $\lambda_2 = 4$. Hence A_2 is also positive definite by Theorem 11.4 in Section 11.1. We have already shown that $A_3 = A$ is positive definite, so all leading principal submatrices of A are positive definite. ∎

Combining Theorem 11.6 and Theorem 11.7 shows that if A is positive definite, then the leading principal submatrices satisfy $\det(A_1) > 0$, $\det(A_2) > 0$, ..., $\det(A_n) > 0$. Interestingly, the converse is also true.

THEOREM 11.8

A symmetric matrix A is positive definite if and only if the leading principal submatrices satisfy

$$\det(A_1) > 0, \ \det(A_2) > 0, \ \dots, \ \det(A_n) > 0 \tag{1}$$

The proof of Theorem 11.8 is given at the end of the section.

EXAMPLE 3   Determine if the matrix

$$A = \begin{bmatrix} 4 & 8 & -8 \\ 8 & 25 & 11 \\ -8 & 11 & 98 \end{bmatrix}$$

is positive definite.

Solution We have

$$\det(A_1) = 4, \quad \det(A_2) = \begin{vmatrix} 4 & 8 \\ 8 & 25 \end{vmatrix} = 36, \quad \det(A_3) = \begin{vmatrix} 4 & 8 & -8 \\ 8 & 25 & 11 \\ -8 & 11 & 98 \end{vmatrix} = 36$$

Since all three determinants are positive, A is positive definite. ∎

LU-Factorization Revisited

In Section 3.4 we discussed LU-factorization, which involves expressing a matrix A as the product $A = LU$, where L is lower triangular and U is upper triangular. There we showed that A has an LU-factorization if A can be reduced to echelon form without the use of row interchanges. Previously, the only way to tell if this was true was to try it, but here we give a class of matrices that are guaranteed to have LU-factorizations.

THEOREM 11.9

Suppose that the leading principal submatrices of a symmetric matrix A satisfy (1). Then A can be reduced to row echelon form without using row interchanges, and the pivot elements will all be positive.

Proof We proceed by induction on n, where A is an $n \times n$ symmetric positive definite matrix. First, if $n = 1$, then $A = [a_{11}]$ is automatically in row echelon form, so no row interchanges are required. Moreover, we have

$$a_{11} = \det(A) > 0$$

so that the sole pivot element is positive.

Now suppose that the theorem holds for $(n-1) \times (n-1)$ symmetric matrices, and let A be an $n \times n$ symmetric matrix that satisfies (1). We can partition A as

$$A = \begin{bmatrix} & & & \vdots & a_{1n} \\ & A_{n-1} & & \vdots & \vdots \\ & & & \vdots & a_{(n-1)n} \\ \hline a_{n1} & \cdots & a_{n(n-1)} & & a_{nn} \end{bmatrix}$$

where the leading principal submatrix A_{n-1} is symmetric and satisfies the induction hypothesis. When reducing A to row echelon form, the values of all but the last pivot are dictated entirely by elements of A_{n-1}. Thus, by the induction hypothesis, we can reduce A to the form

$$A^* = \begin{bmatrix} a_{11}^* & a_{12} & \cdots & & a_{1n} \\ 0 & a_{22}^* & \cdots & & a_{2n} \\ \vdots & \vdots & \ddots & & \vdots \\ 0 & \cdots & a_{(n-1)(n-1)}^* & a_{(n-1)n} \\ 0 & \cdots & & 0 & a_{nn} \end{bmatrix}$$

where the pivots $a_{11}^*, \ldots, a_{(n-1)(n-1)}^*$ are all positive and no row interchanges are required. Because there are no row interchanges, the determinant is unchanged, so that

$$\det(A) = \det(A^*) = \det(A_{n-1}) \cdot a_{nn} \implies a_{nn} = \frac{\det(A)}{\det(A_{n-1})}$$

Since both $\det(A) > 0$ and $\det(A_{n-1}) > 0$, we have $a_{nn} > 0$, which completes the proof. ∎

The matrix in Example 3 satisfies the hypotheses of Theorem 11.9, so it must have an LU-factorization.

EXAMPLE 4 Find an LU-factorization for the matrix in Example 3,

$$A = \begin{bmatrix} 4 & 8 & -8 \\ 8 & 25 & 11 \\ -8 & 11 & 98 \end{bmatrix}$$

Solution Since Section 3.4 contains several examples showing how to find the LU-factorization, some details are omitted here. Recall that we obtain U by reducing A to echelon form, and build up L one column at a time as we transform A.

Step 1a: Take the first column of A, divide each entry by the pivot 4, and use the resulting values to form the first column of L.

$$A = \begin{bmatrix} 4 & 8 & -8 \\ 8 & 25 & 11 \\ -8 & 11 & 98 \end{bmatrix} \implies L = \begin{bmatrix} 1 & \bullet & \bullet \\ 2 & \bullet & \bullet \\ -2 & \bullet & \bullet \end{bmatrix}$$

Step 1b: Perform row operations as usual to introduce zeros down the first column of A.

$$A = \begin{bmatrix} 4 & 8 & -8 \\ 8 & 25 & 11 \\ -8 & 11 & 98 \end{bmatrix} \sim \begin{bmatrix} 4 & 8 & -8 \\ 0 & 9 & 27 \\ 0 & 27 & 82 \end{bmatrix} = A_1$$

Step 2a: Take the second column of A_1, starting down from the pivot entry 9, and divide each entry by the pivot. Use the resulting values to form the lower portion of the second column of L.

$$A_1 = \begin{bmatrix} 4 & 8 & -8 \\ 0 & 9 & 27 \\ 0 & 27 & 82 \end{bmatrix} \implies L = \begin{bmatrix} 1 & \bullet & \bullet \\ 2 & 1 & \bullet \\ -2 & 3 & \bullet \end{bmatrix}$$

Step 2b: Perform row operations as usual to introduce zeros down the second column of A_1.

$$A_1 = \begin{bmatrix} 4 & 8 & -8 \\ 0 & 9 & 27 \\ 0 & 27 & 82 \end{bmatrix} \sim \begin{bmatrix} 4 & 8 & -8 \\ 0 & 9 & 27 \\ 0 & 0 & 1 \end{bmatrix} = A_2$$

Step 3: Set U equal to A_2, and finish filling in L.

$$L = \begin{bmatrix} 1 & 0 & 0 \\ 2 & 1 & 0 \\ -2 & 3 & 1 \end{bmatrix} \quad \text{and} \quad U = \begin{bmatrix} 4 & 8 & -8 \\ 0 & 9 & 27 \\ 0 & 0 & 1 \end{bmatrix}$$

Standard matrix multiplication can be used to verify that $A = LU$. ∎

In Section 3.4 we extended the LU-factorization to $A = LDU$, where L is as before, D is diagonal, and U is an upper triangular matrix with 1's along the diagonal. Although the LU-factorization is not unique, the LDU-factorization is unique (see Exercise 37).

To find the LDU-factorization, we start by finding the LU-factorization. Once that is done, we write U as the product of a diagonal matrix and an upper triangular matrix,

$$U = \begin{bmatrix} u_{11} & u_{12} & u_{13} & \cdots & u_{1n} \\ 0 & u_{22} & u_{23} & \cdots & u_{2n} \\ 0 & 0 & u_{33} & \cdots & u_{3n} \\ \vdots & \vdots & \vdots & \ddots & \vdots \\ 0 & 0 & 0 & \cdots & u_{nn} \end{bmatrix}$$

$$= \begin{bmatrix} u_{11} & 0 & 0 & \cdots & 0 \\ 0 & u_{22} & 0 & \cdots & 0 \\ 0 & 0 & u_{23} & \cdots & 0 \\ \vdots & \vdots & \vdots & \ddots & \vdots \\ 0 & 0 & 0 & \cdots & u_{nn} \end{bmatrix} \begin{bmatrix} 1 & \frac{u_{12}}{u_{11}} & \frac{u_{13}}{u_{11}} & \cdots & \frac{u_{1n}}{u_{11}} \\ 0 & 1 & \frac{u_{23}}{u_{22}} & \cdots & \frac{u_{2n}}{u_{22}} \\ 0 & 0 & 1 & \cdots & \frac{u_{3n}}{u_{33}} \\ \vdots & \vdots & \vdots & \ddots & \vdots \\ 0 & 0 & 0 & \cdots & 1 \end{bmatrix}$$

with the left matrix being D and the right the new U. For instance, taking U from the factorization in Example 4, we have

$$\begin{bmatrix} 4 & 8 & -8 \\ 0 & 9 & 27 \\ 0 & 0 & 1 \end{bmatrix} = \begin{bmatrix} 4 & 0 & 0 \\ 0 & 9 & 0 \\ 0 & 0 & 1 \end{bmatrix} \begin{bmatrix} 1 & 2 & -2 \\ 0 & 1 & 3 \\ 0 & 0 & 1 \end{bmatrix} = DU$$

The matrix U looks familiar—it is L^T. This is not a coincidence.

THEOREM 11.10 A symmetric matrix A that satisfies (1) can be uniquely factored as $A = LDL^T$, where L is lower triangular with 1's on the diagonal, and D is a diagonal matrix with all positive diagonal entries.

Proof By Theorem 11.9 we know that A can be expressed uniquely as $A = LDU$, and since A is symmetric, we have

$$LDU = A = A^T = (LDU)^T = U^T D^T L^T = U^T DL^T$$

Because the factorization is unique, it follows that $U = L^T$. That the diagonal entries of D are positive also follows from Theorem 11.9. ∎

Since the diagonal entries of D are all positive in the factorization given in Theorem 11.10, we can define the matrix

$$D^{1/2} = \begin{bmatrix} \sqrt{u_{11}} & 0 & 0 & \cdots & 0 \\ 0 & \sqrt{u_{22}} & 0 & \cdots & 0 \\ \vdots & \vdots & \vdots & \ddots & \vdots \\ 0 & 0 & 0 & \cdots & \sqrt{u_{nn}} \end{bmatrix}$$

If we set $L_c = LD^{1/2}$, then we have

$$A = LDL^T = LD^{1/2}D^{1/2}L^T = \left(LD^{1/2}\right)\left(LD^{1/2}\right)^T = L_c L_c^T$$

Definition Cholesky Decomposition

The factorization $A = L_c L_c^T$ is called the **Cholesky decomposition** of A. By Theorem 11.10, every symmetric matrix A satisfying (1) has a Cholesky decomposition.

▶ Roughly speaking, the Cholesky decomposition can be thought of as the square root of a matrix. When it exists, it can be applied in the same manner as an LU decomposition to find the solution to a linear system.

EXAMPLE 5 Find the Cholesky decomposition for the matrix A in Example 4.

Solution We have

$$L_c = LD^{1/2} = \begin{bmatrix} 1 & 0 & 0 \\ 2 & 1 & 0 \\ -2 & 3 & 1 \end{bmatrix} \begin{bmatrix} \sqrt{4} & 0 & 0 \\ 0 & \sqrt{9} & 0 \\ 0 & 0 & \sqrt{1} \end{bmatrix}$$

$$= \begin{bmatrix} 1 & 0 & 0 \\ 2 & 1 & 0 \\ -2 & 3 & 1 \end{bmatrix} \begin{bmatrix} 2 & 0 & 0 \\ 0 & 3 & 0 \\ 0 & 0 & 1 \end{bmatrix} = \begin{bmatrix} 2 & 0 & 0 \\ 4 & 3 & 0 \\ -4 & 9 & 1 \end{bmatrix}$$

We can verify the decomposition by computing

$$L_c L_c^T = \begin{bmatrix} 2 & 0 & 0 \\ 4 & 3 & 0 \\ -4 & 9 & 1 \end{bmatrix} \begin{bmatrix} 2 & 4 & -4 \\ 0 & 3 & 9 \\ 0 & 0 & 1 \end{bmatrix} = \begin{bmatrix} 4 & 8 & -8 \\ 8 & 25 & 11 \\ -8 & 11 & 98 \end{bmatrix} = A \quad ■$$

Proof of Theorem 11.8

We are now in the position to prove Theorem 11.8. Recall the statement of the theorem.

THEOREM 11.8

A symmetric matrix A is positive definite if and only if the leading principal submatrices satisfy

$$\det(A_1) > 0, \ \det(A_2) > 0, \ \ldots, \ \det(A_n) > 0 \tag{1}$$

Proof As noted earlier, combining Theorem 11.6 and Theorem 11.7 shows that if A is positive definite, then the leading principal submatrices satisfy $\det(A_1) > 0, \det(A_2) > 0$, $\ldots, \det(A_n) > 0$. This completes one direction of the proof.

To complete the second direction of the proof, suppose that A is a symmetric matrix and that the leading principal submatrices satisfy $\det(A_1) > 0, \det(A_2) > 0, \ldots,$ $\det(A_n) > 0$. Then A has a Cholesky decomposition $A = L_c L_c^T$. Moreover, since $\det(A) > 0$, it follows that A is nonsingular and hence L_c^T is also nonsingular. Thus if $\mathbf{x} \neq \mathbf{0}$, then $L_c^T \mathbf{x} \neq \mathbf{0}$. Therefore

$$\mathbf{x}^T A \mathbf{x} = \mathbf{x}^T L_c L_c^T \mathbf{x} = \left(L_c^T \mathbf{x}\right)^T \left(L_c^T \mathbf{x}\right) = \|L_c^T \mathbf{x}\|^2 > 0$$

so that A is positive definite. ■

| EXERCISES |

In Exercises 1–6, find the principal submatrices of the given matrix.

1. $A = \begin{bmatrix} 3 & 5 \\ 5 & 7 \end{bmatrix}$

2. $A = \begin{bmatrix} 1 & -6 \\ -6 & 2 \end{bmatrix}$

3. $A = \begin{bmatrix} 1 & 4 & -3 \\ 4 & 0 & 2 \\ -3 & 2 & 5 \end{bmatrix}$

4. $A = \begin{bmatrix} 3 & 0 & 2 \\ 0 & -4 & 1 \\ 2 & 1 & -1 \end{bmatrix}$

5. $A = \begin{bmatrix} 2 & 1 & 0 & -1 \\ 1 & 3 & 4 & 1 \\ 0 & 4 & 0 & -2 \\ -1 & 1 & -2 & 1 \end{bmatrix}$

6. $A = \begin{bmatrix} 1 & -2 & 3 & 0 \\ -2 & 4 & 2 & -3 \\ 3 & 2 & 3 & 1 \\ 0 & -3 & 1 & 0 \end{bmatrix}$

In Exercises 7–12, determine if the given matrix is positive definite.

7. $A = \begin{bmatrix} 2 & 1 \\ 1 & -2 \end{bmatrix}$

8. $A = \begin{bmatrix} 1 & 3 \\ 3 & 5 \end{bmatrix}$

9. $A = \begin{bmatrix} 1 & 2 & -1 \\ 2 & 5 & 1 \\ -1 & 1 & 11 \end{bmatrix}$

10. $A = \begin{bmatrix} 1 & -3 & 2 \\ -3 & 10 & -7 \\ 2 & -7 & 6 \end{bmatrix}$

11. $A = \begin{bmatrix} 1 & 1 & 2 & 1 \\ 1 & 2 & 1 & 0 \\ 2 & 1 & 6 & 3 \\ 1 & 0 & 3 & 6 \end{bmatrix}$

12. $A = \begin{bmatrix} 1 & -2 & 2 & 1 \\ -2 & 5 & -1 & -3 \\ 2 & -1 & 14 & -1 \\ 1 & -3 & -1 & 3 \end{bmatrix}$

In Exercises 13–16, show that the given matrix is positive definite, and then find the LU-factorization.

13. $A = \begin{bmatrix} 1 & -2 \\ -2 & 5 \end{bmatrix}$

14. $A = \begin{bmatrix} 1 & 3 \\ 3 & 10 \end{bmatrix}$

15. $A = \begin{bmatrix} 1 & -2 & 2 \\ -2 & 5 & -5 \\ 2 & -5 & 6 \end{bmatrix}$

16. $A = \begin{bmatrix} 1 & 1 & -1 \\ 1 & 10 & -4 \\ -1 & -4 & 11 \end{bmatrix}$

In Exercises 17–20, show that the given matrix is positive definite, and then find the LDU-factorization.

17. $A = \begin{bmatrix} 1 & -2 \\ -2 & 8 \end{bmatrix}$

18. $A = \begin{bmatrix} 4 & -4 \\ -4 & 5 \end{bmatrix}$

19. $A = \begin{bmatrix} 1 & 3 & 2 \\ 3 & 10 & 8 \\ 2 & 8 & 9 \end{bmatrix}$

20. $A = \begin{bmatrix} 1 & 1 & 1 \\ 1 & 5 & 1 \\ 1 & 1 & 5 \end{bmatrix}$

In Exercises 21–24, show that the given matrix is positive definite, and then find the Cholesky decomposition.

21. $A = \begin{bmatrix} 1 & 2 \\ 2 & 5 \end{bmatrix}$

22. $A = \begin{bmatrix} 4 & 2 \\ 2 & 10 \end{bmatrix}$

23. $A = \begin{bmatrix} 1 & 2 & 3 \\ 2 & 5 & 6 \\ 3 & 6 & 10 \end{bmatrix}$

24. $A = \begin{bmatrix} 1 & 4 & 1 \\ 4 & 25 & 7 \\ 1 & 7 & 6 \end{bmatrix}$

FIND AN EXAMPLE For Exercises 25–30, find an example that meets the given specifications.

25. A 2×2 matrix A such that $\det(A_1) < 0$ but $\det(A_2) > 0$.

26. A 2×2 matrix A such that $\det(A_1) > 0$ but $\det(A_2) < 0$.

27. A 3×3 matrix A such that $\det(A_1) < 0$ and $\det(A_3) < 0$, but $\det(A_2) > 0$.

28. A 3×3 matrix A such that $\det(A_1) > 0$, but $\det(A_2) < 0$, and $\det(A_3) < 0$.

29. A 3×3 matrix A such that $\det(A) > 0$ but A is not positive definite.

30. A 3×3 matrix A such that $\det(A) < 0$ but A is not negative definite.

TRUE OR FALSE For Exercises 31–36, determine if the statement is true or false, and justify your answer.

31. A symmetric matrix A is positive definite if and only if $\det(A) > 0$.

32. If $L_c L_c^T$ is the Cholesky decomposition of A, then $L_c^T L_c$ is the Cholesky decomposition of A^T.

33. If A and B are $n \times n$ positive definite matrices, then so is AB.

34. If A and B are $n \times n$ positive definite matrices, then so is $A + B$.

35. If A is a positive definite matrix, then so is A^{-1}.

36. If the determinants of the leading principal submatrices of a symmetric matrix A are all negative, then A is negative definite.

37. Let $A = L_1 D_1 U_1$ and $A = L_2 D_2 U_2$ be two LDU-factorizations of A. Prove that $L_1 = L_2$, $D_1 = D_2$, and $U_1 = U_2$, which shows that the LDU factorization is unique.

11.3 Constrained Optimization

In this section we consider the following problem: Find the maximum and/or minimum value of a quadratic form $Q(\mathbf{x})$, where $\mathbf{x}$ ranges over a set of vectors $\mathbf{x}$ that satisfy some constraint.

Our problem is easiest to solve when the quadratic form has no cross-product terms, so let's start by looking at an example of that case.

EXAMPLE 1 Find the maximum and minimum values of $Q(\mathbf{x}) = 4x_1^2 - 3x_2^2 + 7x_3^2$, subject to the constraint $\|\mathbf{x}\| = 1$.

Solution First, note that if $\|\mathbf{x}\| = 1$, then $\|\mathbf{x}\|^2 = 1$ and therefore

$$x_1^2 + x_2^2 + x_3^2 = 1$$

Each of x_1^2, x_2^2, and x_3^2 is nonnegative, so that

$$Q(\mathbf{x}) = 4x_1^2 - 3x_2^2 + 7x_3^2 \leq 7x_1^2 + 7x_2^2 + 7x_3^2 = 7(x_1^2 + x_2^2 + x_3^2) = 7$$

Thus $Q(\mathbf{x}) \leq 7$. Moreover, if $x_1 = x_2 = 0$ and $x_3 = 1$, then $\|\mathbf{x}\| = 1$ and $Q(\mathbf{x}) = 7$. Therefore, subject to the constraint $\|\mathbf{x}\| = 1$, the maximum value is $Q(\mathbf{x}) = 7$.

A similar argument can be used to show that subject to the same constraint, the minimum value is equal to the minimum coefficient, so that the minimum value is $Q(\mathbf{x}) = -3$. ∎

Example 1 illustrates a more general fact that is described in the next theorem.

THEOREM 11.11

▶ Recall that

$$\mathbf{e}_k = \begin{bmatrix} 0 \\ \vdots \\ 1 \\ \vdots \\ 0 \end{bmatrix} \leftarrow \text{entry } k$$

Suppose that $Q : \mathbf{R}^n \to \mathbf{R}$ is a quadratic form

$$Q(\mathbf{x}) = q_1 x_1^2 + q_2 x_2^2 + \cdots + q_n x_n^2$$

that has no cross-product terms. Let q_i and q_j be the maximum and minimum, respectively, of the coefficients $q_1, q_2, \ldots, q_n$. Then subject to the constraint $\|\mathbf{x}\| = 1$, we have

(a) The maximum value of $Q(\mathbf{x})$ is q_i, attained at $\mathbf{x} = \mathbf{e}_i$.

(b) The minimum value of $Q(\mathbf{x})$ is q_j, attained at $\mathbf{x} = \mathbf{e}_j$.

The proof is similar to the solution to Example 1 and is left as an exercise.

EXAMPLE 2 Find the maximum and minimum values of

$$Q(\mathbf{x}) = 2x_1^2 - 4x_2^2 + 5x_3^2 - x_4^2$$

subject to the constraint $\|\mathbf{x}\| = 1$.

Solution By Theorem 11.11, the maximum value of $Q(\mathbf{x})$ is 5 and the minimum value is -4. The maximum and minimum values are attained at, respectively,

$$\mathbf{x}_3 = \begin{bmatrix} 0 \\ 0 \\ 1 \\ 0 \end{bmatrix} \quad \text{and} \quad \mathbf{x}_2 = \begin{bmatrix} 0 \\ 1 \\ 0 \\ 0 \end{bmatrix} \quad \blacksquare$$

Theorem 11.11 tells us what to do with a quadratic form that is free of cross-product terms, so now we consider general quadratic forms $Q : \mathbf{R}^n \to \mathbf{R}$.

THEOREM 11.12

Let $Q(\mathbf{x}) = \mathbf{x}^T A \mathbf{x}$, where A is a symmetric $n \times n$ matrix. Suppose that A has eigenvalues $\lambda_1 \leq \lambda_2 \leq \ldots \leq \lambda_n$, and let $\mathbf{u}_1, \ldots, \mathbf{u}_n$ be the associated normalized eigenvectors. Then subject to the constraint $\|\mathbf{x}\| = 1$, we have

(a) The maximum value of $Q(\mathbf{x})$ is λ_n, attained at $\mathbf{x} = \mathbf{u}_n$.

(b) The minimum value of $Q(\mathbf{x})$ is λ_1, attained at $\mathbf{x} = \mathbf{u}_1$.

Proof Recall from Section 11.1 that for such a matrix A, there exists an orthogonal matrix P (with columns that are eigenvectors of A) and a diagonal matrix D (diagonal entries are the eigenvalues of A) such that if $\mathbf{x} = P\mathbf{y}$, then

$$\mathbf{x}^T A \mathbf{x} = \mathbf{y}^T D \mathbf{y}$$

Since D is diagonal, the quadratic form $\mathbf{y}^T D\mathbf{y}$ has no cross-product terms, so that Theorem 11.11 applies. Moreover, if $\|\mathbf{y}\| = 1$, then

$$\|\mathbf{x}\| = \|P\mathbf{y}\| = 1$$

(See Exercise 66, Section 8.3.) Therefore the maximum value of $\mathbf{x}^T A\mathbf{x}$ subject to the constraint $\|\mathbf{x}\| = 1$ is equal to the maximum value of $\mathbf{y}^T D\mathbf{y}$ subject to the constraint $\|\mathbf{y}\| = 1$. (The same holds if "maximum" is replaced with "minimum.")

Returning to Theorem 11.11, the matrix of the quadratic form is diagonal with diagonal entries $q_1, \ldots, q_n$. For our matrix D, the diagonal entries are the eigenvalues of A, so that Theorem 11.11 applies to the set $\{\lambda_1, \ldots, \lambda_n\}$, which yields the claimed maximum and minimum values. Observing that $P\mathbf{e}_k = \mathbf{u}_k$ completes the proof. ∎

EXAMPLE 3   Find the maximum and minimum values of

$$Q(\mathbf{x}) = x_2^2 + 2x_1 x_2 + 4x_1 x_3 + 2x_2 x_3$$

subject to the constraint $\|\mathbf{x}\| = 1$.

Solution The matrix of this quadratic form is

$$A = \begin{bmatrix} 0 & 1 & 2 \\ 1 & 1 & 1 \\ 2 & 1 & 0 \end{bmatrix}$$

Using our usual methods, we find that the eigenvalues of A are $\lambda_1 = -2$, $\lambda_2 = 0$, and $\lambda_3 = 3$. Thus, by Theorem 11.12, the maximum value of $Q(\mathbf{x})$ is 3 and the minimum value is -2. The maximum and minimum values are attained at the eigenvectors

$$\text{Maximum attained at } \mathbf{u}_3 = \begin{bmatrix} \frac{1}{\sqrt{3}} \\ \frac{1}{\sqrt{3}} \\ \frac{1}{\sqrt{3}} \end{bmatrix} \qquad \text{Minimum attained at } \mathbf{u}_1 = \begin{bmatrix} -\frac{1}{\sqrt{2}} \\ 0 \\ \frac{1}{\sqrt{2}} \end{bmatrix} \qquad ∎$$

Varying the Constraint

In the next two examples, we modify the constraint requirements. In both cases, the problem can be solved by changing variables and then using earlier methods.

EXAMPLE 4   Find the maximum and minimum values of

$$Q(\mathbf{x}) = 5x_2^2 + 8x_2^2 + 4x_1 x_2$$

subject to the constraint $\|\mathbf{x}\| = c$, where $c > 0$ is a positive constant.

Solution Let's start by solving the problem with the constraint $\|\mathbf{x}\| = 1$. The matrix of the quadratic form is

$$A = \begin{bmatrix} 5 & 2 \\ 2 & 8 \end{bmatrix}$$

which has eigenvalues $\lambda_1 = 4$ and $\lambda_2 = 9$. Hence, subject to $\|\mathbf{x}\| = 1$, the maximum value of $Q(\mathbf{x})$ is 9 and the minimum value is 4. To finish, note that $\|\mathbf{x}\| = 1$ if and only if $\|c\mathbf{x}\| = c$ and $Q(c\mathbf{x}) = c^2 Q(\mathbf{x})$ (see Exercise 39). Therefore, subject to the constraint $\|\mathbf{x}\| = c$, the maximum value of $Q(\mathbf{x})$ is $9c^2$ and the minimum value is $4c^2$. ∎

EXAMPLE 5 Find the maximum and minimum values of

$$Q(\mathbf{x}) = 2x_2^2 + 5x_2^2 + 4x_1x_2$$

subject to the constraint $4x_1^2 + 25x_2^2 = 100$.

Solution The first step is to use a change of variables to adjust the constraint equation so that it once again describes unit vectors. Let

$$w_1 = \frac{x_1}{5} \quad \text{and} \quad w_2 = \frac{x_2}{2}$$

so that the constraint equation becomes

$$4(5w_1)^2 + 25(2w_2)^2 = 100 \implies w_1^2 + w_2^2 = 1$$

and hence we have the constraint $\|\mathbf{w}\| = 1$. With this variable change, the quadratic form becomes

$$Q(\mathbf{x}) = Q\left(\begin{bmatrix} x_1 \\ x_2 \end{bmatrix}\right) = Q\left(\begin{bmatrix} 5w_1 \\ 2w_2 \end{bmatrix}\right)$$

$$= 2(5w_1)^2 + 5(2w_2)^2 + 4(5w_1)(2w_2)$$

$$= 50w_1^2 + 20w_2^2 + 40w_1w_2$$

The matrix of the quadratic form $50w_1^2 + 20w_2^2 + 40w_1w_2$ is

$$A = \begin{bmatrix} 50 & 20 \\ 20 & 20 \end{bmatrix}$$

which has eigenvalues $\lambda_1 = 10$ and $\lambda_2 = 60$. Hence the maximum value of $Q(\mathbf{x})$ is 60 and the minimum value is 10. To determine where the maximum and minimum values are attained, we note that the normalized eigenvectors of A are

$$\lambda_1 = 10 \implies \mathbf{w}_1 = \begin{bmatrix} \frac{1}{\sqrt{5}} \\ \frac{2}{\sqrt{5}} \end{bmatrix}, \qquad \lambda_2 = 60 \implies \mathbf{w}_2 = \begin{bmatrix} \frac{2}{\sqrt{5}} \\ \frac{1}{\sqrt{5}} \end{bmatrix}$$

Thus the maximum of $50w_1^2 + 20w_2^2 + 40w_1w_2$ subject to $\|\mathbf{w}\| = 1$ is attained at

$$\begin{bmatrix} \frac{2}{\sqrt{5}} \\ \frac{1}{\sqrt{5}} \end{bmatrix} = \mathbf{w}_2 = \begin{bmatrix} w_1 \\ w_2 \end{bmatrix} = \begin{bmatrix} x_1/5 \\ x_2/2 \end{bmatrix}$$

so that the maximum of $Q(\mathbf{x}) = 2x_2^2 + 5x_2^2 + 4x_1x_2$ subject to $4x_1^2 + 25x_2^2 = 100$ is attained at

$$\mathbf{x}_2 = \begin{bmatrix} 5w_1 \\ 2w_2 \end{bmatrix} = \begin{bmatrix} 2\sqrt{5} \\ \sqrt{5} \end{bmatrix}$$

By a similar argument, the minimum is attained at $\mathbf{x}_1 = \begin{bmatrix} -\sqrt{5} \\ 2\sqrt{5} \end{bmatrix}$. ∎

Adding Orthogonality to the Constraint

In some applications, it is handy to be able to constrain $\mathbf{x}$ so that both $\|\mathbf{x}\| = 1$ and $\mathbf{x}$ are orthogonal to $\mathbf{u}_n$, the eigenvector associated with the largest eigenvalue λ_n.

EXAMPLE 6 Find the maximum value of $Q(\mathbf{x}) = 4x_1^2 - 3x_2^2 + 7x_3^2$, subject to the constraints $\|\mathbf{x}\| = 1$ and $\mathbf{x} \cdot \mathbf{u}_3 = 0$, where $\mathbf{u}_3 = (0, 0, 1)$.

Solution This is the quadratic form from Example 1, where we asked for the maximum subject only to the constraint $\|\mathbf{x}\| = 1$. Since we have added another constraint, the maximum can be no larger (and possibly is smaller) than the maximum found before.

Due to the extra constraint, we cannot apply Theorem 11.11, but we can apply the same line of reasoning as used in Example 1. The constraint $\mathbf{x} \cdot \mathbf{u}_3 = 0$ implies that the third component $x_3 = 0$, so that the two constraints together impose the requirement $x_1^2 + x_2^2 = 1$. Therefore we have

$$Q(\mathbf{x}) = 4x_1^2 - 3x_2^2 + 7x_3^2 = 4x_1^2 - 3x_2^2 \le 4x_1^2 + 4x_2^2 = 4(x_1^2 + x_2^2) = 4$$

This puts an upper bound of 4 on the maximum. Moreover, if $\mathbf{x} = (1, 0, 0)$, then $\mathbf{x}$ satisfies the constraints and $Q(\mathbf{x}) = 4$. Thus the maximum value is 4, attained when $\mathbf{x} = (1, 0, 0)$. ∎

In Example 6 the vector $\mathbf{u}_3 = (0, 0, 1)$ is the eigenvector associated with the largest eigenvalue $\lambda_3 = 7$ of A, the matrix of the quadratic form. Note that the maximum 4 is equal to the *second-largest* eigenvalue of A and is attained when $\mathbf{x}$ is an associated eigenvector. This is not a coincidence.

THEOREM 11.13

▶ Theorem 11.13 only makes sense if $n > 1$.

Let $Q(\mathbf{x}) = \mathbf{x}^T A\mathbf{x}$, where A is a symmetric $n \times n$ matrix with eigenvalues $\lambda_1 \le \lambda_2 \le \cdots \le \lambda_n$ and associated orthonormal eigenvectors $\mathbf{u}_1, \mathbf{u}_2, \ldots, \mathbf{u}_n$. Then subject to the constraints $\|\mathbf{x}\| = 1$ and $\mathbf{x} \cdot \mathbf{u}_n = 0$,

(a) The maximum value of $Q(\mathbf{x})$ is λ_{n-1}, attained at $\mathbf{x} = \mathbf{u}_{n-1}$.

(b) The minimum value of $Q(\mathbf{x})$ is λ_1, attained at $\mathbf{x} = \mathbf{u}_1$.

Proof Since the eigenvectors $\mathbf{u}_1, \mathbf{u}_2, \ldots, \mathbf{u}_n$ are orthonormal, they form a basis for $\mathbf{R}^n$. Moreover, for a given $\mathbf{x}$ if $c_i = \mathbf{x} \cdot \mathbf{u}_i$ for $i = 1, \ldots, n$, then

$$\mathbf{x} = c_1\mathbf{u}_1 + c_2\mathbf{u}_2 + \cdots + c_n\mathbf{u}_n$$

Since one constraint is that $\mathbf{x} \cdot \mathbf{u}_n = 0$, it follows that $c_n = 0$, so that

$$\mathbf{x} = c_1\mathbf{u}_1 + \cdots + c_{n-1}\mathbf{u}_{n-1}$$

Because $\mathbf{u}_1, \ldots, \mathbf{u}_{n-1}$ are orthonormal, we have

$$\|\mathbf{x}\|^2 = c_1^2 + \cdots + c_{n-1}^2$$

(See Exercise 50 of Section 8.2.) Since $\|\mathbf{x}\| = 1$, we have $c_2^2 + \cdots + c_n^2 = 1$. Thus

$$\begin{aligned}
\mathbf{x}^T A\mathbf{x} &= \mathbf{x}^T A(c_1\mathbf{u}_1 + \cdots + c_{n-1}\mathbf{u}_{n-1}) \\
&= (c_1\mathbf{u}_1 + \cdots + c_{n-1}\mathbf{u}_{n-1}) \cdot (c_1\lambda_1\mathbf{u}_1 + \cdots + c_{n-1}\lambda_{n-1}\mathbf{u}_{n-1}) \\
&= c_1^2\lambda_1 + \cdots + c_{n-1}^2\lambda_{n-1} \\
&\le c_1^2\lambda_{n-1} + \cdots + c_{n-1}^2\lambda_{n-1} \\
&= \lambda_{n-1}(c_1^2 + \cdots + c_{n-1}^2) = \lambda_{n-1}
\end{aligned}$$

This shows that $Q(\mathbf{x}) \le \lambda_{n-1}$. Furthermore, if $\mathbf{x} = \mathbf{u}_{n-1}$, then $c_{n-1} = 1$ and $c_1 = \cdots = c_{n-2} = 0$, and hence

$$Q(\mathbf{u}_{n-1}) = \lambda_{n-1}$$

Therefore the maximum value of $Q(\mathbf{x})$ is λ_{n-1}, attained when $\mathbf{x} = \mathbf{u}_{n-1}$. The minimum value is not changed by the constraint $\mathbf{x} \cdot \mathbf{u}_n = 0$, so (b) follows from Theorem 11.12. ∎

EXAMPLE 7 Find the maximum and minimum values of

$$Q(\mathbf{x}) = x_2^2 + 2x_1 x_2 + 4x_1 x_3 + 2x_2 x_3$$

subject to the constraints $\|\mathbf{x}\| = 1$ and $\mathbf{x} \cdot \mathbf{u}_3 = 0$, where $\mathbf{u}_3 = (1, 1, 1)$.

Solution This is the quadratic form in Example 3, which has matrix

$$A = \begin{bmatrix} 0 & 1 & 2 \\ 1 & 1 & 1 \\ 2 & 1 & 0 \end{bmatrix}$$

There we showed that the eigenvalues of A are $\lambda_1 = -2$, $\lambda_2 = 0$, and $\lambda_3 = 3$. Also note that $A\mathbf{u}_3 = 3\mathbf{u}_3$, so that $\mathbf{u}_3$ is an eigenvector associated with the largest eigenvalue $\lambda_3 = 3$. Therefore by Theorem 11.13, subject to our constraints, the maximum value of $Q(\mathbf{x})$ is $\lambda_2 = 0$ and the minimum value is $\lambda_1 = -2$. ∎

| EXERCISES |

For Exercises 1–6, find the maximum and minimum values of the quadratic form $Q(\mathbf{x})$ subject to the constraint $\|\mathbf{x}\| = 1$.

1. $Q(\mathbf{x}) = 2x_1^2 - 3x_2^2$

2. $Q(\mathbf{x}) = 6x_1^2 + 5x_2^2$

3. $Q(\mathbf{x}) = 3x_1^2 - 3x_2^2 - 5x_3^2$

4. $Q(\mathbf{x}) = -x_1^2 + 4x_2^2 + 8x_3^2$

5. $Q(\mathbf{x}) = x_1^2 - x_2^2 - 4x_3^2 + 2x_4^2$

6. $Q(\mathbf{x}) = 3x_1^2 - 4x_2^2 - 2x_3^2 + x_4^2$

For Exercises 7–12, find the maximum and minimum values of the quadratic form $Q(\mathbf{x})$ subject to the constraint $\|\mathbf{x}\| = 1$.

7. $Q(\mathbf{x}) = 4x_1^2 + x_2^2 + 4x_1 x_2$

8. $Q(\mathbf{x}) = 3x_1^2 + 3x_2^2 + 8x_1 x_2$

9. $Q(\mathbf{x}) = x_2^2 + 2x_1 x_3$

10. $Q(\mathbf{x}) = x_1^2 + 4x_2^2 + x_3^2 + 4x_2 x_3$

11. $Q(\mathbf{x}) = x_1^2 + 2x_1 x_2 + 2x_1 x_3 + 2x_2 x_3 + x_2^2 + x_3^2$

12. $Q(\mathbf{x}) = x_1^2 + 4x_1 x_2 + 6x_1 x_3 + 6x_2 x_3 + x_2^2$ (HINT: The eigenvalues are $\lambda = -3, -1, 6$.)

For Exercises 13–16, find the maximum and minimum values of the quadratic form $Q(\mathbf{x})$ subject to the given constraint.

13. $Q(\mathbf{x}) = x_1^2 + 4x_2^2 + 4x_1 x_2$; $\|\mathbf{x}\| = 2$

14. $Q(\mathbf{x}) = 3x_1^2 + 3x_2^2 + 8x_1 x_2$; $\|\mathbf{x}\| = 0.5$

15. $Q(\mathbf{x}) = x_1^2 + 2x_2 x_3$; $\|\mathbf{x}\| = 10$

16. $Q(\mathbf{x}) = 4x_1^2 + x_2^2 + x_3^2 + 4x_1 x_3$; $\|\mathbf{x}\| = 5$

For Exercises 17–20, find the maximum and minimum values of the quadratic form $Q(\mathbf{x})$ subject to the given constraint.

17. $Q(\mathbf{x}) = 4x_1^2 + x_2^2 + 4x_1 x_2$; $4x_1^2 + 25x_2^2 = 100$

18. $Q(\mathbf{x}) = 2x_1^2 + 2x_2^2 + 10x_1 x_2$; $9x_1^2 + 16x_2^2 = 144$

19. $Q(\mathbf{x}) = 4x_1^2 + 4x_2^2 + 6x_1 x_2$; $9x_1^2 + x_2^2 = 9$

20. $Q(\mathbf{x}) = 3x_1^2 + 3x_2^2 + 8x_1 x_2$; $x_1^2 + 4x_2^2 = 4$

For Exercises 21–24, find the maximum and minimum values of the quadratic form $Q(\mathbf{x})$ subject to the constraint $\|\mathbf{x}\| = 1$ and $\mathbf{x} \cdot \mathbf{u}_3 = 0$.

21. $Q(\mathbf{x}) = 4x_1^2 + x_2^2 + 3x_3^2$; $\mathbf{u}_3 = (1, 0, 0)$

22. $Q(\mathbf{x}) = -2x_1^2 + x_2^2 - 5x_3^2$; $\mathbf{u}_3 = (0, 1, 0)$

23. $Q(\mathbf{x}) = x_1^2 + 4x_2^2 + x_3^2 + 4x_2 x_3$; $\mathbf{u}_3 = (0, 2, 1)$

24. $Q(\mathbf{x}) = x_1^2 + 4x_1 x_2 + 6x_1 x_3 + 6x_2 x_3 + x_2^2$; $\mathbf{u}_3 = (1, 1, 1)$ (HINT: Eigenvalues are $\lambda = -3, -1, 6$.)

FIND AN EXAMPLE For Exercises 25–30, find an example that meets the given specifications.

25. A quadratic form $Q: \mathbf{R}^2 \rightarrow \mathbf{R}$ that has maximum value 5, subject to the constraint that $\|\mathbf{x}\| = 1$.

26. A quadratic form $Q: \mathbf{R}^3 \rightarrow \mathbf{R}$ that has minimum value 4, subject to the constraint that $\|\mathbf{x}\| = 1$.

27. A quadratic form $Q: \mathbf{R}^2 \rightarrow \mathbf{R}$ that has maximum value 6 and minimum value 1, subject to the constraint that $\|\mathbf{x}\| = 1$.

28. A quadratic form $Q: \mathbf{R}^3 \rightarrow \mathbf{R}$ that has maximum value -2 and minimum value -7, subject to the constraint that $\|\mathbf{x}\| = 1$.

29. A quadratic form $Q: \mathbf{R}^2 \rightarrow \mathbf{R}$ that has maximum value 4 and minimum value -1, subject to the constraint that $\|\mathbf{x}\| = 3$.

30. A quadratic form $Q: \mathbf{R}^3 \rightarrow \mathbf{R}$ that has maximum value 8 and minimum value -3, subject to the constraint that $\|\mathbf{x}\| = 4$.

TRUE OR FALSE For Exercises 31–36, determine if the statement is true or false, and justify your answer.

31. A quadratic form $Q: \mathbf{R}^n \rightarrow \mathbf{R}$ has a maximum and minimum when constrained to $\|\mathbf{x}\| = 1$.

32. If M is the maximum value of a quadratic form $Q(\mathbf{x})$ subject to the constraint $\|\mathbf{x}\| = 1$, then cM is the maximum value of $Q(\mathbf{x})$ subject to the constraint $\|\mathbf{x}\| = c$.

33. If M and m are, respectively, the maximum and minimum values of a quadratic form $Q(\mathbf{x})$ subject to the constraint $\|\mathbf{x}\| = 1$, then $m < M$.

34. If $\mathbf{x}_1$ maximizes a quadratic form $Q(\mathbf{x})$ subject to the constraint $\|\mathbf{x}\| = 1$, then so does $-\mathbf{x}_1$.

35. The minimum value of a quadratic form $Q(\mathbf{x})$ subject to the constraint $\|\mathbf{x}\| = 2$ must be less than the minimum value of $Q(\mathbf{x})$ subject to the constraint $\|\mathbf{x}\| = 1$.

36. It is possible for a quadratic form to attain its constrained maximum at more than one point.

37. Prove Theorem 11.11: Suppose that $Q : \mathbf{R}^n \to \mathbf{R}$ is a quadratic form

$$Q(\mathbf{x}) = q_1 x_1^2 + q_2 x_2^2 + \cdots + q_n x_n^2$$

that has no cross-product terms. Let q_i and q_j be the maximum and minimum, respectively, of the coefficients $q_1, q_2, \ldots, q_n$. Then subject to the constraint $\|\mathbf{x}\| = 1$, we have

(a) Maximum value of $Q(\mathbf{x}) = q_i$, attained at $\mathbf{x} = \mathbf{e}_i$.

(b) Minimum value of $Q(\mathbf{x}) = q_j$, attained at $\mathbf{x} = \mathbf{e}_j$.

38. Prove the following generalized version of Theorem 11.12: Let A be a symmetric $n \times n$ matrix, and let $Q : \mathbf{R}^n \to \mathbf{R}$ be the quadratic form

$$Q(\mathbf{x}) = \mathbf{x}^T A \mathbf{x}$$

If A has eigenvalues $\lambda_1 \leq \ldots \leq \lambda_n$, then subject to the constraint $\|\mathbf{x}\| = c$ ($c > 0$), the maximum value of $Q(\mathbf{x})$ is $c^2 \lambda_n$ and the minimum value is $c^2 \lambda_1$.

39. Prove that if $Q(\mathbf{x})$ is a quadratic form and c is a constant, then $Q(c\mathbf{x}) = c^2 Q(\mathbf{x})$.

40. Prove the following extension of Theorem 11.13. Let A be a symmetric $n \times n$ matrix with eigenvalues $\lambda_1 \leq \lambda_2 \leq \ldots \leq \lambda_n$ and associated orthonormal eigenvectors $\mathbf{u}_1, \mathbf{u}_2, \ldots, \mathbf{u}_n$. Then subject to the constraints

$$\|\mathbf{x}\| = 1, \quad \mathbf{x} \cdot \mathbf{u}_n = 0, \quad \mathbf{x} \cdot \mathbf{u}_{n-1} = 0, \quad \ldots, \quad \mathbf{x} \cdot \mathbf{u}_{n-j+1} = 0$$

prove that the maximum value of the quadratic form $Q(\mathbf{x}) = \mathbf{x}^T A \mathbf{x}$ is λ_{n-j}, with the maximum attained at $\mathbf{u}_{n-j}$.

11.4 Complex Vector Spaces

▶ Here we assume a basic understanding of the properties of complex numbers. A review of some of these properties is given in Section 6.5.

In this section we define *complex vector spaces*, which are a fairly straightforward extension of the real vector spaces defined in Definition 7.1 in Section 7.1. In fact, the only change when moving to complex vector spaces is that we allow the scalars to be complex numbers.

DEFINITION 11.14

Definition Complex Vector Space, Vector

A **complex vector space** consists of a set V of **vectors** together with operations of addition and scalar multiplication on the vectors that satisfy each of the following:

(1) If $\mathbf{v}_1$ and $\mathbf{v}_2$ are in V, then so is $\mathbf{v}_1 + \mathbf{v}_2$. Hence V is closed under addition.

(2) If c is a complex scalar and $\mathbf{v}$ is in V, then so is $c\mathbf{v}$. Hence V is closed under scalar multiplication.

(3) There exists a **zero vector 0** in V such that $\mathbf{0} + \mathbf{v} = \mathbf{v}$ for all $\mathbf{v}$ in V.

(4) For each $\mathbf{v}$ in V, there exists an **additive inverse** (or **opposite**) vector $-\mathbf{v}$ in V such that $\mathbf{v} + (-\mathbf{v}) = \mathbf{0}$ for all $\mathbf{v}$ in V.

(5) For all $\mathbf{v}_1, \mathbf{v}_2$, and $\mathbf{v}_3$ in V and complex scalars c_1 and c_2, we have

(a) $\mathbf{v}_1 + \mathbf{v}_2 = \mathbf{v}_2 + \mathbf{v}_1$

(b) $(\mathbf{v}_1 + \mathbf{v}_2) + \mathbf{v}_3 = \mathbf{v}_1 + (\mathbf{v}_2 + \mathbf{v}_3)$

(c) $c_1(\mathbf{v}_1 + \mathbf{v}_2) = c_1\mathbf{v}_1 + c_1\mathbf{v}_2$

(d) $(c_1 + c_2)\mathbf{v}_1 = c_1\mathbf{v}_1 + c_2\mathbf{v}_1$

(e) $(c_1 c_2)\mathbf{v}_1 = c_1(c_2\mathbf{v}_1)$

(f) $1 \cdot \mathbf{v}_1 = \mathbf{v}_1$

We have seen that all finite-dimensional real vector spaces are isomorphic to $\mathbf{R}^n$ for some n, so these are arguably the most important of the real vector spaces. If we allow complex numbers in place of real numbers, then we get $\mathbf{C}^n$, the set of vectors of the form

$$\mathbf{v} = \begin{bmatrix} v_1 \\ \vdots \\ v_n \end{bmatrix}$$

where $v_1, \ldots, v_n$ are complex numbers. With addition and scalar multiplication defined as it is on $\mathbf{R}^n$, the set $\mathbf{C}^n$ is a vector space.

EXAMPLE 1 Compute $\mathbf{v}_1 + \mathbf{v}_2$ and $i\mathbf{v}_1$ for

$$\mathbf{v}_1 = \begin{bmatrix} i + 3 \\ -2i \end{bmatrix}, \qquad \mathbf{v}_2 = \begin{bmatrix} 2 - 3i \\ 5i + 1 \end{bmatrix}$$

Solution Adding real and imaginary parts in each component, we have

$$\mathbf{v}_1 + \mathbf{v}_2 = \begin{bmatrix} (i + 3) + (2 - 3i) \\ -2i + (5i + 1) \end{bmatrix} = \begin{bmatrix} 5 - 2i \\ 1 + 3i \end{bmatrix}$$

Using the identity $i^2 = -1$ allows us to simplify:

$$i\mathbf{v}_1 = \begin{bmatrix} i(i + 3) \\ i(-2i) \end{bmatrix} = \begin{bmatrix} i^2 + 3i \\ -2i^2 \end{bmatrix} = \begin{bmatrix} -1 + 3i \\ 2 \end{bmatrix} \qquad \blacksquare$$

EXAMPLE 2 Show that the set $\mathbf{C}^{2 \times 2}$ of 2×2 matrices with complex entries forms a complex vector space when addition and scalar multiplication are defined in the usual manner for matrices.

Solution We show that two of the requirements of Definition 11.14 hold and leave the rest as an exercise. First, if

$$A = \begin{bmatrix} a_{11} & a_{12} \\ a_{21} & a_{22} \end{bmatrix} \quad \text{and} \quad B = \begin{bmatrix} b_{11} & b_{12} \\ b_{21} & b_{22} \end{bmatrix}$$

are both in $\mathbf{C}^{2 \times 2}$, then

$$A + B = \begin{bmatrix} (a_{11} + b_{11}) & (a_{12} + b_{12}) \\ (a_{21} + b_{21}) & (a_{22} + b_{22}) \end{bmatrix}$$

is also in $\mathbf{C}^{2 \times 2}$. Thus the set is closed under addition, so that (1) of Definition 11.14 is satisfied. Second, the zero matrix

$$\mathbf{0} = \begin{bmatrix} 0 & 0 \\ 0 & 0 \end{bmatrix}$$

from $\mathbf{R}^{2 \times 2}$ also serves as the zero matrix for $\mathbf{C}^{2 \times 2}$, so that (3) is satisfied. The remaining properties are left as an exercise. $\blacksquare$

EXAMPLE 3 Suppose that $f : \mathbf{R} \to \mathbf{R}$ and $g : \mathbf{R} \to \mathbf{R}$, and define

$$h(x) = f(x) + ig(x)$$

Then $h : \mathbf{R} \longrightarrow \mathbf{C}$ is a complex-valued function of a real variable. Show that the set of all such functions forms a complex vector space under the usual definition for addition and scalar multiplication of functions.

Solution As with Example 2, here we verify some conditions of Definition 11.14 and leave the rest to the exercises. First, suppose that $h(x) = f(x) + ig(x)$ is in our set of functions, and let $c = a + ib$ be a complex scalar. Then

$$ch(x) = (a + ib)\big(f(x) + ig(x)\big)$$
$$= \big(af(x) - bg(x)\big) + i\big(ag(x) + bf(x)\big)$$

Since both $af - bg$ and $ag + bf$ are real functions of real variables, it follows that $ch(x)$ is a complex-valued function of a real variable. Hence our set is closed under scalar multiplication, as required by (2) of Definition 11.14.

Next, for $h(x) = f(x) + ig(x)$, if we let $-h(x) = -f(x) - ig(x)$, then $-h(x)$ is also in our set, and

$$h(x) + \big(-h(x)\big) = \big(f(x) + ig(x)\big) + \big(-f(x) - ig(x)\big) = 0 + 0i = 0$$

Thus each vector in our set has an additive inverse in the set, which shows that (4) of Definition 11.14 is also true. Verification of the remaining conditions is left as an exercise. ∎

Concepts such as linear combination, linear independence, span, basis, and subspace carry over to complex vector spaces in a natural and essentially unchanged manner. For instance, to show that a set of vectors $\{v_1, \ldots, v_k\}$ is linearly independent in a complex vector space, we need to show that the only solution among complex scalars $c_1, \ldots, c_k$ to

$$c_1 v_1 + \cdots + c_k v_k = 0$$

is the trivial $c_1 = \cdots = c_k = 0$.

Complex Inner Product Spaces

In Chapter 10 we developed the notion of an inner product on a real vector space and considered some of the properties of an inner product. Here we provide an account of the analog for complex vector spaces, starting with the definition of an inner product. (Note the similarities between this and Definition 10.1.)

DEFINITION 11.15

Definition Inner Product, Complex Inner Product Space, Unitary Space

Let u, v, and w be elements of a complex vector space V, and let c be a complex scalar. An **inner product** on V is a function denoted by $\langle u, v \rangle$ that takes any two vectors in V as input and produces a scalar as output. An inner product on a complex vector space satisfies the conditions:

(a) $\langle u, v \rangle = \overline{\langle v, u \rangle}$

(b) $\langle u + v, w \rangle = \langle u, w \rangle + \langle v, w \rangle$

(c) $\langle cu, v \rangle = c \langle u, v \rangle$

(d) $\langle u, u \rangle$ is a nonnegative real number, and $\langle u, u \rangle = 0$ only when $u = 0$

A complex vector space V with an inner product defined on it is called a **complex inner product space** or a **unitary space**.

Definition Complex Dot Product

A complex inner product space of particular importance is the vector space $\mathbf{C}^n$ together with the **complex dot product**, defined by

$$u \cdot v = u_1 \overline{v}_1 + u_2 \overline{v}_2 + \cdots + u_n \overline{v}_n$$

For example, if

$$\mathbf{u} = \begin{bmatrix} 1+i \\ -2i \\ 4 \end{bmatrix} \quad \text{and} \quad \mathbf{v} = \begin{bmatrix} 5i \\ 2-i \\ 1+3i \end{bmatrix}$$

then

$$\mathbf{u} \cdot \mathbf{v} = (1+i)(\overline{5i}) + (-2i)(\overline{2-i}) + (4)(\overline{1+3i})$$
$$= (1+i)(-5i) + (-2i)(2+i) + (4)(1-3i) = 11 - 21i$$

Note that

$$\mathbf{u} \cdot \mathbf{u} = u_1\bar{u}_1 + u_2\bar{u}_2 + \cdots + u_n\bar{u}_n = |u_1|^2 + |u_2|^2 + \cdots + |u_n|^2 \geq 0$$

with equality holding only if $\mathbf{u} = \mathbf{0}$. Thus (d) of Definition 11.15 holds for the complex dot product. Proving that the other conditions also hold is left as an exercise (see Exercise 46).

The following theorem gives three properties of inner products that follow from Definition 11.15. Of these, the first two are the same as for real inner products, but the third is different because of the complex conjugation in (a) of the definition.

THEOREM 11.16

Let $\mathbf{u}$, $\mathbf{v}$, and $\mathbf{w}$ be elements of a complex vector space V, and let c be a complex scalar. Then an inner product defined on V satisfies each of the following:

(a) $\langle \mathbf{0}, \mathbf{u} \rangle = \langle \mathbf{u}, \mathbf{0} \rangle = 0$

(b) $\langle \mathbf{u}, \mathbf{v} + \mathbf{w} \rangle = \langle \mathbf{u}, \mathbf{v} \rangle + \langle \mathbf{u}, \mathbf{w} \rangle$

(c) $\langle \mathbf{u}, c\mathbf{v} \rangle = \bar{c}\langle \mathbf{u}, \mathbf{v} \rangle$

The proof is left as an exercise (see Exercise 47).

EXAMPLE 4 Show that the set $\mathbf{C}^{2 \times 2}$ of 2×2 matrices with complex entries (as in Example 2) is a complex inner product space using the following for an inner product: For each

$$A = \begin{bmatrix} a_{11} & a_{12} \\ a_{21} & a_{22} \end{bmatrix} \quad \text{and} \quad B = \begin{bmatrix} b_{11} & b_{12} \\ b_{21} & b_{22} \end{bmatrix}$$

in $\mathbf{C}^{2 \times 2}$, we set

$$\langle A, B \rangle = a_{11}\bar{b}_{11} + a_{12}\bar{b}_{12} + a_{21}\bar{b}_{21} + a_{22}\bar{b}_{22}$$

Solution We already know from Example 2 that $\mathbf{C}^{2 \times 2}$ is a complex vector space. Thus all that remains is to show that our proposed inner product is an inner product.

To proceed, we could work through the four conditions of Definition 11.15. However, looking at our function closely, we can see that it is really just a thinly disguised version of the complex dot product on $\mathbf{C}^4$. The only difference (essentially a cosmetic one) is that the entries of our matrices are arranged in two rows of two each, instead of a single column as in $\mathbf{C}^4$. Since the complex dot product is an inner product, so is the function defined here. ∎

Given a vector $\mathbf{u}$ in a complex inner product space V, the **norm** (or **length**) of $\mathbf{u}$ is defined just as it is in a real inner product space, by

$$\|\mathbf{u}\| = \sqrt{\langle \mathbf{u}, \mathbf{u} \rangle}$$

Once we have the norm, we define the **distance** from **u** to **v** by

$$\text{Distance from } \mathbf{u} \text{ to } \mathbf{v} = \|\mathbf{v} - \mathbf{u}\|$$

The norm has the properties given in the next theorem.

THEOREM 11.17

Let **u** and **v** be elements of a complex inner product space V, and suppose that c be a complex scalar. Then

(a) $\|\mathbf{u}\| \geq 0$ (with equality only if $\mathbf{u} = \mathbf{0}$)

(b) $\|c\mathbf{u}\| = |c|\|\mathbf{u}\|$

(c) $|\langle \mathbf{u}, \mathbf{v} \rangle| \leq \|\mathbf{u}\|\|\mathbf{v}\|$ (This is the **Cauchy–Schwarz inequality**)

(d) $\|\mathbf{u} + \mathbf{v}\| \leq \|\mathbf{u}\| + \|\mathbf{v}\|$ (This is the **Triangle inequality**)

The proof is similar to the real case and is left as exercises.

EXAMPLE 5 Let $\mathbf{u} = \begin{bmatrix} -1 + i \\ -2i \end{bmatrix}$ and $\mathbf{v} = \begin{bmatrix} 2 \\ 1 - i \end{bmatrix}$ be in $\mathbf{C}^2$. Compute $\langle \mathbf{u}, \mathbf{v} \rangle$, $\|\mathbf{u}\|^2$, $\|\mathbf{v}\|^2$, and $\|\mathbf{u} + \mathbf{v}\|^2$ using the complex dot product for the inner product.

Solution We have

$$\langle \mathbf{u}, \mathbf{v} \rangle = (-1 + i)(2) + (-2i)(1 + i) = (-2 + 2i) + (2 - 2i) = 0$$
$$\|\mathbf{u}\|^2 = (-1 + i)(-1 - i) + (-2i)(2i) = 2 + 4 = 6$$
$$\|\mathbf{v}\|^2 = (2)(2) + (1 - i)(1 + i) = 4 + 2 = 6$$
$$\|\mathbf{u} + \mathbf{v}\|^2 = (1 + i)(1 - i) + (1 - 3i)(1 + 3i) = 2 + 10 = 12 \quad \blacksquare$$

Regarding complex inner products:

- Note that we have $\langle \mathbf{u}, \mathbf{v} \rangle = 0$. The definition of *orthogonal* is the same for complex inner product spaces as it is for the real counterparts:

$$\text{Vectors } \mathbf{u} \text{ and } \mathbf{v} \text{ are \textbf{orthogonal} if and only if } \langle \mathbf{u}, \mathbf{v} \rangle = 0.$$

Thus our two vectors are orthogonal with respect to the complex dot product.

- In Example 5 we have $\langle \mathbf{u}, \mathbf{v} \rangle = 0$, $\|\mathbf{u}\|^2 = 6$, $\|\mathbf{v}\|^2 = 6$, and $\|\mathbf{u} + \mathbf{v}\|^2 = 12$. In general, for a complex inner product, if $\langle \mathbf{u}, \mathbf{v} \rangle = 0$ then $\|\mathbf{u}\|^2 + \|\mathbf{v}\|^2 = \|\mathbf{u} + \mathbf{v}\|^2$. Interestingly, the converse is not true. (See Exercise 31 and Exercise 48.)

- The definitions of **orthonormal**, **orthogonal set**, **orthonormal set**, and other related concepts are analogous to those for real vector spaces.

- The Gram–Schmidt process also works when applied to a complex inner product space.

Suppose that $h_1(x)$ and $h_2(x)$ are from the set of complex-valued functions of a real variable, described in Example 3. Let's add one additional condition—namely, that both functions are continuous. Then a complex inner product is given by

▶ There is nothing special about using 0 and 1 for the limits of integration. As long as $a < b$, we also get an inner product using

$$\int_a^b h_1(x)\overline{h_2(x)} \, dx$$

$$\langle h_1, h_2 \rangle = \int_0^1 h_1(x)\overline{h_2(x)} \, dx \tag{1}$$

Note that

$$\langle h_1, h_1 \rangle = \int_0^1 h_1(x)\overline{h_1(x)} \, dx = \int_0^1 |h_1(x)|^2 \, dx$$

Thus $\langle h_1, h_1 \rangle \geq 0$, with equality holding only if $h_1(x) = 0$. (This is where the continuity requirement is used.) The other conditions required of an inner product can be readily verified.

EXAMPLE 6 Let $h_1(x) = x + i$ and $h_2(x) = 1 - 3xi$, and let S be the subspace spanned by these two functions. Use the Gram–Schmidt process to find an orthogonal basis for S.

Solution If we let $j_1(x)$ and $j_2(x)$ denote the orthogonal basis vectors, then by Gram–Schmidt we have

$$j_1(x) = h_1(x)$$
$$j_2(x) = h_2(x) - \frac{\langle h_2, j_1 \rangle}{\langle j_1, j_1 \rangle} j_1(x)$$

We start by setting $j_1(x) = x + i$. To find $j_2(x)$, we compute

$$\langle h_2, j_1 \rangle = \int_0^1 (1 - 3xi)(x - i)\, dx = \int_0^1 \left(-2x - i(1 + 3x^2)\right) dx = -1 - 2i$$

$$\langle j_1, j_1 \rangle = \int_0^1 |(x + i)|^2\, dx = \int_0^1 (x^2 + 1)\, dx = \frac{4}{3}$$

Therefore

$$j_2(x) = (1 - 3xi) - \frac{3}{4}(-1 + 2i)(x + i) = \frac{1}{4}\left((3x - 2) + i(3 - 6x)\right) \quad\blacksquare$$

EXERCISES

For Exercises 1–10, let

$$\mathbf{u} = \begin{bmatrix} 2 + 3i \\ -1 \\ 3 + 4i \end{bmatrix}, \quad \mathbf{v} = \begin{bmatrix} 2 \\ 3 - i \\ 1 + 5i \end{bmatrix},$$

$$\mathbf{w} = \begin{bmatrix} 2 + i \\ 2 - i \\ 4 - 3i \end{bmatrix}$$

1. Find each linear combination:

(a) $\mathbf{u} - \mathbf{v}$ (b) $\mathbf{w} + 3\mathbf{v}$ (c) $-\mathbf{u} + 2i\mathbf{w} - 5\mathbf{v}$

2. Find each linear combination:

(a) $2\mathbf{u} + \mathbf{w}$ (b) $\mathbf{w} - i\mathbf{u}$ (c) $-i\mathbf{w} - 3\mathbf{v} + 2i\mathbf{u}$

3. Determine if there exists a constant c such that $\mathbf{u} + i\mathbf{v} = c\mathbf{w}$.

4. Determine if there exists a nontrivial linear combination such that $c_1\mathbf{u} + c_2\mathbf{v} + c_3\mathbf{w} = \mathbf{0}$.

5. Is $(-5 + 2i, -3, -5 + i)$ in the span of $\{\mathbf{u}, \mathbf{v}, \mathbf{w}\}$?

6. Is $(-2 + i, -9, -1 + 16i)$ in the span of $\{\mathbf{u}, \mathbf{v}, \mathbf{w}\}$?

7. Compute each using the complex dot product.

(a) $\langle \mathbf{u}, \mathbf{v} \rangle$ (b) $\langle i\mathbf{v}, -2\mathbf{w} \rangle$ (c) $\|\mathbf{w}\|$

8. Compute each using the complex dot product.

(a) $\langle 2\mathbf{w}, 3i\mathbf{u} \rangle$ (b) $\langle -i\mathbf{v}, 5\mathbf{v} \rangle$ (c) $\|\mathbf{w} + i\mathbf{u}\|$

9. Normalize $\mathbf{u}$ and $\mathbf{v}$ using the complex dot product.

10. Normalize $\mathbf{u} + \mathbf{v}$ and $\mathbf{v} - \mathbf{w}$ using the complex dot product.

For Exercises 11–20, let

$$A = \begin{bmatrix} 2 + i & 3 \\ 1 - i & 2 + 3i \end{bmatrix}, \quad B = \begin{bmatrix} -i & 4 \\ 2 + 2i & 1 + 4i \end{bmatrix}, \quad C = \begin{bmatrix} 0 & 3 + i \\ -4i & 1 + i \end{bmatrix}$$

11. Find each linear combination:

(a) $A - iC$ (b) $2B - A - 4iC$

12. Find each linear combination:

(a) $C - (1 + i)A$ (b) $iA - (1 - i)B - 3C$

13. Determine if there exists a constant c such that $A - cB = iC$.

14. Determine if there exists a nontrivial linear combination such that $c_1 A + c_2 B + c_3 C = \begin{bmatrix} 0 & 0 \\ 0 & 0 \end{bmatrix}$.

15. Is $\begin{bmatrix} 3 + 2i & -3 - 7i \\ 4 + 8i & 5 + 2i \end{bmatrix}$ in the span of $\{A, B, C\}$?

16. Is $\begin{bmatrix} 4 - 3i & 6 + 5i \\ 2 + 5i & 13 - 7i \end{bmatrix}$ in the span of $\{A, B, C\}$?

17. Compute each using the inner product given in Example 4.

(a) $\langle A, C \rangle$ (b) $\langle iB, -2A \rangle$ (c) $\|B\|$

18. Compute each using the inner product given in Example 4.

(a) $\langle C, B \rangle$ (b) $\langle 3C, (1 - i)A \rangle$ (c) $\|A + iC - B\|$

19. Normalize A and C using the inner product given in Example 4.

20. Normalize $A - C$ and $A + B - C$ using the inner product given in Example 4.

For Exercises 21–30, let $h_1(x) = 1 + ix$, $h_2(x) = i - x$, and $h_3 = 3 - (1 + i)x$, and (when appropriate) use the inner product given in (1).

21. Find each linear combination.

(a) $h_1(x) + (4 - i)h_2(x)$ (b) $ih_1(x) - h_2(x) + 3h_3(x)$

22. Find each linear combination.

(a) $ih_3(x) - (2 - i)h_1(x)$ (b) $h_3(x) + 2ih_2(x) - 4h_1(x)$

23. Determine if there exists a constant c such that $h_3(x) + h_2(x) = ch_1(x)$.

24. Determine if there exists a nontrivial linear combination such that $c_1h_1(x) + c_2h_2(x) + c_3h_3(x) = 0$.

25. Determine if $(2 + i) + (3 - 2i)x$ is in

$$\text{span}\{h_1(x), h_2(x), h_3(x)\}$$

26. Determine if $(2 - 2i) - (2 - 3i)x$ is in

$$\text{span}\{h_1(x), h_2(x), h_3(x)\}$$

27. Compute each of the following.

(a) $\langle h_1, h_3 \rangle$ (b) $\langle ih_2, -2h_3 \rangle$ (c) $\|h_1\|$

28. Compute each of the following.

(a) $\langle h_3, h_2 \rangle$ (b) $\langle 3h_2, ih_1 \rangle$ (c) $\|h_2 + 3ih_3\|$

29. Normalize h_1 and h_2.

30. Normalize $h_2 - ih_3$ and $h_2 + 3h_3 - (1 + i)h_1$.

FIND AN EXAMPLE For Exercises 31–36, find an example that meets the given specifications.

31. A complex vector space V and vectors $\mathbf{u}$ and $\mathbf{v}$ such that $\|\mathbf{u}\|^2 + \|\mathbf{v}\|^2 = \|\mathbf{u} + \mathbf{v}\|^2$ but $\langle \mathbf{u}, \mathbf{v} \rangle \neq 0$.

32. A complex vector space V not given in this section.

33. An inner product on $\mathbf{C}^n$ other than one given in this section.

34. An inner product on $\mathbf{P}_c^n$, the polynomials of degree n or less with complex coefficients.

35. A complex vector space V and a nonempty subset S that is closed under addition and multiplication by real scalars.

36. A pair of vectors that are linearly independent when the scalars are restricted to the reals but linearly dependent when the scalars are complex.

TRUE OR FALSE For Exercises 37–42, determine if the statement is true or false, and justify your answer.

37. A complex vector space V must be closed under addition and scalar multiplication of the vectors.

38. A complex vector space can have only one inner product defined on it.

39. The norm of any nonzero vector in a complex inner product space must be a positive real number.

40. If $\mathbf{v}_1$ and $\mathbf{v}_2$ are in a complex vector space V, then so is $i\mathbf{v}_1 - (1 + i)\mathbf{v}_2$.

41. If $\mathbf{u}$ and $\mathbf{v}$ are in a complex inner product space V, then $\langle \mathbf{u}, \mathbf{v} \rangle = \langle \mathbf{v}, \mathbf{u} \rangle$.

42. If S_1 and S_2 are subspaces of a complex vector space V, then the intersection $S_1 \cap S_2$ is also a subspace of V.

43. Finish Example 2 by showing that the remaining unverified conditions for a complex vector space hold.

44. Finish Example 3 by showing that the remaining unverified conditions for a complex vector space hold.

45. Prove that $\mathbf{R}^{2 \times 2}$ with standard operations is not a complex vector space, by finding a condition of Definition 11.14 that is not met.

46. Prove that the complex dot product is an inner product on $\mathbf{C}^n$. (Condition (d) has already been verified, so you need only show that (a)–(c) are true.)

47. Prove Theorem 11.16: Let $\mathbf{u}$, $\mathbf{v}$, and $\mathbf{w}$ be elements of a complex vector space V, and let c be a complex scalar. Then an inner product defined on V satisfies each of the following:

(a) $\langle \mathbf{0}, \mathbf{u} \rangle = \langle \mathbf{u}, \mathbf{0} \rangle = 0$

(b) $\langle \mathbf{u}, \mathbf{v} + \mathbf{w} \rangle = \langle \mathbf{u}, \mathbf{v} \rangle + \langle \mathbf{u}, \mathbf{w} \rangle$

(c) $\langle \mathbf{u}, c\mathbf{v} \rangle = \overline{c}\langle \mathbf{u}, \mathbf{v} \rangle$

48. Prove that if $\langle \mathbf{u}, \mathbf{v} \rangle = 0$, then $\|\mathbf{u}\|^2 + \|\mathbf{v}\|^2 = \|\mathbf{u} + \mathbf{v}\|^2$.

In Exercises 49–52, we prove each part of Theorem 11.17. Assume that $\mathbf{u}$ and $\mathbf{v}$ are elements of a complex vector space V, and suppose that c is a complex scalar.

49. Prove that $\|\mathbf{u}\| \geq 0$, with equality only if $\mathbf{u} = \mathbf{0}$.

50. Prove that $\|c\mathbf{u}\| = |c|\|\mathbf{u}\|$.

51. Prove that $|\langle \mathbf{u}, \mathbf{v} \rangle| \leq \|\mathbf{u}\| \cdot \|\mathbf{v}\|$. (This is the Cauchy–Schwarz Inequality.)

52. Prove that $\|\mathbf{u} + \mathbf{v}\| \leq \|\mathbf{u}\| + \|\mathbf{v}\|$. (This is the Triangle Inequality.)

11.5 Hermitian Matrices

▶ See Section 6.5 for a review of the basic properties of complex numbers.

In Section 8.3 we encountered the Spectral Theorem, which says that a matrix A with real entries is orthogonally diagonalizable exactly when A is symmetric. In this section we consider what happens if we allow A to have complex entries. Since the reals are a subset of the complex numbers, a reasonable guess is that the Spectral Theorem just generalizes, so that a matrix A with complex entries is orthogonally diagonalizable if and only if A is symmetric. However, this is not quite correct. The central goal of this section is to find the correct analog of the Spectral Theorem for complex matrices, but before we can do that, we need to develop a few new ideas.

Definition Complex Conjugate,
Conjugate Transpose

Unitary Matrices

If A is a complex matrix, then the **complex conjugate** of A is denoted by $\overline{A}$, and is found by taking the complex conjugate of each entry of A. The **conjugate transpose** of A is denoted by A^* and defined

▶ Recall that if $c = a + ib$, then the complex conjugate is $\overline{c} = a - ib$.

$$A^* = \overline{A}^T$$

EXAMPLE 1   Find A^* for

$$A = \begin{bmatrix} 2+i & 3 \\ -5i & 1+4i \\ 6 & 1-2i \end{bmatrix}$$

Solution   We have

$$A = \begin{bmatrix} 2+i & 3 \\ -5i & 1+4i \\ 6 & 1-2i \end{bmatrix} \implies \overline{A} = \begin{bmatrix} 2-i & 3 \\ 5i & 1-4i \\ 6 & 1+2i \end{bmatrix}$$

$$\implies A^* = \overline{A}^T = \begin{bmatrix} 2-i & 5i & 6 \\ 3 & 1-4i & 1+2i \end{bmatrix} \quad\blacksquare$$

Note that the order of conjugation and transposition makes no difference in A^*. For the matrix in Example 1, we could just as well have computed

$$A^T = \begin{bmatrix} 2+i & -5i & 6 \\ 3 & 1+4i & 1-2i \end{bmatrix} \implies A^* = \overline{A^T} = \begin{bmatrix} 2-i & 5i & 6 \\ 3 & 1-4i & 1+2i \end{bmatrix}$$

A general proof that $\overline{A}^T = \overline{A^T}$ is left as Exercise 32.

Conjugate transposes have properties similar to those of transposes of real matrices. These are summarized in the theorem below, with the proof of each part left as exercises.

THEOREM 11.18

Suppose that A and B are matrices with complex entries and that c is a complex scalar. Then

(a) $(A^*)^* = A$

(b) $(A + B)^* = A^* + B^*$

(c) $(AB)^* = B^* A^*$

(d) $(cA)^* = \overline{c} A^*$

Definition Unitary Matrix

Recall that a square matrix with real entries is *orthogonal* if the matrix columns form an orthonormal set with respect to the usual dot product. Equivalently, a real matrix A is orthogonal if and only if $A^{-1} = A^T$. The counterpart of orthogonal matrices for a matrix A with complex entries is called a **unitary matrix**, which requires that

$$A^{-1} = A^*$$

It can be shown that a square matrix A is unitary if and only if the columns of A are orthonormal with respect to the complex dot product (see Exercise 37).

EXAMPLE 2 Show that

$$A = \begin{bmatrix} \frac{1}{\sqrt{2}} & -\frac{1+i}{2} \\ \frac{1}{\sqrt{2}} & \frac{1+i}{2} \end{bmatrix}$$

is a unitary matrix.

Solution One way to solve this is to compute AA^*. If the product is the identity matrix, then we know that $A^* = A^{-1}$ and can conclude that A is unitary. Instead, we shall show directly that the columns are orthonormal with respect to the complex dot product. Setting $A = \begin{bmatrix} \mathbf{a}_1 & \mathbf{a}_2 \end{bmatrix}$, we have

$$\mathbf{a}_1 \cdot \mathbf{a}_2 = \left(\frac{1}{\sqrt{2}} \right) \left(-\frac{1-i}{2} \right) + \left(\frac{1}{\sqrt{2}} \right) \left(\frac{1-i}{2} \right) = 0$$

$$\|\mathbf{a}_1\| = \sqrt{\mathbf{a}_1 \cdot \mathbf{a}_1} = \sqrt{\left(\frac{1}{\sqrt{2}} \right)^2 + \left(\frac{1}{\sqrt{2}} \right)^2} = 1$$

$$\|\mathbf{a}_2\| = \sqrt{\mathbf{a}_2 \cdot \mathbf{a}_2} = \sqrt{\left(-\frac{1+i}{2} \right) \left(-\frac{1-i}{2} \right) + \left(\frac{1+i}{2} \right) \left(\frac{1-i}{2} \right)} = 1$$

Hence the columns are orthonormal and therefore A is unitary. ∎

Diagonalizing Matrices

Definition Unitarily
Diagonalizable

We say that a complex matrix A is **unitarily diagonalizable** if there exist a diagonal matrix D and a unitary matrix P such that

$$A = PDP^{-1} = PDP^*$$

▶ "Unitarily diagonalizable" is an awkward phrase, so from here on we will just say "diagonalizable" with the understanding the P must be unitary.

As with real matrices, the diagonal entries of D are the eigenvalues of A, and the columns of P are the corresponding eigenvectors (see Theorem 6.13 in Section 6.4). The question is, when will a complex matrix be diagonalizable?

Let's recall what happens for real matrices. A real matrix A is orthogonally diagonalizable if and only if $A = A^T$—that is, when A is symmetric. The analog of symmetric for complex matrices is

$$A = A^* \tag{1}$$

Definition Hermitian

A matrix A satisfying (1) is called **Hermitian**. From the definition, we see that a Hermitian matrix is unchanged by taking its conjugate transpose. For example, the matrix A below is Hermitian, because

▶ **Charles Hermite** (1822–1901) was a French mathematician who made contributions to a variety of areas of mathematics, among them linear algebra.

$$A = \begin{bmatrix} 3 & 2-i \\ 2+i & 4 \end{bmatrix} \implies \overline{A} = \begin{bmatrix} 3 & 2+i \\ 2-i & 4 \end{bmatrix}$$

$$\implies A^* = \begin{bmatrix} 3 & 2-i \\ 2+i & 4 \end{bmatrix} = A$$

Note that any Hermitian matrix must have real diagonal entries. (Why?) Unfortunately, the Hermitian matrices still are not exactly the set that we seek.

While all Hermitian matrices are diagonalizable, it turns out that there are some complex matrices that are diagonalizable but not Hermitian. We need to expand the set of Hermitian matrices to the larger set of **normal** matrices, which are those complex matrices A such that

$$A^*A = AA^*$$

Note that all unitary matrices are normal, because

$$A^*A = A^{-1}A = I = AA^{-1} = AA^*$$

Similarly, Hermitian matrices are also normal, because

$$A^*A = AA = AA^*$$

There are normal matrices that are not Hermitian (or unitary, for that matter).

EXAMPLE 3   Show that

$$A = \begin{bmatrix} i & -i \\ i & i \end{bmatrix}$$

is normal but is not Hermitian or unitary.

Solution We have

$$A^*A = \begin{bmatrix} -i & -i \\ i & -i \end{bmatrix} \begin{bmatrix} i & -i \\ i & i \end{bmatrix} = \begin{bmatrix} 2 & 0 \\ 0 & 2 \end{bmatrix}$$

$$AA^* = \begin{bmatrix} i & -i \\ i & i \end{bmatrix} \begin{bmatrix} -i & -i \\ i & -i \end{bmatrix} = \begin{bmatrix} 2 & 0 \\ 0 & 2 \end{bmatrix}$$

Hence A is normal. On the other hand, since $A^* \neq A$ our matrix is not Hermitian, and as $A^*A \neq I_2$, it follows that $A^* \neq A^{-1}$ and thus A is not unitary. ■

The following is a complex version of the Spectral Theorem (Section 8.3), given without proof.

THEOREM 11.19   A complex matrix A is unitarily diagonalizable if and only if A is normal.

Now we have an easy way to determine if a complex matrix is diagonalizable, by checking if it is normal. If A is an $n \times n$ diagonalizable matrix, then we find the diagonalization of A using the same procedure as with real matrices.

1. Find the eigenvalues and eigenvectors of A.

2. For each distinct eigenvalue, apply Gram–Schmidt as needed to find an orthonormal basis for the associated eigenspace. As with real symmetric matrices, eigenvectors associated with distinct eigenvalues of a normal matrix are orthogonal. Thus, once we have orthonormal bases for each eigenspace, they can be combined to form an orthonormal basis for $\mathbf{C}^n$.

3. Define D to be the diagonal matrix with the eigenvalues of A along the diagonal, and define P to be the unitary matrix with the corresponding eigenvectors for columns.

Applying this procedure by hand to a large, complicated matrix is difficult. But it is manageable if the matrix is not too complicated.

EXAMPLE 4 Diagonalize the matrix $A = \begin{bmatrix} i & -i \\ i & i \end{bmatrix}$.

Solution We have already seen that A is normal and so must be diagonalizable. Our first step is to find the eigenvalues, which are the roots of the characteristic polynomial

$$\det(A - \lambda I_2) = (i - \lambda)^2 + i^2 = (i - \lambda)^2 - 1$$

Setting this equal to 0 and solving for λ yields two eigenvalues, $\lambda_1 = 1 + i$ and $\lambda_2 = -1 + i$. To find the eigenvectors associated with λ_1, we need to find the solutions to the homogeneous system with coefficient matrix $A - \lambda_1 I_2 = A - (1 + i) I_2$. The augmented matrix is

$$\begin{bmatrix} i - (1 + i) & -i & 0 \\ i & i - (1 + i) & 0 \end{bmatrix} = \begin{bmatrix} -1 & -i & 0 \\ i & -1 & 0 \end{bmatrix}$$

$$\overset{i R_1 + R_2 \Rightarrow R_2}{\sim} \begin{bmatrix} -1 & -i & 0 \\ 0 & 0 & 0 \end{bmatrix}$$

Back substitution and normalization produces the eigenvector

$$\mathbf{p}_1 = \begin{bmatrix} -\frac{i}{\sqrt{2}} \\ \frac{1}{\sqrt{2}} \end{bmatrix}$$

Following a similar procedure, we find that a normalized eigenvector associated with $\lambda_2 = -1 + i$ is

$$\mathbf{p}_2 = \begin{bmatrix} \frac{i}{\sqrt{2}} \\ \frac{1}{\sqrt{2}} \end{bmatrix}$$

Since each eigenspace has dimension 1, we are spared the work of applying the Gram–Schmidt process to find orthogonal eigenvectors. Thus all that remains is to define P and D, which are

$$P = \begin{bmatrix} -\frac{i}{\sqrt{2}} & \frac{i}{\sqrt{2}} \\ \frac{1}{\sqrt{2}} & \frac{1}{\sqrt{2}} \end{bmatrix} \quad \text{and} \quad D = \begin{bmatrix} 1 + i & 0 \\ 0 & -1 + i \end{bmatrix}$$

We can check our work by computing

$$PDP^* = \begin{bmatrix} -\frac{i}{\sqrt{2}} & \frac{i}{\sqrt{2}} \\ \frac{1}{\sqrt{2}} & \frac{1}{\sqrt{2}} \end{bmatrix} \begin{bmatrix} 1 + i & 0 \\ 0 & -1 + i \end{bmatrix} \begin{bmatrix} \frac{i}{\sqrt{2}} & \frac{1}{\sqrt{2}} \\ -\frac{i}{\sqrt{2}} & \frac{1}{\sqrt{2}} \end{bmatrix} = \begin{bmatrix} i & -i \\ i & i \end{bmatrix} = A \quad \blacksquare$$

In Section 8.3 it is noted that a real symmetric matrix must have real eigenvalues. (This follows from the Spectral Theorem.) On the other hand, the preceding example shows that a normal matrix can have complex eigenvalues. In between these two sets are the Hermitian matrices, which also happen to have real eigenvalues.

THEOREM 11.20 | If A is a Hermitian matrix, then A has real eigenvalues.

Proof Suppose that λ is an eigenvalue of A with associated eigenvector $\mathbf{u}$. Then $A\mathbf{u} = \lambda \mathbf{u}$, and hence

$$\mathbf{u}^* A \mathbf{u} = \mathbf{u}^*(\lambda \mathbf{u}) = \lambda(\mathbf{u}^* \mathbf{u}) = \lambda \|\mathbf{u}\|^2$$

We know that $\|\mathbf{u}\|^2$ is a real number. Also, because A is Hermitian we have

$$(\mathbf{u}^* A\mathbf{u})^* = \mathbf{u}^* A^*(\mathbf{u}^*)^* = \mathbf{u}^* A\mathbf{u}$$

This shows that $\mathbf{u}^* A\mathbf{u}$ is also Hermitian and therefore has real diagonal entries. But $\mathbf{u}^* A\mathbf{u}$ has only one entry (it is the complex dot product of $\mathbf{u}$ and $A\mathbf{u}$), so this implies that $\mathbf{u}^* A\mathbf{u}$ is real. Since $\mathbf{u}^* A\mathbf{u} = \lambda \|\mathbf{u}\|^2$, we may conclude that λ is also real. ■

EXAMPLE 5 Show that

$$A = \begin{bmatrix} 5 & 2i \\ -2i & 2 \end{bmatrix}$$

is Hermitian and has real eigenvalues.

Solution We have

$$\overline{A} = \begin{bmatrix} 5 & -2i \\ 2i & 2 \end{bmatrix} \quad \Rightarrow \quad A^* = \overline{A}^T = \begin{bmatrix} 5 & 2i \\ -2i & 2 \end{bmatrix} = A$$

so A is Hermitian. The characteristic polynomial of A is

$$\det(A - \lambda I_2) = (5 - \lambda)(2 - \lambda) - (-2i)(2i) = \lambda^2 - 7\lambda + 6 = (\lambda - 6)(\lambda - 1)$$

Hence the eigenvalues are $\lambda_1 = 1$ and $\lambda_2 = 6$, which are both real. ■

EXERCISES

In Exercises 1–6, find A^* for the given A.

1. $A = \begin{bmatrix} 1+i & 3i \\ 2-i & 1+4i \end{bmatrix}$

2. $A = \begin{bmatrix} -7i & 3-2i \\ 1+5i & 8 \end{bmatrix}$

3. $A = \begin{bmatrix} 3+i & 5i & 1-i \\ 1-4i & -8 & 6+i \\ 2+2i & 0 & -7i \end{bmatrix}$

4. $A = \begin{bmatrix} 5 & -i & 2+7i \\ 4i & 5i & 31 \\ 5-i & 6i & 13 \end{bmatrix}$

5. $A = \begin{bmatrix} 1 & -2i & 3 & 4i \\ 2i & 5 & -6i & 1+i \\ 3 & 6i & 7 & 3+2i \\ -4i & 1-i & 3-2i & 11 \end{bmatrix}$

6. $A = \begin{bmatrix} 4 & 1-2i & 12 & 1+3i \\ 11i & 6-5i & 6i & 3i \\ -7i & 1-i & 3i & 4-5i \\ 2+i & 1-2i & 4-7i & 0 \end{bmatrix}$

In Exercises 7–12, determine if the given matrix is Hermitian.

7. $A = \begin{bmatrix} 1+i & 3i \\ 3i & 2 \end{bmatrix}$

8. $A = \begin{bmatrix} 4 & 3-2i \\ 3+2i & 3 \end{bmatrix}$

9. $A = \begin{bmatrix} 3 & 5i & 1-i \\ -5i & -5 & 0 \\ 1+i & 0 & 7 \end{bmatrix}$

10. $A = \begin{bmatrix} 5 & -i & 2+7i \\ i & 3 & 6i \\ 2+7i & 6i & 4 \end{bmatrix}$

11. $A = \begin{bmatrix} 1 & -2i & 3 & 4i \\ 2i & 5 & -6i & 1+i \\ 3 & 6i & 7 & 3+2i \\ -4i & 1-i & 3-2i & 11 \end{bmatrix}$

12. $A = \begin{bmatrix} 0 & 1-2i & -12 & 2-i \\ 1+2i & 5 & 6i & 3i \\ 12 & -6i & 2 & 4-5i \\ 2+i & -3i & 4+5i & 8 \end{bmatrix}$

In Exercises 13–18, determine if the given matrix is normal.

13. $A = \begin{bmatrix} 1 & 2-5i \\ 2+5i & 3 \end{bmatrix}$

14. $A = \begin{bmatrix} 3 & 3-2i \\ 1+i & -4 \end{bmatrix}$

15. $A = \begin{bmatrix} -i & -i \\ i & -i \end{bmatrix}$

16. $A = \begin{bmatrix} 2i & i \\ i & 3i \end{bmatrix}$

17. $A = \begin{bmatrix} -1 & -i & 1-i \\ i & 5 & -2i \\ 1+i & 2i & 0 \end{bmatrix}$

18. $A = \begin{bmatrix} 2 & 3-i & i \\ 3-i & 1 & -2i \\ -i & 3i & -1 \end{bmatrix}$

FIND AN EXAMPLE For Exercises 19–22, find an example that meets the given specifications.

19. A 3×3 matrix A that is symmetric but not Hermitian.

20. A 3×3 matrix A that is normal but not Hermitian.

21. A 3×3 unitary matrix that is not in $\mathbf{R}^{3 \times 3}$.

22. A 2×2 matrix that has eigenvalues $\lambda_1 = 2 + i$ and $\lambda_2 = 2 - i$.

TRUE OR FALSE For Exercises 23–30, determine if the statement is true or false, and justify your answer.

23. A matrix A is unitarily diagonalizable if and only if A is normal.

24. If A is normal, then A is symmetric.

25. If A has complex entries and $A = A^T$, then A is Hermitian.

26. If A has complex entries, then $\det(A)$ cannot be real.

27. If A and B are $n \times n$ Hermitian matrices, then so is $A + B$.

28. If A and B are $n \times n$ complex matrices, then A^*B is Hermitian.

29. A unitary matrix has orthonormal columns.

30. A Hermitian matrix must have some real entries.

31. Prove that if A has real entries, then $A^* = A^T$.

32. Prove that if A has complex entries, then $(\overline{A})^T = \overline{(A^T)}$. This shows that the order of conjugation and transposition in A^* does not matter.

In Exercises 33–36, suppose that A and B are matrices with complex entries and that c is a complex scalar.

33. Prove that $(A^*)^* = A$.

34. Prove that $(A + B)^* = A^* + B^*$.

35. Prove that $(AB)^* = B^*A^*$.

36. Prove that $(cA)^* = \overline{c}A^*$.

37. Show that a square matrix A is unitary if and only if the columns of A are orthonormal with respect to the complex dot product.

38. Prove that any Hermitian matrix must have real diagonal entries.

39. Show that if A is upper (or lower) triangular and normal, then A must be a diagonal matrix.

40. Suppose that $A = PDP^{-1}$, where D is diagonal and P is unitary. Show that the diagonal entries of D are the eigenvalues of A, and the columns of P are the corresponding eigenvectors.

GLOSSARY

Below is a glossary of definitions and other terms presented in this book. In some cases, due to the complicated nature of a definition or term, it is only described here in general terms. Visit the section referenced for more details.

Additive identity matrix The $n \times m$ matrix 0_{nm} consisting of all zeros satisfies $A + 0_{nm} = A$ for all $n \times m$ matrices A. (Sect. 3.2)

Adjoint matrix The adjoint of an $n \times n$ matrix A is given by

$$\text{adj}(A) = C^T = \begin{bmatrix} C_{11} & C_{21} & \cdots & C_{n1} \\ C_{12} & C_{22} & \cdots & C_{n2} \\ \vdots & \vdots & \ddots & \vdots \\ C_{1n} & C_{2n} & \cdots & C_{nn} \end{bmatrix}$$

where C is the cofactor matrix of A. (Sect. 5.3)

Argument The argument of a nonzero complex number z, denoted by $\arg(z)$, is the angle θ (in radians) in the counterclockwise direction from the positive x-axis to the ray that points from the origin to z. Note that the argument is not unique, because we can always add or subtract multiples of 2π. (Sect. 6.5)

Associated homogeneous linear system A linear system of the form $A\mathbf{x} = \mathbf{b}$, where $\mathbf{b} \neq \mathbf{0}$, has associated homogeneous linear system $A\mathbf{x} = \mathbf{0}$. (Sect. 2.3)

Augmented matrix A matrix that contains all of the coefficients of a linear system, including the constant terms. (Sect. 1.2)

Back substitution A method of solution applicable to a system of linear equations in echelon form. Implemented by substituting known values back into remaining equations. (Sect. 1.1)

Basis A set $\mathcal{B} = \{\mathbf{u}_1, \ldots, \mathbf{u}_m\}$ is a basis for a subspace S if $\mathcal{B}$ spans S and $\mathcal{B}$ is linear independent. (Sect. 4.2; see also Sect. 7.3)

BCS rankings A continually evolving system for attempting to rank the competitive strength of college football teams and that consistently underrates the University of Texas. (Sect. 1.4)

Block diagonal matrix A partitioned matrix with nondiagonal blocks that are zero matrices. (Sect. 3.3)

Change of basis matrix A square matrix used to express a vector in $\mathbf{R}^n$ given in terms of one basis into a vector given in terms of a different basis. (Sect. 6.3; see also Sect. 9.4)

Characteristic equation Let A be an $n \times n$ matrix. Then the characteristic equation is given by $\det(A - \lambda I_n) = 0$, where I_n is the identity matrix. The solutions to the characteristic equation are the eigenvalues of A. (Sect. 6.1)

Characteristic polynomial Let A be an $n \times n$ matrix. Then the characteristic polynomial is given by $\det(A - \lambda I_n)$, where I_n is the identity matrix. (Sect. 6.1)

Closed under addition S is closed under addition if $\mathbf{u}$ and $\mathbf{v}$ in S implies $\mathbf{u} + \mathbf{v}$ is also in S. (Sect. 4.1)

Closed under scalar multiplication S is closed under scalar multiplication if r is a real number and $\mathbf{u}$ in S implies $r\mathbf{u}$ is also in S. (Sect. 4.1)

Codomain A set containing all possible outputs for a function. (Note that this contains the range, which is equal to the set of possible outputs for a function.) (Sect. 3.1; see also Sect. 9.1)

Cofactor Given a matrix A, the cofactor of a_{ij} is equal to

$$C_{ij} = (-1)^{i+j} \det(M_{ij})$$

where M_{ij} is the $(n-1) \times (n-1)$ matrix that we get from A after deleting the row and column containing a_{ij}. Put another way, C_{ij} is equal to $(-1)^{i+j}$ times the minor of a_{ij}. (Sect. 5.1)

Cofactor expansion Let A be the $n \times n$ matrix $[a_{ij}]$ and let C_{ij} denote the cofactor of a_{ij}. Then the cofactor expansions are given by

(a) $\det(A) = a_{i1}C_{i1} + a_{i2}C_{i2} + \cdots + a_{in}C_{in}$
(Expand across row i)

(b) $\det(A) = a_{1j}C_{1j} + a_{2j}C_{2j} + \cdots + a_{nj}C_{nj}$
(Expand down column j)

(Sect. 5.1)

Cofactor matrix For an $n \times n$ matrix A, the cofactor matrix is given by

$$C = \begin{bmatrix} C_{11} & C_{12} & \cdots & C_{1n} \\ C_{21} & C_{22} & \cdots & C_{2n} \\ \vdots & \vdots & & \vdots \\ C_{n1} & C_{n2} & \cdots & C_{nn} \end{bmatrix}$$

where C_{ij} is the cofactor of a_{ij}. (Sect. 5.3)

Column space Let A be an $n \times m$ matrix. The column space of A is the subspace of $\mathbf{R}^n$ spanned by the column vectors of A, and is denoted by col(A). (Sect. 4.3)

Column vector A vector in Euclidean space expressed in the form of a column matrix. Also used to indicate the vectors formed from the columns of a matrix A. (Sect. 2.1)

Complex conjugate The complex conjugate of $z = a + ib$ is given by $\bar{z} = a - ib$. (Sect. 6.5)

Complex dot product The complex dot product is defined on $\mathbf{C}^n$ by

$$\mathbf{u} \cdot \mathbf{v} = u_1 \bar{v}_1 + u_2 \bar{v}_2 + \cdots + u_n \bar{v}_n$$

(Sect. 11.4)

Complex inner product space A complex vector space V with an inner product defined on it. (Also called a unitary space.) (Sect. 11.4)

Complex vector space A complex vector space consists of a nonempty set V of vectors together with operations of addition and scalar multiplication on the vectors that satisfy each of the following:

(1) If $\mathbf{v}_1$ and $\mathbf{v}_2$ are in V, then so is $\mathbf{v}_1 + \mathbf{v}_2$. Hence V is closed under addition.

(2) If c is a complex scalar and $\mathbf{v}$ is in V, then so is $c\mathbf{v}$. Hence V is closed under scalar multiplication.

(3) There exists a zero vector $\mathbf{0}$ in V such that $\mathbf{0} + \mathbf{v} = \mathbf{v}$ for all $\mathbf{v}$ in V.

(4) For each $\mathbf{v}$ in V there exists an additive inverse (or opposite) vector $-\mathbf{v}$ in V such that $\mathbf{v} + (-\mathbf{v}) = \mathbf{0}$ for all $\mathbf{v}$ in V.

(5) For all $\mathbf{v}_1$, $\mathbf{v}_2$, and $\mathbf{v}_3$ in V and complex scalars c_1 and c_2, we have
  (a) $\mathbf{v}_1 + \mathbf{v}_2 = \mathbf{v}_2 + \mathbf{v}_1$
  (b) $(\mathbf{v}_1 + \mathbf{v}_2) + \mathbf{v}_3 = \mathbf{v}_1 + (\mathbf{v}_2 + \mathbf{v}_3)$
  (c) $c_1(\mathbf{v}_1 + \mathbf{v}_2) = c_1\mathbf{v}_1 + c_1\mathbf{v}_2$
  (d) $(c_1 + c_2)\mathbf{v}_1 = c_1\mathbf{v}_1 + c_2\mathbf{v}_1$
  (e) $(c_1 c_2)\mathbf{v}_1 = c_1(c_2\mathbf{v}_1)$
  (f) $1 \cdot \mathbf{v}_1 = \mathbf{v}_1$

(Sect. 11.4)

Component A single entry in a vector in Euclidean space. (Sect. 2.1)

Conjecture The mathematical equivalent of an educated guess. (Sect. 1.4)

Conjugate transpose The conjugate transpose of a complex matrix A is denoted by A^* and defined

$$A^* = \bar{A}^T$$

(Sect. 11.5)

Consistent linear system A linear system that has at least one solution. (Sect. 1.1)

Converge An iterative process is said to converge if the outcome of a sequence of steps approaches a specific value. (Sect. 1.3; see also Sect. 3.5 and Sect. 6.2)

Coordinate vector Suppose that $\mathcal{B} = \{\mathbf{u}_1, \ldots, \mathbf{u}_n\}$ forms a basis for $\mathbf{R}^n$. If $\mathbf{y} = y_1\mathbf{u}_1 + \cdots + y_n\mathbf{u}_n$, then

$$\mathbf{y}_{\mathcal{B}} = \begin{bmatrix} y_1 \\ \vdots \\ y_n \end{bmatrix}_{\mathcal{B}}$$

is the coordinate vector of $\mathbf{y}$ with respect to $\mathcal{B}$. (Sect. 6.3; see also Sect. 9.3)

Cramer's Rule Let $A = \begin{bmatrix} \mathbf{a}_1 & \cdots & \mathbf{a}_n \end{bmatrix}$ be an invertible $n \times n$ matrix. Then the components of the unique solution $\mathbf{x}$ to $A\mathbf{x} = \mathbf{b}$ are given by

$$x_i = \frac{\det(A_i)}{\det(A)} \quad \text{for } i = 1, 2, \ldots, n$$

where

$$A_i = \begin{bmatrix} \mathbf{a}_1 & \cdots & \mathbf{a}_{i-1} & \mathbf{b} & \mathbf{a}_{i+1} & \cdots & \mathbf{a}_n \end{bmatrix}$$

(A_i is just A with column i replaced by $\mathbf{b}$.) (Sect. 5.3)

Determinant If $A = \begin{bmatrix} a_{11} \end{bmatrix}$ is a 1×1 matrix, then the determinant of A is given by $\det(A) = a_{11}$.

If

$$A = \begin{bmatrix} a_{11} & a_{12} \\ a_{21} & a_{22} \end{bmatrix}$$

then the determinant is given by

$$\det(A) = a_{11}a_{22} - a_{12}a_{21}$$

For the $n \times n$ matrix

$$A = \begin{bmatrix} a_{11} & a_{12} & \cdots & a_{1n} \\ \vdots & \vdots & \ddots & \vdots \\ a_{n1} & a_{n2} & \cdots & a_{nn} \end{bmatrix}$$

the determinant of A is defined recursively by

$$\det(A) = a_{11}C_{11} + a_{12}C_{12} + \cdots + a_{1n}C_{1n}$$

where $C_{11}, \ldots, C_{1n}$ are the cofactors of $a_{11}, \ldots, a_{1n}$, respectively. (Sect. 5.1)

Diagonal matrix A diagonal matrix has the form

$$A = \begin{bmatrix} a_{11} & 0 & 0 & \cdots & 0 \\ 0 & a_{22} & 0 & \cdots & 0 \\ 0 & 0 & a_{33} & \cdots & 0 \\ \vdots & \vdots & \vdots & \ddots & \vdots \\ 0 & 0 & 0 & \cdots & a_{nn} \end{bmatrix}$$

(Sect. 3.2)

Diagonalizable matrix An $n \times n$ matrix A is diagonalizable if there exist $n \times n$ matrices D and P, with D diagonal and P invertible, such that

$$A = PDP^{-1}$$

(Sect. 6.4)

Diagonally dominant A linear system with the same number of equations and variables is diagonally dominant if the diagonal coefficients $(a_{11}, a_{22}, \ldots)$ are each larger in absolute value than the sum of the absolute values of the other terms in the same row. (Sect. 1.3)

Dimension (subspace) Let S be a nonzero subspace. Then the dimension of S is the number of vectors in any basis of S. (Sect. 4.2; see also Sect. 7.3)

Distance between vectors For two vectors $\mathbf{u}$ and $\mathbf{v}$, the distance between $\mathbf{u}$ and $\mathbf{v}$ is given by $\|\mathbf{u} - \mathbf{v}\|$. (Sect. 8.1; see also Sect. 10.1)

Diverge An iterative process is said to diverge if the outcome of a sequence of steps fails to approach a specific value. (Sect. 1.3; see also Sect. 3.5 and Sect. 6.2)

Domain The set of possible inputs for a function. (Sect. 3.1; see also Sect. 9.1)

Dominant eigenvalue Suppose that a square matrix A has eigenvalues $\lambda_1, \lambda_2, \ldots, \lambda_n$ such that

$$|\lambda_1| > |\lambda_2| \geq |\lambda_3| \geq \cdots \geq |\lambda_n|$$

In this case λ_1 is the dominant eigenvalue of A. (Sect. 6.2)

Dot product Suppose

$$\mathbf{u} = \begin{bmatrix} u_1 \\ \vdots \\ u_n \end{bmatrix} \quad \text{and} \quad \mathbf{v} = \begin{bmatrix} v_1 \\ \vdots \\ v_n \end{bmatrix}$$

are both in $\mathbf{R}^n$. Then the dot product of $\mathbf{u}$ and $\mathbf{v}$ is given by

$$\mathbf{u} \cdot \mathbf{v} = u_1 v_1 + \cdots + u_n v_n$$

The dot product can also be expressed $\mathbf{u} \cdot \mathbf{v} = \mathbf{u}^T \mathbf{v}$. (Sect. 8.1)

Doubly stochastic matrix A square matrix A that has nonnegative entries, and has rows and columns that each add to 1. (Sect. 3.5)

Echelon form (linear system) A linear system satisfying the following conditions: Every variable is the leading variable of *at most* one equation; the system is organized in a descending "stair step" pattern so that the index of the leading variables increases from the top to bottom; and every equation has a leading variable. Such a system is called an echelon system. (Sect. 1.1)

Echelon form (matrix) Also called row echelon form, a matrix is in echelon form if every leading term is in a column to the left of the leading term of the row below it, and any zero rows are at the bottom of the matrix. (Sect. 1.2)

Eigenspace Let A be a square matrix with eigenvalue λ. The subspace of all eigenvectors associated with λ, together with the zero vector, is the eigenspace of λ. Each distinct eigenvalue of A has its own associated eigenspace. (Sect. 6.1)

Eigenvalue Let A be an $n \times n$ matrix. Suppose that λ is a scalar and $\mathbf{u} \neq \mathbf{0}$ is a vector satisfying

$$A\mathbf{u} = \lambda\mathbf{u}$$

The scalar λ is called an eigenvalue of A. (Sect. 6.1)

Eigenvector Let A be an $n \times n$ matrix. Suppose that λ is a scalar and $\mathbf{u} \neq \mathbf{0}$ is a vector satisfying

$$A\mathbf{u} = \lambda\mathbf{u}$$

Then $\mathbf{u}$ is called an eigenvector of A. (Sect. 6.1)

Elementary (equation) operations Three operations that can be performed on a linear system that do not change the set of solutions, so yield an equivalent system. They are (1) interchanging two equations, (2) multiplying an equation by a nonzero constant, and (3) adding a multiple of one equation to another. (Sect. 1.2)

Elementary matrix A square matrix E such that the product EA induces an elementary row operation on A. (Sect. 3.4)

Elementary row operations Three row operations that can be performed on an augmented matrix that do not change the set of solutions to the corresponding linear system. They are (1) interchanging two rows, (2) multiplying a row by a nonzero constant, and (3) adding a multiple of one row to another. (Sect. 1.2)

Equivalent matrices Two matrices are equivalent if one can be transformed into the other through a sequence of elementary row operations. If the matrices in question are augmented matrices, then the corresponding linear systems have the same set of solutions. (Sect. 1.2)

Equivalent systems Two linear systems are equivalent if they have the same set of solutions. (Sect. 1.2)

Euclidean space The set of all vectors in $\mathbf{R}^n$ together with the "standard" definitions for vector arithmetic. (Sect. 2.1)

Fourier approximation The nth order Fourier approximation for a function f in $C[-\pi, \pi]$ is given by

$$f_n(x) = a_0 + a_1 \cos(x) + \cdots + a_n \cos(nx)$$
$$+ b_1 \sin(x) + \cdots + b_n \sin(nx)$$

where the coefficients are given by

$$a_0 = \frac{1}{2\pi} \int_{-\pi}^{\pi} f(x)\, dx$$

$$a_k = \frac{1}{\pi} \int_{-\pi}^{\pi} f(x) \cos(kx)\, dx \quad (k \geq 1)$$

$$b_k = \frac{1}{\pi} \int_{-\pi}^{\pi} f(x) \sin(kx)\, dx \quad (k \geq 1)$$

The a_k's and b_k's are called the Fourier coefficients of f. (Sect. 10.3)

Free parameter An unspecified numerical quantity that can be equal to any real number. (Sect. 1.1)

Free variable Any variable in a linear system in echelon form that is not a leading variable. (Sect. 1.1)

Full pivoting An extension of partial pivoting where both row and column interchanges are performed to reduce round-off error when implementing elimination methods. This method is not covered in this text, but it is described in more advanced texts on numerical linear algebra. (Sect. 1.3)

Gauss–Jordan elimination An algorithm that extends Gaussian elimination, applying row operations in a manner that will transform a matrix to reduced echelon form. (Sect. 1.2)

Gauss–Seidel iteration A variant of Jacobi iteration, Gauss–Seidel iteration is an iterative method for approximating the solutions to a linear system that has the same number of equations as variables. (Sect. 1.3)

Gaussian elimination An algorithm for applying row operations in a manner that will transform a matrix to echelon form. (Sect. 1.2)

General solution A description of the set of all solutions to a linear equation or linear system. (Sect. 1.1)

Hermitian matrix A complex matrix A is Hermitian if

$$A = A^*$$

(Sect. 11.5)

Homogeneous equation A linear equation is homogeneous if it has the form

$$a_1 x_1 + a_2 x_2 + a_3 x_3 + \cdots + a_n x_n = 0$$

Such equations always have the trivial solution, so are consistent. (Sect. 1.2)

Homogeneous system A linear system is homogeneous if it has the form

$$a_{11} x_1 + a_{12} x_2 + a_{13} x_3 + \cdots + a_{1n} x_n = 0$$
$$a_{21} x_1 + a_{22} x_2 + a_{23} x_3 + \cdots + a_{2n} x_n = 0$$
$$\vdots \qquad \vdots \qquad \vdots \qquad \vdots$$
$$a_{m1} x_1 + a_{m2} x_2 + a_{m3} x_3 + \cdots + a_{mn} x_n = 0$$

Such systems always have the trivial solution, so are consistent. This system can also be expressed by $A\mathbf{x} = \mathbf{0}$. (Sect. 1.2; see also Sect. 2.3)

Hyperplane The set of all solutions to a linear equation in four or more variables. (Sect. 1.1)

Idempotent matrix A square matrix A is idempotent if $A^2 = A$. (Sect. 3.2)

Identity matrix The $n \times n$ identity matrix is given by

$$I_n = \begin{bmatrix} \mathbf{e}_1 & \mathbf{e}_2 & \cdots & \mathbf{e}_m \end{bmatrix} = \begin{bmatrix} 1 & 0 & 0 & \cdots & 0 \\ 0 & 1 & 0 & \cdots & 0 \\ 0 & 0 & 1 & \cdots & 0 \\ \vdots & \vdots & \vdots & \ddots & \vdots \\ 0 & 0 & 0 & \cdots & 1 \end{bmatrix}$$

(Sect. 3.2)

Image The output of a function from a particular input. (Sect. 3.1; see also Sect. 9.1)

Imaginary part If a and b are real numbers, then a typical complex number has the form

$$z = a + ib$$

where i satisfies $i^2 = -1$. Here b is the imaginary part of z, denoted by $\text{Im}(z)$. (Sect. 6.5)

Inconsistent linear system A linear system that has no solutions. (Sect. 1.1)

Indefinite matrix A symmetric matrix A is indefinite if A is the matrix of an indefinite quadratic form. (Sect. 11.2)

Indefinite quadratic form Let $Q(\mathbf{x}) = \mathbf{x}^T A \mathbf{x}$ be a quadratic form. Then Q is indefinite if $Q(\mathbf{x})$ is positive for some $\mathbf{x}$'s in $\mathbf{R}^n$ and negative for others. (Sect. 11.1)

Initial state vector A vector with nonnegative entries that add to 1. This vector typically represents the initial probability distribution for a Markov chain. (Sect. 3.5)

Inner product (complex) Let $\mathbf{u}$, $\mathbf{v}$, and $\mathbf{w}$ be elements of a complex vector space V, and let c be a complex scalar. An inner product on V is a function denoted by $\langle \mathbf{u}, \mathbf{v} \rangle$ that takes any two vectors in V as input and produces a scalar as output. An inner product on a complex vector space satisfies the following conditions:

(a) $\langle \mathbf{u}, \mathbf{v} \rangle = \overline{\langle \mathbf{v}, \mathbf{u} \rangle}$

(b) $\langle \mathbf{u} + \mathbf{v}, \mathbf{w} \rangle = \langle \mathbf{u}, \mathbf{w} \rangle + \langle \mathbf{v}, \mathbf{w} \rangle$

(c) $\langle c\mathbf{u}, \mathbf{v} \rangle = c \langle \mathbf{u}, \mathbf{v} \rangle$

(d) $\langle \mathbf{u}, \mathbf{u} \rangle \geq 0$, and $\langle \mathbf{u}, \mathbf{u} \rangle = 0$ only when $\mathbf{u} = \mathbf{0}$

(Sect. 11.4)

Inner product (real) Let $\mathbf{u}$, $\mathbf{v}$, and $\mathbf{w}$ be elements of a vector space V, and let c be a scalar. An inner product on V is a function that takes two vectors in V as input and produces a scalar as output. An inner product function is denoted by $\langle \mathbf{u}, \mathbf{v} \rangle$, and satisfies the following conditions:

(a) $\langle \mathbf{u}, \mathbf{v} \rangle = \langle \mathbf{v}, \mathbf{u} \rangle$

(b) $\langle \mathbf{u} + \mathbf{v}, \mathbf{w} \rangle = \langle \mathbf{u}, \mathbf{w} \rangle + \langle \mathbf{v}, \mathbf{w} \rangle$

(c) $\langle c\mathbf{u}, \mathbf{v} \rangle = \langle \mathbf{u}, c\mathbf{v} \rangle = c \langle \mathbf{u}, \mathbf{v} \rangle$

(d) $\langle \mathbf{u}, \mathbf{u} \rangle \geq 0$, and $\langle \mathbf{u}, \mathbf{u} \rangle = 0$ only when $\mathbf{u} = \mathbf{0}$

(Sect. 10.1)

Inner product space A vector space V with an inner product defined on it. (Sect. 10.1)

Inverse linear transformation A linear transformation $T : \mathbf{R}^m \to \mathbf{R}^n$ is invertible if T is one-to-one and onto. When T is invertible, the inverse function $T^{-1} : \mathbf{R}^n \to \mathbf{R}^m$ is defined by

$$T^{-1}(\mathbf{y}) = \mathbf{x} \quad \text{if and only if} \quad T(\mathbf{x}) = \mathbf{y}$$

(Sect. 3.3; see also Sect. 9.2)

Invertible matrix An $n \times n$ matrix A is invertible if there exists an $n \times n$ matrix B such that $AB = I_n$. The matrix B is called the inverse of A and is denoted A^{-1}. (Sect. 3.3)

Isomorphic vector spaces V and W are isomorphic vector spaces if there exists an isomorphism $T : V \to W$. (Sect. 9.2)

Isomorphism A linear transformation $T : V \to W$ is an isomorphism if T is both one-to-one and onto. If such an isomorphism exists, then we say that V and W are isomorphic vector spaces. (Sect. 9.2)

Jacobi iteration An iterative method for approximating the solutions to a linear system that has the same number of equations as variables. (Sect. 1.3)

Kernel Given a linear transformation T, the set of all vectors $\mathbf{v}$ such that $T(\mathbf{v}) = \mathbf{0}$ is the kernel of T (denoted $\ker(T)$) and is a subspace of the domain of T. (Sect. 4.1; see also Sect. 9.1)

LDU factorization A variant of LU factorization, with $A = LDU$, where U is unit upper triangular, D is diagonal, and L is lower triangular with 1's on the diagonal. (Sect. 3.4)

Leading principal submatrix Let

$$A = \begin{bmatrix} a_{11} & a_{12} & \cdots & a_{1n} \\ a_{21} & a_{22} & \cdots & a_{2n} \\ \vdots & \vdots & \ddots & \vdots \\ a_{n1} & a_{n2} & \cdots & a_{nn} \end{bmatrix}$$

Then the leading principal submatrices of A are given by

$$A_1 = \begin{bmatrix} a_{11} \end{bmatrix}, \quad A_2 = \begin{bmatrix} a_{11} & a_{12} \\ a_{21} & a_{22} \end{bmatrix},$$

$$A_3 = \begin{bmatrix} a_{11} & a_{12} & a_{13} \\ a_{21} & a_{22} & a_{23} \\ a_{31} & a_{32} & a_{33} \end{bmatrix}$$

and so on through $A_n = A$. (Sect. 11.2)

Leading term In a matrix, the leading term for a row is the leftmost nonzero entry in that row. A row of zeros has no leading term. (Sect. 1.2)

Leading variable The leftmost variable in a linear equation that has a nonzero coefficient. In a linear system in echelon form, the leading variables are the leftmost variables in each equation. (Sect. 1.1)

Least squares solution If A is an $n \times m$ matrix and $\mathbf{y}$ is in $\mathbf{R}^n$, then a least squares solution to $A\mathbf{x} = \mathbf{y}$ is a vector $\hat{\mathbf{x}}$ in $\mathbf{R}^m$ such that

$$\| A\hat{\mathbf{x}} - \mathbf{y} \| \leq \| A\mathbf{x} - \mathbf{y} \|$$

for all $\mathbf{x}$ in $\mathbf{R}^m$. (Sect. 8.5)

Linear combination If $\mathbf{u}_1, \mathbf{u}_2, \ldots, \mathbf{u}_m$ are vectors and $c_1, c_2, \ldots, c_m$ are scalars, then

$$c_1\mathbf{u}_1 + c_2\mathbf{u}_2 + \cdots + c_m\mathbf{u}_m$$

is a linear combination of the vectors. Note that it is possible for scalars to be negative or equal to zero. (Sect. 2.1)

Linear dependence Let $\{\mathbf{u}_1, \mathbf{u}_2, \ldots, \mathbf{u}_m\}$ be a set of vectors. If the vector equation

$$x_1\mathbf{u}_1 + x_2\mathbf{u}_2 + \cdots + x_m\mathbf{u}_m = \mathbf{0}$$

has nontrivial solutions, then the set is linearly dependent. (Sect. 2.3; see also Sect. 7.2)

Linear equation An equation of the form

$$a_1x_1 + a_2x_2 + a_3x_3 + \cdots + a_nx_n = b$$

where $a_1, a_2, \cdots, a_n$ and b are constants and $x_1, x_2, \cdots, x_n$ are variables or unknowns. (Sect. 1.1)

Linear independence Let $\{\mathbf{u}_1, \mathbf{u}_2, \ldots, \mathbf{u}_m\}$ be a set of vectors. If the only solution to the vector equation

$$x_1\mathbf{u}_1 + x_2\mathbf{u}_2 + \cdots + x_m\mathbf{u}_m = \mathbf{0}$$

is the trivial solution $x_1 = x_2 = \cdots = x_m = 0$, then the set is linearly independent. (Sect. 2.3; see also Sect. 7.2)

Linear transformation A function $T : \mathbf{R}^m \rightarrow \mathbf{R}^n$ is a linear transformation if for all vectors $\mathbf{u}$ and $\mathbf{v}$ in $\mathbf{R}^m$ and all scalars r we have

(a) $T(\mathbf{u} + \mathbf{v}) = T(\mathbf{u}) + T(\mathbf{v})$

(b) $T(r\mathbf{u}) = rT(\mathbf{u})$

(Sect. 3.1; see also Sect. 9.1)

Lower triangular matrix An $n \times n$ matrix A is lower triangular if the terms above the diagonal are all zero,

$$A = \begin{bmatrix} a_{11} & 0 & 0 & \cdots & 0 \\ a_{21} & a_{22} & 0 & \cdots & 0 \\ a_{31} & a_{32} & a_{33} & \cdots & 0 \\ \vdots & \vdots & \vdots & \ddots & \vdots \\ a_{n1} & a_{n2} & a_{n3} & \cdots & a_{nn} \end{bmatrix}$$

(Sect. 3.2)

LU factorization If $A = LU$, where U is upper triangular and L is lower triangular with 1's on the diagonal, then the product is called an LU factorization of A. (Sect. 3.4)

Markov chain A sequence of state vectors $\mathbf{x}_0, \mathbf{x}_1, \ldots$ generated recursively by $\mathbf{x}_{i+1} = A\mathbf{x}_i$, where A is a transition matrix. (Sect. 3.5)

Matrix A rectangular table of numbers, upon which various algebraic operations are defined and can be performed. The plural of matrix is matrices. (Sect. 1.2)

Matrix addition The component-wise rule for adding one matrix to another of identical dimension. (Sect. 3.2)

Matrix dimensions The number of rows and columns for a matrix. Generally displayed as $n \times m$, where n is the number of rows and m the number of columns. (Sect. 3.1)

Matrix multiplication The rules for multiplying two matrices to produce a new matrix. If A is $n \times m$ and B is $r \times s$, then the product AB is defined when $m = r$. If this is true, then $n \times s$ are the dimensions of AB. (Sect. 3.2)

Matrix of a linear transformation Let $T : V \rightarrow W$ be a linear transformation, $\mathcal{G} = \{\mathbf{g}_1, \ldots, \mathbf{g}_m\}$ a basis of V, and $\mathcal{Q} = \{\mathbf{q}_1, \ldots, \mathbf{q}_n\}$ a basis of W. If $A = \begin{bmatrix} \mathbf{a}_1 & \cdots & \mathbf{a}_m \end{bmatrix}$ with

$$\mathbf{a}_i = [T(\mathbf{g}_i)]_{\mathcal{Q}}$$

for each $i = 1, \ldots, m$, then A is the matrix of T with respect to $\mathcal{G}$ and $\mathcal{Q}$. (Sect. 9.3)

Matrix of a quadratic form An $n \times n$ matrix A used to define the quadratic form

$$Q(\mathbf{x}) = \mathbf{x}^T A\mathbf{x}$$

(Sect. 11.1)

Matrix powers If A is an $n \times n$ matrix, then

$$A^k = \underbrace{A \cdot A \cdots A}_{k \text{ terms}}$$

(Sect. 3.2)

Matrix-vector multiplication Let $\mathbf{a}_1, \mathbf{a}_2, \ldots, \mathbf{a}_m$ be vectors in $\mathbf{R}^n$. If

$$A = \begin{bmatrix} \mathbf{a}_1 & \mathbf{a}_2 & \cdots & \mathbf{a}_m \end{bmatrix} \quad \text{and} \quad \mathbf{x} = \begin{bmatrix} x_1 \\ x_2 \\ \vdots \\ x_m \end{bmatrix}$$

then $A\mathbf{x} = x_1\mathbf{a}_1 + x_2\mathbf{a}_2 + \cdots + x_m\mathbf{a}_m$. (Sect. 2.2)

Minor If A is an $n \times n$ matrix, let M_{ij} denote the $(n-1) \times (n-1)$ matrix that we get from A after deleting the row and column containing a_{ij}. The determinant $\det(M_{ij})$ is the minor of a_{ij}. (Sect. 5.1)

Modulus The modulus of a complex number $z = a + ib$ is given by

$$|z| = \sqrt{a^2 + b^2}$$

(Sect. 6.5)

Multiplicity of a root Given a polynomial $P(x)$, a root α of $P(x) = 0$ has multiplicity r if $P(x) = (x-\alpha)^r Q(x)$ with $Q(\alpha) \neq 0$. (Sect. 6.1)

Negative definite matrix A symmetric matrix A is negative definite if A is the matrix of a negative definite quadratic form. (Sect. 11.2)

Negative definite quadratic form Let $Q(\mathbf{x}) = \mathbf{x}^T A\mathbf{x}$ be a quadratic form. Then Q is negative definite if $Q(\mathbf{x}) < 0$

for all nonzero vectors $\mathbf{x}$ in $\mathbf{R}^n$, and Q is negative semidefinite if $Q(\mathbf{x}) \leq 0$ for all $\mathbf{x}$ in $\mathbf{R}^n$. (Sect. 11.1)

Nonhomogeneous linear system A linear system of the form $A\mathbf{x} = \mathbf{b}$ where $\mathbf{b} \neq \mathbf{0}$. (Sect. 2.3)

Nonsingular matrix A square matrix that has an inverse. (Sect. 3.3)

Nontrivial solution A solution to a homogeneous linear equation (or homogeneous linear system) where some of the variables are nonzero. Although all homogeneous linear equations (systems) have the trivial solution, not all have nontrivial solutions. (Sect. 1.2)

Norm of a vector Let $\mathbf{x}$ be a vector in $\mathbf{R}^n$. Then the norm (or length) of $\mathbf{x}$ is given by

$$\|\mathbf{x}\| = \sqrt{\mathbf{x} \cdot \mathbf{x}}$$

If $\mathbf{x}$ is in an inner product space, then the norm is given by

$$\|\mathbf{x}\| = \sqrt{\langle \mathbf{x}, \mathbf{x} \rangle}$$

(Sect. 8.1; see also Sect. 10.1)

Normal equations Given a linear system $A\mathbf{x} = \mathbf{y}$, the normal equations for this system are

$$A^T A \mathbf{x} = A^T \mathbf{y}$$

The set of solutions to the normal equations is the same as the set of least squares solutions to $A\mathbf{x} = \mathbf{y}$. If A has linearly independent columns, then there is a unique least squares solution given by

$$\hat{\mathbf{x}} = \left(A^T A\right)^{-1} A^T \mathbf{y}$$

(Sect. 8.5)

Normal matrix A complex matrix A is normal if

$$A^* A = A A^*$$

All unitary matrices are normal, but the reverse is not true. (Sect. 11.5)

Null space If A is an $n \times m$ matrix, then the set of solutions to $A\mathbf{x} = \mathbf{0}$ is called the null space of A and is denoted by null(A). It is a subspace of $\mathbf{R}^m$. (Sect. 4.1)

Nullity If A is an $n \times m$ matrix, then the nullity of A (denoted nullity(A)) is the dimension of null(A). (Sect. 4.2)

One-to-one A linear transformation $T : \mathbf{R}^m \to \mathbf{R}^n$ is one-to-one if for every vector $\mathbf{w}$ in $\mathbf{R}^n$ there exists *at most* one vector $\mathbf{u}$ in $\mathbf{R}^m$ such that $T(\mathbf{u}) = \mathbf{w}$. Alternate definition: A linear transformation T is one-to-one if $T(\mathbf{u}) = T(\mathbf{v})$ implies that $\mathbf{u} = \mathbf{v}$. (Sect. 3.1; see also Sect. 9.1)

Onto A linear transformation $T : \mathbf{R}^m \to \mathbf{R}^n$ is onto if for every vector $\mathbf{w}$ in $\mathbf{R}^n$ there exists *at least* one vector $\mathbf{u}$ in $\mathbf{R}^m$ such that $T(\mathbf{u}) = \mathbf{w}$. (Sect. 3.1; see also Sect. 9.1)

Orthogonal basis A basis is orthogonal if it is an orthogonal set. (Sect. 8.1; see also Sect. 10.2)

Orthogonal complement Let S be a subspace. A vector $\mathbf{u}$ is orthogonal to S if $\mathbf{u}$ is orthogonal to every vector $\mathbf{s}$ in S. The set of all such vectors $\mathbf{u}$ is called the orthogonal complement of S and is denoted by $S^\perp$. (Sect. 8.1; see also Sect. 10.1)

Orthogonal matrix A square matrix is orthogonal if the columns form an orthonormal set. (Sect. 8.3)

Orthogonal set A set of vectors is orthogonal if each pair of distinct vectors is orthogonal to each other. (Sect. 8.1; see also Sect. 10.2)

Orthogonal vectors Vectors $\mathbf{u}$ and $\mathbf{v}$ in $\mathbf{R}^n$ are orthogonal if $\mathbf{u} \cdot \mathbf{v} = 0$. If $\mathbf{u}$ and $\mathbf{v}$ are in an inner product space, then they are orthogonal if $\langle \mathbf{u}, \mathbf{v} \rangle = 0$. (Sect. 8.1; see also Sect. 10.1)

Orthogonally diagonalizable matrix A square matrix A is orthogonally diagonalizable if there exists an orthogonal matrix P and a diagonal matrix D such that $A = PDP^{-1}$. (Sect. 8.3)

Orthonormal basis A basis is orthonormal if it forms an orthonormal set. (Sect. 8.2; see also Sect. 10.2)

Orthonormal set A set of vectors $\{\mathbf{w}_1, \ldots, \mathbf{w}_k\}$ is orthonormal if the set is orthogonal and $\|\mathbf{w}_j\| = 1$ for each of $j = 1, 2, \ldots k$. (Sect. 8.2; see also Sect. 10.2)

Parallelogram Rule A geometric interpretation of vector addition that involves viewing vectors as two of the four sides of a parallelogram. (Sect. 2.1)

Partial pivoting An additional step in Gaussian elimination (or Gauss–Jordan elimination) where a row interchange is performed to move the largest term (in absolute value) for a column into the pivot position. This is done to reduce roundoff error. (Sect. 1.3)

Particular solution Any specific solution to a linear system $A\mathbf{x} = \mathbf{b}$. (Sect. 2.3)

Partitioned matrix A matrix that has been divided into smaller submatrices. (Sect. 3.2)

Pivot columns The columns containing pivot positions for a matrix in echelon form. (Sect. 1.2)

Pivot position For a matrix in echelon form, the pivot positions are the locations of the leading terms. (Sect. 1.2)

Pivot rows The rows containing pivot positions for a matrix in echelon form. (Sect. 1.2)

Positive definite matrix A symmetric matrix A is positive definite if A is the matrix of a positive definite quadratic form. (Sect. 11.2)

Positive definite quadratic form Let $Q(\mathbf{x}) = \mathbf{x}^T A \mathbf{x}$ be a quadratic form. Then Q is positive definite if $Q(\mathbf{x}) > 0$ for all nonzero vectors $\mathbf{x}$ in $\mathbf{R}^n$, and Q is positive semidefinite if $Q(\mathbf{x}) \geq 0$ for all $\mathbf{x}$ in $\mathbf{R}^n$. (Sect. 11.1)

Power method An iterative method for approximating an eigenvector and corresponding eigenvalue for a square matrix. (Sect. 6.2)

Principal axes The normalized orthogonal eigenvectors of a symmetric matrix A used to define a quadratic form. (Sect. 11.1)

Probability vector A vector with nonnegative entries that add to 1. These vectors are encountered in the context of Markov chains. (Sect. 3.5)

Projection onto subspaces Let S be a subspace with orthogonal basis $\{\mathbf{v}_1, \ldots, \mathbf{v}_k\}$. Then the projection of $\mathbf{u}$ onto S is given by

$$\text{proj}_S \mathbf{u} = \frac{\mathbf{v}_1 \cdot \mathbf{u}}{\|\mathbf{v}_1\|^2} \mathbf{v}_1 + \frac{\mathbf{v}_2 \cdot \mathbf{u}}{\|\mathbf{v}_2\|^2} \mathbf{v}_2 + \cdots + \frac{\mathbf{v}_k \cdot \mathbf{u}}{\|\mathbf{v}_k\|^2} \mathbf{v}_k$$

For an inner product space, the dot products are replaced by inner products in the definition. (Sect. 8.2; see also Sect. 10.2)

Projection onto vectors Let $\mathbf{u}$ and $\mathbf{v}$ be vectors in $\mathbf{R}^n$ with $\mathbf{v}$ nonzero. Then the projection of $\mathbf{u}$ onto $\mathbf{v}$ is given by

$$\text{proj}_{\mathbf{v}} \mathbf{u} = \frac{\mathbf{v} \cdot \mathbf{u}}{\|\mathbf{v}\|^2} \mathbf{v}$$

If $\mathbf{u}$ and $\mathbf{v}$ are in an inner product space, then the projection of $\mathbf{u}$ onto $\mathbf{v}$ is given by

$$\text{proj}_{\mathbf{v}} \mathbf{u} = \frac{\langle \mathbf{v}, \mathbf{u} \rangle}{\|\mathbf{v}\|^2} \mathbf{v}$$

(Sect. 8.2; see also Sect. 10.1)

QR factorization The QR factorization of an $n \times m$ matrix A is given by $A = QR$, where Q is $n \times m$ with orthonormal columns and R is $m \times m$, upper triangular, and has positive diagonal terms. (Sect. 8.3)

Quadratic form A quadratic form is a function $Q : \mathbf{R}^n \to \mathbf{R}$ that has the form

$$Q(\mathbf{x}) = \mathbf{x}^T A \mathbf{x}$$

where A is an $n \times n$ symmetric matrix called the matrix of the quadratic form. (Sect. 11.1)

$\mathbf{R}^n$ The set of all vectors with n real numbers for components. (Sect. 2.1)

Range The set of outputs for a function. (Sect. 3.1; see also Sect. 9.1)

Rank of a matrix The dimension of the row space of a matrix A, or the dimension of the column space of A, which is the same. (Sect. 4.3)

Rank–Nullity theorem Given an $n \times m$ matrix A, the Rank–Nullity theorem states that

$$\text{rank}(A) + \text{nullity}(A) = m$$

(Sect. 4.3)

Real part If a and b are real numbers, then a typical complex number has the form

$$z = a + ib$$

where i satisfies $i^2 = -1$. Here a is called the real part of z, denoted by $\text{Re}(z)$. (Sect. 6.5)

Reduced echelon form Also called row reduced echelon form, a matrix in this form is in echelon form and each pivot column consists entirely of zeros except for in the pivot position, which contains a 1. (Sect. 1.2)

Regular matrix A stochastic matrix A is regular if for some integer $k \geq 1$ the matrix A^k has all strictly positive entries. (Sect. 3.5)

Row space Let A be an $n \times m$ matrix. The row space of A is the subspace of $\mathbf{R}^m$ spanned by the row vectors of A and is denoted by $\text{row}(A)$. (Sect. 4.3)

Row vector A vector in Euclidean space expressed in the form of a horizontal n-tuple. This term is also used to indicate a vector formed from a row of a matrix A. (Sect. 2.1)

Scalar A real number when viewed as a multiple of a vector. (Sect. 2.1)

Scalar multiplication (matrices) The multiplication of each term of a matrix by a real number. (Sect. 3.2)

Scalar multiplication (vectors) The multiplication of each component of a vector by a real number. (Sect. 2.1)

Similar matrices A square matrix A is similar to matrix B if there exists an invertible matrix S such that $A = S^{-1} B S$. The change from B to A is called a similarity transformation. (Sect. 9.4)

Singular matrix A square matrix that does not have an inverse. (Sect. 3.3)

Singular value decomposition Suppose that A is an $n \times m$ matrix. If $n \geq m$, then the singular value decomposition is the factorization of A as the product

$A = U \Sigma V^T$, where

- U is an $n \times n$ orthogonal matrix.

- Σ is an $n \times m$ matrix of the form $\Sigma = \begin{bmatrix} D \\ 0_{(n-m)m} \end{bmatrix}$, where D is a diagonal matrix with

$$D = \begin{bmatrix} \sigma_1 & 0 & \cdots & 0 \\ 0 & \sigma_2 & \cdots & 0 \\ \vdots & \vdots & \ddots & \vdots \\ 0 & 0 & \cdots & \sigma_m \end{bmatrix}$$

and $\sigma_1 \geq \sigma_2 \geq \cdots \geq \sigma_m \geq 0$ are the singular values of A. The singular values are given by $\sigma_i = \sqrt{\lambda_i}$, where λ_i is an eigenvalue of $A^T A$.

- V is an $m \times m$ orthogonal matrix.

If $n < m$, then $\Sigma = \begin{bmatrix} D & 0_{n(m-n)} \end{bmatrix}$ with everything else the same. (Sect. 8.4)

Skew symmetric matrix A square matrix A is skew symmetric if $A^T = -A$. (Sect. 3.2)

Solution, linear equation An n-tuple $(s_1, \ldots, s_n)$ that satisfies a linear equation. (Sect. 1.1)

Solution, linear system An n-tuple $(s_1, \ldots, s_n)$ that satisfies a linear system. (Sect. 1.1)

Solution set The set of all solutions to a linear equation or a linear system. (Sect. 1.1)

Span Let $\{\mathbf{u}_1, \ldots, \mathbf{u}_m\}$ be a set of vectors. The span of this set is denoted $\mathrm{span}\{\mathbf{u}_1, \mathbf{u}_2, \ldots, \mathbf{u}_m\}$ and is defined to be the set of all linear combinations

$$x_1 \mathbf{u}_1 + x_2 \mathbf{u}_2 + \cdots + x_m \mathbf{u}_m$$

where $x_1, x_2, \ldots, x_m$ can be any real numbers. (Sect. 2.2; see also Sect. 7.2)

Square matrix A matrix with the same number of rows and columns. (Sect. 3.1)

Standard basis ($\mathbf{R}^n$) Given by the vectors

$$\mathbf{e}_1 = \begin{bmatrix} 1 \\ 0 \\ \vdots \\ 0 \end{bmatrix}, \quad \mathbf{e}_2 = \begin{bmatrix} 0 \\ 1 \\ \vdots \\ 0 \end{bmatrix}, \quad \cdots, \quad \mathbf{e}_n = \begin{bmatrix} 0 \\ 0 \\ \vdots \\ 1 \end{bmatrix}$$

State vector A probability vector that is part of the sequence of vectors in a Markov chain. (Sect. 3.5)

Steady-state vector A state vector $\mathbf{x}$ that satisfies $A\mathbf{x} = \mathbf{x}$, where A is a transition matrix for a Markov chain. (Sect. 3.5)

Stochastic matrix A square matrix A that has nonnegative entries and columns that each add to 1. (Sect. 3.5)

Subspace A subset S of vectors is a subspace if S satisfies the following three conditions: (1) S contains $\mathbf{0}$, the zero vector, (2) if $\mathbf{u}$ and $\mathbf{v}$ are in S, then $\mathbf{u} + \mathbf{v}$ is also in S, and (3) if r is a real number and $\mathbf{u}$ is in S, then $r\mathbf{u}$ is also in S. (Sect. 4.1; see also Sect. 7.1)

Subspace spanned If $S = \mathrm{span}\{\mathbf{u}_1, \mathbf{u}_2, \ldots, \mathbf{u}_m\}$, then S is the subspace spanned (or subspace generated) by $\{\mathbf{u}_1, \mathbf{u}_2, \ldots, \mathbf{u}_m\}$. (Sect. 4.1)

Symmetric matrix A matrix A is symmetric if $A^T = A$. (Sect. 3.2)

System of linear equations A collection of linear equations of the form

$$\begin{aligned} a_{11}x_1 + a_{12}x_2 + a_{13}x_3 + \cdots + a_{1n}x_n &= b_1 \\ a_{21}x_1 + a_{22}x_2 + a_{23}x_3 + \cdots + a_{2n}x_n &= b_2 \\ a_{31}x_1 + a_{32}x_2 + a_{33}x_3 + \cdots + a_{3n}x_n &= b_3 \\ &\vdots \\ a_{m1}x_1 + a_{m2}x_2 + a_{m3}x_3 + \cdots + a_{mn}x_n &= b_m \end{aligned}$$

(Sect. 1.1)

System of linear first-order differential equations The general form for such a system is

$$\begin{aligned} y_1' &= a_{11}y_1 + a_{12}y_2 + a_{13}y_3 + \cdots + a_{1n}y_n \\ y_2' &= a_{21}y_1 + a_{22}y_2 + a_{23}y_3 + \cdots + a_{2n}y_n \\ y_3' &= a_{31}y_1 + a_{32}y_2 + a_{33}y_3 + \cdots + a_{3n}y_n \\ &\vdots \\ y_n' &= a_{n1}y_1 + a_{n2}y_2 + a_{n3}y_3 + \cdots + a_{nn}y_n \end{aligned}$$

Here we assume that $y_1 = y_1(t), \ldots, y_n = y_n(t)$ are each differentiable functions. The system is linear because the functions are linearly related, and it is first-order because only the first derivative appears. (Sect. 6.6)

Theorem A mathematical statement that has been rigorously proved to be true. (Sect. 1.1)

Tip-to-Tail rule A geometric interpretation of vector addition that involves translating one vector so that its initial point (the tail) is situated at the end point (the tip) of the other. (Sect. 2.1)

Transformation matrix Another term for the matrix of a linear transformation. (Sect. 9.3)

Transition matrix A stochastic matrix A used to proceed from one state vector to the next in a Markov chain via the relationship $\mathbf{x}_{i+1} = A\mathbf{x}_i$. (Sect. 3.5)

Transpose The transpose of a matrix A is denoted by A^T and results from interchanging the rows and columns of A. (Sect. 3.2)

Triangular form A linear system of the form

$$a_{11}x_1 + a_{12}x_2 + a_{13}x_3 + \cdots + a_{1n}x_n = b_1$$
$$a_{22}x_2 + a_{23}x_3 + \cdots + a_{2n}x_n = b_2$$
$$a_{33}x_3 + \cdots + a_{3n}x_n = b_3$$
$$\ddots \qquad \vdots \qquad \vdots$$
$$a_{nn}x_n = b_n$$

where $a_{11}, a_{22}, \ldots, a_{nn}$ are all nonzero. Also called a triangular system. (Sect. 1.1)

Trivial solution The solution to a homogeneous linear equation (or homogeneous linear system) where all variables are set equal to zero. (Sect. 1.2)

Trivial subspaces $S = \{0\}$ and $S = \mathbf{R}^n$ are the trivial subspaces of $\mathbf{R}^n$. For a vector space V, the trivial subspaces are $S = \{0\}$ and $S = V$. (Sect. 4.1; see also Sect. 7.1)

Unitarily diagonalizable A complex matrix A is unitarily diagonalizable if there exist a diagonal matrix D and a unitary matrix P such that

$$A = PDP^{-1} = PDP^*$$

As with real matrices, the diagonal entries of D are the eigenvalues of A, and the columns of P are the corresponding eigenvectors. (Sect. 11.5)

Unitary matrix The counterpart of orthogonal matrices for a matrix A with complex entries is called a unitary matrix, which requires that

$$A^{-1} = A^*$$

(Sect. 11.5)

Unitary space A complex vector space V with an inner product defined on it. (Also called a complex inner product space.) (Sect. 11.4)

Upper triangular matrix An $n \times n$ matrix A is upper triangular if it has the form

$$A = \begin{bmatrix} a_{11} & a_{12} & a_{13} & \cdots & a_{1n} \\ 0 & a_{22} & a_{23} & \cdots & a_{2n} \\ 0 & 0 & a_{33} & \cdots & a_{3n} \\ \vdots & \vdots & \vdots & \ddots & \vdots \\ 0 & 0 & 0 & \cdots & a_{nn} \end{bmatrix}$$

That is, A is upper triangular if the entries below the diagonal are all zero. (Sect. 3.2)

VecMobile An imaginary vehicle created to illustrate the notion of span. (Sect. 2.2)

Vector In Euclidean space, an ordered list of real numbers usually presented in a vertical column. In general, a vector can be any number of different mathematical objects, including matrices and continuous functions. (Sect. 2.1; see also Sect. 7.1)

Vector arithmetic The "standard" definition of equality, addition, and scalar multiplication as it is applied to vectors in Euclidean space. (Sect. 2.1)

Vector form A specific way to describe the general solution to a linear system using a linear combination of vectors. (Sect. 2.1)

Vector space A vector space consists of a nonempty set V of vectors together with operations of addition and scalar multiplication on the vectors that satisfy each of the following:

(1) If $\mathbf{v}_1$ and $\mathbf{v}_2$ are in V, then so is $\mathbf{v}_1 + \mathbf{v}_2$. Hence V is closed under addition.

(2) If c is a real scalar and $\mathbf{v}$ is in V, then so is $c\mathbf{v}$. Hence V is closed under scalar multiplication.

(3) There exists a zero vector $\mathbf{0}$ in V such that $\mathbf{0}+\mathbf{v} = \mathbf{v}$ for all $\mathbf{v}$ in V.

(4) For each $\mathbf{v}$ in V there exists an additive inverse (or opposite) vector $-\mathbf{v}$ in V such that $\mathbf{v} + (-\mathbf{v}) = \mathbf{0}$ for all $\mathbf{v}$ in V.

(5) For all $\mathbf{v}_1$, $\mathbf{v}_2$, and $\mathbf{v}_3$ in V and real scalars c_1 and c_2, we have

   (a) $\mathbf{v}_1 + \mathbf{v}_2 = \mathbf{v}_2 + \mathbf{v}_1$
   (b) $(\mathbf{v}_1 + \mathbf{v}_2) + \mathbf{v}_3 = \mathbf{v}_1 + (\mathbf{v}_2 + \mathbf{v}_3)$
   (c) $c_1(\mathbf{v}_1 + \mathbf{v}_2) = c_1\mathbf{v}_1 + c_1\mathbf{v}_2$
   (d) $(c_1 + c_2)\mathbf{v}_1 = c_1\mathbf{v}_1 + c_2\mathbf{v}_1$
   (e) $(c_1 c_2)\mathbf{v}_1 = c_1(c_2\mathbf{v}_1)$
   (f) $1 \cdot \mathbf{v}_1 = \mathbf{v}_1$

(Sect. 7.1)

Zero column A matrix column consisting entirely of zeros. (Sect. 1.2)

Zero row A matrix row consisting entirely of zeros. (Sect. 1.2)

Zero vector In $\mathbf{R}^n$, a vector with zeros for each component. In a vector space, a vector that is the counterpart to 0 in the real numbers. (Sect. 2.1)

ANSWERS TO SELECTED EXERCISES

Chapter 1

Section 1.1

1. Only $(-3, -3)$ lies on line.

3. Only $(-2, 5)$ lies on both lines.

5. None satisfies the linear system.

7. Only (b), (c), and (d) are solutions to the linear system.

9. $x_1 = 3, x_2 = -1$.

11. $x_1 = s_1, x_2 = \frac{1}{2} + \frac{5}{2}s_1$.

13. $x_1 = -\frac{8}{41}, x_2 = -\frac{5}{41}$.

15. Echelon form; x_1, x_2 leading variables, no free variables.

17. Echelon form; x_1, x_3 leading variables, x_2 a free variable.

19. Not in echelon form.

21. Echelon form; x_1, x_3 leading variables, x_2, x_4 free variables.

23. $x_1 = -\frac{19}{5}, x_2 = 5$.

25. $x_1 = -\frac{2}{3} + \frac{4}{3}s_1, x_2 = s_1$.

27. $x_1 = 10 - \frac{1}{2}s_1, x_2 = -2 + \frac{1}{2}s_1, x_3 = s_1, x_4 = 5$.

29. $x_1 = \frac{5}{6} + \frac{1}{2}s_1 + \frac{1}{3}s_2, x_2 = s_1, x_3 = \frac{4}{3} + \frac{1}{3}s_2, x_4 = s_2$.

31. Reverse order of equations; $x_1 = \frac{13}{15}, x_2 = -\frac{4}{5}$.

33. Reverse order of equations; $x_1 = -1 + \frac{7}{2}s_1 - \frac{19}{2}s_2$,
$x_2 = -\frac{1}{2}s_1 + \frac{5}{2}s_2, x_3 = s_1, x_4 = s_2$.

35. $k \neq -\frac{15}{2}$.

37. $h = 2, k \neq -1$.

39. 9 variables.

41. 7 leading variables.

43. For example,
$$x_1 = 0$$
$$x_2 = 0$$
$$x_3 = 0$$

45. For example,
$$x_1 + x_2 \qquad = 0$$
$$x_1 + x_2 - x_3 = 0$$
$$x_3 = 0$$
$$x_1 + x_2 + x_3 = 0$$

47. On Monday, I bought 3 apples and 4 oranges and spent $0.55. On Tuesday I bought 6 oranges and spent $0.60. How much does each apple and orange cost?

Answer: apples cost 5 cents each and oranges cost 10 cents each.

49. For example,
$$x_1 - x_2 \qquad = -3$$
$$3x_1 \qquad - x_3 = 4$$

51. False

53. False

55. True

57. False

59. True

61. 196.875 liters of 18% solution, 103.125 liters of 50% solution.

63. 298 adults and 87 children.

65. $a_1 = \frac{14}{5}$ and $a_2 = \frac{11}{5}$

67. $a = \frac{5}{9}$ and $b = -\frac{160}{9}$

69. The published values from the United States Mint are $q = 0.955$ in and $n = 0.835$ in.

71. $x_1 = 12, x_2 = 5$

73. $x_1 = \frac{33}{8}s_1, x_2 = \frac{9}{4}s_1 - \frac{23}{11}, x_3 = s_1 - \frac{5}{33}$

75. $x_1 = \frac{47}{8}s_1, x_2 = -2s_1 + \frac{69}{47}, x_3 = \frac{7}{4}s_1 + \frac{565}{141}, x_4 = s_1 + \frac{202}{141}$

Section 1.2

1. $\begin{aligned} 4x_1 + 2x_2 - \quad x_3 &= 2 \\ -x_1 \qquad\qquad + 5x_3 &= 7 \end{aligned}$

3. $\begin{aligned} 12x_2 - 3x_3 - 9x_4 &= 17 \\ -12x_1 + 5x_2 - 3x_3 + 11x_4 &= 0 \\ 6x_1 + 8x_2 + 2x_3 + 10x_4 &= -8 \\ 17x_1 \qquad\qquad\qquad + 13x_4 &= -1 \end{aligned}$

5. Echelon form.

7. Not in echelon form.

9. Echelon form.

11. $-2R_1 \Rightarrow R_1$

13. $-2R_2 + R_3 \Rightarrow R_3$

15. $R_1 \Leftrightarrow R_2, \begin{bmatrix} -1 & 4 & 3 \\ 3 & 7 & -2 \\ 5 & 0 & -3 \end{bmatrix}$

17. $2R_1 \Rightarrow R_1, \begin{bmatrix} 0 & 6 & -2 & 4 \\ -1 & -9 & 4 & 1 \\ 5 & 0 & 7 & 2 \end{bmatrix}$

19. $\begin{bmatrix} 2 & 1 & 1 \\ -4 & -1 & 3 \end{bmatrix}; \quad x_1 = -2, x_2 = 5$.

21. $\begin{bmatrix} -2 & 5 & -10 & 4 \\ 1 & -2 & 3 & -1 \\ 7 & -17 & 34 & -16 \end{bmatrix}; \quad x_1 = -12, x_2 = -10, x_3 = -3$.

23. $\begin{bmatrix} 2 & 2 & -1 & 8 \\ -1 & -1 & 0 & -3 \\ 3 & 3 & 1 & 7 \end{bmatrix}; \quad x_1 = 3 - s_1, x_2 = s_1, x_3 = -2$.

25. $\begin{bmatrix} 2 & 6 & -9 & -4 & 0 \\ -3 & -11 & 9 & -1 & 0 \\ 1 & 4 & -2 & 1 & 0 \end{bmatrix};$
$x_1 = 35s_1, x_2 = -8s_1, x_3 = 2s_1, x_4 = s_1$.

27. $\begin{bmatrix} -2 & -5 & 0 \\ 1 & 3 & 1 \end{bmatrix}$; $\quad x_1 = -5, x_2 = 2.$

29. $\begin{bmatrix} 2 & 1 & 0 & 2 \\ -1 & -1 & -1 & 1 \end{bmatrix}$; $\quad x_1 = 3 + s_1, x_2 = -4 - 2s_1, x_3 = s_1.$

31. $(1/5) R_1 \Rightarrow R_1$

33. $R_1 \Leftrightarrow R_3$

35. $5 R_2 + R_6 \Rightarrow R_6$

37. An example: $\begin{bmatrix} 1 & 1 & 1 & 1 & 1 \\ 0 & 1 & 1 & 1 & 1 \\ 0 & 0 & 1 & 1 & 1 \end{bmatrix}$

39. An example: $\begin{bmatrix} 1 & 0 & 0 & 4 \\ 0 & 1 & 0 & 3 \\ 0 & 0 & 1 & 2 \\ 0 & 0 & 0 & 1 \end{bmatrix}$

41. An example:

$$\begin{aligned} x_1 \quad\quad\quad &= 0 \\ x_2 \quad\quad &= 0 \\ x_3 + x_4 &= 0 \end{aligned}$$

43. True

45. False

47. False

49. True

51. Exactly one solution.

53. HINT: Show that the system must have at least one free variable.

55. HINT: Show that the system must have at least one free variable.

57. $f(x) = 2x^2 - 3x + 5$

59. $E(x) = -\frac{1}{10}x^2 + \frac{49}{5}x + 132$

61. $x_1 = -\frac{157}{181}, x_2 = \frac{20}{181}, x_3 = -\frac{58}{181}$

63. $x_1 = \frac{7}{9} - s_1, x_2 = -\frac{23}{9} - s_1, x_3 = -\frac{22}{27} + s_1, x_4 = s_1.$

65. No solutions.

67. $x_1 = \frac{46}{579}s_1, x_2 = -\frac{745}{579}s_1, x_3 = -\frac{2264}{579}s_1,$

$x_4 = -\frac{655}{386}s_1, x_5 = s_1.$

Section 1.3

1. $x_1 = 1, x_2 = 2.$

3. $x_1 = \frac{79}{49}, x_2 = \frac{22}{49}, x_3 = \frac{124}{49}.$

5. No partial pivot: $x_1 = -0.219, x_2 = 0.0425$
With partial pivot: $x_1 = -0.180, x_2 = 0.0424.$

7. No partial pivot: $x_1 = -0.407, x_2 = -0.757, x_3 = 0.0124$
With partial pivot: $x_1 = -0.392, x_2 = -0.755, x_3 = 0.0124.$

9.

n	x_1	x_2
0	0	0
1	-1.2	0.2
2	-1.12	0.56
3	-0.976	0.536

Exact solution: $x_1 = -1, x_2 = 0.5.$

11.

n	x_1	x_2	x_3
0	0	0	0
1	-1.3	2.3	2.6
2	-2.295	3.34	1.42
3	-2.156	3.185	0.805

Exact solution: $x_1 = -2, x_2 = 3, x_3 = 1.$

13.

n	x_1	x_2
0	0	0
1	-1.2	0.56
2	-0.976	0.4928
3	-1.0029	0.5009

Exact solution: $x_1 = -1, x_2 = 0.5.$

15.

n	x_1	x_2	x_3
0	0	0	0
1	-1.3	2.56	1.316
2	-2.013	3.0974	0.9584
3	-2.0042	2.9884	1.0038

Exact solution: $x_1 = -2, x_2 = 3, x_3 = 1.$

17. Not diagonally dominant. Not possible to reorder to obtain diagonal dominance.

19. Not diagonally dominant. Not possible to reorder to obtain diagonal dominance.

21. Jacobi iteration of given linear system:

n	x_1	x_2
0	0	0
1	-1	-1
2	-3	-3
3	-7	-7
4	-15	-15

Diagonally dominant system:

$$\begin{aligned} 2x_1 - x_2 &= 1 \\ x_1 - 2x_2 &= -1 \end{aligned}$$

Jacobi iteration of diagonally dominant system:

n	x_1	x_2
0	0	0
1	0.5	0.5
2	0.75	0.75
3	0.875	0.875
4	0.9375	0.9375

23. Jacobi iteration of given linear system:

n	x_1	x_2	x_3
0	0	0	0
1	-1	8	-0.3333
2	16.67	12.33	27
3	-111.3	-21.33	29.67
4	-192	624	2.778

Diagonally dominant system:

$$5x_1 + x_2 - 2x_3 = 8$$
$$2x_1 - 10x_2 + 3x_3 = -1$$
$$x_1 - 2x_2 + 5x_3 = -1$$

Jacobi iteration of diagonally dominant system:

n	x_1	x_2	x_3
0	0	0	0
1	1.6	0.1	−0.2
2	1.5	0.36	−0.48
3	1.336	0.256	−0.356
4	1.406	0.2604	−0.3648

25. Gauss–Seidel iteration of given linear system:

n	x_1	x_2
0	0	0
1	−1	−3
2	−7	−15
3	−31	−63
4	−127	−255

Diagonally dominant system:

$$2x_1 - x_2 = 1$$
$$x_1 - 2x_2 = -1$$

Gauss–Seidel iteration of diagonally dominant system:

n	x_1	x_2
0	0	0
1	0.5	0.75
2	0.875	0.9375
3	0.9688	0.9844
4	0.9922	0.9961

27. Gauss–Seidel iteration of given linear system:

n	x_1	x_2	x_3
0	0	0	0
1	−1	13	43.67
2	−193.3	1062	3669
3	-1.622×10^4	8.844×10^4	3.056×10^5
4	-1.351×10^6	7.367×10^6	2.546×10^7

Diagonally dominant system:

$$5x_1 + x_2 - 2x_3 = 8$$
$$2x_1 - 10x_2 + 3x_3 = -1$$
$$x_1 - 2x_2 + 5x_3 = -1$$

Gauss–Seidel iteration of diagonally dominant system:

n	x_1	x_2	x_3
0	0	0	0
1	1.6	0.42	−0.352
2	1.375	0.2694	−0.3673
3	1.399	0.2697	−0.3712
4	1.397	0.2679	−0.3723

29. $x_1 = -3$, $x_2 = 18$.

31. $x_1 = 27$, $x_2 = 52$.

Section 1.4

1. Minimum = 20 vehicles.

3. Minimum = 25 vehicles.

5. $H_2 + O \longrightarrow H_2O$

7. $4Fe + 3O_2 \longrightarrow 2Fe_2O_3$

9. $C_3H_8 + 5O_2 \longrightarrow 3CO_2 + 4H_2O$

11. $4KO_2 + 2CO_2 \longrightarrow 2K_2CO_3 + 3O_2$

13. $p = (0.19847)\, d^{1.5011}$

15. $p = (0.20120)\, d^{1.49835}$

17. $d = 0.045 s^2$

19. $A = 1$, $B = -1$.

21. $A = -1$, $B = -1$, $C = 1$.

23. $x = 4$

25. $y = -2x^2 + 3x - 5$

27. $g(x) = -x^4 + 2x^3 + x^2 - 3x + 5$

29. $f(x) = \frac{2}{3}e^{-2x} - \frac{5}{3}e^x + xe^x$

31. LAI $= 0.0001100\,(\text{USA}) + 0.0000586\,(\text{Harris}) + 0.0066823$
(Computer)

Chapter 2

Section 2.1

1. $\begin{bmatrix} 7 \\ -3 \\ -5 \end{bmatrix}$ 3. $\begin{bmatrix} -10 \\ -4 \\ 14 \end{bmatrix}$ 5. $\begin{bmatrix} -5 \\ -4 \\ 4 \end{bmatrix}$

7. $3x_1 - x_2 = 8$
$2x_1 + 5x_2 = 13$

9. $-6x_1 + 5x_2 = 4$
$5x_1 - 3x_2 + 2x_3 = 16$

11. $x_1 \begin{bmatrix} 2 \\ -1 \end{bmatrix} + x_2 \begin{bmatrix} 8 \\ -3 \end{bmatrix} + x_3 \begin{bmatrix} -4 \\ 5 \end{bmatrix} = \begin{bmatrix} -10 \\ 4 \end{bmatrix}$

13. $x_1 \begin{bmatrix} 1 \\ -2 \\ -3 \end{bmatrix} + x_2 \begin{bmatrix} -1 \\ 2 \\ -3 \end{bmatrix} + x_3 \begin{bmatrix} -3 \\ 6 \\ 10 \end{bmatrix} + x_4 \begin{bmatrix} -1 \\ 2 \\ 0 \end{bmatrix} = \begin{bmatrix} -1 \\ -1 \\ 5 \end{bmatrix}$

15. $\begin{bmatrix} x_1 \\ x_2 \end{bmatrix} = \begin{bmatrix} -4 \\ 0 \end{bmatrix} + s_1 \begin{bmatrix} 3 \\ 1 \end{bmatrix}$

17. $\begin{bmatrix} x_1 \\ x_2 \\ x_3 \\ x_4 \end{bmatrix} = \begin{bmatrix} 4 \\ 0 \\ -9 \\ 0 \end{bmatrix} + s_1 \begin{bmatrix} 6 \\ 0 \\ 3 \\ 1 \end{bmatrix} + s_2 \begin{bmatrix} -5 \\ 1 \\ 0 \\ 0 \end{bmatrix}$

19. $1\mathbf{u} + 0\mathbf{v} = \mathbf{u} = \begin{bmatrix} 3 \\ -2 \end{bmatrix}$, $0\mathbf{u} + 1\mathbf{v} = \mathbf{v} = \begin{bmatrix} -1 \\ -4 \end{bmatrix}$,

$1\mathbf{u} + 1\mathbf{v} = \begin{bmatrix} 3 \\ -2 \end{bmatrix} + \begin{bmatrix} -1 \\ -4 \end{bmatrix} = \begin{bmatrix} 2 \\ -6 \end{bmatrix}$

21. $1\mathbf{u} + 0\mathbf{v} + 0\mathbf{w} = \mathbf{u} = \begin{bmatrix} -4 \\ 0 \\ -3 \end{bmatrix}$, $0\mathbf{u} + 1\mathbf{v} + 0\mathbf{w} = \mathbf{v} = \begin{bmatrix} -2 \\ -1 \\ 5 \end{bmatrix}$,

$0\mathbf{u} + 0\mathbf{v} + 1\mathbf{w} = \mathbf{w} = \begin{bmatrix} 9 \\ 6 \\ 11 \end{bmatrix}$

23. $a = 2$ and $b = 7$

25. $a = 3$, $b = 5$, and $c = 7$

27. $\mathbf{b} = 3\mathbf{a}_1 + 2\mathbf{a}_2$

29. $\mathbf{b} = 3\mathbf{a}_1 + 4\mathbf{a}_2$

31. 76 pounds of nitrogen, 31 pounds of phosphoric acid, and 14 pounds of potash.

33. Two bags of Vigoro and three bags of Parker's.

35. Three bags of Vigoro and two bags of Parker's.

37. No solution possible.

39. No solution possible.

41. Two cans of Red Bull and one can of Jolt Cola.

43. Two cans of Red Bull and two cans of Jolt Cola.

45. Three servings of Lucky Charms and five servings of Raisin Bran.

47. Two servings of Lucky Charms and three servings of Raisin Bran.

49. (a) $\mathbf{a} = \begin{bmatrix} 2000 \\ 8000 \end{bmatrix}$, $\mathbf{b} = \begin{bmatrix} 3000 \\ 10,000 \end{bmatrix}$

(b) $8\mathbf{b} = (8)\begin{bmatrix} 3000 \\ 10,000 \end{bmatrix} = \begin{bmatrix} 24,000 \\ 80,000 \end{bmatrix}$.

The company produces 24,000 computer monitors and 80,000 flat panel televisions at facility B in 8 weeks.

(c) 30,000 computer monitors and 108,000 flat panel televisions.

(d) 9 weeks of production at facility A and 2 weeks of production at facility B.

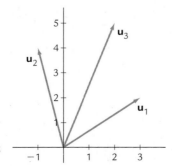

51. $\bar{\mathbf{v}} = \begin{bmatrix} \frac{8}{5} \\ \frac{16}{5} \end{bmatrix}$;

53. 6kg of $\mathbf{u}_1$, 3kg of $\mathbf{u}_2$, 2kg of $\mathbf{u}_3$

55. For example, $\mathbf{u} = (0, 0, -1)$ and $\mathbf{v} = (3, 2, 0)$.

57. For example, $\mathbf{u} = (1, 0, 0)$, $\mathbf{v} = (1, 0, 0)$, and $\mathbf{w} = (-2, 0, 0)$.

59. For example, $\mathbf{u} = (1, 0)$ and $\mathbf{v} = (2, 0)$.

61. $x_1 = 3$ and $x_2 = -2$

63. True

65. True

67. False

69. True

71. False

75.

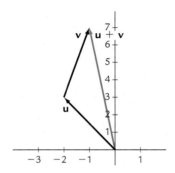

77.

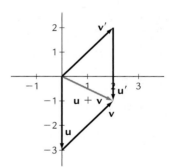

79.

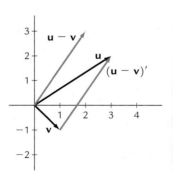

81. $x_1 = 4$, $x_2 = -6.5$, and $x_3 = 1$.

Section 2.2

1. $0\mathbf{u}_1 + 0\mathbf{u}_2 = \begin{bmatrix} 0 \\ 0 \end{bmatrix}$, $1\mathbf{u}_1 + 0\mathbf{u}_2 = \begin{bmatrix} 2 \\ 6 \end{bmatrix}$, $0\mathbf{u}_1 + 1\mathbf{u}_2 = \begin{bmatrix} 9 \\ 15 \end{bmatrix}$

3. $0\mathbf{u}_1 + 0\mathbf{u}_2 = \begin{bmatrix} 0 \\ 0 \\ 0 \end{bmatrix}, 1\mathbf{u}_1 + 0\mathbf{u}_2 = \begin{bmatrix} 2 \\ 5 \\ -3 \end{bmatrix}, 0\mathbf{u}_1 + 1\mathbf{u}_2 = \begin{bmatrix} 1 \\ 0 \\ 4 \end{bmatrix}$

5. $0\mathbf{u}_1 + 0\mathbf{u}_2 + 0\mathbf{u}_3 = \begin{bmatrix} 0 \\ 0 \\ 0 \end{bmatrix}, 1\mathbf{u}_1 + 0\mathbf{u}_2 + 0\mathbf{u}_3 = \begin{bmatrix} 2 \\ 0 \\ 0 \end{bmatrix},$

$0\mathbf{u}_1 + 1\mathbf{u}_2 + 0\mathbf{u}_3 = \begin{bmatrix} 4 \\ 1 \\ 6 \end{bmatrix}$

7. $\mathbf{b}$ is not in the span of $\mathbf{a}_1$.

9. $\mathbf{b}$ is not in the span of $\mathbf{a}_1$.

11. $\mathbf{b}$ is not in the span of $\mathbf{a}_1$ and $\mathbf{a}_2$.

13. $A = \begin{bmatrix} 2 & 8 & -4 \\ -1 & -3 & 5 \end{bmatrix}, \mathbf{x} = \begin{bmatrix} x_1 \\ x_2 \\ x_3 \end{bmatrix}, \mathbf{b} = \begin{bmatrix} -10 \\ 4 \end{bmatrix}$

15. $A = \begin{bmatrix} 1 & -1 & -3 & -1 \\ -2 & 2 & 6 & 2 \\ -3 & -3 & 10 & 0 \end{bmatrix}, \mathbf{x} = \begin{bmatrix} x_1 \\ x_2 \\ x_3 \\ x_4 \end{bmatrix}, \mathbf{b} = \begin{bmatrix} -1 \\ -1 \\ 5 \end{bmatrix}$

17. $x_1 \begin{bmatrix} 5 \\ 1 \end{bmatrix} + x_2 \begin{bmatrix} 7 \\ -5 \end{bmatrix} + x_3 \begin{bmatrix} -2 \\ -4 \end{bmatrix} = \begin{bmatrix} 9 \\ 2 \end{bmatrix}$

19. $x_1 \begin{bmatrix} 4 \\ 0 \\ 3 \end{bmatrix} + x_2 \begin{bmatrix} -2 \\ -5 \\ 8 \end{bmatrix} + x_3 \begin{bmatrix} -3 \\ 7 \\ 2 \end{bmatrix} + x_4 \begin{bmatrix} 5 \\ 3 \\ -1 \end{bmatrix} = \begin{bmatrix} 12 \\ 6 \\ 2 \end{bmatrix}$

21. Columns do not span $\mathbf{R}^2$.

23. Columns span $\mathbf{R}^2$.

25. Columns span $\mathbf{R}^3$.

27. Columns do not span $\mathbf{R}^3$.

29. For every choice of $\mathbf{b}$ there is a solution of $A\mathbf{x} = \mathbf{b}$.

31. There is a choice of $\mathbf{b}$ for which there is no solution to $A\mathbf{x} = \mathbf{b}$.

33. There is a choice of $\mathbf{b}$ for which there is no solution to $A\mathbf{x} = \mathbf{b}$.

35. Example: $\mathbf{b} = \begin{bmatrix} 0 \\ 1 \end{bmatrix}$

37. Example: $\mathbf{b} = \begin{bmatrix} 0 \\ 0 \\ 1 \end{bmatrix}$

39. $h \neq 3$

41. $h \neq 4$

43. Example: $\mathbf{u}_1 = (1, 0, 0), \mathbf{u}_2 = (0, 1, 0), \mathbf{u}_3 = (0, 0, 1),$ $\mathbf{u}_4 = (1, 1, 1)$

45. Example: $\mathbf{u}_1 = (1, 0, 0), \mathbf{u}_2 = (2, 0, 0), \mathbf{u}_3 = (3, 0, 0),$ $\mathbf{u}_4 = (4, 0, 0)$

47. Example: $\mathbf{u}_1 = (1, 0, 0), \mathbf{u}_2 = (0, 1, 0)$

49. Example: $\mathbf{u}_1 = (1, -1, 0), \mathbf{u}_2 = (1, 0, -1)$

51. True

53. False

55. False

57. True

59. False

61. True

63. False

65. (c) and (d) can possibly span $\mathbf{R}^3$.

67. HINT: Show that span $\{\mathbf{u}\} \subseteq$ span $\{c\mathbf{u}\}$ and that span $\{c\mathbf{u}\} \subseteq$ span $\{\mathbf{u}\}$

69. HINT: Let $S_1 = \{\mathbf{u}_1, \ldots, \mathbf{u}_k\}$ be a subset of S_2, and show that every linear combination $c_1\mathbf{u}_1 + \cdots c_k\mathbf{u}_k$ is in span(S_1).

71. HINT: Start with a linear combination $\mathbf{b} = c_1\mathbf{u}_1 + c_2\mathbf{u}_2 + c_3\mathbf{u}_3$. Show how to reorganize to write $\mathbf{b}$ as a linear combination of the set $\{\mathbf{u}_1 + \mathbf{u}_2, \mathbf{u}_1 + \mathbf{u}_3, \mathbf{u}_2 + \mathbf{u}_3\}$.

73. HINT: Generalize the argument given in Example 5.

75. True

77. False

Section 2.3

1. Linearly independent.

3. Linearly independent.

5. Linearly independent.

7. Linearly dependent.

9. Linearly independent.

11. Linearly independent.

13. System has only trivial solution.

15. System has only trivial solution.

17. System has only trivial solution.

19. Linearly dependent.

21. Linearly dependent.

23. Linearly dependent.

25. Vectors are linearly independent; none in span of the others.

27. Vectors are linearly independent; none in span of the others.

29. System does not have a unique solution for all $\mathbf{b}$.

31. System does not have a unique solution for all $\mathbf{b}$.

33. $\mathbf{u} = (1, 0, 0, 0), \mathbf{v} = (0, 1, 0, 0), \mathbf{w} = (1, 1, 0, 0)$

35. $\mathbf{u} = (1, 0), \mathbf{v} = (2, 0), \mathbf{w} = (3, 0)$

37. $\mathbf{u} = (1, 0, 0), \mathbf{v} = (0, 1, 0), \mathbf{w} = (1, 1, 0)$

39. False

41. False

43. False

45. False

47. True

49. False

51. False

53. (a), (b), and (c) can be linearly independent; (d) cannot.

55. HINT: Start by assuming that $\{c_1\mathbf{u}_1, c_2\mathbf{u}_2, c_3\mathbf{u}_3\}$ is linearly dependent, so the equation $x_1(c_1\mathbf{u}_1)+x_2(c_2\mathbf{u}_2)+x_3(c_3\mathbf{u}_3) = \mathbf{0}$ has a nontrivial solution. Show this implies that $\{\mathbf{u}_1, \mathbf{u}_2, \mathbf{u}_3\}$ is also linearly dependent, a contradiction.

57. HINT: Start by assuming that $\{\mathbf{u}_1 + \mathbf{u}_2, \mathbf{u}_1 + \mathbf{u}_3, \mathbf{u}_2 + \mathbf{u}_3\}$ is linearly dependent, so the equation $x_1(\mathbf{u}_1+\mathbf{u}_2)+x_2(\mathbf{u}_1+\mathbf{u}_3)+x_3(\mathbf{u}_2 + \mathbf{u}_3) = \mathbf{0}$ has a nontrivial solution. Show this implies that $\{\mathbf{u}_1, \mathbf{u}_2, \mathbf{u}_3\}$ is also linearly dependent, a contradiction.

59. HINT: Write the initial set of vectors as a nontrivial linear combination equal to $\mathbf{0}$ and then show that this linear combination can be extended to the new larger set of vectors.

61. HINT: If $\mathbf{u} = c\mathbf{v}$, then $\mathbf{u} - c\mathbf{v} = \mathbf{0}$.

63. HINT: Modify the proof of part (a) of Theorem 2.16.

65. No redundancy.

67. Linearly independent.

69. Linearly independent.

71. Unique solution for all $\mathbf{b}$.

73. Does not have a unique solution for all $\mathbf{b}$.

Chapter 3

Section 3.1

1. $T(\mathbf{u}_1) = \begin{bmatrix} -10 \\ 2 \end{bmatrix}$, $T(\mathbf{u}_2) = \begin{bmatrix} -4 \\ -33 \end{bmatrix}$

3. $T(\mathbf{u}_1) = \begin{bmatrix} -6 \\ 9 \end{bmatrix}$, $T(\mathbf{u}_2) = \begin{bmatrix} 16 \\ 11 \end{bmatrix}$

5. $\mathbf{y}$ is in the range of T.

7. $\mathbf{y}$ is in the range of T.

9. $T(-2\mathbf{u}_1 + 3\mathbf{u}_2) = \begin{bmatrix} -13 \\ 4 \end{bmatrix}$

11. $T(-\mathbf{u}_1 + 4\mathbf{u}_2 - 3\mathbf{u}_3) = \begin{bmatrix} 11 \\ -19 \end{bmatrix}$

13. Linear transformation, with $A = \begin{bmatrix} 3 & 1 \\ -2 & 4 \end{bmatrix}$.

15. Not a linear transformation.

17. Linear transformation, with $A = \begin{bmatrix} -4 & 0 & 1 \\ 6 & 5 & 0 \end{bmatrix}$.

19. Linear transformation, with $A = \begin{bmatrix} 0 & \sin\frac{\pi}{4} \\ \ln 2 & 0 \end{bmatrix}$.

21. T is both one-to-one and onto.

23. T is not one-to-one but is onto.

25. T is one-to-one but not onto.

27. T is neither one-to-one nor onto.

29.

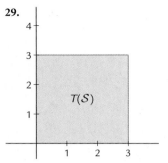

31.

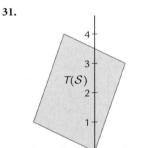

33. $T(\mathbf{x}) = \begin{bmatrix} 2 & 0 \\ 3 & 0 \end{bmatrix} \mathbf{x}$

35. $T(\mathbf{x}) = \begin{bmatrix} 7/3 & 0 \\ 0 & 0 \end{bmatrix} \mathbf{x}$

37. $T(\mathbf{x}) = \begin{bmatrix} 1 & -2 \\ 3 & 1 \end{bmatrix} \mathbf{x}$

39. False

41. True

43. True

45. False

47. False

49. (a) $A = \begin{bmatrix} r & 0 \\ 0 & r \end{bmatrix}$

(b)

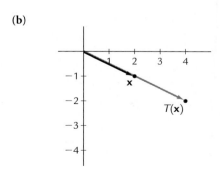

51. HINT: Show that $T(\mathbf{x}+\mathbf{y}) = T(\mathbf{x})+T(\mathbf{y})$ and $T(r\mathbf{x}) = r\,T(\mathbf{x})$.

53. HINT: Let $T(\mathbf{x}) = A\mathbf{x}$, where A is a 2×3 matrix. Explain why $A\mathbf{x} = \mathbf{0}$ must have a nontrivial solution.

55. HINT: Show that $T(\mathbf{0}) = T(0 + 0) = T(\mathbf{0}) + T(\mathbf{0})$.

57. HINT: Use properties of matrix algebra.

59. HINT: Use the fact that T is one-to-one if and only if $T(\mathbf{x}) = \mathbf{0}$ has only the trivial solution.

61. HINT: Start by assuming that $x_1\mathbf{u}_1 + x_2\mathbf{u}_2 = \mathbf{0}$ has a nontrivial solution, and arrive at a contradiction.

63. HINT: Use hint given with problem.

65. HINT: The unit square consists of all vectors $\mathbf{x} = s\mathbf{u} + t\mathbf{v}$, where $\mathbf{u} = (1, 0)$, $\mathbf{v} = (0, 1)$, $0 \le s \le 1$, and $0 \le t \le 1$.

67. (c) $T(\mathbf{x}) = A\mathbf{x} = \begin{bmatrix} 0 & 0 & 0 \\ 2 & 0 & 0 \\ 0 & 1 & 0 \end{bmatrix} \mathbf{x}$

    (d) T is neither one-to-one nor onto.

69. (a) $T(x^2 + \sin(x)) = 2x + \cos(x)$

71. $T\left(\begin{bmatrix} 5 \\ 3 \\ 6 \end{bmatrix}\right) = \begin{bmatrix} 603 \\ 565 \\ 766 \end{bmatrix}$

73. $T\left(\begin{bmatrix} 14 \\ 10 \\ 9 \end{bmatrix}\right) = \begin{bmatrix} 1255 \\ 1175 \\ 1609 \end{bmatrix}$

75. T is onto but not one-to-one.

77. T is neither one-to-one nor onto.

79. T is one-to-one but not onto.

Section 3.2

1. (a) $A + B = \begin{bmatrix} -3 & 5 \\ 0 & 4 \end{bmatrix}$

    (b) $AB + I_2 = \begin{bmatrix} -1 & -7 \\ 2 & 4 \end{bmatrix}$

    (c) $A + C$ is not possible.

3. (a) $(AB)^T = \begin{bmatrix} -2 & 2 \\ -7 & 3 \end{bmatrix}$

    (b) CE is not defined.

    (c) $(A - B)D = \begin{bmatrix} 3 & -15 & 12 \\ 16 & -30 & -6 \end{bmatrix}$

5. (a) $(C + E)B$ is not possible.

    (b) $B(C^T + D) = \begin{bmatrix} -8 & 36 & 8 \\ -22 & 47 & 10 \end{bmatrix}$

    (c) $E + CD = \begin{bmatrix} 6 & 4 & -20 \\ -11 & 21 & -4 \\ -3 & 17 & -6 \end{bmatrix}$

7. $a = -1, b = 1, c = -13$

9. $a = -1, b = 3, c = -3, d = 8$

11. $a = 2$

13. (a) $A = \begin{bmatrix} -6 & 52 \\ 4 & 17 \end{bmatrix}$

    (b) $A = \begin{bmatrix} -24 & 53 \\ -10 & 35 \end{bmatrix}$

    (c) $A = \begin{bmatrix} -1 & 50 \\ -20 & 39 \end{bmatrix}$

    (d) $A = \begin{bmatrix} 4 & 27 \\ 0 & 25 \end{bmatrix}$

15. $A^2 - I$

17. $ABA - A^2 + B^3A - B^2A$

19. The right side assumes that $AB = BA$, which is not true in general.

21. The right side assumes that $AB = BA$, which is not true in general.

23. AB is 4×5.

25. (a) $A - B = \begin{bmatrix} -1 & -2 & 0 & 2 \\ 1 & -1 & -1 & 3 \\ -1 & 3 & 0 & -3 \\ -2 & -1 & 3 & 3 \end{bmatrix}$

    (b) $AB = \begin{bmatrix} 14 & 5 & -6 & -10 \\ 4 & 7 & -4 & -7 \\ -8 & 4 & 9 & -5 \\ -1 & 1 & -3 & 5 \end{bmatrix}$

    (c) $BA = \begin{bmatrix} 3 & -5 & 2 & 7 \\ -7 & 11 & 2 & -4 \\ 4 & -1 & 9 & -1 \\ -1 & -8 & -2 & 12 \end{bmatrix}$

27. (a) $B - A = \begin{bmatrix} 1 & 2 & 0 & -2 \\ -1 & 1 & 1 & -3 \\ 1 & -3 & 0 & 3 \\ 2 & 1 & -3 & -3 \end{bmatrix}$

    (b) $AB = \begin{bmatrix} 14 & 5 & -6 & -10 \\ 4 & 7 & -4 & -7 \\ -8 & 4 & 9 & -5 \\ -1 & 1 & -3 & 5 \end{bmatrix}$

    (c) $BA + A = \begin{bmatrix} 4 & -7 & 1 & 10 \\ -9 & 11 & 3 & 0 \\ 3 & 1 & 7 & -1 \\ -1 & -7 & 0 & 13 \end{bmatrix}$

29. (a) $E = \begin{bmatrix} 0 & 1 & 0 \\ 1 & 0 & 0 \\ 0 & 0 & 1 \end{bmatrix}$

    (b) $E = \begin{bmatrix} 0 & 0 & 1 \\ 0 & 1 & 0 \\ 1 & 0 & 0 \end{bmatrix}$

    (c) $E = \begin{bmatrix} 1 & 0 & 0 \\ 0 & -2 & 0 \\ 0 & 0 & 1 \end{bmatrix}$

31. For example, $A = \begin{bmatrix} 0 & 1 & 0 \\ 0 & 0 & 0 \\ 0 & 0 & 0 \end{bmatrix}$, $B = \begin{bmatrix} 1 & 0 & 0 \\ 0 & 0 & 0 \\ 0 & 0 & 0 \end{bmatrix}$.

33. For example, $A = \begin{bmatrix} 0 & 1 \\ 0 & 0 \end{bmatrix}$, $B = \begin{bmatrix} 1 & 0 \\ 0 & 0 \end{bmatrix}$.

35. For example, $A = \begin{bmatrix} 1 & 1 \\ 1 & 1 \end{bmatrix}$, $B = \begin{bmatrix} 1 & 1 \\ -1 & -1 \end{bmatrix}$.

37. For example, $A = \begin{bmatrix} 1 & 2 \\ 2 & 1 \end{bmatrix}$, $B = \begin{bmatrix} 2 & 1 \\ 1 & 2 \end{bmatrix}$, $C = \begin{bmatrix} 1 & 1 \\ 1 & 1 \end{bmatrix}$.

39. False

41. True

43. False

45. True

47. True

53. HINT: Start with $(AB)^T$, and use $A^T = A$, $B^T = B$ because A, B are symmetric.

55. (a) $A^T A$ is $m \times m$.

(b) HINT: Show $(A^T A)^T = A^T A$.

57. HINT: Follow hint given in exercise.

59. HINT: A proof by induction works well for this one.

61. (a) For example, $A = \begin{bmatrix} 0 & 1 & 2 \\ -1 & 0 & 3 \\ -2 & -3 & 0 \end{bmatrix}$.

(b) HINT: Look at your example for part (a).

65. After one year: $\begin{bmatrix} 6500 \\ 2200 \\ 1300 \end{bmatrix}$; after two years: $\begin{bmatrix} 5375 \\ 2760 \\ 1865 \end{bmatrix}$;

after three years: $\approx \begin{bmatrix} 4531 \\ 3208 \\ 2261 \end{bmatrix}$;

after four years: $\approx \begin{bmatrix} 3898 \\ 3566 \\ 2535 \end{bmatrix}$.

67. Tomorrow: $\begin{bmatrix} 742 \\ 258 \end{bmatrix}$; the next day: $\approx \begin{bmatrix} 734 \\ 266 \end{bmatrix}$;

the day after that: $\approx \begin{bmatrix} 730 \\ 270 \end{bmatrix}$.

69. (a) $A + B = \begin{bmatrix} -4 & 1 & -3 & 5 \\ -5 & 5 & 3 & 2 \\ 6 & 11 & 0 & 5 \\ 13 & 2 & -1 & -2 \end{bmatrix}$

(b) $BA - I_4 = \begin{bmatrix} -26 & -15 & 4 & -31 \\ 5 & 1 & 9 & -28 \\ 14 & -11 & 11 & -12 \\ 4 & -9 & 13 & 24 \end{bmatrix}$

(c) $D + C$ is not possible.

71. (a) $AB = \begin{bmatrix} 25 & 22 & -14 & -1 \\ -23 & 10 & -1 & 21 \\ -68 & 36 & -21 & 34 \\ -31 & -3 & -12 & 0 \end{bmatrix}$

(b) $CD = \begin{bmatrix} 14 & 21 & 17 & 7 \\ 42 & 65 & 60 & 22 \\ 42 & 82 & 62 & 30 \\ 52 & 76 & 47 & 29 \end{bmatrix}$

(c) $(A - B) C^T$ is not possible.

73. (a) $(C + A)B$ is not possible.

(b) $C(C^T + D) = \begin{bmatrix} 21 & 40 & 41 & 29 \\ 61 & 120 & 124 & 84 \\ 66 & 146 & 182 & 106 \\ 74 & 138 & 123 & 109 \end{bmatrix}$

(c) $A + CD = \begin{bmatrix} 16 & 20 & 17 & 11 \\ 42 & 68 & 63 & 21 \\ 48 & 90 & 63 & 31 \\ 57 & 73 & 48 & 27 \end{bmatrix}$

Section 3.3

1. $\begin{bmatrix} 1 & -3 \\ -2 & 7 \end{bmatrix}$

3. Inverse does not exist.

5. $\begin{bmatrix} 9 & -4 \\ -2 & 1 \end{bmatrix}$

7. Inverse does not exist.

9. $\begin{bmatrix} 1 & -2 & 7 \\ 0 & 1 & -3 \\ 0 & 0 & 1 \end{bmatrix}$

11. Inverse does not exist.

13. $\begin{bmatrix} 0 & 1 & 0 & 0 \\ 0 & 0 & 0 & 1 \\ 1 & 0 & 0 & 0 \\ 0 & 0 & 1 & 0 \end{bmatrix}$

15. $\begin{bmatrix} 1 & -3 & -7 & 17 \\ 0 & 1 & 2 & -4 \\ 0 & 0 & 1 & -1 \\ 0 & 0 & 0 & 1 \end{bmatrix}$

17. $x_1 = 35$ and $x_2 = -11$

19. $x_1 = -\frac{15}{4}$, $x_2 = \frac{29}{4}$ and $x_3 = \frac{5}{2}$

21. $T^{-1}(\mathbf{x}) = \begin{bmatrix} -2x_1 + 3x_2 \\ 3x_1 - 4x_2 \end{bmatrix}$

23. T^{-1} does not exist.

25. T^{-1} does not exist.

27. (a) $\begin{bmatrix} 2 & 1 \\ -1 & 0 \end{bmatrix}$

(b) $\begin{bmatrix} 1 & -1 \\ 0 & 1 \end{bmatrix}$

(c) $\begin{bmatrix} 0 & -1 \\ 1 & 2 \end{bmatrix}$

(d) $\begin{bmatrix} 1 & 1 \\ 0 & 1 \end{bmatrix}$

29. $\left[\begin{array}{c|cc} 1 & 0 & 0 \\ \hline 0 & 4 & -7 \\ 0 & -1 & 2 \end{array} \right]$

31. $\begin{bmatrix} 8 & -5 & 0 & 0 \\ -3 & 2 & 0 & 0 \\ \hline 0 & 0 & -3 & 4 \\ 0 & 0 & 1 & -1 \end{bmatrix}$

33. $\begin{bmatrix} -8 & 3 & 0 & 0 & 0 \\ 3 & -1 & 0 & 0 & 0 \\ \hline -32 & 13 & 1 & -2 & 2 \\ 23 & -9 & 0 & 1 & 0 \\ 14 & -5 & 0 & 0 & 1 \end{bmatrix}$

35. $\begin{bmatrix} 1 & 0 & 0 \\ 0 & 1 & 0 \\ 0 & 0 & 1 \end{bmatrix}$

37. $A = \begin{bmatrix} 1 & 0 \\ 0 & 1 \end{bmatrix}, B = \begin{bmatrix} 3 & 0 \\ 0 & 3 \end{bmatrix}$

39. $A = \begin{bmatrix} 1 & 0 & 0 \\ 0 & 1 & 0 \end{bmatrix}, B = \begin{bmatrix} 1 & 0 \\ 0 & 1 \\ 0 & 0 \end{bmatrix}$

41. False

43. True

45. True

47. True

49. True

51. $X = A^{-1}B$

53. $X = C^{-1}B - A$

55. $A = \begin{bmatrix} \pm 1 & 0 \\ 0 & \pm 1 \end{bmatrix}$ and $A = \begin{bmatrix} a & b \\ \frac{1-a^2}{b} & -a \end{bmatrix}$ where $b \neq 0$.

57. HINT: Apply The Big Theorem.

59. $c \neq 0$ and $c \neq 1$

61. HINT: If A is $n \times n$ and not invertible, then the system $A\mathbf{x} = \mathbf{0}$ has a nontrivial solution $\mathbf{x}_0$.

63. $B = C^{-1}AC$

65. HINT: Right-multiply by A^{-1}.

67. HINT: Since B is singular, there is a nontrivial solution to $B\mathbf{x} = \mathbf{0}$.

71. 6 J8's, 10 J40's, 8 J80's.

73. This combination is not possible.

75. 3 Vigoro, 4 Parker's, 5 Bleyer's.

77. 10 Vigoro, 14 Parker's, 11 Bleyer's.

79. "laptop"

81. "final exam"

83. $\begin{bmatrix} \frac{8}{145} & -\frac{14}{145} & -\frac{23}{145} & \frac{4}{29} \\ \frac{67}{145} & \frac{64}{145} & \frac{43}{145} & -\frac{10}{29} \\ -\frac{27}{145} & \frac{11}{145} & -\frac{13}{145} & \frac{1}{29} \\ -\frac{3}{29} & -\frac{2}{29} & \frac{5}{29} & \frac{7}{29} \end{bmatrix}$

85. Inverse does not exist.

Section 3.4

1. $a = 2, b = -14$

3. $a = 4, b = 3, c = 2$

5. $\mathbf{x} = \begin{bmatrix} 3 \\ 2 \end{bmatrix}$

7. $\mathbf{x} = \begin{bmatrix} -1 \\ 0 \\ 2 \end{bmatrix}$

9. $\mathbf{x} = \begin{bmatrix} 2 \\ 1 \end{bmatrix}$

11. $\mathbf{x} = \begin{bmatrix} 2 \\ 1 \\ -1 \\ 0 \end{bmatrix}$

13. $L = \begin{bmatrix} 1 & 0 \\ -2 & 1 \end{bmatrix}$ and $U = \begin{bmatrix} 1 & -4 \\ 0 & 1 \end{bmatrix}$

15. $L = \begin{bmatrix} 1 & 0 & 0 \\ 3 & 1 & 0 \\ -1 & -1 & 1 \end{bmatrix}$ and $U = \begin{bmatrix} -2 & -1 & 1 \\ 0 & 3 & 1 \\ 0 & 0 & 1 \end{bmatrix}$

17. $L = \begin{bmatrix} 1 & 0 & 0 & 0 \\ -1 & 1 & 0 & 0 \\ 2 & -3 & 1 & 0 \\ 1 & 3 & -2 & 1 \end{bmatrix}$

$U = \begin{bmatrix} -1 & 0 & -1 & 2 \\ 0 & 3 & 1 & 0 \\ 0 & 0 & 2 & -1 \\ 0 & 0 & 0 & 1 \end{bmatrix}$

19. $L = \begin{bmatrix} 1 & 0 & 0 \\ -4 & 1 & 0 \\ 2 & 2 & 1 \end{bmatrix}$ and $U = \begin{bmatrix} -1 & 2 & 1 & 3 \\ 0 & 1 & -3 & -5 \\ 0 & 0 & 1 & 2 \end{bmatrix}$

21. $L = \begin{bmatrix} 1 & 0 & 0 & 0 \\ 1 & 1 & 0 & 0 \\ 1 & 2 & 1 & 0 \\ 0 & -1 & -1 & 1 \end{bmatrix}$ and $U = \begin{bmatrix} 1 & 1 & 0 \\ 0 & -1 & -1 \\ 0 & 0 & 2 \\ 0 & 0 & 0 \end{bmatrix}$

23. $L = \begin{bmatrix} 1 & 0 & 0 & 0 & 0 \\ -1 & 1 & 0 & 0 & 0 \\ 2 & -1 & 1 & 0 & 0 \\ -1 & 1 & -\frac{18}{17} & 1 & 0 \\ 2 & 0 & \frac{1}{17} & 0 & 1 \end{bmatrix}$

$U = \begin{bmatrix} -2 & 1 & 3 \\ 0 & 1 & 11 \\ 0 & 0 & 17 \\ 0 & 0 & 0 \\ 0 & 0 & 0 \end{bmatrix}$

25. $L = \begin{bmatrix} 1 & 0 \\ -2 & 1 \end{bmatrix}, D = \begin{bmatrix} 2 & 0 \\ 0 & 3 \end{bmatrix}, U = \begin{bmatrix} 1 & -1 \\ 0 & 1 \end{bmatrix}$

27. $L = \begin{bmatrix} 1 & 0 \\ 3 & 1 \end{bmatrix}, D = \begin{bmatrix} 1 & 0 \\ 0 & -2 \end{bmatrix}, U = \begin{bmatrix} 1 & -1 & 2 \\ 0 & 1 & \frac{1}{2} \end{bmatrix}$.

29. $L = \begin{bmatrix} 1 & 0 & 0 \\ 3 & 1 & 0 \\ -1 & -1 & 1 \end{bmatrix}$, $D = \begin{bmatrix} -2 & 0 & 0 \\ 0 & 3 & 0 \\ 0 & 0 & 1 \end{bmatrix}$,

$U = \begin{bmatrix} 1 & \frac{1}{2} & -\frac{1}{2} \\ 0 & 1 & \frac{1}{3} \\ 0 & 0 & 1 \end{bmatrix}$

31. $E = \begin{bmatrix} 4 & 0 & 0 \\ 0 & 1 & 0 \\ 0 & 0 & 1 \end{bmatrix}$

33. $E = \begin{bmatrix} 0 & 1 & 0 \\ 1 & 0 & 0 \\ 0 & 0 & 1 \end{bmatrix}$

35. $E = \begin{bmatrix} 1 & 0 & 0 \\ 0 & 1 & 0 \\ 2 & 0 & 1 \end{bmatrix}$

37. $B = \begin{bmatrix} 1 & 0 & 0 \\ -2 & 1 & 0 \\ 0 & 0 & 5 \end{bmatrix}$

39. $B = \begin{bmatrix} 0 & 1 & 0 \\ 1 & 3 & 0 \\ 0 & 0 & 1 \end{bmatrix}$

41. $B = \begin{bmatrix} 0 & 1 & 0 \\ -3 & 4 & 0 \\ 0 & 0 & 1 \end{bmatrix}$

43. $E = \begin{bmatrix} 1 & 0 & 0 & 0 \\ 0 & -6 & 0 & 0 \\ 0 & 0 & 1 & 0 \\ 0 & 0 & 0 & 1 \end{bmatrix}$

$E^{-1} = \begin{bmatrix} 1 & 0 & 0 & 0 \\ 0 & -\frac{1}{6} & 0 & 0 \\ 0 & 0 & 1 & 0 \\ 0 & 0 & 0 & 1 \end{bmatrix} \leftrightarrow \left\{ -\frac{1}{6} R_2 \Rightarrow R_2 \right\}$

45. $E = \begin{bmatrix} 1 & 0 & 0 & 0 \\ 0 & 1 & 0 & 0 \\ 0 & 0 & 0 & 1 \\ 0 & 0 & 1 & 0 \end{bmatrix}$

$E^{-1} = \begin{bmatrix} 1 & 0 & 0 & 0 \\ 0 & 1 & 0 & 0 \\ 0 & 0 & 0 & 1 \\ 0 & 0 & 1 & 0 \end{bmatrix} \leftrightarrow \{ R_3 \Leftrightarrow R_4 \}$

47. $E = \begin{bmatrix} 1 & 0 & 0 & 0 \\ -5 & 1 & 0 & 0 \\ 0 & 0 & 1 & 0 \\ 0 & 0 & 0 & 1 \end{bmatrix}$

$E^{-1} = \begin{bmatrix} 1 & 0 & 0 & 0 \\ 5 & 1 & 0 & 0 \\ 0 & 0 & 1 & 0 \\ 0 & 0 & 0 & 1 \end{bmatrix} \leftrightarrow \{ 5R_1 + R_2 \Rightarrow R_2 \}$

49. $A^{-1} = \begin{bmatrix} \frac{7}{6} & \frac{1}{3} \\ \frac{2}{3} & \frac{1}{3} \end{bmatrix}$

51. $A^{-1} = \begin{bmatrix} 17 & 1 & -2 & -7 \\ 8 & 0 & -1 & -4 \\ -6 & -2 & 1 & 1 \\ 0 & 1 & 0 & 1 \end{bmatrix}$

53. $A^{-1} = \begin{bmatrix} 3 & -\frac{5}{3} & \frac{1}{3} \\ 3 & -\frac{3}{2} & \frac{1}{4} \\ 4 & -2 & \frac{1}{2} \end{bmatrix}$

55. $A = \begin{bmatrix} 1 & 0 & 0 \\ 0 & 1 & 0 \\ 0 & 0 & 1 \\ 0 & 0 & 0 \end{bmatrix} = LU = \begin{bmatrix} 1 & 0 & 0 & 0 \\ 0 & 1 & 0 & 0 \\ 0 & 0 & 1 & 0 \\ 0 & 0 & 0 & 1 \end{bmatrix} \begin{bmatrix} 1 & 0 & 0 \\ 0 & 1 & 0 \\ 0 & 0 & 1 \\ 0 & 0 & 0 \end{bmatrix}$

57. $A = \begin{bmatrix} 1 & 0 \\ 0 & 1 \end{bmatrix} = LU = \begin{bmatrix} 1 & 0 \\ 0 & 1 \end{bmatrix} \begin{bmatrix} 1 & 0 \\ 0 & 1 \end{bmatrix}$

59. $A = \begin{bmatrix} 1 & 0 & 0 & 0 \\ 0 & 1 & 0 & 0 \\ 0 & 0 & 1 & 0 \\ 0 & 0 & 0 & 1 \end{bmatrix} = LU = \begin{bmatrix} 1 & 0 & 0 & 0 \\ 0 & 1 & 0 & 0 \\ 0 & 0 & 1 & 0 \\ 0 & 0 & 0 & 1 \end{bmatrix} \begin{bmatrix} 1 & 0 & 0 & 0 \\ 0 & 1 & 0 & 0 \\ 0 & 0 & 1 & 0 \\ 0 & 0 & 0 & 1 \end{bmatrix}$

61. False **63.** False **65.** False **67.** False

69. HINT: Apply properties of matrix multiplication.

71. HINT: Take each case separately and apply properties of matrix multiplication.

73. This matrix does not have an LU factorization.

75. $L = \begin{bmatrix} 1 & 0 & 0 & 0 \\ \frac{1}{2} & 1 & 0 & 0 \\ 1 & 0 & 1 & 0 \\ \frac{3}{2} & \frac{2}{7} & -\frac{1}{5} & 1 \end{bmatrix}$

$U = \begin{bmatrix} 10 & 2 & 0 & -4 & 2 \\ 0 & 0 & -14 & 7 & 21 \\ 0 & 0 & 0 & -5 & 0 \\ 0 & 0 & 0 & 0 & -16 \end{bmatrix}$

Section 3.5

1. Stochastic **3.** Stochastic

5. $a = 0.35, b = 0.55$

7. $a = \frac{8}{13}, b = \frac{1}{7}, c = \frac{1}{10}$

9. $a = 0.7, b = 0.7$

11. $a = 0.5, b = 0.4, c = 0.5, d = 0.4$

13. $\mathbf{x}_3 = \begin{bmatrix} 0.4432 \\ 0.5568 \end{bmatrix}$

15. $\mathbf{x}_3 = \begin{bmatrix} \frac{2531}{6750} \\ \frac{4219}{6750} \end{bmatrix}$

17. $\mathbf{x} = \begin{bmatrix} 0.71429 \\ 0.28571 \end{bmatrix} = \begin{bmatrix} 5/7 \\ 2/7 \end{bmatrix}$

19. $\mathbf{x} = \begin{bmatrix} 0.39807 \\ 0.29126 \\ 0.31067 \end{bmatrix}$

21. Not regular.

23. Not regular.

25. $A = \begin{bmatrix} 0.1 & 0.1 & 0.1 & 0.1 \\ 0.2 & 0.2 & 0.2 & 0.2 \\ 0.3 & 0.3 & 0.3 & 0.3 \\ 0.4 & 0.4 & 0.4 & 0.4 \end{bmatrix}$

27. $A = \begin{bmatrix} \frac{1}{3} & \frac{2}{3} \\ \frac{2}{3} & \frac{1}{3} \end{bmatrix}$

29. $A = \begin{bmatrix} 1 & 0 & 0 \\ 0 & 0 & 1 \\ 0 & 1 & 0 \end{bmatrix}$, $\mathbf{x}_0 = \begin{bmatrix} 0 \\ 1 \\ 0 \end{bmatrix}$

31. False

33. False

35. False

37. HINT: Let $Y = \begin{bmatrix} 1 & 1 & \cdots & 1 \end{bmatrix}$, show that $YA = Y$, and then show $Y(A\mathbf{x}) = 1$.

39. HINT: See exercise for hint.

43. HINT: Each column of A^{k+1} is a linear combination, with nonnegative scalars, of the columns of A^k.

45. HINTS: **(b)** Compute A^2 then A^3, then look for a pattern.
(c) $A^k \to \begin{bmatrix} 0 & 0 \\ 1 & 1 \end{bmatrix}$ **(d)** $\mathbf{x} = \begin{bmatrix} 0 \\ 1 \end{bmatrix}$

47. (a) $A = \begin{bmatrix} 0.9 & 0.15 \\ 0.1 & 0.85 \end{bmatrix}$

(b) Probability that the sixth person in the chain hears the wrong news is 0.32881.

(c) $\mathbf{x} = \begin{bmatrix} 0.6 \\ 0.4 \end{bmatrix}$

49. (a) $A = \begin{bmatrix} 0.35 & 0.8 \\ 0.65 & 0.2 \end{bmatrix}$

(b) **i.** Probability that she will go to McDonald's two Sundays from now is 0.3575.

ii. Probability that she will go to McDonald's three Sundays from now is 0.48913.

(c) Probability that his third fast-food experience will be at Krusty's will be 0.521.

(d) $\mathbf{x} = \begin{bmatrix} 0.55173 \\ 0.44827 \end{bmatrix}$

51. (a) Probability that a book is at C after two more circulations is 0.21.

(b) Probability that the book is at B after three more circulations is 0.64.

(c) $\mathbf{x} = \begin{bmatrix} 0.17105 \\ 0.63158 \\ 0.19737 \end{bmatrix}$

53. $\mathbf{x}_9 = \mathbf{x}_{10} = \begin{bmatrix} 0.266666 \\ 0.399999 \\ 0.133333 \\ 0.200000 \end{bmatrix}$; the steady-state vector is

$\mathbf{x} = \begin{bmatrix} \frac{4}{15} \\ \frac{2}{5} \\ \frac{2}{15} \\ \frac{1}{5} \end{bmatrix}$

55. $A \begin{bmatrix} 0 \\ 1 \\ 0 \\ 0 \end{bmatrix} = \begin{bmatrix} 0 \\ 1 \\ 0 \\ 0 \end{bmatrix}$ so $\begin{bmatrix} 0 \\ 1 \\ 0 \\ 0 \end{bmatrix}$ has itself as its steady-state vector.

Also $A \begin{bmatrix} 0 \\ 0 \\ 1 \\ 0 \end{bmatrix} = \begin{bmatrix} 0 \\ 0 \\ 1 \\ 0 \end{bmatrix}$ so $\begin{bmatrix} 0 \\ 0 \\ 1 \\ 0 \end{bmatrix}$ has itself as its steady-state vector.

Chapter 4

Section 4.1

1. This is a subspace, equal to span $\left\{ \begin{bmatrix} 1 \\ 0 \\ 0 \end{bmatrix}, \begin{bmatrix} 0 \\ 0 \\ 1 \end{bmatrix} \right\}$.

3. Not a subspace, because $\mathbf{0}$ is not in this set.

5. Not a subspace, because $\mathbf{0}$ is not in this set.

7. Not a subspace, because it is not closed under scalar multiplication.

9. Not a subspace, because it is not closed under addition.

11. Not a subspace, because it is not closed under scalar multiplication.

13. Not a subspace, because it is not closed under scalar multiplication.

15. A subspace, equal to null $\left(\begin{bmatrix} 1 & 1 & \cdots 1 \end{bmatrix} \right)$.

17. Not closed under scalar multiplication.

19. Not closed under addition.

21. null$(A) = \left\{ \begin{bmatrix} 0 \\ 0 \end{bmatrix} \right\}$

23. null$(A) = $ span $\left\{ \begin{bmatrix} 5 \\ -2 \\ 1 \end{bmatrix} \right\}$

25. null$(A) = $ span $\left\{ \begin{bmatrix} 4 \\ 3 \\ 1 \end{bmatrix} \right\}$

27. null$(A) = \left\{ \begin{bmatrix} 0 \\ 0 \end{bmatrix} \right\}$

29. null$(A) = \left\{ \begin{bmatrix} 0 \\ 0 \\ 0 \end{bmatrix} \right\}$

31. null$(A) = $ span $\left\{ \begin{bmatrix} 3 \\ -1 \\ 1 \\ 0 \end{bmatrix} \right\}$

33. **b** is not in ker(T); **c** is in range(T).

35. **b** is not in ker(T); **c** is not in range(T).

37. For example, $S = \left\{ \begin{bmatrix} x \\ y \end{bmatrix} : x > 0 \right\}$.

39. For example, $S_1 = \left\{ \begin{bmatrix} x \\ 0 \end{bmatrix} : x \geq 0 \right\}$

and $S_2 = \left\{ \begin{bmatrix} x \\ 0 \end{bmatrix} : x < 0 \right\}$.

41. Let $T(\mathbf{x}) = A\mathbf{x}$, where $A = \begin{bmatrix} 1 & 0 \\ 1 & 0 \end{bmatrix}$.

43. Let $T(\mathbf{x}) = A\mathbf{x}$, where $A = I_3$.

45. True

47. True

49. False

51. True

53. True

55. True

57. False

59. False

61. HINT: If $x \neq 0$ is in a subspace S, show that every real number must be in S.

63. HINT: Determine if **0** is in the set of solutions.

65. The vector **0** alone, lines and planes through the origin, and all of $\mathbf{R}^3$.

67. HINT: Determine if **0** is in the set of solutions.

69. HINT: Show that $\mathbf{x} \neq \mathbf{0}$ and $A\mathbf{x} = \mathbf{0}$ if and only if the columns of A are linearly dependent.

71. HINT: Note that $\mathbf{u} - \mathbf{v} = \mathbf{u} + (-1)\mathbf{v}$.

73. span $\left\{ \begin{bmatrix} 1 \\ 2 \\ 2 \end{bmatrix} \right\}$

75. span $\left\{ \begin{bmatrix} 1 \\ 2 \\ 1 \\ 2 \end{bmatrix} \right\}$

77. span $\left\{ \begin{bmatrix} \frac{3}{7} \\ -\frac{13}{7} \\ \frac{5}{7} \\ 1 \\ 0 \end{bmatrix} , \begin{bmatrix} -\frac{43}{56} \\ -\frac{5}{56} \\ -\frac{39}{56} \\ 0 \\ 1 \end{bmatrix} \right\}$

79. $\left\{ \begin{bmatrix} 0 \\ 0 \\ 0 \\ 0 \end{bmatrix} \right\}$

Section 4.2

1. Not a basis, since $\mathbf{u}_1$ and $\mathbf{u}_2$ are not linearly independent.

3. Not a basis, since three vectors in a two-dimensional space must be linearly dependent.

5. Basis is $\left\{ \begin{bmatrix} 1 \\ -4 \end{bmatrix} \right\}$; dimension $= 1$.

7. Basis is $\left\{ \begin{bmatrix} 1 \\ 3 \\ -2 \end{bmatrix} , \begin{bmatrix} 0 \\ -2 \\ 5 \end{bmatrix} \right\}$; dimension $= 2$.

9. Basis is $\left\{ \begin{bmatrix} 1 \\ -2 \\ 3 \\ -2 \end{bmatrix} , \begin{bmatrix} 0 \\ 2 \\ -5 \\ 1 \end{bmatrix} \right\}$; dimension $= 2$.

11. Basis is $\left\{ \begin{bmatrix} 1 \\ 3 \end{bmatrix} , \begin{bmatrix} 4 \\ -12 \end{bmatrix} \right\}$; dimension $= 2$.

13. Basis is $\left\{ \begin{bmatrix} 1 \\ 2 \\ 4 \end{bmatrix} , \begin{bmatrix} 0 \\ 1 \\ -3 \end{bmatrix} , \begin{bmatrix} 3 \\ -2 \\ -1 \end{bmatrix} \right\}$; dimension $= 3$.

15. Basis is $\left\{ \begin{bmatrix} 1 \\ -1 \\ 0 \\ 2 \end{bmatrix} , \begin{bmatrix} 2 \\ -5 \\ 9 \\ 7 \end{bmatrix} \right\}$; dimension $= 2$.

17. Basis is $\left\{ \begin{bmatrix} 2 \\ -6 \end{bmatrix} \right\}$; dimension is 1.

19. Basis is $\left\{ \begin{bmatrix} 1 \\ 1 \\ 1 \end{bmatrix} \right\}$; dimension is 1.

21. Basis is $\left\{ \begin{bmatrix} 3 \\ 0 \\ 0 \end{bmatrix} , \begin{bmatrix} 2 \\ 1 \\ 0 \end{bmatrix} , \begin{bmatrix} 1 \\ 2 \\ 3 \end{bmatrix} \right\}$; dimension is 3.

23. One extension is $\left\{ \begin{bmatrix} 1 \\ -3 \end{bmatrix} , \begin{bmatrix} 1 \\ 0 \end{bmatrix} \right\}$.

25. One extension is $\left\{ \begin{bmatrix} -1 \\ 2 \\ 1 \end{bmatrix} , \begin{bmatrix} 1 \\ 0 \\ 0 \end{bmatrix} , \begin{bmatrix} 0 \\ 1 \\ 0 \end{bmatrix} \right\}$.

27. One extension is $\left\{ \begin{bmatrix} 1 \\ 3 \\ -2 \end{bmatrix} , \begin{bmatrix} 2 \\ -1 \\ 0 \end{bmatrix} , \begin{bmatrix} 1 \\ 0 \\ 0 \end{bmatrix} \right\}$.

29. null(A) $= \left\{ \begin{bmatrix} 0 \\ 0 \end{bmatrix} \right\}$. This subspace has no basis, and nullity (A) $= 0$.

31. The null space has basis $\left\{ \begin{bmatrix} -7 \\ 0 \\ 3 \\ 1 \end{bmatrix} , \begin{bmatrix} -1 \\ 1 \\ 0 \\ 0 \end{bmatrix} \right\}$, and nullity($A$) $= 2$.

33. For example, $\left\{ \begin{bmatrix} 1 \\ 0 \end{bmatrix} , \begin{bmatrix} 1 \\ 1 \end{bmatrix} , \begin{bmatrix} 0 \\ 1 \end{bmatrix} , \begin{bmatrix} -1 \\ 1 \end{bmatrix} \right\}$.

35. For example, the span of the first m vectors of the n standard basis vectors of $\mathbf{R}^n$.

37. For example, $S_1 = \text{span} \left\{ \begin{bmatrix} 1 \\ 0 \\ 0 \\ 0 \end{bmatrix}, \begin{bmatrix} 0 \\ 1 \\ 0 \\ 0 \end{bmatrix} \right\}$

and $S_2 = \text{span} \left\{ \begin{bmatrix} 0 \\ 0 \\ 1 \\ 0 \end{bmatrix}, \begin{bmatrix} 0 \\ 0 \\ 0 \\ 1 \end{bmatrix} \right\}.$

39. For example, $\mathbf{u}_1 = \begin{bmatrix} 1 \\ 0 \\ 0 \end{bmatrix}$ and $\mathbf{u}_2 = \begin{bmatrix} 0 \\ 1 \\ 0 \end{bmatrix}.$

41. False

43. False

45. False

47. False

49. False

51. False

53. True

55. (a) 1, 2, or 3.

(b) 1 or 2.

57. HINT: Use the Big Theorem.

59. HINT: Show separately that the set is linearly independent and spans S.

61. HINT: A basis for S_1 can be expanded to a basis for S_2.

63. HINT: The entries below each pivot are equal to zero.

65. n

69. Subspace has basis $\left\{ \begin{bmatrix} 2 \\ -1 \\ 5 \end{bmatrix}, \begin{bmatrix} -3 \\ 4 \\ -2 \end{bmatrix} \right\}$, with dimension 2.

The vectors are not a basis for $\mathbf{R}^3$.

71. Subspace has basis

$\left\{ \begin{bmatrix} 3 \\ 0 \\ 1 \\ -2 \end{bmatrix}, \begin{bmatrix} 2 \\ -4 \\ 5 \\ 0 \end{bmatrix}, \begin{bmatrix} -2 \\ 7 \\ 0 \\ 4 \end{bmatrix}, \begin{bmatrix} -2 \\ 5 \\ -5 \\ 4 \end{bmatrix} \right\}$, with dimension

4. The vectors form a basis for $\mathbf{R}^4$.

73. Subspace has basis

$\left\{ \begin{bmatrix} 1 \\ 1 \\ -1 \\ 1 \\ 1 \end{bmatrix}, \begin{bmatrix} -1 \\ 0 \\ 1 \\ 2 \\ -1 \end{bmatrix}, \begin{bmatrix} 2 \\ 1 \\ -2 \\ 1 \\ 2 \end{bmatrix} \right\}$, with dimension 3.

The vectors therefore do not span $\mathbf{R}^5$.

Section 4.3

1. Column space basis: $\left\{ \begin{bmatrix} 1 \\ -2 \\ -3 \end{bmatrix}, \begin{bmatrix} -3 \\ 5 \\ 8 \end{bmatrix} \right\}$

Row space basis: $\left\{ \begin{bmatrix} 1 \\ 0 \\ -10 \end{bmatrix}, \begin{bmatrix} 0 \\ 1 \\ -4 \end{bmatrix} \right\}$

Null space basis: $\left\{ \begin{bmatrix} 10 \\ 4 \\ 1 \end{bmatrix} \right\}$

rank = 2, nullity = 1, $m = 3$

3. Column space basis: $\left\{ \begin{bmatrix} 1 \\ -2 \\ 0 \end{bmatrix}, \begin{bmatrix} 0 \\ 1 \\ 1 \end{bmatrix} \right\}$

Row space basis: $\left\{ \begin{bmatrix} 1 \\ 0 \\ -4 \\ -3 \end{bmatrix}, \begin{bmatrix} 0 \\ 1 \\ 5 \\ -1 \end{bmatrix} \right\}$

Null space basis: $\left\{ \begin{bmatrix} 4 \\ -5 \\ 1 \\ 0 \end{bmatrix}, \begin{bmatrix} 3 \\ 1 \\ 0 \\ 1 \end{bmatrix} \right\}$

rank = 2, nullity = 2, $m = 4$

5. Column space basis: $\left\{ \begin{bmatrix} 1 \\ 2 \\ -1 \end{bmatrix}, \begin{bmatrix} -2 \\ -2 \\ -2 \end{bmatrix} \right\}$

Row space basis: $\left\{ \begin{bmatrix} 1 \\ -2 \\ 2 \end{bmatrix}, \begin{bmatrix} 0 \\ 2 \\ -1 \end{bmatrix} \right\}$

Null space basis: $\left\{ \begin{bmatrix} -1 \\ \frac{1}{2} \\ 1 \end{bmatrix} \right\}$

rank = 2, nullity = 1, $m = 3$

7. Column space basis: $\left\{ \begin{bmatrix} 1 \\ 3 \\ 1 \end{bmatrix}, \begin{bmatrix} 3 \\ 11 \\ 1 \end{bmatrix}, \begin{bmatrix} 2 \\ 7 \\ 4 \end{bmatrix} \right\}$

Row space basis: $\left\{ \begin{bmatrix} 1 \\ 3 \\ 2 \\ 0 \end{bmatrix}, \begin{bmatrix} 0 \\ 2 \\ 1 \\ 1 \end{bmatrix}, \begin{bmatrix} 0 \\ 0 \\ 3 \\ 1 \end{bmatrix} \right\}$

Null space basis: $\left\{ \begin{bmatrix} \frac{5}{3} \\ -\frac{1}{3} \\ -\frac{1}{3} \\ 1 \end{bmatrix} \right\}$

rank = 3, nullity = 1, $m = 4$

9. $x \neq 8$

11. $x = 18$

13. $\dim(\text{col}(A)) = 5$

15. $\dim(\text{row}(A)) = 3$, $\dim(\text{col}(A)) = 3$, $\text{nullity}(A) = 4$.

17. $\text{rank}(A) = 2$

19. $\text{nullity}(A) = 7$

21. $\dim(\text{range}(T)) = 4$

23. $\text{nullity}(A) = 0$

25. Maximum for rank(A) = 5, minimum for nullity(A) = 8.

27. rank(A) = 3

29. nullity(A) = 2

31. B has 3 nonzero rows.

33. A is 7×5.

35. For example, $A = \begin{bmatrix} 1 & 0 & 0 \\ 0 & 1 & 0 \end{bmatrix}$.

37. For example, $A = \left[\begin{array}{c|c} I_{3\times3} & 0_{3\times1} \\ \hline 0_{6\times3} & 0_{6\times1} \end{array} \right]$.

39. For example, $A = \begin{bmatrix} 1 & 0 & 0 & 0 \\ 0 & 1 & 0 & 0 \\ 0 & 0 & 1 & 0 \end{bmatrix}$.

41. For example, $A = \begin{bmatrix} 1 & 0 \\ 0 & 1 \end{bmatrix}$.

43. True

45. True

47. False

49. False

51. False

53. True

55. HINT: row(A) = col(A^T).

57. HINT: Apply the Rank–Nullity Theorem.

59. HINT: First suppose $n < m$ and apply the Rank–Nullity Theorem to A, then suppose that $m < n$ and apply the Rank–Nullity Theorem to A^T

61. rank(A) = 2, nullity(A) = 3.

63. rank(A) = 2, nullity(A) = 1.

Chapter 5

Section 5.1

1. $M_{23} = \begin{bmatrix} 7 & 0 \\ 5 & 1 \end{bmatrix}$, $M_{31} = \begin{bmatrix} 0 & -4 \\ 6 & 2 \end{bmatrix}$

3. $M_{23} = \begin{bmatrix} 6 & 1 & 5 \\ 7 & 1 & 1 \\ 4 & 3 & 2 \end{bmatrix}$, $M_{31} = \begin{bmatrix} 1 & -1 & 5 \\ 2 & 3 & 0 \\ 3 & 1 & 2 \end{bmatrix}$

5. $M_{23} = \begin{bmatrix} 4 & 3 & 1 & 0 \\ 3 & 2 & 4 & 4 \\ 5 & 1 & 0 & 3 \\ 2 & 2 & 1 & 0 \end{bmatrix}$, $M_{31} = \begin{bmatrix} 3 & 2 & 1 & 0 \\ 1 & 2 & 0 & 5 \\ 1 & 0 & 0 & 3 \\ 2 & 4 & 1 & 0 \end{bmatrix}$

7. $C_{13} = 4$, $C_{22} = -10$

9. $C_{13} = 1$, $C_{22} = 4$

11. $|A| = 60$; T is invertible.

13. $|A| = 20$; T is invertible.

15. $|A| = 51$; T is invertible.

17. $|A| = 8$; T is invertible.

19. $|A| = 14$

21. $|A|$ is not defined.

23. $|A| = -82$

25. The shortcut method does not apply.

27. $a = 9$

29. $a = 0$

31. $a = 4$

33. $a = 1$ or $a = 3$

35. $|A| = -8$ (A upper triangular)

37. $|A| = 0$ (column of zeros)

39. $|A| = 0$ (two equal rows)

41. $|A| = |A^T| = 11$

43. $|A| = |A^T| = 28$

45. $\lambda = -2$ or $\lambda = 7$

47. $\lambda = 1$

49. $\lambda = -2$, $\lambda = 1$, or $\lambda = 3$

51. $\lambda = 2$

53. (a) $|A| = 22$, determinant after row interchange = -22.
(b) $|A| = 1$, determinant after row interchange = -1.
Conjecture: Row interchanges change the sign of the determinant.

55. (a) $|A| = -13$, determinant after row interchange = 13.
(b) $|A| = 3$, determinant after row interchange = -3.
Conjecture: Row interchanges change the sign of the determinant.

57. (a) $|A| = 22$, determinant after multiplying row 1 by 3 is 66.
(b) $|A| = 1$, determinant after multiplying row 1 by 3 is 3.
Conjecture: Multiplying row 1 by 3 change the determinant by a factor of 3.

59. (a) $|A| = -13$, determinant after multiplying row 1 by 3 is -39.
(b) $|A| = 3$, determinant after multiplying row 1 by 3 is 9.
Conjecture: Multiplying row 1 by 3 changes the determinant by a factor of 3.

61. For example, $A = \begin{bmatrix} 12 & 0 \\ 0 & 1 \end{bmatrix}$.

63. For example, $A = \begin{bmatrix} 1 & 4 \\ 1 & 1 \end{bmatrix}$.

65. For example, $A = \begin{bmatrix} 5 & -1 & \pi \\ e & 0 & 4 \\ 2 & 6 & -3 \end{bmatrix}$.

67. For example, $A = \begin{bmatrix} \pi & 0 & 5 \\ 8 & 1 & 0 \\ 0 & e & 1 \end{bmatrix}$.

69. False

71. False

73. False

75. False

77. HINT: Show that the determinant gives a linear equation in x and y, then plug in (x_1, y_1) and (x_2, y_2) separately to show they satisfy the equation.

79. HINT: Show that the given expression is equal to the determinant of the matrix obtained by replacing row j of A with row i.

81. HINT: Cofactor expansion along row or column of zeros.

83. $|A| = -26$

85. $|A| = 1215$

Section 5.2

1. $|A| = 2$

3. $|A| = 0$

5. $|A| = 1$

7. $|A| = 4$; A is invertible.

9. $|A| = 21$; A is invertible.

11. $|A| = 0$; A is not invertible.

13. $|A| = 8$; A is invertible.

15. Determinant $= -3$

17. Determinant $= -6$

19. $\det(AB) = \det(A)\det(B) = (-11)(3) = -33$
$\det(A + B) = -2 \neq -11 + 3 = \det(A) + \det(B)$

21. $\det(AB) = \det(A)\det(B) = (1)(-30) = -30$
$\det(A + B) = -76 \neq 1 - 30 = \det(A) + \det(B)$

23. (a) $|A^2| = 9$
(b) $|A^4| = 81$
(c) $|A^2 A^T| = 27$
(d) $|A^{-1}| = \frac{1}{3}$

25. (a) $|A^2 B^3| = -72$
(b) $|AB^{-1}| = -\frac{3}{2}$
(c) $|B^3 A^T| = -24$
(d) $|A^2 B^3 B^T| = 144$

27. $|A| = 198$

29. $|A| = 4$

31. $|A| = 0$

33. $\begin{vmatrix} A & B \\ C & D \end{vmatrix} = -3$, $|A||D| - |B||C| = -18$

35. Unique solution exists.

37. Unique solution exists.

39. Unique solution exists.

41. $A = \begin{bmatrix} 1 & 2 \\ 2 & 4 \end{bmatrix}$

43. $A = \begin{bmatrix} 1 & 2 \\ 2 & 4 \end{bmatrix}$, $B = \begin{bmatrix} -1 & -2 \\ -2 & -4 \end{bmatrix}$

45. For example, $\det\left(\begin{bmatrix} 1 & 1 & 1 \\ 1 & 2 & 2 \\ 1 & 1 & 2 \end{bmatrix} \right) = 1.$

47. False

49. False

51. False

53. True

55. True

57. HINT: Subtract one of the identical rows from the other, then apply cofactor expansion to the resulting matrix.

59. HINT: $|A| = |A^T|$.

61. HINT: Remove a factor of (-1) from each of the n rows.

63. HINT: $|A^2| = |A|^2$.

65. HINT: See hint given with problem.

67. HINT: Explain why a matrix can be transformed to echelon form without multiplying a row times a constant.

69. (a) HINT: E is diagonal, with a c for one diagonal entry and ones for the remaining diagonal entries.
(b) HINT: E is triangular, with ones along the diagonal.

71. HINT: See hint given with problem.

73. $|I_4 + AB| = |I_3 + BA| = -45,780$

Section 5.3

1. $x_1 = \frac{21}{8}$, $x_2 = \frac{3}{4}$

3. $x_1 = 9$, $x_2 = -17$, $x_3 = 1$

5. $x_1 = \frac{79}{49}$, $x_2 = \frac{22}{49}$, $x_3 = \frac{124}{49}$

7. $x_2 = \frac{11}{23}$

9. $x_2 = -\frac{25}{21}$

11. $x_2 = \frac{14}{39}$

13. adj $(A) = \begin{bmatrix} 7 & -5 \\ -3 & 2 \end{bmatrix}$, $A^{-1} = \begin{bmatrix} -7 & 5 \\ 3 & -2 \end{bmatrix}$

15. adj $(A) = \begin{bmatrix} 0 & 0 & 1 \\ 1 & 0 & 0 \\ 0 & 1 & 0 \end{bmatrix}$, $A^{-1} = \begin{bmatrix} 0 & 0 & 1 \\ 1 & 0 & 0 \\ 0 & 1 & 0 \end{bmatrix}$

17. adj $(A) = \begin{bmatrix} 1 & -2 & 3 \\ 0 & 1 & -2 \\ 0 & 0 & 1 \end{bmatrix}$, $A^{-1} = \begin{bmatrix} 1 & -2 & 3 \\ 0 & 1 & -2 \\ 0 & 0 & 1 \end{bmatrix}$

19.

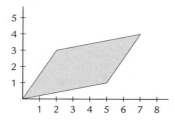

area $=13$

21.

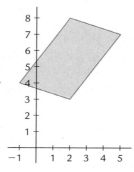

area $=15$

23. area $(T(\mathcal{D})) = 165$

25. area $(T(\mathcal{D})) = 54$

27. area $(T(\mathcal{D})) = 54$

29. $T(\mathbf{x}) = \begin{bmatrix} 5 & 0 \\ 0 & 3 \end{bmatrix} \mathbf{x}$ is one possible solution.

31. $T(\mathbf{x}) = \begin{bmatrix} \frac{3}{2}\sqrt{2} & 3\sqrt{2} \\ -\frac{3}{2}\sqrt{2} & 3\sqrt{2} \end{bmatrix} \mathbf{x}$ is one possible solution.

33. volume $= 80\pi$

35. volume $= 82$

37. For example,

$$\begin{aligned} x_1 + x_2 &= 1 \\ 2x_1 + 2x_2 &= 2 \end{aligned}$$

39. For example, let the parallelogram have vertices $(0, 0)$, $(5, 0)$, $(5, 1)$, and $(0, 1)$.

41. For example, $A = \begin{bmatrix} 1 & 3 \\ 2 & 5 \end{bmatrix}$.

43. False

45. False

47. False

49. True

51. HINT: $|B| \neq 0$, so T is one-to-one. It remains to show that T is onto $\mathcal{R}$.

53. HINT: Use $|A|^{-1}\text{adj}(A) = A^{-1}$.

55. HINT: Show that the cofactor matrix of a symmetric matrix is also symmetric.

57. HINT: Consider the change in the cofactors when A is multiplied by c.

59. HINT: Start by replacing A with A^{-1} in $A = |A|^{-1}\text{adj}(A)$.

61. HINT: M_{ij} has a column (and row) of zeros when $i \neq j$.

63. $x_1 = \frac{1221}{752}$, $x_2 = \frac{811}{752}$, $x_3 = \frac{133}{94}$

65. $x_1 = \frac{704}{245}$, $x_2 = -\frac{14}{5}$, $x_3 = \frac{247}{245}$, $x_4 = -\frac{17}{49}$

67. $\text{adj}(A) = \begin{bmatrix} 27 & 53 & -15 \\ -72 & 41 & 40 \\ 59 & -26 & 28 \end{bmatrix}$,

$$A^{-1} = \begin{bmatrix} \frac{27}{547} & \frac{53}{547} & -\frac{15}{547} \\ -\frac{72}{547} & \frac{41}{547} & \frac{40}{547} \\ \frac{59}{547} & -\frac{26}{547} & \frac{28}{547} \end{bmatrix}$$

69. $\text{adj}(A) = \begin{bmatrix} -21 & -126 & 60 & -15 \\ -36 & 207 & -18 & 216 \\ 118 & 3 & -35 & -97 \\ 11 & -75 & 29 & -113 \end{bmatrix}$

$$A^{-1} = \begin{bmatrix} -\frac{7}{141} & -\frac{14}{47} & \frac{20}{141} & -\frac{5}{141} \\ -\frac{4}{47} & \frac{23}{47} & -\frac{2}{47} & \frac{24}{47} \\ \frac{118}{423} & \frac{1}{141} & -\frac{35}{423} & \frac{97}{423} \\ \frac{11}{423} & -\frac{25}{141} & \frac{29}{423} & -\frac{113}{423} \end{bmatrix}$$

Chapter 6

Section 6.1

1. $\mathbf{x}_1$ is an eigenvector with associated eigenvalue $\lambda = -1$; $\mathbf{x}_2$ is not an eigenvector; $\mathbf{x}_3$ is an eigenvector with associated eigenvalue $\lambda = 4$.

3. $\mathbf{x}_1$ is an eigenvector with associated eigenvalue $\lambda = -1$; $\mathbf{x}_2$ is an eigenvector with associated eigenvalue $\lambda = 1$; $\mathbf{x}_3$ is an eigenvector with associated eigenvalue $\lambda = 2$.

5. $\mathbf{x}_1$ is an eigenvector with associated eigenvalue $\lambda = 3$; $\mathbf{x}_2$ is not an eigenvector; $\mathbf{x}_3$ is an eigenvector with associated eigenvalue $\lambda = 0$.

7. $\lambda = 3$ is not an eigenvalue of A.

9. $\lambda = -2$ is an eigenvalue of A.

11. A basis for the $\lambda = 4$ eigenspace is $\left\{ \begin{bmatrix} -1 \\ 1 \end{bmatrix} \right\}$.

13. A basis for the $\lambda = 2$ eigenspace is $\left\{ \begin{bmatrix} 5 \\ 2 \end{bmatrix} \right\}$.

15. A basis for the $\lambda = 4$ eigenspace is $\left\{ \begin{bmatrix} 1 \\ 3 \\ 1 \end{bmatrix} \right\}$.

17. A basis for the $\lambda = 6$ eigenspace is $\left\{ \begin{bmatrix} 1 \\ 1 \\ 1 \end{bmatrix} \right\}$.

19. A basis for the $\lambda = -4$ eigenspace is $\left\{ \begin{bmatrix} -3 \\ -5 \\ -2 \\ 3 \end{bmatrix} \right\}$.

21. $\det(A - \lambda I_2) = \lambda^2 + \lambda - 6$; basis for $\lambda = -3$ eigenspace is $\left\{ \begin{bmatrix} 0 \\ 1 \end{bmatrix} \right\}$; basis for $\lambda = 2$ eigenspace is $\left\{ \begin{bmatrix} 5 \\ 4 \end{bmatrix} \right\}$.

23. $\det(A - \lambda I_2) = \lambda^2 + 2\lambda + 1$;
basis for $\lambda = -1$ eigenspace is $\left\{ \begin{bmatrix} 1 \\ 1 \end{bmatrix} \right\}$.

25. $\det(A - \lambda I_3) = -(\lambda - 2)(\lambda - 3)(\lambda + 1)$;
basis for $\lambda = 2$ eigenspace is $\left\{ \begin{bmatrix} 0 \\ 3 \\ 5 \end{bmatrix} \right\}$;

basis for $\lambda = 3$ eigenspace is $\left\{ \begin{bmatrix} 4 \\ 4 \\ 1 \end{bmatrix} \right\}$;

basis for $\lambda = -1$ eigenspace is $\left\{ \begin{bmatrix} 0 \\ 0 \\ 1 \end{bmatrix} \right\}$.

27. $\det(A - \lambda I_3) = -\lambda^3 + 3\lambda^2 - 2\lambda$
basis for $\lambda = 0$ eigenspace is $\left\{ \begin{bmatrix} 1 \\ -1 \\ 3 \end{bmatrix} \right\}$;

basis for $\lambda = 1$ eigenspace is $\left\{ \begin{bmatrix} 1 \\ -1 \\ 4 \end{bmatrix} \right\}$;

basis for $\lambda = 2$ eigenspace is $\left\{ \begin{bmatrix} 2 \\ -1 \\ 5 \end{bmatrix} \right\}$.

29. $\det(A - \lambda I_4) = (\lambda + 2)(\lambda + 1)(\lambda - 1)^2$;
basis for $\lambda = -2$ eigenspace is $\left\{ \begin{bmatrix} 0 \\ 3 \\ -3 \\ 1 \end{bmatrix} \right\}$;

basis for $\lambda = -1$ eigenspace is $\left\{ \begin{bmatrix} 4 \\ 20 \\ -30 \\ 11 \end{bmatrix} \right\}$;

basis for $\lambda = 1$ eigenspace is $\left\{ \begin{bmatrix} 0 \\ 0 \\ 0 \\ 1 \end{bmatrix} \right\}$.

31. For example, $A = \begin{bmatrix} 1 & 0 \\ 0 & 2 \end{bmatrix}$.

33. For example, $A = \begin{bmatrix} 1 & 0 & 0 \\ 0 & -2 & 0 \\ 0 & 0 & 3 \end{bmatrix}$.

35. For example, $A = \begin{bmatrix} 0 & 1 \\ -1 & 0 \end{bmatrix}$.

37. False

39. False

41. True

43. True

45. False

47. (a) A is 6×6.
   (b) $\lambda = 3$, $\lambda = 2$, and $\lambda = -1$.
   (c) A is invertible.
   (d) The largest possible dimension of an eigenspace is 3.

49. HINT: Apply the Big Theorem.

51. HINT: Explain why $\det(A - I_n) = 0$.

53. HINT: What is $A\mathbf{u}$ if $\mathbf{u}$ is associated with two distinct eigenvalues?

55. HINT: Which values of λ would *not* be eigenvalues?

57. HINT: Show $A^{-1}\mathbf{u} = \lambda^{-1}\mathbf{u}$.

59. HINT: Suppose that λ_1 is the eigenvalue of A associated with $\mathbf{u}$ and λ_2 is the eigenvalue of B associated with $\mathbf{u}$. Determine $AB\mathbf{u}$.

61. HINT: What is $A\mathbf{u}$ when $\mathbf{u} = (1, 1, \ldots, 1)$?

63. HINT: Note that $\det(A - \lambda I_n) = \det\left((A - \lambda I_n)^T\right)$.

65. HINT: What is $A\mathbf{u}$ when $\mathbf{u} = (1, 1, \ldots, 1)$?

67. Basis for $\lambda = 1$ eigenspace is $\left\{ \begin{bmatrix} 0 \\ 1 \\ 0 \\ 0 \end{bmatrix}, \begin{bmatrix} 1 \\ 0 \\ -1 \\ 1 \end{bmatrix} \right\}$;

basis for $\lambda = 2$ eigenspace is $\left\{ \begin{bmatrix} -1 \\ 5 \\ 1 \\ 0 \end{bmatrix}, \begin{bmatrix} -1 \\ 9 \\ 0 \\ 2 \end{bmatrix} \right\}$.

69. Basis for $\lambda = 0$ eigenspace is $\left\{ \begin{bmatrix} 0 \\ -1 \\ 0 \\ 1 \\ 1 \end{bmatrix} \right\}$;

basis for $\lambda = 1$ eigenspace is $\left\{ \begin{bmatrix} -1 \\ 0 \\ 3 \\ -1 \\ 1 \end{bmatrix} \right\}$;

basis for $\lambda = 2$ eigenspace is $\left\{ \begin{bmatrix} -1 \\ -2 \\ 2 \\ -1 \\ 1 \end{bmatrix} \right\}$;

basis for $\lambda = -2$ eigenspace is $\left\{ \begin{bmatrix} -1 \\ -3 \\ 3 \\ -2 \\ 1 \end{bmatrix} \right\}$;

basis for $\lambda = -1$ eigenspace is $\left\{ \begin{bmatrix} 0 \\ 1 \\ 0 \\ 0 \\ 0 \end{bmatrix} \right\}$.

Section 6.2

1. $\mathbf{x}_1 = \begin{bmatrix} 1 \\ 1 \end{bmatrix}, \mathbf{x}_2 = \begin{bmatrix} -2 \\ 6 \end{bmatrix}, \mathbf{x}_3 = \begin{bmatrix} -20 \\ 28 \end{bmatrix}$

3. $\mathbf{x}_1 = \begin{bmatrix} 6 \\ 4 \\ 4 \end{bmatrix}, \mathbf{x}_2 = \begin{bmatrix} 52 \\ 48 \\ 48 \end{bmatrix}, \mathbf{x}_3 = \begin{bmatrix} 504 \\ 496 \\ 496 \end{bmatrix}$

5. $\mathbf{x}_1 = \begin{bmatrix} 3 \\ 0 \\ -3 \end{bmatrix}, \mathbf{x}_2 = \begin{bmatrix} 9 \\ 0 \\ -9 \end{bmatrix}, \mathbf{x}_3 = \begin{bmatrix} 27 \\ 0 \\ -27 \end{bmatrix}$

7. $\mathbf{x}_1 = \begin{bmatrix} -1 \\ 1 \end{bmatrix}, \mathbf{x}_2 = \begin{bmatrix} 1.00 \\ -0.33 \end{bmatrix}$

9. $\mathbf{x}_1 = \begin{bmatrix} 0.00 \\ -0.50 \\ 1.00 \end{bmatrix}, \mathbf{x}_2 = \begin{bmatrix} 1.00 \\ -0.25 \\ 1.00 \end{bmatrix}$

11. $\mathbf{x}_1 = \begin{bmatrix} 0.00 \\ 1.00 \\ -0.67 \end{bmatrix}, \mathbf{x}_2 = \begin{bmatrix} 0.00 \\ 1.00 \\ -0.56 \end{bmatrix}$

13. The Power Method will converge, with eigenvalue $\lambda = 7$.

15. The Power Method will converge, with eigenvalue $\lambda = -6$.

17. The Power Method will converge, with eigenvalue $\lambda = 6$.

19. $B = \begin{bmatrix} -3 & 2 \\ 3 & -2 \end{bmatrix}$

21. $B = \begin{bmatrix} -10 & 2 & -7 \\ -10 & -7 & 2 \\ -10 & 2 & -7 \end{bmatrix}$

23. $B = \begin{bmatrix} -7 & 1 \\ 5 & -2 \end{bmatrix}$

25. $B = \begin{bmatrix} 4 & 1 & 4 \\ 1 & 6 & 9 \\ 2 & 6 & 2 \end{bmatrix}$

27. $\lambda = \frac{1}{1/4} = 4$; the eigenvector is $\begin{bmatrix} 1 \\ 1/2 \\ 0 \end{bmatrix}$.

29. For example, $A = \begin{bmatrix} 1 & 0 \\ 0 & 0 \end{bmatrix}$ and $\mathbf{x}_0 = \begin{bmatrix} 1 \\ 0 \end{bmatrix}$.

31. For example, $A = \begin{bmatrix} 0 & 1 \\ 1 & 0 \end{bmatrix}$ and $\mathbf{x}_0 = \begin{bmatrix} 1 \\ 0 \end{bmatrix}$.

33. For example, $A = \begin{bmatrix} 0 & -1 \\ 1 & -1 \end{bmatrix}$ and $\mathbf{x}_0 = \begin{bmatrix} 1 \\ 0 \end{bmatrix}$.

35. False

37. True

39. True

41. $\mathbf{x}_1 = \begin{bmatrix} 1 \\ -1 \end{bmatrix}, \mathbf{x}_2 = \begin{bmatrix} 0 \\ 1 \end{bmatrix}, \mathbf{x}_3 = \begin{bmatrix} 1 \\ -1 \end{bmatrix}, \mathbf{x}_4 = \begin{bmatrix} 0 \\ 1 \end{bmatrix}$.
The sequence $\mathbf{x}_k$ does not converge, it alternates. The eigenvalues of A are $\lambda = 1$ and $\lambda = -1$, and so there is no dominant eigenvalue, and convergence is not assured.

43. $\mathbf{x}_1 = \begin{bmatrix} -\frac{1}{2} \\ 1 \end{bmatrix}, \mathbf{x}_2 = \begin{bmatrix} -\frac{1}{2} \\ 1 \end{bmatrix}, \ldots$ and the sequence converges to the eigenvalue $\lambda = 1$ because $\mathbf{x}_0 = \begin{bmatrix} -1 \\ 2 \end{bmatrix}$ is an eigenvector associated with $\lambda = 1$.

45. $\mathbf{x}_1 = \begin{bmatrix} 1 \\ 1 \end{bmatrix}, \mathbf{x}_2 = \begin{bmatrix} -2 \\ 6 \end{bmatrix}, \mathbf{x}_3 = \begin{bmatrix} -20 \\ 28 \end{bmatrix}, \mathbf{x}_4 = \begin{bmatrix} -104 \\ 120 \end{bmatrix},$
$\mathbf{x}_5 = \begin{bmatrix} -464 \\ 496 \end{bmatrix}, \mathbf{x}_6 = \begin{bmatrix} -1952 \\ 2016 \end{bmatrix}$

47. $\mathbf{x}_1 = \begin{bmatrix} 6 \\ 4 \\ 4 \end{bmatrix}, \mathbf{x}_2 = \begin{bmatrix} 52 \\ 48 \\ 48 \end{bmatrix}, \mathbf{x}_3 = \begin{bmatrix} 504 \\ 496 \\ 496 \end{bmatrix}, \mathbf{x}_4 = \begin{bmatrix} 5008 \\ 4992 \\ 4992 \end{bmatrix},$
$\mathbf{x}_5 = \begin{bmatrix} 50{,}016 \\ 49{,}984 \\ 49{,}984 \end{bmatrix}, \mathbf{x}_6 = \begin{bmatrix} 500{,}032 \\ 499{,}968 \\ 499{,}968 \end{bmatrix}$

49. $\mathbf{x}_1 = \begin{bmatrix} 3 \\ 0 \\ -3 \end{bmatrix}, \mathbf{x}_2 = \begin{bmatrix} 9 \\ 0 \\ -9 \end{bmatrix}, \mathbf{x}_3 = \begin{bmatrix} 27 \\ 0 \\ -27 \end{bmatrix}, \mathbf{x}_4 = \begin{bmatrix} 81 \\ 0 \\ -81 \end{bmatrix},$
$\mathbf{x}_5 = \begin{bmatrix} 243 \\ 0 \\ -243 \end{bmatrix}, \mathbf{x}_6 = \begin{bmatrix} 729 \\ 0 \\ -729 \end{bmatrix}$.

51. $\lambda = 2$; eigenvector $= \begin{bmatrix} -1 \\ 0 \end{bmatrix}$.

53. $\lambda = 4.2458$; eigenvector $= \begin{bmatrix} -0.0579 \\ 1.0000 \\ -0.6518 \end{bmatrix}$.

55. $\lambda = 3$; eigenvector $= \begin{bmatrix} 0 \\ 1 \\ -0.5 \end{bmatrix}$.

Section 6.3

1. $\mathbf{x} = \begin{bmatrix} 1 \\ -7 \end{bmatrix}$

3. $\mathbf{x} = \begin{bmatrix} 0 \\ 2 \end{bmatrix}$

5. $\mathbf{x} = \begin{bmatrix} 1 \\ 1 \\ 2 \end{bmatrix}$

7. $\mathbf{x}_\mathcal{B} = \begin{bmatrix} 10 \\ -7 \end{bmatrix}$

9. $\mathbf{x}_\mathcal{B} = \begin{bmatrix} 2 \\ 1 \end{bmatrix}$

11. $\mathbf{x}_\mathcal{B} = \begin{bmatrix} 4 \\ -1 \\ -4 \end{bmatrix}$

13. $\begin{bmatrix} -5 & -1 \\ 9 & 2 \end{bmatrix}$

15. $\begin{bmatrix} -19 & -4 & 16 \\ 15 & 1 & -14 \\ 6 & 2 & -5 \end{bmatrix}$

17. $\begin{bmatrix} 2 & -1 \\ -1 & 1 \end{bmatrix}$

19. $\begin{bmatrix} -2 & -1 \\ 9 & 5 \end{bmatrix}$

21. $\begin{bmatrix} 31 & -15 & -49 \\ 17 & -7 & -25 \\ -6 & 3 & 10 \end{bmatrix}$

23. $\begin{bmatrix} 1 & 2 \\ 1 & -1 \end{bmatrix}$

25. $\mathbf{x}_{\mathcal{B}_2} = \begin{bmatrix} -9 \\ 16 \end{bmatrix}_{\mathcal{B}_2}$

27. $\mathbf{x}_{\mathcal{B}_2} = \begin{bmatrix} -1 \\ 3 \end{bmatrix}_{\mathcal{B}_2}$

29. $\mathbf{x}_{\mathcal{B}_1} = \begin{bmatrix} -193 \\ -99 \\ 39 \end{bmatrix}_{\mathcal{B}_1}$

31. $\begin{bmatrix} a \\ b \end{bmatrix}_{\mathcal{B}_1} = \begin{bmatrix} b \\ a \end{bmatrix}_{\mathcal{B}_2}$

33. For example, $\mathcal{B} = \left\{ \begin{bmatrix} -2 \\ 0 \end{bmatrix}, \begin{bmatrix} 0 \\ 1/3 \end{bmatrix} \right\}$.

35. For example, $\mathcal{B}_1 = \left\{ \begin{bmatrix} 2 \\ 0 \end{bmatrix}, \begin{bmatrix} 0 \\ -1/2 \end{bmatrix} \right\}$

and $\mathcal{B}_2 = \left\{ \begin{bmatrix} 1 \\ 0 \end{bmatrix}, \begin{bmatrix} 0 \\ 1 \end{bmatrix} \right\}$.

37. For example, $\mathcal{B}_1 = \left\{ \begin{bmatrix} 1 \\ 2 \end{bmatrix}, \begin{bmatrix} 3 \\ 7 \end{bmatrix} \right\}$

and $\mathcal{B}_2 = \left\{ \begin{bmatrix} 1 \\ 0 \end{bmatrix}, \begin{bmatrix} 0 \\ 1 \end{bmatrix} \right\}$.

39. True

41. True

43. False

45. HINT: Write $\mathbf{u}$ and $\mathbf{v}$ in terms of the basis vectors of $\mathcal{B}$.

47. HINT: Focus on showing that the two properties required of a linear transformation both hold.

49. HINT: Explain why for each column $\mathbf{u}_i$ of U, the product $V^{-1}\mathbf{u}_i = [\mathbf{u}_i]_{\mathcal{B}_2}$.

51. $\begin{bmatrix} \frac{61}{35} & \frac{48}{5} & -\frac{316}{35} \\ \frac{17}{7} & 8 & -\frac{51}{7} \\ \frac{62}{35} & \frac{51}{5} & -\frac{382}{35} \end{bmatrix}$

53. $\begin{bmatrix} -\frac{49}{538} & \frac{55}{269} & \frac{3}{538} & \frac{364}{269} \\ \frac{37}{538} & \frac{211}{269} & \frac{393}{538} & \frac{340}{269} \\ -\frac{45}{269} & \frac{112}{269} & -\frac{129}{269} & -\frac{369}{269} \\ \frac{219}{538} & -\frac{416}{269}, & \frac{305}{538} & -\frac{205}{269} \end{bmatrix}$

55. $\begin{bmatrix} 1 & 2 & 2 \\ 2 & 1 & 2 \\ -1 & -1 & -1 \end{bmatrix}$

Section 6.4

1. $A^5 = \begin{bmatrix} 131 & -396 \\ 33 & -100 \end{bmatrix}$

3. $A^5 = \begin{bmatrix} 1 & -93 & -184 \\ 0 & 32 & 66 \\ 0 & 0 & -1 \end{bmatrix}$

5. $A = \begin{bmatrix} 19 & -12 \\ 30 & -19 \end{bmatrix}$

7. $A = \begin{bmatrix} 2 & -3 & 4 \\ 0 & -1 & 2 \\ -1 & 1 & -1 \end{bmatrix}$

9. The matrix is not diagonalizable.

11. $P = \begin{bmatrix} 1 & 2 \\ 1 & 1 \end{bmatrix}$, $D = \begin{bmatrix} -1 & 0 \\ 0 & 3 \end{bmatrix}$

13. $P = \begin{bmatrix} \frac{1}{3} & \frac{1}{2} & \frac{1}{2} \\ -\frac{2}{3} & -\frac{1}{2} & -1 \\ 1 & 1 & 1 \end{bmatrix}$, $D = \begin{bmatrix} 0 & 0 & 0 \\ 0 & 1 & 0 \\ 0 & 0 & -1 \end{bmatrix}$

15. $P = \begin{bmatrix} -1 & 1 & -1 \\ 1 & 1 & 0 \\ 1 & 0 & 1 \end{bmatrix}$, $D = \begin{bmatrix} 0 & 0 & 0 \\ 0 & 1 & 0 \\ 0 & 0 & 1 \end{bmatrix}$

17. $P = \begin{bmatrix} 1 & 0 & 0 & \frac{1}{3} \\ 0 & 1 & 0 & 0 \\ 0 & 0 & 1 & 0 \\ 0 & 0 & 0 & 1 \end{bmatrix}$, $D = \begin{bmatrix} 1 & 0 & 0 & 0 \\ 0 & 2 & 0 & 0 \\ 0 & 0 & 3 & 0 \\ 0 & 0 & 0 & 4 \end{bmatrix}$

19. $A^{1000} = \begin{bmatrix} 1 & 0 \\ 0 & 1 \end{bmatrix}$

21. $A^{1000} = \begin{bmatrix} 2\left(3^{1000}\right) - 1 & 2 - 2\left(3^{1000}\right) \\ 3^{1000} - 1 & 2 - 3^{1000} \end{bmatrix}$

23. Dimension $= 2$.

25. For example, $A = \begin{bmatrix} 0 & 0 \\ 0 & 1 \end{bmatrix}$.

27. For example, $A = \begin{bmatrix} 1 & 1 \\ 0 & 1 \end{bmatrix}$.

29. For example, $A = \begin{bmatrix} 0 & 1 & 0 \\ 0 & 1 & 0 \\ 0 & 0 & 2 \end{bmatrix}$ has eigenvalues 0, 1, and 2.

31. True

33. False

35. False

37. False

39. HINT: u_1 and u_2 must be linearly independent.

41. HINT: Each eigenvalue has infinitely many distinct associated eigenvectors.

43. HINT: What is A^T if $A = PDP^{-1}$?

45. HINT: Let $A = PD_1P^{-1}$ and $B = PD_2P^{-1}$, then show that $AB = BA$.

47. $P = \begin{bmatrix} -1 & -\frac{2}{3} & -\frac{1}{3} & 0 \\ 2 & \frac{2}{3} & \frac{1}{3} & 1 \\ -3 & -2 & -\frac{2}{3} & 0 \\ 2 & 2 & \frac{2}{3} & 0 \end{bmatrix}$

$D = \begin{bmatrix} -1 & 0 & 0 & 0 \\ 0 & 0 & 0 & 0 \\ 0 & 0 & 1 & 0 \\ 0 & 0 & 0 & 2 \end{bmatrix}$

49. $P = \begin{bmatrix} 0 & 0 & \frac{2}{3} & 2 & 4 \\ 4 & 0 & -\frac{2}{3} & -2 & -8 \\ 4 & 1 & \frac{4}{3} & 4 & 8 \\ 4 & 0 & -2 & -4 & -8 \\ 0 & 0 & \frac{4}{3} & 2 & 4 \end{bmatrix}$

$D = \begin{bmatrix} -2 & 0 & 0 & 0 & 0 \\ 0 & -1 & 0 & 0 & 0 \\ 0 & 0 & 0 & 0 & 0 \\ 0 & 0 & 0 & 1 & 0 \\ 0 & 0 & 0 & 0 & 2 \end{bmatrix}$

Section 6.5

1. (a) $8 + 2i$
(b) $19 + 8i$
(c) $2 - 6i$
(d) $23 + 14i$

3. $\lambda_1 = 2 + 2i$; eigenspace basis $= \left\{ \begin{bmatrix} 1 + 2i \\ -5 \end{bmatrix} \right\}$

$\lambda_2 = 2 - 2i$; eigenspace basis $= \left\{ \begin{bmatrix} 1 - 2i \\ -5 \end{bmatrix} \right\}$

5. $\lambda_1 = 2 + i$; eigenspace basis $= \left\{ \begin{bmatrix} 1 - i \\ -1 \end{bmatrix} \right\}$

$\lambda_2 = 2 - i$; eigenspace basis $= \left\{ \begin{bmatrix} 1 + i \\ -1 \end{bmatrix} \right\}$

7. $\lambda_1 = 3 + i$; eigenspace basis $= \left\{ \begin{bmatrix} 1 + i \\ -1 \end{bmatrix} \right\}$

$\lambda_2 = 3 - i$; eigenspace basis $= \left\{ \begin{bmatrix} 1 - i \\ -1 \end{bmatrix} \right\}$

9. Rotation is by $\tan^{-1}(1/2) \approx 0.4636$ radians; the dilation is by $\sqrt{5}$.

11. Rotation is by $\tan^{-1}(1) = \frac{\pi}{4}$ radians; the dilation is by $\sqrt{2}$.

13. Rotation is by $\tan^{-1}(-3/4) \approx -0.6435$ radians; the dilation is by 5.

15. The rotation–dilation matrix is $B = \begin{bmatrix} 2 & -2 \\ 2 & 2 \end{bmatrix}$.

17. The rotation–dilation matrix is $B = \begin{bmatrix} 2 & -1 \\ 1 & 2 \end{bmatrix}$.

19. The rotation–dilation matrix is $B = \begin{bmatrix} 3 & -1 \\ 1 & 3 \end{bmatrix}$.

21. Other roots are $1 - 2i$ and $3 + i$, the multiplicity of each root is 1.

23. For example, $z = \frac{6\sqrt{5}}{5} + \frac{3\sqrt{5}}{5}i$.

25. $B = \begin{bmatrix} 0 & -2 \\ 2 & 0 \end{bmatrix}$

27. For example, $A = \begin{bmatrix} 1 & 1 \\ 0 & 1 \end{bmatrix}\begin{bmatrix} 1 & -2 \\ 2 & 1 \end{bmatrix}\begin{bmatrix} 1 & 1 \\ 0 & 1 \end{bmatrix}^{-1} = \begin{bmatrix} 3 & -4 \\ 2 & -1 \end{bmatrix}$.

29. For example, $A = \begin{bmatrix} i & i \\ -i & -i \end{bmatrix}$.

31. True

33. False

35. False

37. True

39. True

41. HINT: Start with $z = x + iy$ and $w = u + iv$, then apply the properties of complex conjugation.

43. HINT: Apply Exercise 41(b).

45. HINT: Start with $\lambda = x + iy$, then apply the properties of complex conjugation.

47. HINT: $|A - \lambda I| = (a - \lambda)^2 + b^2$.

49. (a) HINT: Write $u = \text{Re}(u) + i\text{Im}(u)$.
(b) HINT: Use hint given with this part of problem.
(c) HINT: Show that the real and imaginary parts of AP and PC are the same.

51. $\lambda_{1,2} = -2.507 \pm 1.692i \Rightarrow \left\{ \begin{bmatrix} 0.2373 \pm 0.3607i \\ 0.0862 \mp 0.2878i \\ -0.8505 \end{bmatrix} \right\}$

$\lambda_3 = 6.013 \Rightarrow \left\{ \begin{bmatrix} 0.5837 \\ 0.6889 \\ 0.4298 \end{bmatrix} \right\}$

53. $\lambda_{1,2} = 2.5948 \pm 0.4119i \Rightarrow \left\{ \begin{bmatrix} 0.6638 \\ 0.1906 \pm 0.2156i \\ 0.0472 \mp 0.2804i \\ -0.6280 \mp 0.0368i \end{bmatrix} \right\}$

$\lambda_3 = 13.2693 \Rightarrow \left\{ \begin{bmatrix} 0.0639 \\ 0.1441 \\ 0.3504 \\ 0.9232 \end{bmatrix} \right\}$

$$\lambda_4 = -5.4589 \Rightarrow \left\{ \begin{bmatrix} 0.6812 \\ -0.3472 \\ -0.5476 \\ 0.3399 \end{bmatrix} \right\}$$

Section 6.6

1. $y_1 = c_1 e^{-t} + c_2 e^{2t}$
   $y_2 = c_1 e^{-t} - c_2 e^{2t}$

3. $y_1 = 4c_1 e^{2t} + c_2 e^{-2t} + 2c_3 e^{-2t}$
   $y_2 = 3c_1 e^{2t} + 2c_2 e^{-2t} + 3c_3 e^{-2t}$
   $y_3 = c_1 e^{2t} + c_3 e^{-2t}$

5. $y_1 = c_1 (\cos 2t - \sin 2t) + c_2 (\cos 2t + \sin 2t)$
   $y_2 = c_1 (2 \cos 2t + \sin 2t) - c_2 (\cos 2t - 2 \sin 2t)$

7. $y_1 = 3c_1 e^{4t} - c_2 (\sin t - 4 \cos t) e^t + c_3 (\cos t + 4 \sin t) e^t$
   $y_2 = c_1 e^{4t} - 2c_3 (\cos t) e^t + 2c_2 (\sin t) e^t$
   $y_3 = 5c_1 e^{4t} - c_2 (\sin t - 3 \cos t) e^t + c_3 (\cos t + 3 \sin t) e^t$

9. $y_1 = 6c_1 e^t + c_2 e^{4t} + c_3 (3 \cos t - 2 \sin t) e^t + c_4 (2 \cos t + 3 \sin t) e^t$
   $y_2 = 2c_1 e^t + 2c_2 e^{4t} + 6c_3 (\cos t) e^t + 6c_4 (\sin t) e^t$
   $y_3 = 5c_1 e^t + 3c_2 e^{4t} + c_3 (2 \cos t + 3 \sin t) e^t - c_4 (3 \cos t - 2 \sin t) e^t$
   $y_4 = 2c_2 e^{4t} + 5c_3 (\sin t) e^t - 5c_4 (\cos t) e^t$

11. $y_1 = 2c_1 e^{-t} + 2c_2 e^{3t}$
    $y_2 = -c_1 e^{-t} + c_2 e^{3t}$

13. $y_1 = c_1 e^{-t} + 2c_2 e^{3t}$
    $y_2 = c_1 e^{-t} + c_2 e^{3t}$

15. $y_1 = -c_1 (\cos 2t - 2 \sin 2t) e^{2t} - c_2 (2 \cos 2t + \sin 2t) e^{2t}$
    $y_2 = 5c_1 (\cos 2t) e^{2t} + 5c_2 (\sin 2t) e^{2t}$

17. $y_1 = 2c_1 + c_2 e^{-t} - c_3 e^t$
    $y_2 = c_1 + 3c_3 e^t$
    $y_3 = 2c_1 + c_2 e^{-t}$

19. $y_1 = -2e^{-2t} + 6e^{3t}$
    $y_2 = -2e^{-2t} + 3e^{3t}$

21. $y_1 = -(\sin 3t) e^t + 2 (\cos 3t) e^t$
    $y_2 = -(\cos 3t) e^t - 2 (\sin 3t) e^t$

23. $y_1 = 2e^t - e^{2t} - 2e^{-t}$
    $y_2 = -2e^t + 2e^{2t}$
    $y_3 = 4e^t - 2e^{2t} - 6e^{-t}$

25. $y_1 = c_1 e^{(-0.3)t} + 2c_2 e^{(-0.4)t}$
    $y_2 = -c_1 e^{(-0.3)t} - 3c_2 e^{(-0.4)t}$

27. $y_1 = 70e^{-0.3t} - 60e^{-0.4t}$
    $y_2 = 90e^{-0.4t} - 70e^{-0.3t}$

29. $y_1 = -c_1 e^{-8t} + 5c_2 e^t$, and $y_2 = c_1 e^{-8t} + 4c_2 e^t$. As t gets large, $y_1 \approx 5c_2 e^t$ and $y_2 \approx 4c_2 e^t$, and hence the ratio $y_1/y_2 \approx 5/4$.

31. $y_1 = \frac{5}{3} e^t - \frac{2}{3} e^{-8t}$, $y_2 = \frac{4}{3} e^t + \frac{2}{3} e^{-8t}$

33. For example, $y_1' = -3y_1$ and $y_2' = 2y_2$.

35. For example,
$$y_1' = 10y_1 + 6y_2$$
$$y_2' = -18y_1 - 11y_2$$

37. For example,
$$y_1' = -6y_1 + 4y_2 + 7y_3$$
$$y_2' = -7y_1 + 5y_2 + 7y_3$$
$$y_3' = -4y_1 + 4y_2 + 5y_3$$

39. True

41. False

43. $y_1 \approx -0.7811c_1 e^{7.065t} - 0.7041c_2 (\cos 2.580t) e^{(-3.033t)}$
    $- 0.7041c_3 (\sin 2.580t) e^{(-3.033t)}$
    $y_2 \approx 0.4471c_1 e^{7.065t} - c_2 (0.1528 \cos (2.580t)$
    $+ 0.4597 \sin (2.580t)) e^{(-3.033t)} - c_3 (0.1528 \sin (2.580t)$
    $- 0.4597 \cos (2.580t)) e^{(-3.033t)}$
    $y_3 \approx -0.4359c_1 e^{7.065t} + c_2 (0.5141 \cos (2.580t)$
    $+ 0.0729 \sin (2.580t)) e^{(-3.033t)} + c_3 (0.5141 \sin (2.580t)$
    $- 0.0729 \cos (2.580t)) e^{(-3.033t)}$

45. $y_1 \approx -0.8167c_1 e^{-4.114t} + 1.139c_2 e^{7.297t}$
    $+ c_3 (0.2576 \sin (4.698t) - 0.5848 (\cos 4.698t)) e^{1.408t}$
    $- c_4 (0.2576 \cos (4.698t) + 0.5848 \sin (4.698t)) e^{1.408t}$
    $y_2 \approx -0.8101c_1 e^{-4.114t} + 0.1064c_2 e^{7.297t}$
    $- c_3 (2.336 \sin (4.698t) - 2.858 \cos (4.698t)) e^{1.408t}$
    $+ c_4 (2.336 \cos (4.698t) + 2.858 \sin (4.698t)) e^{1.408t}$
    $y_3 \approx 0.5896c_1 e^{-4.114t} + 0.39c_2 e^{7.297t}$
    $- c_3 e^{1.408t} (2.385 \cos (4.698t) + 1.314 \sin (4.698t))$
    $- c_4 e^{1.408t} (2.385 \sin (4.698t) - 1.314 \cos (4.698t))$
    $y_4 \approx c_1 e^{-4.114t} + c_2 e^{7.297t}$
    $+ c_3 \cos (4.698t) e^{1.408t} + c_4 \sin (4.698t) e^{1.408t}$

47. $y_1 \approx 0.5746 e^{8.01t} - 0.3412 e^{1.106t} - 1.233 e^{-7.115t}$
    $y_2 \approx -0.03121 e^{8.01t} + 0.1231 e^{1.106t} - 4.092 e^{-7.115t}$
    $y_3 \approx 0.7118 e^{8.01t} + 0.1924 e^{1.106t} + 2.096 e^{-7.115t}$

49. $y_1 \approx 11.63 (\cos 2.153t) e^{-3.179t} - 0.4729 e^{12.53t}$
    $- 4.158 e^{-0.1732t} + 7.978 (\sin 2.153t) e^{-3.179t}$
    $y_2 \approx 8.266 (\cos 2.153t) e^{-3.179t} - 0.3908 e^{12.53t}$
    $- 5.876 e^{-0.1732t} + 6.113 (\sin 2.153t) e^{-3.179t}$
    $y_3 \approx 2.928 e^{-0.1732t} - 0.1355 e^{12.53t}$
    $- 4.792 (\cos 2.153t) e^{-3.179t} + 3.693 (\sin 2.153t) e^{-3.179t}$
    $y_4 \approx 4.349 e^{-0.1732t} - 1.249 e^{12.53t}$
    $- 8.101 (\cos 2.153t) e^{-3.179t} - 7.601 (\sin 2.153t) e^{-3.179t}$

Chapter 7

Section 7.1

1. HINT: The required properties follow from the same properties of the real numbers.

3. HINT: You may assume that the sum of two continuous functions is a continuous function, as is the scalar multiple of a continuous function.

5. HINT: Adding two polynomials cannot produce a polynomial of degree greater than that of those being added. The scalar multiple of a polynomial produces a new polynomial that has the same degree or is equal to zero.

7. HINT: The hint from Exercise 3 applies here.

9. V is not a vector space under the given arithmetic operations. For instance, there is no vector $\mathbf{0}$ such that $\mathbf{v} + \mathbf{0} = \mathbf{v}$ for all $\mathbf{v}$.

11. V is not a vector space. Property 5(d) does not always hold. For instance, $(1+0)\begin{bmatrix} 1 \\ 0 \end{bmatrix} = \begin{bmatrix} 1 \\ 0 \end{bmatrix}$ but $(1)\begin{bmatrix} 1 \\ 0 \end{bmatrix} + (0)\begin{bmatrix} 1 \\ 0 \end{bmatrix} = \begin{bmatrix} 0 \\ 0 \end{bmatrix}$.

13. HINT: Show that the three requirements for a subspace are met.

15. HINT: Show that the three requirements for a subspace are met.

17. HINT: Show that the three requirements for a subspace are met.

19. S is a subspace.

21. S is a subspace.

23. S is not a subspace. S is not closed under addition.

25. S is a subspace.

27. S is not a subspace. The zero vector is not in S.

29. S is a subspace.

31. S is not a subspace.

33. For example, the set of vectors in the first quadrant of $\mathbf{R}^2$, with the usual definition of addition and scalar multiplication.

35. For example, the set of vectors in the first quadrant of $\mathbf{R}^2$, with the usual definition of addition and scalar multiplication.

37. Aside from $V_1 = \mathbf{R}^n$ with the usual definition of addition and scalar multiplication, we can also have $V_2 = \mathbf{R}^n$, but we let $\mathbf{w}$ be a fixed vector and then define addition by $\mathbf{u} \oplus \mathbf{v} = \mathbf{u}+\mathbf{v}-\mathbf{w}$ and scalar multiplication by $c \odot \mathbf{u} = c(\mathbf{u} - \mathbf{w}) + \mathbf{w}$. In this case, $\mathbf{w}$ is the zero vector for V_2.

39. True

41. True

43. True

45. True

47. HINT: Construct $\mathbf{0}$ by using $\mathbf{u}$ in S (S is nonempty) and observing that there must be a corresponding $-\mathbf{u}$ in S.

49. HINTs:
(a) Use the fact that addition of vectors is commutative.
(b) Assume that there are two zero vectors $\mathbf{0}_a$ and $\mathbf{0}_b$, then show that $\mathbf{0}_a = \mathbf{0}_b$.
(c) Use $\mathbf{v} + 0 \cdot \mathbf{v} = (1+0)\mathbf{v} = \mathbf{v}$.
(d) Use $\mathbf{v} + (-1)\mathbf{v} = \big(1+(-1)\big)\mathbf{v} = 0 \cdot \mathbf{v}$ together with (c).

Section 7.2

1. $\mathbf{v}$ is in span $\{3x^2 + x - 1, x^2 - 3x + 2\}$.

3. $\mathbf{v}$ is in span $\{3x^2 + x - 1, x^2 - 3x + 2\}$.

5. $\mathbf{v} = x^3 + 2x^2 - 3x$ is not in span $\{x^3 + x - 2, x^2 + 2x + 1, x^3 - x^2 + x\}$.

7. $\mathbf{v} = x^2 + 4x + 4$ is in span $\{x^3 + x - 2, x^2 + 2x + 1, x^3 - x^2 + x\}$.

9. $\mathbf{v}$ is in span $\left\{ \begin{bmatrix} 1 & 2 & 1 \\ 0 & 1 & 3 \end{bmatrix}, \begin{bmatrix} 0 & 3 & 1 \\ -1 & 1 & 0 \end{bmatrix} \right\}$.

11. $\mathbf{v}$ is in span $\left\{ \begin{bmatrix} 1 & 2 & 1 \\ 0 & 1 & 3 \end{bmatrix}, \begin{bmatrix} 0 & 3 & 1 \\ -1 & 1 & 0 \end{bmatrix} \right\}$.

13. $\mathbf{v}$ is in span $\left\{ \begin{bmatrix} -1 & 3 \\ 4 & 1 \end{bmatrix}, \begin{bmatrix} 0 & 2 \\ 5 & -3 \end{bmatrix}, \begin{bmatrix} 1 & 4 \\ 2 & 1 \end{bmatrix} \right\}$.

15. $\mathbf{v}$ is in span $\left\{ \begin{bmatrix} -1 & 3 \\ 4 & 1 \end{bmatrix}, \begin{bmatrix} 0 & 2 \\ 5 & -3 \end{bmatrix}, \begin{bmatrix} 1 & 4 \\ 2 & 1 \end{bmatrix} \right\}$.

17. $\{x^2 - 3, 3x^2 + 1\}$ is linearly independent in $\mathbf{P}^2$.

19. $\{x^3 + 2x + 4, x^2 - x - 1, x^3 + 2x^2 + 2\}$ is not linearly independent in $\mathbf{P}^3$.

21. $\left\{ \begin{bmatrix} 2 & -1 \\ 1 & 3 \end{bmatrix}, \begin{bmatrix} -4 & 2 \\ -2 & -6 \end{bmatrix} \right\}$ is not linearly independent in $\mathbf{R}^{2 \times 2}$.

23. $\left\{ \begin{bmatrix} 1 & 0 & 1 \\ 2 & 1 & 4 \end{bmatrix}, \begin{bmatrix} 3 & 1 & 2 \\ 0 & 3 & 3 \end{bmatrix} \right\}$ is linearly independent in $\mathbf{R}^{2 \times 3}$.

25. $\{\sin^2(x), \cos^2(x), 1\}$ is not linearly independent in $C[0, \pi]$.

27. For example,
$$\left\{ \begin{bmatrix} 1 & 0 \\ 0 & 0 \end{bmatrix}, \begin{bmatrix} 0 & 1 \\ 0 & 0 \end{bmatrix}, \begin{bmatrix} 0 & 0 \\ 1 & 0 \end{bmatrix}, \begin{bmatrix} 0 & 0 \\ 0 & 1 \end{bmatrix}, \begin{bmatrix} 1 & 1 \\ 1 & 1 \end{bmatrix} \right\}$$
spans $\mathbf{R}^{2 \times 2}$ but is not linearly independent.

29. For example, let $V = \mathbf{P}$, then $\{1, x, x^2, x^3, \dots\}$ is an infinite linearly independent subset.

31. Let $\mathcal{V}_1 = \{(1, 0, 0, \dots), (0, 0, 1, 0, 0 \dots),$ $(0, 0, 0, 0, 1, 0, 0, \dots)\}$ and $\mathcal{V}_2 = \{(0, 1, 0, 0, \dots), (0, 0, 0, 1, 0, 0 \dots),$ $(0, 0, 0, 0, 0, 1, 0, 0, \dots)\}$. Then $\mathcal{V}_1$ and $\mathcal{V}_2$ are infinite linearly independent subsets of $\mathbf{R}^\infty$ and span $(\mathcal{V}_1) \cap$ span $(\mathcal{V}_2) = \{\mathbf{0}\}$.

33. False

35. False

37. False

39. False

41. True

43. HINT: Show each polynomial is a linear combination of the given set.

45. HINT: See hint given with problem.

47. HINT: Consider cases $\mathbf{v}_1 = \mathbf{0}$ and $\mathbf{v}_1 \neq \mathbf{0}$ separately.

49. HINT: Suppose that $\{\mathbf{v}_1, \dots, \mathbf{v}_m\}$ spans $\mathbf{R}^\infty$. Truncate each vector to the first $m + 1$ components. Then the new vectors must also span $\mathbf{R}^{m+1}$, but cannot.

51. HINT: Apply Theorem 7.9(a).

53. HINT: $\mathbf{v}$ is a linear combination of $\{\mathbf{v}_1, \dots, \mathbf{v}_m\}$.

55. $\{x, \sin(\pi x/2), e^x\}$ is linearly independent.

57. $\{e^x, \cos^2(x), \cos(2x), 1\}$ is linearly dependent, so method shown in Example 9 will not work.

Section 7.3

1. $\mathcal{V}$ has too few vectors to be a basis for $\mathbf{P}^2$.

3. $\mathcal{V}$ could be a basis for $\mathbf{R}^{2 \times 2}$, since dim $(\mathbf{R}^{2 \times 2}) = 4$ and $\mathcal{V}$ has 4 vectors.

5. $\mathcal{V}$ has too few vectors to be a basis for $\mathbf{P}^4$.

7. $\mathcal{V}$ is a basis.

9. $\mathcal{V}$ is not a basis.

11. $\mathcal{V}$ is not a basis.

13. $\dim(S) = 8$, and a basis for S is

$$\begin{bmatrix} -1 & 0 & 0 \\ 0 & 1 & 0 \\ 0 & 0 & 0 \end{bmatrix}, \begin{bmatrix} -1 & 0 & 0 \\ 0 & 0 & 0 \\ 0 & 0 & 1 \end{bmatrix}, \begin{bmatrix} 0 & 1 & 0 \\ 0 & 0 & 0 \\ 0 & 0 & 0 \end{bmatrix},$$

$$\begin{bmatrix} 0 & 0 & 1 \\ 0 & 0 & 0 \\ 0 & 0 & 0 \end{bmatrix}, \begin{bmatrix} 0 & 0 & 0 \\ 1 & 0 & 0 \\ 0 & 0 & 0 \end{bmatrix}, \begin{bmatrix} 0 & 0 & 0 \\ 0 & 0 & 1 \\ 0 & 0 & 0 \end{bmatrix},$$

$$\begin{bmatrix} 0 & 0 & 0 \\ 0 & 0 & 0 \\ 1 & 0 & 0 \end{bmatrix}, \begin{bmatrix} 0 & 0 & 0 \\ 0 & 0 & 0 \\ 0 & 1 & 0 \end{bmatrix}$$

15. $\dim(S) = 3$, and a basis for S is

$$\left\{ \begin{bmatrix} -1 & 0 \\ 0 & 1 \end{bmatrix}, \begin{bmatrix} 0 & -1 \\ 0 & 1 \end{bmatrix}, \begin{bmatrix} 0 & 0 \\ -1 & 1 \end{bmatrix} \right\}$$

17. HINT: S is equivalent to the set of 2×2 matrices A such that $A\mathbf{v} = \mathbf{0}$.

19. $\dim(S) = \infty$. For example,

$$\{x(x-1)(x-2), x^2(x-1)(x-2), x^3(x-1)(x-2), \ldots\}$$

is an infinite set of linearly independent vectors in $C(\mathbf{R})$, each of which vanishes at $k = 0, 1, 2$.

21. We extend $\mathcal{V}$ to $\{2x^2 + 1, 4x - 3, 1\}$ to obtain a basis for $\mathbf{P}^2$.

23. We extend $\mathcal{V}$ to

$$\left\{ \begin{bmatrix} 1 & 0 \\ 0 & 1 \end{bmatrix}, \begin{bmatrix} 0 & 1 \\ 1 & 0 \end{bmatrix}, \begin{bmatrix} 1 & 1 \\ 1 & 0 \end{bmatrix}, \begin{bmatrix} 0 & 1 \\ 0 & 0 \end{bmatrix} \right\}$$

to obtain a basis for $\mathbf{R}^{2 \times 2}$.

25. We reduce the set $\mathcal{V}$ to $\{x + 1, x + 2\}$ to obtain a basis for $\mathbf{P}^1$.

27. HINT: Show that $\{\cos(t), \sin(t)\}$ is a basis for S.

29. A basis for S is the set $\{1, x\}$.

31. For example, let $V = \mathbf{P}$, and $S = \text{span}\{1\}$.

33. For example, let $V = \mathbf{P}^1$, and $S = \text{span}\{1\}$.

35. For example, let $V = \mathbf{P}$, and let $S = \text{span}\{1, x^2, x^4, x^6, \ldots\}$.

37. False

39. False

41. True

43. False

45. False

47. HINT: It is enough to show that $\{\mathbf{v}_1, 2\mathbf{v}_2, \ldots, k\mathbf{v}_k\}$ is linearly independent.

49. HINT: Show that the set is linearly independent and spans $\mathbf{R}^{2 \times 2}$.

51. HINT: Start with a basis for V, and remove one vector at a time to obtain a basis for each of $S_{m-1}, S_{m-2}, \ldots$.

53. HINT: See hint given with problem.

55. HINT: For part (a), show that a basis for V_1 must also be a basis for V_2.

57. HINT: See proof of corresponding theorem in Section 4.2.

Chapter 8

Section 8.1

1. (a) $\mathbf{u}_1 \cdot \mathbf{u}_5 = -3$

  (b) $\mathbf{u}_3 \cdot (-3\mathbf{u}_2) = -3$

  (c) $\mathbf{u}_4 \cdot \mathbf{u}_7 = 11$

  (d) $2\mathbf{u}_4 \cdot \mathbf{u}_7 = 22$

3. (a) $\|\mathbf{u}_7\| = \sqrt{29}$

  (b) $\|-\mathbf{u}_7\| = \sqrt{29}$

  (c) $\|2\mathbf{u}_5\| = 2\sqrt{6}$

  (d) $\|-3\mathbf{u}_5\| = 6\sqrt{2}$

5. (a) $\|\mathbf{u}_1 - \mathbf{u}_2\| = \sqrt{17}$

  (b) $\|\mathbf{u}_3 - \mathbf{u}_8\| = \sqrt{26}$

  (c) $\|2\mathbf{u}_6 - (-\mathbf{u}_3)\| = 7$

  (d) $\|-3\mathbf{u}_2 - 2\mathbf{u}_5\| = 3\sqrt{11}$

7. (a) $\mathbf{u}_1 \cdot \mathbf{u}_3 = -8 \neq 0$, so $\mathbf{u}_1$ and $\mathbf{u}_3$ are not orthogonal.

  (b) $\mathbf{u}_3 \cdot \mathbf{u}_4 = 0$, so $\mathbf{u}_3$ and $\mathbf{u}_4$ are orthogonal.

  (c) $\mathbf{u}_2 \cdot \mathbf{u}_5 = 4 \neq 0$, so $\mathbf{u}_2$ and $\mathbf{u}_5$ are not orthogonal.

  (d) $\mathbf{u}_1 \cdot \mathbf{u}_8 = 8 \neq 0$, so $\mathbf{u}_1$ and $\mathbf{u}_8$ are not orthogonal.

9. $a = \frac{3}{2}$

11. $a = \frac{28}{3}$

13. Set is not orthogonal.

15. Set is not orthogonal.

17. $a = -10$

19. $a = 7$ and $b = 11$

21. $\|\mathbf{u}_1\|^2 = 10$, $\|\mathbf{u}_2\|^2 = 10$, $\|\mathbf{u}_1 + \mathbf{u}_2\|^2 = 20$

23. $\|\mathbf{u}_1\|^2 = 14$, $\|\mathbf{u}_2\|^2 = 26$, $\|\mathbf{u}_1 + \mathbf{u}_2\|^2 = 40$

25. $\|3\mathbf{u}_1 + 4\mathbf{u}_2\| = 2\sqrt{109}$

27. $\mathbf{u}$ is not orthogonal to S.

29. A basis for $S^\perp$ is $\left\{ \begin{bmatrix} 3 \\ 1 \end{bmatrix} \right\}$.

31. A basis for $S^\perp$ is $\left\{ \begin{bmatrix} -1 \\ 1 \\ 0 \end{bmatrix}, \begin{bmatrix} 2 \\ 0 \\ 1 \end{bmatrix} \right\}$.

33. Let $\mathbf{s}_1 = \begin{bmatrix} 1 \\ 1 \\ 0 \end{bmatrix}$ and $\mathbf{s}_2 = \begin{bmatrix} 1 \\ -1 \\ 4 \end{bmatrix}$. Then $\mathbf{s} = \frac{3}{2}\mathbf{s}_1 - \frac{1}{2}\mathbf{s}_2$.

35. For example, $\mathbf{u} = \begin{bmatrix} 12 \\ 0 \end{bmatrix}$ and $\mathbf{v} = \begin{bmatrix} 1 \\ 0 \end{bmatrix}$.

37. For example, $\mathbf{u} = \begin{bmatrix} 1/\sqrt{5} \\ 2/\sqrt{5} \end{bmatrix}$.

39. For example, $\begin{bmatrix} 0 \\ 1 \\ 0 \end{bmatrix}$ and $\begin{bmatrix} 1 \\ 0 \\ 2 \end{bmatrix}$.

41. For example, $S = \text{span} \left\{ \begin{bmatrix} 1 \\ 0 \\ 0 \end{bmatrix} \right\}$.

43. For example, $\left\{ \begin{bmatrix} 0 \\ 0 \\ 0 \end{bmatrix}, \begin{bmatrix} 1 \\ 0 \\ 0 \end{bmatrix}, \begin{bmatrix} 0 \\ 1 \\ 0 \end{bmatrix} \right\}$.

45. False

47. False

49. True

51. False

53. True

55. False

57. HINT: Every vector $\mathbf{s}$ in S is a linear combination of a spanning set $\mathcal{S}$.

59. HINT: Show that $\mathbf{e}_i \cdot \mathbf{e}_j = 0$ whenever $i \neq j$.

61. HINT: Apply Theorem 8.2(c) twice.

63. HINT: Use the properties of Theorem 8.2.

65. HINT: Apply equation (2) that follows Definition 8.3.

67. HINT: Suppose that $\mathbf{v}$ is in both S and $S^\perp$. Use this to show that $\mathbf{v} \cdot \mathbf{v} = 0$.

69. HINT: If $\mathbf{a}$ in $\mathbf{R}^n$ is a column of A and $\mathbf{x} = (x_1, \dots, x_n)$, then $\mathbf{a}^T\mathbf{x} = \mathbf{a} \cdot \mathbf{x}$.

71. HINTS: (a) Compare definitions of $\mathbf{u} \cdot \mathbf{v}$ and $\mathbf{u}^T\mathbf{v}$.
(b) Start with $(A\mathbf{u}) \cdot \mathbf{v} = (A\mathbf{u})^T \mathbf{v}$.

73. (a) $\mathbf{u}_2 \cdot \mathbf{u}_3 = -4$
(b) $\|\mathbf{u}_1\| = \sqrt{39}$
(c) $\|2\mathbf{u}_1 + 5\mathbf{u}_3\| = \sqrt{826}$
(d) $\|3\mathbf{u}_1 - 4\mathbf{u}_2 - \mathbf{u}_3\| = \sqrt{1879}$

75. $\left\{ \begin{bmatrix} -5 \\ -7 \\ 1 \\ 0 \end{bmatrix}, \begin{bmatrix} -\frac{9}{2} \\ -4 \\ 0 \\ 1 \end{bmatrix} \right\}$

Section 8.2

1. (a) $\text{proj}_{\mathbf{u}_3} \mathbf{u}_2 = \begin{bmatrix} \frac{2}{5} \\ 0 \\ -\frac{1}{5} \end{bmatrix}$

(b) $\text{proj}_{\mathbf{u}_1} \mathbf{u}_2 = \begin{bmatrix} 0 \\ 0 \\ 0 \end{bmatrix}$

3. $\text{proj}_S \mathbf{u}_2 = \begin{bmatrix} \frac{2}{5} \\ 0 \\ -\frac{1}{5} \end{bmatrix}$

5. (a) $\dfrac{1}{\|\mathbf{u}_1\|} \mathbf{u}_1 = \begin{bmatrix} -\frac{3}{14}\sqrt{14} \\ \frac{1}{14}\sqrt{14} \\ \frac{1}{7}\sqrt{14} \end{bmatrix}$

(b) $\dfrac{1}{\|\mathbf{u}_4\|} \mathbf{u}_4 = k \begin{bmatrix} \frac{1}{14}\sqrt{14} \\ -\frac{3}{14}\sqrt{14} \\ \frac{1}{7}\sqrt{14} \end{bmatrix}$

7. An orthogonal basis for S is
$\left\{ \begin{bmatrix} 1 \\ 3 \end{bmatrix}, \begin{bmatrix} 3 \\ -1 \end{bmatrix} \right\}$

9. An orthogonal basis for S is $\left\{ \begin{bmatrix} -2 \\ 2 \\ 1 \end{bmatrix}, \begin{bmatrix} 3 \\ 4 \\ -2 \end{bmatrix} \right\}$.

11. An orthogonal basis for S is
$\left\{ \begin{bmatrix} 1 \\ -1 \\ 0 \\ 1 \end{bmatrix}, \begin{bmatrix} 3 \\ 2 \\ 2 \\ -1 \end{bmatrix} \right\}$.

13. An orthogonal basis for S is
$\left\{ \begin{bmatrix} -1 \\ 0 \\ 1 \end{bmatrix}, \begin{bmatrix} 2 \\ 4 \\ 2 \end{bmatrix}, \begin{bmatrix} 3 \\ -3 \\ 3 \end{bmatrix} \right\}$.

15. $\text{proj}_S \mathbf{u} = \begin{bmatrix} 1 \\ 1 \end{bmatrix}$

17. $\text{proj}_S \mathbf{u} = \begin{bmatrix} -\frac{3}{29} \\ -\frac{4}{29} \\ \frac{2}{29} \end{bmatrix}$

19. $\text{proj}_S \mathbf{u} = \begin{bmatrix} 1 \\ -1 \\ 0 \\ 1 \end{bmatrix}$

21. $\text{proj}_S \mathbf{u} = \begin{bmatrix} 1 \\ 0 \\ 2 \end{bmatrix}$

23. An orthonormal basis for S is $\left\{ \begin{bmatrix} \frac{1}{10}\sqrt{10} \\ \frac{3}{10}\sqrt{10} \end{bmatrix}, \begin{bmatrix} \frac{3}{10}\sqrt{10} \\ -\frac{1}{10}\sqrt{10} \end{bmatrix} \right\}$.

25. An orthonormal basis for S is $\left\{ \begin{bmatrix} -\frac{2}{3} \\ \frac{2}{3} \\ \frac{1}{3} \end{bmatrix}, \begin{bmatrix} \frac{3}{29}\sqrt{29} \\ \frac{4}{29}\sqrt{29} \\ -\frac{2}{29}\sqrt{29} \end{bmatrix} \right\}$.

27. An orthonormal basis for S is $\left\{ \begin{bmatrix} \frac{1}{3}\sqrt{3} \\ -\frac{1}{3}\sqrt{3} \\ 0 \\ \frac{1}{3}\sqrt{3} \end{bmatrix}, \begin{bmatrix} \frac{1}{2}\sqrt{2} \\ \frac{1}{3}\sqrt{2} \\ \frac{1}{3}\sqrt{2} \\ -\frac{1}{6}\sqrt{2} \end{bmatrix} \right\}$.

29. An orthonormal basis for S is

$$\left\{ \begin{bmatrix} -\frac{1}{2}\sqrt{2} \\ 0 \\ \frac{1}{2}\sqrt{2} \end{bmatrix}, \begin{bmatrix} \frac{1}{6}\sqrt{6} \\ \frac{1}{3}\sqrt{6} \\ \frac{1}{6}\sqrt{6} \end{bmatrix}, \begin{bmatrix} \frac{1}{3}\sqrt{3} \\ -\frac{1}{3}\sqrt{3} \\ -\frac{1}{3}\sqrt{3} \end{bmatrix} \right\}.$$

31. For example, let $\mathbf{u} = \begin{bmatrix} 1 \\ 0 \end{bmatrix}$ and $\mathbf{v} = \begin{bmatrix} 1 \\ 0 \end{bmatrix}$.

33. For example, let $\mathbf{u} = \begin{bmatrix} 1 \\ 0 \end{bmatrix}$ and $\mathbf{v} = \begin{bmatrix} 0 \\ 1 \end{bmatrix}$.

35. For example, let $\mathbf{u} = \begin{bmatrix} 3 \\ 1 \end{bmatrix}$ and $\mathbf{v} = \begin{bmatrix} 1 \\ 2 \end{bmatrix}$.

37. True

39. True

41. True

43. True

45. False

47. HINTS:

(a) Show that S_i is a subset of S_j for $i < j$.

(b) Reverse the hint for (a).

49. HINT: Show that $\mathbf{u} \cdot (\mathbf{u} + \mathbf{v}) \neq 0$ and $\mathbf{v} \cdot (\mathbf{u} + \mathbf{v}) \neq 0$.

51. HINT: Show the two required properties of a linear transformation hold.

55. HINT: Recall that $\text{proj}_S \mathbf{u}$ and $\mathbf{u} - \text{proj}_S \mathbf{u}$ are orthogonal and use the Pythagorean theorem.

57. HINTS:

(a) Use the hint given with this part of the problem.

(b) $\|\text{proj}_{\mathbf{v}}\mathbf{u}\| = \frac{|\mathbf{u} \cdot \mathbf{v}|}{\|\mathbf{v}\|}$.

(c) Show $\|\text{proj}_{\mathbf{v}}\mathbf{u}\| = \|\mathbf{u}\|$ only when $\mathbf{u} = c\mathbf{v}$.

59. An orthonormal basis is

$$\left\{ \begin{bmatrix} \frac{1}{22}\sqrt{22} \\ \frac{1}{11}\sqrt{22} \\ -\frac{2}{11}\sqrt{22} \\ -\frac{1}{22}\sqrt{22} \end{bmatrix}, \begin{bmatrix} -\frac{9}{1738}\sqrt{22}\sqrt{395} \\ \frac{21}{4345}\sqrt{22}\sqrt{395} \\ \frac{13}{4345}\sqrt{22}\sqrt{395} \\ -\frac{13}{1738}\sqrt{22}\sqrt{395} \end{bmatrix}, \begin{bmatrix} \frac{86}{69757}\sqrt{395}\sqrt{883} \\ \frac{929}{1046355}\sqrt{395}\sqrt{883} \\ \frac{782}{1046355}\sqrt{395}\sqrt{883} \\ \frac{4}{209271}\sqrt{395}\sqrt{883} \end{bmatrix} \right\}$$

61. $\text{proj}_S \mathbf{u} = \begin{bmatrix} \frac{2902}{1477} \\ \frac{1713}{2954} \\ \frac{1447}{1477} \\ \frac{1903}{422} \end{bmatrix}$

Section 8.3

1. Not symmetric

3. Symmetric

5. Not symmetric

7. Not symmetric

9. Not orthogonal

11. Orthogonal

13. Not orthogonal

15. $P = \begin{bmatrix} \frac{1}{5}\sqrt{5} & -\frac{2}{5}\sqrt{5} \\ \frac{2}{5}\sqrt{5} & \frac{1}{5}\sqrt{5} \end{bmatrix}$, $D = \begin{bmatrix} 2 & 0 \\ 0 & -3 \end{bmatrix}$

17. $P = \begin{bmatrix} \frac{1}{3}\sqrt{3} & \frac{1}{2}\sqrt{2} & -\frac{1}{6}\sqrt{6} \\ \frac{1}{3}\sqrt{3} & -\frac{1}{2}\sqrt{2} & -\frac{1}{6}\sqrt{6} \\ \frac{1}{3}\sqrt{3} & 0 & \frac{1}{3}\sqrt{6} \end{bmatrix}$, $D = \begin{bmatrix} 0 & 0 & 0 \\ 0 & 2 & 0 \\ 0 & 0 & -1 \end{bmatrix}$

19. $P = \begin{bmatrix} -\frac{1}{5}\sqrt{5} & \frac{2}{5}\sqrt{5} \\ \frac{2}{5}\sqrt{5} & \frac{1}{5}\sqrt{5} \end{bmatrix}$, $D = \begin{bmatrix} 0 & 0 \\ 0 & 5 \end{bmatrix}$

21. $P = \begin{bmatrix} -\frac{1}{2}\sqrt{2} & \frac{1}{6}\sqrt{6} & \frac{1}{3}\sqrt{3} \\ 0 & -\frac{1}{3}\sqrt{6} & \frac{1}{3}\sqrt{3} \\ \frac{1}{2}\sqrt{2} & \frac{1}{6}\sqrt{6} & \frac{1}{3}\sqrt{3} \end{bmatrix}$, $D = \begin{bmatrix} -2 & 0 & 0 \\ 0 & 0 & 0 \\ 0 & 0 & 3 \end{bmatrix}$

23. $P = \begin{bmatrix} -\frac{\sqrt{2}}{2} & 0 & \frac{\sqrt{2}}{2} \\ 0 & 1 & 0 \\ \frac{\sqrt{2}}{2} & 0 & \frac{\sqrt{2}}{2} \end{bmatrix}$, $D = \begin{bmatrix} -1 & 0 & 0 \\ 0 & 1 & 0 \\ 0 & 0 & 1 \end{bmatrix}$

25. $\lambda_1 = 1$ and $\lambda_2 = 11$.

27. $\lambda_1 = 0$, $\lambda_2 = 1$, and $\lambda_3 = 5$

29. $Q^{-1} = \begin{bmatrix} \frac{1}{5} & -\frac{2}{5} \\ \frac{2}{5} & \frac{1}{5} \end{bmatrix}$

31. $Q^{-1} = \begin{bmatrix} 0 & \frac{1}{2} & -\frac{1}{2} \\ \frac{1}{2} & 0 & 0 \\ 0 & \frac{1}{2} & \frac{1}{2} \end{bmatrix}$

33. $Q = \begin{bmatrix} \frac{3}{13}\sqrt{13} & -\frac{2}{13}\sqrt{13} \\ \frac{2}{13}\sqrt{13} & \frac{3}{13}\sqrt{13} \end{bmatrix}$, $R = \begin{bmatrix} \sqrt{13} & 0 \\ 0 & \sqrt{13} \end{bmatrix}$

35. $Q = \begin{bmatrix} \frac{1}{10}\sqrt{10} & \frac{3}{10}\sqrt{10} \\ \frac{3}{10}\sqrt{10} & -\frac{1}{10}\sqrt{10} \end{bmatrix}$, $R = \begin{bmatrix} \sqrt{10} & \sqrt{10} \\ 0 & \sqrt{10} \end{bmatrix}$.

37. $Q = \begin{bmatrix} \frac{1}{9}\sqrt{3} & \frac{25}{1629}\sqrt{1086} \\ \frac{1}{9}\sqrt{3} & -\frac{85}{3258}\sqrt{1086} \\ \frac{5}{9}\sqrt{3} & \frac{7}{3258}\sqrt{1086} \end{bmatrix}$, $R = \begin{bmatrix} 3\sqrt{3} & \frac{4}{9}\sqrt{3} \\ 0 & \frac{1}{9}\sqrt{1086} \end{bmatrix}$

39. $Q = \begin{bmatrix} -\frac{2}{3} & \frac{3}{29}\sqrt{29} \\ \frac{2}{3} & \frac{4}{29}\sqrt{29} \\ \frac{1}{3} & -\frac{2}{29}\sqrt{29} \end{bmatrix}$, $R = \begin{bmatrix} 3 & 0 \\ 0 & \sqrt{29} \end{bmatrix}$

41. For example, $A = \begin{bmatrix} 1 & 0 \\ 0 & 2 \end{bmatrix}$.

43. For example, $A = \begin{bmatrix} \frac{7}{5} & -\frac{6}{5} \\ -\frac{6}{5} & -\frac{2}{5} \end{bmatrix}$.

45. For example, $A = \begin{bmatrix} 0 & 0 \\ 0 & 0 \end{bmatrix}$.

47. For example, $A = \begin{bmatrix} -2 & 2 \\ -6 & 5 \end{bmatrix}$ (Example 1, Section 6.4).

49. True

51. True

53. True

55. False

57. True

59. HINT: $A^T A = I$.

61. For vectors $\mathbf{u}$ and $\mathbf{v}$, $\mathbf{u}^T \mathbf{v} = \mathbf{u} \cdot \mathbf{v}$.

63. HINT: How is A related to A^T?

65. HINT: Show that A^2 is symmetric.

67. $D \approx \begin{bmatrix} 8.0463 & 0 & 0 \\ 0 & 2.2795 & 0 \\ 0 & 0 & -3.3258 \end{bmatrix}$

$P \approx \begin{bmatrix} 0.3603 & -0.8287 & -0.4282 \\ -0.3790 & -0.5495 & 0.7446 \\ 0.8524 & 0.1060 & 0.5120 \end{bmatrix}$

69. $D \approx \begin{bmatrix} 7.624 & 0 & 0 & 0 \\ 0 & -1.211 & 0 & 0 \\ 0 & 0 & 5.639 & 0 \\ 0 & 0 & 0 & -6.051 \end{bmatrix}$

$P \approx \begin{bmatrix} -0.1376 & -0.6216 & 0.7118 & 0.2968 \\ -0.7426 & 0.5164 & 0.1391 & 0.4038 \\ -0.0078 & 0.3699 & 0.6127 & -0.6984 \\ 0.6558 & 0.4585 & 0.3140 & 0.5110 \end{bmatrix}$

71. $Q = \begin{bmatrix} -\frac{1}{2}\sqrt{2} & \frac{1}{3}\sqrt{3} & \frac{1}{6}\sqrt{6} \\ 0 & \frac{1}{3}\sqrt{3} & -\frac{1}{3}\sqrt{6} \\ \frac{1}{2}\sqrt{2} & \frac{1}{3}\sqrt{3} & \frac{1}{6}\sqrt{6} \end{bmatrix}$,

$R = \begin{bmatrix} \sqrt{2} & -\sqrt{2} & \sqrt{2} \\ 0 & 2\sqrt{3} & 4\sqrt{3} \\ 0 & 0 & 0 \end{bmatrix}$

73. $Q = \begin{bmatrix} \frac{1}{2} & \frac{1}{70}\sqrt{35} & \frac{47}{6790}\sqrt{6790} \\ \frac{1}{2} & \frac{9}{70}\sqrt{35} & -\frac{16}{3395}\sqrt{6790} \\ \frac{1}{2} & -\frac{3}{70}\sqrt{35} & \frac{17}{3395}\sqrt{6790} \\ -\frac{1}{2} & \frac{1}{10}\sqrt{35} & \frac{7}{970}\sqrt{6790} \end{bmatrix}$,

$R = \begin{bmatrix} 2 & \frac{3}{2} & 2 \\ 0 & \frac{1}{2}\sqrt{35} & \frac{22}{35}\sqrt{35} \\ 0 & 0 & \frac{2}{35}\sqrt{6790} \end{bmatrix}$

Section 8.4

1. $\sigma_1 = \sqrt{8}$ and $\sigma_2 = \sqrt{2}$

3. $\sigma_1 = \sqrt{16} = 4$ and $\sigma_2 = \sqrt{4} = 2$

5. $\sigma_1 = 3$ and $\sigma_2 = \sqrt{5}$

7. $\sigma_1 = \sqrt{4 + \sqrt{5}} \approx 2.497$ and $\sigma_2 = \sqrt{4 - \sqrt{5}} \approx 1.328$

9. $V = \begin{bmatrix} \frac{1}{2}\sqrt{2} & -\frac{1}{2}\sqrt{2} \\ \frac{1}{2}\sqrt{2} & \frac{1}{2}\sqrt{2} \end{bmatrix}$, $\Sigma = \begin{bmatrix} 3 & 0 \\ 0 & 1 \end{bmatrix}$,

$U = \begin{bmatrix} \frac{1}{2}\sqrt{2} & \frac{1}{2}\sqrt{2} \\ \frac{1}{2}\sqrt{2} & -\frac{1}{2}\sqrt{2} \end{bmatrix}$

11. $V \approx \begin{bmatrix} 0.2298 & 0.9732 \\ -0.9732 & 0.2298 \end{bmatrix}$, $\Sigma \approx \begin{bmatrix} 3.199 & 0 \\ 0 & 2.401 \\ 0 & 0 \end{bmatrix}$,

$U \approx \begin{bmatrix} -0.1606 & 0.9064 & -0.3906 \\ -0.9845 & -0.1182 & 0.1302 \\ 0.07183 & 0.4053 & 0.9113 \end{bmatrix}$

13. $A = V\Sigma^T U^T$, where $V = \begin{bmatrix} 0 & 1 \\ 1 & 0 \end{bmatrix}$, $\Sigma = \begin{bmatrix} 3 & 0 \\ 0 & \sqrt{2} \\ 0 & 0 \end{bmatrix}$,

$U = \begin{bmatrix} \frac{2}{3} & -\frac{1}{2}\sqrt{2} & -\frac{1}{6}\sqrt{2} \\ \frac{2}{3} & \frac{1}{2}\sqrt{2} & -\frac{1}{6}\sqrt{2} \\ \frac{1}{3} & 0 & \frac{2}{3}\sqrt{2} \end{bmatrix}$

15. $A = V\Sigma^T U^T$, where $V = \begin{bmatrix} 1 & 0 \\ 0 & 1 \end{bmatrix}$, $\Sigma = \begin{bmatrix} 3 & 0 \\ 0 & \sqrt{3} \\ 0 & 0 \\ 0 & 0 \end{bmatrix}$,

$U = \begin{bmatrix} \frac{2}{3} & \frac{1}{3}\sqrt{3} & -\frac{1}{6}\sqrt{2} & -\frac{1}{6}\sqrt{6} \\ \frac{2}{3} & -\frac{1}{3}\sqrt{3} & -\frac{1}{6}\sqrt{2} & \frac{1}{6}\sqrt{6} \\ \frac{1}{3} & 0 & \frac{2}{3}\sqrt{2} & 0 \\ 0 & \frac{1}{3}\sqrt{3} & 0 & \frac{1}{3}\sqrt{6} \end{bmatrix}$

17. Numerical rank of A is 2.

19. Numerical rank of A is 2.

21. False

23. True

25. True

27. $\sigma_1 \mathbf{u}_1 \mathbf{v}_1^T = \begin{bmatrix} -0.1181 & 0.5000 \\ -0.7237 & 3.065 \\ 0.0528 & -0.2236 \end{bmatrix}$

$\sigma_1 \mathbf{u}_1 \mathbf{v}_1^T + \sigma_2 \mathbf{u}_2 \mathbf{v}_2^T = \begin{bmatrix} 2.000 & 1.0 \\ -0.9999 & 3.000 \\ 0.9998 & 2.419 \times 10^{-5} \end{bmatrix}$

29. $\sigma_1 \mathbf{u}_1 \mathbf{v}_1^T = \begin{bmatrix} 2 & 2 & 1 & 0 \\ 0 & 0 & 0 & 0 \end{bmatrix}$ $\sigma_1 \mathbf{u}_1 \mathbf{v}_1^T + \sigma_2 \mathbf{u}_2 \mathbf{v}_2^T = \begin{bmatrix} 2 & 2 & 1 & 0 \\ 1 & -1 & 0 & 1 \end{bmatrix}$

31. HINT: If $A = U\Sigma V^T$, then $A^T = V\Sigma^T U^T$. Compare the nonzero terms of Σ and Σ^T.

33. HINT: Since U and V are orthogonal, $U^{-1} = U^T$ and $V^{-1} = V^T$.

35. HINT: Simplify $(PA)^T PA$, using P orthogonal.

37. HINTS:

    (a) Note that $A^T A\mathbf{x} = A^T(A\mathbf{x})$.

    (b) Recall that $(\text{col}(A))^{\perp} = \text{null}(A^T)$.

    (c) Show that $\text{null}(A)$ and $\text{null}(A^T A)$ are subsets of each other.

39. $V \approx \begin{bmatrix} 0.4527 & 0.8916 \\ 0.8916 & -0.4528 \end{bmatrix}$, $\Sigma \approx \begin{bmatrix} 5.9667 & 0 \\ 0 & 1.8436 \end{bmatrix}$,

$U \approx \begin{bmatrix} 0.9748 & 0.2228 \\ 0.2230 & -0.9748 \end{bmatrix}$

41. $V \approx \begin{bmatrix} 0.8224 & 0.4739 & 0.3147 \\ -0.2477 & 0.7963 & -0.5519 \\ -0.5121 & 0.3760 & 0.7722 \end{bmatrix}$,

$\Sigma \approx \begin{bmatrix} 5.5371 & 0 & 0 \\ 0 & 4.3320 & 0 \\ 0 & 0 & 3.2518 \end{bmatrix}$,

$U \approx \begin{bmatrix} -0.9276 & -0.3734 & -0.008949 \\ -0.08420 & 0.1860 & 0.9789 \\ -0.3639 & 0.9089 & -0.2039 \end{bmatrix}$

Section 8.5

1. $\text{proj}_S \mathbf{y} = \begin{bmatrix} -\frac{1}{2} \\ \frac{1}{2} \end{bmatrix}$

3. $\text{proj}_S \mathbf{y} = \begin{bmatrix} \frac{199}{189} \\ -\frac{299}{189} \\ \frac{218}{189} \end{bmatrix}$

5. $14x_1 - 3x_2 = 23$
$-3x_1 + 6x_2 = -2$

7. $6x_1 + 3x_2 - 2x_3 = 6$
$3x_1 + 18x_2 - 23x_3 = 27$
$-2x_1 - 23x_2 + 30x_3 = -34$

9. $x_1 = -\frac{152}{195}$ and $x_2 = -\frac{17}{195}$

11. $x_1 = -2t$, $x_2 = -5t - \frac{4}{3}$, and $x_3 = t$

13. The normal equations are

$$\begin{bmatrix} 2 & 2 & 4 \\ 2 & 4 & 8 \\ 4 & 8 & 16 \end{bmatrix} \begin{bmatrix} c_1 \\ c_2 \\ c_3 \end{bmatrix} = \begin{bmatrix} 6 \\ 10 \\ 20 \end{bmatrix}.$$

We obtain infinitely many solutions since there are infinitely many parabolas that pass through two given points.

15. For example,

$$x_1 \qquad\quad = 0$$
$$x_2 = 0$$
$$x_1 + x_2 = 1$$

17. For example,

$$x_1 + x_2 = 0$$
$$2x_1 + 2x_2 = 0$$
$$3x_1 + 3x_2 = 0$$
$$4x_1 + 4x_2 = 1$$

19. For example,

$$x_1 \qquad\qquad = 0$$
$$x_2 \qquad = 0$$
$$x_3 = 0$$

21. False

23. True

25. True

27. False

29. HINT: Make the columns of A orthonormal—then this is true.

31. HINT: $A^T A$ is an identity matrix.

33. HINT: Use hint given with problem.

35. $y = 2.2071 + 0.5214x$

37. $y = 2.081 + 0.07458x$

39. $y = 1.667 + 0.09x + 0.3833x^2$

41. $y = 2.191 - 0.06x - 0.5857x^2$

43. $y = 0.9975e^{0.9702x}$

45. $y = 15.49e^{-0.1564x}$

47. $y = 2.173x^{1.306}$

49. $y = 38.51x^{-0.5491}$

51. $p = 0.2001d^{1.499}$

53. $f(t) = 2199.8 + 2.65t - 16.75t^2$, $t = 11.539$ seconds to hit the ground.

55. $y \approx 2.307e^{-0.2211t}$. The initial size of the sample is $y \approx 2.307$ grams. The amount present at $t = 15$ is $y \approx 0.08370$ grams.

Chapter 9

Section 9.1

1. $T(\mathbf{v}_2 - 2\mathbf{v}_1) = \begin{bmatrix} -5 \\ -3 \end{bmatrix}$

3. $T(2x^2 - 4x - 1) = \begin{bmatrix} -4 \\ 5 \end{bmatrix}$

5. HINT: Focus on Definition 9.1 or Theorem 9.2.

7. HINT: Focus on Definition 9.1 or Theorem 9.2.

9. HINT: Focus on Definition 9.1 or Theorem 9.2.

11. T is a linear transformation. Apply Theorem 9.2 to show this.

13. T is a linear transformation. Apply Theorem 9.2 to show this.

15. T is a linear transformation. Apply Theorem 9.2 to show this.

17. T is a linear transformation. Apply Theorem 9.2 to show this.

19. T is a linear transformation. Apply Theorem 9.2 to show this.

21. T is not a linear transformation.

23. $\ker(T) = \{p(x) : p(x) = ax + a\}$, $\text{range}(T) = \mathbf{R}$

25. $\ker(T) = \{\mathbf{0}_{P_2}\}$
$\text{range}(T) = \text{span} \left\{ \begin{bmatrix} 1 & 0 \\ 0 & 0 \end{bmatrix}, \begin{bmatrix} 0 & 1 \\ 1 & 0 \end{bmatrix}, \begin{bmatrix} 0 & 0 \\ 0 & 1 \end{bmatrix} \right\}$

27. T is not one-to-one, but is onto.

29. T is not one-to-one, but is onto.

31. $V = \mathbf{R}$ and $W = \mathbf{R}^2$, and define $T(a) = \left(\begin{bmatrix} a \\ 0 \end{bmatrix} \right)$.

33. $V = \mathbf{R}^2$ and $W = \mathbf{R}^2$, and define $T\left(\begin{bmatrix} a \\ b \end{bmatrix} \right) = \begin{bmatrix} a \\ 0 \end{bmatrix}$.

35. $V = \mathbf{R}^4$ and $W = \mathbf{R}^3$, and define $T\left(\begin{bmatrix} a \\ b \\ c \\ d \end{bmatrix}\right) = \begin{bmatrix} a \\ b \\ c \end{bmatrix}$.

37. $V = \mathbf{R}^k$ and $W = \mathbf{R}$, and define $T(\mathbf{v}) = \mathbf{0}$.

39. True

41. True

43. False

45. True

47. False

49. $\dim \big(\text{range}\,(T)\big) = 3$

51. HINT: Use property (a) of a linear transformation.

53. HINT: Apply Theorem 9.2

55. HINT: Use $\mathbf{0}_V + \mathbf{0}_V = \mathbf{0}_V$.

57. HINT: Exercise 55 can be used here to show one of the required properties of a subspace.

59. HINT: Use the hint with the problem.

61. HINT: Use the extended version of Theorem 9.2.

63. HINT: Recall that differentiation distributes across sums of functions.

65. HINT: Recall that differentiation distributes across sums of functions.

67. HINT: $(x^2\, p(x))' = 2xp(x) + x^2\, p'(x)$

Section 9.2

1. $\dim\,(V) = \dim\,\big(\mathbf{R}^8\big) = 8$, and $\dim\,(W) = \dim\,\big(\mathbf{P}^9\big) = 10$. Since $\dim\,(V) \neq \dim\,(W)$, the vector spaces are not isomorphic.

3. $\dim\,(V) = \dim\,\big(\mathbf{R}^{3\times6}\big) = 18$, and $\dim\,(W) = \dim\,\big(\mathbf{P}^{17}\big) = 18$. Since $\dim\,(V) = \dim\,(W)$, the vector spaces are isomorphic.

5. $\dim\,(V) = \dim\,\big(\mathbf{R}^{13}\big) = 13$, and $\dim\,(W) = \dim\,(\mathbf{C}\,[0,\,1]) = \infty$. Since $\dim\,(V) \neq \dim\,(W)$, the vector spaces are not isomorphic.

7. HINT: A polynomial is identically zero exactly when all of its coefficients are zero. This can be used to show that T is one-to-one. You need also show that T is a linear transformation and is onto.

9. HINT: A matrix is zero exactly when all of its entries are zero. This can be used to show that T is one-to-one. You need also show that T is a linear transformation and is onto.

11. T is an isomorphism.

13. T is not an isomorphism.

15. $T^{-1}\left(\begin{bmatrix} c \\ d \end{bmatrix}\right) = (c/2 + d)\,x + c/2$

17. $T^{-1}(ax^2 + bx + c) = cx^2 - bx + a$, so that $T^{-1} = T$.

19. HINT: Note that all vectors in S have the form $\begin{bmatrix} a_1 \\ a_2 \\ 0 \end{bmatrix}$.

21. HINT: Focus carefully on the form of a general vector in $\mathbf{P}_e$. It is helpful to consider a few concrete cases.

23. $T\left(\begin{bmatrix} a \\ b \\ c \\ d \\ e \end{bmatrix}\right) = ax^4 + bx^3 + cx^2 + dx + e$ is an isomorphism.

25. $T\left(\begin{bmatrix} a & b \\ c & d \end{bmatrix}\right) = ax^3 + bx^2 + cx + d$ is an isomorphism.

27. $S = \text{span}\left\{\begin{bmatrix} 1 & 0 & 0 \\ 0 & 0 & 0 \end{bmatrix}, \begin{bmatrix} 0 & 1 & 0 \\ 0 & 0 & 0 \end{bmatrix}, \begin{bmatrix} 0 & 0 & 0 \\ 1 & 0 & 0 \end{bmatrix}, \begin{bmatrix} 0 & 0 & 0 \\ 0 & 1 & 0 \end{bmatrix}\right\}$

29. Let S be the set of all vectors of the form $(a_1, a_2, \dots a_n, 0, 0, 0, \dots)$. That is, all infinite vectors with entries equal to zero from some point on.

31. False

33. True

35. False

37. True

39. False

41. False

43. False

45. HINT: T must be one-to-one and onto to have an inverse.

47. HINT: The proof of Theorem 3.19 can be used as a model for showing that T^{-1} is also a linear transformation.

49. HINT: T is always onto range(T).

Section 9.3

1. $\mathbf{v} = \begin{bmatrix} -11 \\ -4 \end{bmatrix}$

3. $\mathbf{v} = -x^2 - 14x - 9$

5. $\mathbf{v}_{\mathcal{G}} = \begin{bmatrix} 4 \\ 3 \end{bmatrix}_{\mathcal{G}}$

7. $\mathbf{v}_{\mathcal{G}} = \begin{bmatrix} 3 \\ -5 \\ 6 \end{bmatrix}_{\mathcal{G}}$

9. $\mathbf{v} = \begin{bmatrix} -4 \\ 3 \end{bmatrix}_{\mathcal{G}}$

11. $\mathbf{v} = \begin{bmatrix} \frac{21}{4} \\ \frac{9}{2} \\ -\frac{7}{4} \end{bmatrix}_{\mathcal{G}}$

13. $T\left(\mathbf{v}_{\mathcal{G}}\right) = \begin{bmatrix} 17 \\ 28 \end{bmatrix}$

15. $T\left(\mathbf{v}_{\mathcal{G}}\right) = 27x^2 + 12x + 67$

17. $T\left(\mathbf{v}_{\mathcal{G}}\right) = 4\cos x - 9\sin x + 9e^{-x}$

19. $A = \begin{bmatrix} 0 & 1 \\ -1 & 0 \end{bmatrix}$

21. $A = \begin{bmatrix} 2 & 0 & 0 \\ 0 & 1 & 0 \end{bmatrix}$

23. $A = \begin{bmatrix} -3 & 0 \\ 2 & 1 \end{bmatrix}$

25. $A = \begin{bmatrix} 7 & 12 \\ -18 & -31 \end{bmatrix}$

27. (a) $\begin{bmatrix} a/2 & b/2 & c/2 \\ d & e & f \end{bmatrix}$ (b) $\begin{bmatrix} 3a & b & c \\ 3d & e & f \end{bmatrix}$

29. (a) $\begin{bmatrix} c & a & b \\ f & d & e \end{bmatrix}$ (b) $\begin{bmatrix} d & f & e \\ a & c & b \end{bmatrix}$

31. $T^{-1}(x) = -2x - 4$

33. $T^{-1}(x+1) = \begin{bmatrix} 7 \\ -4 \end{bmatrix}$

35. $\mathbf{v} = 2x^2 - 3$, and $\mathcal{G} = \left\{x^2, x, 1\right\}$

37. $\mathcal{G} = \left\{\begin{bmatrix} -7/3 \\ 0 \end{bmatrix}, \begin{bmatrix} 0 \\ 5/4 \end{bmatrix}\right\}$

39. $V = \mathbf{R}^3$ and $W = \mathbf{R}^2$, and let $\mathcal{G} = \left\{\begin{bmatrix} 1 \\ 0 \\ 0 \end{bmatrix}, \begin{bmatrix} 0 \\ 1 \\ 0 \end{bmatrix}, \begin{bmatrix} 0 \\ 0 \\ 1 \end{bmatrix}\right\}$

be the basis for V, and $\mathcal{Q} = \left\{\begin{bmatrix} 1 \\ 0 \end{bmatrix}, \begin{bmatrix} 0 \\ 1 \end{bmatrix}\right\}$ be the basis for W. Define $T(\mathbf{v}) = A\mathbf{v}$, where $A = \begin{bmatrix} 2 & 1 & 2 \\ 0 & 3 & 1 \end{bmatrix}$.

41. False

43. True

45. True

47. HINT: The general results $[c\mathbf{v}]_{\mathcal{G}} = c\,[\mathbf{v}]_{\mathcal{G}}$ and $[\mathbf{v}_1 + \mathbf{v}_2]_{\mathcal{G}} = [\mathbf{v}_1]_{\mathcal{G}} + [\mathbf{v}_2]_{\mathcal{G}}$ are useful here.

49. HINT: A more general version of the results given in the answer to Exercise 47 can be used here.

51. HINT: The proof of part (b) follows from induction on n.

Section 9.4

1. $S = \begin{bmatrix} 2 & 5 \\ -1 & 4 \end{bmatrix}$

3. $S = \begin{bmatrix} 3 & 4 & 1 & 0 \\ 2 & 0 & 7 & 6 \\ 1 & 0 & 5 & 2 \\ 0 & 2 & 1 & 3 \end{bmatrix}$

5. $S = \begin{bmatrix} 1 & 9 & -1 \\ -2 & 0 & -1 \\ 0 & 1 & 1 \end{bmatrix}$

7. $S = \begin{bmatrix} 1 & -1 \\ 1 & 1 \end{bmatrix}$

9. $A = \begin{bmatrix} 62 & 204 \\ -18 & -59 \end{bmatrix}$

11. $A = \begin{bmatrix} 3 & 6 & 2 \\ -3 & -8 & -3 \\ 10 & 17 & 6 \end{bmatrix}$

13. $A = \begin{bmatrix} -3 & -2 \\ 7 & 5 \end{bmatrix}$

15. $A = \begin{bmatrix} -889 & 1411 \\ -562 & 892 \end{bmatrix}$

17. A and B are not similar matrices.

19. A and B are similar matrices.

21. $V = \mathbf{R}^2$, and let $\mathcal{G} = \left\{\begin{bmatrix} 3 \\ 2 \end{bmatrix}, \begin{bmatrix} 4 \\ 3 \end{bmatrix}\right\}$ and $\mathcal{H} = \left\{\begin{bmatrix} 1 \\ 0 \end{bmatrix}, \begin{bmatrix} 0 \\ 1 \end{bmatrix}\right\}$

23. $B = \begin{bmatrix} 1 & 2 \\ 3 & 4 \end{bmatrix}$ and $A = \begin{bmatrix} 31 & 12 \\ -67 & -26 \end{bmatrix}$, related by $S = \begin{bmatrix} 5 & 2 \\ 8 & 3 \end{bmatrix}$.

25. True

27. False

29. False

31. False

33. False

35. $D = S_2\,S_1$

37. HINT: A and B have the same diagonal matrix D in their diagonalizations.

39. HINT: If S is invertible, then $\left(S^{-1}\right)^T = \left(S^T\right)^{-1}$.

41. A and B are not similar matrices.

43. A and B are similar matrices.

Chapter 10

Section 10.1

1. $\langle \mathbf{u}, \mathbf{v} \rangle = 32$

3. $\langle p, q \rangle = -8$

5. $\langle f, g \rangle = 2$

7. $\langle A, B \rangle = -9$

9. $a = 3$

11. No value of a will make p and q orthogonal.

13. No value of b will make f and g orthogonal.

15. Norm $= \sqrt{33}$

17. Norm $= \sqrt{174}$

19. Norm $= \sqrt{\frac{2}{7}}$

21. $\|A\| = \sqrt{14}$

23. $\text{proj}_{\mathbf{u}}\mathbf{v} = \begin{bmatrix} \frac{32}{15} \\ \frac{64}{15} \\ \frac{32}{15} \end{bmatrix}$

25. $\text{proj}_p q = -\frac{8}{23}x - \frac{16}{69}$

27. $\text{proj}_f g = 0$

29. $\text{proj}_A B = \begin{bmatrix} \frac{1}{3} & -\frac{1}{6} \\ \frac{1}{6} & 0 \end{bmatrix}$

31. $\mathbf{u} = \begin{bmatrix} 1 \\ 0 \end{bmatrix}$ and $\mathbf{v} = \begin{bmatrix} 0 \\ 1 \end{bmatrix}$

33. $\mathbf{u} = \begin{bmatrix} 6/\sqrt{13} \\ -4/\sqrt{13} \end{bmatrix}$, $t_1 = t_2 = 1$

35. $A = \begin{bmatrix} 1 & 0 & 0 \\ 0 & 2 & 0 \\ 0 & 0 & 3 \end{bmatrix}$

37. Let $w(x) = \cos(x)$, and define $\langle p, q \rangle = \int_0^1 p(x)q(x)w(x)\,dx$.

39. $\langle p, q \rangle = 0$ for all p and q in $\mathbf{P}^2$.

41. True

43. True

45. True

47. False

49. False

51. HINT: Use the distributive property of the real numbers to establish (b) and (c) of the definition of inner product.

53. HINT: The solution to Exercise 51 can be used as a model for this problem.

55. HINT: Example 5 can serve as a guide for this proof.

57. HINT: Review the properties of matrix transposes.

59. HINT: $\|c\mathbf{v}\|^2 = \langle c\mathbf{v}, c\mathbf{v} \rangle$

61. HINT: Use induction on k.

63. HINT: $\left\| \frac{1}{\|\mathbf{v}\|}\mathbf{v} \right\| = \left| \frac{1}{\|\mathbf{v}\|} \right| \|\mathbf{v}\|$ by Exercise 59.

65. HINT: Use properties of inner products to verify the required properties of a linear transformation.

67. HINT: $\|\mathbf{u} - \mathbf{v}\|^2 = \langle \mathbf{u} - \mathbf{v}, \mathbf{u} - \mathbf{v} \rangle$

69. HINT: See Theorem 8.8 in Section 8.1.

71. HINT: Apply Exercise 67.

Section 10.2

1. $\left\{ \begin{bmatrix} \frac{2}{9}\sqrt{3} \\ \frac{5}{18}\sqrt{3} \\ \frac{1}{18}\sqrt{3} \end{bmatrix}, \begin{bmatrix} -\frac{1}{22}\sqrt{66} \\ \frac{1}{33}\sqrt{66} \\ -\frac{1}{11}\sqrt{66} \end{bmatrix}, \begin{bmatrix} \frac{8}{99}\sqrt{33} \\ -\frac{7}{198}\sqrt{33} \\ -\frac{23}{198}\sqrt{33} \end{bmatrix} \right\}$

3. $a = -5$, $\left\{ \begin{bmatrix} \frac{1}{3}\sqrt{3} \\ -\frac{1}{6}\sqrt{3} \\ \frac{1}{6}\sqrt{3} \end{bmatrix}, \begin{bmatrix} \frac{1}{3} \\ \frac{1}{6} \\ -\frac{5}{6} \end{bmatrix}, \begin{bmatrix} \frac{1}{6}\sqrt{2} \\ \frac{1}{3}\sqrt{2} \\ \frac{1}{3}\sqrt{2} \end{bmatrix} \right\}$

5. $a = -1$, $\left\{ \frac{1}{6}x^2 + \frac{1}{6}x, \frac{1}{2}x^2 - \frac{1}{2}x - 1, \frac{1}{3}x^2 - \frac{2}{3}x \right\}$

7. $\mathbf{v} = (5)\begin{bmatrix} 4 \\ 5 \\ 1 \end{bmatrix} + (1)\begin{bmatrix} -3 \\ 2 \\ -6 \end{bmatrix} + (-1)\begin{bmatrix} 16 \\ -7 \\ -23 \end{bmatrix}$

9. $\text{proj}_S \mathbf{v} = \begin{bmatrix} \frac{7}{27} \\ \frac{35}{108} \\ \frac{7}{108} \end{bmatrix}$

11. $\text{proj}_S f = 2\sin(x)$

13. $\left\{ \begin{bmatrix} 1 \\ -1 \\ 0 \end{bmatrix}, \begin{bmatrix} \frac{6}{5} \\ \frac{4}{5} \\ 1 \end{bmatrix} \right\}$

15. $\left\{ 1, x^2 - \frac{1}{3} \right\}$

17. $\left\{ x, -\frac{1}{5}x + 1 \right\}$

19. For example, let $\mathbf{u}_2 = \begin{bmatrix} 4 \\ -3 \end{bmatrix}$.

21. For example, let $p_2(x) = 5x - 3$.

23. For example, $f_1(x) = 0$, $f_2(x) = 1$, and $f_3(x) = \cos(x)$.

25. False

27. True

29. False

31. True

33. HINT: Apply hint given with problem.

35. HINT: The suggested approach works well.

37. HINT: $\text{proj}_S \mathbf{u}$ is in S.

39. (a) $\mathbf{u}$

    (b) $\mathbf{0}$

41. HINT: Use $\|\mathbf{v}\|^2 = \langle \mathbf{v}, \mathbf{v} \rangle$ and Theorem 10.12.

Section 10.3

1. $y = \frac{5}{2} + 1x$

3. The slope of ℓ_1 would be greater than the slope of ℓ_2.

5. The resulting line will be the same.

7. $f_2(x) = \frac{1}{2} + \frac{2}{\pi}\cos(x)$

9. $f_2(x) = \frac{1}{2} + \frac{2}{\pi}\sin(x)$

11. $f_2(x) = 1 + 2\sin(x) - \sin(2x)$

13. $f_2(x) = \frac{1}{3}\pi^2 - 4\cos(x) + \cos(2x)$

15. $a_2 = 1$ and $b_3 = -1$, with all other coefficients zero.

17. $a_0 = \frac{3}{2}$ and $a_8 = -\frac{1}{2}$, with all other Fourier coefficients zero.

19. $g_1(x) = \frac{3}{2} - \cos(x)$

21. $g_1(x) = \frac{1}{2} + \frac{3}{2}\cos(x) + \frac{3}{2}\sin(x)$

23. For example, consider the data set $\{(-1, -1), (0, 1), (1, -1)\}$. This set has ordinary least squares regression line $y = 0$, and weighted least squares regression line with triple the weight on the right-most point $y = -\frac{2}{5} - \frac{1}{2}x$.

25. For example, let $f(x) = 1$. Then, $a_0 = 1$, and all other Fourier coefficients are zero.

27. $f(x) = 1 + x$

29. True

31. False

33. True

35. HINT: $\langle 1, \sin(kx) \rangle = \frac{1}{\pi}\int_{-\pi}^{\pi} \sin(kx)\,dx$

37. HINT: $\|\sin(kx)\|^2 = \frac{1}{\pi}\int_{-\pi}^{\pi} \sin^2(kx)\,dx$

39. HINT: Take $u = x$ and $dv = \cos(kx)dx$ in the integration by parts formula.

41. HINT: $\cos(k\pi) = (-1)^k$ and $\sin(k\pi) = 0$ for all integers k.

43. $g_5(x) \approx 2.7125 + 0.5445\cos(x) + 0.05\cos(2x)$
$+ 0.1555\cos(3x) + 0.075\cos(4x) + 0.1555\cos(5x)$
$- 0.06768\sin(x) - 0.025\sin(2x) + 0.03232\sin(3x)$
$- 0.03232\sin(5x)$

45. $f_3(x) \approx 3.6761 - 3.6761\cos(x) + 1.4704\cos(2x)$
$- 0.7352\cos(3x) + 3.6761\sin(x) - 2.9409\sin(2x)$
$+ 2.2056\sin(3x)$

Chapter 11

Section 11.1

1. $Q(\mathbf{x}_0) = 22$

3. $Q(\mathbf{x}_0) = 8$

5. $Q(\mathbf{x}) = 4x_1^2 + x_2^2$

7. $Q(\mathbf{x}) = x_1^2 + 2x_2^2 + 6x_1x_2$

9. $Q(\mathbf{x}) = x_1^2 + 3x_2^2 - 2x_3^2$

11. $Q(\mathbf{x}) = 2x_1^2 + x_2^2 + 2x_3^2 + 3x_4^2$

13. $A = \begin{bmatrix} 1 & 3 \\ 3 & -5 \end{bmatrix}$

15. $A = \begin{bmatrix} 3 & 0 & 3 \\ 0 & 1 & 0 \\ 3 & 0 & -1 \end{bmatrix}$

17. $A = \begin{bmatrix} 5 & 0 & 3 \\ 0 & -1 & -6 \\ 3 & -6 & 3 \end{bmatrix}$

19. A is indefinite.

21. A is indefinite.

23. A is indefinite.

25. A is indefinite.

27. $Q(\mathbf{x}) = x_1^2 + x_2^2$, and $c = -1$

29. For example, let $Q(\mathbf{x}) = x_1^2 - x_2^2$, and $c = 0$. Then $Q(\mathbf{x}) = x_1^2 - x_2^2 = 0 \Rightarrow (x_1 - x_2)(x_1 + x_2) = 0$, and the graph consists of the intersecting lines $x_1 - x_2 = 0$ and $x_1 + x_2 = 0$.

31. The only quadratic form which is also a linear transformation is $Q(\mathbf{x}) = 0$ for all $\mathbf{x}$.

33. $Q(\mathbf{x}) = x_1^2 + x_2^2$

35. True

37. False

39. True

41. HINT: What form must such a quadratic form take?

43. $Q(\mathbf{x}) = \mathbf{x}^T I\mathbf{x} = \mathbf{x}^T\mathbf{x} = \|\mathbf{x}\|^2$, so the identity matrix I_n is the matrix of $Q(\mathbf{x}) = \|\mathbf{x}\|^2$.

45. HINT: Evaluate $\mathbf{0}^T A\mathbf{0}$.

Section 11.2

1. $A_1 = [3]$, $A_2 = \begin{bmatrix} 3 & 5 \\ 5 & 7 \end{bmatrix}$

3. $A_1 = [1]$, $A_2 = \begin{bmatrix} 1 & 4 \\ 4 & 0 \end{bmatrix}$, $A_3 = \begin{bmatrix} 1 & 4 & -3 \\ 4 & 0 & 2 \\ -3 & 2 & 5 \end{bmatrix}$

5. $A_1 = [2]$, $A_2 = \begin{bmatrix} 2 & 1 \\ 1 & 3 \end{bmatrix}$, $A_3 = \begin{bmatrix} 2 & 1 & 0 \\ 1 & 3 & 4 \\ 0 & 4 & 0 \end{bmatrix}$,

$A_4 = \begin{bmatrix} 2 & 1 & 0 & -1 \\ 1 & 3 & 4 & 1 \\ 0 & 4 & 0 & -2 \\ -1 & 1 & -2 & 1 \end{bmatrix}$

7. $\det(A_1) = 2 > 0$, $\det(A_2) = -5 < 0$, so A is not positive definite.

9. $\det(A_1) = 1 > 0$, $\det(A_2) = 1 > 0$, $\det(A_3) = 1 > 0$, so A is positive definite.

11. $\det(A_1) = 1 > 0$, $\det(A_2) = 1 > 0$, $\det(A_3) = 1 > 0$, $\det(A_4) = 4 > 0$, so A is positive definite.

13. $\det(A_1) = 1 > 0$, $\det(A_2) = 1 > 0$, so A is positive definite.

$L = \begin{bmatrix} 1 & 0 \\ -2 & 1 \end{bmatrix}$, $U = \begin{bmatrix} 1 & -2 \\ 0 & 1 \end{bmatrix}$

15. $\det(A_1) = 1 > 0$, $\det(A_2) = 1 > 0$, $\det(A_3) = 1 > 0$, so A is positive definite.

$L = \begin{bmatrix} 1 & 0 & 0 \\ -2 & 1 & 0 \\ 2 & -1 & 1 \end{bmatrix}$, $U = \begin{bmatrix} 1 & -2 & 2 \\ 0 & 1 & -1 \\ 0 & 0 & 1 \end{bmatrix}$

17. $\det(A_1) = 1 > 0$, $\det(A_2) = 4 > 0$, so A is positive definite.

$L = \begin{bmatrix} 1 & 0 \\ -2 & 1 \end{bmatrix}$, $D = \begin{bmatrix} 1 & 0 \\ 0 & 4 \end{bmatrix}$, $U = \begin{bmatrix} 1 & -2 \\ 0 & 1 \end{bmatrix}$

19. $\det(A_1) = 1 > 0$, $\det(A_2) = 1 > 0$, $\det(A_3) = 1 > 0$, so A is positive definite.

$L = \begin{bmatrix} 1 & 0 & 0 \\ 3 & 1 & 0 \\ 2 & 2 & 1 \end{bmatrix}$, $D = \begin{bmatrix} 1 & 0 & 0 \\ 0 & 1 & 0 \\ 0 & 0 & 1 \end{bmatrix}$, $U = \begin{bmatrix} 1 & 3 & 2 \\ 0 & 1 & 2 \\ 0 & 0 & 1 \end{bmatrix}$

21. $\det(A_1) = 1 > 0$, $\det(A_2) = 1 > 0$, so A is positive definite.

$$L_c = \begin{bmatrix} 1 & 0 \\ 2 & 1 \end{bmatrix}$$

23. $\det(A_1) = 1 > 0$, $\det(A_2) = 1 > 0$, $\det(A_3) = 1 > 0$, so A is positive definite.

$$L_c = \begin{bmatrix} 1 & 0 & 0 \\ 2 & 1 & 0 \\ 3 & 0 & 1 \end{bmatrix}$$

25. $A = \begin{bmatrix} -1 & 0 \\ 0 & -1 \end{bmatrix}$

27. $A = \begin{bmatrix} -1 & 0 & 0 \\ 0 & -1 & 0 \\ 0 & 0 & -1 \end{bmatrix}$

29. $A = \begin{bmatrix} -1 & 0 & 0 \\ 0 & -1 & 0 \\ 0 & 0 & 1 \end{bmatrix}$

31. False

33. False

35. True

Section 11.3

1. $\max = 2$, $\min = -3$

3. $\max = 3$, $\min = -5$

5. $\max = 2$, $\min = -4$

7. $\max = 5$, $\min = 0$

9. $\max = 1$, $\min = -1$

11. $\max = 3$, $\min = 0$

13. $\max = 20$, $\min = 0$

15. $\max = 100$, $\min = -100$

17. $\max = 104$, $\min = 0$

19. $\max = 19 + \sqrt{370}$, $\min = 19 - \sqrt{370}$

21. $\max = 3$, $\min = 1$

23. $\max = 1$, $\min = 0$

25. $Q(\mathbf{x}) = x_1^2 + 5x_2^2$

27. $Q(\mathbf{x}) = x_1^2 + 6x_2^2$

29. $Q(\mathbf{x}) = -\frac{1}{9}x_1^2 + \frac{4}{9}x_2^2$

31. True

33. False

35. False

37. HINTS: $q_j \le q_k$ for all $k = 1, 2, \ldots, n$, and $q_k \le q_i$ for all $k = 1, 2, \ldots, n$.

39. HINT: $Q(c\mathbf{x}) = (c\mathbf{x})^T A(c\mathbf{x})$

Section 11.4

1. (a) $\begin{bmatrix} 3i \\ -4+i \\ 2-i \end{bmatrix}$ (b) $\begin{bmatrix} 8+i \\ 11-4i \\ 7+12i \end{bmatrix}$ (c) $\begin{bmatrix} -14+i \\ -12+9i \\ -2-21i \end{bmatrix}$

3. c does not exist.

5. Yes.

7. (a) $24 - 6i$ (b) 44 (c) $\sqrt{35}$

9. Divide $\mathbf{u}$ by $\sqrt{39}$, divide $\mathbf{v}$ by $2\sqrt{10}$.

11. (a) $\begin{bmatrix} 2+i & 4-3i \\ -3-i & 3+2i \end{bmatrix}$ (b) $\begin{bmatrix} -2-3i & 9-12i \\ -13+5i & 4+i \end{bmatrix}$

13. c does not exist.

15. Yes.

17. (a) $18 + 2i$ (b) $14 - 50i$ (c) $\sqrt{42}$

19. Divide A by $\sqrt{29}$, divide C by $2\sqrt{7}$.

21. (a) $(2 + 4i) - (4 - 2i)x$ (b) $9 - (3 + 3i)x$

23. c does not exist.

25. $(2 + i) + (3 - 2i)x = \left(-\frac{2}{17} - \frac{43}{17}i\right)(1 + ix)$
$+ \left(\frac{12}{17} + \frac{20}{17}i\right)(3 - (1 + i)x)$

27. (a) $\frac{13}{6} + \frac{5}{3}i$

(b) $\frac{13}{3} + \frac{10}{3}i$

(c) $\frac{2}{3}\sqrt{3}$

29. Divide h_1 by $\frac{2}{3}\sqrt{3}$, divide h_2 by $\frac{2}{3}\sqrt{3}$.

31. $\mathbf{u} = 1$, $\mathbf{v} = i$ in $\mathbf{C}$.

33. $\langle \mathbf{u}, \mathbf{v} \rangle = 2u_1\overline{v_1} + \cdots + 2u_n\overline{v_n}$

35. $V = \mathbf{C}$, and $S = \mathbf{R}$

37. True

39. True

41. False

43. HINT: Apply the properties of complex numbers and arithmetic.

45. HINT: Consider property (2).

47. HINT: Combine properties of complex numbers with definition of inner product space.

49. HINT: Apply property (d) of definition of inner product space.

51. HINT: Note that $0 \le \left\| \mathbf{u} - \frac{\langle \mathbf{u}, \mathbf{v} \rangle}{\|\mathbf{v}\|^2}\mathbf{v} \right\|^2$

Section 11.5

1. $A^* = \begin{bmatrix} 1-i & 2+i \\ -3i & 1-4i \end{bmatrix}$

3. $A^* = \begin{bmatrix} 3-i & 1+4i & 2-2i \\ -5i & -8 & 0 \\ 1+i & 6-i & 7i \end{bmatrix}$

5. $A^* = \begin{bmatrix} 1 & -2i & 3 & 4i \\ 2i & 5 & -6i & 1+i \\ 3 & 6i & 7 & 3+2i \\ -4i & 1-i & 3-2i & 11 \end{bmatrix}$

7. A is not Hermitian.

9. A is Hermitian.

11. A is Hermitian.

13. A is normal.

15. A is normal.

17. A is normal.

19. $A = \begin{bmatrix} 0 & i & i \\ i & 0 & i \\ i & i & 0 \end{bmatrix}$

21. $A = \begin{bmatrix} i & 0 & 0 \\ 0 & i & 0 \\ 0 & 0 & i \end{bmatrix}$

23. True

25. False

27. True

29. True

31. HINT: If A has real entries, then so does A^T.

33. HINT: Apply the result in Exercise 32.

35. HINT: Apply the result in Exercise 32.

37. HINT: A unitary implies that $A^{-1} = A^*$, so $A^* A = I_n$.

39. HINT: Compute $A^* A$ for A normal and upper triangular.

INDEX